Classic Futures

**LESSONS FROM THE PAST
FOR THE ELECTRONIC AGE**

Classic Futures
LESSONS FROM THE PAST
FOR THE ELECTRONIC AGE

Published by Risk Books, a division of Risk Publications.

Haymarket House
28–29 Haymarket
London SW1Y 4RX
Tel: +44 (0)20 7484 9700
Fax: +44 (0)20 7930 2238
E-mail: books@risk.co.uk
Home Page: http://www.riskpublications.com

Monadnock Building
53 West Jackson Boulevard
Suite 252
Chicago IL 60604
USA
Tel: 001 312 554 0556
Fax: 001 312 554 0558

Every effort has been made to secure the permission of individual copyright holders for inclusion.

Introduction © Financial Engineering Ltd 2000
This compilation © Financial Engineering Ltd 2000

ISBN 1 899332 92 8

British Library Cataloguing in Publication Data
A catalogue record for this book is available from the British Library

Risk Books Commissioning Editor: William Falloon
Project Editor: Martin Llewellyn
Typeset by Keyword Publishing Services

Printed and bound in Great Britain by Antony Rowe, Bumper's Farm, Chippenham, Wiltshire

Conditions of sale
All rights reserved. No part of this publication may be reproduced in any material form whether by photocopying or storing in any medium by electronic means whether or not transiently or incidentally to some other use for this publication without the prior written consent of the copyright owner except in accordance with the provisions of the Copyright, Designs and Patents Act 1988 or under the terms of a licence issued by the Copyright Licensing Agency Limited of 90, Tottenham Court Road, London W1P 0LP.

Warning: the doing of any unauthorised act in relation to this work may result in both civil and criminal liability.

Every effort has been made to ensure the accuracy of the text at the time of publication. However, no responsibility for loss occasioned to any person acting or refraining from acting as a result of the material contained in this publication will be accepted by Financial Engineering Ltd.

Many of the product names contained in this publication are registered trade marks, and Risk Books has made every effort to print them with the capitalisation and punctuation used by the trademark owner. For reasons of textual clarity, it is not our house style to use symbols such as TM, ©, etc. However, the absence of such symbols should not be taken to indicate absence of trademark protection; anyone wishing to use product names in the public domain should first clear such use with the product owner.

Lester G. Telser

Lester G. Telser has been a professor at the University of Chicago since 1958, first in the Graduate School of Business and then in the Department of Economics from 1965 to 1998, when he became emeritus. He is a Fellow of the American Statistical Association and the Econometric Society. He has published more than 65 articles on economics of which 25 are on futures, the first in 1955. His books include *Competition, Collusion and Game Theory* 1972, *Functional Analysis in Mathematical Economics* co-authored with Robert L. Graves 1972, *Economic Theory and the Core* 1978, *A Theory of Efficient Cooperation and Competition* 1987, and *Joint Ventures of Labor and Capital* 1998. His most recent research focuses on the Great Depression.

Contents

Introduction – Electronic Trading   xi
Lester G. Telser

Section 1 – Introduction   1

1. Structures and Methods for Meeting Uncertainty   5
Frank H. Knight

2. Constructive Speculation   27
Alfred Marshall

Section 2 – Theory of Storage, Hedging and Futures Markets   43

3. Speculation and Economic Stability   53
Nicholas Kaldor

4. Theory of the Inverse Carrying Charge in Futures Markets   89
Holbrook Working

5. Professor Vaile and the Theory of Inverse Carrying Charges   113
Holbrook Working

6. Futures Trading and the Storage of Cotton and Wheat   119
Lester G. Telser

7. Speculation on Hedging Markets   149
Holbrook Working

8. New Concepts Concerning Futures Markets and Prices   191
Holbrook Working

9. The Pricing of Commodity Contracts   221
Fischer Black

Section 3 – The Rationale, Structure and Performance of Futures Markets   235

10. The Technique of Mediaeval and Modern Produce Markets   241
Abbott Payson Usher

11. Futures and Actual Markets: How they are Related   261
Lester G. Telser

Contents

12 **Origins of the Modern Exchange Clearinghouse: A History of Early Clearing and Settlement Methods at Futures Exchanges** 277
James T. Moser

13 **Organised Futures Markets: Costs and Benefits** 321
Lester G. Telser and Harlow N. Higinbotham

14 **Futures Markets: Their Purpose, Their History, Their Growth, Their Successes and Failures** 355
Dennis W. Carlton

15 **Monopoly, Manipulation, and the Regulation of Futures Markets** 395
Frank H. Easterbrook

Section 4 – Effects of Speculation 421

16 **The Suspension of the Berlin Produce Exchange and its Effect upon Corn Prices** 427
R. H. Hooker

17 **The State of Long-Term Expectation** 467
John Maynard Keynes

18 **Speculation, Profitability, and Stability** 479
William J. Baumol

19 **A Theory of Speculation Relating Profitability and Stability** 495
Lester G. Telser

20 **Profitable Speculation** 509
Michael J. Farrell

21 **Price Destabilising Speculation** 521
Oliver D. Hart and David Kreps

Section 5 – Returns to Speculators and the Costs of Hedging 547

22 **Can Speculators Forecast Prices?** 553
H. S. Houthakker

23 **Normal Backwardation, Forecasting, and the Returns to Commodity Futures Traders** 569
Charles S. Rockwell

24 **Futures Trading and Investor Returns: An Investigation of Commodity Market Risk Premiums** 597
Katherine Dusak

25 **Luck versus Forecast Ability: Determinants of Trader Performance in Futures Markets** 619
Michael L. Hartzmark

Section 6 – New Results on the Random Properties of Futures Prices 643

26 **The Variation of Certain Speculative Prices** 649
Benoit Mandelbrot

27 **The Variation of Some Other Speculative Prices** 685
Benoit Mandelbrot

Notes on Illustrations 709

Index 719

Introduction
Electronic Trading
Lester G. Telser

Without computers there could be no electronic trading. But even without electronic trading, transactions do not require the physical presence of the transactors. Therefore, the immediate appeal of electronic trading does not lie in its potential to dispense with physical contact among the traders. Instead it lies in its promise of lowering the cost of transactions, provided suitable rules are devised for electronic trading. By fulfilling this promise, electronic trading would have several important consequences. First, it would raise the volume of trading in those existing organised futures markets that now use traditional open-outcry bilateral auction trading. Second, it could stimulate the creation of new futures markets in which, at present, costs are no high relative to the benefits that organised trading cannot occur. Third, it could lead to more public participation in the present and in new organised futures markets. However, such developments would not be beneficial to everybody. There would be losers as well as gainers among the participants in futures markets. In addition, the extension of trading owing to the lower cost of electronic trading can pose some dangers as well as benefits to the public interest.

Although electronic trading is new, it must be recognised as merely the latest in a long sequence of innovations in futures markets going back more than a century and a half. We must keep this in mind because the past, carefully studied, can help us anticipate the likely pitfalls that will come with the new developments. The articles in this anthology anticipate and discuss many of the problems that electronic trade will encounter although such was not their manifest intention. It is useful to remember, for example, that in the 1870s, traders in a Des Moines broker's office could send an order to the floor of the Chicago Board of Trade and receive confirmation of the execution of their trades in minutes by telegraph. Plainly, the high cost of this rapid mode of trade confined its use to a few for whom the benefit was high enough.

I stress the adjective "organised" for these markets to distinguish them from other markets in which trades are subject only to the standard provisions for commercial dealings but not to the special rules applied to organised futures markets. Futures trading began in the United States as a spontaneous response to meet the needs of grain merchants. Even at the beginning of organised futures trade, it was understood that trade should be confined to the members of the exchange because trading in futures contracts pose special problems. These rules originate from the enlightened self-interest of the traders themselves, who have learned by experience the need for them. Many of these rules focus on making sure that traders fulfil their obligations. It must be borne in mind that futures contracts, of necessity, have two phases separated in time. In the first a buyer and seller reach agreement on price and quantity. Consummation occurs later at or before the expiration date of the futures contract. In the second phase the buyer and seller must take suitable actions to fulfil the obligations they undertook in the first phase. It is inherent in a futures contract not settled by actual delivery that one of the parties must have suffered a loss at the time of the second phase because the prevailing price of the futures contract must then be different from when the parties entered their contract. Therefore either the buyer, where the price has fallen, or the seller, where the price has risen, will have a loss. The loser has the tempting prospect of walking away from the obligation, that is, reneging on the contract. It would not take many such instances for sufficient loss of confidence as could destroy the futures market itself. Finally, even if the futures contract were fulfilled by delivery, a loss on one side is inevitable because it is exceedingly unlikely that the spot price prevailing at the maturity date will equal the price of the futures contract in the first phase when the initial trade took place. The certainty that one of the parties to a futures contract must have a loss is fully understood by both parties at the outset when they make their original trade. Nevertheless, each must be confident that the contract will be honoured. All of this applies as much to electronic trading as it does to the more traditional forms.

RULES COMBAT POTENTIAL CONFLICT

In the beginning futures markets' trade was mostly among the merchants who were members of the exchange, trading among themselves for their own account. Before long, before merchants recognised the advantage of delegating their futures activities to speciality merchants who had become highly skilled practitioners acting as agents on behalf of their clients. Because some traders on the floor of the exchange were trading on behalf of others, some were trading for themselves and some were doing both, potential conflicts of interests were present. This led to new rules being implemented to protect the principals and thereby encourage them to continue using the exchange. Some examples of these rules are revealing: for instance, a floor trader had to execute a customer's order before his own. Also, a broker had to execute a customer's order at a price emerging from open outcry among competing traders on the floor but he could not trade with his own customer. There were also rules to protect the broker. Customers had to deposit good-faith sums called "margin" with their brokers. If prices turn against the customer, then the broker could demand more margin as protection against loss. Should this not be forthcoming rapidly enough, the broker could close the customer's position and use the proceeds plus the remaining margin to cover the loss. It was evidently in the interest of the broker to require a customer to have enough margin, depending on the volatility of prices, to be reasonably confident that the customer, not the broker, suffered the consequences of a loss. Therefore, futures exchanges have adopted a rule which is sometimes interpreted as an impediment to competition. This is the rule that stops trading when the price change reaches a prescribed limit. However, once it is seen that much trade is undertaken by brokers trading under instructions from their customers, the benefit of the limit price rule is easy to understand. By stopping trade temporarily, brokers gain some time to communicate with their customers and obtain new instructions. Without this rule, outside traders, being at a disadvantage relative to the floor traders, would be reluctant to trade at all; trading volume would shrink, liquidity would diminish, and the organised market in futures would stop altogether.

Futures contracts became highly standardised fungible instruments capable of being traded rapidly on the basis of relatively simple instructions confined to price and quantity. Consequently, because the clearing houses of the exchanges guarantee the validity of each trade, the identity of the parties in a trade on the floor of the exchange are not relevant to its integrity. A special class of exchange members (the clearing members of the exchanges) back up the clearing house with some of their capital, which is at the disposal of the clearing house to ensure that the terms of each trade will be fulfilled. The open positions of the traders are marked to the closing market price at the end of each trading session. This ensures that there is not sufficient time to allow traders to keep their margins current to the condition of the market. These rules are designed to reduce the chance of somebody running up so big a loss that it could threaten a wide circle of creditors and cause widespread defaults throughout the financial system.

These rules also apply to trades among members of the exchange. Members' obligations to the exchange impel them to guard themselves against hazards to which they are exposed by the trades and positions of their customers, who are not exchange members. Each broker must know enough about his customer's financial standing and must use the powerful force of self-interest to demand enough margin from them. Although a broker is usually willing to accept business from anyone who can satisfy his financial requirements and who is regarded as a suitable trader, these criteria have become more stringent because customers who have incurred losses occasionally go to court in an attempt to recover their losses on the ground that their broker gave them bad advice or failed to stop them from making bad trades. To protect themselves, brokers vet their customers, perhaps more carefully now than they had done in the past. Nevertheless, individuals who are outside any business related to the futures contract also trade these futures. These traders may be well or ill informed about the fundamental determinants of prices in these markets. Often successful amateurs measure their ability by their success, while unsuccessful amateurs exhibit their ingenuity by the reasons they give for their failures. The presence of these non-commercial traders in a futures market can move prices away from the path they would take in their absence. Such departures of prices determined by fundamental factors tempt traders who specialise in studying the emotions of amateurs and who can make profits commensurate with the skill with which they understand the psychology of the amateurs. If electronic trading lowers the cost of trading and thereby attracts more non-professionals, then it may also increase the difference between the market price path and that path which would be followed under the exclusive influence of commercial forces. A self-correcting factor mitigates the effects of non-professional trading – the losses typically incurred by amateurs. A group of unsuccessful traders cannot survive in the face of losses unless they can obtain resources from other activities to subsidise their futures activities. But this is not all: even if there is rapid turnover of unsuccessful traders so that the losers depart and are replaced by new aspirants, the potential number of these aspirants may be big enough to feed a steady stream of optimistic non-professionals into the market. The best available estimates of the gains and losses of traders show that few, if any, non-commercial traders encounter success. Therefore, one can hardly expect electronic trading to alter this picture.

SECURITY AND REGULATION

Eventually, electronic trading could use computer algorithms to calculate market clearing prices and thereby dispense with floor traders. It could also accommodate large volumes of trade and keep accurate records. Hence, it could enhance the advantages of more liquidity in a single market. However, these advances would not forgo the need to police traders as electronic trading will no doubt inspire new and ingenious ways to commit fraud that must be encountered by even more ingenious safeguards. While computers can lower the costs of trading, they do not guarantee honesty.

At present, each member of an organised exchange has a financial obligation to the exchange, the extent of which depends directly on the scrupulous behaviour of the other members. Even so, Federal agencies can and do enforce Federal laws and regulations that apply to these exchanges. However, the magnitude of Federal intervention would mushroom if the incentive for self-enforcement by the exchange were to diminish. Owing to the members' obligations to the exchange and especially the clearing members, the exchange does not allow a member to sell his seat to just anyone willing to pay the price. Potential members must be approved by the exchange or its agents. By converting an exchange from its present form to a corporation in which owners would have limited liability, this would inevitably result in more Federal regulation and supervision to replace the deficient self-interest inherent in a corporate form of organisation.

The intrusion of government into the business of the exchange is not a far-fetched possibility. An existing example of this are the rules that limit the size of a trader's position and the additional rules that require traders with positions above a prescribed level at the close of a trading session, to report details about these positions each day that they exceed this level to the Commodity Futures Trading Commission (the Federal agency that regulates futures trading). Futures traders and the exchanges were compelled to accept these laws which actually carry consequences contrary to public interest insofar as they impair liquidity and are irrelevant to their intended purpose, ie, prevention of corners.

ECONOMIES OF SCALE

Also pertinent is another important feature of futures markets – the distribution of the open interest by maturity of the futures contract. These figures cast light on the needs of traders and permit conclusions about how well futures prices can forecast spot prices. The concentration of the open interest in the nearest term contracts is especially striking in Treasury Bond futures and the S&P 500 Index futures. For these two futures well over 90% of the total open interest is always in the nearest term contract. Many other futures also tend to show high concentration in a handful of maturities. This holds for most agricultural commodities in which the open interest clusters around contracts maturing close to the time of harvest, namely, foods and fibre. A similar explanation is apposite for livestock and meat futures. The open interest for metals such as copper and gold also tends to cluster around the nearest term contracts. The main apparent exceptions are the various oil futures contracts for which there is substantial open interest in some contracts going as far as about seven months into the future.

These facts have several implications. First, they demonstrate the presence of economies of scale in futures trading which appear from the effects of the volume of trading on the dispersion of prices. Given the underlying determinants of supply and demand, market prices can be described by a probability distribution such that the dispersion of prices decreases as the volume of trade increases. Consequently, a customer can be confident that a trade can be executed at a price close to that of the last trade. This means the market in that contract is highly liquid. Second, the weakness of the market in the more distant futures suggests the absence of firmly held opinions about the future. It implies that the forecast error of futures prices decreases as the maturity date approaches. Third, the shape of the distribution of the open interest demonstrates the importance of hedging in futures markets. Therefore, we can infer where the main hedging interest lies by observing where the open interest concentrates.

A given volume of trade can be handled more cheaply when carried out electronically rather than manually. Therefore, a small electronic market can offer traders the same terms as a larger, manually traded one. However, it does not follow that the smaller electronic market can prevail over the larger manual market. Economies of scale are inherent in markets owing to the effects of liquidity and the advantages of a single price and this holds no less for electronically rather than for manually executed trades. While several electronic markets have sprung up, differing in some of their features, the forces of competition will select the best combination of those pertinent to electronic trading. In the future perhaps one of these new markets will grow and dominate the scene or perhaps a traditional market will adapt to the new circumstances and retain its dominant position. However, there is no reason to believe that electronic trading can destroy the benefits of concentrating trade in one market. In fact the reverse is more likely. With electronic trading a market can handle a larger volume of trade thereby obtaining the advantages that accrue at a lower cost than with manual trading.

HOW PRICES ARE DETERMINED

Prices on US organised exchanges are determined by means of a "bilateral auction". The term "bilateral" refers to the buyers being on one side and the sellers on the other. Presently traders in a given futures exchange congregate in a small space on the trading-floor called the pit. They reveal their bids and offers showing price and quantity by open outcry and hand signals. But these are valid only for the instant they are made unless the trader repeats the same bid or offer. Despite the seemingly chaotic tumult this presents to an outside observer, the process is actually very efficient. Transactions take place rapidly with few mistakes, honest or otherwise, and they are consummated as soon as a seller signals acceptance of a bid or a buyer signals acceptance of an offer. The mutually agreed-upon price and quantity are rapidly recorded electronically by an exchange employee and are quickly sent to various data banks including the clearing house of the exchange. The information embodied in a trade, ie price, quantity and the contract, is made visible immediately to all traders on the floor and to members elsewhere. This information is the valuable property of the exchange and is not freely available to others. However, the traders on the floor also possess a visible sense of the market as they can see the bids, the offers and who made them. Thus, the traders on the floor can see the unconsummated bids and offers as well as the consummated trades. Consequently, because many of the bids and offers come from brokers acting on behalf of their customers, the floor traders may not know the identity of the parties behind these bids and offers. Indeed, the larger traders may wish to conceal their activities by using agents to act on their behalf.

By its nature, electronic trading would be different. Traders would not congregate in one location where they would have physical contact with each other. Instead they would sit in front of their computers and watch the bids and offers on their computer screens and therefore transmit their bids and offers by computer. To this extent electronic traders would know as much as floor traders do now, but electronic traders would not necessarily know who is making these bids and offers. Even if the computer did reveal this – and it could, albeit at the cost of slowing the trading process – the information available to these traders would still be more desiccated than that which floor traders now have in their hands. Among other things, the waves of excitement would be missing and so would other cues to the floor traders. However, a different type of person is likely to be a successful electronic trader from the type currently successful on the floor. A successful floor trader needs a commanding physical presence. A successful electronic trader would resemble a skilful player of computer games. Therefore, it would not be surprising if floor traders were opposed to electronic trading.

Electronic traders would have a formidable advantage over floor traders in one major area – multi-commodity trading. It would be easier for an electronic trader to acquire and use information about many different contracts from a computer than for a floor trader to obtain similar information about other contracts in different markets. Furthermore, it would be equally simple to send bids and offers very quickly for many different contracts by computer. Floor traders cannot make such trades by themselves. Should they wish to operate across different contracts, they would need the help of other floor traders. Although not out of the question, this practice would be fraught with problems arising from occasionally garbled messages and instructions resulting in misunderstanding, delays and disagreements – a problem unfamiliar to an electronic trader who would alone be able to do at least as much as a team of floor traders. For example, program trading in stocks, a familiar example of electronic trading, would be prohibitively costly without computers.

Table 1	
buyer$_1$ = 100	seller$_1$ = 90
buyer$_2$ = 98	seller$_2$ = 92
buyer$_3$ = 94	seller$_3$ = 95

The hurly-burly of trading obscures the underlying determinants of the market prices and quantities. A simple model can clarify the nature of the trading process and bring some important principles to light. Suppose each trader were willing to trade at most one contract and had a reservation price in mind for that contract. A buyer would have a maximal acceptable price and a seller would have a minimal acceptable price. To be specific, suppose buyer 1 was willing to pay up to 100 for the contract and seller 1 would accept anything not below 90 (see Table 1). The two traders could reach a mutually acceptable bargain for one contract at a price between 90 and 100. For the gap to narrow the appearance of more traders would be required. So let another pair appear, buyer 2 with a maximally acceptable price of 98 and seller 2 with a minimally acceptable price of 92. This pair could reach an agreement at a price between 92 and 98. Therefore, there could be a sequence of two prices, the first between 90 and 100 and the second between 92 and 98. It is possible that by chance, both prices would be the same since the second price interval is contained in the first. However, if the four trades occurred simultaneously instead of sequentially the picture would be different. The simultaneous presence of all four traders would narrow the price range. First, let us suppose the price was between 90 and 92. Buyer 2 would compete with buyer 1 for the purchase from seller 1 driving the price up to at least 92, at which point competition from seller 2 would become effective. Therefore 92 would become the lower limit on the price. Next, suppose the price were between 98 and 100. Now seller 2 would compete with seller 1 for the sale to buyer 1. This would drive the price down to 98 where competition from buyer 2 would set a lower limit on the price. Therefore, to satisfy all four traders requires the same price of between 92 and 98 to apply to both trades. The forces of competition would not be able to produce a narrower range than this one as it is narrower than the trade between only buyer 1 and seller 1. It would not be a matter of what particular trades took place among the four, buyer 1 would be equally happy to buy from seller 1 or seller 2 at the same price in the range 92 to 98. By narrowing the price range competition produces a better price for both the buyers and the sellers.

We can explore this example further by introducing a third seller, seller

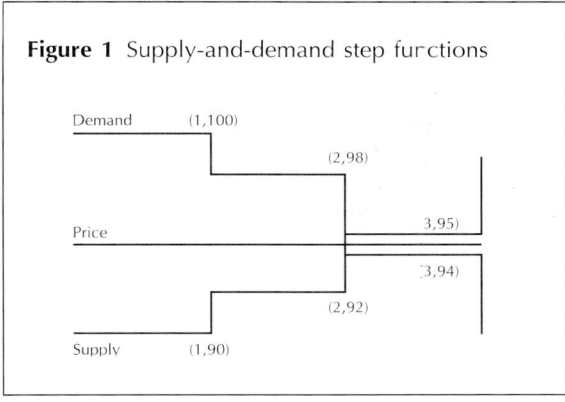

Figure 1 Supply-and-demand step functions

3, with a minimally acceptable price of 95 (see Table 1). Seller 3 could make a deal with buyer 2, who is willing to pay up to 98, or with buyer 1, who is willing to pay up to 100, but there are only two units of demand. Even though seller 3 could make a mutually satisfactory trade with either buyer 1 or buyer 2, competition from sellers 1 and 2, who both have lower minimally acceptable prices, would prevent a trade between seller 3 and either buyer 1 or buyer 2. Nevertheless, although seller 3 cannot make a trade with either of these two buyers, seller 3 does have an effect on the outcome. It would shorten the interval within which the market price must lie from 92 to 98, to 95 to 98. Similarly, the appearance of buyer 3 with a maximally acceptable price of 94 would reduce the range in which the market price must lie from 95 to 98, to 94 to 95. This reduction occurs because buyer 3 could make a trade with seller 2, whose minimally acceptable price is 92, or with seller 1, whose minimally acceptable price is 90, but not with seller 3, whose minimally acceptable price being 95 exceeds the maximal price that buyer 3 would be willing to pay. Therefore, despite the presence of two more traders and the fact that the quantity traded remains at two units, the price range narrows. If buyer 3 had a maximal acceptable price of 95 and seller 3 a minimal acceptable price of 94, then 3 units could be traded at a price between 94 and 95. The common price in this interval would allow any pair of simultaneous trades among the three buyers and the three sellers such that they would be indifferent with respect to the identity of their trading partner.

Figure 1 illustrates the procedure using step functions to show the supply and demand. The demand curve is the descending step function going through the points labelled (1,100), (2,98) and (3,94). The supply curve is the ascending step function through the points (1,90), (2,92) and (3,94). The market clearing price is shown by the horizontal line labelled "Price" which is above the point (3,94) on the demand curve and below the point (3,95) on the supply curve. This simple diagram illustrates an important principle. The market clearing price can satisfy the terms required by any submarket of the active buyers and sellers. This means that buyers 1 and 2 with sellers 1 and 2 could not improve their individual gains by withdrawing from the whole market and trading among themselves. To express it another way, the pairs (buyer 1, seller 1) and (buyer 2, seller 2) reveal price ranges that include the price that all pairs would be willing to accept. However, it does not follow that their individual gains in the submarket are smaller than their individual gains in the whole market. For example, a price between 98 and 100 would be acceptable to the pair (buyer 1, seller 1) if trade were confined to these two, but the buyer has a bigger gain and the seller a smaller one at a price between 94 and 95.

The market clearing prices determine the amounts per unit paid by each buyer and received by each seller, but these prices are not independent of the quantities traded. Traders cannot take these prices as given. Changes in the quantities offered or demanded by the trades, even without changes of their reservations prices, typically affect the market clearing prices.

The situation would change if the trades were sequential instead of being simultaneous. Sequential trading would make the outcomes depend on the order with which the trades were executed. There could be different prices for each pair of trades, and prices could differ more from one trade to the next. Table 1 above can illustrate these effects if the numbers in Table 1 give the order of appearance of the six traders and we assume the price paid by a buyer becomes the minimally acceptable price at which the buyer would become a seller if the first trade took place at a price of 97, then buyer 1 would be a seller at any price not less than 97. If the buyer remained in the market when the second pair (buyer 2, seller 2) came into the market, it could affect the outcome. Plainly, sequential market clearing enlarges the set of mutually acceptable trades and implies more price variability from one trade to the next.

This example demonstrates several important principles. First, while mutually beneficial trades between certain pairs of traders could occur, the traders' desire for gain drives them to seek the best possible trades. Although only mutually beneficial trades can take place, so that a maximally acceptable buyer's price must exceed the minimally acceptable price of a seller, this condition is only necessary but is not sufficient for the trade to take place. Second, the presence of traders who cannot consummate trades still has an effect on the market clearing prices. They serve to narrow the range within which the market clearing price must lie.

Trading is continuous in futures markets at present, ie, whenever a bid above an ask appears so that a trade could occur, it does. As a result there is an almost continuous stream of market prices and quantities given by these trades. There are no pauses to evaluate a price capable of clearing a block of bids and offers. Block trades would leave intervals during which there would be no trades, but the sequence of market prices for the blocks would differ less than the presently generated stream of pairwise trades. Block trades with the use of computers and suitable programs could achieve the same speed as pairwise trades. There would be the additional advantage that such programs could calculate market clearing prices for many different commodities at the same time. Traders could thereby pursue elaborate strategies involving simultaneous trades among many commodities.

ELECTRONIC TRADING OPENS NEW VISTAS

Electronic trading could enable a radical departure from current practice that would also bring trading closer to the theoretical ideal. Ideally, as much trade as possible would be consummated at the same price. To the extent this can be done, the outcome approaches the optimum of placing the commodity into the hands of those who value it the most. This ideal meets another important condition: it maximises the traders' total net gain; proofs of these assertions appear below.

To attain these desirable results requires several factors. First, orders containing quantities and limit prices should be accumulated in a batch and entered into a computer without revealing these individual orders to anybody. Secrecy is essential to prevent manipulation. Second, there must be a rule determining the minimal size of a batch. When the minimal size is reached, the computer temporarily closes its files to new orders and calculates the trades with their associated tentative market clearing prices. Third, having made these prices public and revealed the trades privately to the pertinent individuals, the computer can accept new orders and allow old ones to be withdrawn or modified. Fourth, the computer recalculates the transactions and the new tentative market clearing prices. This process continues until no new orders appear or old orders are withdrawn or revised. At this stage the transactions and market clearing prices determined by the program are final and the trading session ends. The end of the process can be ensured in two different ways: first, in the course of events as just described or, second, by a rule that closes the session after the lapse of a prescribed time. According to this procedure, traders could observe a nearly continuous sequence of tentative market clearing prices. Consequently, individual traders would be informed privately by the computer what they have bought and sold at these tentative prices. (These prices are tentative because they may change in response to the entry of new orders and the removal or revision of old ones.) However, unlike the present system, consummation is not instantaneous. This is because in the present system trades take place as soon as there is a bid above an offer or an offer below a bid, so that a mutually acceptable trade is possible. A suitable program could handle many contracts in many markets simultaneously. Traders could enter complicated schedules of reservation prices and maximal quantities at these prices in many different contracts all at the same time. The vistas opened up by the marriage of computers with trade go far beyond the capabilities of traditional open outcry methods of futures trading.

Several questions naturally arise. First, how much can an individual trader affect the outcome of the trade? This question refers to the effect on the tentative market clearing prices of the entry of a new trader and the exit or revision of the schedule of an existing trader. These effects depend both on the number of traders included in the computer's calculation of the best trades and the accompanying market clearing prices, and on the distance between a trader's schedule and the current solution. Traders fall into two classes in this calculation. In the first class are those traders whose schedules are so far from the current solution that they do not affect it. In the second class are those traders near enough to the solution that changes in their schedule could affect the outcome. In either class, even if each trader were armed with his own powerful computer and knew the program used by the central computer, it would not be possible for an individual trader to predict the effect of changes in his own schedule on the solution without knowing the database of the central computer that contains the schedules of all the traders in the market. To put it another way, no individual trader would have computable power to affect the outcome if they lacked the information in the database of the central computer used to compute the trades and prices. This conclusion assumes the database in the central computer is kept secret from individual traders.

A second question concerns the stability of the process. Because traders would be allowed to insert, revise or remove their trading schedules in the database which determines the tentative outcome, the current outcomes could affect the final result. There would be a well-defined solution at any moment based on the current schedules in the computer's memory. Yet it seems possible that convergence could not be assured. The proper question is whether the *procedure* for calculating the outcomes is inherently conducive to instability, not whether changes in the minds of traders as reflected by the schedules they insert into the computer's memory cause destabilising changes. Consequently, because the proposed program faithfully mimics an ideal market, the program itself cannot affect the stability of the market. The true question is whether even an ideal market is stable. Are there forces inherent in a speculative market inimical to convergence and stability? This important topic is the subject of Section 4.

Third, we must consider whether traders would be inclined to furnish the computer with false limit prices in an attempt to move the market clearing prices favourably. If a seller or buyer submits reservation prices above or below what he would truly be willing to accept or pay, then he forgoes the chance of profitable sales or purchases owing to competition from sellers or buyers. Also, because the computer posts tentative market clearing prices and gives traders opportunities to revise their order, another tactic is feasible. A trader may move his reservation prices towards the posted, tentative market clearing prices in an attempt to raise his gain. However, it can be shown that success in such an endeavour is doubtful. It is even possible that a trader with gains at the posted prices would have none at the market clearing prices that would emerge in response to such changes. Therefore, all of these tactics turn against those who try them.

DECREASING UNIT VALUATIONS ILLUSTRATED WITH SOME NUMERICAL EXAMPLES

Complications arise when traders' reservation prices depend on the quantities they wish to trade. The most plausible hypothesis says that the unit valuations of commodities vary inversely with their quantities. Consequently, their marginal valuations are below their unit valuations. In this case the best trades could be supported by constant-unit prices only if there are at least two active sellers in the group. If there is only one active seller, then it may be necessary to have two-part prices so that buyers would pay a lump-sum amount in addition to an amount proportional to the quantities they buy, and the seller would receive the total lump sum in addition to receipts proportional to the quantities sold. Numerical examples can illustrate some of these complications.

The purpose of the following numerical examples is to illustrate some general propositions about exchange that are relevant to open outcry and electronic trading. In the first example there is one potential seller (S) and one potential buyer (B), and there is one commodity. The seller is willing to offer up to 6 units of this commodity and attaches a valuation to s units, $0 < s \leq 6$, shown by the following formula:

$$V(S) = 47.9 + 3.1 * s$$

Where the seller has none of the commodities the valuation is 0. That is:

$$s = 0 \text{ which implies } V(S) = 0$$

Therefore, the seller would be willing to sell 6 units for a sum not less than $66.5 = 47.9 + 3.1 * 6$. The marginal valuation is 3.1. The presence of the term 47.9 means that the unit valuation to the seller varies inversely with quantity. The unit valuation is $V(S)/s = 3.1 + 47.9/2$. Thus the unit variation is never below the marginal valuation, 3.1, and it approaches 3.1 as s increases.

Next we turn to the potential buyer. The buyer derives a valuation from q units according to the formula:

$$V(B) = 49.1 + 5.95*q \qquad \text{if } 0 < q \leq 10$$

For the buyer, the marginal valuation per unit is 5.95 and the valuation per unit is $V(B)/q = 49.1/q + 5.95$. Like the seller above, the buyer's valuation per unit decreases as the quantity increases and the marginal valuation is always below the unit valuation. If the buyer obtains none of the goods so that $q = 0$, then the valuation is zero. Thus:

$$V(B) = 0 \qquad \text{if } q = 0$$

Since purchases must equal sales, the maximal quantity that could be sold is the smaller of the two numbers, 6 for the seller and 10 for the buyer. If the buyer could obtain 6 units, the valuation would be $84.8 = 49.1 + 5.95*6$. Because 6 units are more valuable to the buyer than to the seller, there is a trade on suitable terms that both would accept. The total valuation from a trade of 6 units would be higher by the amount equal to the valuation to the buyer of 6 units minus the valuation to the seller of 6 units, namely, $84.8 - 66.5 = 18.3$. Therefore, the buyer could pay the seller an amount $66.5 + R$ where $0 < R < 18.3$, and both would be better off: the seller would obtain an amount not less than the valuation of 6 units to him and the buyer would pay an amount not more than the buyer's valuation of 6 units. Could such a mutually beneficial trade occur at a price acceptable to the buyer and the seller? Given the marginal valuations of the two, the seller would not take less than 3.1 per unit and the buyer would not pay more than 5.95 per unit. Moreover, even if the buyer did pay 5.95 per unit, the seller's receipts would be only 35.7, which is below his total valuation of 6 units, 66.5. Therefore, a trade confined to a unit price of the commodity between 3.1 and 5.95 would be refused by the seller. To deny the relevance of this example by saying that if the seller were willing to increase sales by 3.5, then the receipts would be ample enough to cover the gap would miss the point. If the seller accepted the unit price 3.1, then he would have to receive in addition a lump-sum payment of 47.9 to break even. At the price 3.1, the buyer's gain would be $66.2 = 49.1 + (5.95 - 3.1)*6$. This is enough to cover the seller's cost of 47.9 and leave the buyer with a gain of 18.3. If the unit price were at the upper end, 5.95, then the seller would have receipts of 17.1 from this source and would need a lump-sum payment of 30.8 to break even. Since $30.8 > 49.1$, there is still room for a bargain between the two and the gain to the buyer would remain 18.3 even if he paid the higher price.

As there is only one seller in these examples, they show that there may be no constant unit price which could accommodate both the buyer and the seller, even though there is the prospect of a mutually beneficial trade between them. There are two necessary conditions for the intervention of a lump-sum payment: first, the presence of the constant terms in the total valuations that imply an inverse relation between unit value and quantity is required and, second, only one seller should be active.

The next example illustrates another complication when there is more than one buyer but only one seller. The simplest case has two buyers and one seller. The new complication introduced here is how to divide the cost of the seller between the two buyers when this cost would not be covered by the receives from sales at a constant unit price. As above, assume there is one commodity to trade and the seller is the same as in the preceding example. Buyer 1 wants, at most, 4 units and buyer 2 at most 3 units. The valuation of q_1 units by buyer 1 is given by:

$$V(B_1) = 24.5 + 5.95 * q_1 \quad \text{where } 0 \leq 4$$

and for buyer 2, the valuation of q_2 units is given by:

$$V(B_2) = 24.6 + 6.3 * q_2 \quad \text{where } 0 < q_2 \leq 3$$

For both, the valuation of zero units is zero.

It is easy to verify from the above that the best trades put 3 units in the hands of each buyer so that the seller exhausts his whole supply. The valuation of this outcome for buyer 1 is $42.35 = 24.5 + 3*5.95$, and for buyer 2, it is $43.50 = 24.6 + 3*6.3$. The total, $85.85 = 42.35 + 43.5$, exceeds the seller's valuation of 6 units, which is 66.5. Hence, these trades would generate the surplus, 19.35. There is room for a bargain in which all three traders would be better off. However, the highest price compatible with the assumed limits is 5.95. At this price the seller would have a loss of $30.8 = 47.9 - (5.95 - 3.1)*6$. Buyer 1 would have a gain of 24.5 and buyer 2 a gain of 25.65. There would need to be some agreement among the three traders on how to share the remaining cost of 30.8 between the two buyers. At one extreme, buyer 1 could make the lump-sum payment of 24.5 and buyer 2 would pay 6.3. This would leave buyer 1 with a net gain of zero, buyer 2 with a net gain of 19.35 and the seller with a net gain of zero. At the other extreme, buyer 2 could pay $24.6 + 1.05 = 25.65$, leaving the balance $30.8 - 25.65 = 5.15$ to be paid by buyer 1. At this extreme buyer 2 has a net gain of zero and buyer 1 a net gain of $19.35 = 24.5 - 5.15$. This range of outcomes is inevitable given the bounds assumed by the valuations. Indeterminacy would appear even if the valuations had no constant terms owing to differences among the traders' marginal valuations.

However, the situation changes when there is more than one active seller, as there is no longer a need for two-part prices. Simple constant-unit prices are capable of clearing the markets. For multi-commodity trading, blocks of transactions could be cleared by these constant-unit prices. However, the blocks would vary in size and the corresponding sets of market clearing prices would also vary from one block to the next.

More elaborate examples involving many commodities and traders could not be handled using simple arithmetic. However, a computer, suitably programmed, could find the unique set of trades that maximises the net valuations of the traders, but the terms of these trades typically would not be uniquely determined. This is not to say that the program could not find all these terms. Of course, it could; but the different terms of trade would be better for some and worse for other traders as the two examples above show. The indeterminacy is not due to the presence of the constant terms – it would be present even if there were no constant terms. Indeterminacy is inescapable as long as buyers and seller have different reservation prices. Constant terms do implicate the calculations but pose no major obstacle to electronic trading.

Consequently, before traders would be willing to accept the answers given by a computer program for the more elaborate possibilities now within reach, they would need to study the results from many solutions for a variety of trading schedules, if it is anticipated that in the process they would become more willing to fulfil the promise of electronic trading.

An Algorithm for a Bilateral Auction with Avoidable Fixed Charges

Lester G. Telser

The following material presents a model of a multi-commodity market, that is, one in which traders may buy or sell many commodities. It gives a program written in *Mathematica* 4.0 to calculate the trades and the market clearing prices. It thereby fulfills the promise in the last section of "Electronic Trading", to describe in detail how to compute the trades and prices when traders attribute decreasing unit valuations to their holdings of the commodities.

DEFINITIONS OF THE PRIMAL VARIABLES

Seller i $\{g^i, \{a_j^i\}, \{c_j^i\}\}$; Buyer k $\{f^k, \{b_j^k\}, \{d_j^k\}\}$

- j = 1, 2, ..., n commodity index
- i = 1, 2, ..., m seller index
- k = 1, 2, ..., l buyer index
- g = m-tuple of seller specific avoidable cost
- f = l-tuple of buyer specific avoidable valuation
- a = m.n matrix of marginal valuations for sellers
- b = l.n matrix of marginal valuations for buyers
- c = m.n matrix of sellers' capacities
- d = l.n matrix of buyers' demands
- s_j^i = quantity of commodity j sold by seller i
- s^i = n-tuple whose jth component is s_j^i
- q_j^k = quantity of commodity j bought by buyer k
- q^k = n-tuple whose jth component is q_j^k
- u^i = 1 if seller i is active and = 0 otherwise
- v^k = 1 if buyer k is active and = otherwise
- x_j^i = 1 if seller i sells a positive quantity of commodity j and = 0 otherwise

$y_j^k = 1$ if buyer k buys a positive quantity of commodity j and = 0 otherwise

The binary variables are u, v, x and y.

RATIONALE FOR THE FORM OF THE OBJECTIVE

Each trader has a valuation function describing the maximal value that the trader charges to his holdings. The identity of potential buyers and sellers is given at the outset and remains fixed throughout. This means that every potential trader can trade, but not that he necessarily does trade. The total *valuation* of $\{s_j^i\}$ to potential seller i is denoted by $V(S^i)$ and the total valuation of $\{q_j^k\}$ to potential buyer j is denoted by $V(B^i)$. These valuations are defined as follows:

$$V(S^i) = u^i g^i + \Sigma_{j=1}^n a_j^i s_j^i$$
$$V(B^k) = v^k f^k + \Sigma_{j=1}^n b_j^k q_j^k$$

The seller of $s^i \geq 0 \Rightarrow u^i = 1$ gives up the valuation $V(S^i)$ and the buyer of q^k obtains the valuation $V(B^k)$. We may regard $V(S^i)$ as the cost and $V(B^k)$ as the benefit. The net benefit from all these transactions is

$$\Sigma_k V(B^k) - \Sigma_i V(S^i)$$

The best trades are those values of s and q which maximise the net benefit.

The marginal valuations are defined as follows:

$$\beta_j^k = b_j^k \quad \text{if} \quad 0 < q_j^k \leq d_j^k \quad \text{and} \quad \beta_j^k = 0 \quad \text{if} \quad q_j^k > d_j^k$$
$$\alpha_j^i = a_j^i \quad \text{if} \quad 0 < s_j^i \leq c_j^i \quad \text{and} \quad \alpha_j^i = \infty \quad \text{if} \quad s_j^i > c_j^i$$

Note that marginal valuations for a commodity may differ among buyers and sellers.

Let λ_j denote the unit price of commodity j. The gains of seller i and buyer k are defined as follows:

$$G(S^i) = -u^i g^i + \Sigma_j (\lambda_j - a_j^i) s_j^i$$
$$G(B^k) = v^k f^k + \Sigma_j (b_j^k - \lambda_j) q_j^k$$

The term g^i is the avoidable cost to seller i. The term $u^i g^i$ enters the seller's gain $G(S^i)$ with a minus sign because g^i is regarded as a cost incurred only by an active seller. Note that

$$G(B^k) = V(B^k) - \lambda q^k \quad \text{and} \quad G(S^i) = \lambda s^k - V(S^k)$$

in which λ, q^k and s^i are n-tuples and λq^k is the scalar product of two n-tuples, λ and q^k.

The following theorem is matched in simplicity only by its importance. Note: replacing an index with · denotes summation over the replaced index. Thus, $q^{\cdot}_j = \Sigma^i_{k=1} q^k_j$.

Theorem 1
For any $\{s^i_j\}$ and $\{q^k_j\}$ such that $s^{\cdot}_j = q^{\cdot}_j$, the sum of the gains to the buyers and sellers equals the sum of their valuations

$$\Sigma_k G(B^k) + \Sigma_i G(S^i) = \Sigma_k V(B^k) - \Sigma_i V(S^i)$$

Proof
From the preceding relations between gains and valuations,

$$\Sigma_k [V(B^k) - \lambda q^k] + \Sigma_i [\lambda s^i - V(S^i)] = \Sigma_k G(B^k) + \Sigma_i G(S^i)$$

Because $s^{\cdot}_j = q^{\cdot}_j$ by hypothesis, $\Sigma_j \lambda_j (q^{\cdot}_j - s^{\cdot}_j) = 0$. The desired conclusion follows. □

Owing to this theorem, the program seeks the maximal net benefit derived by putting the commodities into the hands of those traders who will value them the most. The dual problem finds the simple or two-part prices that correspond to the best allocation. It also finds the gains to the individual traders induced by these prices. Theorem 3 below is also pertinent. Two-part prices may be necessary only if there is just one active seller.

PURCHASE AND SALE CONSTRAINTS
Purchases and sales are constrained by the following inequalities:

$$q^k_j \leq y^k_j \quad \text{and} \quad s^i_j \leq x^i_j c^i_j$$

However, the effect on q and s of the manner in which the algorithm changes d and c is such that there is always equality for q and s. Therefore, the two continuous variables, purchases and sales, can be removed from the system and the problem becomes a pure binary programming problem.

MARKET CLEARING PRICES
The following theorem and discussion relates market clearing prices (the λ's from the dual problem) to the traders' reservation prices.

Theorem 2

$$x_j^i > 0 \Rightarrow (\lambda_j - a_j^i)c_j^i = \xi_j^i \geq 0$$
$$y_k^j > 0 \Rightarrow (b_j^k - \lambda_j)d_j^k = \eta_j^k \geq 0$$

Proof
Immediate from inequalities (13) and (15) below. □

Theorem 2 says that active sellers get at least their marginal valuations and active buyers pay at most their marginal valuations. Note that ξ_λ^i shows how much commodity j contributes to the avoidable cost of seller i and η_j^i shows how much the payment for the purchases of commodity j reduces the avoidable benefit to buyer k.

It is important to recognise that these market clearing prices are the amounts per unit paid by active buyers and the amounts per unit received by active sellers, and that they depend on the quantities exchanged. No individual trader may take these prices as given independently of his quantities. It is possible, however, that small changes in the c's or d's would not affect the λ's given by the solution of the dual problem. This would depend on whether it was a change of a nonbinding c or d. If the change were for a binding c or d, then the λ's would change.

While the best trades are unique, the prices are typically determinate only within a range. The reason is in the nature of the valuation functions that imply a range between the marginal valuations of the buyers and sellers. It must also not escape attention that the prices cannot be calculated until the best trades have been found.

DESCRIPTION OF THE PRIMAL CONSTRAINTS

In (1)–(6) the constraints are shown on the left and the dual variables on the right:

$$x_j^i \leq u^i, \qquad \xi_j^i \qquad (1)$$
$$u^i \leq 1, \qquad \mu^i \qquad (2)$$
$$y_k^j \leq v^k, \qquad \eta_j^k \qquad (3)$$
$$v^k \leq 1, \qquad \nu^k \qquad (4)$$
$$v^k \leq \Sigma_i u^i \qquad r^k \qquad (5)$$

Inequality (5) prevents a situation in which nothing is either sold or bought, but buyers could obtain a gain. Without inequality (5), the program would allow v's = 1, although all the u's would be 0, and buyers' gains would come from their f's without buying anything. Neither individual sellers nor individual buyers incur losses as is shown in the dual problem.

The equilibrium condition for each commodity j is

$$\sum_{i=1}^{m} c_j^i x_j^i = \sum_{k=1}^{i} d_j^k y_j^k, \quad \lambda j \tag{6}$$

so that the quantity sold of each commodity equals the quantity bought. Equation (6) may also be written $s_j^* = q_j^*$.

The traders' valuations in terms of the binary variables u, v, x, and y are as follows:

$$V(B^k) = \sum_j y_j^k b_j^k d_j^k + v^k f^k \tag{7}$$

$$V(S^i) = \sum_j x_j^i a_j^i c_j^i + u^i g^i \tag{8}$$

The problem seeks the values of the binary variables v^k, u^i, x_j^i, and y_j^k that will maximise

$$\sum_{k=1}^{i} V(B^k) - \sum_{i=1}^{m} V(S^i) \tag{9}$$

subject to (1)–(6).

The Lagrangian for the primal is the following expression

$$\sum_k V(B^k) - \sum_i V(S^i) + \sum_{i,j} \xi_j^i(u^i - x_j^i) + \sum_i \mu^i(1 - u^i) + \sum_{j,k} \eta_j^k(v^k - y_j^k)$$
$$+ \sum_k v^k(1 - v^k) + \sum_j \lambda_j(\sum_j c_j^i x_j^i - \sum_k y_j^k d_j^k) + \sum_k \tau^k(\sum_i u^i - v^k) \tag{10}$$

A trader specific term is positive if the trader is active and is zero otherwise. These binary terms furnish a simple approximation to a concave function for each trader that implies decreasing unit valuations for each commodity. Note that these functions are discontinuous at the origin. (See Figures 1 and 2 below). In a more elaborate model there would also be commodity specific terms. Thus, the valuation function for seller i in that model would be

$$V(S^i) = u^i g^i + \sum_j (x_j^i h_j^i + s_j^i a_j^i)$$

in which h_j^i denotes the commodity specific term for commodity j, seller i. If seller i did sell a quantity s_j^i of commodity j, then its valuation would be $h_j^i + s_j^i a_j^i$ and the valuation of a zero quantity of commodity j would be zero. For buyer k, the valuation function would be

$$V(B^k) = v^k f^k + \sum_j (y_j^k r_j^k + q_j^k b_j^k)$$

so that the valuation of a purchase of q_j^k units of commodity j by buyer k would be $y_j^k r_j^k + q_j^k b_j^k$ and would be zero if none of that commodity were bought. These valuation functions are concave like the simpler versions but have more elaborate discontinuities at the boundaries. Commodity specific terms would complicate the algorithm for finding the trades giving the optimal net valuation, and they would add little of interest to the economics of the problem.

EFFECTS OF BINARY VARIABLES

In view of the complications from the trader specific terms f^k and g^i, what are the advantages? The main one is that they yield inverse relations between unit valuations of the commodities and quantities. Consider the following implications for buyer k.

$$f^k = 0 \Rightarrow \text{unit valuation} \begin{array}{l} > \\ = \\ < \end{array} \begin{array}{l} \text{decreasing} \\ \text{constant} \\ \text{increasing} \end{array}$$

Figure 1 shows two things; first, the valuation function is a discontinuous, linear, concave approximation to a strictly concave valuation function and, second, the valuation function gives decreasing unit valuations. The unit valuation is the slope of a ray drawn from the origin to the line $f + bq$. The slopes of these rays are decreasing as the end point of the ray on the line goes from left to right. The simplest case would have no term f^k so that $f^k = 0$, unit valuations would be constant and would equal marginal valuations.

In Figure 2, the slope of OP is the price, λ, of the commodity. The break-even level of sales is the x-coordinate of the point on the intersection of the line OP and the line $G + as$. The segment $OS(\lambda)$ shows the size of the break-even sales.

$$\text{sales} = OS(\lambda) \Rightarrow \begin{array}{l} > \\ = \\ < \end{array} \begin{array}{l} \text{loss} \\ \text{\{break-even\}} \\ \text{gain} \end{array}$$

Break-even sales depend on the price, and $OS(\lambda) \to 0$ as $\lambda \to \infty$ if and only

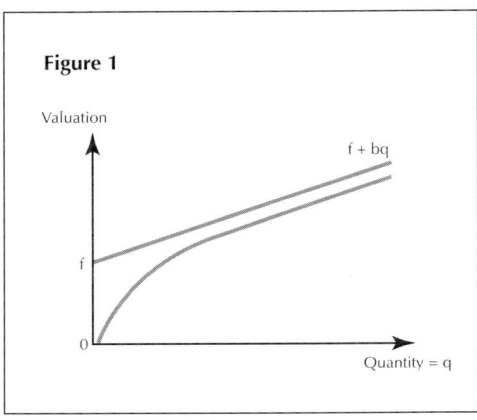

Figure 1

AN ALGORITHM FOR A BILATERAL AUCTION WITH AVOIDABLE FIXED CHARGES

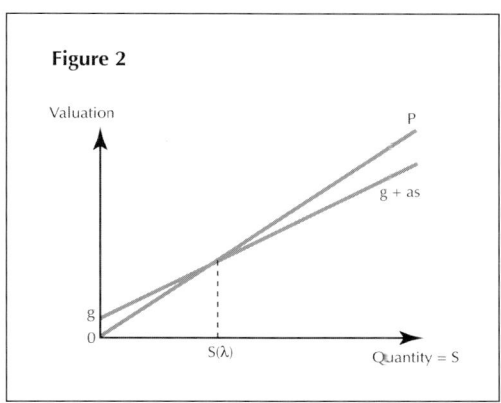

Figure 2

if $g > 0$. $OS(\lambda) = 0$ for all prices if and only if $g = 0$. Positive break-even sales cause most of the complications.

Reconsider inequality (5). When the binary solution has been found the u's and v's are either 0 or 1. If more than one seller is active, then the right-hand side of inequality (5) is at least 2 but the left-hand side is never more than one. Hence if more than one seller is active, τ must equal zero. This completes the proof the following important result.

Theorem 3
The dual variable τ^k can be positive only if one seller is an active trader. If more than one seller is active then all the τ's are zero.

This theorem says that two-part prices are necessary to solve the problem only if there is only one active seller. In this case there could be payments from the buyers to that seller who would incur losses at the prices given by the solution of the BP dual problem. The dual problem determines the two-part prices that some or all of the active buyers must pay to the single active seller. Even so, not every active buyer must pay a positive tau when only one seller is active.

THE DUAL PROBLEM

The solution of the dual problem gives the prices. The dual problem minimises the objective

$$\Sigma_i \mu^i + \Sigma_k \nu^k \tag{11}$$

with respect to nonnegative η_j^k, ν^k, λ_j, ξ_j^i, μ^i and τ^k subject to the following dual constraints that are the necessary conditions for the primal problem:

$$f^k + \Sigma_j \eta_j^k - \nu^k - \tau^k \leq 0, \quad \nu^k \quad (12)$$

$$d_j^k b_j^k - \eta_j^k - \lambda_j d_j^k \leq 0, \quad y_j^k \quad (13)$$

$$-g^i + \Sigma_j \xi_j^i - \mu^i + \Sigma_k \tau^k \leq 0, \quad u^i \quad (14)$$

$$\lambda_j c_j^i - \xi_j^i - a_j^i c_j^i \leq 0, \quad x_j^i \quad (15)$$

The net return to seller i is μ^i and to buyer k is ν^k. The dual constraints (12)–(15) assert that

$$y_j^k > 0 \Rightarrow \eta_j^k = d_j^k(b_j^k - \lambda_j) \geq 0 \Rightarrow b_j^k - \lambda_j \geq 0$$

$$x_j^i > 0 \Rightarrow \xi_j^i = c_j^i(\lambda_j - a_j^i) \geq 0 \Rightarrow \lambda_j - a_j^i \geq 0$$

$$u^i > 0 \Rightarrow \mu^i = \Sigma_j \xi_j^i - g^i + \Sigma_k \tau^k \geq 0 \Rightarrow \Sigma_j \xi_j^i + \Sigma_k \tau^k \geq g^i > 0$$

$$\nu^k > 0 \Rightarrow \nu^k = \Sigma_j \eta_j^k + f^k - \tau^k \geq 0$$

The role of τ^k is especially noteworthy. It is the amount an active buyer k pays in order to cover any deficiency between the receipts of a seller from sales of the commodities at the constant unit prices, the λ's, and his avoidable cost, g. Hence it is the fixed component of the two-part price paid by buyer k when only one seller is active. Note that different buyers may pay different τ's.

Theorem 4
No active trader incurs a loss.

Proof
The return to seller i is μ^i. If $\xi_.^i - g^i + \tau^{\cdot} < 0$, then $u^i = 0$ so seller i must be idle and his return must be zero. The return to buyer k, ν^k, must be nonnegative. Hence $f^k + \eta^{\cdot} - \tau^k < 0 \Rightarrow \nu^k = 0$ so buyer k is idle and $\nu^k = 0$. □

Theorem 5
If there is an active buyer, then there must be an active seller.

Proof
If buyer k is active, then $\nu^k = 1$, so by inequality (5) at least one $u^i = 1$. □

FEASIBILITY
It remains to show that the primal and dual constraints have finite, nonnegative solutions. This is obvious for the primal constraints as follows:

$$0 \leq x_j^i \leq u^i \leq 1,\ 0 \leq y_j^k \leq \nu^k \leq 1,\ \nu^k \leq \Sigma_j \mu^i \leq m \text{ and } \Sigma_j c_j^i x_j^i = \Sigma_k d_j^k y_j^k$$

Because the primal objective is a continuous linear function on a closed bounded set, the Weierstrass Extremum Theorem applies and guarantees the primal has a maximum.

The analysis for the dual constraints (12)–(13) is more difficult. As we shall see, the dual constraints do have finite nonnegative solutions, but when no trade can occur, the solution, while algebraically correct, is economic nonsense. There are two cases.

1. No trade can occur if $\min_{ik}\{a_j^i\} > \max_k\{b_j^k\} \ \forall\ j$.
2. Trade can occur only if there is a j such that $\min_i\{a_j^i\} \leq \max_k\{b_j^k\}$.

We study the feasibility of the dual constraints starting with the first case. Let us focus attention on the λ's. There are four alternatives:

1.1 $\qquad\qquad\qquad \lambda_j \geq b_j^k$
1.2 $\qquad\qquad\qquad \lambda_j < b_j^k < a_j^i$
1.3 $\qquad\qquad\qquad \lambda j \leq a_j^i$
1.4 $\qquad\qquad\qquad \lambda j > a_j^i > b_j^k$

Note that $1.4 \Rightarrow 1.1$ but 1.1 does not imply 1.4. Also, $1.2 \Rightarrow 1.3$ but 1.3 does not imply 1.2. Hence it suffices to show that the dual constraints are feasible for 1.2 and 1.4 since feasibility follows from these for the rest.

1.2 $\qquad\qquad\qquad 0 \leq \lambda_j < b_j^k < a_j^i$

In dual constraint (13), $d_j^k(b_j^k - \lambda_j) > 0$, so let $\eta_j^k = d_j^k(b_j^k - \lambda_j)$. In constraint (15), $c_j^i(\lambda_j - a_j^i) < 0$, so let $\xi_j^i = 0 \Rightarrow \xi_j^* = 0$. Now we need nonnegative solutions for (12) and (14). That is,

$$f^k + \eta_\cdot^k \leq v^k + \tau^k \quad \text{and} \quad \tau^* \leq g^* + \mu^i$$

Equivalently, we need

$$f^* + \eta_\cdot^* - v^* \leq \tau^* \leq g^i + \mu$$

We can satisfy these inequalities even if $f^* > g^i$ with suitable choices of $\eta_\cdot^*\ \&\ v^*$, if necessary setting $\eta_\cdot^* - v^* < 0$ and small enough to satisfy these inequalities. However, although these results are algebraically correct, they are economic nonsense because they say that buyers can get a benefit without buying anything.

1.4 $\qquad\qquad\qquad \lambda_j > a_j^i > b_j^k$

Dual constraint (15) asserts $0 < c_j^i(\lambda_j - a_j^i) \leq \xi_j^i$. Hence let ξ_j^i equal its positive lower bound. In dual constraint (13) we have $d_j^k(b_j^k - \lambda_j) < 0$ so set $\eta_j^k = 0 \Rightarrow \eta_\cdot^k = 0$. Now we must satisfy inequalities (12) and (14) in the following shape:

$$\tau^\cdot + \xi_\cdot^i \leq g^i + \mu^i \quad \text{and} \quad f^k \leq v^k + \tau^k$$

These reduce to

$$g^i + \mu^i - \xi_\cdot^i \geq \tau^\cdot \geq f^\cdot - v^k$$

One can satisfy these inequalities even if $f^\cdot > g^i$ by means of suitable choices for v^k, μ^i and λ_j. While this would give an algebraically correct answer, since it would imply that sellers would derive a benefit even when they sell nothing, an economic absurdity, it remains true that the dual constraints are feasible even in this case.

In order to avoid producing economically meaningless results, the program does not compute the benefit unless the necessary conditions for mutually advantageous trade, case 2, hold.

The dual constraints do give finite nonnegative solutions even when no trade would occur. It is easy to verify this also for case 2, when the necessary conditions for trade, hold. Because the primal and dual constraints are both feasible and give finite solutions, the duality theorem of linear programming applies and we may conclude that

$$\min(\mu^\cdot + v^\cdot) = \max \Sigma_j V(S^i) + \Sigma_k V(B^k)$$

STATUS OF THE CORE

Returns to the traders are in the core of the market if and only if they are undominated. Call G the grand coalition of all the traders in the market and call S a subcoalition of G. Equivalently, S is a submarket of the whole market G. For each S and G compute the best trades that determine the returns to the traders in S and G, call these $\mu^i(S)$ and $v^k(S)$ for seller i and buyer k. These are given by the solution of the dual problem for S and G. Let $\Phi(C)$ denote the indicator function for a set C. It is an $(1 + m)$-tuple such that coordinate $h = 1$, if trader h is in C and $= 0$ if trader h is not in C. For the grand coalition, G, every coordinate of $\Phi(G)$ is 1. The biggest return that the traders in the submarket S can assure themselves under the worst conditions, namely, those that arise when they must confine their trades to themselves, is $\Phi[S][\mu(S) + v(S)]$. Similarly, the best outcome for the grand market is given by $\Phi(G)[\mu(G) + v(G)]$. The n-tuples of returns, $\mu(G)$ and $v(G)$, given by the solution of the dual problem for G are in the core if the following inequality holds for all S in G.

$$\Phi(S)[\mu(G) + v(G)] \geq \Phi(S)[\mu(S) + v(S)] \qquad (16)$$

Seller i gets a bigger return by trading in the grand market G than in any submarket S if

$$\mu^i(G) \geq \mu^i(S) \quad \text{for all} \quad S \subset G \tag{17}$$

Buyer k gets a bigger return by trading in the grand market G than in any submarket S if

$$\nu^k(G) \geq \nu^k(S) \quad \text{for all} \quad S \subset G \tag{18}$$

These inequalities do not say that the μ's and ν's are increasing; they say that the individual returns are bigger in the grand market than in any submarket. The returns generated by the solutions of the dual problem are in the core of G if they can satisfy these inequalities (17) and (18) for all traders in G. These returns are calculated using the prices that satisfy the dual constraints (12)–(14). Note that (17) and (18) are sufficient but not necessary conditions for (16). It is easy to make examples in which the returns to some traders do not satisfy inequalities (17) and (18) if there are more than two potential buyers and two potential sellers. When the returns do not satisfy (17) and (18) for a submarket S, traders in S would withdraw from the grand market because it would be worse for them to trade in G than in their submarket S. Therefore, the grand market G would not be stable. The presence of avoidable costs and valuations are necessary for this conclusion. If these were zero, then the returns given by the solution of the dual problem for the grand market would satisfy inequalities (17) and (18).

Reconsider inequality (16). Because this inequality says the *sum* of the returns that the traders in S would get by trading in the grand market would exceed the *sum* of their returns confined to the submarket S, transfers among the traders, adding to or subtracting from the returns given by the solution of the dual problem, would be necessary to carry out the terms of inequality (16). Thus the constraints given by (16) are weaker than those given by (17) and (18). Nevertheless, we should not expect the solutions of the dual problems to satisfy even these weaker conditions save for the simplest case when one potential buyer meets one potential seller. While inequalities (17) and (18) imply (16), even if inequality (16) were satisfied, it need not imply (17) and (18). Therefore, even if the core constraints in additive form, (18), could be satisfied, the individualistic form shown by (17) and (18) need not be satisfied. This means that the returns produced by the prices may require additional monetary transfers among the traders. Although the solution of the primal problem does give the best trades, and the solution of the dual problem does give the corresponding prices, these prices do not generate gains that can satisfy the core constraints. It would be possible but impractical to satisfy inequalities (17) and (18) if traders in each submarket were forced to trade the same quantities as would be best for the grand market.

The largest set of traders admitting a core determines the size of the block of trades that can clear the market. Each trader in this block gets a return satisfying (17) (seller) and (18) (buyer). The market clearing price and the size of the blocks of trades can differ from one block or another.

PROGRAM

```
Needs [''Utilities''MemoryConserve''']
Off [General :: spell1]
Off [General :: spell]
dx : = 10 ^ - 6
add [thing_List] : = Apply[Plus, thing]
m : = Length [g]
l : = Length [f]
n : = Dimensions [b] [[2]]
zero [num_Integer] : = Table [0, {num}]

CheckDeal : = (nodeal = 0; Tra = Transpose [a]; Trb = Transpose [b];
Do [If[Min[Tra[[j]]] > Max [Trb[[j]]], nodeal ++], {j, n}];
If [nodeal > = n, Print [StringForm[''No deals are possible'']]; Abort[]])
```

PRIMAL CONSTRAINTS

```
Up [stuff_List] : = Thread[Flatten[stuff] < = 1

(*UpU : = Thread[U ≤ 1]
 UpV : = Thread[V ≤ 1]*)

bnd [1st1_List, 1st2_List] : = Thread[Flatten[1st1 - 1st2] < = 0]
```

The preceding is a substitute for the following. It is neater but perhaps slower.

```
(*UXTop : = Thread[Flatten[X-U] ≤ 0]
 VYTop : = Thread[Flatten[Y - V] ≤ 0]*)

Makevbnd : =
(vbnd = {}; Do [vbnd = {vbnd, Sum[u[i], {i, m}] - v[k] > = 0}, {k, 1}]; vbnd = Flatten[vbnd])

(*Makeubnd : =
(ubnd = {}; Do [ubnd = {ubnd, Sum [v[k], {k, 1}] - u[i] > = 0}, {i, m};: ubnd = Flatten [ubnd])*)

walras : = (equil = {}
Do[equil = {equil, Sum[x[1, j] * c [[1, j]], {i, m}] -
    Sum[y[k, j] * d[[k, j]], {k, 1}] = = 0}, {j, n}]; equil = Flatten[equil])

MakConstraints : = (walras; Makevbnd; Makeubnd)

constraints : = Flatten [{Up[U], Up[V], bnd[X, U], bnd{Y, V], equil, vbnd}]
```

PRIMAL LP PROBLEM; VARIABLES AND OBJECTIVE

```
S[i_] : = u[i] * g[[i]] + Sum [a[[i, j]] * c[[i, j]] * x[i, j], {j, t,}]
B[k_] : = v[k] * f[[k]] + Sum[b[[k, j]] * d[[k, j]] * y[k, j], {j, n}]

obj : = Sum[B[k], {k, l}] - Sum[S[i], {i, m}]

vars : = Flatten[[X, Y, U, V}]

prob : = (X : = Array[x, {m, b}]; Y : = Array[y, {l, n}]; U : = Array[u, m]; V : = Array[v, l];
MakeConstraints;
primal = ConstrainedMax[obj, constraints, vars];
{X, Y, U, V} = {X, Y, U, V} /. primal [[2]])
```

ALGORITHM FOR PRIMAL

The algorithm employs a sequence of linear programs. It replaces the binary constraints on u, v, x and y with nonnegative fractions so that $0 \le u^i \le 1$, $0 \le v^k \le 1$, $0 \le x_j^i \le 1$, and $0 \le y_j^k \le 1$. Given j, if there are several sellers whose x's are positive fractions, then the algorithm finds the biggest a[[i,j]]*c[[i,j]] and reduces its c[[i,j]] so that the new c equals x[[i,j]]*c[[i,j]]. Hence the old c[[i,j]] remains feasible on the next iteration. Similarly, if there are several buyers of commodity j with positive fractional y's, it finds the biggest b[[k,j]]*d[[k,j]] and makes the new d equal y[[k,j]]*d[[k,j]] so the old d[[k,j]] is feasible on the next iteration.

For any commodity j at an iteration for which there is a positive fraction x (or y) at most one c (or d) changes. All the buys and sells from the preceding iteration for that commodity j are feasible at the next iteration so that the value of the objective can stay the same. What cannot occur in this procedure is an *increase* of any c or d.

```
MakeSellmat[X_] : = (Do[If[dx < = X[[i, j]] < = 1 -
 dx, e[j, i] = 1, e[j, i] = 0], {j, n}, {i, m}];
Sellmat = Table[e[k, i], {j, n}, {l, m}])

MakeBuymat[Y_] : = (Do[If[dx < = Y[[k, j]] < = 1 - dx, h[j, k] = 1, h[j, k] = 0], {j, n}, {k, l}];
Buymat = Table[h[j, k], {j, n}, {k, l}])

sellsum : = add[Flatten[Sellmat]];

FixSell[X_] : = (MakeSellmat[X]; sellsum;
If[sellsum > 0, Do[If[add[Sellmat[[j]]] > 0, Do[If[Sellmat[[j, i]] > 0,
sellchk[i] = {i, a[[i, j]] *c[[i, j]]}, sellchk[i] = {i, 0}], {i, m}];
sellpick = Table[sellchk[i], {i, m}]; sellpick = Flatten[sellpick];
sellpick = Partition [sellpick, 2]; sellavail = Table[sellpick[[γ, 2]], {?, m}];
sellinx = First[Flatten[Position[sellavail, Max[sellavail]]]];
ind = sellpick[[sellinx, 1]]; c[[ind, j]] = c[[ind, j]] * X[[ind, j]]], {j, n}]])

buysum : = add[Flatten[Buymat]]

FixBuy[Y_] : = (MakeBuymat[Y]; buysum;
If[buysum > 0, Do[If[add[Buymat[[j]]] > 0, Do[If[Buymat[[j, k]] > 0,
buychk[k] = {k, b[[k, j]] *d [[k, l]]}; buychk[k] = {k, 0}], {k, l}];
buypick = Table[buychk[k], {k, l}]; buypick = Flatten[buypick];
```

```
buypick = Partition[buypick, 2]; buyavail = Table[buypick[[γ, 2]], {?, 1}];
buyinx = First[Flatten[Position[buyavail, Max[buyavail]]]];
knd = buypick[[buyinx, 1]]; d[[knd, j]] = d[[knd, j]] *Y[[knd, j]]], {j, n}]]

repair[Y_] : = Do[If[add[Y[[k]]] < = 0, f[[k]] 0], {k, 1}]
```

Comments
1. Repair{Y} is essential for the correct answer with respect to V.
2. Because the primal satisfies the equilibrium conditions, the dual must use quantities in the primal solution. Therefore, it is wrong to return to Ur values of c and d. The value of the dual objective must equal the value of the primal objective. Otherwise, the dual is wrong.

```
work : = (prob; FixSell[X]; FixBuy[Y]; repair[Y])

figure[maxiter_Integer] : = (X =.; Y =.; U =.; V =.; CheckDeal;
iternum = 0; work;
While[(buysum > 0||sellsum > 0) && iternum < maxiter, work; iternum ++];
Print[StringForm["Iteration = ", U = ", V = ", X = ", Y = ", Net Valuation = ",
    Purchases = ", Sales = "",
iternum, U, V, X, Y, NumberForm[primal[[1]], 4], purchases, sales]]; dualprob]
```

TRADES
```
sales : = Table [X[[i, j]] *c[[i, j]], {i, m}, {j, n}]

purchases : = Table [Y[[k, j]] *d[[k, j]], {k, 1}, {j, n}]
```

DUAL PROBLEM
```
MakeVconstr : = (vconstr = {};
Do[vconstr = {vconstr, f[[k]] + Sum[η[k, j], {j, n}] - ν[k] - τ[k] < = 0}, {k, 1}];
vconstr = Flatten[vconstr])

MakeUconstr : = (uconstr = {};
Do[uconstr = {uconstr, - g[[i]] + Sum[ξ[i, j], {j, n}] -μ[i] + Sum[τ[k],
    {k, 1}] < = 0, {i, m}]; uconstr = Flatten[uconstr])

MakeXconstr : =
(Xconstr = {}; Do[Xconstr = {Xconstr, λ[j] *c[[i, j]] -c[[i, j]]
    *a[[i, j]] - ξ[i, j] < = 0, {i, m}, {j, n}]; Xconstr = Flatten[Xconstr])

MakeYconstr : =
(Yconstr = {}; Do[Yconstr = {Yconstr, b[[k, j]] *d[[k, j]] - λ[j] *d[[k, j]]
    -η[k, j] < =0}, {k, 1}, {j, n}]; Yconstr = Flatten[Yconstr])

dualobj : = add[mu] + add[nu]

dualvars : = Flatten[{mu, nu, ksi, eta, lambda, tau}]

MakeDualconstraints : = (MakeVconstr; MakeUconstr; MakeXconstr;
MakeYconstr; dualconstraints = Flatten[{vconstr, uconstr, Xconstr, Yconstr}])

dualprob : = (mu =.; nu =.; ksi =.; eta =.; lambda =.;
tau =.; mu : = Array[μ, m]; nu : = Array [ν, 1]; ksi; + Array[ξ, {m, n}];
eta = Array[η, {1, n}]; lambda : = Array[λ, n]; tau : = Array[τ, 1];
```

```
If[primal[[1]] <= 0, f = zero[1]];
MakeDualconstraints; dual = ConstrainedMin[dualobj, dualconstraints, dualvars];
{mu, nu, ksi, eta, lambda, tau} = {mu, nu, ksi, eta, lambda, tau} /. dual[[2]];
Print[StringForm["Min Obj = ', Prices = ", Sellers' Commodity Gains = ",
Buyers' Commodity Gains = ", Sellers' Total Gains = ", Buyers' Total
      Gains = ", tau = "", NumberForm[dual[[1]], 4], lambda, ksi, eta, mu, nu, tau]])
```

INTRODUCTION

Introduction

Lester G. Telser

The first chapter is from *Risk, Uncertainty and Profit* by Frank H. Knight, his first and most important work in economics. published in 1921. The second selection is from *Industry and Trade* by Alfred Marshall, his last major book on economics.

Knight's book, based on his doctoral dissertation, is perhaps one of the five most important books on economics in the 20th century. Because risk and uncertainty is the essence of futures markets, reading this chapter should whet the appetite for the whole book to anyone interested in the theory of futures markets. "Wise" is the most accurate adjective that describes Knight. His writings have nuances, subtleties and deep insights to be found only among the greatest thinkers Points of departure by lesser scholars are grist for Knight's critical mill. This section contains Knight's famous distinctions between risk and uncertainty and between reflective and objective probability. It gives several applications the distinction between risk and uncertainty. The chapter begins with a profound analysis of insurance. Why health insurance poses special problems does not elude Knight's attention and he uses his theory of uncertainty to analyse corporations. Knight starts his analysis of speculation and hedging by emphasising the great principle of specialisation, especially as it explains the function of speculators. However, it would be unjust and simplistic to Knight to say that he describes speculators as akin to sellers of price insurance to hedgers. Knight draws analogies between speculators and hedgers on one side, and, on the other side, owners and renters of certain kinds of capital goods. An owner of these capital goods incurs risks that the renter avoids. Rental resembles a short hedge because the rent may be regarded as the difference between the purchase price of an asset and the subsequent sales price.

The second selection by Alfred Marshall operates on a very different level to Knight's. Marshall's purpose was to explain economics to the curious businessman as well as to the student. His last book, *Industry and*

Trade, shows this perfectly because it treats the subject in a less abstract fashion than Knight, the concrete details in Marshall illuminate the principles that Knight elucidates. Marshall does much more than this. Nearly every important issue that has arisen in the formal study of futures markets can be found in this chapter. Moreover, his discussion of hedging is more sophisticated than Kaldor's (in Section 2). Marshall recognises the advantages of long hedging as well as short hedging and shows that an organised exchange furnishes a place for both the trade and the offset of their opposite risks. Marshall makes no pronouncements about the bias towards the futures price and the expected spot price. Nor is this all. Marshall refers to US evidence against the complaint of farmers that spot prices are unduly depressed at the time of harvest by the actions of speculators. He also refers to Hooker's study (in Section 4), that showed how speculation on the Berlin Produce Exchange raised prices at harvest times and lowered them subsequently. Thus speculation smoothed prices. Yet Marshall is no apologist for speculators and points out circumstances when it may harm the public interests.

1
*Structures and Methods for Meeting Uncertainty**

Frank H. Knight[†]

To preserve the distinction which has been drawn between the measurable uncertainty and an unmeasurable one we may use the term "risk" to designate the former and the term "uncertainty" for the latter. The word "risk" is ordinarily used in a loose way to refer to any sort of uncertainty viewed from the standpoint of the unfavourable contingency, and the term "uncertainty" similarly with reference to the favourable outcome; we speak of the "risk" of a loss, the "uncertainty" of a gain. But if our reasoning so far is at all correct, there is a fatal ambiguity in these terms, which must be gotten rid of, and the use of the term "risk" in connection with the measurable uncertainties or probabilities of insurance gives some justification for specialising the terms as just indicated. We can also employ the terms "objective" and "subjective" probability to designate the risk and uncertainty respectively, as these expressions are already in general use with a signification akin to that proposed.

 The practical difference between the two categories, risk and uncertainty, is that in the former the distribution of the outcome in a group of instances is known (either through calculation *a priori* or from statistics of past experience), while in the case of uncertainty this is not true, the reason being in general that it is impossible to form a group of instances, because the situation dealt with is in a high degree unique. The best example of uncertainty is in connection with the exercise of judgement or the formation of those opinions as to the future course of events, which opinions (and not scientific knowledge) actually guide most of our conduct. Now if the distribution of the different possible outcomes in a group of instances is known, it is possible to get rid of any real uncertainty

---

*This paper was first published in *Risk, Uncertainty and Profit*, 1921, (Boston: Houghton Mifflin & Co).
†Frank Knight, 1895–1973.

by the expedient of grouping or "consolidating" instances. But that it *is possible* does not necessarily mean that it *will be done*, and we must observe at the outset that when an individual instance only is at issue, there is no difference for conduct between a measurable risk and an unmeasurable uncertainty. The individual, as already observed, throws his estimate of the value of an opinion into the probability form of "*a* success in *b* trials" (*a/b* being a proper fraction) and "feels" toward it as toward any other probability situation.

As so commonly in this subject fraught with logical difficulty and paradox, reservations must be made to the above statement. In the first place, it does not matter how unique the instance, if a real probability can be calculated, if we can know with certainty how many successes there *would be* in (say) one hundred trials *if* the one hundred trials could be made. If we know the odds against us it does not matter in the least whether we place all our wagers in one kind of game or in as many different games as there are wagers; the laws of probability hold in the second case just as well as in the first. But in business situations it so rarely happens that a probability can be computed for a single unique instance that this qualification has less weight than might be supposed. However, in so far as objective probability enters into a calculation, it is hard to imagine an intelligent individual considering any single case as absolutely isolated. The only exception would be a decision in which one's whole fortune (or his life) were at stake. The importance of the contingency and probable frequency of recurrence in the individual lifetime of situations similar in the magnitude of the issues involved should make a difference in the attitude assumed toward any one case as well as the mathematical probability of success or failure.

A second reservation of more importance is connected with the possibility referred to in the preceding chapter, of forming classes of cases by grouping the decisions of a given person. That is, even though we do not get a quantitative probability by the process of grouping, still there is some tendency for fluctuations to cancel out and for the result to approach constancy in some degree. There appear to be in the making of judgements the same two kinds of elements that we find in probability situations proper; ie, (a) determinate factors (the quality of the judging faculty, which is more or less stable) and (b) truly accidental factors varying from one decision to another according to a principle of indifference. The difference between the uncertainty of an opinion and a true probability is that we have no means of separating the two and evaluating them, either by calculation *a priori* or by empirical sorting. But in the second case the difference is not absolute; the sorting method does apply to some extent, though within narrow limits. Life is mostly made up of uncertainties, and the conditions under which an error or loss in one case may be compensated by other cases are bafflingly complex. We can only say that "in so far as" one confronts a situation involving uncertainty and deals with it on its merits as an isolated case, it is a matter of practical indifference whether the uncertainty is measurable or not.

The problem of the human attitude toward uncertainty (not for the present purpose distinguishing kinds) is as beset with difficulties as that of uncertainty itself. Not merely is the human reaction to situations of this character apt to be erratic and extremely various from one individual to another, but the "normal" reaction is subject to well-recognised deviations from the conduct which sound logic would dictate. Thus it is a familiar fact, well discussed by Adam Smith, that men will readily risk a small amount in the hope of winning a large when the adverse probability (known or estimated) against winning is much in excess of the ratio of the two amounts, while they commonly will refuse to incur a small chance of losing a larger amount for a virtual certainty of winning a smaller, even though the actuarial value of the chance is in their favour. To this bias must be added an inveterate belief on the part of the typical individual in his own "luck", especially strong when the basis of the uncertainty is the quality of his own judgement. The man in the street has little more sense of the real value of this opinion than he has knowledge of the "logic" (if such it may be called) on which they rest. In addition, we must consider the almost universal prevalence of superstitions. Any coincidence that strikes attention is likely to be elevated into a law of nature, giving rise to a belief in an unerring "sign". Even a mere "hunch" or "something tells me", with no real or imaginary basis in the mind of the person himself, may readily be accepted as valid ground for action and treated as an unquestionable verity.

Doubtless in the long run of history there is a tendency toward rationality even in men's whims or impulses. And if for no other reason than the impossibility of intelligently dealing with conduct on any other hypothesis, we seem justified in limiting our discussion to rational grounds of action. We shall assume, then, that if a man is undergoing a sacrifice for the sake of a future benefit, the expected reward must be larger in order to evoke the sacrifice if it is viewed as contingent than if it is considered certain, and that it will have to be larger in at least some general proportion to the degree of felt uncertainty in the anticipation.[1] It is clearly the subjective uncertainty which is decisive in such a case, what the man believes the chances to be, whether his degree of confidence is based upon an objective probability in the situation itself or in an estimate of his own powers of prediction. We hold also that both the objective and subjective types may be involved at the same time, though no doubt most men do not carry their deliberations so far; the man's opinion or prediction may be an estimate of an objective probability, and the estimate itself be recognised as having a certain degree of validity, so that the degree of felt uncertainty is a product of two probability ratios. It is to be emphasised again that practically all decisions as to conduct in real life rest upon opinions, and doubtless the greater part rest upon opinions which on scrutiny easily resolve themselves into an opinion of a probability – though as noted this "scrutiny" many not in most cases be given to the judgement by the individual making it.

The normal economic situation is of this character: the adventurer has an opinion as to the outcome, within more or less narrow limits. If he is inclined to make the venture, this opinion is either an expectation of a certain definite gain or a belief in the real probability of a larger one. Outside the limits of the anticipation any other result becomes more and more improbable in his mind as the amount thought of diverges either way. Hence it is correct to treat all instances of economic uncertainty as cases of choice, between a smaller reward more confidently, and a larger one less confidently anticipated.

At the bottom of the uncertainty problem in economics is the forward-looking character of the economic process itself. Goods are produced to satisfy wants; the production of goods requires time, and two elements of uncertainty are introduced, corresponding to two different kinds of foresight which must be exercised. First, the end of productive operations must be estimated from the beginning. It is notoriously impossible to tell accurately when entering upon productive activity what will be its results in physical terms, what (a) quantities and (b) qualities of goods will result from the expenditure of given resources. Second, the wants which the goods are to satisfy are also, of course, in the future to the same extent, and their prediction involves uncertainty in the same way. The producer, then, must *estimate* (1) the future demand which he is striving to satisfy and (2) the future results of his operations in attempting to satisfy that demand.

It goes without saying that rational conduct strives to reduce to a minimum the uncertainties involved in adapting means to ends. This does not mean, be it emphasised, that uncertainty as such is abhorrent to the human species, which probably is not true. We should not really prefer to live in a world where everything was "cut and dried", which is merely to say that we should not want our activity to be all perfectly rational. But in attempting to act "intelligently" we *are* attempting to secure adaptation, which means foresight, as perfect as possible. There is, as already noted, an element of paradox in conduct which is not to be ignored. We find ourselves compelled to strive after things, which in a "calm, cool hour" we admit we do not want, at least not in fullness and perfection. Perhaps it is the manifest impossibility of reaching the end, which makes it interesting to strive after it. In any case we *do* strive to reduce uncertainty, even though we should not want it eliminated from our lives.

The possibility of reducing uncertainty depends again on two fundamental sets of conditions: first, uncertainties are less in groups of cases than in single instances. In the case of *a priori* probability the uncertainty tends to disappear altogether, as the group increases in inclusiveness; with statistical probabilities the same tendency is manifest in a lesser degree, being limited by defectiveness of classification. And even the third type, true uncertainties, show some tendency toward regularity when grouped on the basis of nearly any similarity or common element. The second fact or set of facts making for a reduction of uncertainty is the differences among human individuals in regard to it. These differences are of many kinds and an enumeration of them will be undertaken presently. We may note here that they may be differences in the men themselves or differences in their position in relation to the problem. We may call the two fundamental methods of dealing with uncertainty, based respectively upon reduction by grouping and upon selection of men to "bear" it, "consolidation"[2] and "specialisation", respectively. To these two methods we must add two others which are so obvious as hardly to call for discussion: (3) control of the future, and (4) increased power of prediction. These are closely inter-related, since the chief practical significance of knowledge is control, and both are closely identified with the general progress of civilisation, the improvement of technology and the increase of knowledge. Possibly a fifth method should be named, the "diffusion" of the consequences of untoward contingencies. Other things equal, it is a gain to have an event cause a loss of a thousand dollars each to a hundred persons rather than a hundred thousand to one person; it is better for two men to lose one eye than for one to lose two, and a system of production which wounds a larger number of workers and kills a smaller number is to be regarded as an improvement. In practice this diffusion is perhaps always associated with consolidation, but there is a logical distinction between the two and they may be practically separable in some cases. We must observe also that consolidation and specialisation are intimately connected, a fact which will call for repeated emphasis as we proceed. In addition to these methods of dealing with uncertainty there is (6) the possibility of directing industrial activity more or less along lines in which a minimal amount of uncertainty is involved and avoiding those involving a greater degree.

One of the most immediate and most important consequences of uncertainty in economics may be disposed of as a preliminary to a detailed technical discussion. The essence of organised economic activity is the production by certain persons of goods which will be used to satisfy the wants of other persons. The first question which arises then is, which of these groups in any particular case, producers or consumers, shall do the foreseeing as to the future wants to be satisfied. It is perhaps obvious that the function of prediction in the technological side of production itself inevitably devolves upon the producer. At first sight it would appear that the consumer should be in a better position to anticipate his own wants than the producer to anticipate them for him, but we notice at once that this is not what takes place. The primary phase of economic organisation is the production of goods for a general market, not upon direct order of the consumer. With uncertainty absent it would be immaterial whether the exchange of goods preceded or followed actual production. With uncertainty (in the two fields, production and wants) present it is still conceivable that men might exchange productive services instead of products, but the fact of uncertainty operates to bring about a different result. To begin with, modern society is organised on the theory (whatever the facts, about which some doubt may be expressed) that men predict the future and adapt their conduct to it more effectively when the results accrue to themselves than when they accrue to others. The responsibilities of controlling production thus devolve upon the producer.

But the consumer does not even contract for his goods in advance, generally speaking. A part of the reason might be the consumer's uncertainty as to his ability to pay at the end of the period, but this does not seem to be important in fact. The main reason is that he does not know what he will want, and how much, and how badly; consequently he leaves it to producers to create goods and hold them ready for his decision when the time comes. The clue to the apparent paradox is, of course, in the "law of large numbers", the consolidation of risks (or uncertainties). The consumer is, to himself, only one; to the producer he is a mere multitude in which individuality is lost. It turns out that an outsider can foresee the wants of a multitude with more ease and accuracy than an individual can attain with respect to his own. This phenomenon gives us the most fundamental feature of the economic system, *production for a market*, and hence also the general character of the environment in relation to which the effects of uncertainty are to be further investigated. Before continuing the inquiry into other phases and methods of the consolidation of risks, we shall turn briefly to consider the differences among individuals in their attitudes and reactions toward measurable or unmeasurable uncertainty.

We assume, as already observed, that although life is no doubt more interesting when conduct involves a certain amount of uncertainty – the proper amount varying with individuals and circumstances – yet that men do actually strive to anticipate the future accurately and adapt their conduct to it. In this respect we may distinguish at least five variable elements in individual attributes and capacities. (1) Men differ in their capacity by perception and inference to *form correct judgements* as to the future course of events in the environment. This capacity, furthermore, is far from homogeneous, some persons excelling in foresight in one kind of problem situations, others in other kinds, in almost endless variety.

Of especial importance is the variation in the power of reading human nature, of forecasting the conduct of other men, as contrasted with scientific judgement in regard to natural phenomena. (2) Another, though related, difference is found in men's capacities to *judge means* and discern and plan the steps and adjustments necessary to meet the anticipated future situation. (3) There is a similar variation in the power to *execute* the plans and adjustments believed to be requisite and desirable. (4) In addition there is diversity in conduct in situations involving uncertainty due to differences in the amount of *confidence* which individuals feel in their judgements when formed and in their powers of execution; this degree of confidence is in large measure independent of the "true value" of the judgements and powers themselves. (5) Distinct from confidence felt is the *conative attitude* to a situation upon which judgement is passed with a given degree of confidence. It is a familiar fact that some individuals want to be sure and will hardly "take chances" at all, while others like to work on original hypotheses and seem to prefer rather than to shun uncertainty. It is common to see people act on assumptions in ways which their own opinions of the value of the assumption do not warrant; there is a disposition to "trust in one's luck".

The amount of uncertainty effective in a conduct situation is the degree of subjective confidence felt in the contemplated act as a correct adaptation to the future – (4) above. It is clear that we may speak in some sense of the "true value" of judgement and of capacity to act, but it is the person's own opinion of these values which controls his activities. Hence the five variables are, from the standpoint of the person concerned, reduced to two, the (subjective or felt) uncertainty and his conative feeling toward it. For completeness we should perhaps add a sixth uncertainty factor, in the shape of occurrences so revolutionary and unexpected by any one as hardly to be brought under the category of an error in judgement at all.

In addition to the above enumeration of five or six distinct elements in the uncertainty situation we must point out that the first three variables named are themselves not simple. Judgement or foresight and the capacity for planning and the ability to execute action are each the product of at least four distinguishable factors, in regard to which the faculties in question may vary independently. These are (a) accuracy, (b) promptness or speed, (c) time range, and (d) space range, of the capacity or action. The first two of these require no explanation; it is evident that accuracy and rapidity of judgement and execution are more or less independent endowments. The third refers to the length of time in the future to which conduct is or may be adjusted, and the fourth to the scope or magnitude of the situation envisaged and the operations planned. Familiar also is the difference between individuals who have a mind for detail and those who confine their attention to the larger outlines of a situation. Even this rather complex outline is extremely simplified as compared with the facts of life in that it compasses only a rigidly "static" view of the problem. Quite as important as differences obtaining at any moment among individuals in regard to the attributes mentioned are their differences in capacity for change or development along the various lines. Knowledge is more a matter of learning than of the exercise of absolute judgement. Learning requires time, and in time the situation dealt with, as well as the learner, undergoes change.

We have classified the possible reactions to uncertainty under some half-dozen heads, each of which gives rise to special problems, though the social structures for dealing with these problems overlap a good deal. The most fundamental facts regarding uncertainty from our point of view are; first, the possibility of reducing it in amount by grouping instances; and, second, the differences in individuals in relation to uncertainty, giving rise to a tendency to specialise the function of meeting it in the hands of certain individuals and classes. The most fundamental effect of uncertainty on the social-economic organisation – production for a general market on the producer's responsibility – has already been taken up; it is primarily a case of reduction of uncertainty by consolidation or grouping of cases. In the mere fact of production for a market, there is little specialisation of uncertainty-bearing, and what there is is on a basis of the producer's position in relation to the problem, not his peculiar characteristics as a man. To isolate the phenomenon of production for a market from other considerations we must picture a pure "handicraft stage" of social organisation. In such a system every individual would be an independent producer of some one finished commodity, and a consumer of a great variety of products. The late Middle Ages afford a picture of an approximation to such a state of affairs in a part of the industrial field.

The approximation is rather remote, however. A handicraft organisation shows an irresistible tendency to pass over, even before well established, into a very different system, and thus further development is also a consequence of the presence of uncertainty. The second system is that of "free enterprise" which we find dominant today. The difference between free enterprise and mere production for a market represents the addition of specialisation of uncertainty-bearing to the grouping of uncertainties, and takes place under pressure of the same problem, the anticipation of wants and control of production with reference to the future. Under free enterprise the solution of this problem, already removed from the consumer himself, is further taken out of the hands of the great mass of producers as well and placed in charge of a limited class of "entrepreneurs" or "business men". The bulk of the producing population cease to exercise responsible control over production and take up the subsidiary role of furnishing productive resources (labour, land,

and capital) to the entrepreneur, placing them under his sole direction for a fixed contract price.

We shall take up this phenomenon of free enterprise for detailed discussion in the next chapter, though we may note here two further facts regarding it; first, the "specialisation" of uncertainty-bearing in the hands of entrepreneurs involves also a further consolidation; and, second, it is closely connected with changes in technological methods which (a) increase the time length of the production process and correspondingly increase the uncertainty involved, and (b) form producers into large groups working together in a single establishment or productive enterprise and hence necessitates concentration of control. The remainder of the present chapter will be devoted to a survey of the social structures evolved for dealing with uncertainty. Some of the phenomena will thus be finally disposed of, so far as the present work is concerned, especially those which already have a literature of their own and whose general bearings and place in a systematic treatment of uncertainty alone call for notice here. Other problems will be merely sketched in outline and reserved for fuller treatment in subsequent chapters, as has just been done with the subject of entrepreneurship.

Following the order of the classification already given of methods of dealing with uncertainty, the first subject for discussion is the institutions or special phenomena arising from the tendency to deal with uncertainty by *consolidation*. The most obvious and best known of these devices is, of course, *insurance*, which has already been repeatedly used as an illustration of the principle of eliminating uncertainty by dealing with groups of cases instead of individual cases. In our discussion of the theory of uncertainty in the foregoing chapter and at other points in the study we have emphasised the radical difference between a measurable and an unmeasurable uncertainty. Now measurability depends on the possibility of assimilating a given situation to a group of similars and finding the proportions of the members of the group which may be expected to exhibit the various possible outcomes. This assimilation of cases into classes may be exceedingly accurate, and the proportions of the various outcomes *may* be computable on *a priori* grounds by the application of the theory of permutations and combinations to determine the possible groupings of *equally probable* alternatives; but this rarely if ever happens in a practical business situation. The classification will be of all degrees of precision, but the ascertainment of proportions must be empirical. The application of the insurance principle, converting a larger contingent loss into a smaller fixed charge, depends upon the measurement of probability on the basis of a fairly accurate grouping into classes. It is in general not enough that the insurer who takes the "risk" of a large number of cases, be able to predict his aggregate losses with sufficient accuracy to quote premiums which will keep his business solvent while at the same time imposing a burden on the insurer, which is not too large a fraction of his contingent loss. In addition he must be able to present a fairly plausible contention that the particular insured is contributing to the total fund out of which losses are paid as they accrue in an amount corresponding reasonably well with his real probability of loss; ie, that he is bearing his fair share of the burden.

The difficulty of a satisfactory logical discussion of the questions we are dealing with has repeatedly been emphasised, due to the fact that distinctions of the greatest importance tend to run together through intermediate degrees and become blurred. This is conspicuously the case with the measurability of uncertainty through classification of instances. We hardly find in practice really homogeneous classifications (in the sense in which mathematical probability implies, as in the case of successive throws of a perfect die) and at the other extreme it is hard to find cases which do not admit of some possibility of assimilation into groups and hence of measurement. Indeed, the very concept of contingency seems to preclude absolute uniqueness (as for that matter there is doubtless nothing absolutely unique in the universe). For to say that a certain event is contingent or "possible" or "may happen" appears to be equivalent to saying that "*such*" things" have been known to happen before, and the "such things" manifestly constitute a class of cases formed on some ground or other. The principal subject for investigation is thus the *degree* of assimilability, or the amount of homogeneity of classes securable, or, stated inversely, the *degree of uniqueness* of various kinds of business contingencies. Insurance deals with those which are "fairly" classifiable or show a relatively low degree of uniqueness, but the different branches of insurance show a wide range of variation in the accuracy of measurement of probability which they secure.

Before taking up various types of insurance we may note in passing a point which it is superfluous to elaborate in this connection, namely, that different forms of organisation in the insurance field all operate on the same principle. It matters not at all whether the persons liable to a given contingency organise among themselves into a fraternal or mutual society or whether they separately contract with an outside party to bear their losses as they fall in. Under competitive conditions and assuming that the probabilities involved are accurately known, an outside insurer will make no clear profit and the premiums will under either system be equal to the administrative costs of carrying on the business.

The branch of insurance which is most highly developed, meaning that its contingencies are most accurately measured because its classifications are most perfect, and which is thus on the most nearly "mathematical" basis is, of course, what is called "life insurance". (In so far as it is "insurance" at all, and not a mere investment proposition, it is clear that it is insurance against "premature" loss of earning power, and not against death). It is possible, on the basis of medical examinations, and taking into account age, sex, place of residence, occupation, and habits of life, to select "risks" which closely approximate the ideal of mechanical probability. The chance of death of two healthy individuals similarly circumstanced in the above regards seems to be about as near an objective equality, the life or death of one rather than the other about as nearly really indeterminate, as anything in nature. To be sure, when we pass outside the relatively narrow circle of "normal" individuals, difficulties are encountered, but the extension of life insurance outside this circle has also been restricted. Some development has taken place in the insurance of sub-standard lives at higher rates, but it is limited in amount and could be characterised as exceptional.[3]

The very opposite situation from life insurance is found in insurance against sickness and accident. Here an objective description and classification of cases is impossible, the business is fraught with great difficulties and susceptible of only a limited development. It is notorious that such policies cost vastly more than they should; indeed, the companies find it profitable to adopt a generous attitude in the adjustment of claims, raising the premium rates accordingly, it is needless to say. Accident compensation for workingmen, under social control, is on a somewhat better footing, but only on condition that the payments are restricted to not too large a fraction of the actual economic loss to the individual, with nothing for discomfort, pain, or inconvenience. In the whole field of personal, physical contingencies, however, there is nothing that is strictly of the nature of a "business risk", unless it is the now happily obsolescent phenomenon of commercial employers' liability insurance.

The typical application of insurance to business hazards is in the protection against loss by fire, and the theory of fire insurance rates forms an interesting contrast with the actuarial mathematics of life assurance. The latter, as we have observed, is a fairly close approximation to objective probability; it is in fact so close to this ideal that life insurance problems are worked by the formulae derived from the binomial law, in the same way as problems in mechanical probability. Fire insurance rating is a very different proposition; only in rather recent years has any approach been made to the formation of fairly homogeneous classes of risks and the measurement of real probability in a particular case. At best there is a large field for the exercise of "judgement" even after literally thousands of classes of risks have been more or less accurately defined.[4] More important is the fact that, in consequence, insurance does not take care of the whole risk against loss by fire. On account of the "moral hazard" and practical difficulties, it is necessary to restrict the amount of insurance to the "direct loss or damage" or even to a part of that, while of course there are usually large indirect losses due to the interruption of business and dislocation of business plans which are entirely unprovided for.

Thus there is a large margin of uncertainty both to insurer and insured, in consequence of the impossibility of objectively homogeneous groupings and accurate measurement of the chance of loss. Corresponding to this margin of uncertainty in the calculations there is a chance for a profit or loss to either party, in connection with the fire hazard. The probabilities in the case of fire are, of course, complicated by the fact that risks are not entirely independent. A fire once started is likely to spread and there is a tendency for losses to occur in groups. In so far, however, as fire losses in the aggregate are calculable in advance, they are or may be converted into fixed costs by every individual exposed to the possibility of loss, and in so far no profit, positive or negative, will be realised by any one on account of this uncertainty in his business.

The principle of insurance has also been utilised to provide against a great variety of business hazards other than fire – the loss of ships and cargoes at sea, destruction of crops by storms, theft and burglary, embezzlement by employees (indirectly through bonding, the employee doing the insuring), payment of damages to injured employees, excessive losses through credit extension, etc. The unusual forms of policies issued by some of the Lloyd's underwriters have attained a certain amount of publicity as popular curiosities. These various types of contingencies offer widely divergent possibilities for "scientific" rate-making, from something like the statistical certainty of life insurance at one extreme to almost pure guesswork at the other, as when Lloyd's insures the business interests concerned that a royal coronation will take place as scheduled, or guarantees the weather in some place having no records to base calculations upon. Even in these extreme cases, however, there is a certain vague grouping of cases on the basis of intuition or judgement; only in this way can we imagine any estimate of a probability being arrived at.

It is therefore seen that the insurance principle can be applied even in the almost complete absence of scientific data for the computation of rates. If the estimates are conservative and competent, it turns out that the premiums received for insuring the most unique contingencies cover the losses; that there is an offsetting of losses and gains from one venture to another, even when there is no discoverable kinship among the ventures themselves. The point seems to be, as already noticed, that the mere fact that judgement is being exercised in regard to the situations forms a fairly valid basis for assimilating them into groups. Various instances of the exercise of (fairly competent) judgement even in regard to the most heterogeneous problems, show a tendency to approach a constancy and predictability of result when aggregated into groups.

The fact which limits the application of the insurance principle to business risks generally is not therefore their inherent uniqueness alone, and the subject calls for further examination. This task will be undertaken in detail in the next chapter, which deals with entrepreneurship. At this point we may anticipate to the extent of making two observations: first, the typical uninsurable (because unmeasurable and this because unclassifiable) business risk relates to the exercise of judgement in the making of decisions by the business man; second, although such estimates do tend to fall into groups within which fluctuations cancel out and hence to approach constancy and measurability, this happens only *after the fact* and, especially in view of the brevity of a man's active life, can only to a limited extent be made the basis of prediction. Furthermore, the classification or grouping can only to a limited extent be carried out by any agency outside the person himself who makes the decisions, because of the peculiarly obstinate connection of a *moral hazard* with this sort of risk. The decisive factors in the case are so largely on the inside of the person making the decisions that the "instances" are not amenable to objective description and external control.

Manifestly these difficulties, insuperable when the "consolidation" is to be carried out by an external agency such as an insurance company or association, fall away in so far as consolidation can be effected within the scale of operations of a single individual; and the same will be true of an organisation if responsibility can be adequately centralised and unity of interest secured. The possibility of thus reducing uncertainty by transforming it into a measurable risk through grouping constitutes a strong incentive to extend the scale of operations of a business establishment. This fact must constitute one of the important causes of the phenomenal growth in the average size of industrial establishments which is a familiar characteristic of modern economic life. In so far as a single business man, by borrowing capital or otherwise, can extend the scope of his exercise of judgement over a greater number of decisions or estimates, there is a greater probability that bad guesses will be offset by good ones and that a degree of constancy and dependability in the total results will be achieved. In so far uncertainty is eliminated and the desideratum of rational activity realised.

Not less important is the incentive to substitute more effective and intimate forms of association for insurance, so as to eliminate or reduce the moral hazard and make possible the application of the insurance principle of consolidation to groups of ventures too broad in scope to be "swung" by a single enterpriser. Since it is capital which is especially at risk in operations based on opinions and estimates, the form of organisation centres around the provisions relating to capital. It is undoubtedly true that the reduction of risk to borrowed capital is the principal desideratum leading to the displacement of individual enterprise by the partnership and the same fact with reference to both owned and borrowed capital explains the substitution of corporate organisation for the partnership. The superiority of the higher form of organisation over the lower from this point of view consists both in the extension of the scope of operations to include a larger number of individual decisions, ventures, or "instances", and in the more effective unification of interest which reduces the moral hazard connected with the assumption by one person of the consequences of another person's decisions.

The close connection between these two considerations is manifest. It is the special "risk" to which large amounts of capital loaned to a single enterpriser are subject which limits the scope of operations of this form of business unit by making it impossible to secure the necessary property resources. On the other hand, it is the inefficiency of organisation, the failure to secure effective unity of interest, and the consequent large risk due to moral hazard when a partnership grows to considerable size, which in turn limit its extension to still larger magnitudes and bring about the substitution of the corporate form of organisation. With the growth of large fortunes it becomes possible for a limited number of persons to carry on enterprises of greater and greater magnitude, and today we find many very large businesses organised as partnerships. Modifications of partnership law giving this form more of the flexibility of the corporation with reference to the distribution of rights of control, of participation in income, and of title to assets in case of dissolution have also contributed to this change.

With reference to the first of our two points above mentioned, the extension of the scope of operations, the corporation may be said to have solved the organisation problem. There appears to be hardly any limit to the magnitude of enterprise which it is possible to organise in this form, so far as mere ability to get the public to buy the securities is concerned. On the second score, however, the effective unification of interests, though the corporation has accomplished much in comparison with other forms of organisation, there is still much to be desired. Doubtless the task is impossible, in any absolute sense; nothing but a revolutionary transformation in human nature itself can apparently solve this problem finally, and such a change would, of course, obliterate all moral hazards at once, without organisation. In the meanwhile the internal problems of the corporation, the protection of its various types of members and adherents against each other's predatory propensities, are quite as vital as the external problem of safeguarding the public interests against exploitation by the corporation as a unit.[5]

Another important aspect of the relations of corporate organisation to risk involves what we have called "diffusion" as well as consolidation. The minute divisibility of ownership and ease of transfer of shares enables an investor to distribute his holdings over a large number of enterprises in addition to increasing the size of a single enterprise. The effect of this distribution on risk is evidently twofold. In the first place, there is to the investor a further offsetting through consolidation; the losses and gains in different corporations in which he owns stock must tend to cancel out in large measure and provide a higher degree of regularity and predictability in his total returns. And again, the chance of loss of a small fraction of his total resources is of less moment even proportionally than a chance of losing a larger part.

There are other aspects of the question which must be passed over in this summary view. Doubtless a significant fact is the greater publicity attendant upon the organisation, resources, and operations of a corporation, due to its being a creature of the State and to legal safeguards. It must be emphasised that this type of organisation actually reduces risks, and does not merely transfer them from one party to another, as might seem at first glance to be the case. Superficial discussions of limited liability tend to give the impression, or at least leave the way open to the conclusion, that this is the main advantage over the partnership. But it must be evident that the mere fact of limited liability only serves to transfer losses in excess of invested resources from the owners of the concern to its creditors; and if this were the only effect of incorporation, the loss in credit standing should offset the gain in security to the owners. The vital facts are the twofold consolidation of risks, together with greater publicity, and diffusion in a minor role, not really separable from the fact of consolidation.

It is particularly noteworthy that large-scale organisation has shown a tendency to grow in fields where division of labour is absent and consolidation or grouping of uncertainties is the principal incentive. Occupations in which the work is of an occasional and intermittent character tend to run into partnerships and even corporations where there is no capital investment, or relatively little, and the members work independently at identical tasks. Examples are the syndicating of detectives, stenographers, and even lawyers and doctors.

The second of the two main principles for dealing with uncertainty is specialisation. The most important instrument in modern economic society for the specialisation of uncertainty, after the institution of free enterprise itself, is *Speculation*. This phenomenon also combines different principles, and the mere specialisation of uncertainty-bearing in the hands of persons most willing to assume the function is probably among the lesser rather than the greater sources of gain. It seems best to postpone for the present a detailed theoretical analysis of the factors of specialisation of uncertainty-bearing in the light of the many ways in which individuals differ in their relations to uncertainty; this discussion will be taken up in the next chapter, in connection with the treatment of enterprise and entrepreneurship. At this point we wish merely to emphasise the association in several ways between specialisation and actual reduction of uncertainty.

Most fundamental among these effects in reducing uncertainty is its conversion into a measured risk or elimination by grouping which is implied in the very fact of specialisation. The typical illustration to show the advantage of organised speculation to business at large is the use of the hedging contract. By this simple device the industrial producer is enabled to eliminate the chance of loss or gain due to changes in the value of materials used in his operations during the interval between the time he purchases them as raw materials and the time he disposes of them as finished product, "shifting" this risk to the professional speculator. It is manifest at once that even aside from any superior judgement or foresight or better information possessed by such a professional speculator, he gains an enormous advantage from the sheer magnitude or breadth of the scope of his operations. Where a single flour miller or cotton spinner would be in the market once, the speculator enters it hundreds or thousands of times, and his errors in judgement must show a correspondingly stronger tendency to cancel out and leave him a constant and predictable return on his operations.

The same reasoning holds good for any method of specialising uncertainty-bearing. Specialisation implies concentration, and concentration involves consolidation; and no matter how heterogeneous the "cases" the gains and losses neutralise each other in the aggregate to an extent increasing as the number of cases thrown together is larger. Specialisation itself is primarily an application of the insurance principle; but, like large-scale enterprise, it grows up to meet uncertainty situations where, on account of the impossibility of objective definition and external control of the individual ventures or uncertainties, a "moral hazard" prevents insurance by an external agency or a loose association of ventures for this single purpose.

Besides organised speculation as carried on in connection with produce and security exchanges, the principle of specialisation is exemplified in the tendency for the highly uncertain or speculative aspects of industry to become separated from the stable and predictable aspects and be taken over by different establishments. This is, of course, what has really taken place in the ordinary form of speculation already noticed, namely, the separation of the *marketing* function from the technological side of production, the former being much more speculative than the latter. A separation perhaps equally significant in modern economic life is that which so commonly takes place between the *establishment* or *founding* of new enterprises and their *operation* after they are set going. To be sure, by no means all the business of *promotion* comes under this head, but still the tendency is manifest. A part of the investors' commitment in promoted concerns look to the future earnings from regular operations for their return, but a large part expect to sell out at a profit after the business is established, and to devote their capital to some new venture of the same sort. A considerable and increasing number of individual promoters and corporations give their exclusive attention to the launching of new enterprises, withdrawing entirely as soon as the prospects of the business become fairly determinate. The gain from arrangements of this sort arises largely from the consolidation of uncertainties, their conversion by grouping into measured risks, which are for the group of cases not uncertainties at all. Such a promoter takes it as a matter of course that a certain proportion of his ventures will be failures and involve heavy losses, while a larger proportion will be relatively unprofitable, and counts on making his gains from the occasional conspicuous successes. That is – to face frankly that paradoxical element which is really involved in such calculations – he does not "expect" to have his "expectations" verified by the results in every case; the expectations on which he really counts are based on an average, on an "estimate" of the long-run value of his "estimates". The specialisation in the speculative phase of the business enables a single man or firm to deal with a larger number of ventures, and is clearly a mode of applying the same principle which underlies ordinary insurance.

Other illustrations of the same phenomenon will come to the reader's mind. Industries which utilise land whose value is largely speculative are more likely to rent rather than own their sites where the nature of the utilisation makes such a procedure practicable. Even expensive machines and articles of equipment of other sorts, ownership of which involves heavy risks to a small concern, may be rented instead of bought outright. The owner of leased land or equipment is presumably a specialist in that sort of business and his risks are reduced by the grouping of a larger number of ventures.

Other advantages of specialisation of speculative functions in addition to the reduction of uncertainty through consolidation are manifest, and no intention of belittling or concealing them is implied in the separation of the latter aspect of the case in the foregoing discussion. It is apparent in particular that the specialist in any line of risk-taking naturally knows more about the problem with which he deals than would a venturer who dealt with them only occasionally. Hence, since most of these uncertainties relate chiefly to the exercise of judgement, the uncertainty itself is reduced by this fact also. There is in this respect a fundamental difference between the speculator or promoter and the insurer, which must be kept clearly in view. The insurer knows more about the risk in a particular case – say of a building burning – but the *real risk* is no less because he assumes it in that particular case. His risk is less only because he assumes a large number. But the transfer of the "risk" of an error in judgement is a very different matter. The "insurer" (entrepreneur, speculator, or promoter) now substitutes his own judgement for the judgement of the man who is getting rid of the uncertainty through transferring it to the specialist. In so far as his knowledge and judgement are better, which they almost certainly will be from the mere fact that he is a specialist, the individual risk is less likely to become a loss, in addition to the gain from grouping. There is better management, greater economy in the use of economic resources, as well as a mere transformation of uncertainty into certainty.

The problem of meeting uncertainty thus passes inevitably into the general problem of management, of economic control. The fundamental uncertainties of economic life are the errors in predicting the future and in making present adjustments to fit future conditions. In so far as ignorance of the future is due to practical indeterminateness in nature itself we can only appeal to the law of large numbers to distribute the losses, and make them calculable, not to reduce them in amount, and this is only possible in so far as the contingencies to be dealt with admit of assimilation into homogeneous groups; ie, in so far as they repeat themselves. When our ignorance of the future is only partial ignorance, incomplete knowledge and imperfect inference, it becomes impossible to classify instances objectively, and any changes brought about in the conditions surrounding the formation of an opinion are nearly sure to affect the intrinsic value of the opinion itself. This is true even of the method of grouping by extending the scale of operations of a single entrepreneur, for the quality of his estimates will not be independent of the number he has to make and the mass of the data involved. But it is especially true of grouping by specialisation, as we have seen. The inseparability of the uncertainty problem and the managerial problem will be especially important in the discussion (in the next chapter) of entrepreneurship, which is the characteristic phenomenon of modern economic organisation and is essentially a device for specialising uncertainty-bearing or the improvement of economic control. The relation between management, which consists of making decisions, and taking the consequences of decisions, which is the most fundamental form of risk-taking in industry, will be found to be a very intricate as well as intimate one. When the sequence of control is followed through to the end, it will be found that from the standpoint of the ultimately responsible manager, the two functions are always inseparable.

We are thus brought naturally around to a discussion of the most thoroughgoing methods of dealing with uncertainty; ie, by securing better knowledge of and control over the future. As previously observed, however, these methods represent merely the objective of all rational conduct from the outset, and they call for discussion in such a work as the present only in so far as they affect the general outline of the social economic structure. Thus it is fundamental to the entrepreneur system that it tends to promote better management in addition to consolidating risks and throwing them into the hands of those most disposed to assume them. The only further comment here called for is to point out the existence of highly specialised industrial structures performing the functions of furnishing knowledge and guidance.

One of the principal gains through organised speculation is the provision of information on business conditions, making possible more intelligent forecasting of market changes. Not merely do the market associations or exchanges and their members engage in this work on their own account. Its importance to society at large is so well recognised that vast sums of public money are annually expended in securing and disseminating information as to the output of various industries, crop conditions, and the like. Great investments of capital and elaborate organisations are also devoted to the work as a private enterprise, on a profit-seeking basis, and the importance of trade journals and statistical bureaux and services tends to increase, as does that of the activities of the Government in this field. The collection, digestion, and dissemination in usable form of economic information is one of the staggering problems connected with our modern large-scale social organisation. It goes without saying that no very satisfactory solution of this problem has been achieved, and it is safe to predict that none will be found in the near future. But all these specialised agencies for the supply of information help to bridge the wide gap between what the individual business manager knows or can find out by the use of his own resources and what he would have to know to conduct his business in a perfectly intelligent fashion. Their output increases the value of the intuitive "judgements" on the basis of which his decisions are finally made after all, and greatly extends the scope of the environment in relation to which he can more or less intelligently react.

The foregoing relates chiefly to the production side of the problem of economic information. In the field of information for consumers, we have the still more staggering development of advertising. This complex phenomenon cannot be discussed in detail here, beyond pointing out its connection with the fact of ignorance and the necessity of knowledge to guide conduct. Only a part of advertising is in any proper sense of the term informative. A larger part is devoted to persuasion, which is a different thing from conviction, and perhaps the stimulation or creation of new wants is a function distinguishable from either. In addition to advertising, most of the social outlay for education is connected with informing the population about the means of satisfying wants, the education of taste. The outstanding fact is that the ubiquitous presence of uncertainty permeating every relation of life has brought it about that information is one of the principal commodities that the economic organisation is engaged in supplying. From this point of view it is not material whether the "information" is false or true, or whether it is merely hypnotic suggestion. As in all other spheres of competitive economic activity, the consumer is the final judge. If people are willing to pay for "Sunny Jim" poetry and "It Floats" when they buy cereals and soap, then these wares are economic goods. If a certain name on a fountain pen or safety razor enables it to sell at a 50% higher price than the same article would otherwise fetch, then the name represents one third of the economic utility in the article, and is economically no different from its colour or design or the quality of the point or cutting edge, or any other quality which makes it useful or appealing. The morally fastidious (and naïve) may protest that there is a distinction between "real" and "nominal" utilities; but they will find it very dangerous to their optimism to attempt to follow the distinction very far. On scrutiny it will be found that most of the things we spend our incomes for and agonise over, and notably practically all the higher "spiritual" values, gravitate swiftly into the second class.

Somewhat different from the production and sale of information is the dealing in actual instructions for the guidance of conduct directly. Modern society is characterised by the rapid growth of this line of industry also. There have always been a few professions whose activities consisted essentially of the sale of guidance, notably medicine and the law, and more or less the preaching and teaching professions. Recent years, however, have witnessed a veritable swarming of experts and consultants in nearly every department of industrial life. The difference from dealing in information is that these people do not stop at diagnosis; in addition they prescribe. They are equally conspicuous in the fields of business organisation, accounting, the treatment of labour, the lay-out of plants, and the processing of materials; they are the scientific managers of the managers of business; and though they by no means serve business or its managers for naught, and in spite of a large amount of quackery, they probably pay their way and more on the whole in increasing the efficiency of production. Certainly they do a useful work in forcing the intelligent, critical consideration of business problems instead of a blind following of tradition or the use of guesswork methods.[6]

The last of the alternatives named for meeting uncertainty relates to the problem of a tendency to prefer relatively predictable lines of activity to more speculative operations. It is common to assume[7] that society pays for the assumption of risk in the form of higher prices for commodities whose production involves uncertainty and a deficient supply of these in comparison with goods of an opposite character. This subject will come up again in connection with the closely related question of a tendency of profit to zero, and it seems best to postpone discussion of it for the present.[8] We shall find reasons for being very sceptical as to the reality of any such abhorrence of uncertainty as to decrease productivity in any line below the level that an equivalent fixed cost would bring about.

1 The chief limitation in fact relates less to the proposition as stated than to the dogma of "conduct" or activity exclusively in order to a future reward. Means and end seem to be the form in which we think about our behaviour rather than the actual form of the behaviour

itself. The literature of ethics is one long record of failure to find any absolute end; in life every end becomes a means to some new and farther goal. The attempt to rationalise human behaviour seems to be a perpetual chase after one's own shadow, and the conclusion forces itself upon us that the *summum bonum* or any other objective *bonum* is an *ignis fatuus*. We are compelled to believe that in a great proportion of cases we take more interest in action whose fruition is only probable than we would if it were certain.

2 Professor Irving Fisher's term (*The Nature of Capital and Income*, p. 288). I should prefer simply "grouping" as both shorter and more descriptive.

3 It would be out of place here to go into the social aspects of life insurance, but one observation may be worth making. From the social point of view it is arguable that all classification of risks is a bad thing, except in so far as the special hazard is purely occupational and the cost of carrying it can be transferred to the consumer of the product. It is hard to discover any good reason why the unfortunate should be especially burdened because of their handicaps. It would, therefore, be better if all were insured at a uniform rate. Indeed, we may go father and contend that the rate should be graduated inversely with the risk (occupational risks excepted, as noted). It goes without saying that only a state compulsory insurance scheme could operate on any such principles; under private profit incentives, competition will compel any insurance agency to classify its risks as accurately and minutely as practicable.

4 Cf. Huebner, *Property Insurance*, chapters XVI and XVII.

5 Haney (*Business Organization and Combination*, chapter XXIII) uses the terms "The Corporation Problem" and "The Trust Problem" to designate what I have called the "internal" and "external" problems respectively. He properly emphasises the importance of the former in view of the tendency of the evils of monopoly, etc, to overshadow it in the popular mind and in much of the literature of the subject.

6 On the production and sale of "guidance" see J. M. Clark, *Journal of Political Economy*, 26 (1 and 2).

7 Cf. Willett, *Economic Theory of Risk and Insurance*, chapter III.

8 Cf. chapter XII.

2
*Constructive Speculation**
Alfred Marshall[†]

A good deal has already been said incidentally as to the ever increasing complexity of marketing, and as to the aids which its operations derive from modern facilities for transport and for the transmission of intelligence. The purpose of this chapter is to consider the influences, which these changes are exerting on the costs and methods of marketing in various classes of industries, and on the structure of business on them.

It will appear that the influences, exerted by developments of the means of communication, operate much as do those exerted by changes in the technique of production: on the whole they strengthen the strong producer and dealer relatively to his weaker competitor; but in some directions their aid is of greater service to the weak than to the strong.

Good marketing has always provided or helped in providing: (a) a supply, both steady and elastic, of everything for which there is a considerable demand (measured in terms of purchasing power) in any place; and (b) a fairly steady supply of employment for the labour, skill, plant and managing faculty required for turning them to effective account. Most of these tasks have been comparatively simple where people have been satisfied with a few products, mostly of local origin; and where, as has often happened, the handicraftsman has been also an agriculturist, and divided his time between his two occupations so as to meet fairly well the more pressing demands of nature and man: but many of them have gradually become very complex.

The advance of knowledge and wealth has brought so great and ever-changing a variety of goods into ordinary consumption, that if the resources, on which the local dealer could draw, had remained without

*This paper was first published in *Industry and Trade*, 1920, Third edition (London: Macmillan).
[†]Alfred Marshall, 1842–1924.

great change, the stocks needed to meet the impatient demand of the modern western consumer would have been larger in proportion to population than the whole movable wealth of, say, the sixteenth century. And, on the same supposition, the employment found in any place for the highly specialised skill and plant of modern industry would have been so fitful and irregular, that modern knowledge and wealth would scarcely have sufficed to make modern industry commercially possible. But the same causes, that have brought new difficulties, have brought also new methods of overcoming them.

Nearly all western markets are now united by so many various connections, that a need for any common product almost anywhere can be filled in a couple of days, if not in a few hours, from a large reservoir; which can be replenished quickly from still larger reservoirs near or far. In return for these services, which central reservoirs render to local consumption, linked-up local demand provides a fairly steady market in time of peace for nearly all commodities, however highly specialised, the consumption of which is not greatly varied by such widespread influences as changes in the season or in fashion.[1]

Thus technical improvements in transport and marketing are ever overcoming old difficulties; but they are also ever stimulating further developments, which open out new difficulties. One of the most prominent of these, the increase in the distances over which food and other goods travel before arriving at their final resting place, has perhaps nearly reached its full force already. For, as backward countries and districts gain on those which had an earlier start in the industrial race, the average distance travelled by ordinary manufactures will diminish. Long distance transport will be increasingly concerned with fine industrial specialities; and with such crude agricultural and mineral products, as nature yields in greater relative abundance, or on easier terms, in some parts of the world than in others.

MODERN ORGANISATION TENDS SO TO DISTRIBUTE THE RISKS INHERENT IN MAKING AND MARKETING THAT THEY FALL INCREASINGLY ON THE SHOULDERS BEST FITTED TO BEAR THEM

The intensest risks of early times and of some backward countries even now, have been "consumers' risks" connected with harvest failures. But cheap and speedy transport by land and sea, aided by the telegraph and the specialised activities of dealers in harvest products, have very greatly diminished such risks in the Western World: and there is some tendency to think of "speculation" as a thing, with which no one need be concerned who does not set out to seek it. But in fact the great risks of business have much in common with the many small risks, which must be faced by every responsible citizen.

For when a traveller is in doubt which of two roads to take, he speculates. He must ultimately take one risk or the other: he has no choice but to speculate as to their relative advantages and act according to his speculative judgement: and, in the more weighty affairs of ordinary life, everyone is frequently at a crossing of roads. If a man is in a hired house which he much likes, but is not sure how long he will stay in the neighbourhood, he takes a risk, whether he accepts or refuses a long lease of it: and similarly when a man decides for what occupation to prepare his son, and to what school to send him, he must take some of many risks. But the *looking into the future*, which such risks involve, is seldom recognised as speculation: and even business transactions which follow the ordinary routine are not commonly regarded as "speculative": that term being almost confined to dealings in things the future prices of which are eminently uncertain.

The speculative taking of high risks has many varieties. Some are in effect mere reckless gambling. Others are shrewd business ventures, aimed at gains that must be balanced by losses to traders who are concerned in the same affairs. Others tend to improve the general application of efforts to the attainment of desirable ends: these last alone are entitled to be called "constructive" in the full sense of the term.

Aristotle's doctrine, that neither party to trade can gain except at the expense of the other, is true only of that particular form of trade which is classed as gambling, a class to which many varieties of trade speculation belong. But genuine trade commonly benefits both parties to it: because, though each receives only what the other gives up, what he receives is more desired by him than that which he gives up. In gambling, when conducted fairly and on equal terms, every transaction is an exchange of equal risks.[2]

When a man, having superior knowledge as to horses, lays a wager about them on advantageous terms to himself, he effects an immediate increase of his property; but without advantage to the world. On the other hand, when a man has superior knowledge that the supply of anything is likely to run short in any particular country or in the world generally; and buys it either outright or for future delivery; then, on the assumption that his judgement is right, his action is to be regarded as constructive speculation. Such work adds to the world's wealth, just as diverting a stream to work a watermill does, for it tends to increase the supply of things where and when they are likely to be most wanted, and to check the supply of things where and when they are likely to be in less urgent demand. This is its most conspicuous service.

But it also renders another service; which, though less conspicuous, is not much less important: for it often enables a man whose whole energies are needed for the internal work of his business, to insure himself against the risk that the materials which he will need in his business will not need to be purchased at an enhanced price. The risk is governed by broad causes over which he has scarcely any control, and the study of which requires knowledge and faculties other than his own.

As a rule the manufacturer who has not contracted to deliver, but works for the general market, desires to be insured against a fall and not against a rise in the price of his material: he stands to lose by such a fall; since purchasers, unless specially pressed for time, will decline to buy his finished product at a price much above that which corresponds to the current price of the material. There is no simple means of insuring against this risk, which corresponds to a contract for delivery of material at a fixed price; but the two sets of risks are in opposite directions, and it is obvious that much economy might be effected by setting these to neutralise one another. In spite of the abuses connected with them, organised markets for dealing in standardised produce, render many services to business men and to the world at large; and perhaps the chief, though not the most prominent of these is their indirect effect in so concentrating risks, that those of them which, like those just considered, are in opposite directions, will tend to extinguish each other.

But, before entering on this matter, it is worthwhile to note that insurance against a risk may be so important a benefit to anyone whose capital is not large relatively to the risk, that it may be worth his while to pay far more than the actuarial value of it. For instance, suppose that the chance that a building of a certain class worth £10,000 will be destroyed by fire in the course of a year, is one in ten thousand: then the actuarial value of that risk will be £1. And yet, if the destruction of the building would ruin its owner, and the chance that he would be ruined in that way would materially increase the difficulty of his borrowing additional capital, it may be worth his while to pay £10 (or even more if it were not to be had at a lower rate), for a secure insurance against that risk. But an insurance company, with an income from insurance premiums of £1,000,000, would scarcely feel such a loss; and it is therefore able to reckon the risk at its actuarial value. Of course allowance must be made for the company's heavy expenses of administration, for risks of fraud, etc. but even so it might make good profits by charging only £2 against the risk.[3]

This illustration shows how a bank or any other powerful business might benefit capitalists of smaller means, if a method could be devised by which it could bear their risks, supposed to run all in the same direction: for it would be to their advantage to insure at more than the actuarial value of their risks. But a characteristic of those risks of a business, which are governed by causes external to it, is that they turn on price movements which throw about equal risks in the opposite direction on other businesses. If an insurance company could contrive that the risks, which it undertook in regard to a particular price movement in the two directions up and down were equal and opposite, the aggregate burden of its risk would be nothing at all: it would need compensation only for its expenses of administration.

A balance of this kind is seldom attempted under the name of insurance. But miscellaneous risks are shared out in various ways, some of which are commonly described as insurance; while many others, not so described, amount in effect to indirect insurance. For instance, bad weather on the day set for a coronation procession is a risk against which an insurance rate is definitely quoted at Lloyds; and it is of great benefit to caterers for public entertainment whose risks are exceptionally large. Insurance against the danger of fine weather is not of frequent occurrence: but it is notorious that fine weather on a Bank holiday causes indoor entertainment's to be deserted for outdoor entertainment's and excursions. Now, if it were possible to insure simultaneously indoor entertainment's against fine weather and outdoor entertainment"s (together with railways, etc.) against bad weather, for the same day and to equivalent amounts, the insurers might take many grievous risks off the shoulders of others: they might reap a goodly profit for themselves, and yet bear little net risk themselves: for opposite risks would have partly extinguished one another. We are now to see how a chief function of organised markets is to accomplish what is in effect a double insurance of this kind, though its manner is rather that of wagering than of insurance.

CHARACTERISTICS OF ORGANISED MARKETS

"An organised market" is one the proceedings of which are formally regulated. As a rule those who deal on it are in effect a corporation: they elect new members and also the executive of their body and appoint the committee by which their own regulations are enforced. In some countries their status is fixed and their actions are superintended by government. Their regulations generally provide, implicitly or explicitly, for the completion of a contract to buy or sell a quantity of a few words on the one side, and by a brief response, sometimes a mere nod, on the other. They generally prescribe a rather large unit as that to which the contract refers, at all events in the absence of any specific statement to the contrary.

The most highly organised exchanges are the Stock Exchanges of the chief industrial countries. But they can profitably be considered only in connection with the Money Markets to which they belong: and their special problems have but little direct connection with those of the present group of chapters. They may therefore be left on one side for the present.

The chief conditions needed for rendering any class of products suitable to be handled in an organised market are, (1) that it be not quickly perishable; (2) that the quantity of each thing can be expressed by number, weight or measure; (3) that its quality can be determined by tests that yield almost identical results when applied by different officials, assumed to be expert and honest; and (4) that the class is important enough to occupy large bodies of buyers and sellers.

These conditions are sufficient to render organised marketing practicable. But a fifth condition is required to make it attractive: it is that the class of things dealt in should be generally liable to considerable fluctuations in price. For otherwise the dealings would be confined almost exclusively to producers, consumers, and merchants: there would be little scope for those professional dealers who make a living by speculative purchases and sales; and who, as we shall presently see, in some cases render great public services by carrying risks that would otherwise need to be borne by people whose special aptitudes lie in other directions. It is true that this beneficent work is often marred, and sometimes over-borne, by evil practices which intensify fluctuations and mislead honest dealers: but, for the present at least, that evil has to be taken with the good. An organised market generally gives scope for purchases and sales for immediate delivery; and for dealings in "futures", that is in goods to be delivered at specified future times.[4]

This fifth condition implies that the things in question are not of such a nature, that their supply can be varied by rapid and extensive changes in the rate of production; so that their price is prevented from fluctuating rapidly, and remains always close to normal cost of production. There are few material things which satisfy all these conditions in very high degree: the chief among them are various grains, especially wheat; and raw cotton. The authorities of each organised produce market define the standard, or standards, in which dealings may be made; and all produce, which comes for delivery in these dealings, is inspected and certified as being truly up to the standard which it claims.

Comparatively few transactions in futures lead to the actual delivery of the produce. In most cases the buyer pays to the seller any amount, by which the official price of the quantity sold may have fallen below that at which the sale was made; or receives from him any amount, by which it may have risen. Either side may insist on completion: but that is generally effected through the organisation of the exchange, by bringing together those who wish actually to deliver with those who wish actually to receive; the rest being "rung out". The practical effect of this is that anyone can as a rule buy a future, without being called upon to pay its price either at the time of making the contract, or afterwards. Each part is required to put in a "margin", which will cover a small movement of the price against him: and, as soon as the price moves considerably against him, he is likely to be required to make a corresponding addition to his margin.

Thus by far the greater part of the transactions are in substance merely wagers to the effect that the price of produce will rise or fall. Of these wagers some are, as we have seen, careful, deliberate business operations, sometimes classed as "legitimate" speculation: others are the almost random guesses of foolish gamblers; and others again are parts of large manipulative policy, which is in the main evil economically and morally.

THE SERVICES RENDERED BY CONSTRUCTIVE SPECULATION ON A WHEAT EXCHANGE, ILLUSTRATED BY ITS EFFICACY IN LIGHTENING THE BURDEN OF RISKS BORNE BY GRAIN MERCHANTS AND MILLERS

In early times the population bore economic risks, arising out of the uncertainties of the harvests, incomparably more grievous than any that fall on any large class of persons in the present age: and the machinery of modern grain markets cannot be adequately judged without some recognition of the evils from which mankind has been delivered by the gradual development of organised trading in grain. Milling is now a subtle industry, requiring a highly technical knowledge of machinery and of the many variations of grain. Flour is not now made of any grain that comes first to hand: it is worked to definite standards, for particular districts and classes of consumers, by appropriate blendings of different sorts of wheat; and the miller therefore has more work to do outside of speculation than before. On the other hand, prices in his local market are now so closely bound to those of the world markets, that to form a good opinion about them requires the undivided energy of an able man. The miller is therefore often glad to insure himself against those risks of his business, which arise, not out of the local conditions of which he has special knowledge, but out of world movements.

A produce exchange can best undertake such risks as these, because many minds of first rate ability and many large capitals are occupied there in dealing with just these risks: and because many of the risks are in opposite directions and cancel one another. The broad shoulders of an exchange can carry without effort the intense risk, relatively to his financial strength, which the chance of a rise in price has imposed on one man; and can generally neutralise it by carrying the equal risk, which the chance of a fall in price has imposed on another.

It is of course impossible to recognise officially contracts in every sort of wheat. But a standard sort and quality is set for official dealing: and when actual delivery is being arranged, sorts differing in some specified small degrees are allowed to be substituted for the standard grade, under special adjustments as to price, which are ordered by the officials after inspection.[5]

A British miller may bring shiploads of wheat of various descriptions alongside his own elevator, and mix them by automatic flow in various proportions to make different sorts of flour. He often buys direct from far-off farmers or local elevators, through agents on the spot, who know his requirements exactly: but he can, if need be, send special instructions in return to telegraphic reports; basing himself on the last records that have been received in Liverpool, or other centre of the wheat trade, of the prices of standard grades in all the chief markets.

Having ordered the purchase of a certain quantity of what he needs, he "hedges", by selling at once in a central market an equal quantity of standard wheat for delivery at about the time at which he expects that the wheat, which he has just bought, will be in his elevator ready to be made quickly into flour. If wheat falls in the interval, his flour has to compete with that made from cheaper wheat; but, what he loses through that fall, is returned to him almost exactly by his gain on the "future" which he has sold. Conversely, if wheat rises in the interval, he has to pay on the sale of his "future" about as much as he gains from the corresponding upward movement of his flour. By buying a future he does *not* speculate; he throws on the shoulders of the general market the risks and the chances of gain that would otherwise have come to him through general movements external to his own business. The elements of risk, that stay with him, are only the chances of some divergence between the price movement of the standard grade which he has undertaken to sell, and that of flour of the sort which he is preparing to make: and experience shows that that divergence is seldom large. He does not speculate: he insures.[6]

Meanwhile many millers have made contracts to deliver flour to bakers and others in specified quantities and at specified times and places: and prefer to buy their grain as they need it. These millers run opposite risks to those of the set just considered: for a rise in the price of wheat might cause them heavy losses. So they insure themselves by buying futures on the exchange. In so far as the sales of futures by the first set, and the purchases by the second, are for equal amounts and like times, the resulting risks cancel one another: whatever excess of risk there is on the one side or the other remains to be borne by the dealers on the exchange: and their shoulders are very strong for the work. Of course the miller who buys a future, can demand delivery only in a standard grade: and if he is producing fine qualities he sells his right to that grade, and buys direct those sorts which he wants.[7]

All great wheat exchanges are in close touch with one another; their movements are now reported at short intervals to the local markets of the districts which they severally dominate; and though passing fluctuations are ignored, a great change from one month to another tends to influence the breadth of sowing and the assiduity of weeding, etc, of any who may be thoughtfully inclined. It is to be remembered that there is no month in the year in which there are not many cultivators in some part of the northern or southern hemisphere, who are making preparation for, or actually sowing, winter or spring wheat.

Long ago English farmers complained that corn-factors arranged to keep down the price of wheat just after harvest when the farmers "had their payments to make", and were compelled to sell; and to raise the price against the public later on.[8] The modern version of that complaint in America is that futures are sold down on the exchanges immediately after harvest, in order to lower the prices of wheat for immediate delivery; and that the prices are raised afterwards so that the consumer does not gain what the producer lost. And, while farmers' organisations complain that speculation in futures lowers prices, millers' organisations complain that they have the opposite effect: but analysis and statistics seem to show conclusively that neither contention can be sustained.[9]

On the whole it seems safe to conclude that, since those who buy because their investigations lead them to think that the supply is likely to run short, or sell because they are convinced it has been underrated, will gain if they are right and lose if they are wrong; therefore they are in their own interest contributing to the public the best judgement of minds that are generally alert, well-informed and capable. Their influence certainly tends to lessen the amplitude of price variations from place to place and from year to year. But let us turn to look at the evil side of such speculations.

DEALINGS IN ORGANISED MARKETS ARE LIABLE TO ABUSE BY UNSCRUPULOUS MEN, AIDED AS THEY OFTEN ARE BY THE FOLLY OF ILL-INFORMED SPECULATORS: BUT THE POWER OF SELLING THE FUTURE COMMAND OF A THING NOT YET IN POSSESSION HAS IMPORTANT USES

Manipulative speculation has many forms and many degrees. Its chief method is to create false opinions as to the general conditions of demand and supply. A clique will lead the market generally to believe that they are working for a fall, when really they are buying quietly and by indirect means much more largely than they are selling; and conversely they will buy openly, when they are really speculating for a fall. To publish definite false news is an extreme measure, bringing so prompt a punishment that it is generally avoided by shrewd manipulators. But false suggestion is a chief weapon and it has so many shades, some of which seem trivial, that men of fairly upright character are apt to be drawn on insensibly to condoning and even practising it.[10]

When a large command over wheat (or other produce) has been quietly obtained, a clique will sometimes go on buying, till all that can be made available before the time of settlement is already sold; and then go on buying, till the price has been forced up to an exorbitant level. This process is of course facilitated by the practice, that prevails on

American exchanges, of buying and selling in a sort of auction open to the whole market. The clique, if successful, finally let out every one who has over-sold, at prices varying with his means: for it is against their interest to make people bankrupt. Some of the corners thus made have a considerable place in history. But the very forces of the modern money-market and modern means of communication, which strengthen the attack, strengthen also and in a greater degree the defence offered by the community. The clique may plot in secret: but their dealings, however disguised, are soon interpreted by operators about as shrewd as themselves. And the larger the plot, the more surely will energy and ability be directed to the inquiry, whether the movement which is on foot is really justified by the general relations of demand and supply in world markets. If that appears not to be the case, a hostile clique well financed will enter the field: and it will be secure of victory, if its calculations are right. For if two teams of nearly equal strength are pulling in opposite directions, that one which is pulling with the slope of the hill must surely prevail.[11]

In all such cases a powerful clique reckons on obtaining great, though unwilling assistance, at all events in the earlier stages of its campaign, from the folly of amateur speculators. For such men do not understand that the affairs of a great speculation require thorough equipment with knowledge that is beyond the reach of the general public: they do not speculate altogether at random; but they act more mischievously and disastrously to themselves than if they did. For when a man decides, without any bias whatever, on which side of an uncertain event he will wager, he will of course lose any charges that may be levied on his wagering, and these will accrue to the members of the exchange: but he will, as a rule, not lose any more; for, if he acts absolutely at random, he is about as likely to go in the right direction as in the wrong. The fees which he pays help to finance constructive speculation and trade, and contribute a little also towards the expenses of malign strategy.

But ill-informed speculators generally suppose themselves to be basing their action on the most recent news. Now, the latest information accessible to outsiders has nearly always been acted on by well-informed persons, and has exerted the full influence, belonging to it, before it reaches the public. They are therefore likely to buy, when a fall is more probable than a rise, and *vice versa* ; and in the long run they would make losses by which better informed dealers would profit, even if all the news, which comes in their way, were designed to lead them aright. But in fact many of the statements and suggestions, by which they are guided, have been specially prepared with the purpose either of inducing the unwary to buy, because an unscrupulous speculator or clique wishes to unload; or of inducing them to sell for the opposite reason. In other words they are, to their own great loss, a powerful force on the side of evil manipulations of the market.[12]

Thus the power of selling the future command of a thing, not yet in possession, is liable to abuse. But, when used by able and honest men, it is beneficial: as is shown by the havoc caused by epidemics of unorganised speculation in the value of land, such as are not infrequent in new countries. Speculation for a rise in the value of land is always easy: anyone, who has considerable means and believes that the land is being sold under its true value, or turned to uses below those of which it is capable, can move its price upwards by buying, and using his purchases as security for loans with which he buys again. But no one can speculate for a fall in the value of land, except to the extent of selling any that he happens to hold. Those who have knowledge, but no land to sell, are unable to turn it to effect in checking an excessive rise, in the same way as they could if it referred to a thing for which there is an organised market. Of course in all such matters opinion is under influence from a public press, in which wise and honest counsels have the upper hand in ordinary times: but this influence is apt to fail at critical times.

For instance, the land boom of 1887–1890 in Melbourne began in a shrewd anticipation by able men that the business of the city would require a great extension of the areas used for wholesale trade, domestic and foreign; and for general purposes. They bought, and were able to sell at higher prices; for their success hastened a general appreciation of the great possibilities of Melbourne: and they were joined by others, not all of whom were capable business men like themselves. A little later, a large part of the population had bought land to the full extent of their own capital, if not beyond. No effective note was sounded, or could well be sounded, to warn them that the rise in price had far outrun all reasonable expectations; and that when the bubble was pricked, business would be so crippled that the value of land would fall fast and far. Consequently Melbourne passed through a period of grievous distress.[13]

SOME MARKETING RISKS RELATING TO "ORDINARY" PRODUCTS CAN BE TRANSFERRED BY FORWARD CONTRACTS: BUT THE MAJORITY CAN BE DELEGATED ONLY AS INCIDENTS OF THE DELEGATION OF CORRESPONDING FUNCTIONS

The methods of transferring risk from the shoulders of one set of men to those of others, more fitted to bear it, which have now been considered, are not applicable to *ordinary* products, that is those for which no highly organised market is available. But something can be done in that direction by other methods; and the increasing complexity of business often inclines the producer to delegate all such responsibilities as he safely can.

When two men are in partnership on equal terms, it is frequently arranged that one gives nearly his whole energies to making, and the other to marketing. Again a very large business, whether in joint stock or not, commonly entrusts to each of several heads of departments responsibility for a group of details, some of which are considerable, relating to the whole affairs of a small business: each of them discharges a share of the functions of the business; but, as a rule, he bears little or no share of its risks. Function can indeed be delegated easily without associated risks: but the class of risks which can be delegated without any corresponding function is narrow. The risks, which can be transferred without function, relate almost exclusively to definite particular transactions; and the chief instance of these has just been considered. The associated functions and risks with which the next chapter is chiefly concerned are incidental to some stage in the journal of various products on the way from manufacturers, or other producers, to ultimate consumers.

We have seen how an organised market enables a producer to secure in advance an adequate supply of certain materials: and how it also enables him to insure against a loss, that might result to him indirectly from a fall in the market value of his material while he is working it up. So far as future supplies of material go, the same security can be had in many classes of work by a contract for future delivery: though it cannot be undone or modified to suit altered circumstances by a compensatory sale, as easily as it could if an organised market offered facilities for a sale of futures. For instance, shipbuilders and other users of half-finished steel products commonly buy their materials in advance as far as possible, when building under contract; in which course they are aided indirectly at least by partially organised markets for some products. Coal, a chief material for many industries, is too various in character to be handled in a highly organised market: otherwise the dealings in such a market might at times surpass all others in volume and excitement.[14]

Thus by private contract or otherwise it is sometimes practicable to insure a business against loss by definite changes in recognised prices, as thoroughly as against losses by fire or other specific accident. But it is not possible, it is scarcely even conceivable, that insurance should be effected against the results of slackness in action or errors of judgement. Such risks must remain with those who control the business and appoint its officers. They may delegate some of their functions, and yet bear these risks either in whole or in part: but it is generally impracticable to transfer such risks without transferring the functions to which they are related. A producer can indeed transfer to middlemen some of the risks of marketing, which he must otherwise bear himself: but he can do so only because that transference is incidental to a transference of some functions to them.[15]

To this fact and another which is closely allied to it may be traced many of the chief characteristics of modern marketing. The second fact is that though each bargain on an open market, whether an organised cotton market, a cattle market, or a fish market, stands very much by itself; the dealings of a mercantile house with its customers cannot generally be isolated, either as regards their costs or as regards their rewards.

1 On this side of its work good marketing may be compared to a fly-wheel, rather than to a reservoir. The engine which drives a single machine, liable to short sudden increase of strain, must have a fairly massive fly-wheel. If it drives a score of such machines it needs a mass perhaps only six times as great: and if it drives 500 of them, the wheel may be perhaps less than a thirtieth part as massive in proportion to the work done, as if it drove only one.
2 This statement is not inconsistent with the fact that if a man stakes half of his fortune of £10,000 on an equal risk, his prospects of material well-being are lowered. Of the increase of well-being derived from an addition of say £10 to property or income, is in accordance with fundamental laws of human nature, generally the less, the greater the amount already possessed: and therefore his welfare stands to be diminished by his exchanging a certainty of £10,000 for two equal chances of £15,000 and £5,000. But almost every general rule has its exceptions: and if he could with £15,000 get something that would make him permanently happy, while a loss of half his £10,000 would do him no special injury, the bargain might be wise.
3 The actuarial calculations are of the same kind, but much more intricate in practice: for partial losses by fire, which are very much more frequent than total losses, are commonly included. In consequence much higher rates than that suggested above are charged: but the general principle by which they are governed is not affected by this.
4 A contract in relation to a future often takes the form of an "option" by which the payment of a certain sum secures the right to demand certain things (or to sell them) within a given period at a specified price: these two options may be combined, an option to buy at a stated price being coupled with one to sell at a stated higher price. There are a few cases in which dealings in options are part of legitimate trade. But there appears to be more force in the arguments for prohibiting them by law, than for prohibiting a simple buying or selling of futures; for they are relatively more serviceable to the gambler and the manipulator than to the straightforward dealer.
5 Grading by American elevator companies, of grain sent to them by farmers, is not always without reproach: though in the main it seems to be fairly correct. Official grading is generally careful though it varies a little from one place to another. A long rod, with a row of small boxes along its whole length, is pushed to the bottom of the truck or other receptacle in which the grain is. The boxes are then opened and filled by a single movement; and old fraudulent methods of making the grain appear better than it is, by concentrating good qualities in those parts of the receptacle from which a sample was most likely to be drawn, are stopped by grading the whole as of the lowest quality shown in any box, where there is any different. The grading is based chiefly on weight per bushel, colour, condition and cleanliness: the last three elements being decided by individual judgement, which appears to be liable to perceptible though small variations; and it takes little or no direct account of strength in gluten, starch or other matters which are important to the miller. See the *Report of the Industrial Commission*, 4, p. 432. That Commission, appointed in 1898, issued its nineteenth volume in 1902; and is the authority for many statements as to American conditions made in the present work.
6 Exception must indeed be made for the improbable case in which his sale of a certain grade for future delivery comes at a time when the price of that grade is being forced up by a campaign on the Exchange to "corner" those who have sold futures.
7 The general position is set out clearly by Professor Emery in his work on *Stock and Produce Exchanges*, 1896, and in his paper read to the American Economic Association in 1899. It may be noted that the miller who contracts in advance to deliver flour of a specially fine

quality is put in a better position by buying a future on the Exchange than he would probably be by contracting with individual dealers for the delivery to him of the sorts that he expects to need. Such a contract would be difficult to make, and it would be liable to uncertainty and friction in the execution. While buying a future on the Exchange he puts himself in command of funds apportioned to the prices of the time at which he needs them; he then buys after inspection, and possibly readjusts his combinations of various sorts according to their several conditions and current prices. The British miller, especially on the Mersey, the Clyde, or other chief waterway, has a choice of wheat from a great number of soils and climates in all parts of the world, and has generally at command several alternative combinations of diverse qualities by which he can reach any desired result.

8 See, for example, evidence of a Sussex farmer before H. of L. Committee on Agriculture, 1836, Q. 34, 25, 6.

9 The Industrial Commission after a full hearing of the arguments of farmers and others in the opposite direction, concluded (*Report*, 4, p. 223) that "prices prevailing at the time when producers dispose of the greater part of their products are greater in comparison to the rest of the year than they were before the advent of modern speculation".

In regard to German experience, see a chart leading to a similar conclusion in Conrad's *Grundriss der Pol. Oek* . I. p. 220. And Table VII, in Mr Hooker's paper on "the suspension of the Berlin Produce Exchange", *Statistical Journal*, 1900, indicates that the German prices fell below the American exceptionally just after harvest, as a result of the suspension.

10 A rise in price at (say) Chicago, is often started by ordering the purchase of some millions of bushels in Liverpool on a certain morning; so as to suggest to the Chicago market, when it opens a few hours later, that information in possession of the English wheat trade points to some scarcity in supply relatively to demand. Again, when a powerful speculator wishes to buy, he often makes large sales openly in his own market; and a little later causes to be bought by agents, acting secretly on his account, much larger amounts, partly in distant markets, to which the news of his sales has of course been telegraphed.

11 Thus Mr Partridge in 1891–1892 sold vast quantities of wheat at Chicago, in the belief that the market's estimates of the crop were too low and that the current high price was too high. He was right, so he was able to buy back the wheat at a lower price than he paid for it. The market is not often thus wrong, but it was wrong again in 1897 when Mr Leiter thought that the current estimates of the crop were too high, and the price too low. Having large funds at his back, and aided by others, he bought great quantities, adopting subtle devices for preventing wheat that was not under his control from reaching Chicago; and was thus able to make many of those, who had sold to him, pay very high prices for default of delivery when the time for settlement came. But having won one battle, he attempted another: and, partly because he had ranged strong enmities against him, he failed grievously. His second venture indeed illustrated the general rule that sensational success in great speculation tends to strengthen the nervous, confident temper in which it originated. It engenders rashness in venturing, and an even more dangerous inability to recognise defeat. In fact a great speculator has scarcely ever rested on his victory: he has nearly always persisted till overtaken by disaster.

12 A full explanation of the folly of such men, written many years ago by Crump, *The Theory of Stock Exchange Speculation*, is in the main applicable to present conditions.

13 While the boom was yet far from its climax, a Melbourne business man told me the price of land in a part which was indeed well suited for commerce and industry, but was adjoined by much neglected land with almost equal natural advantages. The price was in fact higher than in any part of London, except two or three acres near the Royal Exchange. He said that prudent men knew that a terrible catastrophe was near: but having sold all the land which they controlled, they had no direct means of influencing prices; and their opinions were unheeded. If a great quantity of futures could have been sold by such men, as soon as prices had gone a little beyond their reasonable level, the sellers would have enriched themselves,

and conferred on Melbourne as a whole a benefit many times as large as their own gains.

14 There are many local coal markets in Britain, each having trade terms and usage's peculiar to itself: and, though the broad dealings of the London Coal Exchange in almost every variety have introduced some uniformity in the movements of prices, no approach to a satisfactory reduction of all varieties to their equivalents in several standard sorts is yet in sight. Comparatively few consumers make their contracts in advance with mine-owners: but the middlemen's dealing in such contracts are so vast as to bear some comparison with those of an organised market. They even occasionally bear the market, in order to obtain favourable terms from coal-owners for large contracts for the coming season: but for this purpose they do not sell futures; they merely withhold their customary purchases for a time. (See the account of "the marketing of coal" by Professor H. Stanley Jevons and Mr David Evans in *Practical Coal Mining*).

15 Some "profit and loss sharing" schemes may seem to point in another direction: but the general trend of such schemes does not. The ordinary shareholders of a large joint-stock company bear its chief risks, but delegate nearly the whole of the control.

2

THEORY OF STORAGE, HEDGING AND FUTURES MARKETS

Introduction
Lester G. Telser

Nicholas Kaldor's *Speculation and Economic Stability* is a landmark in the theory of futures markets. Most notably, it explains why stocks are held even when the spot price is above the futures price. It does so by introducing the concept of the convenience yield of stocks that measures the sources of gain from inventories. Second, he presents an argument to show why the futures price is above the expected spot price and, even more remarkable, in a revision of his article published 50 years after the original appeared, he shows what is wrong with his original argument. Nevertheless, his original treatment remains influential for some economists. A summary of his original argument follows.

Let i be the marginal interest cost, r the marginal risk premium, q the marginal convenience yield of stocks, and c the marginal carrying cost. Let FP denote the futures price, CP the current price and EP the expected spot price. A positive quantity of stocks will be held unhedged up to the level where

$$EP - CP = i + c - q + r \qquad (1)$$

The presence of the positive marginal risk premium indicates, according to Kaldor, the compensation a holder of unhedged stocks requires for the risk incurred by so doing. A short hedger, one who sells futures contracts equal in quantity to his inventory, bears no risk because delivery of the physical inventory to settle the futures commitment is a feasible and riskless alternative. Therefore, a positive amount of hedged stocks will be held up to the point where

$$FP - CP = i + c - q \qquad (2)$$

Note that there is no marginal risk premium r in Equation (2). Now, Kaldor takes a crucial step without stating the underlying implicit assumptions that are needed to justify it. If someone holds a positive quantity of *both* hedged and unhedged stocks, then it follows from simple algebra applied to Equations (1) and (2) that

$$EP - FP = r \tag{3}$$

Kaldor concludes that the expected spot price exceeds the futures price by the marginal risk premium, r. This has led many researchers to search for the risk premium. To put it another way, Kaldor's argument implies that the futures price is a downward-biased estimate of the spot price.

Before explaining Kaldor's later criticism of his own argument, let us note two qualifications. First, if it is best to hold hedged stocks but not to hold any unhedged stocks, then, instead of (1), there is the following inequality

$$EP - CP < i + c - q + r \tag{4}$$

However, (2) is still a valid equation for holding a positive quantity of hedged stocks. It is an implication of Equation (2) and inequality (4) that

$$EP < FP \tag{5}$$

On the other hand, if it is best to hold unhedged stocks and not hold any hedged stocks, then Equation (2) is replaced by the inequality

$$FP - CP < i + c - q \tag{6}$$

Equation (1) remains a valid condition for holding a positive amount of unhedged stocks. It follows from Equation (1) and inequality (6) that

$$EP > FP \tag{7}$$

Inequality (5) implies that the futures price is an upward-biased estimate of the spot price and inequality (7) that it is a downward-biased estimate. Therefore, even with Kaldor's original framework his conclusion follows if and only if it is best for a firm to hold *both* hedged and unhedged stocks. Nor is this all. If firms act differently because they have different views about the future, then they will not use the same hedging strategy. If some firms hold only hedged stocks, others only unhedged stocks and some both, then there is no unambiguous relation between the futures price and the expected spot price. Moreover, even if, as Kaldor assumes, all firms are alike, his conclusion does not follow unless firms hold positive quantities of *both* hedged and unhedged stocks.

Kaldor's criticism of his formulas, for which he gives credit to Hawtrey, takes a different line. If there are many speculators, both long and short in the market, then, even if short hedgers would be willing to pay an insurance premium to long speculators (as an inducement for them to shoulder risk) this may not be necessary. As Kaldor puts it, there may be bull speculators who are buyers of futures and for them the demand price is $EP - r$, and there may be bear speculators for whom the supply price is $EP + r$. Therefore, we are no longer entitled to conclude there is a bias one way or another in the relation between the spot and the future price.

This section includes three chapters by Holbrook Working, a prominent student of futures markets. Working immersed himself in detailed studies of futures markets throughout his long career as an academic economist at a research institute affiliated with Stanford University. These three chapters summarise his main results, including his criticisms of the views on futures markets of Keynes, Hicks and Kaldor, among others.

Working's article explaining his theory of inverse carrying charge was published in 1948 in the *Journal of Farm Economics* where, not surprisingly, it did not attract as much attention among general economists as it deserved. However, it eventually became his best known and most influential contribution to the subject. Reading his article is essential for understanding futures markets notwithstanding its primary focus on seasonally produced agricultural commodities. Working relegates some of his most important points to footnotes. Note 10 for example, claims that organised futures markets cannot survive without a substantial hedging interest. Working's reasoning, cast in terms of differences of opinion among speculators, is not without interest. Speculators are attracted to commodities in which prices change unexpectedly by large amounts. The volume of trade due to speculators goes up when prices display such behaviour. When prices are relatively stable, speculation diminishes and trade languishes. However, there is a difficulty with the implications of this argument because the same conditions that are conducive to speculation, namely differences of opinion leading to price variability, are also conducive to hedging. How can one distinguish between these two effects? The second of Working's tracts tries to answer this question and introduces an important concept, the supply of storage. It also begins the empirical hunt for bias in the futures price by studying whether there is a seasonal in the futures price of a contract. Such a seasonal could be explained as remuneration to speculators for selling price insurance to hedgers.

The next chapter is a logical successor to Working's inverse carrying charges article that has a graph showing a positive relation between the price of storage and the amount of stock. Working called this positive relation the supply of storage. Working also claimed to have statistical estimates of this graph, but he did not explain why the data show a positive relation between these two variables. The demand for storage was missing from Working's explanation. My chapter contains a more complete model, the first formal model of a futures market, with both the demand and the supply of storage that explains why we observe a positive relation between the amount of stocks held and the spread between prices of adjacent futures contracts. The question whether a seasonal appears in futures prices is answered in the negative. In this chapter, I assert that the futures price is an unbiased estimate of the expected spot price. It also shows that when the price level is falling, there is a downward seasonal, and when the price level is approximately constant (taken to be changes of less than 5% per year) no seasonal is visible in the futures price. This relationship between the price level and the futures price may indicate that people tend to underestimate changes in the general price level. It need not cast doubt on the proposition that the futures price is an unbiased estimate of the spot price. (I must note here that Cowles Commission Monograph 17 referred to in note 13 was never completed and does not exist.)

The third chapter by Working has the provocative title "Speculation on Hedging Markets" to emphasise Working's contention that hedging is essential to the survival of an organised futures market, but this is not his most important idea. The key idea is to regard a futures contract as a temporary substitute for an actuals contract. That is, lacking a customer, the holder of an inventory substitutes sales of a futures contract, a short hedge. Similarly, a firm, having made a commitment to produce a commodity such as flour or export wheat without having the raw materials on hand, buys futures as a temporary substitute for the actual commodity. This is a long hedge against a forward sale. Working also introduces an anticipatory hedge. Such a hedge does not match an actual commitment but a potential one expected to take place in a short time. These ideas that regard a futures contract as a temporary abode in place of an actuals contract lead to an isomorphism between the theory of futures contracts and monetary theory, that is explained at length in my article on actual and futures markets, in the second chapter of Section 3.

Working argues that official statistics, the classification of open interest into hedging and speculation by those with large enough positions requiring them to report to regulatory authorities, are misleading as they do not furnish information about the positions of non-reporting traders.

The latter are often, mistakenly, taken to be speculators because it is thought that hedging positions are of a size that require reporting to the regulators.

Working attempts to allocate the non-reporting commitments between hedging and speculation by assuming that ratios between long and short positions for the reporting traders are the same as for non-reporting traders.

This is to say that the ratios are independent of the size of the traders' positions. Working does not state explicitly as he should have, his assumption that the ratios are the same for traders regardless of the size of their commitments. Working can check his assumption only in those cases in which the authorities did a complete survey of all traders' positions, large and small. He claims that his estimates are tolerably close to the actual figures for these surveys, while admitting that the surveys are usually carried out when unusual conditions prevail. There is another difficulty that Working does not mention: the hedging and speculative positions differ markedly by futures contract. Long hedging is far more important than short hedging towards the end of the crop year than at the beginning. The ratios ignore these differences among futures contracts, since they rely on the total commitments in all contracts. The reader can judge whether to accept Working's estimates. In my opinion, the division into commercial and non-commercial traders, as explained in Hartzmark's chapter in Section 5, is better.

The official statistics on open contracts classified by position are subject to another problem Working does not mention but surely deserves attention. Federal regulations place limits on the size of speculative positions but not on hedging positions. This gives large traders an incentive to classify their positions as hedging, if they can. A leading example is anticipatory hedging that lies on a murky dividing line between hedging and speculation. While anticipatory hedging may indeed be a legitimate form of hedging as Working claims, it is not always so. The position limits on speculation and their absence on hedging cast doubt on the validity of the classifications in the government's publications, a point that Working overlooks in his efforts to allocate non-reporting positions by long, short, hedging and speculation.

Chapter 5 is one of Working's last publications It was published in the *American Economic Review*, which has the biggest circulation, and presumably, the largest potential audience among economists. Here, Working summarises the results of his lifetime of studies of futures markets. For this reason alone it deserves our respectful attention. Three points about this article deserve a mention. First, it refers to the relatively new research showing that futures prices take a random walk, a notion far more familiar today than when Working, a pioneer in these studies, emphasised its importance. Second, Working refers to the research on seasonals in futures prices and their failure to find a systematic pattern application to all futures. Third, he refers to Blair Stewart's study of the gains and losses of futures traders and expresses the opinion it "was singularly unproductive of positive conclusions". Nothing could be further from the truth. Stewart's study is the first rigorous attempt to see what the facts show about gains and losses of commodity futures traders. Because it runs counter to prevailing opinions about what should be found, it should have been taken very seriously and occupy a prominent place in the minds of students of futures markets. Instead it tended to be ignored, an especially ironic fate given Working's pronouncements about scientific evidence. What is noteworthy is Working's failure to relate the findings about the absence of systematic seasonals among futures to Stewart's findings that speculators, on average, incur losses. This should have alerted Working to the deficiencies in the received theory about the role of speculators as sellers of price insurance to hedgers. Working also failed to connect research on the random character of futures prices to the absence of a systematic pattern of seasonals. If there were a seasonal, then the random walk should apply to the difference between the actual futures price and the seasonal, not to the futures prices themselves.

The last chapter by Fischer Black, already famous for the Black–Scholes formula for pricing stock options (1973), is a crisp, spare, logical analysis of futures prices. Black derives and publishes his formula for pricing commodity options in this chapter. An essential feature of a futures contract is that it is marked to the market daily. This means that the holder of a futures contract can withdraw profits (if any) and must make up losses (if any) on a daily basis. These payments and receipts change the balance sheet of the clearinghouse that credits or debits the account of the holder or his agent, the latter being a clearinghouse member. Therefore, the net value of a futures contract is zero at the end of each day. While the market value of the futures contract is zero, the futures price is not zero. Black is among the first to recognise the distinction between a forward and a futures contract. A forward contract is not marked to the market continuously. Therefore, a forward contract can have a market value that is positive, zero or negative. A spot contract is a special case of a forward contract because a spot contract has an immediate expiration date while the expiration date of a forward contract is some later date. A futures contract is like a bet but a forward contract refers to a transaction in an actual commodity. Among the many important ideas in this article is its analysis of risk based on modern portfolio theory. Black shows that the risk, if any, to holders of futures contracts must be measured with respect to their whole portfolio. Consequently, it does not necessarily follow that a speculative position in futures incurs risk, and so on this account no theoretical reason for a bias in futures contracts. Thus, Black disposes of decades of confusion on this important aspect of the theory of futures markets. Dusak (1973), (in the third chapter of Section 5) had completed her research showing the absence of a risk premium in futures prices several years before Black published his article. Experts in futures markets are universal in their admiration of this article by Black, a landmark in the subject.

However, Black does not discuss the important point that risk and uncertainty remain for the economy as a whole although individuals can hold suitably diversified portfolios. The invention of financial paper and the organisation of exchanges to trade financial paper facilitates specialisation. This great principle, explained by Adam Smith in the first three chapters of the *Wealth of Nations* and emphasised by Knight, is a major source of growing economic wealth, not the least of which is the harnessing of the skills of those best able and temperamentally suited to trade financial paper.

3

Speculation and Economic Stability*
Nicholas Kaldor[†]

The purpose of the following paper is to examine, in the light of recent doctrines, the effects of speculation on economic stability. Speculation, for the purposes of this paper, may be defined as the purchase (or sale) of goods with a view to re-sale (re-purchase) at a later date, where the motive behind such action is the expectation of a change in the relevant prices relatively to the ruling price and not a gain accruing through their use, or any kind of transformation effected in them or their transfer between different markets. Thus, while merchants and other dealers do make purchases and sales, which might be termed "speculative", their ordinary transactions do not fall within this category. What distinguishes speculative purchases and sales from other kinds of purchases and sales is the expectation of an impending change in the ruling market price as the sole motive of action. Hence "speculative stocks" of anything may be defined as the difference between the amount actually held and the amount that would be held if, other things being the same, the price of that thing were expected to remain unchanged; and they can be either positive or negative.[1]

The traditional theory of speculation viewed the economic function of speculation as the evening out of price-fluctuations due to changes in the conditions of demand or supply. It assumed that speculators are people of better than average foresight who step in as buyers whenever there is a temporary excess of supply over demand, and thereby moderate the price-fall; they step in as sellers, whenever there is a temporary deficiency of supply, and thereby moderate the price-rise. By thus stabilising prices, or at any rate, moderating the range of price-fluctuations, they also automatically act in a way which leads to the transfer of goods from uses

---
*This paper is taken from *The Essential, Kaldor*, 1989, F. Targeth and A. P. Thirbwall (eds.) and is reprinted with the permission of Gerald Duckworth & Co. Ltd.
[†]Nicholas Kaldor, 1908–1986.

where they have a lower utility to uses where they yield a higher utility. (If future conditions of demand and supply were generally foreseen, this transfer would no doubt be effected without the agency of speculators. But in a world of perfect foresight nobody could make a speculative gain; speculators would be non-existent. In a world of imperfect foresight, the existence of speculators enables the system to behave with more foresight than the average individual in the system possesses). Hence speculative gains are very much of the same order as other kinds of entrepreneurial gains; they are earned, similarly to the profits of wholesalers or retailers, as a result of the transference of goods from less important to more important uses.

The possibility that speculative activity might cause the range of price fluctuations to become greater rather than narrower, and that it might lead to the transfer of goods from more or less important uses, was not seriously contemplated in traditional theory. For this would imply that the speculators' foresight, instead of being better than the average, is worse than the average; such speculative activity would be speedily eliminated. Only the speculator with better than average foresight can hope to remain permanently in the market. And this implies that the effect of speculative activity must be price-stabilising, and in the above sense, wholly beneficial.

This argument, however, implies a state of affairs where speculative demand or supply amounts only to a small proportion of total demand to supply, so that speculative activity, while it can influence the magnitude of the price-change, cannot at any time change the direction of the price-change. If this condition is not satisfied, the argument breaks down. It still remains true that the speculator, in order to be permanently successful, must possess better than average foresight. But it will be quite sufficient for him to forecast correctly (or more correctly) the degree of foresight of other speculators, rather than the future course of the underlying non-speculative factors in the market.[2] If the proportion of speculative transactions in the total is large, it may become, in fact, more profitable for the individual speculator to concentrate on forecasting the psychology of other speculators, rather than the trend of the non-speculative elements. In such circumstances, even if speculation as a whole is attended by a net loss, rather than a net gain, this will not prove, even in the long-run, self-corrective. For the losses of a floating population of unsuccessful speculators; and the existence of this body of successful speculators will be a sufficient attraction to secure a permanent supply of this floating population. So long as the speculators differ in their own degree of foresight, and so long as they are numerous, they need not prove successful in forecasting events outside; they can live on each other.

But the traditional theory can also be criticised from another point of view. It ignored the effect of speculation on the general level of activity – or rather, it concentrated its attention on price-stability and assumed (implicitly perhaps, rather than explicitly) that if speculation can be shown to exert a stabilising influence upon price, it will *ipso facto* have a stabilising influence on activity. This, however, will only be true under certain special assumptions regarding monetary management which are certainly not fulfilled in the real world. In the absence of those assumptions, as will be shown below, speculation, in so far as it succeeds in eliminating price fluctuations will, in many cases, generate fluctuations in the level of incomes. Its stabilising influence on price will be accompanied by a destabilising influence on activity. Hence the question of the effect of speculation on price-stability and its effect on the stability of employment ought to be treated, not as part of the same problem, but as separate problems.

In the subsequent sections of this paper we shall deal first with the conditions under which speculation can take place, secondly, with the effect of speculation on price-stability, and finally, with the influence of speculation on economic stability in general.

THE PRE-REQUISITES OF SPECULATION

Not all economic goods are the objects of speculative activity; in fact the range of things in which speculation, on any significant scale, is possible is rather limited. The two main conditions, which must be present in normal circumstances in order that a particular good or asset should be the object of speculation, is the existence of a *perfect, or semi-perfect, market* and *low carrying cost*. For if carrying costs are large and/or the market is imperfect, and thus the difference between buying price and selling price is large, speculation becomes far too expensive to be undertaken – except perhaps sporadically, or under the stress of violent changes.[3]

The presence of these two conditions presupposes, on the other hand, a number of attributes which only a limited number of goods possess simultaneously. These attributes are: (1) The good must be fully standardised, or capable of full standardisation; (2) It must be an article of general demand; (3) It must be durable; (4) It must be valuable in proportion to bulk.

The first two conditions are indispensable for anything resembling a perfect market to develop in exchange.[4] The last two ensure low carrying cost. For the greater the durability the less is the wastage due to mere passage of time, the greater the value in proportion to bulk the less the cost of storage.

These last two factors make up what might be called carrying costs *proper*. But *net* carrying cost also depends on a third factor: the yield of goods. In normal circumstances, stocks of all goods possess a yield, measured in terms of themselves,[5] and this yield which is a compensation to the holder of stocks, must be deducted from carrying costs proper in calculating net carrying cost.[6] The latter can, therefore, be negative or positive.

From the point of view of yield, it is important to distinguish between two categories of goods: those which are used in production and those which are used up in production. (There is no convenient English equivalent for the German distinction between *Gebrauchsgüter* and *Verbrauchsgüter*, both of which can refer to durable goods). Stocks of goods of the latter category also have a yield, *qua* stocks, by enabling the producer to lay hands on them the moment they are wanted, and thus saving the cost and trouble of ordering frequent deliveries, or of waiting for deliveries.[7] But the important difference is that with this latter category, the amount of stock which can be thus "useful" is, in given circumstances, strictly limited; their marginal yield falls sharply with an increase in stock above "requirements" and may rise very sharply with a reduction of stocks below "requirements".[8] When redundant stocks exist, the marginal yield is zero.[9] With the other category of goods, items of fixed capital, the yield declines much more slowly with an increase in stock, and it is normally always positive. Hence, as we defined "speculative stocks" as the *excess* of stocks over normal requirements (ie, that part of stocks which is only held in the expectation of a price-rise and would not be held otherwise) we may say that with working-capital-goods (*Verbrauchsgüter*) carrying costs are likely to be positive, when speculative stocks are positive, and negative when they are negative; with fixed-capital-goods (*Gebrauchsgüter*), carrying costs are normally negative, irrespective of whether "speculative stocks" are positive or negative.

It would follow from this that fixed-capital-goods like machines or buildings, whose carrying cost is negative and invariant with respect to the size of speculative stocks, ought to be much better objects of speculation than raw materials, whose carrying costs are so variable. The reason why they are not, is because the condition of high standardisation, necessary for a perfect market, is not satisfied, and hence the gap between buying price and selling price is large. It is not that machines, etc., by being used, become "second-hand", and thereby lose value, since the depreciation due to use is already allowed for in calculating their net yield. The reason is that all second-hand machines are to some extent de-standardised; it is very difficult to conceive a perfect market in such objects.[10,11] The same lack of standardisation accounts for the comparative absence of speculation in land and buildings.

This explains, I think, why in the real world there are only two classes of assets which satisfy the conditions necessary for large-scale speculation. The first consists of certain raw materials, dealt in at organised produce exchanges. The second consists of standardised future claims or titles to property, ie, bonds and shares. It is also obvious that the suitability of the second class for speculative purposes is much greater than that of the first. Bonds and shares are perfect objects for speculation; they possess all the necessary attributes to a maximum degree. They are perfectly standardised (one particular share of a company is just as good as any other); perfectly durable (if the paper they are written on goes bad it can be easily replaced); their value is very high in proportion to bulk (storage cost is zero or a nominal amount); and in addition they (normally) have a yield, which is invariant (in the short period at any rate) with respect to the size of speculative commitments. Hence their net carrying cost can never be positive, and in the majority of cases, is negative.[12]

If expectations were quite certain, speculative activity would so adjust the current price that the difference between expected price and current price would be equal to the sum of interest cost and carrying cost. For if the difference is greater than this, it would pay speculators to enlarge their commitments; in the converse case, to reduce them. The interest rate relevant in calculating the interest cost is always the short term rate of interest, since speculation is essentially a short period commitment.[13] The carrying cost, as mentioned above, is equal to the sum of storage cost and "primary depreciation",[14] minus the yield.

If expectations are uncertain, the difference between expected price and current price must cover, in addition, a certain risk premium, which will be the greater (1) the greater the dispersion of expectations from the mean (the less the standard probability); (2) the greater the size of commitments. Given the degree of uncertainty, marginal risk premium is an increasing function of the size of speculative stocks.

It will be useful to re-state the relationship between expected price and current price in algebraical form. We denote marginal interest costs by i, the marginal risk premium by r, the marginal yield by q, the marginal carrying cost proper by c (so that carrying cost = $c - q$); and the current price and expected price by CP and EP respectively. The following relationship must then be satisfied:

$$EP - CP = i + c - q + r$$

For certain problems in the theory of speculation, the concept of the "representative expectations" is perfectly legitimate; it enables us, as do other similar constructions, to simplify the problem without materially affecting the result. From the point of view of the theory of the forward market, however, it may not be legitimate; for the determination of the futures price, and in particular, the relation of the futures price to the expected price, will not be the same in the case where everybody's expectations are the same as in the case where the "representative expectation" is an average of divergent individual expectations. It will be convenient therefore to divide the analysis into two stages: in the first stage, to assume that all individuals have the *same* expectations at any one time, and to deal afterwards with the consequences of differences in individual expectations.

In both cases, individuals participating in the forward market can perform three different functions: "hedging", "speculation" and "arbitrage". "Hedgers" are those who have certain commitments independent of any transactions in the forward market, either because they hold stocks of the commodity, or are committed to produce, in the future, something else for which the commodity is required as a raw material; and who enter the forward market in order to reduce the risks arising out of these commitments. "Speculators", in general, have no commitments[15] apart from those entered into in connection with forward transactions; they assume risks by entering the market. Both hedgers and speculators can be, in particular circumstances, buyers or sellers of "futures",[16] but in both cases, it is the speculators who assume the risks and the hedgers who get rid of them.

The possibility of arbitrage, ie, buying spot and selling futures simultaneously and holding the stock until the date of delivery, arises when the relationship between the futures prices and the current price ensures a riskless profit. An arbitrage operation differs from an ordinary hedging operation only in that the ordinary hedger enters the futures market in order to reduce a risk arising out of a commitment which occurs independently of the existence of the forward market; whereas the arbitrageur assumes risks which he would not have assumed if the facilities of the forward market did not enable him to pass them on, on advantageous terms. Hence, any ordinary holder of stocks of a commodity becomes an "arbitrageur" in so far as the existence of the futures market tempts him not only to hedge the stocks he would ordinarily hold, but to enlarge his stocks in relation to turnover owing to the advantageous terms on which they can be "hedged".

The possibility of arbitrage sets an upper limit to the futures price in relation to the spot price. While there is no limit, apart from expectations, to backwardation, ie, to the extent to which the futures price may fall short of the current price, there is a limit to contango, in that the futures price cannot exceed the current price by more than the cost of arbitrage, ie, by more than the sum of interest plus carrying costs. Since, as explained above, ordinary holders of stock automatically become arbitrageurs whenever the relation between the futures price and the spot price tempts them to do so, the carrying cost consists of the costs of storage and wastage *minus the yield*, so that the net cost of arbitrage is $i + c - q$.[17] Denoting the futures price by *FP*, we have

$$FP - CP = i + c - q$$

and since, in all cases

$$EP - CP = i + c - q + r$$

it follows that

$$FP = EP - r$$

Thus, arbitrage prevents the futures price from rising above $EP - r$, while speculation (see below) prevents it from falling below this amount. This conclusion is quite general,[18] and it is consistent with both contango and backwardation, ie, with the futures price being either above or below the current price. When the yield is one of convenience, the marginal yield (q) varies inversely with the size of stocks in relation to turnover. When speculative stocks are zero, the expected price equals the current price (the stocks of ordinary holders are "normal"), and therefore $-(c - q) = i + r$, ie, the negative of carrying cost must be equal to the sum of interest cost and risk premiums. Since i and r are always positive, the carrying cost must be negative – the yield must exceed the sum of storage cost and primary depreciation by the required amount. In this case, $FP = CP - r$, and the futures price falls short of the current price by the amount Mr. Keynes called "normal backwardation".[19]

If stocks are sufficiently large in relation to turnover, q declines to zero, though it cannot become negative. It follows that, given an expectation of a rise in price, the upper limit to contango is $i + c$. In this case $FP - CP = i + c$, but as $q = 0$, $EP - CP = i + c + r$; FP again equals $EP - r$. This upper limit to contango is likely to appear at times when the market is unduly depressed by the prevalence of excessive stocks in relation to current consumption; and when, as a result of this, the spot price has fallen to such low levels that there is a definite expectation in the market that the price will recover in the future, with the gradual absorption of excessive stocks. This is another way of saying that when the stocks carried are excessive in relation to turnover (so that q is zero), the current price must be below the expected price, since only the expectation of a rise in price can induce the market to carry the outstanding volume of stocks.

On the other hand, when stocks are scarce, the marginal yield is large so that abnormal backwardation must appear, ie, $q > i + c + r$. In this case, the high marginal convenience yield ensures that the spot price rises to abnormally high levels and, as a result, the market expects a fall in price in the future as stocks are restored to more normal levels. Just as it is impossible for the market to expect a rise in prices in the (foreseeable) future when stocks are scarce in relation to current consumption (since the spot price will have to rise sufficiently to eliminate any such expectation), so in the absence of an expectation of a change in price (ie, when stocks are normal and the underlying factors are considered stable) there must always be backwardation.[20]

We have seen that arbitrage prevents the futures price from rising above $EP - r$. It remains to be shown that speculation prevents the futures price from falling below $EP - r$. Hedgers, in principle, could be either sellers or buyers of futures, though normally, hedging will be predominantly on the selling side.[21] Since transactions between hedgers who are sellers and hedgers who are buyers cancel out, it is the net sale of futures by hedgers (ie, the excess of hedging on the selling side) which requires to be taken up by speculators. The hedgers will *sell* futures if the futures price is *equal to or higher than* $EP - r$; whilst speculators will *buy* futures if the futures price is *equal to or lower than* $EP - r$ (where r is the individual marginal risk premium in both cases).[22] If everybody's expectations are the same, futures transactions between hedgers and speculators can only arise owing to differences in the marginal risk premium, ie, in the marginal willingness to bear risks among the different individuals in the two groups, and would proceed until these differences are eliminated. Since the marginal risk premium varies, not only with individual psychological propensities to bear risks, but also with the size of commitments, ie, with the amount of the possible loss relatively to the individual's total assets, we need not assume any variation in psychological propensities in order to account for such differences. When hedgers are predominantly sellers of futures, the buying of futures by speculators therefore prevents the futures price from falling below $EP - r$, whilst arbitrage, as we have seen, prevents it from rising above this level.

The above theory of "normal backwardation" (according to which the futures price must be below the spot price when stocks in relation to turnover are normal, while contango can only develop in times of excess stocks and abnormally low spot prices) is subject however to an important qualification when we allow for the fact that the expectations of different individuals comprising the market are not uniform.[23] Transactions will now take place not only between hedgers and arbitrageurs, but also between speculators and speculators; and transactions of the latter type may swamp all others. We now have to divide speculators into two groups: bulls and bears. Bull speculators will be *buyers of futures*, and their demand price is $EP - r$ (where both EP and r are subjective terms and refer to the mean value of the individual speculator's expectation, and his individual risk premium, respectively); bear speculators will be *sellers of futures* and their supply price is $EP + r$.

It should be clear at once that if by "expected price" we mean not the expectations either of bulls or of bears, but some kind of average expectation for the market as a whole, the price which will tend to get established as the outcome of transactions between bulls and bears is neither $EP + r$, nor $EP - r$, but something in between the two. We cannot say that the futures price will *correspond* to this average "expected price"; this would only be true if the marginal risk premia of different speculators were equal; and when the expectations themselves are different, there is no reason to assume that the marginal risk premia will be the same. But it is clear that the opposite risks assumed by bulls and bears will tend to cancel each other out – leaving the futures price if not equal to, at any rate fairly near, the "expected" price.[24] In a market where bulls predominate, the futures price will tend to exceed the "expected" price and vice versa, when bears predominate. Hedging on the selling side and arbitrage will act in the same fashion as a strengthening of bearish sentiment, while hedging on the buying side will act in the same fashion as a strengthening of bullish sentiment.

Thus, in addition to the factors mentioned above, the determination of the futures price will also depend, in the real world, on *divergence of opinion*; this factor will be all the more important the greater the degree of divergence and the more equally bulls and bears are divided. In markets where this divergence is important, and where transactions between speculators dominate over hedging transactions, we cannot say that the futures price will either be above, or below, the expected price, but simply that it will reflect the "expected" price; always subject, of course, to the provision that it cannot exceed the current price by more than the cost of arbitrage.

The elasticity of speculative stocks may be defined as the proportionate change in the amount of speculative stocks held as a result of a given percentage change in the *ratio* of the expected price to the current price.[25] This elasticity will obviously depend on the variations in the terms i, c and r, which are associated with a change in speculative stocks, in other words on the elasticity of marginal interest cost, marginal carrying cost, and the marginal risk premium, with respect to a change in speculative commitments.

Of these three factors the marginal interest cost, as we have seen, may be subject to discontinuous variation if the "marginal speculator", in a particular market turns from a lender of money into a borrower, or vice versa, but apart from this its elasticity is likely to be fairly high, if not infinite.[26] The marginal risk premium is normally rising, and its elasticity probably differs greatly between different markets. The more numerous are speculators in a particular market, and the more steady the price on the basis of past experience[27] the higher this elasticity is likely to be. Finally, the marginal carrying cost, as we have seen, can be assumed to be constant in the case of securities, while it will rise sharply (at any rate over a certain range) in the case of raw materials and primary products. Hence, taking all factors together, the elasticity of speculative stocks is likely to be much higher in the case of long-term securities than in the case of raw materials.

The higher the elasticity of speculative stocks, the greater the dependence of the current price on the expected price. In the limiting case when this elasticity is infinite, the current price may be said to be entirely determined by the expected price; changes in the conditions of non-speculative demand or supply can then have no direct influence on the current price at all (since speculative stocks will immediately be so adjusted as to leave the price unchanged); any change in the current price must be the result of a change in price-expectations.[28]

In the opposite limiting case, when the elasticity of speculative stocks is zero, changes in the expected price have no influence upon the current price; the latter is entirely determined by the non-speculative factors.

SPECULATION AND PRICE STABILITY

We can now attempt to analyse the question of the effect of speculation on price-stability. We have seen above that the argument that speculation *must* exert its influence in a price-stabilising direction in order to be successful presupposes that speculative purchases (or sales) make up only a small fraction of total transactions, and does not hold otherwise. But this still leaves the question of whether the effect of speculation is price stabilising or destabilising open.

We have seen that in all circumstances speculation must have the effect of narrowing the range of fluctuations of the current price *relatively to the expected price*. Hence, if the expected price is taken as given, speculation must necessarily exert a stabilising influence: a rise in the current price will be followed by a fall in speculative stocks, and vice versa.

In order that speculation should be price destabilising, therefore, we must assume one of two things: either (a) that changes in the current price lead to a change in the expected price by *more* (in proportion) than the current price has changed; or (b) that there are spontaneous changes in the expected price which are speculative in origin and are not justified or not fully justified in the movement of non-speculative factors.

The second of these needs further elucidation. It may be that an impending change in data is foreseen by speculators, but exaggerated in importance and, therefore, creates price movements greater than those which would have occurred in the absence of speculation; eg, a good harvest which, in the absence of speculative activity would have caused a 10% reduction in price, will involve a price movement which is both smoother and smaller in extent, if it is correctly foreseen by speculators. But if speculators foresee a 50% price reduction, in consequence of the expected harvest, the resultant price oscillation will be much greater than if they had not foreseen it at all. It may be also that the speculators expect that a certain event will react, favourably or unfavourably, upon a particular price, though in the absence of the expectation no such reaction would have occurred at all. The day-to-day movements on the stock exchange, where considerable changes in prices occur in accordance with the day's political news, could hardly be accounted for on any other ground but on the attempt of speculators to forecast the psychology of other speculators. If one is selling this is because he thinks the other would do likewise. If speculators combined and formed a monopoly, these price-movements could not take place at all.

The first of these factors – ie, the reaction of expectations to *changes* in the current price – may be measured by Professor Hicks' concept of the "elasticity of expectations".[29] This elasticity is defined as unity when a change in the current price causes an equi-proportionate change in the expected price. Hence, if the elasticity of expectations is positive, but less than unity, speculation will still have a stabilising influence, though, of course, a weaker one than if the elasticity is zero. The case of unit elasticity of expectations is, as Professor Hicks has said, on the borderline between stability and instability.

Thus the elasticity of expectations and the elasticity of speculative stocks *together* determine what may be termed "the degree of price-stabilising influence of speculation": the extent to which price-variations, due to outside causes, are eliminated by speculation. This may be measured by the proportionate change in stocks in response to a given change in the *current* price, since the larger this change, the smaller the extent to which any given change in outside factors (a shift in demand or supply) can affect the price. If we denote the degree of price-stabilising influence by σ, the elasticity of speculative stocks (as above defined) by e, and the elasticity of expectations by η, their relation is as follows:

$$\sigma = -e(\eta - 1)$$

Since e cannot be negative, the expression is negative or positive according as η is greater or less than 1.

It is not possible, however, to express the behaviour of expectations at any given moment in terms of a single elasticity. For what this elasticity will be, on a particular day, will depend on the magnitude of the price-change on that day, the price-history of previous days, and on whether the price expectation refers to next day, next month, or next year. This elasticity is thus likely to be both large and small, at the same time, according as the price-change has been large or small, and according as the expectation refers to the near future or the more distant future. It will vary, moreover, with the *cause* of the price-change. For it is permissible to assume that in most markets, speculators regard price-changes merely as indicators of certain forces at work, and that they attempt to form some idea as to the nature of these forces before adjusting their expectations. A given change in price will react differently on speculators' expectations according as they regard it as the result of speculative forces, or of "outside" demand or "outside" supply and so on.

If any generalisation can be made it is that expectations are likely to be less elastic as regards the more distant future than as regards the near future, and as regards larger changes in price than as regards smaller changes. These two factors moreover are not independent of each other. For the expectations as regards the more distant future are likely to be more and more influenced by the speculators' idea as to the "normal price" – this "normal price" is determined by different factors in different markets, but as we shall see below, it is likely to function in most – and *the larger the deviation of the current price from the normal, the longer must the period be which speculators expect to elapse before the price reverts to normality.* Beyond a certain point, therefore (whether the word "point" is understood to be a distance of time in the future or a magnitude of price-change), expectations become insensitive; the elasticity of expectations becomes zero or may even become negative.

Speculation, therefore, is much more likely to operate in a destabilising direction when we consider price-fluctuations within smaller ranges, than larger ranges; and when we consider the movements over a shorter period, than over a longer period. This is so not only because it is the short-period expectations which are most *elastic* (show the strongest reaction to price-changes), but also because it is these short-period expectations which are most *flexible* (are most liable to spontaneous changes). Those changes in expectations which are caused by the speculators' own attempts to anticipate each other's reactions, and thus set up purely spurious price-movements, are essentially short-period expectations; and they are responsible for short-period movements.

Hence to our question: does speculation exert a price-stabilising influence, or the opposite? The most likely answer is that it is neither, or rather that it is both simultaneously. It is probable that in every market there is a certain range of price-oscillation within which speculation works in a destabilising direction while *outside* that range it has a stabilising effect. Where markets differ, is in the magnitude of this critical range of price-oscillation. In some markets (among which the market for long-term bonds is conspicuous) this range is in normal circumstances relatively small, the stabilising forces of speculation dominate over the others. In other cases (and here one may count perhaps the markets for certain classes of ordinary shares) the range is large and the stabilising forces are relatively weak. The explanation of these differences lies in the varying degree of influence exerted by the idea of the "normal price" in the different markets. It is to this factor that we must now turn.[30]

In the case of many commodities, and in particular, non-agricultural raw materials, the elasticity of supply is rather high if the period of adjustment is allowed for, though supply is inelastic for periods shorter than this period of adjustment.[31] The upper limit of the expected price, for periods longer than this period of adjustment, is given by this supply price. The stability, therefore, depends on the general belief that the normal supply price in the future will not be very different from the normal supply price of the past; it is ultimately a belief in the stability of money wages.[32] It is in this way that the rigidity of money wages contributes to the stability of the economic system, by inducing the forces of speculation to operate in a much more stabilising fashion than they would do if money wages were flexible.

If the current price is in excess of the normal supply price, the future date at which the price is expected to return to the normal is given by the period of adjustment. If the current price is below the normal supply price, the date at which the price is expected to return to the normal is determined by the speculators' expectations as to the period of absorption of excessive stocks. Hence in markets where the belief in the existence of a normal supply price is strong, and where speculators are in a position to form a reasonable opinion as to the size of redundant stocks, we should expect speculation to work mainly in a stabilising direction.

Such is the case with the main industrial raw materials; and yet we find that here price-fluctuations are more violent than in most other markets.[33] The explanation for this, however, lies in the sudden changes to which the yield of stocks of such raw materials is liable, combined with the low elasticity of speculative stocks. Traders' requirements of such stocks, as stated before, are a fairly fixed proportion of the expected turnover; and a relatively slight reduction in the expected turnover is sufficient to bring their marginal yield down to zero. But this necessitates a very sharp reduction in the current price, since net carrying costs, which were negative before, have now become positive. Assuming that previously the price was equal to the normal supply price, so that the forward price fell short of the current price by the amount of the "normal backwardation", the price must fall far enough for the backwardation to turn into a contango, sufficient to cover interest plus carrying costs. If the "normal backwardation" (ie, the marginal risk premium) was equal to 10% per annum[34] and the interest cost plus carrying cost (when the yield is zero) is also equal to 10% per annum, the current price must fall by 20% if the excess stocks are expected to be absorbed in two years.[35] In the converse case, where owing to a rise in turnover, stocks fall below the normal proportion, there might be an equally sharp rise in price due to the rapid rise in the yield of stocks. According to Mr. Keynes,[36] it is easy to quote cases where the backwardation amounted to 30% per annum; and this implies that the marginal yield of stocks must have risen to 40%. Hence the explanation for the wide fluctuations in raw material prices need not be sought in any inelasticity of supply (except, of course, in the short period), or in the destabilising influence of speculative activities, but simply in instability of demand and the low elasticity of speculative stocks.

In the case of agricultural crops, the supply curve is much less elastic and is subject to frequent and unpredictable shifts, due to the weather. It is impossible to foretell the size of the crop a year ahead, and it is not possible to say when prices will revert to normal. Hence, in this case, the effect of speculation on price-stability is much more doubtful. If, nevertheless, the recorded range of price-fluctuations is no greater than in the case of industrial raw materials, the explanation might be sought in the fact, as an article in *The Economist* recently suggested,[37] that these goods are mostly foodstuffs the demand for which is much more stable than the demand for raw materials.[38] The fluctuations in the size of stocks are caused by changes in the supply side; and since the southern hemisphere became an important source of supply, the extent of the annual fluctuation in world output is much less considerable. It would probably be found, however, that in the case of agricultural crops the *same* percentage changes in stocks are associated with greater variations of prices than in the case of individual raw materials (not because the elasticity of speculative stocks is less, but the elasticity of expectations is probably greater).

In the market for securities (bonds and shares) there is no such "external" determinant of normal price as the producers' supply price in the case of commodities. Yet in the market for long-term bonds the notion of a "normal price" operates very strongly; only thus can the remarkable stability of the long-term rate, relatively to the short-term rate, be explained. Since the elasticity of speculative stocks, in this case, is very large, the current price is largely determined by the expected price; if the current price is stable, this must be because the expected price is insensitive to short-period variations of the current price. But how is this expected price determined?

The simplest explanation which suggests itself is that the expected price is determined by some average of past prices. The longer the period in the past, which enters into the calculation of this average, the less sensitive is the expected price to movements in the current price.

This hypothesis, however, taken by itself, is not very satisfactory. In the first place, as Professor Robertson has pointed out, it appears to leave the long-term rate of interest "hanging by its own bootstraps". If the current price of consols is determined by the expected price, and the expected price by an average of past prices, how were these past prices determined? How did the rate of interest come to settle at one particular level rather than at some other level? In the second place, why should expectations be so inelastic in the long-term bond market in particular? The fact that the price of many industrial shares moves fairly closely with fluctuations in current earnings, does not suggest that this is a characteristic attribute of security markets in general.

These difficulties can be resolved, however, if account is taken of the dependence of the current long-term rate on the expected future short-term rates. A loan for a particular duration, say for two years, can be regarded, as Professor Hicks has shown,[39] as being made up of a "spot" short-term loan (say for three months) *plus* a series of "forward" short-term loans; a short-term loan renewed so many times, and contracted forward. Hence the existence of a long-term loan market *implies* the existence of a series of forward markets in short-term loans. And since the forward price of anything, if uncertainty is present, must be below the expected price, this implies that the current rate of interest on loans of any particular duration must be above the average of expected future short-term rates, over the same period.

This explains why the yield of bonds with a currency of, say, twenty years or over is so stable. For it depends on the expected average of short-term rates over the next twenty years and we need not suppose that the elasticity of expectations with respect to the short-term rate is small in order that this average should be stable, eg, if the current short-term rate is half the "normal rate", and is expected to remain in operation for as long as five years, this will only reduce the average over the next 20 years by one-eighth. Thus the explanation is not that interest-expectations are particularly inelastic; the expectations regarding the short-term rate would need to be extremely elastic in order that the expected long-term rate should be responsive to changes in the current rate.[40] This also answers the objection that the expectation-theory leaves the structure of interest rates, current and expected, hanging in the air by its own bootstraps. For while the current long rate depends on the expected short rates, the current short rate is *not* dependent either on the expected short rates or the expected long rate.[41] The latter is not dependent on expectations at all (or only to a very minor extent) but only on the current demand for cash balances (for transaction purposes) and the current supply. And since the elasticity of supply of cash with respect to the short-term rate is normally much larger than the elasticity of demand, the current short-term rate can be treated simply as a datum, determined by the policy of the central bank.[42]

How is this related to Mr. Keynes' theory of the long-term rate of interest being determined by liquidity-preference? It leads to much the same result (ie, that the rate, in the short period, is not determined by savings and investment), but the route by which it is reached is rather different. The insensitivity of the long-term rate to "outside" influences (ie, the supply and demand for "savings") is not due to any "liquidity premium" attached to money or short-term bills; in fact, the notion of a "liquidity premium" appears entirely absent. It may be objected that it is merely replaced by the notion of a marginal risk premium, and that Mr. Keynes' "liquidity premium" on the holding of short-term assets is merely the negative of our marginal risk premium on the holding of long-term bonds.

But in this case, the peculiar behaviour of the long-term interest rate is certainly not explained by the existence of this risk-premium.[43] For let us suppose that subjective expectations are quite certain, so that this risk premium is completely absent. In this analysis the current long-term rate, instead of being above the average of expected short-term rates, will be equal to that average. But "the elasticity of demand for ideal balances with respect to the long-term interest rate", in the sense in which Mr. Keynes employs this concept, would certainly not be removed thereby; on the contrary, it would be rendered infinite. For in the absence of uncertainty, when the marginal risk premium is zero, the elasticity of speculative stocks in the bond market would be infinite (since the other factors determining the elasticity of speculative stocks, ie, interest costs and carrying costs, are, in this case, also infinitely elastic). The current price of bonds would be entirely governed by the expected price; if we assume the expectations concerning future interest rates as given, the current long-term rate would be uniquely determined.[44] Changes in the rate of savings or in the marginal efficiency of capital could not affect the long-term rate at all, except through the slow and dubious channel of changing the level of incomes, and, hence, the demand for cash; this reacting on the short-term rate of interest, and this, in turn, gradually affecting the interest expectations for the future.

It is therefore not so much the *uncertainty* concerning future interest rates as the *inelasticity* of interest expectations which is responsible for Mr. Keynes' "liquidity preference function", ie, which causes the demand for short-term funds to be elastic with respect to the long-term rate.[45] The uncertainty of expectations works rather in the opposite direction. For when uncertainty is high not only is the amount of the marginal risk premium high, but the elasticity of the marginal risk premium curve (and hence the elasticity of speculative stocks) is low; speculators are much less willing to extend their commitments or to reduce them. It is in such periods that variations in the "outside" demand or supply can be expected to have the most marked effect on bond prices.[46]

Nor is it quite accurate to represent the forces determining the long-term rate as the demand for liquid funds (money + bills) *vis-à-vis* the supply of such funds. We have seen that the proposition that "the rate of interest is the agency which equates the supply and demand for money" is true if under "rate of interest" is meant the short rate of interest, and under "money" the sum total of things which mainly serve as a means of payment (ie, bank notes and current deposits).[47] But in the case of the long-term rate, we are not really confronted with a demand curve for liquid funds but with a supply curve of long-term assets by speculators;[48] and the two methods of representation do not always yield the same conclusion. It is not clear, for example, why an increase in the aggregate amount of liquid funds should have any direct effect on the speculators' willingness to hold long-term assets, other than through the change in the short-term rate. An increase in the supply of money will directly reduce the short-term rate (unless it is already so low that it cannot fall further); this will have a minor effect on the long-term rate (for the current price of bonds will now be higher relatively to the expected price); a major effect will only come about slowly as expectations are affected.[49]

To sum up. In the market for long-term bonds, the elasticity of expectations is normally small, and the elasticity of speculative stocks is large, both absolutely and relatively to the elasticity of the non-speculative demand and supply (ie, the elasticity of supply of savings, and the elasticity of the producers' demand for long-term funds, as a function of the rate of interest). Hence the price in the short period is largely determined by speculative influences. It would not be correct to say that these outside factors have *no* influence upon price;[50] only that this influence is largely overshadowed by speculative forces. And for reasons given in the next section, the inability of these forces to influence the price in the short period will much weaken their price-determining influence in the long period as well.[51]

There remains to examine the situation in the market for shares; and after what has been said above, it should not be difficult to account for the apparent contrast between the stability in the gilt-edged market and the instability of the share market. Shares are also subject to the same influences as bond prices; changes in the expectations regarding future interest rates should affect the one just as much as the other. But in addition they are also subject to changes in the expectations concerning the expected level of profit. And here experience suggests that the elasticity of expectations concerning the future level of profits is fairly high; share prices correlate fairly well with fluctuations in current earnings.

There are several explanations for this. In the first place, the course of profits of any particular company is likely to exhibit a *trend*, in addition to fluctuations; and this trend may make it impossible for the expectation of a long-run "normal level" to develop such as is formed in the case of the short-term interest rate where there is no such trend. In the second place, owing to the absence of any means of information concerning the more distant future, the profits relating to the more distant periods are likely to be heavily discounted for uncertainty; thus the profits relating to the mere fact of discounting at compound interest. And, lastly, just because the short-period price fluctuations are large, investors are likely to concentrate attention on the short-period expectations concerning capital value, rather than the long-period prospects of enterprise; speculative motives get inseparably mixed up with other motives and it is no longer possible to distinguish between speculative demand and non-speculative demand. It is in such circumstances that the destabilising influence of speculation will dominate over a wide range of price oscillation.

SPECULATION AND INCOME-STABILITY

There remains, finally, to examine the effects of speculation on the stability of the general level of economic activity. Speculation affects the level of activity through the variations in the size of speculative stocks held. We have seen above that if there is speculation, any change in the conditions of the outside demand or supply will be followed by a change in the amount of stock held; if speculation is price-stabilising (the elasticity of expectations less than unity), a reduction in demand or an increase in supply will be followed by an increase in stocks; if speculation is price-destabilising (elasticity of expectations greater than unity) a reduction in demand or increase in supply will be followed by a decrease in stocks, and vice versa.

An increase in speculative stocks of any commodity implies an increase of investment in that commodity, or, to use Mr. Hawtrey's expression, a "release of cash" by the market. (Conversely, a reduction of stocks implies an absorption of cash.) Unless this increase is simultaneously compensated by a decrease in the amount of stocks held in other commodities, it must imply a corresponding change in the level of investment for the system as a whole. Such compensation can take place in two ways. (i) If there are several goods whose markets are speculative, an increase in speculative stocks in one market may, in certain cases which we shall analyse below, involve a consequential decrease in speculative stocks in other markets. (ii) If the monetary authorities raise the short-term rate of interest they can induce a reduction in stocks in the system as a whole. Without making some assumptions about monetary policy, it is impossible, therefore, to make any generalisations about the effects of changes in the amount of speculative stocks. In the present section we shall assume that the monetary authorities maintain the short-term rate of interest constant. In the last section we shall deal with the question of how far the instabilities due to speculation can be counteracted if a different monetary policy is adopted.

Before proceeding further we must introduce the distinction between "income goods" and "capital goods". There are certain categories of goods in the case of which it can be assumed that *spontaneous* changes in the amount of money spent on them (per unit period) will be associated with equivalent and opposite changes in the amount of money spent on other goods. There are other categories of goods for which this is not (or not necessarily) the case. The first category may be said to consist of goods which are ordinarily bought "out of income", the second category, of goods bought "out of capital" (ie, whose purchase is charged to a capital account, and not to an income account). For under the assumption that *all* the income of individuals is spent one way or another,[52] and the level of income is given,[53] any change in the amount of money spent on anything must be associated by an opposite change in the amount of money spent on something else. In the case of goods purchased on a capital account this is not so, for variations in the amount of money spent on them may represent variations in the total volume of borrowing. Thus our distinction between "income goods" and "capital goods" comes close to the usual distinction between "consumption goods" and "capital goods", though we must remember that in our case, it is not the character or the destination of the goods which is at the basis of the distinction but simply whether spontaneous changes in the amount of money spent on them imply simultaneous changes in the total amount of expenditure or not.

In the case of income goods, a change in speculative stocks, which has a stabilising effect on price, must exert a destabilising influence on the level of economic activity, irrespective of whether the change was due to a change in the conditions of demand or supply. For an increase in the demand (a shift in the outside demand curve to the right), by causing a reduction in speculative stocks, also entails a reduction in the level of incomes, since the increase in the demand is associated with a reduction in the amount spent, and thus of incomes earned, in the production of other things, without any compensating increase in incomes earned in the particular commodity in question. Conversely, a reduction in demand will involve an increase in the level of incomes, for the same reason. An increase in supply on the other hand (a shift in the outside supply curve to the right) will involve an increase in speculative stocks and thus an increase in the level of incomes; a reduction in supply a reduction in incomes. None of these changes in incomes could have taken place in the absence of speculation; and the extent of the change in incomes, following upon a given shift in supply and demand, will be the greater the larger the price-stabilising influence of speculation as above defined.[54]

We can conclude, therefore, that in the case of income goods, an increase in demand or a reduction of supply will cause a reduction in incomes, and vice versa, the magnitude of which depends, *inter alia*, on the degree of price-stabilising influence. When this is negative (speculation is price destabilising) a reduction in demand or an increase in supply will cause a reduction in incomes, and an increase in demand or a reduction in supply an increase in incomes. Only when the degree of price-stabilising influence is zero will a shift in demand or supply leave the level of incomes unchanged.

It might be objected that this effect of speculative activity on incomes will be purely temporary; for while a discrepancy between "outside" demand and "outside" supply lasts, speculative stocks must increase (or decrease, as the case may be) continuously and this process cannot go on forever. There is no market which has either resources or the inclination to go on "absorbing cash" or "releasing cash" indefinitely; sooner or later, the elasticity of speculative stocks and, hence, the degree of price-stabilising influence, must become zero. This is true, but it does not follow from this that either the price-stabilising or the income-raising (or income-lowering) effect of speculation is purely temporary. For in certain cases the change in incomes itself provides a mechanism which brings the discrepancy between demand and supply gradually to an end. For commodities whose income-elasticity of demand is greater than zero (or to use another terminology, where the marginal propensity to consume is positive) an increase in incomes will involve an expansion in demand, and vice versa. And since we have seen that if the price-stabilising influence of speculation is positive, an excess of supply over demand must lead to an increase in incomes, and an excess of demand over supply to a reduction in incomes, in both cases the discrepancy between demand and supply will tend to get adjusted through the variations in demand caused by the change in incomes.[55]

This is nothing else but the well-known doctrine of the Multiplier put forward in a generalised form. The latter, strictly speaking, assumes that the degree of price-stabilising influence is infinite, so that the price can be regarded as constant. In that case the increase in incomes created by a given increase in supply (or a given reduction in demand) will be (ultimately) as many times the value of the original increase in supply (or reduction in demand) as the reciprocal of the proportion of marginal income spend on the commodity in question. This is so because so long as the excess of supply over demand continues, stocks are accumulated and entrepreneurs' receipts continue to rise. When income has risen in the ratio stated above, the discrepancy between demand and supply disappears (owing to the adjustment in demand) and the accumulation of stocks ceases.

The doctrine of the Multiplier could therefore be restated as follows. The demand and supply of anything always tend towards equality. In normal cases this equality is secured by adjustments in the price of that commodity. But if changes in the price of a particular commodity are attended by variations in the amount of stocks held, the equality may be secured – to a greater or lesser extent – by adjustments in the level of incomes. Only if the stabilising influence of speculation is infinite and the income-elasticity of demand greater than zero can the adjustment come about entirely through a change in incomes. In general, following upon a change in the demand or the supply schedule, the mechanism of adjustment is all the more likely to operate through a change in incomes, rather than a change in price: (i) the greater the degree of price-stabilising influence of speculation; (ii) the greater the proportion of marginal income spent on the good; (iii) the less is the elasticity of the (outside) demand and supply schedules.

This doctrine is subject, however, to an important qualification. If there is more than one good whose price is stabilised, or quasi-stabilised, through speculative activities, the mechanism of the demand-and-supply equalisation through a change in incomes will not operate in the same way as in the case of only a single good. Suppose that out of a large number of things on which consumers spend their incomes, coffee is the sole good whose price is stabilised by compensatory stock-variation. In that case, whenever there is, say, an increase in the supply of coffee, stocks of coffee will increase, incomes will rise, and this process will go on until the demand for coffee, in consequence, has risen in the same proportion in which the supply has increased. But if the price of sugar is also subject to the same influences, and consumers spend part of their increased incomes on that commodity, then whenever there is an increase in the stocks of coffee and a consequent increase in incomes there will also be a decrease in the stocks of sugar, which will act as a brake on the rise in incomes. In this case the level of incomes can never rise sufficiently (as a consequence of the increase in the supply of coffee) to adjust demand fully to the increase in supply. Hence, the existence of more than one "multiplier" weakens the operation of any one of them.

In the real world the most important case of "income goods" whose prices are stabilised through speculation are long-term bonds dealt in the capital market. On our definition the things bought and sold in the capital market are not "capital goods", but "income goods", for the outside or non-speculative demand for securities consists of savings, savings are "paid" out of income; a change in the rate of savings out of a given income must imply an opposite change in the amount spent on other income goods. And since both the proportion of marginal income going to savings and the elasticity of speculative stocks for bonds is large, it is in the case of the long-term investment market that one would expect disequilibria due to changes in supply or demand (ie, in the propensity to save or in the investment-demand schedule) to be adjusted through variation in the level of incomes, rather than in the long-term rate of interest.

This, however, cannot be entirely true, except for a closed system. In the case of an individual country which is on the Gold Standard, or where the price of foreign exchange is stabilised through the operations of an Exchange Equalisation Fund, the price of foreign exchange, and, hence, the price of imported goods, behaves in much the same manner as the prices of goods stabilised by speculation. Hence, in addition to the "savings-investment multiplier" there exists a "foreign trade multiplier", and the operation of the latter must weaken the price-stabilising forces operating in the former (and vice versa).

This can be best elucidated by an example. Let us suppose that for the community as a whole, 25% of additional income is saved, and that there is an increase in the rate of long-term investment (financed through the sale of securities) by £1 million per week. This will immediately increase incomes in the investment goods industries by £1 million and savings by £250,000. Thus, while in the first "week" the investment market was required to furnish the whole of the additional expenditure of £1 million out of its speculative funds, in the second week it only has to furnish three-quarters of that amount; one quarter will be furnished through the additional savings.[56] If we suppose that all the income which is not saved is spent on home-produced goods, there will be a further increase in incomes by £750,000 in the second week, and £562,500 in the third week, and so on.[57] Similarly, the outside demand for securities will expand by £187,500 in the third week, £140,625 in the fourth week, and so on. As a result, after a certain number of "weeks", total incomes will have expanded by £4 millions per week, and total savings by £1 million; the outside demand for securities will ultimately have increased in the same rate as the outside supply. The size of speculative commitments has increased, of course, during this process, but this increase will come to a halt once the increase in outside demand has caught up with the increase in the rate of supply; the contribution which the speculative resources of the market have to make is limited. Provided that the total required increase in the size of speculative stocks is not too large relatively to the resources of the market (ie, provided it does not impair the degree of price-stabilising influence) there will be no pressure on the price of securities (ie, no tendency for the rate of interest to rise) either in the long run, or in the short run.

Let us suppose now, however, that not three-quarters, but only a half of the marginal income is spent on home-produced goods; and while 25% is saved, another 25% is spent on additional imports. In that case the "multiplier" will not be four, but only two; the ultimate increase in incomes, following upon a £1 million increase in the rate of investment, will be £2 millions (per week) and of savings £500,000. Hence, even after incomes have been fully adjusted to the change in the level of investment the rate of outside demand for securities will have only increased by one-half of the increase in the rate of supply; if the rate of interest is to remain unchanged, speculators will have to furnish £500,000 per week, *indefinitely*.

It is true that in this case the increase in speculative stocks in the securities market is attended by a decrease in stocks (of gold and foreign exchange) in the foreign exchange market; while in the first market speculators continuously borrow (on short term) in the second market they will continuously lend; and since the one must be numerically equal to the other, there can be no tendency for the short-term interest rate to change.[58] In Keynesian terminology one might also say that investment and savings will be equal; for if the export surplus is added to investment, the net increase in the rate of investment will not be £1,000,000, but only £500,000. Yet the situation will obviously not be one of equilibrium: it could not be indefinitely maintained. As the stocks of gold and foreign exchange continuously decrease, and the speculative commitments in the long-term investment market increase, one of two things must happen: either the price of gold (and foreign exchange) must rise, thus raising the prices of imported goods and thereby the home-investment multiplier, or the price of securities must fall (the long-term rate of interest rise) and thus the level of home investment will be reduced. Whether equilibrium is finally reached through a rise in the price of foreign exchange, or of the rate of interest, will depend on the relative elasticity of speculative stocks in the two markets (or rather on the range over which, in the two markets, speculative stocks are elastic). Assuming that it takes place through the second and not the first, it would be wrong to attribute the rise in the interest rate to any shortage of *cash*. In the situation contemplated there is no shortage of cash, and no pressure on the short-term rate of interest. The long-term rate rises *relatively* to the short-term rate simply because, owing to a shortage of savings, speculators are required to expand continuously the size of their commitments: and there are limits to the extent to which this is possible.

It is not quite accurate therefore, when Mr. Keynes says that "the investment market can be congested on account of a shortage of cash. It can never be congested on account of a shortage of savings".[59] This proposition would only hold if it were assumed, in addition to the degree of price-stabilising influence being infinite, that the long-term investment market is the only market which is subject to such speculative influences; in other words, that dC/dY (the marginal propensity to consume) which is relevant for calculating the home-investment multiplier, is necessarily equal to $1 - (dS/dY)$ (where dS/dY is the marginal propensity to save). This, as we have seen, need not be the case.[60]

Mr. Keynes' General Theory, therefore, in so far as it concerns his theory of the rate of interest and the theory of the multiplier, is rather in the nature of a "special case"; strictly it only holds if σ is infinite in the case of long-term bonds and zero in the case of everything else. But if we abstract from the case of foreign exchange under a gold standard or quasi-gold-standard regime, it is true that the degree of price-stabilising influence, though not perhaps infinite, is very much larger in the case of long-term bonds than for any other commodity; and this means that the Keynesian theory, though a "special case", gives, nevertheless, a fair approximation to reality.

In the case of "capital goods", a change in speculative stocks which has a stabilising effect on price will also have a stabilising influence on activity, if the change in stocks was due to a shift in demand, and a destabilising influence, if it was due to a change in supply. This difference is due to the fact that changes in the demand for capital goods are not (normally) associated with opposite change in the demand for other goods; they represent variations in the investment-demand schedule (in the marginal efficiency of capital) and such variations, in a regime where the long-term interest rate is itself subject to the price-stabilising influence of speculation, and/or where the short-term rate of interest is held constant, must imply variations in the total scale of expenditures. Hence, an increase in the demand for capital goods involves an increase in the level of incomes, when σ is zero (in the absence of speculation); in so far as σ is positive (the increase in demand is associated with a reduction in stocks), the increase in incomes will be less than it would be otherwise. Thus, the existence of speculation in capital goods, if price-stabilising, damps down the multiplier and hence the range of income fluctuations, in much the same way as the existence of a stabilised price for foreign exchange in our previous example. But whereas in the case of foreign exchange, there can be "spontaneous" changes in demand which themselves set up income-variations (when, eg, there is a spontaneous change in the propensity to consume foreign goods) the changes in demand for capital goods are always "induced"; they reduce the magnitude of other multipliers but do not have a multiplier-effect of their own.

It follows from this that in a world where the long-term rate of interest is a "conventional phenomenon" (to use Mr. Hawtrey's phrase) the existence of price-stabilisation schemes for raw materials which are mainly used for the production of capital goods, and operated by means of compensatory stock-variation[61] (and *not* by the restriction of supply) will have a beneficial effect on economic stability; it will reduce the instability arising from the operation of the savings-investment market. We must remember however that this conclusion only holds if the price-variation which the stabilisation scheme is designed to eliminate, arises wholly or mainly from fluctuations in demand, rather than fluctuations in supply. In the case of agricultural crops, where it is supply, rather than demand, which is unstable, such price-stabilisation schemes must have a destabilising influence on activity.[62]

We may now summarise the argument of this section. Speculation in any good, by creating uncompensated variations in the amount of stocks held, is responsible for variations in the general level of activity. If the effect of speculation is price-destabilising (a rise in price being associated with an increase in speculative stocks, and vice versa) an increase in demand or a decrease in supply will be followed by an increase in the level of activity, and vice versa. If the effect of speculation is price-stabilising (a rise in price is associated with a decrease in speculative stocks) an increase in demand or a decrease in supply will be followed by a reduction in activity, a fall in demand or an increase in supply by a rise in activity.

If there is more than one good which is subject to speculation of the price-stabilising type, and if the income elasticity of demand of such goods is positive, changes in the amount of speculative stocks in any one market will set up compensatory changes in the amount of stocks held in other markets, and thereby make the variations in the level of activity smaller than they would be otherwise. This compensatory effect will be all the stronger the more alike is the degree of price-stabilising influence in the different markets. Hence, while the introduction of speculation in any particular market in a system which is otherwise free from speculation is bound to have an unsettling effect on economic activity (quite irrespective of whether it is price-stabilising or price-destabilising) the extension of speculation over *further markets*, if price-stabilising, may reduce the instability in the level of incomes. It is not so much therefore the existence of speculation as such but its widely differing incidence between the different markets – the fact that while in some markets speculation plays such a dominating role, in others it has only an insignificant influence – which is responsible for the instability of our economic system.

MONETARY POLICY AND STABILITY

The analysis of the previous section was based on the assumption that the monetary authorities hold the short-term interest rate constant, despite the fluctuations in the level of money incomes. Suppose that the monetary authorities raise the short-term rate whenever there is an expansion of activity, and lower it when there is a contraction, would this not secure stability of incomes, despite the destabilising influences emanating from speculative markets? Mr. Hawtrey, who more than anyone else regards economic stability as a matter of bank rate policy, himself admits that the *sources* of inflationary and deflationary tendencies may lie in markets other than the market for short-term credit.[63] But he argues that it is the business of the monetary authorities to counteract such tendencies by appropriate adjustments of the bank rate.

There are, of course, serious practical difficulties confronting a policy of this kind: the difficulty of determining the correct timing and the appropriate magnitude of the changes that are required. For if the changes in the bank rate are incorrectly timed, or if they are greater than necessary, the instrument of the bank rate becomes a source of further instability and not a stabiliser. But apart from these practical difficulties, there are also certain theoretical limitations to the extent to which stability can be secured by monetary policy; and the remaining part of this paper may be devoted to a discussion of some of these limitations.

The instrument of the bank rate operates through its effect on the amount of stocks held of commodities in general. Hence, if any particular market becomes a source of inflation (ie, begins to accumulate stocks) it is never possible to eliminate the source of the disturbance *directly* through the changes in the short-term rate; all that can be achieved is to offset the accumulation of stocks in the one market by enforcing a reduction in stocks held in all the other markets. There is, of course, always *some* rate of interest which, if enforced, would be sufficient to stop the inflationary tendency emanating from any particular market. But unless this rate of interest is confined to lenders and borrowers in that particular market, its effect would be clearly deflationary; for in addition to preventing the accumulation of stocks in the "inflationary" market, it would involve a decumulation of stocks in all the others. Hence, the "appropriate" short-term rate (the one which leaves the level of incomes unchanged) is one which balances the accumulation of stocks by a decumulation elsewhere, not one which prevents it.

The effect of any change in the short rate of interest on activity is purely temporary. It operates by causing an adjustment in the size of stocks to the new interest rate; once stocks have been adjusted to the new rate, its effect is exhausted. A 6% bank rate, provided it has been in operation long enough, is no more "deflationary" than a 4% rate or a 2% rate (apart from its effect on the long-term rate, which we shall discuss below). It is not the absolute level of the short-term rate, therefore, but its change from some previous level which acts as a stimulant (or the reverse). Hence, if a particular market becomes the source of an inflationary tendency, and continues to do so, even after the effect of the rise in the bank rate has exhausted itself, a further rise in the bank rate becomes necessary, and so on. An expanding or contracting tendency in the level of incomes may not be counteracted, therefore, by a single change of the bank rate, but only by a series of changes.

The instrument of the bank rate is subject, moreover, to a great constitutional weakness: it cannot operate equally freely in both directions. It should always be possible to stop an expansion of activity by a rise in the short-term rate of interest, whatever the strength of the forces of expansion. As Mr. Hawtrey says: "If a rate of, say, 6% were found to be negligible, the rate could be raised indefinitely. No one would regard a rate of 60% as negligible."[64] But it is not equally possible to prevent a contraction of activity through a reduction in the rate. For the rate can never be below zero; and as we have seen, there is a certain level, above zero, at which the power of the monetary authorities to determine the rate (by varying the quantity of money) ceases to function.[65] In order that the bank rate mechanism should be efficacious in the downward direction, as well as in the upward direction, its *average level* must be kept high so as to leave sufficient "elbow room" for reductions, as well as increases. For if, after a period of stable activity, a deflationary process develops, it is much easier to counteract it by monetary measures if the short-term rate stood previously at 10% or 6% than if it stood at 4% or

3%. The effectiveness of a reduction in the bank rate from 4% to 2% is no greater than that of a reduction from 10% to 8%; and so far, Central Banks have never found it practicable to reduce the official rate of discount below 2%.

We have seen above that as far as the short-term rate of interest is concerned, it is the change in the rate, and not its absolute level, which has an influence on activity. Hence, if the average long-run level of the short-term rate is kept at 10% this is no more detrimental to activity than if the average rate is 3%. This is not so, however, when the effect on the long-term rate is taken into account. In the short period, as we have seen, changes in the short-term rate have only a minor effect on the long-term rate, because the latter depends on the *average* of short-term rates expected over a long period in the future,[66] and this average is but little affected by the changes in the current rate. But if the monetary authorities, in their long-period policy, regulate the short-term rate in such a way as to keep its average level high, then it is this average level which will come to be regarded as the "long period normal" and thus govern the level of the long-term rate. If over considerable periods in the past the monetary authorities varied the bank rate between 2% and 10% and maintained it on the average at 6% (thus allowing for a 4% elbow room in each direction) the long-term rate will settle at a level which is *above* 6%.[67] This in turn, by restricting long-term investment, will have a considerably depressing effect on the *average* level of activity.

In the long run, therefore, the monetary authorities are not free to vary the short-term rate as they like; if they want to maintain activity at a satisfactory level, they must keep the mean level of the short-term rate sufficiently low so as to secure a long-term rate which permits a sufficient amount of long-term investment. Alternatively, if they want to secure stability by means of monetary policy, they must allow the average level of employment to fall to a low enough level to permit the mean level of the short-term rate to be sufficiently high. Thus the two main aims of monetary policy, to secure a satisfactory level of incomes, and to secure stability of incomes, may prove incompatible: the one may only be achieved by sacrificing the other.

Assuming that the monetary authorities regard the achievement of a satisfactory level of activity as their paramount consideration, we must expect the bank rate mechanism, as an instrument of economic policy, to become increasingly ineffectual. As real incomes increase, savings rise, and the available investment opportunities become smaller, the bank rate, though still available for dealing with an occasional boom, becomes more and more ineffective as a safeguard against the ravages of deflation.

Mr. Hawtrey is well aware of this constitutional weakness of the bank rate mechanism; and he calls the state of affairs where the mechanism is put out of action through the bank rate having reached its minimum level a "credit deadlock". But he is too much inclined, I think, to attribute the emergence of a "credit deadlock" to past mistakes in banking policy – to the central bank not having lowered the rate sufficiently soon, or sufficiently suddenly – rather than to the inherent causes connected with the long-term rate.[68] If Professor Hicks' calculations are right,[69] and 2% is now to be regarded as the necessary marginal risk premium by which the current long-term rate exceeds the average savings deposit rate, then a 3% rate on consols presupposes a 1% average on deposits; and a 1% rate on deposits is perilously near its absolute minimum level. If "full employment" requires a long-term rate which is below 3%, and this is what the monetary authorities aim at, the "credit deadlock" becomes a more or less permanent state of affairs.

In a world of perpetual or semi-perpetual credit deadlock, stability cannot be achieved by monetary policy (in the sense in which this term is ordinarily understood); nor can fluctuations in the level of activity be regarded as a "purely monetary phenomenon". For in those circumstances monetary factors can neither be said to have caused the fluctuations, nor have they the power to prevent them.

1 The expectations of different individuals composing the market are normally different, of course. But it is permissible to speak of a single expectation for the market as a whole, since *cet. par.* there is always a definite amount of any good that would be held, at any particular expectation, if all individuals' expectations were the same.

2 Cf. Keynes, *General Theory*, Chapter 12 on "Long Term Expectations". "We have reached the third degree [in the share markets] where we devote our intelligences to anticipating what average opinion expects average opinion to be." (p. 156.)

3 In conditions of hyper-inflation – as in Germany in 1923 – the range of goods in which speculation takes place is, of course, very much extended.

4 In particular the degree of standardisation required is very high. For the difference between the simultaneous buying price and selling price can only be small when there is a large and steady volume of transactions, in the same article, per unit of time, so that it pays individuals to undertake purchases and sales through the agency of an organised market. It is necessary for this that the commodity in question should have but few attributes of quality (that it should be simple) so that a specification of standard qualities ("grades") can be drawn up without difficulty. In the organised exchanges of the world this standardisation is carried so far that buyers rarely see the article they are actually buying – contracts are made between buyers and sellers with reference to standard grades and places of delivery. It is also necessary that the amount bought by the representative individual in a single transaction should represent considerable value. Nobody would go to the trouble of buying, say, cigarettes, through the agency of a central market, even if this would imply some saving.

5 By defining yield as that return which goods obtain when measured in terms of themselves, we exclude here any return due to appreciation of value (in terms of some standard) whether expected or unexpected.

6 Our definition of net carrying cost is, therefore, the negative of Mr. Keynes' "own-rate" of own-interest" in Chapter 17 of the *General Theory* – except that no allowance is made here for the factor termed "liquidity premium". Our reason for deducting the yield from the carrying cost, and not the other way round, is because (as will be clear below) from the point of view of speculation, net carrying cost is the significant concept rather than the own-rate of interest.

7 There is, of course, in addition, the stock of goods in the course of production (goods in process) which depends on the length of the production process, but with this we are not here concerned (since they are not standardised).
8 This, as we shall see below, is equally true of stocks of money as of other commodities.
9 Mr. Keynes, in the *Treatise on Money*, uses the term "working capital" for stocks which have a positive yield, and "liquid capital" for those which have a zero yield. (vol. 2, p. 130.)
10 That is, the seller of a second-hand machine must not only allow for a reduction of value due to depreciation, but also an extra loss due to the fact that he is selling the machine and not buying it.
11 Mr. Keynes, in certain parts of the *General Theory*, appears to use the term "liquidity" in a sense which comes very close to our concept of "perfect marketability"; ie, goods which can be sold at any time for the same price, or nearly the same price, at which they can be bought. Yet it is obvious that the attribute of goods is not the same thing as what Mr. Keynes really wants to mean by "liquidity". Certain gilt-edged securities can be bought on the stock exchange at a price which is only a small fraction higher than the price at which they can be sold; on this definition, therefore, they should have to be regarded as highly liquid assets. In fact it is very difficult to find a satisfactory definition of what constitutes "liquidity" – a difficulty, I think, which is inherent in the concept itself. As will be argued below, what appears to be the result of a preference for "liquidity" may be explained as the consequence of certain speculative activities in which "liquidity preference" in any private sense, plays a very small part.
12 (This section is a revised amalgam of the one published in the original version of the article, *Review of Economic Studies*, 1939–1940, pp. 5–7, and of my contribution to the Symposium, "A Note on the Theory of the Forward Market", *Review of Economic Studies*, 1939–1940, pp. 196–201).
13 If there is a difference between the speculators' borrowing rate and their lending rate, one or the other will be relevant, according as the marginal unit of commitment is made with borrowed funds or not. Hence when speculators reduce their commitments, interest costs may fall, and this has (even if the market rate is unchanged) a similar effect as a fall in carrying cost (rise in yield).
14 The term "primary depreciation" is Mr. Hawtrey's (*Capital and Employment*, p. 272), and is used to denote that part of depreciation which inevitably arises through the mere passage of time.
15 Ordinary stockholders and producers of the commodity may of course also indulge in speculation, in so far as they carry extra stocks in the expectation of a rise in price.
16 I use the term "futures" here as equivalent to a forward purchase or sale; in other words, I assume that the commodity dealt in is completely standardised – as in the case of the foreign exchange market – and hence there is no difference in the obligations assumed in a spot contract or a futures contract. If there are differences then as Mr. Dow has shown, they can be introduced as an additional element causing some further deviation in the futures price from the spot price. Cf. J. C. R. Dow, "A Theoretical Account of Futures Markets", *Review of Economic Studies*, 7, 1939–1940, pp. 185–95.
17 We must bear in mind however that the marginal convenience yield of stocks varies (in some cases fairly rapidly) with changes in the size of stocks in relation to turnover, so that even a moderate enlargement of stocks for arbitrage purposes might be sufficient to reduce q to zero.
18 (In the original article and again in the Symposium, *Review of Economic Studies*, 1939–1940, I held that where the yield was one of convenience, arbitrageurs may not obtain it (since the convenience yield accrues only to those who hold stocks as part of their normal course of business), so that the upper limit to contango is $i + c$ even in cases where q is positive, and $FP = EP - r + q$. I ignored, however, the fact that when the futures price exceeds the current price by more than $i + c - q$, it will pay ordinary holders to enlarge their stocks and

hedge by selling futures at the same time).
19 J. M. Keynes, *Treatise on Money* (vol. 2, pp. 142–4).
20 The reason why this is not the case in the forward transactions of the stock exchange is due to the convention that the yield is credited to the forward buyer (from the date of the contract) and not to the forward seller who actually holds the stock. Hence the forward price is equal to the current price plus interest cost and there is contango (since in this case $c = 0$, $FP = EP - r + q = CP + i$). Backwardation can only arise on the stock exchange if a fall in price is expected, and this expected fall, on account of a shortage of stock for immediate delivery which is known to be purely temporary, cannot be adequately reflected in the current price (eg, in the case of trans-Atlantic stocks, when arbitrageurs run out of stock and have to wait for fresh supplies to be sent across the Atlantic). It is always a sign, therefore, of the current price not being in equilibrium in relation to the expected price.
21 This is because the risk normally hedged against is the risk of a fall in the prices of the commodities actually carried by the manufacturer or trader, ie, the risk of an inventory loss. The opposite risk of a future rise in the prices of the materials which the manufacturer uses, with its attendant possibility of a lower profit on fabrication (or at least the sacrifice of the inventory profit he would have made if he had bought earlier) is not normally compensated by buying futures, simply because manufacturers are less afraid of making losses on account of rising raw material prices than on account of falling prices (since in times of rising raw material prices the increase in cost can normally be passed on to the buyer). A failure to make a profit on account of the failure to make an investment is not considered as contributing the same kind of "risk" as that of a loss on the capital actually invested. In exceptional cases, however, eg, where the manufacturer works on long-term contracts, and does not expect to be able to pass on the higher prices of materials by raising his selling price, or where the futures price is abnormally low in relation to the spot price (due to temporary shortages of stock), manufacturers may well allow their stocks to be run down below normal levels, and compensate for the risks arising therefrom by becoming buyers in the futures market.
22 If hedgers are buyers of futures on balance, they will buy futures if FP is equal to or lower than $EP + r$; whilst the speculators will sell futures if FP is equal to or higher than $EP + r$. However, with uniform expectations, this will not establish the futures price at $EP + r$ on account of the fact that it will be the arbitrageurs and not the speculators who provide the supply of futures to match the excess hedging on the buying side. Arbitrage, as we have seen, will always bring the futures prices to $CP + i + c - q$ which, on the assumption of unanimous expectations, equals $EP - r$. Speculation on the selling side can only appear as a result of differences in expectations.
23 (I am indebted to Mr. R. G. Hawtrey's criticism of the original version of this paper for the conclusions that follow).
24 The more favourable relationship between the futures price and the expected price resulting from this will mean however that the market will carry larger stocks than it would have carried otherwise. For, as we have seen, when expectations are uniform and the expected price is the same as the current price, the futures price will necessarily be below the current price. In a market where the bullish sentiment expecting a rise in price and the bearish sentiment expecting a fall are equally strong, the futures price may well establish itself in the neighbourhood of the current price, as being the only price capable of bringing speculative supply and demand into equilibrium. In that case, however, ordinary traders will enlarge their stocks in connection with arbitrage operations until the marginal convenience yield falls sufficiently to eliminate the profitability of such operations. This action by arbitrageurs may not depress the futures price significantly, when the speculative demand and supply is large in relation to the non-speculative demand for holding stocks.
25 (In the original version this elasticity was defined with reference to the percentage change

in the *difference* $EP - CP$. It was pointed out by E. Rothbarth in 1942 and recently by Mr. Streeten that to be consistent with the formula on p. 33, the elasticity should be defined in terms of the *ratio* of the two prices. The latter definition is correct in cases where the items $(i + c + r - q)$ are proportional to CP, whilst the former definition is appropriate when $(i + c + r - q)$ is fixed in money terms and hence varies, as a percentage, with a change in CP. The correct formula corresponding to the former definition is that given by Streeten in equation (3a). (*Review of Economic Studies*, October, 1958, p. 67.)

26 In certain markets the lending rate only is normally relevant, and not the borrowing rate. In this case the elasticity of interest cost can be taken as infinite.

27 In other words, the elasticity of the marginal risk premium is likely to vary inversely with the amount of the risk premium. When the risk premium is low, its elasticity is also likely to be high.

28 This does not imply, of course, that the current price must be *equal* to the expected price; this would only be the case if in addition, the sum of $i + c - q + r$ were zero. Nor does it imply that changes in the non-speculative factors can have no influence upon price at all, for changes in these factors may influence price-expectations.

29 *Value and Capital*, p. 205.

30 The following owes much to Hicks, *Value and Capital*, pp. 270–2.

31 This period of adjustment varies, I believe, for different commodities from six months to anything up to two years, though it is likely to be between six and twelve months in the majority of cases. The case of rubber and tin, where the period is several years, is exceptional. But in these last two cases, the short-period elasticity (through more or less intensive "tapping" or "plucking") is considerable.

32 For it is the level of money wages which governs money supply prices if the elasticity of supply is high.

33 Mr. Keynes stated in the *Economic Journal*, September, 1938, p. 451, that the difference between the highest and lowest price, in the same year, in the case of rubber, cotton, wheat and lead amounted in the average to 67% over the last ten years.

34 In markets where the range of price-fluctuations is large, the marginal risk premium is also likely to be large, thus making the range of fluctuations still larger. 10% is a "normal" figure in the case of seasonal crops; 1 to 2% may be regarded as the appropriate figure in the case of long-term government bonds.

35 Keynes, *Treatise on Money*, (vol. 2, p. 144).

36 *Ibid*, p. 143.

37 *The Economist*, August 19, 1939, p. 350.

38 Industrial crops, such as cotton, have shown in the past more violent price-fluctuation than either foodstuffs or industrial raw materials.

39 *Value and Capital*, pp. 144–5. Cf. also Pigou, *Industrial Fluctuations*, pp. 230–2.

40 We can regard the current long-term rate as being determined either by the expected future long-term rate, or the average of expected future short-term rates. The two come to the same thing since the expected long-term rate also depends on the average of short rates.

If R_1 is the current yield of a bond repayable in ten years' time and R_2 is its expected yield next year, r_1 is the short-term rate this year, r_2 ... etc., are the "forward" short-term rates in subsequent years (ie, the expected short-term rates plus the risk premium), their relation is given by the following equations:

$$(1 + R_1)^{10} = (1 + r_1)(1 + r_2) \ldots (1 + r_{10})$$

$$(1 + R_2)^9 = (1 + r_2)(1 + r_3) \ldots (1 + r_{10})$$

Hence,

$$(1 + R_2)^9 = (1 + R_1)^{10}/(1 + r_1)$$

Thus the expected long-term rate will exceed the current long-term rate if the current short-term rate is below its expected average, and vice versa. (Cf. Hicks, op. cit., p. 152; also Hicks, "Mr. Hawtrey on Bank Rate and the Long-Term Rate of Interest", *Manchester School*, 1939).

41 The current short rate is not dependent on the expected *short rates*, simply because the life-time of short-term bills is much too short for expectations to have much influence. The expected rate on bills next year can have no influence in determining the rate on a three-months' bill today; while the expected rates for the next three months are very largely determined by the current rate. It is only in exceptional circumstances that the market expects a definite change in the short-term discount rates within the next few months. It is just because the elasticity of expectations for very short periods is generally so near to unity, that the short-term interest market is largely non-speculative.

Similarly, a change in the long-term rate (either the current or the expected rate) cannot react back on the short-term rate except perhaps indirectly by causing a change in the level of income and, hence, in the demand for cash. For, supposing the change in the long rate causes speculators to sell long-term investments, this could only affect the short rate if they substituted the holding of cash for the holding of long-term bonds; it cannot affect the short rate if the substitution takes place in favour of short-term investments other than cash (savings, deposits, etc.). But there is no reason to expect, in normal circumstances at any rate, that the substitution will be in favour of cash. "Idle balances" – ie, that part of short-term holdings which the owner does not require for transaction purposes – can be kept in forms such as savings deposits, which offer the same advantages as cash (as far as the preservation of capital value is concerned) and yield a return in addition. It is only when the short rate is so low that investment in savings deposits is no longer considered worthwhile (see footnote below) that there can be a net substitution in favour of cash; but precisely in those circumstances the elasticity of substitution between cash and savings deposits is likely to be so high that this cannot have any appreciable effect on the short-term rate.

Thus, while the current short rate does determine the relation between the current long rate and the expected long rate, this is not true the other way round.

42 The nature of the equilibrium in the short-term interest market is shown in the accompanying diagram, where the quantity of cash is measured along Ox, the short-term interest rate along Oy. DD and SS stand for the demand and supply of money, respectively. The demand curve is drawn on the assumption that the volume of money transactions (ie, the level of income) is given. This demand curve is inelastic, since the marginal yield of money declines fairly rapidly with an increase in the proportion of the stock to turnover. Below a certain

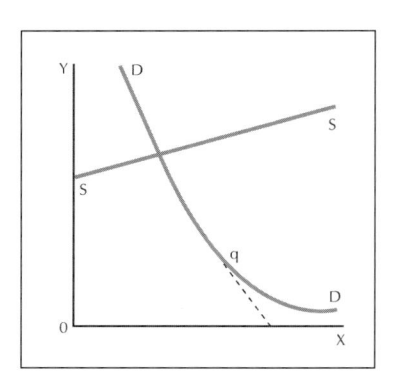

point (q) the demand curve becomes elastic, however, since the holding of short-term rate is lower than the necessary compensation for this. There is a certain minimum, therefore, below which the short-term rate cannot fall, though this minimum might be very low. (The dotted line shows what the demand curve would be if this risk were entirely absent). Hence, when the short-term rate is very low, it can be said to be determined by the risk premium attaching to the holding of the safest short-term asset; otherwise it is determined by the supply-price of money (ie, banking policy) and changes in the short-term rate are best regarded as due to shifts in this supply price. (The elasticity of the supply of money in a modern banking system is ensured partly by the open market operations of the central bank, partly by the commercial banks not holding to a strict reserve ratio in the face of fluctuation in the demand for loans, and partly it is a consequence of the fact that under present banking practices a switch-over from current deposits to savings deposits automatically reduces the amount of deposit money in existence, and vice versa.)

43 Moreover, the long-term rate is never *equal* to this risk premium (or liquidity premium); this only accounts for the difference between the long-term rate and the expected average of short-term rates. Professor Hicks has calculated this risk premium to have been about 1% in Great Britain before the last war, and 2% after the war (*Manchester School*, 10(1), p. 31.)

44 To assume that subjective expectations are *certain* is not the same thing, of course, as assuming "perfect foresight" or assuming that the expected prices are the same as current prices. Suppose eg, that, despite the current short-term rate being 2%, speculators are absolutely certain that a year hence, and forever afterwards, the rate will be 4%. The existence of this expectation will not cause any change in the current short-term rate; but it will invariably fix the level of the current long-term rate at 3.92%.

45 Mr. Keynes appears to be aware of this, when he speaks of the influence of the notion of the "safe rate" of interest in determining the current rate. (*General Theory*, p. 201.) But in the general statement on the liquidity preference function he relies on the factor of uncertainty of which money is free and, hence, commands a "premium".

46 It is well known that in periods of great political uncertainty it becomes impossible either to buy or to sell gilt-edged in large amounts without thereby directly affecting the price.

47 The proposition is true, partly because both the demand and the supply of money vary with the rate of interest and also because any factor influencing the short-term rate can do so only by affecting the demand or the supply of cash. As mentioned before, emphasis should be placed on the elasticity of the supply of money (with respect to the short-term rate) rather than on the elasticity of demand. The latter, except when the rate is very low, is relatively small; which is shown by the fact that the short-term rate can be varied within fairly wide limits without these changes being associated with equally significant variations in the quantity of money in existence or in the level of incomes.

48 If the price of bonds is regarded as determined by the supply and demand for liquid funds, the price of any other thing which is subject to the price stabilising influence of speculation could be equally regarded as such. In all these cases it is the speculators' willingness to undertake commitments in a particular direction, and not their preference for avoiding commitments in general (ie, their preference for "liquidity") which is relevant.

The "liquidity premium" on money and savings deposits, etc., in the sense in which Mr. Keynes employs this concept, may be regarded as 2% per annum if the market for consols is chosen as the standard with reference to which this liquidity premium is measured. But it will be 10% per annum if, say, the market for wheat is chosen as a standard, and something else again if the tin market is chosen. But is there any particular reason for selecting the gilt-edged market as the standard?

49 It might be argued that an increase in the proportion of short-term assets in the total of assets will reduce the marginal risk premium on long-term assets and, hence, lower their yield. But it is doubtful whether this effect could be appreciable, unless the change in this

proportion is large; and since short-term assets amount in any case only to a small proportion of total wealth, a *large* change in their amount can only cause a *small* change in this proportion. In the case of banks and other financial institutions, however, which keep a *large* proportion of their assets on short term, the effect may be considerable.

50 Here I am not thinking of the day-to-day fluctuations in gilt-edged prices which are, of course, also due to speculative influences – they either reflect genuine variations in the degree of certainty with which expectations are held, or the expectation of such variations on behalf of others – but changes which sometimes follow from the pressure of new issues.

51 Cf. pp. 71–75 below.

52 Income *saved* is also spent on something in so far as it is spent on the purchase of securities or other income-yielding assets. What we are assuming here, therefore, is that there is no attempt to "hoard"; that changes in the deposition of incomes between the different lines of expenditure do not involve changes in the demand for money.

53 This is the meaning of the word "spontaneous" in the above definition. If there is a change in the amount spent as a result of a change in the level of incomes, there need not be, of course, any reduction in the amount spent on other things.

54 Cf. pp. 62–64 above.

55 These equalising forces operate if (a) σ (the price-stabilising influence) is positive and (b) the income-elasticity of demand is positive. If the income-elasticity of demand is negative, the equalising forces operate σ if is negative.

56 On the assumption, of course, that all increase in "genuine savings" is directed at the purchase of long-term assets.

57 Abstracting from any involuntary reduction in stocks in the hands of retailers and wholesalers, which is purely temporary; this will merely delay adjustment not prevent it.

58 There will be an initial increase in the demand for short-term funds with the increase in incomes; but provided this demand is satisfied by the banking system, no further contribution is required to keep the short-term rate stable. The continuous increase in the short-term indebtedness of the speculators in securities will be exactly offset by the decrease in the short-term borrowings of the foreign exchange market.

59 "The Ex-Ante Theory of the Rate of Interest", *Economic Journal*, December, 1937, p. 669.

60 Mr. Keynes' own proof of this proposition is that there must "always be *exactly* enough ex-post saving to take up the ex-post investment and so release the finance which the latter had been previously employing" (*ibid*. p. 669). *Ex-post* investment and *ex-post* saving will always be equal if "ex-post investment" is so defined as to include consequential changes in stocks (in our example, foreign exchange balances). But in order that the funds released through the reduction in stocks should be available for the finance of long-term investment somebody must perform the "arbitrage operation" of borrowing on short term and re-lending it on long term; and the willingness to perform this arbitrage must necessarily depend on the total amount of such transactions already undertaken. Sooner or later this willingness must fall off; which in our own terminology, is expressed by saying that the marginal risk premium by which the current long-term rate exceeds the average of expected short rates, must rise. But no stretching of the meaning of words could describe this situation as one of a "shortage of cash" or one which could be remedied through an increase in cash. Temporarily, the banks might prevent the long-term rate from rising by *reducing* the short-term rate; but however much they inundate the system with cash, sooner or later the long rate must rise.

61 As recommended by Mr. Keynes in the *Economic Journal*, September, 1938, pp. 449–60.

62 Apart from the case, of course, where the price fluctuations are themselves created and enhanced by speculation (ie, when σ is negative) in which case the elimination of such price variations is always income-stabilising.

63 *Capital and Employment*, p. 66. *A Century of Bank Rate*, p. 38. He still regards the fluctuations in short-term investment as the prime mover of the trade cycle, and the inflationary (or

deflationary) influences emanating from the capital market as subsidiary. But he would concede, I think, that this need not *necessarily* be the case.
64 *A Century of Bank Rate*, p. 195.
65 Cf. the diagram on p. 84. When the rate falls below the point q the demand for cash balances becomes elastic and this makes further reductions in the rate difficult or impossible. It is only when the short-term rate is at this point that "idle balances", in the sense of ideal *cash* balances, really exist; and it is only then that the short-term rate reflects "liquidity preference", in any significant sense. (When the rate is above this level, the marginal demand price for cash is determined by the marginal yield of money-stocks in terms of convenience, and not by the risk premiums attached to alternative forms of investment.)
66 We need not suppose, of course, that people have any definite expectations as to what the short-term rate is going to be at any particular date in the future, other than the immediate future. If the theory depended on there being such a definite expectation, it would clearly be wrong. But all that is necessary to suppose is that people have certain ideas as to what constitutes the "normal" level of the short-term rate of interest, and that they consider deviations from this normal level as temporary. Their idea as to the normal is based, of course, on past experience, and reflects the average level which ruled over some period in the past.
67 Ignoring the lag between the official discount rate and the market rates.
68 This is because he is sceptical of any influence of the short-term rate on the long-term rate (either the current short rate or the average of past short rates) and regards the two rates as independently determined. But it is difficult to see how the long-term rate can remain below the short-term rate, once the prevailing level of the short-term rate comes to be accepted as "normal". The supply of long-term funds comes from savers, and why should savers place any money in long-term investments if they expect a higher return on savings deposits?
69 *Manchester School* 10(1), p. 31.

4

*Theory of the Inverse Carrying Charge in Futures Markets**

Holbrook Working[†]

In the technical language of American futures markets, *carrying charge* refers to a difference at a given time between prices of a commodity for two different dates of delivery.[1] If prices in Chicago wheat futures, for example, were quoted as, September, US$1.30, December, US$1.38, and May, US$1.48, it would be said that the market showed a carrying charge of 8 cents from September to December, of 10 cents from December to May, and of 18 cents from September to May.

Carrying charges in the futures market sometimes approximate, or even exceed, the cost of holding the wheat in public storage at the regular rates (*storage charges*) fixed by the grain exchanges for such service, but more often they are less. Not infrequently prices of the deferred futures are below that of the near future; the market is then said to show an *inverse carrying charge*, or sometimes an "inverted" or a "reversed" carrying charge. In British usage, "contango" and "backwardation" refer to positive and inverse carrying charges respectively.

There seems to be substantial agreement among writers on futures markets that positive carrying charges tend to reflect marginal net costs of storage, and that when carrying charges are positive, prices of near and distant futures must respond about equally to any causes of price change. With regard to inverse carrying charges there is more difference of opinion and much reason for the student of futures markets to be dissatisfied with the present state of theory. How shall one explain a large inverse carrying charge between December and May in a United States wheat market? Do inverse carrying charges reliably forecast price declines? When prices of deferred futures fall below the spot price, does the futures market tend to lose effective connection with the cash[2] market?

---

*This paper was first published in the *Journal of Farm Economics* 30(1), pp. 1–28 (1948) and is reprinted with the permission of the Food Research Institute.
[†]Holbrook Working, 1895–1985.

Table 1 Temporal price structure of Chicago wheat market on selected dates*

Quotation	September 5, 1929[a] Price	Difference	September 1, 1934 Price	Difference	September 1, 1925 Price	Difference	September 3, 1946[b] Price	Difference
"Cash" wheat	126 7/8		103 3/8		156		194	
September future	132 1/8	(+5 1/4)	102 5/8	(−3/4)	154	(−2)	191 3/8	(−2 5/8)
December future	140 3/8	+8 1/4	103 5/8	+1	152 5/8	−1 3/8	189 3/4	−1 5/8
May future	150	+9 5/8	105	+1 3/8	157 1/8	+4 1/2	183 1/2	−6 1/4
	May 1, 1930 Price	Difference	May 1, 1935 Price	Difference	May 1, 1926 Price	Difference	May 1, 1947[b] Price	Difference
"Cash" wheat	98 1/2		98 3/4		166 1/4		267	
May future	101 1/2	(+3)	97 3/4	(−1)	164 1/4	(−2)	258	(−9)
July future	104 1/4	+2 3/4	98 1/8	+3/8	143 3/8	−20 7/8	219 7/8	−38 1/8
September future	107 7/8	+3 5/8	98 3/4	+5/8	137 5/8	−5 3/4	213	−6 7/8
December future	112 5/8	+4 3/4	—		—		210 1/2	−2 1/2

*Based on closing quotations on the first trading day of the month indicated. The cash quotations apply as nearly as possible to wheat of the quality and location that will be delivered on the near future, but the correspondence may not be exact in any instance. The "differences" are in each instance differences between the specified futures price and the cash or future price immediately above it; those between futures are pure carrying charges, while those between cash and futures prices, shown in parentheses, may include a quality or location differential.

[a]The third trading day, taken because quotations on the first trading day and a few neighbouring ones were unrepresentative of the price structure that prevailed through most of August and September.

[b]Kansas City quotations are used for 1946–1947 in order to show the September price structure in terms of quotations for the same delivery months as in other years; trading in Chicago in the autumn of 1946 was for October and January delivery. The price structure at Kansas City was essentially the same as at Chicago throughout 1946–1947, as it usually is.

These are some specific questions that call for answer and to which no satisfactory reply seems to be offered by prevailing theory.

Other questions may be suggested by attempts to explain price relations shown in the table above, which records the wheat price structure in Chicago or in Kansas City early in September and early in the following May of each of four crop years selected with a view to exhibiting a variety of conditions, ranging from that of large positive carrying charges in the column at the left to large inverse carrying charges at the right.

Four different lines of attempt to explain inverse carrying charges and other market conditions which accompany them will be considered in this chapter, as follows:

(1) "Cash and future prices, though related, are not equivalents aside from the time element, at least in the United States wheat market".
(2) "The future, as against the present, is discounted".
(3) Expectations regarding future demand and supply conditions tend to have more effect on prices of deferred futures that on cash prices or on near futures.

(4) An inverse carrying charge is a true negative price of storage, arising from the fact that stocks may have a high marginal "convenience yield".

The first two of these are quoted from a note by Lawrence L. Vance in the *Journal of Farm Economics*,[3] not because they are original with him – others have expressed similar views before – but in recognition of his perception of the importance of inverse carrying charges for the theory of futures markets and his attempt to relate theory to the observable facts. Inverse carrying charges, as Vance has said, are not adequately explained by conventional theories of futures markets. Because they occur frequently and are often larger than any positive carrying charges that ever arise, they cannot be regarded merely as the effects of special influences which do not warrant attention in a general realistic theory of futures markets.

TWO MARKETS OR ONE?

The suggestion that cash and futures prices are not equivalents apart from the time element may be taken either narrowly or broadly. On a narrow interpretation, which seems what Vance intended, it may imply only that cash and futures prices differ for such reasons as differences in quality or location of the commodity or because of uncertainty as to time of delivery on futures contracts.

It appears to me illogical to treat differences in quality or in location as explanations of carrying charges, either positive or inverse, except under extraordinary circumstances. In general a carrying charge, defined as a market difference between prices of a commodity for different dates of delivery, should be a difference between prices for essentially the same quality, in the same location, at the different dates. Most cash price quotations on grades of wheat deliverable on futures contracts in American markets apply to a quality superior to that which will be delivered on the futures contracts. Cash-future price differences therefore usually reflect quality differentials as well as time differentials. But in such cases it should be said that the price difference is a combination of carrying charge and quality differential, not a pure carrying charge.

Chicago cash wheat price quotations which seemed to provide a nearly perfect basis for calculating the market carrying charge were available until 1917.[4] Now, published cash price quotations in terminal markets apply almost invariably to wheat on tract rather than to wheat in store. Therefore cash-future price differences commonly include a location differential, which may amount to a cent or two per bushel, the price on track being below the price in store when the prevailing movement is into storage.

In such an extreme case as is presented by the quotations for September 5, 1929, a price on track may properly be used for calculating a market carrying charge. Wheat on track sold then more than five cents under the price of the September future (which may be regarded as also a spot price, since large deliveries were being made on futures contracts) because elevator space was lacking for all the wheat arriving; two different sets of carrying charges existed simultaneously, one for wheat in store in "regular" elevators and another, not exactly determinate,[5] for wheat "without a home".

Concerning the effect of uncertainty regarding the time when delivery will be made on futures contracts, two facts should be noted: (a) there usually is little or no doubt among interested traders regarding the approximate time when delivery may be expected;[6] and (b) the actual or potential volume of buying of futures by people for whom precise timing of delivery is important is so small that it may be ignored as a price factor.

Taken more broadly, the question whether cash and futures markets are equivalent apart from the time element includes the question whether cash and futures prices may differ because they reflect the opinions of substantially different groups of traders. The view that persistent differences arise for that reason requires the supposition that arbitrage between the cash and futures prices may be ineffective. The prevalent opinion of recent theoretical writers seems to be that "arbitrage can always be relied upon to prevent the forward price from exceeding the spot price by more than net carrying cost ... (but) cannot be equally effective in preventing the forward price from exceeding the spot price by *less* than net carrying cost. In fact, it is possible that the forward price may not exceed the spot price at all or even fall short of the spot price (even when net carrying cost is positive)".[7] "There is no limit ... to the extent to which the futures price may *fall short* of the current (spot) price".[8]

This theory of inverse carrying charges is clearly inapplicable to wheat prices in the United States and may be quite generally untenable. It rests in part on the mistaken assumption that hedging is not arbitrage. The placing of a hedge against a purchase of cash wheat is one part of a double transaction which is arbitrage in fact as well as in form because its occurrence depends on a judgement regarding the relation between two prices. Sometimes the judgement takes the form of a decision to hedge a cash transaction that would be made in any event, but more commonly in United States terminal wheat markets the occurrence of the cash transaction (and, therefore, of the futures transaction also) turns on a judgement that the relation between the cash and the futures prices warrants the double transaction.[9] Since cash prices may be depressed relative to futures quite as effectively by diminution of purchases as by increase of sales in the cash market against opposite transactions in futures, arbitrage between cash and futures markets for wheat remains effective even in the presence of narrow physical limits on the possible sales of cash against purchases of futures.[10]

This is a conclusion that should not be accepted lightly. It requires the abandonment of a pattern of thought regarding futures markets which is extraordinarily difficult to relinquish; the reader who accepts it will almost certainly soon find himself thinking and speaking, when he tries to discuss a wheat price situation involving inverse carrying charges, as though he still regarded cash and futures prices as substantially independent.

Continuous effectiveness of arbitrage between cash and futures prices makes it necessary in most consideration of price influences to regard the two sets of prices as determined in a single market. To treat them as determined in separate markets is to risk implying a degree of independence that does not exist.

In primary wheat markets of the United States the prevalence of hedging by both buyers and sellers and the "breadth" of futures trading (which is, indeed, a requisite for prevalence of hedging) have led to abandonment of even the outward appearance of independence in the cash market.[11] At the cash-grain tables buyers and sellers ordinarily do not discuss prices; they bargain in terms of cents "over" and cents "under". When agreement is reached in these terms the premium or discount settled on is applied to the latest quotation for the "basic" future to arrive at a formal price. The cash market is therefore clearly subsidiary from the standpoint of price formation, as country markets or retail markets are usually subsidiary to large wholesale markets. The function of primary price formulation lies with the futures market.[12]

The fact that wheat markets in the United States offer no opportunity for significant inverse carrying charges to develop from a tendency for the cash and futures market to pull apart, does not by itself imply that such tendencies cannot be effective in markets where there is relatively little hedging, if there be such futures markets. But since large inverse carrying charges do occur in American wheat markets for other reasons, it follows that such other reasons may afford the principal or sole explanation of inverse carrying charges in all markets.

DISCOUNT ON THE FUTURE

Although this section takes its title from a statement by Vance, it must consider also an alternative theory advocated by J. M. Keynes, and an idea treated by the Federal Trade Commission, which is common to the theories of both Vance and Keynes but admits of still other theoretical development.

Vance's theory is expressed in two sentences;

traders have not been willing, at the start of trading in a particular future, to give full effect in their commitments to the forces which existed at the time, and . . .

only after the passage of time and the confirmation of what theory would have predicted at the start were prices revised upward. The mere fact that future events always bear some degree of uncertainty is perhaps sufficient to justify a discounting of expectations.[13]

Keynes' explanation of "normal" inverse carrying charges, commonly referred to as his "theory of normal backwardation", ran as follows:

> If supply and demand are balanced, the spot price must exceed the forward price by the amount which the producer is ready to sacrifice in order to "hedge" himself, ie, to avoid the risk of price fluctuations during his production period. Thus in normal conditions the spot price exceeds the forward price, ie, there is backwardation. In other words, the normal supply price on the spot includes remuneration for the risk of price fluctuation during the period of production, whilst the forward price excludes this. The statistics of organised markets show that 10% per annum is a modest estimate of the amount of this backwardation in the case of seasonal crops which have a production period approaching a year in length and are exposed to all the chances of the weather. In less organised markets the cost is much higher . . .[14]

The Federal Trade Commission was content to discuss and try to measure "downward bias" in future prices.[15] Because both Vance and Keynes relied on supposed statistical proof of existence of downward bias to support their theories, the three theories are closely related. (There is an element of theory in the concept of downward bias, since it assumes the existence of a general tendency measurable by statistical averages). Vance went further than the Federal Trade Commission by suggesting an explanation of the downward bias, and Keynes, offering a different explanation, gave explicit expression to the relation he conceived to exist between downward bias and "normal" inverse carrying charge.

As an explanation of inverse carrying charges, Vance's theory of discount on the future is difficult to discuss on logical grounds because he did not set out the reasoning by which he would connect downward bias and carrying charge. It may hinge on a concept of substantial independence of cash and futures markets such as has been criticised in the foregoing section of this paper. Keynes' theory of normal backwardation has been called in question in later theoretical discussion.[16] That discussion has tended toward the conclusion that the logical expectation depends on what assumptions are made, but it has not produced clear evidence as to what assumptions are appropriate to actual futures markets.

We may as well confine further attention at this point to the statistical evidence. I shall attempt in later pages to show that inverse carrying charges may best be explained by a theory that does not involve explicit consideration of any supposed downward bias of futures prices. The remainder of this section argues that such actual downward bias as may exist cannot contribute significantly to explanation of inverse carrying charges, irrespective of the theory accepted as relating them.

Contrary to Keynes' apparent assumption it is not a simple matter, as we shall see, to determine precisely what "the statistics or organised markets show" with respect to the price tendency he was considering. One of the most critical and painstaking inquiries into the subject was that made by the Federal Trade Commission. It attacked the problem in several different ways. All the methods produced evidence, in price data subsequent to 1896, of some "downward bias" in futures prices of wheat and corn, but not of oats; but for the ten-year period prior to 1896, the indicated bias was in the opposite direction for all grains.[17] The method which the Federal Trade Commission appeared to regard as quantitatively most trustworthy, and the only one from which it drew a value which was discussed as a measure of bias, yielded for wheat 1906–1916, the estimate that it amounted to −2.39 cents (about 2.4%) for a 12-month interval.[18]

An inquiry of mine into the seasonal tendency of wheat futures prices led to the conclusion that I could "find no evidence of a general tendency toward post-harvest depression of prices of Chicago wheat futures".[19] But I should incline to think that a tendency averaging only 2% per year might exist without being capable of statistical proof. Another investigation along different lines led to the conclusion, as a statement of historical fact, that "speculators in wheat futures as a group have in the past carried the risks of price changes on hedged wheat and have received no reward for the service, but paid heavily for the privilege".[20]

With regard to corn futures prices, which Vance cited as showing strong evidence of "discount on the future", some pertinent data are presented in Figure 1. Price averages for the May future, in which trading is usually heaviest and starts farthest in advance of the delivery date, are shown from August to the following May. The record is continued through a 14-month interval by means of averages for the September future.

Section (a) of Figure 1 gives the results of using prices for 32 of the 33 years on which Vance relied for evidence; data for the September 1918 and May 1919 futures are omitted because quotations on each ran for only a few months and cannot be included without greatly decreasing the number of months for which averages can be calculated. The averages give a strong impression of a general tendency for futures prices to rise from month to month; but is it reasonable to base conclusions regarding a general tendency on price data for 1904–1905 to 1936–1937, a period dominated by years in which the general price level was rising, and one which included the abnormal price fluctuations during and just after the first world war? When data are added for 20 earlier years, and one later (the last crop-year before onset of the second world war), a considerably smaller tendency for corn futures prices to advance during the crop year is indicated, see Section (b). Eliminating from the calculations data for

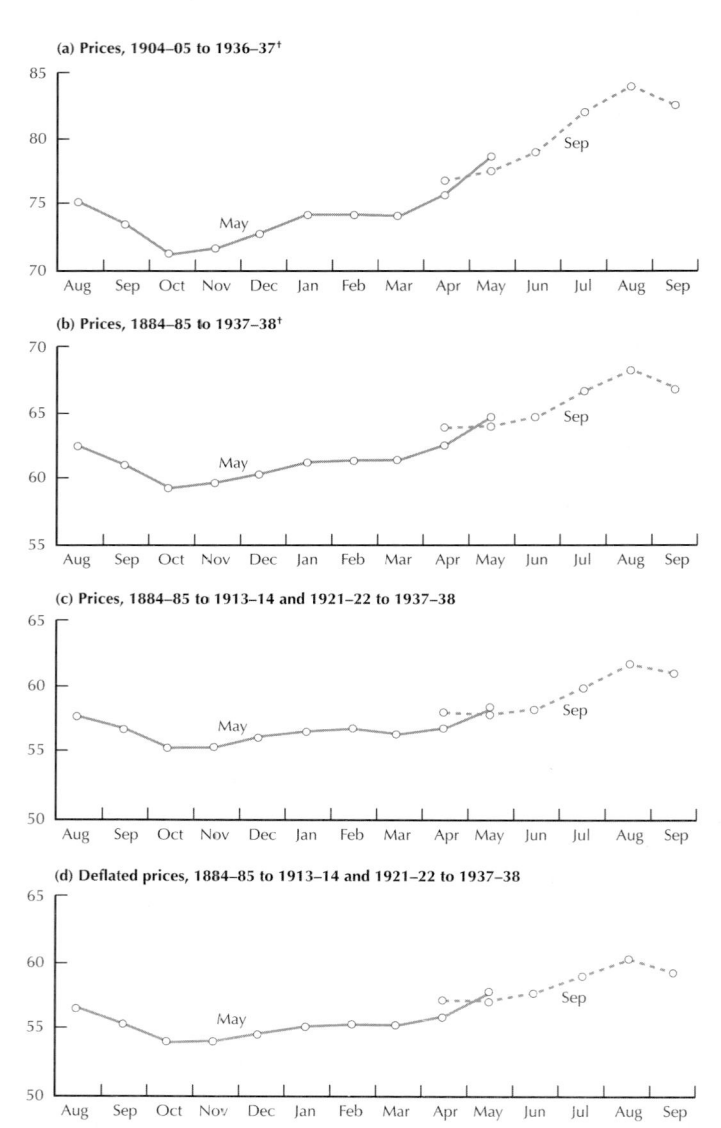

Figure 1 Average seasonal course of prices of Chicago May and September corn futures* *(cents per bushel)*

(a) Prices, 1904–05 to 1936–37†

(b) Prices, 1884–85 to 1937–38†

(c) Prices, 1884–85 to 1913–14 and 1921–22 to 1937–38

(d) Deflated prices, 1884–85 to 1913–14 and 1921–22 to 1937–38

*Computed generally from Friday closing prices; when direct closing quotations have not been available, as occasionally happened for the May future in August (and once in September) and for the September future in April and May, approximations based on other quotations have been substituted. Deflated prices are in terms of the 1910–14 price level.
†Omitting prices for the September 1918 and May 1919 futures; see text.

seven war and post-war years, when price fluctuations were abnormal and unusually difficult to anticipate, results (Section (c)) in still further reduction of the indicated price advance. Deflation of the original price series before averaging (Section (d)) produces little change in the tendencies suggested.[21]

If one takes the data of Sections (c) and (d) of Figure 1 as representative of tendencies in corn futures prices in the United States, it appears that on the average prices tended to rise about two cents per bushel between October and April, with the advance concentrated mainly in December and January. A further rise of the May future in the delivery month means little from the present standpoint. It can be attributed to two tendencies other than a discounting of the future, namely: (1) when stocks are large the expiring future tends to be effectively a spot price during the delivery month and for that reason it rises relative to deferred futures during the course of the month; and (2) when stocks are short, the price of the expiring future is sometimes raised relative to other futures and most cash prices by a corner or squeeze – developments which it is neither practically possible nor theoretically desirable to have affect prices in advance.[22] The September future shows an average rise of over 3 cents per bushel between June and August, which must be considered with the nearly equal average decline in the May future from August to October. Shall we say that the September future tends to underestimate them, or shall we say that the price averages show a three-month seasonal bulge that culminates in August, perhaps because crop scares occur so frequently in July and August?

The evidence of Section (c) may be summarised more simply in another way. Prices of the May corn future during 46 years show an average decline of 1.0 cent per bushel between August, the first month in which that future has usually been traded, and the following April; between April and the next August, the September future rose 3.5 cents per bushel, on the average. The total average change over twelve months, therefore, was an increase of 2.5 cents, or slightly over 4%. On this basis, downward bias in Chicago corn futures seems to have averaged rather less than 0.4% per month.

More evidence might be offered[23] tending to show that prices in major futures markets have little downward bias, but we risk over-emphasis on this issue for present purposes, important as it may be from the standpoint of appraising the economic utility of futures trading. Enough has been said to show that no theory founded on downward bias could explain more than a very small inverse carrying charge.

PRICE EXPECTATIONS

Views which seem to prevail widely regarding the influence of expectations were expressed in the Federal Trade Commission's *Report on the*

Grain Trade. "There may be an active demand for cash wheat, for example", the *Report* says, "at whatever terminal market is under consideration, while the speculators dealing in the more remote futures ... may believe that prices are bound to go down markedly by the time the futures contracts mature – for example, in May. Anticipations of more abundant supplies or of slackened demand in the spring may press down the price of future contracts almost without limit ... there is nothing in the machinery and methods of the grain trade, to limit the discount, provided there are speculators willing to trade, and specifically short sellers willing to sell, at relatively low prices".[24] Furthermore, "July and September (futures) prices ... are both much affected by (anticipated) receipts from the new wheat crops, the one of winter and the other of spring wheat, and by the comparatively unsettled conditions and estimates regarding crop years may relate to entirely different supply conditions and prices".[25]

Opinions such as these are often qualified by the observation that existence of heavy surplus stocks may force spot and futures prices into a pattern of relationship determined by the costs of carrying stocks. The text from which the foregoing quotations are drawn includes a somewhat obscure phrase which was doubtless intended as such a qualification. I should like in the next few pages to advance first the theory, admittedly oversimplified, that in a perfectly functioning futures market continuous existence of any stocks should prevent the emergence of price differences that depend on expectations, and then to examine whether actual price behaviour agrees better with that theory or with the prevalent view.

Consider first a case in which stocks are heavy and the market shows a strong positive carrying charge, as in the quotations for May 1, 1930, in Table 1. The fact that September wheat (the distant future) was quoted then at nearly six cents per bushel above May wheat (the near future) may be regarded in a sense as recording market expectation of a price advance, but the expectation concerns only the consequences of the existing level of wheat supplies. The prices show a "market expectation" that wheat will be worth about US$1.08 per bushel in September, and a consequent opinion that the May value, in view of costs of storage, is about US$1.02. Expectations regarding future supply and demand conditions must affect the price of the May future as much as the price of the September.[26]

It should be particularly noted that this discussion avoids a direct answer to the question: do futures prices reflect expected price change? That query is like the question, have you stopped beating your wife? To state the facts without false implication, it is necessary to refuse to answer the question as put.

An important merit of futures trading is that it helps to make spot prices conform to expectations regarding events to come.[27] It does so partly by affording those who have opinions on prices a chance to deal in contracts applying to a time that suits their thinking, and partly by the mere fact that it shows where prices should stand later in order to justify present cash prices.

The idea that a futures market *should* quote different prices for different future dates in accordance with developments anticipated between them cannot be valid when stocks must be carried from one date to another. It involves supposing that the market should act as a forecasting agency rather than as a medium for rational price formation when it cannot do both. The business of a futures market, so far as it may differ from that of any other, is to anticipate future developments as best it may and to give them due expression in present prices, spot and near futures as well as distant futures.

Now suppose that wheat stocks in May are such that a portion of them will be carried through July, but none to September, will expectations pertinent to the price of the September future deserve to have a bearing on the price of the May future? Logically they should. Stocks of new wheat will be in existence in July which will be carried through September, requiring the price of July wheat to respond fully to expectations pertinent to the price of September; and since the May and July prices are logically connected by stocks to be carried from May through July, expectations that bear on the price of September wheat should bear equally on the price of May wheat.

Extension of this reasoning through whatever number of steps may be required indicates that, in the slightly simplified conditions supposed, expectations which deserve to influence the price of the most distant future quoted should always bear equally on spot prices, unless a period intervenes when stocks from both past and future production are expected to be non-existent.

Stocks of wheat in the United States during over half a century have never fallen much below 60 million bushels except under governmental control in 1917, and during the last quarter-century have not fallen below 80 million bushels. At their lowest level of any crop year, near the first of July, much less than half of the stocks of old wheat may be hard winter, which is usually the class represented directly by the Chicago future, but even stocks of that class have never been negligibly small. Therefore this theory requires that expectations which affect the price of the most distant Chicago wheat future should always be reflected equally in the price of the near future, and that such expectations should have no influence on price differences.

Looking now to the facts, we may hope to determine as a matter of actual observation whether futures prices do behave about as should be expected on the foregoing theory, or more nearly in accord with the belief that price difference between futures commonly reflect expectations regarding future developments.[28] The clearest case of supposing that near futures may respond weakly to expectations which affect the price of a distant future exists when inverse carrying charges are large; the expectations which are most closely measurable and seem capable of having the greatest differential price effected are those pertaining to size of a crop not yet harvested. It seems, therefore, that a critical test may be made by studying the effects of large changes in crop prospects on price relations between the May wheat future at Chicago and the July future. The May is clearly an "old-crop" future while the July is universally regarded in the grain trade as a "new-crop" future; and the July has often sold at large discounts under the May.

In long and critical study of price relations between old- and new-crop wheat futures at Chicago, the only evidence I have found of effects of crop prospects is of effects that lasted for not more than a week or two. Now and then reports of crop damage have seemed to result at first in a slightly greater price rise in the July future than in the May, but the consequent diminution of the inverse carrying charge has usually been followed within a few days by an increase to about what the carrying charge was before the report of crop damage. Unexpectedly favourable crop reports have had similar effects in the opposite direction, which have tended also to be followed quickly by a return toward the previous price relation. The behaviour in such instances suggests that reports of crop damage, for example, may have an initial effect which is slightly greater on the new-crop futures than on the May because so many traders believe that such should be the case, but that the result is a maladjustment of price relations which cannot persist. Even such wholly temporary correspondence of price behaviour with the theory that new-crop prospects should especially affect the new-crop futures is the exception rather than the rule.[29]

For the Liverpool wheat market, the evidence is a little different. On the supposition that expectations should affect prices for different forward dates differently, the spread between prices of the December and March futures at Liverpool should depend on the expected size of the Southern hemisphere exportable surplus, on which Europe used to rely for much of its imported wheat after January or February of each year. Analysis of the data show that such expectations had no important bearing on the December–March price spread. Yet when *changes* between the earliest and latest per-harvest estimates of southern hemisphere surplus in each crop year are related to concurrent changes in the December–March price spread, a small but statistically significant correlation appears.[30] At Liverpool, unlike Chicago, expectations regarding supplies that would become available between two future months apparently had some influence, small in relation to others, but measurable and not wholly temporary, on the differences between forward prices applying to those two months.

It cannot be said, then, that expectations regarding supplies which will become available later always, in all future markets, have exactly as much effect on prices of near as of distant futures. But if such expectations tend in some markets to have an influence on inverse carrying charges, the effect is small in relation to other influences. An inverse carrying charge may never be regarded as a specific measure of the price effect of expected future developments.

We turn in the next section to a development of the theory that existence of stocks provides a link between spot prices and near and distant futures even when the temporal price structure is "inverted". It will then be found that some influence of price expectations on carrying charges is not excluded as a possibility under certain circumstances.

THE PRICE OF STORAGE

The origin and prevalence of the term "carrying charge" in trade usage reflects the fact that hedgers commonly regard the designated price differences as in fact equivalent to a price for "carrying" the commodity, or what may be called for the purpose of economic analysis a *price of storage*. The market is said to show a "full carrying charge" when the price differences approximate the costs of carrying the commodity in public storage at the fixed rates for such storage plus interest and insurance. One meets also the expression "the market shows a carrying charge", as equivalent to a statement that the carrying charge is positive. This usage implies at least a tendency to associate the concept of price of storage more particularly with positive than with inverse carrying charges.

When the carrying charge is positive, as in the example in our opening paragraph, it is clearly a measure or an index of returns for carrying the commodity hedged. Anyone who is in a position to handle the necessary business transactions may buy September wheat, in such a situation, sell December or May, and take delivery on the September contract with assurance of a gross return of eight cents per bushel for carrying the wheat from the time of delivery to the first of December, or of 18 cents for carrying it to the first of May. For one who already owns hedged wheat on the first of September, the meaning of the price relations may not be quite so precise. If the wheat is of minimum quality for delivery on futures contracts and is eligible for such delivery, the assured returns are the same as when the wheat is acquired by purchase of the September future, unless such wheat is currently selling slightly over the price of the future, as it might well be if the "carrying charges" were much smaller than are here assumed, but could scarcely be under the conditions supposed. If the wheat owned is like most stored wheat at any time, not of minimum deliverable quality, or not in position for delivery, or both, the return actually released from storage will depend on whether the price obtainable for that wheat next December (or May) stands in the same relation to the December (or May) future as it had to the September future on September 1; the returns for storage will probably approximate the carrying charge reflected in the futures market, but may be greater or less.

If it is advantageous to regard a positive carrying charge as a price for storage, as traders do, there may be similar advantages to gain from extending the concept to include inverse carrying charges. Other theories of inverse carrying charges have proved so inadequate and misleading, as we have seen, that a better theory is much to be desired.

There is nothing obvious in the behaviour of market carrying charges to indicate that they take on a different character when they shift from positive to negative or from negative to positive.[31] Market transactions that are directly related to the carrying charge – purchase and storage of the commodity against sales in the futures markets – tend to be on a large scale when the carrying charge is positive and large, and on a smaller scale when the carrying charge is negative and large, but the transition between these extremes is a continuous one; no sharp change in hedging practice occurs when the carrying charge changes sign.

Carrying charges behave like prices of storage as regards their relation to the quantity of stocks held in storage. Graphically represented, the relation should be that of a supply curve, showing small amounts of storage service rendered when the price of storage is low, and increasing amounts as the price of storage advances. The general form of the curves seems to be like that in Figure 2. Statistical analysis, treating carrying charge as a price, has shown such relationships to exist, and to be capable in some instances of fairly precise statistical determination.[32] Correlations between stocks and carrying charge tend to be highest for relationships involving carrying charges that often take on large negative values, like that for wheat between May and July in the Chicago market. Clearly, therefore, the supply-curve relationship between amount of storage and price of storage does not break down when the "price" becomes negative.

The statistical results indicate also that the market carrying charge,

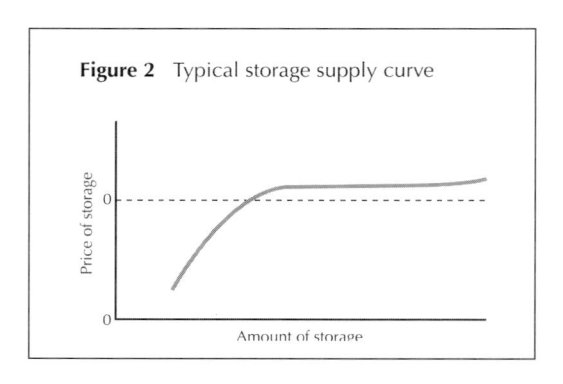

Figure 2 Typical storage supply curve

viewed as a price of storage, is broadly representative. The correlations with the carrying charge in the Chicago market are higher for statistics of all stocks of wheat in the United States than for statistics covering only stocks likely to be hedged, or covering only stocks likely to be hedged in the Chicago market. Carrying charges recorded in the Liverpool wheat market show similar evidence of representatives.[33]

The treatment of inverse carrying charges as prices of storage raises some problems of theory. First is a difficulty arising from the logical presumption that no substantial volume of stocks will be carried without assurance or expectation of at least a small return for carrying it. The presumption is not open to question, but it does not necessarily require that the price of storage be positive. For example, people "store" rented works of art in their homes, paying for the privilege. Storage of goods without direct remuneration and without expectation of price appreciation is to be observed in every retail store. A merchant might adopt the practice of buying today only what he could be sure of selling before tomorrow, or before the next delivery day, but if he did so he would be unlikely to remain long in business; he must carry stocks beyond known immediate needs and take his return in general customer satisfaction. Merchants who deal in goods that are subject to whims of fashion, or to sudden obsolescence for other reasons, must lay in stocks and carry them in expectation that some parts of the stocks will have to be sold at a heavy loss.

These observations illustrate a fact which Nicholas Kaldor has expressed in general terms by saying that "stocks of all goods possess a yield ... and this yield which is a compensation to the holder of stocks, must be deducted from carrying costs proper in calculating net carrying cost. The latter can, therefore, be negative or positive".[34]

Though stocks, as such, may be productive, with the result that some quantity will be carried even when loss from price depreciation is expected, or, if they are hedged, when an inverse carrying charge makes a loss virtually certain, it has not been considered obvious that the marginal yield of stocks can be sufficient to offset a large inverse carrying charge. Kaldor, thinking of producers' stocks, ascribed the yield to "convenience" and held that it is largely lost if the stocks are sold forward – for example, hedged;[35] and Gerda Blau has argued that "the convenience yield" of standard commodities dealt with in Future Exchanges is very small.[36] For the wheat market of the United States these doubts may best be answered by the fact that the largest inverse carrying charges between the May and July futures at Chicago – the largest in terms of monthly rates that occur there – have never prevented the carrying of substantial aggregate quantities of wheat hedged in the Chicago market.[37] The reasons, which have never been carefully explored, doubtless vary with the circumstances and type of business of the individual or firm that carried the stocks. For present purposes it is sufficient to note the fact.

Distinction is sometimes drawn between "surplus stocks" which a merchant or manufacturer will not carry without assurance or expectation of a special return, and "necessary working stocks" which will be carried for other reasons. The concept of working stocks as fixed in amount (necessary) may be warranted for many purposes but is inadmissible in serious consideration of the economics of stock-carrying. The aggregate of "working stocks" of a commodity in a market area may vary through a considerable range. Its upper limit may be defined as the level at which an increase in stocks will produce no increment of "yield", in Kaldor's phrase. Additional stocks, which will be carried only in response to assurance or expectation of a special return at least equal to "carrying costs proper", may be defined as surplus stocks. For wheat stocks in the United States as of June 30, marginal "yield" seems to disappear and additional stocks to be regarded as surplus when the total is somewhere between 150 and 200 million bushels.[38] The transition point is then depressed to about its minimum because producers, merchants and processors are all anticipating large additions to their stocks very shortly.[39] Minimum working stocks, though difficult to define in principle, seem fairly closely determinable for practical purposes at about 80 million bushels as of June 30; reduction below that level, under present conditions, seems to be very difficult.

Having found that the apparent logical obstacles to treating inverse carrying charges as prices of storage disappear on analysis, let us examine somewhat further the practical merits of the concept.

The carrying charge is one of three variables related by the equation

$$P_1 + P_s - P_2 = 0 \tag{1}$$

where P_1 = a spot or forward price,
P_2 = a forward price for a date later than that to which P_1 applies, and
P_s = carrying charge (price of storage).

Equation (1) gives no information by itself regarding the economic influences which determine any one of the variables included, but it affords a basis for explaining any one of them in terms of known explanations of the other two. When so used, it may well be rewritten by transposition in the appropriate one of the following three forms:

$$P_1 = P_2 - P_s \tag{1a}$$

$$P_2 = P_1 + P_s \tag{1b}$$

$$P_s = P_2 - P_1 \tag{1c}$$

Treatment of the carrying charge as a price of storage involves providing a direct explanation of P_s instead of relying on explanation in terms of differences between the explanation of P_1 and that of P_2. Direct explanation is useful for two reasons: (a) it is simpler and more reliable than indirect explanation on the basis of differences between the explanations of P_1 and P_2; and (b) it opens the possibility of explaining either P_1 or P_2 in terms of an explanation of the other and of P_s. These advantages prove substantial in practice. They arise partly because the important influences affecting P_s are usually fewer than those affecting P_1 and different from those affecting P_2; and partly because P_1 and P_2 always have important determining influences in common, with the result that indirect appraisal of the effect of those influences on P_s requires quantitatively precise knowledge of their effects on both P_1 and P_2. I may illustrate with a few examples.

Consider first the effect of reports of damage to the American winter-wheat crop on the difference between prices of May and July wheat at Chicago at a time when the price of the July future is some 10% below the price of the May. The indirect approach has led uniformly, so far as I know, to the opinion that in those circumstances the reports of crop damage should affect the price of July wheat considerably more than the price of May wheat, and therefore should reduce the price difference between them. This is a conclusion, however, that is contrary to the facts, as we have noted in the previous section. If one takes the direct approach permitted by treating the difference as itself a price of storage, it becomes apparent that no substantial effect is to be expected; the crop damage has no influence on the volume of stocks to be carried from May to July, and it is unlikely to have much influence on the "price" at which holders are willing to carry given stocks.[40]

Consider, second, the effect of reports of damage to the southern hemisphere wheat crop on the difference between the prices of the December and the March futures at Liverpool, when the price of the March future was considerably below the December. The indirect approach has tended toward conclusions which were quite erroneous, at least as regards the magnitude of the effect on the price difference. The direct approach leads to the conclusion that the effect might differ according to circumstances. Study of the behaviour of the "price of storage" at Liverpool when international trade in wheat was relatively free indicates that the December–March carrying charge there was causally related to expected total stocks of imported wheat in Europe in December.[41] The level of these stocks could be appreciably affected by southern hemisphere crop damage. One possible chain of effects was that news of the crop damage might produce such a strong price advance in northern hemisphere exporting countries that Europe importers would allow their stocks to run down, hoping to buy more cheaply later, and that the reduction of stocks would be accompanied by increase in the inverse carrying charge (elevation of the December future relative to the March). What more often happened was that southern hemisphere crop damage was viewed rather more seriously by European importers than by most traders in exporting countries of the northern hemisphere; importers tended to avoid letting stocks run so low in December as they would have otherwise, and the inverse carrying charge tended to narrow slightly (the December future declined relative to the March).

Now consider an influence of another sort: the rather prevalent belief in the winter of 1946–1947 that a business recession was in prospect. This was regarded by many as sufficient to explain the large inverse carrying charges which prevailed in many futures markets at that time. Any of several lines of reasoning could be followed to reach that conclusion: it might be thought that the futures markets, being open to short selling as cash markets are not, tended especially to be depressed by selling on the part of those who anticipated business recession; or it might be argued that prices of deferred futures were depressed because the expectation of business recession implied expectation of price decline and that it is the business of futures markets to reflect expectations of price change; or again, it might be reasoned that the futures prices were depressed relative to spot prices because the people who trade only in futures markets more generally expected a business recession than did those who trade only in cash markets.

A direct approach to explanations of inverse carrying charges in 1946–1947, treating them as prices of storage, could lead to conclusions somewhat similar to those from the indirect approach. Though expectation of business recession could scarcely affect the price of storage through producing a significant change in the amount of wheat, for example, that would be held in storage from December to May or from May to July, it might have an influence on the distribution of stocks. If wheat growers tended to ignore or to disagree with the prevalent opinion in commercial circles, they might consider the current market price to be unduly depressed, and be led, therefore, to hold wheat back on the farms. Because the market price of storage, though related to total stocks and not merely to hedged stocks or to stocks in commercial channels, may be especially affected by the level of commercial stocks, strong holding by producers may have some tendency toward development of inverse carrying charges. But the direct approach, permitting this line of explanation, permits also a check on the hypothesis, which in this case compels its rejection. Informed observers were nearly unanimous in the opinion that wheat growers in the winter of 1946–1947 were not showing any notable tendency to hold back their wheat. Commercial stocks of wheat outside of producing areas were indeed relatively low, but they were so principally because of the strenuous efforts that were being made to expand exports, and partly because of shortage of freight cars, which retarded movement to primary markets. The direct approach thus leads to the conclusion that, in the case of wheat at least, the large inverse carrying charges during the winter and spring of 1946–1947 are to be attributed mainly to relative shortage of supplies, aggravated by transportation difficulties.[42]

Direct explanation of inverse carrying charges yield a gain in the forms of better understanding of cash and futures prices that may be as important as the gain in understanding of carrying charges themselves. Because the carrying charge, whether positive or negative, is determined by influences that are largely or wholly independent of those affecting the price of a distant future, the equation

$$P_1 = P_2 - P_s \tag{1a}$$

has special economic significance. Usually the "nearer" price, P_1, may be explained best in terms of the set of influences that affect the more distant price and the almost wholly separate and additional set of influences that affect the price of storage, P_s. Because P_2 and P_s are substantially independent, influences affecting the price of a distant future, P_2, may be regarded as having an approximately equal effect on any nearer price. Similarly, any influence affecting the carrying charge, P_s, must tend to have an equal and opposite effect on the near price. The latter fact deserves notice especially in studies of particular price situations, because it permits significant information, which can often be obtained regarding reasons for a given carrying charge, to be applied in the explanation of price.

CONCLUSIONS

The thread of argument in the foregoing pages has proceeded from a consideration of several attempted explanations of inverse carrying charges, which are found defective, to an explanation which views inverse carrying charges, like positive carrying charges, as market-determined prices of storage. Since the explanations rejected were not figures of straw set up as a mere device of exposition, it should be surprising if they do not carry some elements of truth. I should like in concluding to bring into relation the final explanation and what seem to me the true and useful features of the unsatisfactory attempts.

For some purposes it is necessary to treat cash markets as distinct from futures markets. They are distinguishable in much the same sense that country market may be distinguished from the terminal market which it supplies, or the Kansas City wheat market from that at Chicago. Error arises only if the strength of the influences which connect their prices is underestimated. The foregoing demonstration that inverse carrying charges may be treated as prices of storage is independent evidence that "arbitrage" between cash and futures markets remains effective even when the futures price falls much below the spot price. Though the direct proof has been given here only for the Chicago wheat market, the conclusion may be applicable to all important futures markets since it requires only that hedging continue in spite of inverse carrying charges, and I think it likely that no future market can persist without the continuous presence of a substantial amount of hedging.

The idea that futures prices tend to have a downward bias probably has some validity, but it is not a useful starting point for a theory of carrying charges. The view which Keynes combined with it, that hedgers may accept a sacrifice to avoid risk, can be useful, however, in another context; it seems pertinent as partial explanation of the "supply curve for storage". The belief that expectations concerning future events tend to be discounted is unquestionably sound, but it must be applied with recognition that the possible future events include those which would be price-depressing as well as those which would be price-elevating. Discounting expectations of a price-depressing event must tend toward a present price higher than would prevail without discounting. With that amendment, the theory of discounting has useful applications in interpretation of carrying charges.[43]

Prevalence of the idea that expectations of price change should be reflected in differences between spot and futures prices is significant principally as evidence of the extent to which the actual economic functions of futures markets are misunderstood, even by some people who have much practical knowledge of the markets. There is a good deal of excuse for such misunderstanding because the facts are difficult to express without ambiguity, and can be understood only if clear distinction is drawn between expectations arising from an existing supply situation and expectations regarding future supply or demand developments. There is, moreover, a grain of truth in the idea that expectations regarding future supply and demand may affect spot-future price spreads; but any supposed effect that cannot be explained in terms of the concept of price of storage deserves to be met with suspicion.

Finally, the main positive conclusions that we have reached may be summarised and perhaps illuminated a little by expressing them as answers to three questions propounded in an introductory paragraph above. Taking the questions in reverse order, answers may now be given as follows:

(1) Spot and futures prices for a commodity are intimately connected at all times. Even when the spot price is far above the price of a deferred future, it tends to respond fully to influences that affect the price of the future.

(2) Inverse carrying charges are reliable indications of current shortage; the forecast of price decline which they imply is no more reliable than a forecast which might be made from sufficient knowledge of the current supply situation itself. Inverse carrying charges do not, in general, measure expected consequence of future developments.

(3) An inverse carrying charge between December and May wheat futures in the United States must be explained in somewhat different terms in August, say, than in December; in August it rests on expectations regarding conditions that will exist in December. As of December, it means that supplies of wheat in commercial channels are so small that holders of hedged stocks expect to dispose of most of their stocks before the spot price declines appreciably (relative to the May future), and that they are willing to risk loss on a fraction of the stocks for the sake of assurance against having their merchandising or manufacturing activities handicapped by shortage of supplies.

1 "Carrying charge", like most other economic terms, is used in varying senses. The Federal Trade Commission in its valuable *Report on the Grain Trade*, uses "carrying charge" only in the sense of carrying costs, treated usually as comprising commercial charges for interest, insurance, and storage in public elevators. (*Op. cit* ., VI, 1924, 147, 192 ff.) What are here called carrying charges it refers to always as spreads.

2 The adjectives "cash" and "spot" are often used with the same meaning, but cash has also the sense of "not-future"; the official rules of the Chicago Board of Trade include the definition, "Cash Grain – Spot grain and grain to arrive". In this paper I intend "spot" to designate that cash price which applies to the particular description of the commodity which is expected to be delivered on the future to which its price is related in calculation of the carrying charge; other spot prices I designate as "cash". Thus in Table 1 the adjective is "cash" because the quotations probably do not represent quite the quality to which futures quotations apply.

3 18 November (1946), pp. 1036–40.

4 Holbrook Working, "Price of Cash Wheat and Futures at Chicago since 1883", *Wheat Studies of the Food Research Institute* (hereafter referred to as *Wheat Studies*), 11 November (1934), p. 79. Later references to works by the present writer omit designation of authorship.

5 The length of time that the wheat might have to be held before it could be sold at a parity with wheat in store was uncertain.

6 Deliveries on futures tend to be concentrated either early in the delivery period or late, and traders usually can judge well in advance which to expect. Occasionally such expectations prove mistaken, but even then uncertainty is short-lived; one expectation is quickly replaced by another on the basis of events in the first few days of the delivery period.

7 Gerda Blau, "Some Aspects of the Theory of Futures Trading", *Review of Economic Studies*, (1944–1945), 12, p. 11. The above quotation in the context I have given it reflects Miss Blau's views correctly, I think, though incompletely. Elsewhere she says that "arbitrage can always be relied (upon) to keep prices in the cash market and in the futures market in line" (*loc. cit* ., p. 23); but by this she means, to keep the "cash spot" price in line with the "futures spot" price. More fully expressed, the statement quoted in the text above would read in part, "arbitrage ... cannot be equally effective in preventing the [futures] forward price from exceeding the [futures] spot price by *less* than net carrying cost". It follows that she does not regard arbitrage as reliably effective between the "futures forward" price and the "cash spot" price.

8 Nicholas Kaldor, 1939–1940, "A Symposium on the Theory of the Forward Market", *Review of Economic Studies*, 8, p. 198.
9 There is this possibly significant difference between pure arbitrage and hedging: the arbitrager, concerned only with trading on price differences, might be expected to initiate a transaction only after prices had departed appreciably from the relation which he believed warranted; the hedger, whose arbitrage is incidental to merchandising or processing, tends to be satisfied to take profits from his major operation and to require of the price relations only that they be such as not to threaten him with loss. Arbitrage associated with merchandising or processing (ie, hedging) may thus be expected to operate more promptly and strongly than would pure arbitrage against an unwarranted tendency of futures prices to rise relative to cash.
10 It would not be difficult to produce evidence for several other commodities that hedging provides continuously effective arbitrage between the cash and futures prices, but little would be gained thereby for present purposes; addition of a number of other examples of continuous effectiveness would not exclude the possibility, or greatly reduce the probability, that contrary examples might be found. More profit might be had from a study of the causes of failure of efforts to establish futures markets; there might emerge the conclusion that no futures market can long persist without support from routine hedging. One reason for thinking that such may be the case is that speculative interest, which in some circumstances may provide ample buying and selling orders to support a futures market without any hedging business, tends to dry up when market conditions do not encourage differences of opinion. Merchants and manufactures whose decisions to hedge or not to hedge turn on specific appraisals of price risks, will then tend also to abandon the futures market. Unless it has a large backlog of business from those who hedge on general principles – perhaps to avoid need for making an appraisal of general price prospects, or because their bankers require it – a futures market must find it difficult to survive such a period.
11 The same condition existed in Canada when the Winnipeg futures market was functioning.
12 Bargaining in terms of differences rather than of prices, which is to be observed in markets for several commodities, represents an advanced stage of dependence of one market on another. It is rarely found in country or in retail markets because in them at least one of the parties to each transaction commonly has an active interest in the price and not merely in a price difference.

The practice of trading in terms of price differences has long been recognised in specialised economic literature, though it has received little general attention by economists. The Federal Trade Commission's *Report on the Grain Trade*, III, 1922, pp. 44–46, describing trading at Chicago on August 27, 1918 (the date is not stated but is identifiable from the quotations), indicates that cash sales of oats were being made directly in terms of premiums and discounts while bargaining for corn was directly in terms of prices. Wheat prices were then under governmental control. G. Wright Hoffman, writing 10 years later, speaks of spot transactions in grains on terminal markets as being uniformly in terms of premiums and discounts (*Future Trading upon Organised Commodity Markets in the United States*, Philadelphia, 1932, pp. 260–63).
13 *Journal of Farm Economics*, 18, November (1946), p. 1039.
14 *Treatise on Money*, 1930, Volume II, 143 (New York). The hedgers in wheat and corn markets of the United States are rarely producers, but Keynes' theory may be rendered directly applicable to those markets by substituting "merchant and processor" in place of "producer" in the first sentence quoted above and making a corresponding substitution for "period of production" in the next sentence.
15 Cf. references below.
16 Recent literature includes: J. R. Hicks, 1939, *Value and Capital* (Oxford) pp. 135–39; Nicholas Kaldor, "Speculation and Economic Stability", *Review of Economic Studies*, 7, 1939–1940, pp.

1–27; J. C. R. Dow, "A Theoretical Account of Futures Markets", *ibid.*, 7, 1939–40, pp. 185–95; Nicholas Kaldor, J. C. R. Dow and R. G. Hawtrey, "A Symposium on the Theory of the Futures Market", *ibid.*, 7, 1939–40, pp. 196–205; and Gerda Blau, *ibid.*, 12, 1944–45, pp. 1–30.

17 *Report on the Grain Trade*, 6, 1924, chapters IX, X and appendices.

18 This was stated with a footnote qualification that a slight modification of the procedure, which might be preferred, gave a result of -2.04 cents (*ibid.*, p. 204).

19 "The Post-Harvest Depression of Wheat Prices", *Wheat Studies*, 6 November 1929, cover page summary.

20 "Financial Results of Speculative Holding of Wheat", *Wheat Studies*, 7 July 1931, p. 435.

21 In the long series beginning with 1884–1885, years of declining general price level approximately balance those of rising price level and deflation serves mainly as a weighting device; applied to the series beginning with 1904–1905, deflation would also eliminate part of the upward trend found in that period.

22 The discrepancy between Vance's interpretation of the evidence on corn prices and that presented here rests almost wholly on the selection of years taken as representative, and the treatment of delivery-month prices.

23 For example, intensive study of Winnipeg wheat prices, directed particularly at measurement of the effects of hedging pressure on futures prices, may be found in "Price Effects of Canadian Wheat Marketing", *Wheat Studies*, 14 October 1937, pp. 37–68.

24 Vol. VI, 1924, p. 147. Cf. the quotations on p. 5, above from recent writers.

25 *Ibid.*, p. 204.

26 I leave to one side the interesting theoretical question, in what sense any futures price reflects expectations, because it has no significant bearing on the immediate problem here. For an exchange of views on that subject see Kaldor, Dow and Hawtrey, "Symposium", and papers cited therein. My observations accord with the views expressed by Mr. Hawtrey in the "Symposium".

27 Prices of the current-month future during May 1930 were effectively spot prices because deliveries were heavy from the beginning of the month; the Chicago *Daily Trade Bulletin* reported deliveries of 5,362,000 bushels on September 1. The three-cent discount of cash wheat under the May shown in Table 1 is probably a location differential similar to that observed early in the previous September.

28 In this instance, as usually in the development of scientific concepts, the new theory arose out of observations that were found to conflict with earlier theory, and led in turn to further observations. In summarising the evidence, it is convenient to state the theory first and follow with pertinent observational data, without discriminating between observations that preceded formulation of the theory and those that followed. The original observations that affected my own thinking concerned relations between the July and September futures at Chicago, which do not bear so obviously on the theoretical issue as similar observations made later.

29 Details of the original studies on which this statement is based have not been published. They were made early in the investigation and served to force search for other explanations of inverse carrying charges; the affirmative findings were published in "Price Relations between July and September Wheat Futures at Chicago", *Wheat Studies*, 9 March 1933, pp. 187–238, and "Price Relations between May and New-Crop Wheat Futures at Chicago", *Wheat Studies*, 10 February 1934, pp. 183–228. The published results include correlation analyses which bear directly but not quite so obviously on the present question. Week-to-week changes in prices and price spreads were found to follow a pattern which is the reverse of that predicted by the theory that inverse carrying charges depend largely on expectations concerning future developments. On that theory, a widening, for example, of an inverse carrying charge should occur typically in connection with a decline in price of the distant future accompanied by little or no decline in the price of the near future. The statistical analyses, however, show that widening of an inverse carrying charge occurs

typically in connection with a price advance in the near future, accompanied by a much smaller price advance in the distant future.

30 Sidney Hoos and Holbrook Working, November 1940, "Price Relations of Liverpool Wheat Futures", *Wheat Studies*, 17, p. 114.

31 See, for example, charts of the May–July and July–September inter-option spreads at Chicago, in "Relations between May and New-Crop Futures", *Wheat Studies*, 10 February 1934, pp. 220, 222, and "Relations between July and September Futures", *Wheat Studies*, 9 March 1933, pp. 192, 194.

32 See "Relations between July and September Futures", *Wheat Studies*, 9, March 1933, p. 206; "Relations between May and New-Crop Future", 10 February 1934, pp. 187, 190, 191; Hoos and Working, "Liverpool Wheat Futures", *Wheat Studies*, 17 November 1940, p. 118.

33 Some of the evidence for the Chicago market may be found in *Wheat Studies*, 9, pp. 204–208; for Liverpool, *ibid.*, 17, pp. 124–26. In United States markets, differences between the spot price, or the price of a future soon to expire, and prices of distant futures have often been rendered quite unrepresentative by corners and squeezes.

34 "Speculation and Economic Stability", *Review of Economic Studies*, 7, 1939–40, p. 3.

35 *Op. cit.*, p. 6, and "Symposium on the Theory of the Forward Market", *ibid.*, p. 199 n.

36 "Some Aspects of the Theory of Futures Trading", *Review of Economic Studies*, 12, 1944–45, p. 4.

37 For example, in spite of the extraordinarily large inverse carrying charges that prevailed in the spring of 1947 (see Table 1), the Commodity Exchange Authority found that reporting hedgers on May 31, 1947, were "short" 6,599,000 bushels in Chicago futures. On June 14 the reported short hedging commitments were 9,439,000 bushels (preliminary figures kindly supplied by letter).

38 The transition point is not precisely determinate and doubtless varies considerably from year to year.

39 During autumn and early winter the upper limit of aggregate "working stocks" of wheat for the United States may continuously exceed 500 million bushels. This estimate includes "working stocks" of wheat growers. The producer of a seasonal crop provides storage, among other reasons, to avoid the excessive cost of marketing very rapidly immediately after harvest. The quantity which he is willing to carry at a given time without expectation of price appreciation is at its maximum just after the harvest, when it may approximate his total crop, and tends to decline progressively thereafter. The volume of wheat stocks which a typical merchant regards as just short of surplus for him likewise varies greatly through the crop year, with its maximum probably at about the time of maximum rate of marketing by producers.

40 Some effect is conceivable, principally because the crop damage may affect the distribution of stocks, either directly or through the attendant price change, but no such effects on carrying charges in the United States have been shown to occur.

41 Hoos and Working, November 1940, "Price Relations of Liverpool Wheat Futures", *Wheat Studies*, 17, pp. 114–25.

42 Year-end stocks were drawn down to the remarkably low level of 83 million bushels.

43 For example, inter-option price spreads for wheat at both Chicago and Liverpool have been found to show no important general tendency for the distant future to rise or to decline during the "life" of the spreads; but if the years be classified on a rational basis, some conspicuous tendencies may be found which invite explanation on the ground that advance evidence of extreme conditions in either direction from the average tend to be discounted toward the average. See charts in *Wheat Studies*, 9 February 1934, p. 210 and 17 November 1940, p. 128.

5

*Professor Vaile and the Theory of Inverse Carrying Charges**

Holbrook Working[†]

I am disappointed that my treatment of inverse carrying charges seems to Professor Vaile "inadequate and unrealistic",[1] but such a stricture alone might be accepted with reasonable equanimity. Theory always involves abstraction, and so cannot avoid being in some sense inadequate and unrealistic. Professor Vaile, however, charges specifically that the theory which I advanced "makes no start toward explaining" what it aimed to explain, and such a charge requires at least an attempt to remove misapprehensions on which it may rest.

Let me deal first with two comments of Vaile's which appear to be subsidiary, and then return to the major charge.

Vaile holds my theory to be unrealistic ". . . because whenever there are high inverse carrying charges, terminal grain merchants tend to modify their market transactions so that full contractual storage charges are paid on an increasing proportion of the total grain in store". I readily grant that the average return for storage received by a grain merchant who stores both for himself and for others is not measured by the market carrying charge (price of storage). Nor is it measured by the fixed price for contractual storage. The conclusion to be drawn is not that either the market price or the contractual price is unrealistic, but only that neither price by itself measures the return for storage received by that grain merchant. The case differs in no essential respect from that of the producer of a physical commodity who sells in two markets which have different price patterns, for example, a dairyman who sells part of his product as market milk and part to a cheese factory.[2]

Turning from criticism to suggestion, Vaile holds that "it is not true" (as

---

*This paper was first published in the *Journal of Farm Economics* 31(1), pp. 167–72 (1949) and is reprinted with the permission of the Copyright Clearing Centre.
[†]Holbrook Working, 1895–1985.

I wrote) "that

$$P_1 + P_s - P_2 = 0 \tag{1}$$

but rather

$$P_1 + P_s - P_2 = \pm x \tag{2}$$

in which equation P_s is the constant standard charge for storage".

The significance of Vaile's suggestion becomes clearer if we allow P_s to retain the meaning I gave it, under which $P_s = P_2 - P_1$ by definition. Then we may take C to represent the "constant standard charge for storage" and rewrite Vaile's equation (2) as follows:

$$P_1 + C - P_2 = x = C - P_s \tag{2'}$$

With this clarification, I can accept Vaile's reformulation, though I do not see what it gains; because C is a *known constant*, it makes little difference whether we undertake to explain P_s or $C - P_s$.

We come now to Vaile's charge that my theory "is inadequate because it makes no start toward explaining why the carrying charges are positive at one time and negative at another". I accept this as the main charge because the only merit I claim for my theory is that it *goes farther* than the generally accepted theory in explaining observed price behaviour. Vaile has done his part toward permitting a test of effectiveness by applying the conventional theory to a particular set of price observations; I shall undertake to show that my theory serves better.

The problem Vaile poses is to explain price relations between the "December and the following May, wheat futures in Minneapolis from 1900 to the present". He observes that persistent inverse carrying charges between the December and the May futures occurred "only in 1921 when the post-war foreign demand still was strong and the domestic crop was short, in 1934–1937 when the crop was seriously curtailed by drought, and in the recent years of high post-war foreign demand. In all other years there were positive, although varying carrying charges between these two futures". Then he adds by way of generalisation "Whenever supply is conspicuously short relative to demand, buyers bid up the cash and nearby prices in their efforts to obtain actual grain of desired quality".

If one asks no more than this by way of explanation, the traditional interpretation which Vaile wishes to retain seems to serve fairly well in such circumstances. But suppose we look a little farther into the facts, taking for example 1936–1937, the crop year prior to World War II in which the inverse carrying charges between December and May at Minneapolis were most extreme. From mid-September through early

Week	December future	May future	Carrying charge
September 21–26	129.1	123.9	–5.2
December 7–12	136.9	131.6	–5.3
Price increase	7.8	7.7	–0.1

December the price of the May future remained steadily about five cents under the price of the December future; thereafter the difference fluctuated somewhat erratically between five cents and nearly nine cents until the end of December. Average prices for the last full week of September and the first full week of December, and the spread between them in cents per bushel, were:

Did the price of December wheat increase because demand became more insistent relative to supply? If so, why did the inverse carrying charge fail to widen correspondingly, in accordance with Vaile's principle of explanation?

Note also that stocks of wheat in Minneapolis for the week ending September 26 were reported as 6,408,050 bushels; for December 12, as 5,681,330 bushels; and that on the following May 1 there remained 3,567,427 bushels in Minneapolis elevators. Most of this wheat was hedged in the futures markets and subject, therefore, to a negative carrying charge. How shall one explain this condition?

To answer these questions, I start with the last and explain that the grain merchants and mills of Minneapolis require "goods on their shelves" and working stocks, and will carry large supplies with no return at all for storage as such, and smaller, but still substantial, supplies when the return specifically attributable to storage is strongly negative. Having taken this view that the price relation between the futures should be considered as the return for storage necessary to induce the carrying of certain stocks by hedgers (or to induce a certain restriction of holdings), I find no difficulty in understanding why it should be possible, and indeed common, for the price relation between futures to remain relatively constant from week to week while prices change.

By concentrating attention on a price relation between two futures calling for delivery within the same crop year. Vaile avoids what in other circumstances is a most serious embarrassment to the doctrine that the two prices must be explained separately. If one studies the relation between the July and the September wheat futures at Minneapolis, the one expiring before the spring-wheat harvest and the other after harvest, and takes periods in which spring-wheat crop prospects changed substantially, it appears that the attempt at separate explanation breaks down completely. Consider for example, the period, May–July 1934, when severe drought in the spring–wheat area contributed to a sharp price

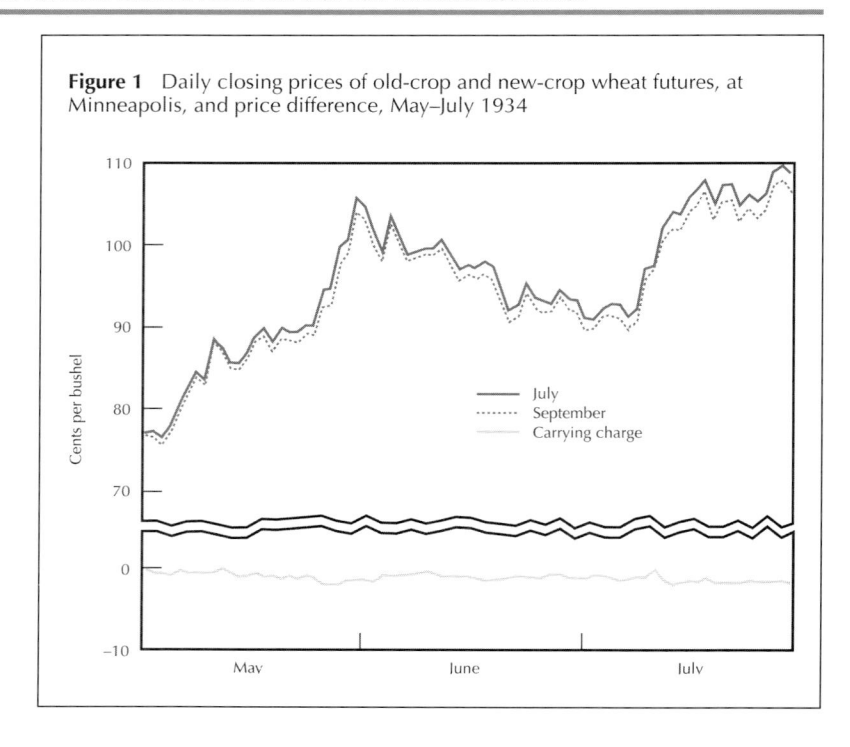

Figure 1 Daily closing prices of old-crop and new-crop wheat futures, at Minneapolis, and price difference, May–July 1934

advance during May; good rains in early June encouraged a severe price reaction; and in July prices rose again in response to renewed drought in the spring–wheat area, together with a crop report which confirmed that much of the damage done in May had been irreparable.[3] Figure 1 tells the detailed story of wide changes in prices of the two futures and negligible changes in the July–September carrying charge during these months.

It was such evidence as this which first convinced me of the futility of attempts at separate explanation of prices of different futures. *If, instead, one regards the influences pertinent to the price of the more distant future as bearing equally on the nearer one, and recognises that the price of the nearer future is affected also by influences which bear specifically on the price difference between the futures, explanation becomes easy.* In this instance we may note that stocks of wheat in Minneapolis elevators, which were 18,672,000 bushels on May 5, declined in normal fashion to 14,647,000 bushels on August 11, and then rose, reaching 16,150,000 bushels on September 1. Most of this wheat was held by merchants and processors who wanted it for convenience and would have held somewhat larger supplies except for the existence of a small inverse carrying charge. Their disposition to hold was virtually unaffected, it appears, by the drastic crop developments of the period.

If Vaile found difficulty in seeing how to apply the ideas which I tried to set out, others must have experienced like difficulty. I hope that the foregoing explanations clarify the main problems.

1 Roland S. Vaile, 1948, "Inverse Carrying Charges in Futures Markets", *Journal of Farm Economics* 30, p. 574. His discussion concerns my paper, "Theory of the Inverse Carrying Charge in Futures Markets", *ibid.*, 1948, 30, pp. 1–28.
2 Vaile may also have underestimated the risk he ran of seriously misleading people unfamiliar with details of the grain trade. The opportunities for giving effect to the "tendency" he asserts are quite limited, for three reasons: (1) Under conditions which have prevailed during most of the last half century the amount of grain stored by "others" in terminal markets has been small relative to the amount stored by the operators of the elevators; (2) the conditions which have led merchants to store little grain in their own elevators have generally reduced similarly the amounts offered for storage by others; and (3) the terminal grain merchant who elects to store grain for others is not free to choose what proportion of his space shall be used by them. He is required by law to operate any elevator used for public storage as a public warehouse, and to accept any grain offered for storage, to the limit of the elevator capacity. In some states, such as Illinois, he is even prohibited from storing his own grain in any elevator which he uses for public storage.
3 For a more detailed account of these developments see M. K. Bennett and Helen C. Farnsworth, "World Wheat Survey and Outlook, September 1934", *Wheat Studies of the Food Research Institute* 11(1), pp. 8–11.

Painting of the Amsterdam Bourse, 1608–1611
Courtesy of William J. Brodsky

2 An Early Painting of the Rice Futures Market in Osaka, Japan, c. 1600–1730
Courtesy of William J. Brodsky

[OCTOBER 31, 1868.] HARPER'S WEEKLY. 701

CHICAGO, ILLINOIS—CORN EXCHANGE AND GRAIN MARKET.

CHICAGO, ILLINOIS—THE CATTLE MARKET.

DRIVING HOGS TO THE CHICAGO MARKET.—[SEE PAGE 702.]

3
Etchings of Chicago, *Harper's Weekly,* October 31, 1868
Courtesy of William J. Brodsky

4 The "First" Octagonal Trading Pit (Symbolic of Open-Outcry Trading) at the Milwaukee Grain Exchange, c. 1870–1880
Courtesy of the Milwaukee County Historical Society

5
Trading Hall of the Liverpool Cotton Exchange
Courtesy of the Liverpool Cotton Association

6 The Interior of the London Metal Exchange on Whittington Avenue in London and home to the LME for 98 years
Courtesy of the London Metal Exchange

7
The Early Gas Oil "Trading Pit", International Petroleum Exchange, London
Courtesy of the International Petroleum Exchange

8
Exterior, Etching of the "New" Chicago Board of Trade on LaSalle Street, *The Graphic News*, March 6, 1886
Courtesy of the Board of Trade Clearing Corporation

9

Interior, Etching of the "New" Chicago Board of Trade, *Harper's Weekly*, May 16, 1885
Courtesy of the Board of Trade Clearing Corporation

10
"Old Hutch" — Futures Market Speculator Benjamin P. Hutchinson, *Harper's Weekly*, May 10, 1890

11 Trading Floor of the former Beijing Commodity Exchange, 1994
Courtesy of William J. Brodsky

12
"Study for Ceres", oil on board, owned by John Norton Garrett
Courtesy of the Illinois State Museum

6

*Futures Trading and the Storage of Cotton and Wheat**

Lester G. Telser
University of Chicago

Commodity futures markets provide the economist with abundant data for studying the behaviour of individuals coping with uncertainty. However, relatively little empirical work has been undertaken to test some of the theories of uncertainty developed for futures markets, and relatively little theoretical explanation of the empirical results can be found in the more factual studies of futures markets.[1] This article combines a theoretical approach with an examination of the evidence, and develops a theory of stockholding and futures markets capable of predicting the relations among some of the data collected in futures markets. The evidence presented is inconsistent with a theory of futures markets advanced by J. M. Keynes and J. R. Hicks.

The first part of the chapter presents the theory of stockholding for a competitive industry. Since most of the commodities traded on futures markets are seasonally produced by farmers, the stockholding theory is developed with this in mind. The intricacies of futures markets are avoided in the first part; we see how the quantity of stocks firms hold at the end of each period depends on their expected price change from

---

*This paper was first published in the *Journal of Political Economy* 66, June, pp. 233–55 (1958) and is reprinted with the permission of the University of Chicago Press.

The research upon which this paper is based was partially financed by a co-operative agreement between the United States Department of Agriculture and the Department of Economics, University of Chicago. The views expressed are mine and do not necessarily reflect those of either institution. For helpful comments and criticisms I am indebted to Milton Friedman, Zvi Griliches, Arnold C. Harberger, H. S. Houthakker, L. D. Howell, D. Gale Johnson, T. W. Schultz, and George Tolley. Holbrook Working pioneered in this subject (see especially "Theory of the Inverse Carrying Charge in Futures Markets", *Journal of Farm Economics* 30, February 1948, and "The Theory of Price of Storage", *American Economic Review*, 39, December 1949.

period to period and what determines the actual price change from one period to another.

Next I discuss a hypothetical futures market in which there are speculators only. Such a futures market can exist side by side with a storage industry none of whose firms trade in futures. Thus there can be two distinct markets: the speculators' futures market and the dealers' and consumers' market in which the physical commodity is traded. This case is not of great interest per se, but it clears the way for the analysis of the real commodity futures market in terms of the interaction between speculators and dealers in the physical commodity. Firms hold stocks determined by the difference between *futures prices* of contracts of successive maturities, whereas in the absence of futures trading the stocks held depend on firms' *expected price changes*. Hence for commodities traded on futures markets we can estimate the relation between the two observable variables, stocks and futures prices, provided we can establish a relation between the expected price and the futures price.

The article next explores the relation between the expected price and the future price. According to the theory of Keynes and Hicks, the expected spot price is higher than the futures price, so that the futures price is a biased estimate of the expected spot price. This implies that futures prices display an upward trend as the delivery date approaches. If there is such a trend in futures prices, the theory of stockholding becomes more complicated because we must then analyse how the risk of holding stocks varies with the amount held. But if futures prices display no trend, so that they can be regarded as an unbiased estimate of the expected spot price, a very simple theory is adequate to relate stocks to futures prices. The evidence tends to support the simpler theory and fails to support the one advanced by Keynes and Hicks.

In the last part of the article I present the data on stocks and futures prices for cotton and wheat and estimate the firm's stockholding schedule. As predicted by the theory, the greater the excess of the more distant futures price over the nearer futures price, the more stocks are held. The problems that arise in measuring the variables and identifying the theoretical relations in the data and those raised by the effect of the government price-support program are also discussed.

THE DERIVED CONSUMERS' STOCK SCHEDULE

In developing a theory of storage for a competitive industry, we need two schedules analogous to the supply-and-demand schedules of elementary price theory. The first schedule is derived from the consumers' demand schedule for consumption, and the second is obtained by considering the behaviour of stockholding firms in the industry. The equilibrium is determined by the intersection of these two schedules, in a sense to be described below. Consumers and firms do not engage in futures trading

Figure 1

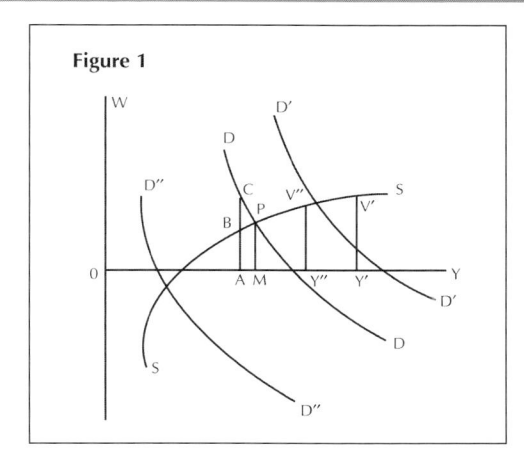

at this stage of the analysis. In this section I derive the consumers' stock schedule.

Consumption during any period[2] equals stocks carried into the period plus current production minus stocks carried out of the period. For an agricultural commodity harvested once a year production is zero during most of the year.[3] For simplicity assume that the entire crop becomes available at the beginning of the crop year. Hence consumption equals stocks carried into the period minus stocks carried out of the period.

The derived consumers' stock schedule relates the change in price between the current period and the preceding period, $p(t) - p(t - 1)$, to the stocks held at the end of period $t - 1$ (which equal stocks carried into period t). This schedule has a negative slope. Suppose that firms decide to carry smaller stocks out of period $t - 1$ into period t and that the quantity of stocks held for all other periods remains the same. More of the commodity is offered for sale in period $t - 1$, and correspondingly less is offered for sale in period t. Consumers are willing to increase their consumption in $t - 1$, provided the price in $t - 1$ falls; and, because a smaller quantity of the commodity is available for consumption in period t, competition among the consumers results in a higher price in that period. It follows that a decrease in stocks carried out of period $t - 1$ into period t decreases the price in period $t - 1$, $p(t - 1)$, relative to the price in period t, $p(t)$. Thus $p(t) - p(t - 1)$ is a decreasing function of stocks carried out of $t - 1$ into t. This function is shown by DD in Figure 1. Stocks are plotted on the horizontal axis, and $p(t) - p(t - 1)$ is plotted on the vertical axis. The curve DD has a negative slope.[4]

THE FIRMS' STOCKHOLDING SCHEDULE

In a competitive industry in an uncertain world a firm maximising expected net revenue holds an amount of stocks such that the net marginal cost of holding these stocks equals the expected change in their price during the time they are held. The net marginal cost of holding a given amount of stocks is the marginal cost of storage minus the marginal convenience yield. The marginal cost of storage is defined in the usual way as the change in the total cost of storage per unit change in the quantity of stocks held. Costs of storage include such things as the rent of the storage facility, interest charges, and loading and unloading charges.

The convenience of stocks is the benefit to the firm of holding stocks, which is derived from two sources. First, the availability of stocks permits a processor or producer to maintain a given *level* of output at a lower cost than would be required without stocks. Similarly, an inventory firm can maintain a given rate of sales at lower cost. Stocks held to realise such economies are often called "pipeline stocks". Second, holding stocks permits the rate of production or sales to be *varied* at a lower cost than would be incurred if the firm attempted to purchase the stocks as they were needed. Immediate purchases are more expensive than leisurely purchases that permit the firm to learn the state of the market and to make the best transaction it can. The marginal convenience yield is the change in the total convenience yield per unit change in stocks held.[5]

The reader may wonder why it is necessary to introduce the concept of convenience yield to explain the holding of stocks. Why is it not sufficient to consider only the marginal cost of storage? The answer lies in the fact that stocks are held even when prices are expected to fall. Hence those holding stocks may expect to suffer a "capital loss" and in addition incur storage charges. That stocks are indeed held under these circumstances may be seen in Figures 3 and 4, which will be discussed in greater detail below. More concretely, a firm may hold a stock of wheat even though the futures price, the price for delivery next month, is US$2.00 per bushel, and the spot price, the price for immediate delivery, is US$2.20 per bushel.

Consider a firm determining the quantity of stocks, y, it will acquire in one period to be sold in the next. Let the convenience derived from these stocks be the function $a(y)$ and the total cost of holding these stocks be the function $b(y)$. The net cost of holding stocks $y, c(y)$ is

$$c(y) = b(y) - a(y) \tag{1}$$

Total costs increase as stocks held increase ($b'[y] > 0$). For a particular firm these costs may increase at either a constant or an increasing rate ($b''[y] \geq 0$). If each firm can store all it wants without affecting the cost per unit of the commodity stored, then its marginal cost of storage is constant ($b''[y] = 0$). If storage is subject to decreasing returns to the scale of the firm, then the marginal cost of storage increases at an increasing rate ($b''[y] > 0$).

The total convenience of stocks is an increasing function of the quantity of stocks held; $a'(y) > 0$. For a given rate of production or sales there is some quantity of stocks so large that the marginal convenience yield of this quantity is zero. Therefore, I assume that the total convenience yield of stocks increases at a decreasing rate; $a''(y) < 0$.

Let us call the supply price of storage w. For a monopolist, w is a decreasing function of the amount of stocks held. For a competitive firm, w is the expected change in the price from period 1 to period 2 and is independent of the stocks held by the particular firm. Since w is the supply price of storage, yw is the gross return from holding a quantity of stocks y. The net return is

$$yw + a(y) - b(y) \qquad (2)$$

The quantity of stocks that maximises the net return can be found by differentiating equation (2) with respect to y and setting the derivative equal to zero. For a competitive firm we obtain

$$w = b'(y) - a'(y) \qquad (3)$$

Equation (3) may be solved for y as a function of w to obtain the firm's stockholding schedule

$$y = s(w) \qquad (4)$$

Equation (3) states that the net return is a maximum when the expected change in the price equals the marginal cost of storage minus the marginal convenience yield. The conditions on the second derivative of $a(y)$ and $b(y)$ given previously insure that the solution is a maximum. The slope of the firm's stockholding schedule is positive. Differentiating equation (3) with respect to y, we obtain

$$\frac{dw}{dy} = b''(y) - a''(y) > 0 \qquad (5)$$

With neither external economies nor diseconomies of scale the industry stock schedule for a given supply price of storage can be found by summing the stocks held by the firms. If there are economies or diseconomies of scale, the net marginal cost of storage for each firm depends on the total quantity of stocks held by the industry. Each firm holds the amount of stocks at which the net marginal cost of storage equals the expected supply price of storage, when the sum of the stocks held equals the total industry stocks used in determining the net marginal cost of storage for each firm. It is conceivable that the supply curve of the industry has a negative slope because of external economies of scale, even though there are internal diseconomies of scale to each firm. However, the empirical evidence to be presented indicates that the aggregate stockholding schedule has a positive slope.

DETERMINATION OF THE EQUILIBRIUM

The equilibrium levels of stocks, consumption, and prices in a two-period model are determined by the firms' stockholding schedule and the derived consumers' stocks schedule. In Figure 1 the derived consumers' stock schedule is *DD* and the firms' stockholding schedule is *SS*. A proof is required to establish that the equilibrium quantity of stocks carried out of the first period and the supply price of storage can be read off the intersection of these schedules at the point *P*.

The derived consumers' stock schedule relates the quantity of stocks carried out of the first period to the realised change in the price. However, the firms' stockholding schedule relates the carry-out to the expected change in the price. How is the realised change in the price related to the expected change in the price? How is account taken of difficulties in expectations among stockholders?

At least two alternative ways of answering the latter question suggest themselves. First, the total quantity of stocks held by all firms in the industry implies a price expected to prevail in the second period. The expected price associated with this total may be defined to be the average of the firms' expected prices. Second, for each expected price change and level of total stocks held by the industry we can find the quantity of stocks held by each firm. To obtain a point on the aggregate stockholding curve, the sum of the stocks held must equal the total industry stocks used in obtaining the amount held by each firm. Deriving an internally consistent schedule in this way permits us to regard each firm in the industry as expecting the same price to prevail in the second period. A reason for the difficulty of obtaining the aggregate schedule is that there may be external economies or diseconomies.

It is quite possible that firms expect the price change to be *AB*, and, accordingly, they hold stocks *OA*. From Figure 1 it is clear that they have miscalculated. When stocks held are *OA*, the realised change in price is *AC*, not *AB*, so the market is not in *ex ante* equilibrium because firms' price expectations are wrong.

However, there are important reasons for believing that, in the kinds of markets studied, the firms' expectations tend to be correct on the average. Suppose, on the contrary, that firms always miscalculated the price change or, in other words, that firms always held a quantity of stocks based on an expected price change that was never realised. Clearly, one of two possibilities holds. Either firms consistently suffer losses (because they overestimate the change in price) and are eliminated from the industry or firms receive large profits which induce new firms to enter the industry. The persistence of firms in the industry, on the one hand, and the stability in the number of firms in the industry, on the other hand, create the presumption that, on the average, neither large losses nor large profits occur. Therefore, on the average, the expected change in the price equals the realised price change. In Figure 1 firms hold OM stocks expecting a price change of MP; and, in fact, the realised price change is MP. The system is in *ex post* as well as *ex ante* equilibrium.

Changes in either the initial supplies or the expected crop alter the *ex ante* equilibrium because they effect the derived consumers' stock schedule. Suppose the demand for consumption in each period is $p(t) = h_t(q[t])$, $t = 1, 2$. Initial stocks are y_0, stocks carried out of the first period are y, and the crop expected at the beginning of the second period is x. Consumption in the first period is $q(1) = y_0 - y$, and consumption in the second period is $q(2) = x + y$. (In a two-period model no stocks are carried out of the second period.) The derived consumers' stock schedule is $p(2) - p(1) = h_2(q[2]) - h_1(q[1])$. Inserting stocks for consumption according to the definition, this becomes

$$p(2) - p(1) = h_2(x + y) - h_1(y_0 - y) \qquad (6)$$

Thus a change in the expected crop, x, or initial supplies, y_0, shifts the derived consumers' stock schedule. The direction of the shift may be formally obtained by partially differentiating (6) with respect to what are here regarded as the exogenous variables, x or y_0. An increase in the initial supplies shifts the derived consumers' stock schedule to the right from DD to $D'D'$. Hence stocks carried into the second period increase, and so does the expected price change. Such a change in the equilibrium is similar in effect to a decrease in the size of the crop expected at the beginning of the second period. A decrease in initial supplies or an increase in the expected crop shifts the derived consumers' stock schedule to the left from DD to $D''D''$, resulting in smaller stocks carried out of the first period and a decrease in the expected price change. Similar conclusions follow when more than two periods are considered. However, the proofs require the use of algebra and are omitted here.[6]

Suppose, however, that the firms' stockholding schedule does not shift seasonally and is represented by SS in Figure 1. If the equilibrium level of stocks carried out of the first period is OY', then the expected price change from the first to the second period is $Y'V'$, which is equal to the realised price change. Stocks carried out of the second period are smaller than stocks carried out of the first period by the amount consumed during the second period. The derived consumers' stock schedule shifts to the left and intersects the firms' stockholding schedule at V''; equilibrium stocks carried out of the second period are OY'', and the expected price change from the second to the third period is $Y''V''$, which equals the realised price change. As stocks decrease during the crop year, the derived consumers' stock schedule shifts to the left and determines equilibria at intersections with SS, which lie progressively farther to the left. The price changes successively decrease during the crop year.

INTRODUCING FUTURES TRADING

So far we have considered the determination of the equilibrium levels of stocks and prices in the absence of futures markets. In this section and the following one I discuss stockholding in the presence of futures trading. First, I analyse a futures market consisting solely of speculators. Second, I analyse a commodity futures market including hedgers as well as speculators; that is, one which more closely resembles a real market. To obtain empirical estimates of the firms' stockholding schedule, we need data for stocks held and the expected price changes. Expected price changes cannot be directly observed, but, when there is futures trading, they are related to differences in prices of futures contracts of successive maturities. In these two sections I develop a theoretical argument and examine futures prices data to show that the price expected in period t equals the futures price of the contract maturing in that period. Hence the firms' stockholding schedule relates the stocks held to futures prices. Since both are variables we can observe, we can obtain empirical estimates of this schedule.

In a futures market we can distinguish two types of participants: hedgers who handle the physical commodity and process or sell it and speculators who do not handle the physical commodity at all.

There are two kinds of handlers in a futures market, short hedgers and long hedgers. In general, a firm that purchases the commodity in a spot market and simultaneously sells an equal quantity of futures is engaged in a short hedge. For example, a typical short hedger in wheat stores wheat purchased from a farmer and simultaneously sells an equal amount of wheat futures. A typical long hedger is a processor of the raw commodity who sells the processed commodity forward and buys futures. For example, a flour miller who sells his flour forward to a baker may prefer to buy wheat futures instead of purchasing an equal amount of wheat in the spot market: such a pair of transactions is a long hedge.

In a futures market consisting solely of speculators, one who thinks the price will rise above its current level purchases futures contracts, and one who expects futures prices to fall sells futures contracts. In the parlance of futures markets the former speculator is "long" and the latter is "short". Since it is sufficient to consider two periods, the expected futures price is the same as the spot price expected in the second period.

If all the speculators expected the same price, there would be no trading. Suppose that each speculator behaves as if there were an excess demand Equation, (7), relating his net purchases of a given futures contract, x_i, to the difference between the current futures price, p, of that contract and the price he expects, p'_i, where i refers to the ith speculator,

$$x_i = F_i(p - p'_i) \qquad (7)$$

If the current futures price exceeds the expected spot price, the speculator sells contracts; $x_i < 0$. Conversely, if the expected spot price exceeds the current futures price, he buys contracts; $x_i > 0$. As $p - p'_i$ increases, x_i decreases. The spot price expected by the ith speculator depends on the current spot price and other variables not explicitly defined, and the exact nature of this relationship is unspecified here.

The equilibrium futures price is such that the quantity of contracts speculators wish to purchase equals the quantity that speculators wish to sell,

$$\Sigma x_i = 0 \qquad (8)$$

Although speculators' expectations differ, the equilibrium futures price is defined to be the expected spot price and is an average of the prices expected by the speculators.[7]

The more speculators there are in the market, whether or not they happen to be holding a commitment at the moment, the closer is the slope of the aggregate excess demand curve to zero.[8] Speculators enter the market whenever they expect it to be profitable, and, since transactions costs are relatively low, the supply of speculators is large enough so that the slope of the excess demand curve of the speculators, SS in Figure 2, does not differ appreciably from zero. In Figure 2 the equilibrium futures price is OB.

Let hedgers enter the futures market. The nature of the risks borne by handlers designated long and short hedgers are different. A short hedger holds stocks and incurs the risk that the price expected in the second period will be lower than anticipated. Such a firm increases its risk by holding stocks. If it can hedge its stocks – sell a quantity of futures contracts equal to the stocks it is holding – then it can avoid the risk of a change in the price of the second period. Therefore selling forward reduces the risk of holding stocks and increases the quantity of stocks a firm is willing to hold for a given supply price of storage.

The existence of futures markets provides a processor who has sold the finished product forward with an alternative way of avoiding price risk. In the absence of a futures market in a raw material, the processor can avoid the risk of a rise in the price of the raw material by purchasing the raw material at the same time the finished product is sold forward. With a futures market, instead of holding a stock of the raw material, he can

Figure 2

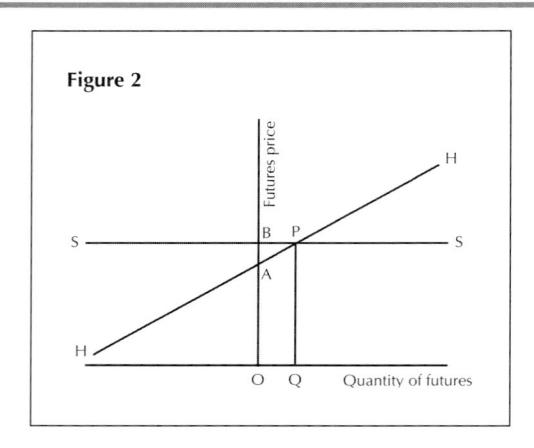

[Figure 2: A graph with "Futures price" on the vertical axis and "Quantity of futures" on the horizontal axis. An upward-sloping line HH intersects a horizontal line SS at point P above point Q on the x-axis. Points B and A are marked on the vertical line at Q, with B on the SS line and P slightly to the right.]

substitute futures contracts for stocks and thereby hold smaller stocks and still avoid a price risk, given the total forward sales of the finished product.

From these remarks we see that the net effect on the firms' stockholding schedule of the introduction of futures trading depends on (1) the increase in stocks held by the short hedgers and on (2) the decrease in stocks held by the long hedgers, both for a given supply price of storage. On a priori grounds we cannot determine which of these two factors predominates and, therefore, what is the net effect on the firms' stockholding schedule of the introduction of futures trading. In the empirical sections to follow I estimate the firms' stockholding schedule in the presence of futures trading. But I am unable to estimate this schedule in the absence of futures trading because I have no evidence with which to determine empirically the net effect of the introduction of futures trading on the firms' stockholding schedule.

Given the current spot price and the expected spot price, the excess supply of futures contracts offered by the handlers of the commodity increases as the futures price increases. An increase in the futures price increases the remuneration to the holders of stocks, the short hedgers, and they are willing to hedge more stocks. In addition, an increase in the futures price increases the cost of a long hedge, which induces the long hedgers to reduce the quantity of futures they buy. The excess supply of futures contracts, being the offers of futures by the short hedgers less the demand for contracts by the long hedgers, has a positive slope because an increase in the futures price increases the net offers of futures contracts by the handlers. This excess supply curve is the schedule *HH* in Figure 2. If the introduction of hedgers into the market does not alter the excess demand for futures contracts by the speculators, then the equilibrium futures price is *OB*, the same as in a market consisting solely of

Table 1 Wheat: annual average commitments of reporting traders in millions of bushels

Year*	Short Hedging	Short Speculating†	Long Hedging	Long Speculating†	Per cent of total open contracts reported Short	Per cent of total open contracts reported Long
1937	48.4	16.8	10.8	15.5	68.5	27.2
1938	57.7	11.2	3.4	17.8	73.4	22.6
1939	55.7	9.6	3.5	17.4	74.7	23.9
1946	7.4	6.8	13.6	7.7	37.4	55.8
1947	38.6	19.2	20.5	26.3	61.3	49.5
1948	32.5	22.6	28.1	23.1	60.6	56.4
1949	34.3	16.3	17.9	27.7	58.9	53.2
1950	50.0	20.8	20.4	29.5	72.5	51.1
1951	52.3	25.4	20.4	35.2	72.5	51.8
1952	55.1	25.7	20.8	36.1	75.7	53.3
1953	42.3	24.1	11.6	33.4	68.5	46.4

Source: United States Department of Agriculture, Commodity Exchange Authority, *Commodity Futures Statistics* (annual), and United States Department of Agriculture, Commodity Exchange Authority, *Grain Futures Statistics, 1921–51* (Statistical Bull. 131 [July, 1953]).

*Prewar years include figures for the Chicago Board of Trade only. Postwar years include figures for all wheat futures markets in the United States. The war years are omitted because there was little trading in futures. Reporting traders hold in excess of 200,000 bushels in one future.

†Speculative commitments include commitments of traders who take opposite positions in different futures (are long in one future and short in another). Such traders are called "spreaders". Many speculative commitments are of this kind.

speculators. The evidence of Tables 1 and 2 indicates that hedgers' commitments are net short and speculators commitments net long, so that SS and HH intersect to the right of O. When hedgers are net long and speculators net short (during the later part of the crop year), the schedules intersect to the left of O.

How is OB related to the spot price realised in the second period, the maturity date of the futures contract? If there were a systematic tendency for the spot price realised in the second period to differ from the futures price quoted in the first period, then it would be possible for one group of speculators to make systematic profits. If, for example, the futures price tended to exceed the realised spot price, the short speculators would systematically profit and the long speculators would systematically lose. Conversely, if the futures price were systematically less than the realised spot price, long speculators would systematically profit and short hedgers would systematically lose on their futures transactions. A glance at Tables 1 and 2 indicates that short and long speculators are both present in the futures market in significant numbers. In view of this it is not possible to obtain systematic profits by always maintaining a long position or by always maintaining a short position. The prospect of such systematic

Table 2 Cotton: annual average commitments of reporting traders in millions of bales

					Per cent of total open contracts reported	
	Short		Long			
Year*	Hedging	Speculating†	Hedging	Speculating†	Short	Long
1937	1.818	0.078	0.315	0.127	68.7	16.4
1938	1.413	0.118	0.292	0.158	61.9	23.2
1939	1.040	0.109	0.610	0.172	68.8	47.1
1940	0.765	0.080	0.447	0.117	68.6	45.4
1946	1.139	0.537	0.718	0.459	50.8	41.1
1947	0.928	0.373	0.589	0.502	48.1	40.3
1948	0.517	0.380	0.590	0.505	37.3	45.6
1949	0.620	0.606	0.565	0.494	42.9	44.6
1950	1.257	0.572	0.927	0.770	55.6	51.6
1951	0.989	0.612	0.970	0.674	45.0	46.1
1952	1.299	0.520	0.737	0.636	49.5	37.4
1953	0.092	0.421	0.476	0.625	54.7	39.8

Source: United States Department of Agriculture, Commodity Exchange Authority, *Commodity Futures Statistics* (annual). Prewar figures are from United States Department of Agriculture, Agricultural Marketing Administration, *Cotton Futures Statistics, October, 1937–July, 1941* (June, 1942).
*Prewar figures are for the New York Cotton Exchange only. Postwar figures include all cotton futures markets in the United States. Reporting traders hold in excess of 5,000 bales of cotton in one future.
†See footnote † to Table 1.

profits induces the entrance into the market of speculators who eliminate this profit by their effect on prices as they either buy or sell. This implies that the excess demand curve of the speculators is highly elastic and that the futures price *OB* tends to equal the realised price in the second period. Professional speculators may earn profits, but this is due to their superior skill in choosing when to be long and when to be short. By hedging, handlers can reduce their price risks at little or no cost to themselves. In the process, however, hedgers substitute a basis risk for a price risk. The basis risk is the risk the hedger incurs that the difference between the spot and futures price may change during the period of the hedge. The entire matter becomes complicated, and these remarks must suffice before we stray too far from the main thread of the argument.[9]

THE KEYNES–HICKS THEORY OF FUTURES MARKETS
Both Keynes and Hicks believe that the expected spot price exceeds the current futures price.[10] They call the excess of the expected spot price over the futures price "normal backwardation". This excess is said to be the insurance premium the hedgers pay the speculators to induce them to bear the risk of a change in the price. Speculators are regarded as people who could receive a certain return on their capital if it were employed in

other ventures. They can only be induced to invest in the risky commodity futures markets if they can receive a normal return on their investment plus a remuneration for bearing risk. If the futures price is below the expected spot price, and speculators buy futures, then at the maturity date of the futures contract the speculators pocket the difference between the futures price and the spot price then prevailing. In this somewhat indirect way they are rewarded for their risk.

Even if the futures price exceeds the current spot price (an eventuality predicted by the convenience yield theory when stocks are large), Keynes maintained that there is still backwardation because the expected spot price exceeds both the current spot price and the futures price.[11]

Whether we accept or reject this theory depends on whether the implications of the theory are in accord with the facts of commodity futures markets. At least two implications are present. The first is that, on average, long speculators should receive profits and short hedgers should suffer losses on their future transactions. The long speculators' profits must be net of transactions costs.[12]

A second implication, the one explored here, is that there is an upward trend in futures prices as the contract approaches maturity. In their theory the futures price is below the expected spot price by the amount of the insurance premium paid the long speculators by the short hedgers. This insurance premium is the remuneration for the risk of price changes. The risk of an unanticipated price change increases the farther away the maturity date of the futures contract is from the current date, assuming that it is more difficult to foresee the distant future. Since the risk premium is the excess of the expected spot price over the futures price, this excess decreases as the futures contract approaches maturity. Under normal conditions, when the expected spot price is not expected to change, this implies that the futures price rises as it approaches maturity. Although we cannot directly observe the expected spot price, the theory that the futures price is a biased estimate of the expected spot price can be tested by observing whether there is an upward trend in the futures price as it approaches maturity.

IS THERE A TREND IN FUTURES PRICES?

Using wheat and cotton futures-prices data, we may try to find out whether futures prices display an upward trend as they approach their maturity dates. Some of the previous studies of the downward bias of futures prices are discussed by Professor Working. He states: "Contrary to Keynes' apparent assumption it is not a simple matter, as we shall see, to determine precisely what 'The statistics of organised markets show' with respect to the price tendency he, [Keynes] was considering."[13] After examining these studies, Working concludes that, if there is such a tendency, it is probably very small and might not be capable of statistical verification.

For each cotton and wheat futures contract during the period 1926–1954 I recorded the sign of the month-to-month change in the futures price; that is, the sign of the first differences of the monthly averages of the daily closing prices. A plus sign indicates a price rise and a minus sign a price fall. If the Keynes-Hicks theory is correct, then the number of plus signs should exceed the number of minus signs. On the other hand, if futures prices display no trend as the delivery date approaches, we expect the same number of plus signs as minus signs. We may test the null hypothesis of no trend against the alternative hypothesis of an upward trend by computing the statistic

$$K = \frac{(S - \frac{1}{2} - M_S)}{\sigma_s}$$

where, for a futures contract traded for n months, S is the number of plus signs, M_S is the expected number of plus signs which equals $(n - 1)/2$, and σ_s is the standard deviation of S, which equals $\sqrt{(n + 1)/(12)}$. The exact distribution of S is known, but, using the given correction for continuity, the distribution of K is approximately normal.[14]

We would not be fair to the Keynes–Hicks theory if we merely counted the number of months the futures prices increased. During a period of generally falling prices (1930) futures prices fell also, and we observe few plus signs. But this excess of minus over plus signs should not discredit the Keynes–Hicks theory, and, similarly, a period of generally rising prices (1947), when we observe many more plus than minus signs, should not confirm their theory. This argument requires elaboration.

Changes in futures prices are not necessarily influenced by changes in the general price level. If changes in the general price level were fully anticipated, the futures prices would reflect the general price level expected on the delivery date. Were this true, changes in the futures prices would be independent of changes in the general price level. But if changes in the general price level are not fully anticipated, then the futures prices tend to move in the same direction as the general price level. During an inflationary period, futures rise because people underestimate the rise in the general price level. Similarly, during a deflationary period, futures tend to fall because people underestimate the fall in the general price level. The hypothesis that changes in the general price level are not fully anticipated together with the hypothesis of the Keynes–Hicks theory, can be tested by counting the number of months futures prices increased for three separate periods: one of generally falling prices, one of generally stable prices, and one of generally rising prices. During a period of generally falling prices we expect less than half the price changes to be positive, and during a period of generally rising prices we expect more than half the price changes to be positive. A fair test of the Keynes–Hicks theory can be made for the periods of generally stable prices. If there is no Keynes–Hicks effect, we expect as many price increases as decreases.

Using the monthly Wholesale Price Index of the Bureau of Labor Statistics (1947–1949 = 100), I classified the years studied into three groups. A year of rising prices is defined as a year in which the price index rose by more than 5%. A year of stable prices is defined as one in which the price index changed by less than 5%. A year of falling prices is one for which the price index fell by more than 5%. The years were classified independently of the movements in futures prices.

Let us first investigate the situation in cotton. Five futures contracts mature in each year: March, May, July, October, and December. Trading in these five contracts begins at least twelve months in advance of their maturity dates for the entire period studied. For example, trading in the July 1928, contract began in August, 1927. In any given month there are always at least five futures being traded. In each month the signs of the first differences of the futures prices tend to be the same for all five contracts. In fact, the signs of all five agree for 85% of the months, and the signs of at least four agree for 10% of the months; thus there is substantial agreement in 95% of the months. Therefore, it suffices to present the results for a single futures contract in detail, and I have chosen the December futures contract. The war years and the years 1939 and 1951 are omitted. I omitted 1939 because a sharp downward movement in the Wholesale Price Index was reversed to a sharp upward movement in September, 1939. Cotton trading was suspended for the early part of 1951.

The results for the December cotton futures are shown in Table 3. In the years of falling prices there are fewer plus signs than we expect, supposing no trend in futures prices; that is, futures prices tend to fall month by month. In the years of rising prices there are many more plus signs than expected, supposing no trend; that is, futures prices tend to rise month by month. In the years of stable prices there are slightly more plus than minus signs. This result suggests there is no trend, since the standard deviation is large enough that the hypothesis of no trend cannot be rejected. A fair test of the Keynes–Hicks theory can be made during those years with a stable general price level. The evidence for years of stable prices is inconsistent with the implication of the Keynes–Hicks theory that futures prices display an upward trend.

Only four wheat futures were traded during the entire period: May, July, September, and December. Trading in these begins from six to twelve months in advance of their delivery dates. Since the trading period for the May future is the longest, this is the one for which I present detailed results. For wheat, as for cotton, when several contracts are being traded during a month, the signs of the first differences of their prices tend to agree, so that little information is lost by presenting detailed data for only

Table 3 Cotton: number of months in which the December futures price increased, 1926–54

During years of falling prices		During years of stable prices*		During years of rising prices	
Year of delivery	No. of months futures price increased	Year of delivery	No. of months futures price increased	Year of delivery	No. of months futures price increased
1926	2	1927	7	1933	7
1930	1	1928	8	1934	6
1931	3	1929	5	1936	6
1932	4	1935	5	1941	8
1937	3	1940	7	1946	8
1938	4	1948	6	1947	7
1949	7	1952	4	1950	9
		1953	4		
		1954	6		
Totals	**24**		**52**		**51**
M_S	38.5	M_S	49.5	M_S	38.5
σ_S	2.753	σ_S	3.158	σ_S	2.753
χ^2	33.23	χ^2	7.38†	χ^2	17.53†
$\chi^2_{0.001}(7)$	24.322	$\chi^2_{0.5}(9)$	8.343	$\chi^2_{0.02}(7)$	16.662
K	−4.35	K	0.73	K	3.63

*The December futures price rose for eight months during 1939.

†$\chi^2 = \sum_i \left(\frac{S_i - M_{Si} - C_i}{\sigma_{Si}}\right)^2$, where c_i is a correction for continuity. When frequencies are greater than expected, they are reduced by 1/2, and, when they are less than expected, they are increased by 1/2.

one futures contract. Trading in the May contract begins in the middle of one calendar year and continues until May of the following year. Instead of studying movements in the price level for calendar years, as for the December cotton contract, I classified the twelve-month periods from June to the following May, using the 5% rule already explained. The results for the May wheat future are shown in Table 4. The wheat data confirm the conclusions reached from the cotton data.[15]

The evidence indicates that changes in the general price level are not fully anticipated. Granting this, the Keynes–Hicks theory implying a downward bias in the futures price can be rejected, and the hypothesis that there is no trend in futures prices can be accepted.

I have argued earlier that in the absence of a futures market the equilibrium established would be such that the expected price change would tend to equal the realised price change. On the assumption that expected profits are the major variable affecting the behaviour of speculators, it was shown that the futures price equals the expected spot price. A rival argument advanced by Keynes and Hicks states on the contrary that the futures price underestimates the expected spot price.

Table 4 Wheat: number of months in which the December futures price increased, 1927–54*

	During years of falling prices			During years of stable prices			During years of rising prices	
Year of delivery	No. of months futures price increased	No. of months contract traded	Year of delivery	No. of months futures price increased	No. of months contract traded	Year of delivery	No. of months futures price increased	No. of months contract traded
1927	3	10	1928	4	9	1934	4	11
1930	1	10	1929	3	9	1937	7	10
1931	3	10	1933	4	10	1941	6	9
1932	4	10	1935	3	10	1946	9	12
1938	2	10	1936	4	10	1947	5	8
1949	6	11	1939	5	10	1948	7	12
			1940	7	10	1950	8	11
			1952	6	11	1951	6	11
			1953	3	10			
			1954	6	12			
Totals	19	61		45	101		52	84
M_S		27.5	M_S		45.5	M_S		38
σ_S		2.347	σ_S		3.041	σ_S		2.770
χ^2		16.61†	χ^2		7.39†	χ^2		21.87†
$\chi^2_{0.01}(6)$		16.812	$\chi^2_{0.70}(10)$		7.267	$\chi^2_{0.01}(8)$		20.090

*Only contracts maturing during 1942–45 are omitted.
†See footnote † to Table 3.

The futures data offer no evidence to contradict the simpler hypothesis that the futures prices are an unbiased estimate of the expected spot price. Hence it is possible to use the theoretical storage model developed earlier to explain the relation between stocks held on various dates and the difference between futures prices of successive maturities. Avoiding all these theoretical questions, the fact remains that a certain empirical relation holds between stocks and futures prices, which it would be the task of any theory to explain. From now on, only the empirical relations concern us.

STATISTICAL ESTIMATES OF THE FIRMS' STOCKHOLDING SCHEDULE IN COTTON AND WHEAT

The vertical axis of Figures 3 and 4 shows the average spread (the difference between futures prices of successive maturities) adjusted for storage costs, the general price level, and the time spanned by the maturity dates of the adjacent futures; the horizontal axis shows the logarithm of stocks on the maturity date of the nearer future. For both

Figure 3

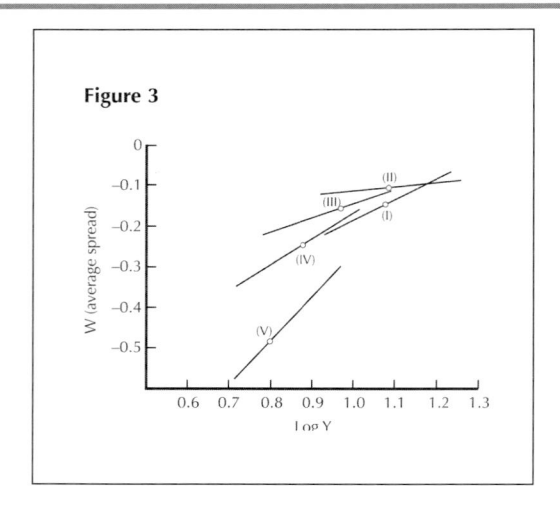

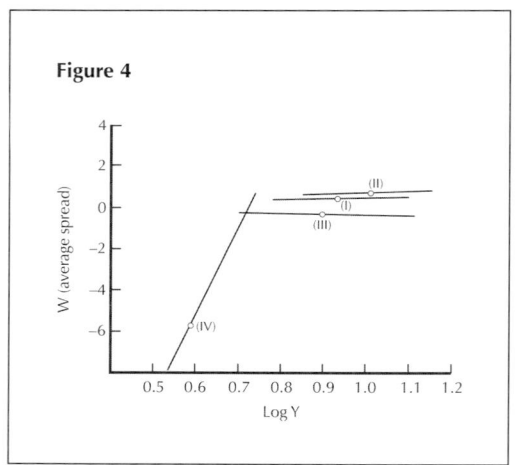

(graph: W (average spread) vs Log Y, with curves labeled (I), (II), (III), (IV), (V))

Figure 4

(graph: W (average spread) vs Log Y, with curves labeled (I), (II), (III), (IV))

cotton and wheat the adjusted spread is greatest in absolute value for those futures prices maturing toward the end of the crop year when stocks are low – the adjusted October–July spread for cotton and the adjusted July–May spread for wheat. The adjusted absolute spreads are smallest for futures maturing in the beginning of the crop year when stocks are large.

In addition, the effect on spreads of a change in stocks is shown. This relationship is derived from regressions to be discussed in greater detail below. An increase in stocks on the appropriate date increases the spread in all cases but one. The shape of the relation between stocks and spreads conforms to that predicted by the convenience-yield theory; that is, schedule *SS* in Figure 1.

Table 5 Wheat: relation between average stocks and average adjusted spreads for sample period*

Average adjusted spreads in cents per bushel		Average stocks in hundreds of millions of bushels on month-end dates	
September–July	0.48	July	8.636
December–September	0.78	September	10.286
May–December	−0.26	December	7.938
July–May	−5.66	May	3.866

*These figures are discussed in detail in the Statistical Appendix.

Table 6 Cotton: relation between average adjusted spreads and average stocks for sample period*

Average adjusted spreads in cents per pound		Average stocks in millions of bales on month-end dates	
December–October	−0.149	October	11.938
March–December	−0.105	December	12.052
May–March	−0.157	March	9.336
July–May	−0.243	May	7.567
October–July	−0.484	July	6.271

*These figures are discussed in detail in the Statistical Appendix.

The slopes of the regressions are larger for spreads maturing toward the end of the crop year, when stocks are low. The slopes are smaller at the beginning of the crop year, when stocks are large. Spreads for futures of various maturities may be ranked according to the quantity of stocks held on the maturity date of the nearer future, the smallest spreads, algebraically, being associated with the smallest quantities of stocks.

Regressions showing the relation between stocks and spreads for cotton and wheat are presented in Tables 7 and 8. These regressions were estimated using least squares. It is easy to see why the relation between stocks and spreads, the firms' stockholding schedule, is identifiable. In Figure 1 the equilibrium relation between stocks and spreads is shown to be determined by the firms' stockholding schedule and the derived consumers' stock schedule DD. Since the harvests at the beginning of each crop year differ, the initial supplies at the beginning of each crop year also differ. Hence the derived consumers' stock schedule for the first period in each crop year intersects SS at a different point in each crop year. From the first period to the last in each crop year, DD shifts to the left, intersecting SS at a series of points such as V' and V''. Thus the schedule SS is traced out by the shifting schedule DD.

Table 7 Cotton: intraseasonal stock spread regressions, 1934–54

	Coefficients of*				Standard	Correlation
Spread	Log y_i	g_i	Log q	Constant	deviation	squared
(I)[†] October–December	0.471	−0.002	0.210	−0.884	0.108	0.240
	(0.389)	(0.241)	(0.571)	(0.828)		
(II) December–March	0.193	−0.065	−0.172	−0.101	0.053	0.120
	(0.189)	(0.112)	(0.278)	(0.359)		
(III) March–May	0.359	−0.143	−0.186	−0.228	0.086	0.244
	(0.209)	(0.149)	(0.463)	(0.567)		
(IV) May–July	0.630	−0.177	−0.022	−0.671	0.172	0.407
	(0.308)	(0.282)	(0.947)	(1.117)		
(V) July–October	1.559	−0.585	−0.940	−1.170	0.269	0.736
	(0.343)	(0.371)	(1.507)	(1.732)		
(VI) Grand regression	0.909	−0.222	0.077	−1.058	0.180	0.607
	(0.079)	(0.079)	(0.424)	(0.495)		

*Stocks, y_i, are measured in millions of bales at the end of October in the first regression, at the end of December in the second, at the end of March in the third, at the end of May in the fourth, and at the end of July in the fifth. The fraction of total stocks held by the government is g. Total consumption during the crop year in millions of bales is q. Figures in parentheses are the standard deviations of the coefficients.

[†]Each of the first five regressions is based on twenty observations. The grand regression is based on a hundred observations.

Table 8 Wheat: intraseasonal stock spread regressions

	Coefficients of*					Standard	Correlation	
Spread	Log y_i	g_i	Log q	Log x	Constant	deviation	squared	Period[‡]
(I) September –July	4.017	−0.627	−5.560	−0.792	4.954[†]	0.876	0.607	1927–53
	(1.130)	(0.896)	(2.578)	(0.930)				
(II) December –September	3.313	1.401	−6.010	...	1.461	0.705	0.486	1934–53
	(4.047)	(2.699)	(3.738)					
(III) May –December	−1.150	5.150	−7.821	...	7.057	1.068	0.541	1935–53
	(4.227)	(2.762)	(3.921)					
(IV) July–May	41.558	−17.075	−43.871	...	12.986	3.938	0.692	1935–54
	(12.740)	(9.019)	(11.587)					

*Stocks, y_i (in hundreds of millions of bushels), as of July 1 are used in (I), as of October 1 in (II), as of January 1 in (III), and as of April 1 in (IV). The fraction of total stocks held by the government is g_i. Total disappearance during the crop year in billions of bushels is q. The visible supply of corn, oats, and barley in hundreds of millions of bushels on July 1 is x. Figures in parentheses are the standard deviations of the coefficients.

[†]The constant for (I) is computed using stocks of old-crop wheat on July 1. This procedure is discussed in the Statistical Appendix.

[‡]1942–46 omitted.

The schedule *DD* can also be estimated, using a model dealing explicitly with the consumers' demand for the commodity. Estimates of this schedule are not presented here because my primary interest is in the behaviour of the stockholding firm.

The appearance of the scatter between stocks and spreads indicates that non-linear regressions would give the best fit. After some experimentation it appeared that the best mathematical form for the curve to be fitted (taking account of computation costs) is linear in the spread w, the logarithm of stocks y, the logarithm of consumption q, and the fraction of total stocks held by the government g, an equation of the form (9),

$$w = a + b \log y + c \log q + dg \qquad (9)$$

(The reasons for including g and q are discussed below.)

Equation (9) represents the firms' stockholding schedule in the presence of futures trading. On the left side of the equation is the spread, w, which equals the marginal cost of storage less the marginal convenience yield of a quantity of stocks y. Factors that affect the marginal storage cost and the marginal convenience yield thereby determine the spread. A variable that increases the marginal convenience yield for a given quantity of stocks decreases the spread, and one that increases the marginal cost of storage for a given quantity of stocks increases the spread. Hence those factors increasing the marginal convenience yield enter the right side of the equation with a negative sign, and those increasing the marginal cost of storage enter the right side of the equation with a positive sign.

Two variables are included on the right side of (9) which have not yet been discussed: consumption and the percentage of total stocks under the government price-support program. Both of these variables increase the marginal convenience yield of a given quantity of stocks and should enter the right side of equation (9) with a negative coefficient.

The marginal convenience yield of a given quantity of stocks increases when consumption increases. This becomes clear if stocks are measured not in physical quantities but in time units, say, months of supply. If stocks are 100 units and consumption during the month is expected to be 50, then 100 units of stock represent two months' supply. But if stocks are 150 units and consumption is expected to be 100, then stocks represent only one and a half months' supply. In the latter case there are "smaller" stocks than in the former, which implies that the marginal convenience yield is greater. Hence consumption is included in the regressions as a separate variable explaining the spread. The sign of this variable is expected to be negative, so that an increase in consumption, stocks remaining the same, will decrease the spread.

The effect of the government price-support program on the spread is more complicated. The effect of large government stocks on spreads differs from that of the same amount of privately held stocks. The convenience yield of government stocks is less than that of an equal amount of privately held stocks because of the way the support program operates. Therefore it seems a plausible hypothesis that the convenience yield of a given total quantity of stocks decreases the greater the proportion under the government price-support program.

Government stocks consist of two parts. One part is the collateral on the price-support loans made to the farmers by the Commodity Credit Corporation (CCC). Only farmers can obtain such loans, which are made up to about the middle of the crop year. The amount of the loan equals the support price per unit of the commodity multiplied by the number of units placed under loan less certain costs such as interest charges and handling costs. Farmers can redeem their collateral by repaying the loans before a certain date, usually the end of the crop year.[16] Such redemptions generally occur when the market price rises above the support price, and the redeemed stocks are then sold to private concerns. If the loans are not redeemed before the deadline, the government can take title to the collateral in full settlement of the loan. Once this has occurred, the stocks are owned by the government and can be sold under conditions which have varied from time to time. At present these are, roughly, that the stocks can be sold at a price not less than the acquisition price plus storage costs and not less than the current support price. These set a lower limit to the government's selling price. However, the government need not sell its stocks even though the market price is well above the current support price. In short, although the government may be holding large inventories these may not be readily available for sale on the free market. Hence the convenience yield of such stocks is less than that of the same amount of stocks privately held. Because stocks under loan are more readily available to the free market, they have a higher convenience yield than an equal amount of government-owned stock. Hence it seems a plausible hypothesis that the spread is lower for a given quantity of price-support stocks the greater the proportion under loan. Unfortunately, this hypothesis cannot be tested because there are no adequate data separating price-support stocks into stocks under loan and government-owned stocks. Therefore, only the total amounts of price-supported stocks are included in the regressions.

In the early years of the price-support program (1933–1939), no differentials in the loan rates were provided for stocks in various locations. Moreover, the differentials for quality were rough and perhaps not in accord with the differentials prevailing in the market. The government tended to acquire stocks of commercially poor quality in regions far removed from consuming centres, thus reducing the convenience yield of government stocks to firms in the industry. We may think of one unit of government stocks as equal to a fraction of a unit of commercial stocks.

In spite of all this, the convenience yield of stocks decreases as the total amount held, public or private, increases, and this causes spreads to increase. However, the spread tends to be lower for a given total of stocks the greater the fraction of this total under loan or owned by the government.

Finally, the convenience yield is affected by other variables. The lower the real cost of communication and transport, the greater is the convenience of a given total of stocks. A relative shortage of a particular quality of the commodity increases its convenience yield for a given total of all other qualities. An increase in the average distance of stocks from consuming centres (owing to regional effects of weather on the harvest) also reduces the convenience of a given total of stocks. However, in the computer regressions such variables are not explicitly included because the data do not permit fine manipulations.

Cotton stocks are largest in the fall and gradually decrease until the end of July. Algebraically, the largest spread is the December minus the October futures price. The smallest spread is the October minus the old-crop July. All the other spreads lie in between. There is a regular relation between the average level of stocks on the appropriate dates and the average adjusted spreads as shown in Figure 3 and Table 6.

Substantial quantities of wheat are harvested during June and July. Stocks reach their peak during the fall and gradually decline toward the end of the crop year. The relation between average stocks on the appropriate dates during the year and the relevant adjusted spreads is shown in Figure 4 and Table 5. For wheat, as for cotton, the spreads are determined by the seasonal movement of stocks.

The correlations for cotton (Table 7) are very low for the first three spreads and are much higher for the last two spreads. However, this should not alarm us. The standard error of the predicted spread (for average values of the explaining variables) is very small for the first three regressions. In fact, the error of a prediction for Table 7, regression II, which has the lowest correlation, is considerably better than for Table 7, regression V, which has the highest correlation. Figure 3 shows the reason for this. During the early part of the crop year, when stocks are large, the curve relating stocks and spreads is very flat, indicating that variation in stocks has a small effect on the spread. In that part of the crop year we could use the average spread to obtain a fairly good prediction of the actual spread during any given year. In the later part of the crop year annual variations in stocks and spreads are much greater, and they are systematically related. That part of the curve relating stocks and spreads for the latter part of the crop year is steeply curved, and the correlation between stocks and spreads is quite high. In the later part of the crop year stocks predict the actual spread better than the average spread does.

The effect of changes in the fraction of stocks held by the government seems to vary during the season. In every case an increase in the fraction of total stocks held by the government decreases the spread. However, only for the October–July spread is the absolute value of the coefficient of this variable greater than its standard error. The effect of cotton consumption on the spread is also slight because in no case does its coefficient exceed its standard error.

The last regression in the series was obtained by pooling all the spreads and stocks into a single regression. This regression summarises the intraseasonal relation between stocks on the various dates and spreads. For example, to predict the average value of the July–May spread for some year, we insert the logarithm of stocks on May 31, the fraction of these stocks held by the government on this date, and the consumption during the year into regression VI of Table 7 and obtain the predicted July–May spread. However, the shape of the relation between stocks and spreads shown in Figure 3 indicates that a prediction made using the regression relevant for the particular spread is better (in the example, regression IV).

Table 8 gives the intraseasonal stock spread regressions for wheat. There, too, the highest correlation is obtained for spreads toward the end of the crop year.[17] Again, in spite of this, the standard error of a prediction is lower in the early part of the crop year than in the later part.

Statistically significant estimates of the effect of consumption on the wheat spread can be obtained because the annual variation in consumption over the sample period is greater for wheat than for cotton. Fluctuations in consumption have a considerable effect on the wheat spreads. For example, an increase in consumption of 1% decreases the July–May spread by 43 cents per bushel in real terms.

The coefficient of stocks is negative in regression III, Table 8. However, the standard error of the estimate is four times greater than the coefficient itself, and the estimated coefficient is consistent with the hypothesis that the "true" slope is positive.

The fraction of total stocks held by the government plays a curious role in these wheat regressions. For both the December–September spread and the May–December spread the coefficient of the fraction of total stocks held by the government is positive, implying that an increase in this fraction increases spreads. For the other two spreads the coefficient is negative. During the middle of the crop year, when a considerable part of the government stocks is held as collateral for price-support loans, it is possible that the convenience yield of such stocks is quite high. Unfortunately, adequate data separating government wheat stocks into their various categories are not available, so that this statement must remain a conjecture.

A grand regression was not computed for wheat because the data are not available for a long enough period for all the stocks and spreads. In any event, the separate cotton regressions are a better approximation to the firms' stockholding schedule than a single regression, and the same is probably true of wheat.[18]

SUMMARY

A model of stockholding behaviour in the absence of future markets began this article. The equilibrium level of stocks and prices are determined by the firms' stockholding schedule and the derived consumers' stock schedule. The firms' stockholding schedule shows the stocks held at the end of one period as determined by the expected price change from that period to the next. Hence empirical estimates of the firms' stockholding schedule require data on expected price changes as well as on stocks held. Although we cannot observe the expected price directly, we can estimate the firms' stockholding schedule for those commodities traded on futures markets because the expected price is related to the futures price.

Next the relation between the expected price and the futures price was analysed. A widely accepted theory advanced by Keynes and Hicks which relates the futures price and the expected spot price regards hedgers as buyers of insurance and speculators as sellers of insurance who must be induced to bear the risk of price changes. When statistical evidence was examined to see whether futures prices display an upward trend as they approach maturity predicted by this theory, it was found instead that futures prices display no trend. Although hedgers may be willing to pay speculators to bear the risks of price changes, they need not do so if speculators are eager to speculate. Firms that hedge can reduce their price risks at little or no cost to themselves. I accepted the hypothesis that the futures price equals the expected spot price.

The last part of the article estimates the firms' stockholding schedule for two commodities, cotton and wheat. The regressions relating stocks to futures prices are justified by the previous argument that stocks are related to the expected price changes, which in turn equal the difference in futures price of successive maturities. These regressions have the form predicted by the marginal convenience-yield theory. The seasonal pattern of stocks determines the spreads.

STATISTICAL APPENDIX

The dependent variable in all the regressions is the supply price of storage or the "spread". The spread is defined as the futures price for contracts maturing in the tth period less the futures price for contracts maturing in the $t - 1$th period. For cotton the basic data used to calculate the spreads are the monthly averages of the daily closing prices of futures traded on the New York Cotton Exchange, the most important cotton futures market in the world measured by the volume of trading. For wheat the basic price data used are the monthly averages of daily closing prices on the Chicago Board of Trade, the most important wheat futures market.

Since the price level varied considerably during the period studied, the spreads are deflated by the Bureau of Labor Statistics Wholesale Price

Index (1947–1949 = 100). This converts the spreads from current to constant dollars.

The intervals between futures contracts vary in length. To obtain comparable spreads using different futures contracts, the deflated spreads are divided by the number of months spanned by the interval between maturity dates of adjacent futures contracts. This adjustment of spreads is carried out for the intraseasonal regressions relating spreads of different futures contracts to appropriate stock figures.

The spread equals the marginal cost of storage minus the marginal convenience yield. Hence changes in the marginal cost of storage affect the spread. The coefficients of two variables affecting the spread via the cost of storage were not directly estimated in the regressions; these are the rate of interest and the average cost of storage. By subtracting the interest and storage costs from the spread, I can account for their effect on the spread without using up degrees of freedom.

The Department of Agriculture has prepared estimates of the average physical costs of storage for cotton beginning in 1932. These figures do not include an allowance for insurance charges except for the last two years of the sample period.[19] From each cotton spread I subtracted the average cost of storage per pound of cotton per month. Since storage costs are not available for wheat, the wheat spreads are not adjusted in this way.

Data on the actual interest cost of holding commodities are not available. In any event, published data on interest rates charged by banks for customers' loans are untrustworthy because they do not take account of variations in conditions under which loans are made. However, the cost of a loan must be highly correlated with the rate of interest in the New York money market. I use the rate on four- to six-month commercial paper at New York as published by the Federal Reserve Board to estimate the interest cost of holding stocks of commodities. The interest cost is the rate of interest per month times the deflated nearer future price used in computing the spread. Both cotton and wheat spreads are adjusted in this way.

Average spreads during the crop year are used in computing the cotton intraseasonal regressions. The average is computed from spreads for each month preceding the expiration of some futures contract. For example, the July–May spread is the average of the July–May spreads for December, February, April, and June.

A slightly different procedure is followed for wheat. The spreads used in the wheat intraseasonal regressions are computed for a single month (the month before the expiration of the nearer futures contract) rather than for the entire crop year. The use of weighted annual averages may improve the results, but they were not used because of the additional expense.

The data on cotton stocks and consumption are from *Statistics on Cotton and Related Data* (United States Department of Agriculture Statistical Bull. 99, June 1951, and supplements). Stocks used are total stocks less mill stocks. The Commodity Credit Corporation supplied me with figures on government holdings of cotton, including collateral for price-support loans and government-owned stocks.

The data on wheat stocks and related variables are from *Grain and Feed Statistics through 1954* (United States Department of Agriculture Statistical Bull. 159, March 1955). Problems arise in measuring wheat stocks. The Department of Agriculture estimates wheat stocks on four dates during the crop year: July 1, October 1, January 1, and April 1. The four important futures contracts are July, September, December and May. The September–July spread is determined by stocks on July 31, the December–September spread by stocks on September 30, the May–December spread by stocks on December 31, and the July–May spread by stocks on May 31. Hence stock figures are unavailable on two of the four relevant dates. Moreover, reported stocks on July 1 (since 1937) do not include new-crop wheat harvested up to July 1.

Stocks on April 1 are generally larger than stocks on May 1, and the July–May spread during April is explained by stocks on April 1. Using stocks on April 1 rather than June 1 does not result in a biased estimate of the stocks' coefficient if the error in underestimating stocks on April 1 is not proportional to the quantity of stocks. If the error is proportional to the quantity of stocks, then the estimated slope is less than the true slope. However, the estimated constant term is definitely in error. To obtain a better estimate of the constant term, I estimate average stocks on June 1 by subtracting two-thirds of the difference between stocks on July 1 and stocks on April 1 from stocks on April 1. The graph in Figure 4 is drawn through the average value of these stocks and the July–May spread during April.

Since almost all winter wheat is harvested during June and July, I estimate wheat stocks on July 1 by adding five-sixths of the winter wheat crop to the published stock figure on July 1, thus allowing for the consumption of the winter wheat crop during the months of June and July, admittedly in a crude way. However, in the regression explaining the September–July spread the published stocks on July 1, which exclude the new-crop wheat, are used. A biased estimate of the stocks' coefficient results only if the error between actual stocks on July 1 and the published figure is proportional to actual stocks. However, the constant in the regression is a biased estimator of the true constant due to the error in estimating stocks. In Figure 4 a less biased constant is obtained by using estimated stocks on July 31 and the September–July spread.

The war years and 1946 are omitted for wheat because almost no trading occurred, and trading was actually suspended for part of 1946. War years are included for cotton, although somewhat better results could be obtained were they omitted. Some months in the early part of 1951 were excluded because cotton futures trading was suspended.

The Federal Farm Board held stocks of wheat and cotton during 1929 and 1930. Although the wheat was soon sold, a considerable amount was still held in 1931 and probably did not have the same convenience yield as the privately held wheat. Unfortunately, data on the quantity of wheat held by the Federal Farm Board are unavailable. The government stock figures for cotton in 1933 (the first year of the sample period) include estimated Federal Farm Board holdings. The Commodity Credit Corporation supplied me with data on the quarterly total of wheat owned by the government and held as collateral for price-support loans.

1 Purely theoretical accounts of futures markets are given by Gerda Blau, "Some Aspects of the Theory of Futures Trading", *Review of Economic Studies*, 20, 1944–45; J. C. R. Dow, "A Theoretical Account of Futures Markets", *Review of Economic Studies*, 7, 1939–40; and N. Kaldor, "Speculation and Economic Stability", *Review of Economic Studies*, 7, 1939–40. Some purely factual studies are: L. D. Howell, *Price Risks for Cotton and Cotton Products and Means of Reducing Them* (United States Department of Agriculture Technical Bulletin 1119, July 1955); G. Wright Hoffman, *Grain Prices and the Futures Markets: A 15 Year Survey, 1923–38* (United States Department of Agriculture Technical Bulletin 747, January 1941); and Holbrook Working, "Price Relations between May and New-Crop Wheat Futures at Chicago since 1885", *Wheat Studies of Food Research Institute*, 10, February 1934, and "Price Relations between July and September Wheat Futures at Chicago since 1885", *ibid.*, 9, March 1933.

2 The length of a period is the time spanned by the maturity dates of two adjacent futures contracts. Thus each crop year is divided into a certain number of discrete periods.

3 Production is treated as exogenous in this article because this involves no loss in generality and simplifies the argument. Even so, in the model to be described the carry-out behaves as a substitute for production. These matters are considered in greater detail by J. B. Williams, 1936, "Speculation and the Carry-over", *Quarterly Journal of Economics*, L, 436–55.

4 Let y_t be stocks carried out of period t, q_t consumption during period t, and $p(t)$ the price during period t. The demand for consumption during the tth period is

$$p(t) = h_t(q_t)$$

But, by the definition of consumption,

$$q_t = y_{t-1} - y_t$$

The derived consumers' stock schedule is

$$p(t) - p(t-1) = h_t(q_t) - h_{t-1}(q_{t-1})$$

Hence

$$p(t) - p(t-1) = h_t(y_{t-1} - y_t) - h_{t-1}(y_{t-2} - y_{t-1})$$

5 A particular convenience yield equation for an inventory firm is derived in K. J. Arrow, T. Harris, and J. Marschak, 1951, "Optimal Inventory Policy", *Econometrica*, XIX, 250–72.

6 See the mathematical appendix to my unpublished doctoral dissertation, "The Supply of Stocks: Cotton and Wheat" (University of Chicago, August, 1956).
7 To illustrate the result in the text that the equilibrium futures price is a kind of average expected price, suppose that the excess demand curve of each speculator is approximated by the linear equation,

$$x(i) = a_i(p - p'[i]), \; a_i < 0$$

The equilibrium condition of the text implies that

$$\sum_i a_i(p - p'[i]) = 0 \quad \text{or} \quad \frac{\Sigma a_i p'(i)}{\Sigma a_i} = p$$

Therefore the equilibrium futures price is a weighted average of the spot prices expected by the speculators. If all speculators expect the same spot price, then $p'(i) = p$ for all i, and, accordingly, $x(i) = 0$ for all i. Disagreement is the essence of speculation.
8 The slope of the aggregate speculators' excess demand curve is the reciprocal of Σa_i. As the number of speculators increases, Σa_i becomes large in absolute value (it approaches minus infinity), and the slope approaches zero.
9 For more detailed accounts of hedging see Holbrook Working, 1953, "Futures Trading and Hedging", *American Economic Review* 43, pp. 314–43 and L. G. Telser, 1955–56, "Safety First and Hedging", *Review of Economic Studies* 23, pp. 1–16.
10 J. M. Keynes, *A Treatise on Money*, Vol. II: *The Applied Theory of Money* (London: Macmillan & Co.), pp. 142–47, and J. R. Hicks, *Value and Capital* Second edition 1930; (Oxford: Clarendon Press), chaps. IX and X, esp. pp. 136–39.
11 *Op. cit.*, p. 144.
12 Gains and losses of traders in cotton, corn, and wheat are estimated in a forthcoming Cowles Foundation Monograph by H. S. Houthakker assisted by L. G. Telser, *Commodity Futures: A Study in the Economics of Uncertainty* (Cowles Foundation Monograph, 17, New York: John Wiley & Sons). In general, the evidence presented there is not conclusive because transaction costs are not deducted from the speculators' income, only nine years are studied (1937–39 and 1946–52), and the method of estimating gains and losses neglects changes in commitments and prices within each month. See also B. Stewart, *An Analysis of Speculative Trading in Grain Futures* (United States Department of Agriculture Technical Bulletin 1001, October 1949).
13 "Theory of the Inverse Carrying Charge in Futures Markets", *op. cit.*, p. 9.
14 This non-parametric statistical test is discussed in W. A. Wallis and H. Roberts, *Statistics: A New Approach* (Glencoe: Free Press, 1956), pp. 575–73.
15 Conceivably, the price increases may be larger than the price decreases even though there are as many price increases as decreases. The price increases would have to be sufficiently larger than the price decreases to cover the transactions costs of the speculators. The size of the price increases seems to be the same as the size of the price decreases for the actual price changes of the December cotton future and the May wheat future. I do not believe that I distort the data by using the signs of the first differences rather than the actual magnitudes. Using actual magnitudes creates problems in obtaining an appropriate price deflator and constructing an appropriate statistical test.
16 During the late 1930s the maximum amount of cotton the government could sell was set by law. A situation could have arisen in which the market price rose considerably above the support price and in which the government could not have sold more than the legal maximum.
17 The apparent exception is regression I, Table 8. However, the July future cannot always be considered a new crop future owing to variations in the date of the harvest.

18 A variable relevant for wheat but not for cotton is the stocks of other commodities that may be stored in the same structure used to store wheat. When stocks of such commodities (shelled corn, oats, barley, and rye) increase, the marginal cost of storing wheat may be expected to increase. In regression I, Table 8, the coefficient of this variable is negative instead of positive, but it is much smaller in absolute value than its standard error. It is possible that error in the estimates of stocks accounts for the poor result. The appropriate stock figures are off-farm stocks (particularly for corn). Since such stock figures are available only since 1943, I use "visible supply" instead. Visible supply includes stocks held in terminal elevators, which form a small part of total off-farm stocks. Since visible supply need not vary proportionately to total off-farm stocks, there is an unknown error of measurement in this variable.

19 *Statistics on Cotton and Related Data* (United States Department of Agriculture Statistical Bulletin 99, June 1951 and supplements).

7
*Speculation on Hedging Markets**
Holbrook Working†

Though statistical evidence, accumulated first by the Grain Futures Administration, predecessor of the present Commodity Exchange Authority, long ago afforded proof to the contrary, it is still rather generally believed that futures markets are primarily speculative markets. They appear so on superficial observation, as the earth appears, from such observation, to be flat. A conspicuous recent result of reliance on superficial appearance was that an administrator of the CEA and a majority of members of the United States Congress were persuaded, mistakenly, that the onion futures market had attracted an excessive amount of speculation, supposedly requiring the prohibition of futures transactions in that commodity. The result was enactment of Public Law 85–839 prohibiting futures trading in onions.[1]

In the present chapter, I seek chiefly to put available official statistics of futures markets into such form as to make clear what they show regarding the relations of speculation and hedging on such markets. First, however, it is necessary to find a definition of *speculation* that can be used consistently and without confusion.

The commercial meaning of "speculation" was undoubtedly derived from earlier use of the verb *speculate* in the sense of *observe* (the meaning of its Latin root, *speculari*), hence *to try to see*, or *try to understand*. In that sense of the word, we speculate on the nature of the universe, on the reasons for a person's actions, or on the probable consequences of a given situation. The verb implies uncertainty, coupled with some reasonable basis for an opinion regarding the subject concerning which we speculate, or meditate. Presumably the present commercial use of the term originated from frequent references to speculation (in the sense of

---

*This paper was first published in *Food Research Institute Studies* 1(2), pp. 185–220 (1960), and is reprinted with the permission of the Food Research Institute.
†Holbrook Working, 1895–1985.

meditation) about future commercial events. In time "speculate" came to mean the *actions taken* on the basis of such meditation rather than the meditation itself.

Then someone, impressed by the hazards of commercial speculation, could speak of a "speculative venture" and have people understand that by *speculative* he meant *risky* – a meaning far removed from the original meaning of *speculate*. "Steal" has taken on a similarly new meaning in the baseball expression, "steal a base".

Speculation, in the commercial sense, appears always to have been criticised by many people; the word seems always to have had a derogatory flavour. Such disapproval, arising in a society that has tended to honour the taking of risks in good causes, must be supposed to have rested on a prevalent belief that commercial speculation tends to be predatory rather than productive.

Economists and businessmen who have seen virtues in commercial speculation have often sought to define speculation as *economically necessary* risk-taking. Thus they have argued that a farmer speculates when he postpones sale of part of his crop for several months after harvest, in the hope of getting a higher price later. By the same argument, a manufacturer may be said to speculate when he contracts the purchase of supplies, well in advance of need for them, in the belief that he can buy more cheaply then than later. Such a definition of speculation amounts to defence by definition. As such, it has been ineffective, doing little or nothing to improve most people's opinions of speculation in general. The main result has been to introduce confusion concerning the meaning of the word. Economic discussion of speculation has thus reached conclusions that tend to be misleading in practical application, because it has considered one thing, and the conclusions are applied to something rather different that goes under the same name.

Scarcely anybody uses the word "speculation" consistently in the artificial sense of "economically necessary risk-taking", while nearly everybody uses it sometimes or always in another, commonly understood, sense. In ordinary usage, speculation in commodities means seeking profit from transactions undertaken especially for that purpose, and not in the normal course of conducting a business of producing, merchandising, or processing a commodity. This definition might be considered to include arbitrage, but in ordinary usage arbitrage is not counted as speculation. Many people are unaware of the existence of arbitrage, and so do not mean to include it as speculation, and people who recognise its existence ordinarily distinguish between speculation and arbitrage.

The distinction ordinarily drawn between speculation and investment in securities follows the same principle that is commonly followed in distinguishing between speculation and other dealings in commodities. Though investors often acquire and hold securities primarily in expectation of appreciation in "value", rather than for current income, this is not regarded as speculation so long as the operations are only those normal to the business of keeping funds invested prudently and profitably. It is more difficult in practice to draw a line between speculation and investment in securities than between speculation and other dealings in commodities, but the principle on which people ordinarily mean to draw the line is the same in both cases. It is that of distinguishing between obviously desirable or appropriate business activities, and activities that, if useful and desirable, do not always appear so on the surface.

By way of formal definition we may say that *speculation in commodities is the holding of a net long or net short position,*[2] *for gain, and not as a normal incident to operating a producing, merchandising, or processing business.*

For our present purposes this definition of speculation, excluding from it all profit-seeking transactions normal to the conduct of production, merchandising, or processing, and excluding arbitrage, has three advantages. It conforms with the commonly understood meaning of *speculation*. It is a logical accompaniment of a good general definition of hedging (use of futures contracts as a *temporary substitute* for contracts intended to transfer ownership of a quantity of the commodity, in the normal course of business).[3] And it accurately describes the principle underlying the classification used in those statistics of commodity futures, published by the CEA, that we shall be using.[4]

There has sometimes been discussion of the question whether speculation is significantly distinguishable from gambling In the United States this question has often come before the courts, because men who have lost money at speculation in commodity futures, thereby incurring debts to brokers, have sought to avoid paying by claiming them to be gambling debts.

Gambling does indeed resemble speculation, and likewise resembles many business undertakings, in that all involve taking risks in the hope of financial gain. Moreover, a man *can* undertake speculation, or a business venture, in a purely gambling spirit. Similarly, a man firing a rifle goes through the same motions whether he is aiming at a target on a rifle range, at a deer, or at a man across the street. And there apparently are some people who can shoot at a man with as little feeling as at a practice target. It is nevertheless profitable for society to distinguish among different uses of a rifle, and among different uses of risk-taking for monetary gain. Nor do we have any real difficulty in drawing these distinctions when we reject sophistry and apply common sense. We call a man an entrepreneur when he takes risks in a clearly useful type of business venture; a gambler when he takes risks of a nature that clearly serve no substantially useful economic purpose; and a speculator when he takes risks of another sort, that some people do not recognise as economically useful, though others regard them as highly useful.

The courts in the United States have tended to draw the line between speculation in futures and wagering according to the criterion of "intent to deliver". The adoption of this criterion by the courts seems to reflect an imperfect understanding of futures markets. The economic usefulness of futures contracts does not arise from their usability for merchandising, but mainly from their use for hedging. The courts can refuse to enforce gambling contracts in many states on the ground that such contracts are illegal, and in any state on the ground that such contracts are frivolous matters with which the courts refuse to concern themselves, as they would refuse to enforce the decisions of an umpire in a ball game. On the other hand, the courts seek to enforce futures contracts, and other contracts related to them, because futures markets are accepted as economically useful institutions. An economist would rather see the usefulness of futures markets affirmed in court on the basis of their true principal merits, rather than on the basis of a technical characteristic of the contracts that is necessary, but that, when emphasised, misrepresents the main function of such markets.

The choice of grounds on which courts in the United States have traditionally sustained futures contracts has had the unfortunate effect of hampering efforts of exchanges in the United States to control corners and squeezes. The exchanges have often felt compelled to countenance recognisably unreasonable demands for delivery, made for manipulative purposes, lest in the process of controlling manipulation they lose the court-recognised ground for distinguishing between economically useful contracts and economically unuseful wagers. In England, where the courts have relied on other criteria than intent to deliver, corners and squeezes have never presented a serious problem. No corner, and no squeeze of consequence, has ever been carried through on the Liverpool wheat futures market (Working and Hoos, 1938, pp. 137–38; Hoos and Working, 1940, p. 104).

RELATIONSHIP BETWEEN SPECULATION AND HEDGING

The first published statement of the conclusion that speculation on a futures market responds to hedging needs, appeared in an article by Irwin (1935).[5] Subsequently, further evidence was published by Hoffman (Hoffman and Duvel, 1941, pp. 33–9) and by Schonberg (1956, pp. 279–88). All of it showed that as commercially owned stocks of the commodity increased or decreased, tending to cause increase or decrease in the volume of short hedging contracts held against such stocks, speculative holdings of futures contracts tended to increase or decrease correspondingly.

Despite the published statements of this conclusion, and the steady appearance year by year of new statistics that always tended to confirm it and never contradicted it, the idea that speculation in futures depends on hedging gained little ground among either economists or members of the exchanges. In 1953 there occurred a striking demonstration of the continued adherence of exchange members to the old concept, and of the truth of Irwin's conclusion, published 18 years earlier, that speculation depends on hedging. In April and early May of 1953, flour mills with long hedges in Kansas City wheat futures (against unfilled flour orders) took substantial losses because soft wheat, unexpectedly drawn to Kansas City for delivery on futures contracts, depressed the price of the May future relative to prices of the hard wheats needed by mills to fill their flour orders. The millers promptly petitioned for a revision of the Kansas City futures contract to make it strictly a hard-wheat contract. It had always previously been so in effect, hedgers and speculators thought of it as such, and many members of the exchange had been surprised to learn that delivery of soft wheat was permitted by the contract.

The members of the exchange, however, seem to have been almost unanimous in the belief that the amount of futures business done on the exchange depended on attracting speculators, and the majority held also the common belief that speculators want a "broad" contract, allowing delivery of more than one class and grade of the commodity.[6] So the exchange refused the plea of the millers for a revision of the contract terms. But in July and August millers took even larger losses, per bushel, on their long hedges, for the same reason as earlier, and this time the losses occurred on a great volume of such hedges, held against recently placed flour orders for milling from the new crop. These new losses caused most millers who had been hedging in Kansas City wheat futures to transfer their hedging business either to Minneapolis, where the hedge was in a hard-wheat contract, or to Chicago, where the hedges, though no more reliable than at Kansas City, could be placed and removed more economically.[7] And speculators apparently deserted the market in about the same large proportion as did hedgers (Working, 1954).

If the Kansas City exchange had persisted in rejecting the pleas of its principal hedgers, the wheat futures market there would very soon have joined the considerable list of such markets that have died because hedgers stopped using them. Mess pork and lard were among the commodities in which futures markets were established early in Chicago, and in the 1880s short rib sides were added to the list. Of these only lard remains, because development of mechanical refrigeration operated to so curtail the accumulation of stocks of cured pork products that hedging of them dwindled. New York City and St. Louis had important wheat futures markets at the beginning of the present century, but their business declined as changes in the wheat trade, and improved communications, reduced the special advantages of hedging in those markets rather than in the more economical Chicago market. Kansas City, Minneapolis and Duluth held their hedging in competition with Chicago because their contracts were distinctive, applying to hard winter wheat,[8] hard spring wheat and durum wheat, respectively; but in the 1930s drought and rust damage so curtailed production of durum wheat that the Duluth market was discontinued for lack of enough hedging business to support it.[9]

Further evidence of the sort cited above might be added in great quantity,[10] but without meeting the major obstacles to recognition of the significance of the evidence. These appear to me to be: (1) the existence of a great amount of evidence that *seems* to indicate that most speculation in futures occurs without any relation to hedging; (2) certain apparently reasonable grounds for doubt whether speculators have reliable means for appraising the hedging needs of a futures market, such as would be required for any close adjustment of the amount of speculation to the amount of hedging; and (3) certain shortcomings of the available statistics that have tended to render them unconvincing, except to people with a good deal of collateral information to aid in interpreting them, and considerable skill in reading the meaning of crude statistical evidence.

The prevalent opinion that much speculation in futures has no significant connection with hedging is a mistaken one that has arisen from a long-established habit of using an available crude indicator of the amount of speculation as though it were a direct and accurate measure of speculation. The only aspect of speculation that is readily observable is the *transaction* by which a speculator initiates or closes out a speculative venture. That is also the only aspect of speculation in futures concerning which any statistics were regularly published prior to July 1923 (US Department of Agriculture, 1937, p. 24).[11]

But these readily observable transactions are only incidental to speculation, as starting and stopping are incidental to driving an automobile across town. In speculation, to pursue the analogy, a new driver takes the wheel at each stop, and each new driver is usually another speculator. But one may observe these changes of "drivers" without learning anything about the question whether most of the trips across town are made on behalf of hedgers. To learn how much speculation is connected with hedging it is necessary to find out what proportion of total speculative open contracts is needed to carry the hedging open contracts. The principal observed facts that have *seemed* to show that speculation in futures does not depend on hedging have been misleading; we need to study the statistics of open contracts in order to learn the extent of connection between hedging and speculation.

Let us defer until later the question whether there exists any reasonably reliable mechanism by which speculators might be led to undertake the holding of futures contracts mainly in response to the offering of such contracts by hedgers. What then, do the statistics of open contracts show regarding the degree of correspondence between amounts of hedging and of speculation in futures?

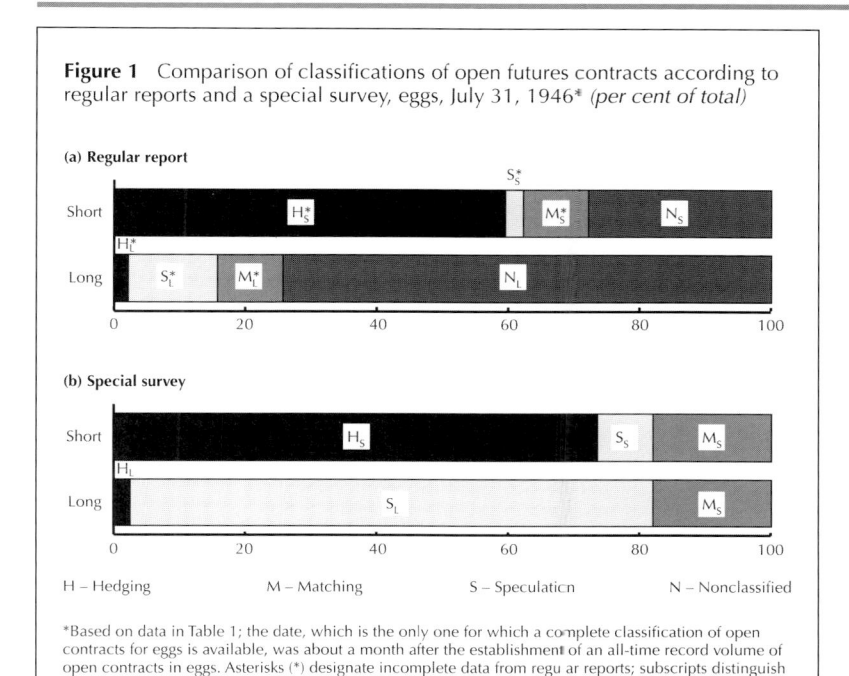

Figure 1 Comparison of classifications of open futures contracts according to regular reports and a special survey, eggs, July 31, 1946* *(per cent of total)*

(a) Regular report

(b) Special survey

H – Hedging   M – Matching   S – Speculation   N – Nonclassified

*Based on data in Table 1; the date, which is the only one for which a complete classification of open contracts for eggs is available, was about a month after the establishment of an all-time record volume of open contracts in eggs. Asterisks (*) designate incomplete data from regular reports; subscripts distinguish short and long contracts.

A principal obstacle to drawing conclusions from the statistics of open contracts is illustrated by Figure 1, which compares the evidence from two sorts of available statistics. Data for two additional commodities are shown in Table 1. The lower bars in the figure compare amounts of speculative contracts (S) and of hedging contracts (H), long and short (indicated by subscripts), for a date for which a complete classification of the open contracts in eggs is available. (M represents matching contracts, presumably arising largely from inter-option spreading, but partly from temporary failure to cancel out offsetting long and short contracts on the books). Such a complete classification of contracts has been published for only a few commodities and a few isolated dates – often dates on which some exceptional conditions existed in the market, rendering the data unrepresentative. The data for eggs, given for the only date for which a complete classification of open contracts has been published, is one on which total open contracts in eggs were near their all-time record level, reached about a month earlier.

The two upper bars in the figure show, for the same date, the sort of statistics of open contracts that are published regularly, as of the middle and end of each month, for all regulated commodities. In these regularly published statistics there is a large "nonclassified" category (N), which

Table 1 Comparison of regularly reported classification of open contracts with complete classification, eggs, cotton and wool tops, available dates* *(Carlots; thousand bales; thousand pounds)*

	Eggs July 31, 1946		Cotton Sept. 28, 1956		Wool tops Dec. 31, 1957	
Class of contracts	Long	Short	Long	Short	Long	Short
A. Regular reports						
Hedging	225	6,203	873	670	1,690	4,290
Speculative	1,416	284	102	3	1,785	0
Matching[a]	1,035	1,035	220	220	205	205
Nonclassified	7,746	2,900	668	970	3,665	2,850
Total	**10,422**	**10,422**	**1,863**	**1,863**	**7,345**	**7,345**
B. Special surveys						
Hedging	261	7,710	1,001	1,140	2,290	5,300
Speculative	8,333[b]	884	268	130	3,385	375
Matching[c]	1,887	1,887	606	606	1,660	1,660
Total	**10,481**	**10,481**	**1,875**[d]	**1,876**	**7,335**	**7,335**

*Data from US Department of Agriculture, *Commodity Futures Statistics* (1947, 1957 and 1958) and from reports on CEA special surveys for commodities and dates indicated.
[a]Contracts explicitly classed as "spreading".
[b]Includes 234 carlots reported without classification by foreign futures commission merchants.
[c]Amounts long (short) that were offset by equal or greater amounts short (long) in individual accounts, not necessarily in different futures.
[d]Discrepancy between long and short totals reflects minor error in compilation of survey data.

absorbed 28% of the total short egg contracts on this date, and 74% of the long contracts. Looking at the major elements in the short and long contracts, we find only 81% of the short hedging contracts and only 17% of the long speculative contracts, explicitly classified as such in the regular reports (percentages calculated from data in Table 1).

The data in the figure are fairly representative of the degrees of which short hedging and long speculation get explicitly classified in the regularly published statistics for most commodities. But for eggs they are highly unrepresentative. We shall see subsequently that, on the average during the last five years, only about 27% (instead of 81%) of total short hedging in eggs has been explicitly classified as such in the regular reports. With egg stocks exceptionally large on July 31, 1946, contract holdings by individual hedgers tended to be large and so an exceptionally large proportion of the hedgers held contracts for 2 carlots or more in a single future – the amount that brought them under the administrative requirement for reporting the classification of their contract holdings.

Despite the large proportions of open contracts that remain unclassified in the regularly published official statistics, it is possible to break down the nonclassified category with a reasonable degree of reliability. The

Table 2 Comparison of results of two estimation methods for handling nonclassified futures contracts* *(per cent of all contracts)*

Class of contracts	Eggs July 31, 1946		Cotton Sept. 28, 1956		Wool tops Dec. 31, 1957	
	Long	Short	Long	Short	Long	Short
A. Conventional estimates from regular reports						
Hedging	2	60	47	36	23	58
Speculative	88	30	41	52	74	39
Matching	10	10	12	12	3	3
B. New estimates from regular reports						
Hedging	3	73	61	50	30	75
Speculative	73	3	27	38	57	12
Matching	24	24	12	12	13	13
C. Special surveys						
Hedging	2	74	54	61	31	72
Speculative	80	8	14	7	46	5
Matching	18	18	32	32	23	23

*Derived from Table 1; new estimates by procedure described in Technical Appendix.

common practice hitherto has been to regard the nonclassified contracts as mainly small-scale speculative contracts and therefore to add them to the reported (large-scale) speculative contracts. This procedure, applied to the data in Table 1, results in the percentage distribution of contracts between hedging, speculation, and spreading and other matching contracts, that is shown in section (A) of Table 2. Section (B) of the table shows the results of applying a simple estimation technique, described in an appendix below, to allocate nonclassified contracts more appropriately. Section (C) shows, for comparison, the correct distribution of contracts among the three categories, as revealed by special surveys made as of the dates for which data are shown.

The estimation procedure, it will be seen, has worked excellently in its apportionment of contracts to the hedging category. For eggs and wool tops the discrepancies exceed one percentage point only in the case of short hedging of wool tops, where the estimation procedure gave 75% short hedging contracts as compared with a true value of 72%. For cotton, the estimation procedure could not correct the false indication of the regular reports that there was nearly one-third more long hedging than short hedging (actually the amount of short hedging exceeded that of long hedging by over one-eighth), but the estimation procedure nevertheless gave nearly the correct total for the sum of the two sorts of hedging.

In its allocation of the remaining contracts between the categories of speculation and of spreading or other matching contracts, the estimation procedure resulted in understatement of the amounts of speculative contracts for eggs, and overstatement of the amount of speculation in cotton and wool tops. But in all cases it produced estimates of the amounts of speculation that were closer to the truth, and usually much closer, than those obtained by merely treating all nonclassified contracts as speculative. The false indication of the regular reports that there was more long hedging than short hedging of cotton necessarily carried on into the estimates of speculative contracts, in the form of a false indication of more short speculation than long speculation, but for the other two commodities, as will be seen, the estimates produced an approximately correct indication of the relation between the amounts of long and of short speculation.

The procedure tested above on data for three commodities, for individual dates on which the reliability of the method can be checked against completely classified statistics of open contracts, has been used to derive the data in Table 3 based on five-year averages (three years for onions) for 11 commodities. In order to provide comparability between commodities, average amounts of open contracts are expressed in dollar values, obtained by multiplying averages in physical units for each commodity by the average price of the commodity for the period of the average.

Simple comparison of the figures in two columns of Table 3 shows that there has been at least a fairly close correspondence between the amounts of long speculation in these eleven commodities (col. 5) and the amounts of short hedging (col. 1). Either the amounts of speculation have been largely determined by the amounts of hedging in the several commodities, or the amounts of hedging have depended largely on the amounts of speculation. A little knowledge of the commodities makes it clear that it has been principally the amount of short hedging that has determined the amount of long speculation, rather than the other way round. For example, consider why bran is at the bottom of the list. The bran market was indeed regarded as a relatively unsatisfactory one for hedging, because it had too little speculation (the market closed November 5, 1957, for the lack of enough business to warrant its continued maintenance). Given more speculation, it would doubtless have attracted more hedging that it did; but no amount of speculation in bran could have raised the amount of hedging in that market above, say, the amount of hedging shown for onions. There simply was not enough bran that might have been hedged. Bran is a minor byproduct of wheat milling, and is produced at a fairly regular rate, so that stocks are never more than a small fraction of the annual production.[13]

Wheat, next to the top of the list in terms of either amount of hedging or amount of speculation, had more than twice as much of both as did corn. That may seem inconsistent with the fact that the average value of

Table 3 Estimated average US dollar values of short hedging and of long speculative open contracts, ratios, and speculative index, eleven commodities, mostly 1954/55–1958/59* *(million US dollars; ratios)*

Commodity	Short hedging contracts			Hedging ratio[c] r_h' (4)	Estimated total long speculative contracts[a] S_L' (5)	Speculative ratio[a] r_s' (6)	Speculative index[d] T' (7)
	Estimated total[a] H_s' (1)	"Reported" (large scale) H_s^* (2)	Reporting ratio[b] H_s'/H_s^* (3)				
Cotton	179.8	109.0	0.61	0.74	130.7	0.73	1.27
Wheat	125.1	88.1	0.70	0.58	97.1	0.78	1.22
Soybeans	102.6	57.9	0.56	0.52	92.7	0.90	1.28
Corn	53.3	40.2	0.75	0.36	45.3	0.85	1.16
Soybean oil	26.0	23.6	0.90	0.46	19.5	0.75	1.14
Eggs	13.8	3.8	0.27	0.09	16.3	1.18	1.25
Wool tops	12.1	9.8	0.81	0.25	10.1	0.84	1.07
Soybean meal	14.0	12.4	0.89	0.61	9.0	0.65	1.16
Potatoes	3.3	2.2	0.67	0.30	3.5	1.05	1.27
Onions[e]	2.0	1.6	0.79	0.03	2.2	1.10	1.12
Bran[f]	0.6	0.6	0.88	0.66	0.3	0.53	1.12

*Five-year averages, 1954/55–1958/59, except as otherwise noted; computed from data *Commodity Futures Statistics*, USDA Statistics Bulletin 256, pp. 57–60, for quantities, and pp. 11–13, (Supplemented, for bran, by similar data from earlier issues) for average prices; cols. (2), (4), calculated directly from data in source. All calculations carried to more places than shown here.

[a]See Technical Appendix below: "speculative ratio" is amount of long speculation divided by amount of short hedging, col. (5) ÷ col. (1).

[b]Col. (2) ÷ col. (1); the reliability of the estimates in cols. (1) and (5) and of the speculative index, T, tends to be highest when this ratio is large.

[c]Ratio of amount of long hedging to amount of short hedging, based on data for "reported" (large-scale) contracts.

[d]Defined as unity plus the ratio, (short speculative contracts) ÷ (short hedging + long hedging contracts); calculated here as $T = (2r_h + r_s)/(1 + r_h)$.

[e]Averages for 1955/56–1957/58; data not available earlier, and market conditions abnormal 1958/59 owing to uncertain legal status.

[f]Averages for 1949/50–1953/54; market shrank rapidly thereafter and was discontinued in 1957.

the annual production of corn is about double the value of the wheat produced annually in the United States. Corn, however, is mainly kept on the farms where it is grown, and there used for feeding, whereas wheat moves quickly into commercial hands, where it tends to be hedged. So there has been in fact only about half as much corn as wheat, in terms of money value, to be hedged. Thus one might continue through the list and find all of the substantial differences in amounts of short hedging explained principally by differences in volume of stocks that holders might reasonably wish to hedge.

One may observe, nevertheless, some irregularity in the relationship between amounts of hedging and of speculation, as shown in Table 3.

An extreme example appears just below the middle of the table: soybean meal, with slightly more short hedging than eggs, had only 55% as much long speculation as eggs. Such an irregularity in relationship, though inconsequential in comparisons between commodities that show numerous examples of one commodity with ten times as much speculation as another, and several commodities with between one hundred and several hundred times as much speculation as the commodity of the bottom of the list, are nevertheless worthy of notice. To bring these small irregularities into prominence for further study, the speculation-hedging rather called for brevity the "speculative ratio", is shown for each commodity in col. 6 of the table.

This speculative ratio is evidently rather closely related to the hedging ratio (col. 4), as should logically be expected. When there is much *long* hedging in a futures market, it serves in part to offset the short hedging, permitting the short hedging to be effectively carried by a smaller amount of long speculations than would be needed to carry an equal amount of short hedging with little long hedging. Figure 2 exhibits graphically the relationship between the speculation ratio and the hedging ratio.

Many economists, myself included, have tended at times to reason that the amount of long speculation needed in a futures market should depend on the *net* amount of hedging – short hedging minus long hedging. Students of the statistics, however, observed early that the amount of long speculation actually present in futures markets depends primarily on the total amount of short hedging rather than on the net short hedging position. It was evident from this that long hedging serves only in part to reduce the need for long speculation. The main reasons that this is so are easily seen when the nature of the long hedging is understood. Though long hedging commonly arises, directly or indirectly, as an expression of price judgement,[14] it is not always an expression of such judgements, and often it expresses somewhat inexpert judgement. Consequently, most long hedging must commonly be absorbed initially by short speculation, instead of serving immediately to absorb simultaneously placed short hedging orders. Long hedges, moreover, tend individually to have a short life. The flour mill that has hedged a large flour order starts very soon, if not immediately, to buy the wheat for use in filling that order, and "lifts" its hedge piecemeal as it does so. The manufacturer of rolled oats who has bought oat futures as an anticipatory hedge, or the fruit-canner who has bought sugar futures similarly, does not wait long before making merchandising contracts for oats or sugar, and lifting the hedge as he does so.

The extent to which long hedging serves to balance short hedging is calculable (as will be explained presently) from the speculative index shown in the final column of Table 3. By calculating thus the amount of short hedging that, on average, was balanced by long hedging for each

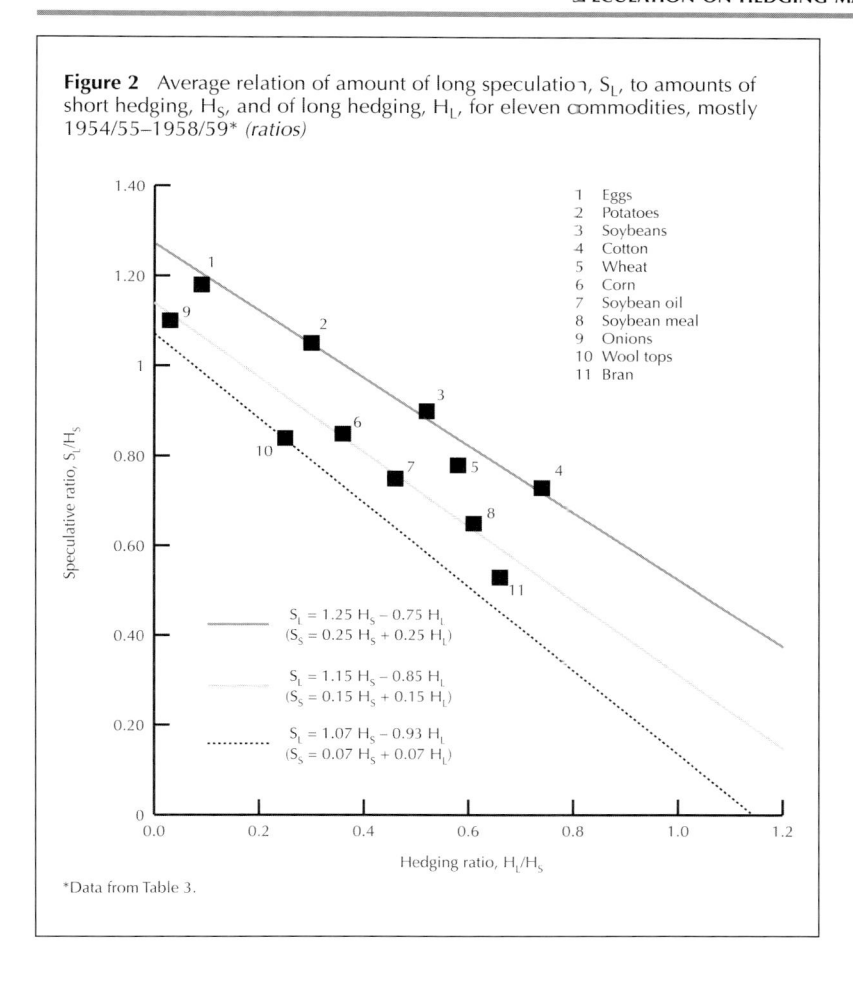

Figure 2 Average relation of amount of long speculation, S_L, to amounts of short hedging, H_S, and of long hedging, H_L, for eleven commodities, mostly 1954/55–1958/59* *(ratios)*

[Graph: Speculative ratio, S_L/H_S (y-axis) vs Hedging ratio, H_L/H_S (x-axis)]

Commodities:
1. Eggs
2. Potatoes
3. Soybeans
4. Cotton
5. Wheat
6. Corn
7. Soybean oil
8. Soybean meal
9. Onions
10. Wool tops
11. Bran

$S_L = 1.25 H_S - 0.75 H_L$
$(S_S = 0.25 H_S + 0.25 H_L)$

$S_L = 1.15 H_S - 0.85 H_L$
$(S_S = 0.15 H_S + 0.15 H_L)$

$S_L = 1.07 H_S - 0.93 H_L$
$(S_S = 0.07 H_S + 0.07 H_L)$

*Data from Table 3.

commodity,[15] and subtracting that from the total amount of short hedging, we arrive at figures for amounts of short hedging that had to be carried by long speculation (Table 4). These are shown graphically in Figure 3, in comparison with the actual average amounts of long speculation in each commodity.

We thus find that each commodity appears to have had somewhat more speculation than was "needed" to carry the unbalanced short hedging in that commodity. Indeed, the speculative index itself is a direct measure of the amount of that "excess". But at least a large part of what may be called technically an "excess" of speculation is *economically necessary*. This can be seen most readily if we imagine a futures market with *no* long hedging (a condition rather closely approached in the egg market). In such a situation the amount of long speculation *could* exactly equal the amount of short

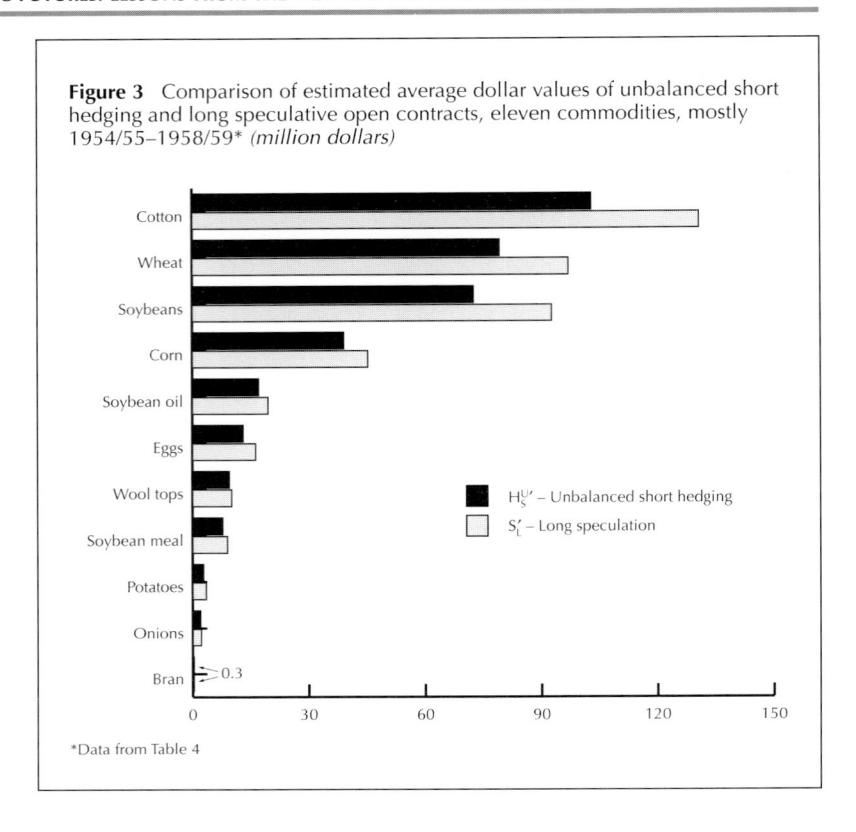

Figure 3 Comparison of estimated average dollar values of unbalanced short hedging and long speculative open contracts, eleven commodities, mostly 1954/55–1958/59* *(million dollars)*

[Bar chart showing commodities: Cotton, Wheat, Soybeans, Corn, Soybean oil, Eggs, Wool tops, Soybean meal, Potatoes, Onions, Bran (≥0.3), with x-axis from 0 to 150. Legend: $H_S^{U'}$ – Unbalanced short hedging; S_L' – Long speculation]

*Data from Table 4

hedging; but it could do so only if there were also *no short speculation*. And that could happen in practice only if the price were so low that no speculator thought the price likely to go lower.

The uncertainties of price appraisal being what they are, a price so low that *no* speculator thought it likely to go lower would assuredly be too low. Any futures market must have more speculation than the minimum technically necessary to carry the hedging, else it will be one in which heavy short hedging causes excessive price depression. How much speculation, beyond an absolute minimum, is needed depends on several circumstances that will be considered presently.

THE SPECULATIVE INDEX AND ITS USE

The speculative index, mentioned above and used in calculation of the amounts of unbalanced short hedging for Figure 3, rests on a simple basic concept that would be most easily applied to a market with no long hedging. In such a case, closely approximated in the egg futures market, the speculative index would be simply the ratio of the amount of long

Table 4 Estimated average US dollar values of long hedging, balanced and unbalanced short hedging, and long speculative open contracts, eleven commodities, mostly 1954/55–1958/59* *(million US dollars)*

	Long hedging H_L'	Short hedging			Long speculation S_L'
Commodity		Total H_S'	Balanced[a] $H_S^{B\prime}$	Unbalanced[b] $H_S^{U\prime}$	
Cotton	132.5	179.8	76.8	103.0	130.7
Wheat	72.2	125.1	45.8	79.3	97.1
Soybeans	52.9	102.6	30.0	72.6	92.7
Corn	19.4	53.3	14.1	39.1	45.3
Soybean oil	11.9	26.0	8.9	17.1	19.5
Eggs	1.2	13.8	0.7	13.1	16.3
Wool tops	3.0	12.1	2.6	9.5	10.1
Soybean meal	8.6	14.0	6.2	7.8	9.0
Potatoes	1.0	3.3	0.6	2.8	3.5
Onions	0.1	2.0	[c]	2.0	2.2
Bran	0.4	0.6	0.3	0.3	0.3

*Data as for Table 3, and partly from it.
[a] Equal amounts long as well as short; computed as $(2/T - 1)H_L'$.
[b] Estimated total short hedging (Table 3) minus balanced short hedging; occasional discrepancy in final digit arises from rounding figures calculated to more places than are shown here.
[c] Less than 0.05.

speculation to the amount of short hedging. In symbols, the speculative index would then be given by the formula,

$$T = \frac{S_L}{H_S} = 1 + \frac{S_S}{H_S} \qquad (H_L = 0)$$

where the parenthetical remark at the right is a reminder of the assumption made as a basis for writing the formula.

If there is a purely logical basis for deducing how to write the formula for the speculative index for markets *with* long hedging, it has escaped me. But it can be shown (see the Technical Appendix below) that the data in Figure 2 reveal how the speculative index should be calculated for markets with long hedging; the previous formulas need to be modified only as follows:

$$T = \frac{S_L + 2H_L}{H_S + H_L} = 1 + \frac{S_S}{H_S + H_L} \qquad (H_S \geq H_L)$$

Application of this formula, as is indicated by the statement in parentheses, is restricted to conditions in which the amount of short hedging exceeds or equals the amount of long hedging. In the reverse condition,

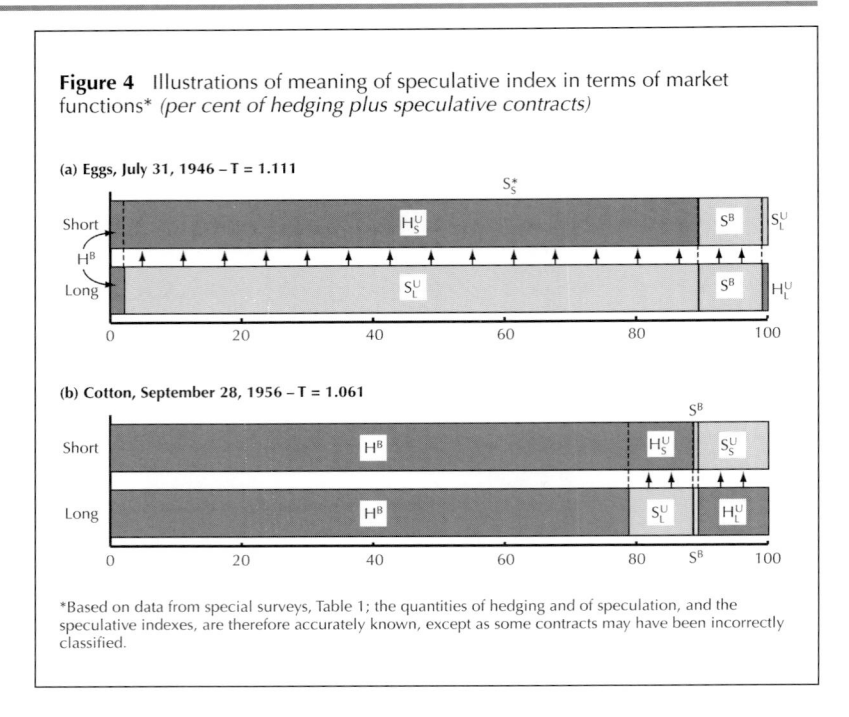

Figure 4 Illustrations of meaning of speculative index in terms of market functions* *(per cent of hedging plus speculative contracts)*

(a) Eggs, July 31, 1946 – T = 1.111

(b) Cotton, September 28, 1956 – T = 1.061

*Based on data from special surveys, Table 1; the quantities of hedging and of speculation, and the speculative indexes, are therefore accurately known, except as some contracts may have been incorrectly classified.

which occurs occasionally in some markets, the formula needs only to be rewritten with subscripts changed, L for S and S for L.

Figure 4 illustrates, for two contrasting commodities, the principles involved in calculation of the speculative index according to the foregoing formula. The data for eggs show very little long hedging, and therefore nearly all of the short hedging must be carried by long speculation. In the data for eggs, the effects of long hedging are inconsequential because there was so little of it.

The data for cotton, however, with nearly four times as much long hedging as there was long speculation, present a situation in which seven-ninths of the short hedging was balanced by long hedging,[16] leaving only two-ninths of the short hedging to be carried by long speculation. And in the cotton market at that time, the greater part of the short speculation was serving to carry a part (one-eighth) of the long hedging, leaving only a small amount of short speculation to be balanced by long speculation.[17]

When we start thus with a definition of the speculative index and proceed to calculate the amounts of hedging and of speculation that balance each other, and the amounts of speculation, both long and short, that serve directly to carry hedging, as seems to me necessary in a non-technical explanation, we seem to have omitted a step in the reasoning. By what justification do we calculate amounts of mathematically determined parts of both speculation and of hedging is implicit in our second formula for the speculative index. When we used the statistical data of Figure 2 to derive a general formula for the speculative index – one not restricted to application in the absence of long hedging – we in fact determined empirically, from statistics for actual markets, the extent to which long hedging *does* balance short hedging and the extent to which short speculation has to be balanced by long speculation. It was lack of knowledge of these facts that prevented our writing, initially, a formula applicable in the presence of long hedging, and the statistical data supplied that information, in the guise of a general formula for the speculative index.

We need next to test the validity of the formula for the speculative index. For that purpose, the accuracy of the mathematics can here be taken for granted, but the adequacy of the data used cannot. Does the formula derived from the data give sensible results, in reasonable agreement with what is known otherwise about the 11 markets for which we have average values of their speculative index?

In order to appraise the reasonableness of the average values of the speculative index, as shown for 11 commodities in Table 3 and graphically in Figure 5, we need to review the major characteristics of the futures markets for those 11 commodities. The commodities were selected from among 21 for which the necessary statistics are available,[18] with a view to obtaining adequate representation, within a list of moderate length, all of major different sorts of market conditions.

The present "big three" among futures markets – cotton, wheat, and soybeans – should of course be included. Addition of corn, which ranks next in size, provides, with wheat, enough representation of grains in the list. Soybean oil and soybean meal differ from all the other commodities in the list except bran and wool tops (both included for a different reason) in that they are processed products rather than primary products; and listing them along with soybeans includes the sole existing example of a group of futures markets that deal in a primary product and in all of its major derivatives.[19]

None of the foregoing commodities is used directly for human consumption, so I add eggs, potatoes, and onions to represent consumer goods in the list of commodities. Their special market characteristic is that, because they are not processed on a large scale, there is relatively little long hedging of these commodities. I take three such commodities in order to include among them a fairly mature futures market, eggs (established in 1922), along with the newcomers, potatoes and onions (1941 and 1942, respectively). Other relatively young markets in the list are soybeans (1936), soybean oil (1940), and soybean meal (1940).[20] The most noteworthy special characteristic likely to be present in a young

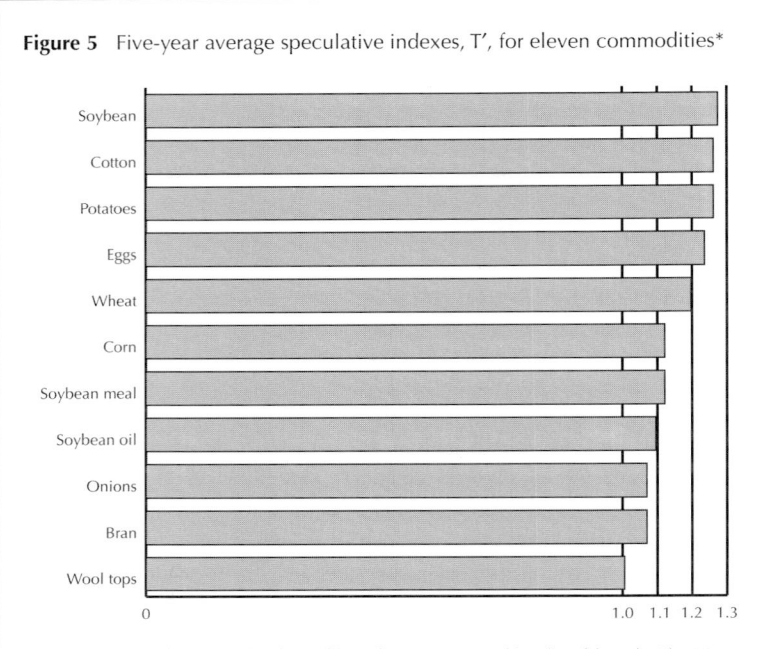

Figure 5 Five-year average speculative indexes, T', for eleven commodities*

Commodity	
Soybean	
Cotton	
Potatoes	
Eggs	
Wheat	
Corn	
Soybean meal	
Soybean oil	
Onions	
Bran	
Wool tops	

0 1.0 1.1 1.2 1.3

*Three-year average for onions; data from Table 2. The minimum possible value of the index T' = 1.0. With one exception, the values shown above appear reasonable in the light of other information about the markets; the index for bran, however, is too high, owing to official classification of anticipatory long hedging as speculation (see text).

futures market is a tendency for short hedging to be done selectively, according to the price expectations of potential hedgers. Dealers in such a commodity tend to persist for a time in making their own appraisals of price prospects, and hence to hedge stocks only when they expect a price decline, rather than to hedge routinely.

Finally, I add to the list two commodities whose futures markets appear to have been struggling against a handicap of neglect by speculators – wool tops struggling successfully, and bran, unsuccessfully. The futures market for the latter commodity, discontinued in 1957, may be truly an example of a market that died for lack of enough speculation rather than for lack of enough hedging to keep it going. The most thorough discussion of the early history of any futures market that has been published in Stewart's study of the wool top futures market (Stewart, 1941).

Because the values of the speculative index shown graphically in Figure 5 have been derived from statistics in which open contracts have been only incompletely classified (as speculative, hedging or matching contracts), it will be necessary to keep in mind that the values shown may be more or less inaccurate. Moreover, the values obtained are dependent on the statutory definition of hedging, which has not entirely corresponded, especially prior to August 1956, with what businessmen regard as hedging.

The principal economic circumstances that tend to influence the value of the speculative index for a commodity are: (1) degree of speculative "interest" in the commodity; and (2) quality of the knowledge and ability of the speculators participating in a market. A third circumstance that may appreciably influence the speculative index is the quality of the price judgements behind the long hedging that enters a market.

Variation in degree of interest in different commodities on the part of skilled speculators presumably occurs principally in the form of a tendency to avoid certain commodities. Skilled speculators shift their attention from one commodity to another with considerable freedom, according to the opportunities that they find in them for speculative profit. By thus shifting from one commodity to another, according to circumstances, they tend to equalise profit opportunities among markets. But certain circumstances can lead speculators to avoid dealing in particular commodities. A commodity that is dealt in only on an obscure market, from which quotations are not widely available, tends to get limited speculative attention. High commission rates, which tend to be necessary in a small market, also discourage speculative use of a market. And informed speculators tend to avoid dealing in a commodity whose price is subject to a substantial degree of control to their disadvantage.[21]

Contrary to an opinion that is sometimes expressed, degree of price variability does not, by itself, appreciably affect speculative interest in a commodity. A speculator simply deals in larger quantities if price variations tend to be narrow than if they tend to be wide.

Unskilled spectators tend to be attracted particularly to the more prominent commodities. And because skilled speculators make profits at the expense of unskilled speculators, as well as by providing a desired service to hedgers, the presence of unskilled speculation in a commodity tends to add to the amount of skilled speculation in that commodity. Thus skilled speculation restrains the vagaries of price movement that unskilled speculation tends to produce, in the same way that it sharply restricts the influence of hedging pressure on a market.

The main variations in the speculative index between commodities that should be expected are, therefore, particularly low values of the index for some commodities that, for one reason or another, have attracted relatively little speculation, and rather high values of the speculative index for those commodities that attract a substantial amount of unskilled speculation. If long hedging in some commodities arises largely from particularly well informed forward buying by processors and manufacturers, either in the form of anticipatory long hedging, or in the form of advance orders which in turn are hedged, such long hedging will tend to partially serve the market function of skilled speculation, and thus make the speculative index of the market a fairly low one.

In the light of these considerations, one should expect the speculative indexes for soybeans, cotton, and wheat to be fairly high, as they are; and it may well be true, as the speculative index for wheat suggests, that that commodity no longer has as much special attraction to unskilled speculators as it used to. The speculative indexes for corn, soybean meal, and soybean oil, falling near the middle of the observed range of speculative indexes, was picked for inclusion in this list of commodities in the belief that it is a commodity with a relative scarcity of speculation.

The speculative indexes for potatoes and eggs are higher, in relation to other commodities, than I expected them to be, but my expectations did not take into account the fact that these are the leading commodities on the New York and the Chicago Mercantile Exchanges, respectively. As such, they would tend especially to attract the attention of such unskilled speculators as do business on those exchanges. Because the amount of hedging and total speculation in those commodities is small (Table 4), a small amount of unskilled speculation in potatoes and eggs can represent as large a *proportion* of total speculation in those commodities as occurs in cotton, wheat, and soybeans.

In the case of bran, it seems clear that the speculative index is distorted upward by a peculiarity of the classification of open contracts. The CEA seeks to have contracts classified according to the statutory definition of hedging, and it was only by amendment of the Commodity Exchange Act, approved July 24, 1956, that anticipatory hedging by processors and manufacturers was legally recognised as hedging rather than speculation (US Statutes, 70 Stat. 630). The data for bran, which had an exceptionally large proportion of anticipatory long hedging, all applies to a time when such hedging, if classed in accordance with the relevant statute, would have been listed as speculation. The data for the other commodities cover years during most of which anticipatory long hedging was legally recognised as hedging. If open contracts in bran had been classed according to the present statutory definition, bran might well have been found to have the lowest speculative index of all.

Potatoes, eggs, and onions, being primarily consumer goods, have little anticipatory hedging by processors and manufacturers, but data from CEA special surveys for potatoes and onions show a large amount of long "speculation", as it is classed, by dealers. A large fraction of the long contracts held by dealers in these commodities appears classifiable, by business standards and on economic grounds, as anticipatory hedging by dealers. If a dealer with storage facilities in the country accumulates stocks of potatoes or onions while growers are selling heavily just after harvest, that is not regarded as speculation, by business standards. If a terminal-market dealer, lacking economical storage facilities, makes contracts directly with country dealers for purchase of potatoes or onions, thus sharing the burden of stock-carrying, that is no more speculative than the holding of stocks by a country dealer. If then, given a futures market, terminal-market dealers accumulate long futures contracts merely as a convenient temporary substitute for making purchase contracts directly with country dealers, is the holding of those futures contracts any more speculative than the holding of contracts directly with country dealers? If long futures contracts held by dealers, in amounts appropriate to normal operation of the business of such dealers, had been classed as hedging contracts, potatoes and onions would have had somewhat lower speculative indexes than those shown in the accompanying figure.[22]

The one complete CEA special survey for eggs that is available gives evidence of no large amount of anticipatory long hedging by dealers in eggs. This is reasonably to be expected, because eggs are held in cold storage, mainly in the larger cities, and the dealers there can accumulate such stocks as they wish in their own hands, instead of relying on purchase contracts with country dealers, or on anticipatory long hedging. There is consequently no reason to suppose that the speculative index for eggs is appreciably distorted upward by classification as speculative of futures contracts regarded, from a business standpoint, as anticipatory hedging.[23]

Having found that a five-year average speculative index of about 1.15 appears to characterise futures markets that have neither a peculiar shortage of speculation, nor any substantial amount of particularly unskilled speculation, we have in the speculative indexes a basis for estimating the proportion of unskilled speculation in, say, the soybean market, which is the one with the highest speculative index, namely 1.28. On the reasonable supposition that the amount of relatively skilled speculation needed to satisfactorily "carry", or balance, a given amount of unskilled speculation, is the same as the amount of skilled speculation necessary to carry a given amount of hedging, the problem is a simple one in arithmetic. Subtracting 1.15 from 1.28, dividing by 1.28, and converting to a percentage, we arrive at the conclusion that about 10% of the speculation in the soybean market was sufficiently unskilled that it had to be offset, like hedging, by better-informed and more skilful speculation.

CONTINUOUS RESPONSE OF SPECULATION TO HEDGING
Next we should examine the degree to which speculation responds to the continuously changing needs for hedge-carrying in a given market. For that purpose I take the wheat market, which is the one in which I am best

Table 5 Estimated total amounts of long and short hedging and speculation, and of balanced and unbalanced hedging, and speculative index, wheat, monthly, July 1956–June 1959* *(million bushels; ratio)*

Date	Hedging		Speculation		Speculative index T' (5)	Balanced hedging[a] $H^{B'}$ (6)	Unbalanced hedging	
	Long H_L' (1)	Short H_S' (2)	Long S_L' (3)	Short S_S' (4)			Long $H_L^{U'}$ (7)	Short $H_S^{U'}$ (8)
1956 July 15	30.7	68.5	51.4	13.6	1.137	23.3	7.4	45.2
Aug. 15	43.8	100.8	70.0	13.0	1.090	36.6	7.2	64.2
Sept. 15	38.5	104.8	77.8	11.5	1.081	32.7	5.8	72.1
Oct. 15	32.4	98.8	77.7	11.2	1.085	27.3	5.1	71.5
Nov. 15	37.5	91.9	68.5	14.1	1.109	30.1	7.4	61.8
Dec. 15	37.9	80.1	59.6	17.4	1.147	28.2	9.7	51.9
1957 Jan. 15	33.7	77.0	62.8	19.4	1.175	23.6	10.1	53.4
Feb. 15	40.3	63.8	48.6	25.0	1.241	24.6	15.7	39.2
Mar. 15	36.8	55.7	44.4	25.5	1.276	20.9	15.9	34.8
Apr. 15	41.9	48.2	37.5	31.2	1.346	20.4	21.5	27.8
May 15	36.4	39.5	31.5	28.4	1.374	16.6	19.8	22.9
Jun. 15	36.4	41.3	34.3	29.4	1.378	16.4	20.0	24.9
July 15	17.8	83.5	74.7	9.0	1.089	14.9	2.9	68.6
Aug. 15	49.2	106.6	76.4	19.0	1.122	38.5	10.7	68.1
Sept. 15	43.7	97.6	74.3	20.4	1.144	32.7	10.9	64.9
Oct. 15	36.4	89.7	71.8	18.5	1.147	27.1	9.3	62.6
Nov. 15	38.7	82.6	70.0	26.1	1.215	25.0	13.7	57.6
Dec. 15	40.4	69.5	57.6	28.5	1.259	23.8	16.6	45.7
1958 Jan. 15	53.4	60.1	47.5	40.7	1.358	25.3	28.1	34.8
Feb. 15	57.8	45.6	32.1	44.3	1.310	24.0	33.8	21.6
Mar. 15	49.1	33.9	24.9	40.1	1.300	18.2	30.9	15.7
Apr. 15	51.4	29.6	22.4	44.2	1.277	16.8	34.6	12.8
May 15	52.0	17.3	13.4	48.1	1.193	11.7	40.3	5.6
Jun. 15	51.5	20.7	14.0	44.8	1.194	14.0	37.5	6.7
July 15	57.4	35.1	21.8	44.1	1.236	27.0	35.7	13.4
Aug. 15	52.5	87.7	59.8	24.6	1.176	36.8	15.7	50.9
Sept. 15	56.9	86.2	54.5	25.2	1.176	39.9	17.0	46.3
Oct. 15	38.7	79.3	58.8	18.2	1.154	29.9	8.8	49.4
Nov. 15	32.6	68.8	54.4	18.2	1.180	22.7	9.9	46.1
Dec. 15	35.4	64.3	50.4	21.4	1.215	22.9	12.5	41.4
1959 Jan. 15	34.8	59.6	47.1	22.4	1.237	21.5	13.3	38.1
Feb. 15	27.4	53.7	45.5	19.1	1.236	16.9	10.5	36.8
Mar. 15	33.6	50.8	42.3	25.1	1.297	18.2	15.4	32.6
Apr. 15	31.8	48.8	42.3	25.3	1.314	16.6	15.2	32.2
May 15	28.2	31.5	27.8	24.5	1.410	11.8	16.4	19.7
Jun. 15	28.8	36.1	30.1	22.8	1.351	13.8	15.0	22.3

*Computed as described in text, from data in US Department of Agriculture, *Commodity Futures Statistics*, Statistics Bulletin 221, 239, 256.
[a]Equal amounts long and short

acquainted with hedging operations and can interpret changes in short and long hedging most reliably.

The available data permit following the course of hedging and speculation, not quite week by week, but as of the middle and end of each month. They are shown in this detail, for three years, in Figures 6 and 7, and as of the middle of each month in Table 5. The second of the two figures serves primarily as an aid to interpreting the first. In Figure 6 interest focuses especially on the course of long speculation, which has to support all of the stock-carrying short hedging that is not balanced by long hedging. Therefore Figure 6 is drawn with the data on long speculation (long open contracts) plotted upward from the base line, and short speculation plotted downward. Short hedging is therefore plotted upward, for ready comparison with the long speculation, and long hedging plotted downward from the base line, for comparison with the short speculation. The speculative index is plotted to a scale with $T = 1.0$, the minimum possible value of the index, at the base line used for the other data.

Figure 7, which shows only hedging data, in comparison again with the speculative index, needs to have hedging plotted as in the previous figure, hence short hedging is plotted upward and long hedging downward, though it would seem more logical to show them otherwise if the figure were not merely a supplement to the previous one. It shows the totals of short and of long hedging, and the portions of each that balance. The difference between the balancing portions of the two sorts of hedging and the total of each, is the "unbalanced" short and long hedging of the previous figure.

The extent to which long hedging balances short hedging (or, when long hedging exceeds short, the extent to which short hedging balances the long) depends both on the amount of long (or short) hedging and on the speculative index, as can readily be seen from the figure. At those times when the amount of short hedging is so large as to strain the carrying capacity of speculation in a market, the price at which hedged stocks will be carried depends relatively heavily on the amount of long hedging present to aid in carrying the short hedging. At such times, a large fraction of the long hedging – up to 89% in these data (July 31, 1957) – serves to directly balance an equal amount of short hedging. At times when the hedging load is light relative to the amount of speculation in a market (the speculative index high), the price is determined mainly by speculative opinion, with little reliance on long hedging to carry part of the short hedging (or little reliance on short hedging to carry part of the long hedging). It is for this reason that the amount of balancing hedging, long and short, is a varying proportion of the total long (or of the total short) hedging. In the data shown in the figure, the amount of balancing hedging fell as low as 41% of the smaller of the two hedging totals (May 31, 1959).

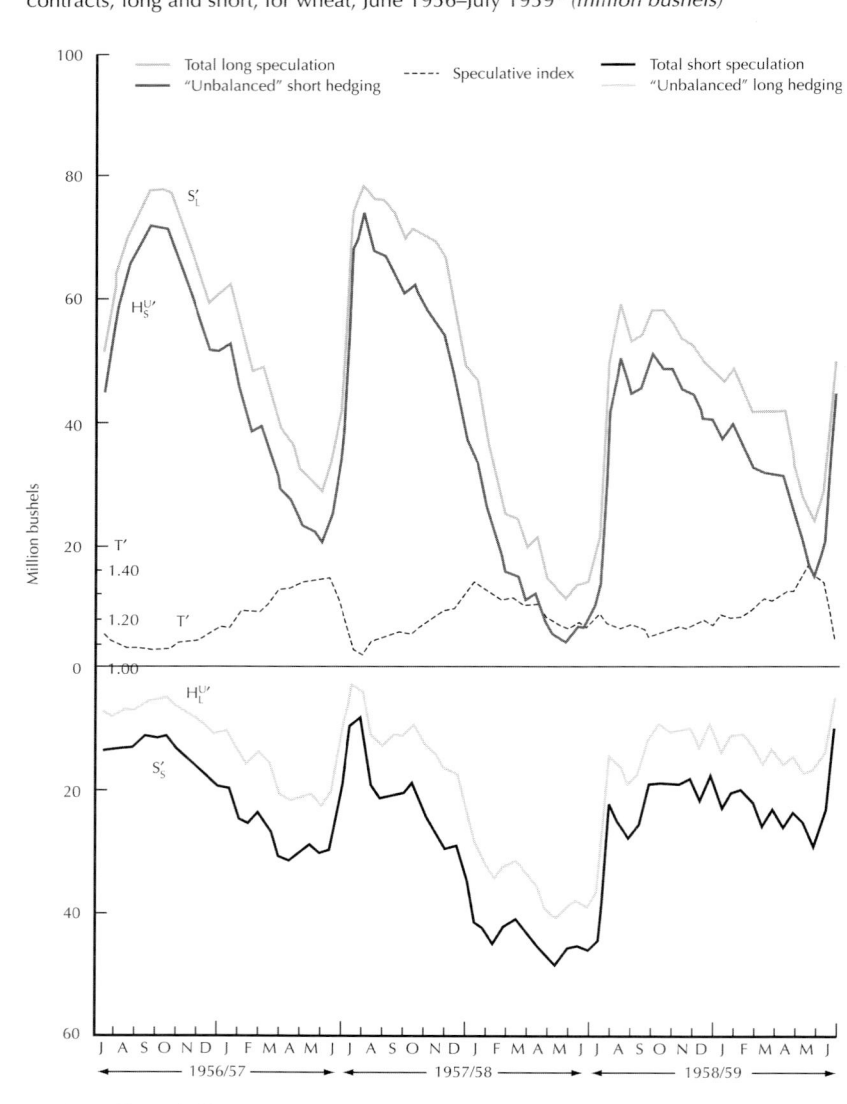

Figure 6 Changes, semi-monthly, in amounts of speculative and of unbalanced hedging contracts, long and short, for wheat, June 1956–July 1959* *(million bushels)*

Legend: Total long speculation; "Unbalanced" short hedging; Speculative index; Total short speculation; "Unbalanced" long hedging

*Data from Table 5 and a like manuscript table including month-end statistics.
The total of long speculative open contracts in wheat during these three years exceeded the amount of "unbalanced" short hedging (see Figure 7) by a nearly constant 8 to 10 million bushels except for a bulge that carried it to 12.8 million on November 30, 1957. "Unbalanced" long hedging was carried by short speculation with the same margin of surplus as for long speculation. The speculative index (T') is close to unity when there is little short speculation relative to the amount of long speculation, as it normally is when hedging is heavy, and increases when either the amount of short speculation increases or the amount of long speculation decreases. (When hedging is net long, as during January–June 1958, speculation must be net short, and the formula for the speculative index is altered accordingly.)

SPECULATION ON HEDGING MARKETS

Figure 7 Changes, semi-monthly, in amounts of hedging contracts, long and short, and in balancing portions of hedging, wheat, June 1956–July 1959* *(million bushels)*

*Data as for Figure 6.
When the pressure of short hedging on a futures market is heavy, requiring a great amount of long speculation to carry it, the speculative index tends to fall close to unity. When it does so, most of the long hedging in the market serves to balance an equal amount of short hedging, thus decreasing the amount of long speculation needed. The amount of short hedging of wheat follows a smooth course, corresponding closely with the volume of commercial stocks to be hedged. The amount of long hedging, on the other hand, tends to change erratically, under the influence of spurts of buying, or the withdrawal of buying, by certain handlers of the commodity (in wheat, chiefly bakers and exporters). Unusual behavior in such buying accounts for the abnormal course of the speculative index for wheat during the latter half of 1957/58.

Two mathematical relationships between curves in Figure 6 deserve comment: (1) the excess of long speculation, S_L', over unbalanced short hedging, $H_S^{U'}$, is an excess that should be measured by the speculative index, according to our definition of that index;[24] and it is $-S_L'/H_S^{U'} = T$. (2) Because total long contracts in a futures market must equal total short contracts, the excess of total short speculation, S_S', over unbalanced long hedging, $H_L^{U'}$, must equal the excess of long speculation over unbalanced short hedging.

The close correspondence between changes in the amount of long speculation and changes in the amount of unbalanced short hedging needing to be carried by long speculation is so obvious as to require no verbal emphasis. The amount of short speculation obviously varied in about equally close correspondence with variations in the amount of unbalanced long hedging. In the latter connection, it should be noted that the amount of long hedging, which depends primarily on the opinions of export buyers and of bakers regarding whether the price of wheat (and hence of flour) is likely to rise or fall, was more erratically variable than was the amount of short hedging (Figure 7).

Less conspicuous than the close correspondence between the two curves in each pair in Figure 6, but at least equally important, is the evidence that the figure gives of an approximately constant margin of difference between the two curves in each pair. This difference represents the amount of "unneeded" speculation, short and long, in the market. It is, of course, "unneeded" only in a physical, or arithmetical, sense. Economically, it is needed because no one can judge *accurately* what price is warranted by known supply and demand conditions at any given time, and price opinions among speculators therefore differ. This margin gives protection against the price, at times of heavy hedging pressure, falling so low that *no* speculator thinks it likely to go lower.

The amount of protective margin of "unneeded" speculation that is *economically needed* in a market depends primarily on the amount of divergence of price judgement that exists among speculators. It tends obviously to be relatively narrow if most speculators are expert and well-informed, and to be relatively wide if there is a considerable group of inexpert or ill-informed speculators in the market. And, what is not so well recognised, the amount of "unneeded" speculation that is present in a market depends on the extent to which speculators specialise in giving attention to particular sorts of market information.

A great amount of economic information concerning current supplies and probable future supplies, and concerning current and probable future consumption demand – the latter influenced by consumer incomes and by prices of other commodities, both in this country and abroad – needs to be considered if the price is to reflect well the available supply and demand information. No one man can get all that information promptly and appraise it wisely. If all speculators give their attention to the same information, they may all form nearly the same price opinions, with the result that there will be little divergent speculation. But in that case there will be times when the price will not conform well to the basic supply and demand conditions, because everybody has overlooked some important information. On the other hand, if different groups of speculators specialise in giving attention to different classes of information, their price opinions will tend to differ considerably, giving rise to a good deal of divergent speculation. And with such specialisation by speculators, there is less likelihood than otherwise that important economic conditions will fail to be reflected properly in the price.

Evidently, then, a widening of the margin of "unneeded" speculation in the market, as in wheat during October–November of 1957, may reflect either of two conditions. It may reflect entry into the market of a considerable group of inexpert or ill-informed speculators; or it may reflect the recognition by one group of expert speculators of significant economic conditions or prospects that are currently being ignored by other, equally expert and generally well-informed speculators. One would like to know to what extent each of these two possible explanations properly accounts for widening or narrowing of the margin at particular times, but presumably the statistics of hedging and of speculation cannot reveal that by themselves. We cannot undertake to fully explore such questions here.

I remarked earlier that there have been three obstacles to acceptance of the evidence that speculation enters a futures market in response to the needs for carrying hedges placed in the market, and enters in an amount fairly appropriate to the needs for carrying the hedging. The common impression that speculation in futures proceeds with little or no regard to hedging has been shown above to rest on a practice of identifying speculation with the readily visible evidences of it, namely the transactions by which it is initiated and terminated, whereas it is properly to be measured, in futures markets, by the volume of open contracts *held* speculatively. The obstacle of unsatisfactory character of the available data on hedging and speculative open contracts, in its original form, has been dealt with above by applying new and improved estimation procedures. There remain for consideration now the questions; what incentives do speculators have for adjusting the amount of speculation to the needs for carrying hedges? And how are speculators able to recognise hedging needs?

The incentive that speculators have for carrying hedges is simply the opportunity to profit by doing so. Speculators, as a group, can make profits only by "buying" from short hedgers at prices lower than those at which they can "sell" later, or by "selling" to long hedgers at prices higher than those at which they can "buy" later. An individual speculator is as willing to make a profit by dealing with another speculator as by dealing with a hedger, but speculators as a group can profit only by rendering a service for which hedgers are willing to pay. Though hedgers forego whatever profits speculators, as a group, make, they do so for the sake of one or another of the advantages obtainable by hedging. The incentive for speculators to seek to carry hedges is, then, simply the profit motive.

There are, broadly, two ways in which speculators recognise the profit opportunities presented to them by the offering of hedging contracts in the market. One of these is by such criteria as are used by scalpers and other quick-turn speculators. They learn, perhaps most importantly, to judge with satisfactory reliability when a small price dip is the result of "selling" pressure that will soon end and be followed by price recovery, and when a small price advance is the result of "buying" that will be similarly short-lived. The dips and advances on which scalpers operate are those very small ones associated with oscillation of the price between what amount to "bid" and "asked" prices. Besides the commonly recognised bid and asked prices that are close together, but are subject to rapid shifts in level, there exist in any market analogous pairs of prices that are farther apart, but more stable. In an *inactive* market it is these more widely spaced limits that provide the quoted bid and asked prices. In an active market, individual hedging orders of moderate size move the price only within the range of the commonly quoted and closely spaced bid and asked prices, while very large hedging orders, or waves of hedging, move the price through wider limits. Scalpers, who normally follow each "purchase" with a "sale" a few minutes later, tend to absorb, initially, the individual hedging orders of moderate size; other traders, willing to take larger market positions, and to hold them for a day, or a week, absorb the larger hedging orders and the waves of hedging, long or short. We noted above, in comment on Figure 7, evidence of the erratic waves of long hedging in wheat that have to be absorbed by that market.

Short hedging tends to enter the market for any annual crop in large quantity once a year, and much of it must be carried for many months. The various sorts of quick-turn traders help to carry this hedging for a time after it enters the market because, while the volume of short hedging is building up, they tend to operate always on the long side. Though it is commonly said that scalpers and other quick-turn traders always stand equally ready to either "buy" or "sell", this is not strictly true. Though they seek their profits on more or less quick turnover of contracts, they always seek to judge the longer-run price prospects, and normally take positions only on one side of the market, either long or short according to their judgement of those longer-run prospects. They do this partly as a means of selecting the larger profit opportunities, but more particularly as a means of limiting risks of loss.

But in the main, short hedges must be carried by speculators who are willing to hold positions for longer periods than the quick-turn traders like to carry their holdings. The incentive for them to carry short hedges must be that they think the price low enough to warrant taking a long position and holding it for a considerable period – not necessarily for the six or eight months or more that much of the short hedging will have to be carried, but at least for a substantial fraction of that time. Do speculators have such good price judgement that many of them are in fact led to "buy", and to hold contracts over fairly long periods, by recognition that the price is only moderately depressed by heavy short hedging? Economists, at least, have found it difficult to credit speculators with such good price judgement as seems necessary in order that speculation should respond sensitively to the hedging needs of a market.

My first clue to the answer to this troubling question was given me by professional speculators. With few exceptions, they have told me that they do not try to judge what the price "ought" to be. They also often say that they disregard the market news, to which economists properly believe that speculators should be alert if they are to serve their economic function well. In order to understand what speculators are trying to say when they make these statements, it is necessary to recognise that neither statement is true in the same sense that most hearers tend to understand it.

Consider the latter of the two statements first, since it is the more readily interpreted. It is a statement that is made characteristically by quick-turn speculators whom one would expect to be especially alert to new market information. Watch their behaviour at close range and you observe that they are indeed eagerly alert to new market information. What they mean when they say that they disregard the market news, is that what passes generally as market "news" is to them like last week's newspaper. And even so, they do not disregard it. It tends to be an explanation of the part rather than an indicator of the future, and most speculators follow it closely as an aid to interpreting what has happened lately in the market.

To understand how speculators are able to act as though they could form highly reliable judgements of what the price "ought" to be, it is particularly necessary to understand what they mean when they say they do not try to form such judgements. My best help on this point came from a speculator from whom the statement was superficially a believable one, because he was a "day trader" specialising in scalping and other quick-turn speculation within the day. He kept careful notes for me for a week, recording the thoughts uppermost in his mind at the beginning of each day's business, and every transaction that he made, and tried to explain the reason for each transaction.[25]

The notes showed that he was giving close attention to a great amount of information pertinent to appraisal of what the price "ought" to be. And quite obviously he was doing so for profit, not out of mere curiosity or general interest.

Why, then, did this man tell me that he did not try to judge what the price "ought" to be; and why have others, of whom it was at least equally untrue, said the same thing? They are all very keenly aware of the unreliability of any individual's judgement of what the price "ought" to be.[26] What they meant was that they never rely heavily on their personal judgement on that score. They all have great respect for the "judgement of the market" as an expression of the pooled opinions of many people – a sort of average of many opinions, in which individual errors of judgement tend to cancel out. They know also, consciously or unconsciously, that the market can be more or less mistaken, and that they must make their profit by taking advantage of the market's "mistakes". Their problem is to find means of striking a delicate balance between relying on their own fallible judgements, but not relying on them too much, and relying on the "market judgement", yet finding opportunities to make profits from disagreeing with the market judgement.

Different speculators find more or less different ways of striking this balance. A typical sort of solution is that found by my friend the "day trader". He formed his opinions on which way the price ought to move;[27] but he gave expression to those opinions indirectly and cautiously, using them as a guide to actions in which other considerations entered also. For example, if he expected a price rise for "fundamental" reasons, his quick-turn trading would tend to be always on the long side ("selling" only to close out long positions taken previously). In such ways skilled speculators make their opinions effective as part of the pool of many opinions that are reflected in the market price, and at the same time use the pooled market opinion as a guide in forming what an economist, if not they themselves, would call their opinions of what the price "ought" to be.

SUMMARY

The main conclusion that flows from the foregoing analysis and evidence is only a more precise and stronger statement of one reached by a few students of futures markets long ago, just a few years after the commencement of regular reporting of open contracts on futures markets. The main statistical evidence provided as a basis for the present version of the conclusion appears graphically in Figures 3 and 6 above. Figure 3 shows that the amount of long speculation, measured in dollar value of open speculative contracts, has differed greatly between commodities, some commodities having 10, 20, or as much as several hundred times as much speculation as others; and that these differences in amounts of speculation depend primarily on the amounts of hedging in the markets. And Figure

6 illustrates the very close correspondence between changes from month to month, and within months, in amounts of speculation and of hedging. The conclusion that follows is obvious: futures markets are primarily *hedging* markets. The amount of speculation in any futures market depends primarily on the amount of hedging that enters the market, and no such market can exist without a sufficient amount of hedging to support it.

These characteristics of futures markets could not be recognised until after statistics on open contracts were collected, because, in the absence of such statistics, people could observe only the transactions by which individual speculators initiated or terminated a period of speculative *holding* of futures contracts. People thus fell into the habit of identifying speculation with the transactions rather than with the *speculative holding* that is its essence, and in consequence were misled.

One obstacle to recognition of the dependence of speculation in hedging has been a belief that speculators have no great incentive to respond to the hedging needs of a market, and no reliable guide for adjusting their speculation to the hedging needs of the market. The final section above explains briefly why speculators try to respond to hedging needs, and how they are enabled to do so both sensitively and reliably.

The basic statistics on hedging and speculation, though published regularly now for a long list of futures markets. are incomplete, and especially so for speculation. In the past if has been usual to deal with this situation very crudely by taking the amount of hedging to be measured by the incomplete record of "reported" hedging, and taking the amount of speculation to be measured by "reported" speculation plus all nonclassified futures contracts. This procedure obviously gave a distorted comparison, but it had to be used until a better one was discovered. Even that crude procedure had sufficed to reveal the existence of a fairly close correspondence between amounts of short hedging and of long speculation in futures markets, but a better procedure was needed before the relation between speculation and hedging could be seen very clearly.

So long as only very crude comparisons could be made between amounts of speculation and of hedging, it was possible to learn only that amounts of long speculation corresponded roughly with amounts of short hedging. By using improved statistical estimates of total amounts of hedging and of speculation, we have gained a more accurate knowledge of the relation between long speculation and short hedging in futures markets and have discovered, additionally, that the amount of short speculation in a market is strongly influenced by the amount of long hedging. This discovery has been aided by solution of the problem of measuring the extent to which long and short hedging directly balance each other. Students of the statistics of futures markets recognised early that long hedging served in part to offset short hedging, but that the offset was only partial. Derivation, through statistical and mathematical analysis, of a *speculative index*, based on the available statistics of hedging and speculation, had led to a formula by which to calculate the amounts of long and short hedging that balance each other. The extent of direct balancing varies with the speculative index, for a reason that has been explained above.

TECHNICAL APPENDIX

1. *Procedure*. In a full statement of the argument that follows, the terms to be used and the formulas derived would differ systematically according to whether $H_S \geq H_L$, as is usually the case in any futures market, or the infrequent opposite condition exists. The argument would be complicated, and nothing would be gained, if the two possible conditions were carried forward simultaneously. Therefore, I proceed for the most part on the assumption that $H_S \geq H_L$, and at the end I note what changes are required to fit the condition, $H_L > H_S$. Under the final heading two parallel sets of formulas are given for computational use. It is mainly in computational work that there is need to take account of whether $H_S \geq H_L$ or not. At the point $H_S = H_L$, the two sets of formulas give identical results.

2. *Estimation of totals of speculation and hedging*. The problem of estimating total amounts of speculation and of hedging, long and short, is one of making an allocation of nonclassified contracts, N, among the categories of hedging, H, speculation, S, and matching contracts, M (the latter being long and short contracts that offset each other in individual accounts). To these letters we add subscripts, $_L$ and $_S$, to designate whether the contracts are long or short. Total short hedging, for example, may then be described in relation to the officially reported statistics as consisting of $H_S = H_S^* + H_S^0$, where the asterisk designates open hedging contracts explicitly classified ("reported") in the published statistics, and the superscript zero designates hedging contracts of "non-reporting" (small-scale) holders of short contracts. *Estimates* of total short hedging are designated H_S'. Statistics and estimates of total contracts in other categories than H are similarly distinguished by superscripts, in addition to the subscripts distinguishing long from short contracts.

Several necessary equalities have to be borne in mind, represented by the following:

$$H_S = H_S^* + H_S^0 \tag{1}$$

$$H_S + S_S = H_L + S_L \tag{2a}$$

$$M_S = M_L \qquad (2b)$$

Equation (1) holds with replacement of S or M for H, and with either subscripts $_S$ or $_L$ throughout. Equations (2) hold either for true values of the quantities, as written above, or for estimates of each, such as H_S', etc. (see Figure 1).

I make only two assumptions that affect estimates of the ratios among the four quantities, H_S', H_L', S_S' and S_L' and these lead directly to the estimation equations for two ratios,

$$r_h' = \frac{H_L'}{H_S'} = \frac{H_L^*}{H_S^*} \qquad (H_S^* \geq H_L^*) \qquad (3a)$$

$$r_s' = \frac{S_L'}{H_S'} = \frac{S_L^* + N_L}{H_S^* + N_S} = \frac{S_L + H_L^0 + M_L^0}{H_S + S_S^0 + M_S^0} \qquad (H_S^* \geq H_L^*) \qquad (3b)$$

From these it follows obviously that $H_L'/S_L' = r_h'/r_s'$, and any other desired ratio between estimates may be obtained with the further aid of the equality, $H_S' + S_S' = H_L' + S_L'$. For example, $S_S'/H_L' = 1 + (r_s - 1)/r_h$. If $H_L^* > H_S^*$, better reliability is obtained by basing estimates on the ratio

$$r_s' = \frac{S_S'}{H_L'} = \frac{S_S^* + N_S}{H^* + N_L} \qquad (H_L^* > H_S^*) \qquad (3c)$$

The same symbol, r_s', is used in both (3b) and (3c) above because the parenthetical remark at the right constitutes a part of each expression. We shall deal throughout with expressions in which subscripts of the variables interchange when $H_L > H_S$. The ratio r_h' might be defined without regard to the relative magnitudes of H_S^* and H_L^*, but to do so would require treating expression (3a) differently than other expressions, and thus tend to make trouble.

One further assumption is needed in order to assign a value to any one or more of the desired *quantities*, such as H_S', and for that purpose I assume that $H_S^0 = H_L^0 + M_L^0$ (if $H_S^* \geq H_L^*$), which leads to the estimation equation,

$$H_S' + S_L' = H_S^* + S_L^* + N_L \qquad (H_S^* \geq H_L^*) \qquad (4)$$

Consequently, this assumption, along with the two previous ones, allows estimation of the six quantities, H_S, H_L, S_S, S_L, M_S and M_L, the latter two of which are of course equal.

Though each of the assumptions made above is subject to error, their usefulness in combination is to be judged in relation to the assumptions implicit in the previously general practice of taking hedging to be satisfactorily measured by H_S^* and H_L^*, and speculation by $S_L^* + N_L$ and $S_S^* + N_S$. That implicitly assumed that $H_S^0 = H_L^0 = M_S^0 = M_L^0 = 0$, which involves three distinct assumptions that together must ordinarily, if not always, be farther from the truth than the three assumptions that we make here.

3. *The speculative index.* Derivation of the speculative index starts from the obvious proposition that, in the absence of long hedging, the ratio, $r_s = S_L/H_S$ would make a good index of the relation of total speculation to the amount of speculation "needed" to carry the hedging in the market. In the absence of long hedging, $S_L = H_S + S_S$, hence for that condition the speculative index, T, may be written $T = 1 + S_S/H_S$. The question then arises: how should the speculative index be calculated in the presence of long hedging?

This problem may be treated empirically by ascertaining how r_s varies with r_h and for that purpose Figure 2 was constructed. If the 11 observations shown in the figure could all be regarded as samples from a common population, an appropriate procedure would be to compute a least-squares regression to express the relation between the two ratios. But I know that the observations came from markets having differing speculative characteristics, making it appropriate to derive more than one line of relationship, each intended to represent the relation between the two ratios in a group of markets with common speculative characteristics. When that was undertaken, it became apparent that each line drawn tended to have an equation that could be written either in the form, $S_L = (1 + \alpha)H_S - (1 - \alpha)H_L$, or as $S_S = \alpha(H_S + H_L)$. Thus the parameter reflected the speculative characteristics common to each group of markets; and because α may be determined from a single observation, that parameter could be used, with these data, to measure the average speculative characteristic of a single market. Used thus, its equation would be, $\alpha = S_S/(H_S + H_L)$; and under the condition for which I know how to derive a useful speculative index, namely with $H_L = 0$, I had $\alpha = S_S/H_S = T - 1$. Hence it was evident that an expression for the speculative index could be written,

$$T = 1 + \frac{S_S}{H_S + H_L} = \frac{S_L + 2H_L}{H_S + H_L} \qquad (H_S \geq H_L) \qquad (5)$$

The second expression for T in (5) above comes from converting the first one to an improper fraction and recognising that $H_S + S_S = S_L + H_L$. Obviously, $T \geq 1$.

4. *Partial balancing of H_S and H_L, and of S_S and S_L.* It is useful to be able to compute the extent to which long hedging balances short hedging,

thereby reducing the amount of net long speculation needed to carry the short hedging. Such balancing, which is accompanied also by partial balancing of long speculation by short speculation, has been illustrated in Figure 4 above.

To deal with the balancing concept, I use the notation of the following equations:

$$H_S = H^B + H_S^U \qquad H_L = H^B + H_L^U$$
$$S_S = S^B + S_S^U \qquad S_L = S^B + S_L^B \qquad (H_S \geq H_L)$$

where the superscripts indicate, respectively, the amounts of short and long hedging (speculation) that balance each other (B) and the remaining, unbalanced (U), portions of both short and long hedging (speculation). Then the *relative* amounts of balancing may be expressed in the form,

$$H^B = p_h H_S$$
$$S^B = p_s S_L \qquad (H_S \geq H_L)$$

where p_h and p_s represent *proportions* of short hedging and of long speculation, respectively, that are balanced by equal amounts of long hedging and short speculation, respectively.

It follows then (see Figure 4) that,

$$S_L - S^B = H_S - H^B \qquad (H_S \geq H_L)$$

and we may readily derive,

$$1 - p_s = (1 - p_h)H_S/S_L \qquad (H_S \geq H_L) \qquad (6)$$

Note that the relation between p_s and p_h thus arrived at is general, within the limits stated parenthetically, and is reached without the aid of any assumption except that balancing *may* occur. We do not require that balancing occur, inasmuch as both p_s and p_h are left free to take zero values. The ranges of possible values of the two p's are given by,

$$0 \leq p_s \leq S_S/S_L; \ 0 \leq p_h \leq H_L/H_S \qquad (H_S \geq H_L)$$

Suppose, now, that we are ignorant of the empirical evidence concerning the general expression for T; let us see how much we can learn about the speculative index, T, from expression (6) and the basic definition,

$$T = S_L/H_S \qquad (H_L = 0)$$

In the special case of this definition, there is no long hedging to balance short hedging, hence $p_h = 0$, and expression (6) becomes, $1 - p_s = H_S/S_L = 1/T$. We see, therefore, that T must satisfy the general condition,

$$(1 - p_s) = 1/T \tag{7}$$

It will become evident later, if it is not at this point, that expression (7) is entirely general, not restricted to the condition $H_S \geq H_L$.

Substituting the definition of p_s in expression (7) leads readily to

$$S^B = [1 - (1/T)]S_L \qquad (H_S \geq H_L) \tag{8}$$

where the parenthetical qualification is required again because of its association with the definition of p_s.

Substituting expression (7) in expression (6), taking reciprocals of both sides, and writing an equivalent for $(1 - p_h) H_S$, gives

$$T = \frac{S_L}{H_S - H^B} = \frac{S_L}{H_S^U} \qquad (H_S \geq H_L) \tag{9}$$

If we now substitute in (9) the empirically derived expression, $T = (S_L + 2H_L)/(H_S + H_L)$, and rearrange terms, we have

$$H_S + H_L = (H_S - H^B)\left(1 + \frac{2H_L}{S_L}\right) = H_S - H^B + 2H_L\frac{H_S - H^B}{S_L} \quad (H_S \geq H_L)$$

Then substituting from expression (9), rearranging and simplifying,

$$H^B = [(2/T) - 1]H_L \qquad (H_S \geq H_L) \tag{10}$$

Expressions (8) and (10) show that T, conceived as a speculative index to express the relation of amount of speculation to amount of hedging in relative terms, with unity representing the minimum amount of speculation capable of carrying the hedging, measures also the extent of direct balancing that occurs between long and short speculation and between long and short hedging. This appears more explicitly if the two expressions are written as

$$p_s = 1 - (1/T)$$
$$p_h = 2/(T - 1)$$

Here, as in expression (7), the relations are entirely general, not restricted to the condition, $H_S \geq H_L$.

To adapt the argument of this appendix to the condition $H_L > H_S$, all that is needed is to interchange subscripts, writing L for S and S for L in all expressions where either appears, and to make corresponding changes in all verbal references to the variables. The lower-case subscripts of r and p remain unchanged.

5. *List of useful formulas.* As an aid to computations, the more widely useful expressions are given below in the two forms in which they may need to be used. My experience suggests that in computing estimates from the published data it is wise to carry at least four significant figures, wherever possible, even though it be unlikely that more than two digits will be meaningful in the final estimates. The symbol ($\approx$) is used to mean "estimated as"; hence the expressions, $r_h \approx H_S^*/H_S^*$ and $r_h' = H_L^*/H_S^*$ are equivalent. The final four pairs of expressions below, written explicitly in terms of known values of the variables, apply also to estimates – that is, with (') added throughout.

When $H_S \geq H_L$	When $H_L \geq H_S$
$r_h = H_L/H_S \approx H_L^*/H_S^*$	$r_h = H_S/H_L \approx H_S^*/H_L^*$
$r_s = S_L/H_S \approx \dfrac{(S_L^* + N_L)}{(H_S^* + N_S)}$	$r_s = S_S/H_L \approx \dfrac{(S_S^* + N_S)}{(H_L^* + N_L)}$
$H_S' = (H_S^* + S_L^* + N_L)/(1 + r_s')$	$H_L' = (H_L^* + S_S^* + N_S)/(1 + r_s')$
$S_L' = r_s' H_S'$	$S_S' = r_s' H_L'$
$H_L' = r_h' H_S'$	$H_S' = r_h' H_L'$
$S_S' = S_L' + H_L' - H_S'$	$S_L' = S_S' + H_S' - H_L'$
$M_L' = (S_L^* + H_L^* + M_L^* + N_L)$ $- (S_L' + H_L') = M_S'$	$M_S' = (S_S^* + H_S^* + M_S^* + N_S)$ $- (S_S' + H_S') = M_L'$
$T = 1 + \dfrac{S_S}{H_S + H_L} = \dfrac{2r_h + r_s}{1 + r_h}$	$T = 1 + \dfrac{S_L}{H_L + S_S} = \dfrac{2r_h + r_s}{1 = r_h}$
$H^B = (2/T - 1)H_L$	$H^B = (2/T - 1)H_S$
$H_S^U = H_S - H^B = S_L/T$	$H_L^U = H_L - H_B = S_S/T$
$H_L^U = H_L - H^B = 2(1 - 1/T)H_L$	$H_S^U = H_S - H^B = 2(1 - 1/T)H_S$

---

1 That action, closing one economically useful market and implying an imminent danger of similar closing of other such markets, emphasises the need for better and more widespread understanding of the economics of futures markets. In a previous article (Working (1960) p.3), I reviewed evidence on the price effects of futures trading – evidence that directly contradicts what the congressional committees had been led to believe; and I promised there a subsequent discussion of reasons why the congressional committees interpreted as they did the evidence put before them. The reasons are not specific to the onion market, but general, influencing most people who are otherwise well informed, and producing misunderstanding and misjudgement of all futures markets. Only part of them can be considered here.

2 This phrase excludes arbitrage, and also any other holdings of matching long and short positions, such as may arise from mere failure to promptly cancel out directly offsetting

contracts on the books. I intentionally avoid specifying that the net position should be calculated for a single commodity. For some purposes holding of long soybean contracts, for example, against short contracts in soybean oil and soybean meal, in appropriate proportions, should be considered arbitrage (and classed, more specifically, as either hedging or spreading, according to whether or not the holder is a soybean crusher, and the holdings of a size appropriate to the size of his crushing business). On the other hand, when considering the relative amounts of speculation and of hedging in, for example, Minneapolis wheat futures, the Minneapolis end of spreads between Chicago and Minneapolis should be counted as speculation in Minneapolis. In short, the line between arbitrage and speculation needs to be drawn according to the purpose of the analysis or discussion involved, hence a general definition of speculation must allow latitude for drawing that line differently in different circumstances.

Hedgers often engage in spreading operations, and sometimes their spreads may well be classed as a special form of hedging, as in an example cited in note 19 below. Usually, however, the holding of matching futures contracts in different markets or different delivery months seems to be best classed as spreading even when it is done by a firm that engages also in hedging.

3 This definition, worded a bit differently, appeared first in Working (1953, p. 560).
4 Until 1956 the Commodity Exchange Act, which the CEA seeks to follow in its statistical classification of futures contracts, was somewhat inconsistent with this principle, in that it failed to recognise anticipatory long hedging as true hedging. An amendment enacted July 24, 1956, recognised such hedging by processors and manufacturers as hedging, but implicitly leaves similar hedging by *dealers* to be classed as speculation. The small apparent inconsistency in definition that thus remains in the Act should perhaps be regarded not as inconsistency with the principle, but as an expression of opinion that forward contracting of supplies by dealers is not "normal merchandising practice".
5 Irwin, then an economist in the Grain Futures Administration, subsequently made an historical study of the origins of futures markets in butter and eggs, and undertook a reinterpretation of historical information on the origins of grain futures (Irwin, 1954). Both indicated that futures markets had grown out of business needs, or wishes, of the sort that hedging meets. One cannot say accurately that it grew out of a desire for a means of hedging, in the modern sense, because only a primitive concept of the usefulness of hedging could emerge until after hedging facilities came into use. Even now it often takes several years following the establishment of a new futures market for handlers of the commodity to learn to use the market effectively for hedging. Irwin's interpretation met such resistance at the time that the results of his study had to be published privately, but it can now be seen as certainly correct, at least in its main outlines.
6 There is no valid evidence, so far as I am aware, that speculators do have such a preference, but the opinion that they do has, for some obscure reason, gained wide acceptance on the exchange.
7 More economically because a hedger, wanting prompt execution of his orders, must expect ordinarily to buy at an "asked" price and sell at a "bid" price, and therefore, can obtain prompt execution at the lower cost in a "broad" active market, where bid and asked price are closer together than in a less active market.
8 The Kansas City contract being distinctive, not by its terms but because, prior to 1953, the location of the market had effectively assured delivery of hard winter wheat on its futures contracts.
9 The course of business on the several wheat futures markets from 1921 through 1935 can be followed conveniently in US Department of Agriculture (1937, Tables 1 and 2) and for earlier years in US Federal Trade Commission (1920–1926, V, pp. 36–40). No statistics of futures transactions in the cured pork products exists, so far as I know, but emergence of irregularity in the recording of price quotations (*Annual Reports*, Chicago Board of Trade)

shows when business had fallen very low.

10 For example, the fluctuations from year to year in average volume of open contracts in wheat at Minneapolis reflect annual fluctuations in size of the spring-wheat crop of Minnesota and the Dakotas; fluctuations in open contracts at Duluth reflected variations in size of the durum crop; fluctuations at Kansas City reflect variations in size of the wheat crop of the Nebraska–Kansas–Oklahoma area. And examples such as the foregoing and those in the text might be cited for any commodity; it is only special familiarity with wheat production and marketing that has led me to pick examples from that commodity.

11 The Grain Futures Administration then began publication of statistics of open contracts in grain futures, first in the form only of totals, later with a partial classification between speculative, hedging, and spreading contracts. The Federal Trade Commission subsequently published with discussion some special compilations of earlier data on open contracts (principally in US Federal Trade Commission, 1920–1926, VII, pp. 124–36).

12 Statistics of long hedging and of short speculation fail to appear explicitly in the table because space that they would occupy is needed for other information, but it should be noticed that estimates of long hedging and of short speculation, for each commodity, do appear *implicitly* in the table. The long hedging is directly calculable by multiplying the short hedging, H_S', by the hedging ratio, r_h' and when that has been done, the short speculation is directly calculable because the total short hedging plus short speculation must equal the total of long hedging plus long speculation ($H_S' + S_S' = H_L' + S_L'$).

13 Because no large stocks of bran are ever accumulated, the maximum hedging potential of the bran futures market must be estimated on the basis of another, larger, source of reason for hedging in bran futures, namely the mills' problem of imperfect offset between unfilled flour orders and either wheat owned or long futures contracts held against them. The bran stocks hedged tend to be mainly because of bran in the form of wheat. Flour constitutes, by weight, only some 70–72% of the product of wheat milling (for some data on variation in the conversion ratio, see Working, 1927, pp. 92–7). A calculation based on unfilled flour orders reported by the Millers National Federation for Sept. 30, 1950, a date of near maximum open contracts in the futures markets for bran and shorts (the two major milling byproducts) indicates a maximum hedging potential for mill byproducts about 19 times as great as the actual volume of open contracts in bran and shorts on the date. I am indebted to Roger W. Gray for this calculation.

But it must not be hastily assumed that there might have been nearly 20 times as much hedging of bran and shorts as there was on September 30, 1950, if only there had been enough speculation in those markets to carry the hedging. Mills sell byproducts forward as well as flour, thus greatly reducing the occasion that they found for hedging byproducts. And, perhaps more important, risk reduction is not the primary reason for mill hedging in any case, but certain other advantages that are obtainable from hedging (see Working, 1953, pp. 549–54, 559) and byproduct hedging contributed little in those respects. So I judge that the amount of bran hedging might possibly have reached four or five times the volume that it did, if speculation had been attracted to the market in sufficient volume.

14 Directly so in the case of anticipatory long hedging; indirectly so for such long hedging as is done by flour mills to offset large forward flour orders placed by bakers. The bakers try to place such orders when wheat, and therefore flour, appears to be priced relatively low.

15 The appropriate formula, as shown in the appendix, is $H_L^B = (2/T - 1)H_L$, where T is the speculative index; H_L is the amount of long hedging; and H_L^B is the amount of "balancing" long hedging, that serves to carry, or "balance", an equal amount of short hedging. The reader may be helped by referring at this point to Figure 7, which illustrates how the amount of balancing between long and short hedging changes as T changes.

16 The formula for calculating the amounts of hedging that balance each other, as shown in the appendix, is $H^B = (2/T - 1)H_L$, assuming $H_S > H_L$. For the data on cotton in Figure 4,

T = 1.061.
17 The amounts of balancing speculation are given by the formula, $S^B = (1 - 1/T)S_L$.
18 The number is 25 if one includes barley, rice, tallow, and middlings, which have been dealt in on futures markets only sporadically and on a very small scale since the war.
19 Emergence of this condition led to the emergence of a new and unique sort of hedging. Soybean crushers often "hedge their processing margins", either by hedging soybean stocks in short futures contracts for soybean oil and soybean meal, in appropriate proportions, or by acquiring long soybean contracts, before stocks of beans have been accumulated, and simultaneously "selling" appropriate amounts of oil and meal contracts (Andreas, 1955, pp. 23, 24).
20 These are dates of the beginning of successful attempts to establish a futures market in each commodity. In at least one instance (potatoes) there was an earlier, but unsuccessful, attempt (Wesson, 1958, pp. 1, 2). Owing partly to the effects of price controls established during the war, and continued for a time afterward, none of the futures markets established in 1936 or subsequently attracted a large volume of either hedging or speculative open contracts until about 1949, roughly a decade ago.
21 This consideration operates to prevent the existence of futures markets in some commodities, but is rarely of importance, so far as I know, in restricting speculation in commodities for which such a market now exists. Belief that the price was subject to a degree of producer control led the Chicago Mercantile Exchange, a few years ago, to reject a proposal for establishment of a futures market in concentrated orange juice, a commodity for which producers desired hedging facilities.
22 According to the classifications of open contracts by occupation of the holder, as given in CEA reports on special surveys, onions appear to have had a good deal more anticipatory long hedging that was classed as speculation than did potatoes, and much more than eggs. Taking the onion data for October 31, 1956, as classified between speculation and hedging by the CEA (Stewart, 1941, p. 5), and eliminating matching contracts, the value of the speculative index for onions on that date was

$$T = 1 + \frac{S_S}{H_S + H_L} = 1 + \frac{587}{2979} = 1.197$$

But of the long contracts classed as speculative, 2,024 carlots were reported as held by persons or firms concerned with the growing, marketing or processing of onions (Stewart, 1941, p. 12). Of these, some 200 carlots were probably matched by offsetting short contracts in individual accounts, and I judge it possible that 1,500 carlots of the remainder might have been classifiable, on business standards, as anticipatory long hedging. Shifting the classification of 1,500 carlots from long speculative to long hedging, however, would reduce the speculative index only to $T = 1 + 587/4479 = 1.131$. This is not a suspiciously low value index for a date of near-maximum short hedging, but a fairly high value.
23 The general outcome of the foregoing appraisal of average values of the speculative index for eleven commodities is rather different from what I expected when I undertook it. The tests of reliability of the estimation procedure made for special survey dates (Table 2) had led me to expect that some average values of the speculative index derived by that procedure would have to be judged appreciably in error, owing to inability of the estimation procedure to cope adequately with the shortcomings of the original data. But I find myself unable to say confidently that the calculated average value of the speculative index is appreciably in error for any commodity except bran, and the error there seems attributable to faulty classification of the contracts classified, not to incompleteness of the official classification. Perhaps those characteristics of the data that produced clear imperfections of the estimates on individual dates are peculiarities that tend to cancel out in long-period averages.
24 This statement and the one that follows must have the words "long" and "short"

interchanged when long hedging exceeds short hedging (in accordance with the definition of the speculative index).

25 Keeping such notes was itself a remarkable performance, because he was doing a bewilderingly complicated business all the while. My debt of gratitude to him must be obvious.

26 There is much reason to believe that the principal explanation of the tendency for unskilled speculators to lose money rapidly is their naïve faith in their own judgements of what the price ought to be. They tend to pit their opinions directly against the almost certainly better "judgement of the market". If they were merely gambling, as a speculator might by merely tossing a coin to determine, initially, whether to "buy" or to 'sell", and next day, whether to hold his previously taken position or close it out, they would tend to break even, apart from commission charges. Actually, they do much worse than that; which must mean that they are using judgement, but bad judgement. The outstanding factual analysis of speculator experience (Stewart, 1941) attributes the losses of unskilled speculators largely to a tendency to "cut profits and let losses run"; but it can be demonstrated that the very real tendency that is commonly expressed in those words cannot produce an overall tendency toward loss except when combined with use of bad market judgement otherwise. To test this assertion, the reader may try to devise a trading system based on coin-tossing that would tend to produce losses. One such system would use the rule: "buy" or "sell" according to the toss of a coin, then close out the trade at the first opportunity to take either a one-cent profit or a ten-cent loss, and repeat indefinitely. The mathematical expectation for such a system is that the "trader" will break even. If that were not so, a reversal of the rule would provide an easy route to wealth.

27 One morning for example, his opening notes included the comment: "I was even and hadn't a clue, so decided to take it very easy today".

BIBLIOGRAPHY

Andreas, D., 1955, "Commodity Markets and the Processor", *Proceedings, Eighth Annual Symposium* (Chicago Board of Trade).

Chicago Board of Trade, *Annual Reports.*

Hoffman, G. W. and J. W. T. Duvel, 1941, *Grain Prices and the Futures Market* (US Department of Agriculture, Technical Bulletin, 747).

Hoos, Sidney and Holbrook Working, 1940, "Price Relations of Liverpool Wheat Futures with Special Reference to the December–March Spread", *Wheat Studies of the Food Research Institute,* November.

Irwin, H. S., 1935, "Seasonal Cycles in Aggregates of Wheat-Futures Contracts", *Journal of Political Economy,* February.

Irwin, H. S., 1954, *Evolution of Futures Trading,* (Madison, Wisconsin).

Schonberg, J. S., 1956, *The Grain Trade* (New York).

Stewart, Blair, 1941, *Trading in Wool Top Futures* (US Department of Agriculture, Circular 604, August).

Stewart, Blair, 1949, *An Analysis of Speculative Trading in Grain Futures* (US Department of Agriculture, Technical Bulletin 1001).

US Department of Agriculture, 1937, *Wheat Futures,* Statistical Bulletin 54.

US Department of Agriculture, 1947, 1955, 1957, 1958, 1959, *Commodity Futures Statistics,* Stat. Bulletins 109, 196, 221, 239, 256.

US Department of Agriculture, 1946, *Classification of Open Contracts in Egg Futures on the Chicago Mercantile Exchange*, July.

US Department of Agriculture, 1956, *Cotton Futures: Survey of Open Contracts on the New York and New Orleans Cotton Exchange*, September.

US Department of Agriculture, 1956, *Onion Futures: Survey of Open Contracts on the Chicago Mercantile Exchange*, October.

US Department of Agriculture, 1957, *Survey of Open Contracts in Wool and Wool Top Futures*, December.

US Federal Trade Commission, 1920–1926, *Report on the Grain Trade* (7 Vols.)

US Statutes, 70 Stat. 630.

Wesson, W. T., 1958, *The Economic Importance of Futures Trading in Potatoes* (US Department of Agriculture, Marketing Research Report 241).

Working, Holbrook, 1927, "Statistics of American Wheat Milling and Flour Disposition Since 1879", *Wheat Studies of the Food Research Institute*, December.

Working, Holbrook and Sydney Hoos, 1938, "Wheat Futures Prices and Trading at Liverpool Since 1886", *Wheat Studies of the Food Research Institute*, November.

Working, Holbrook, 1953, "Hedging Reconsidered", *Journal of Farm Economics*, November.

Working, Holbrook, 1954, "Whose Markets? – Evidence on Some Aspects of Futures Trading", *Journal of Marketing*, July.

Working, Holbrook, 1960, "Price Effects of Futures Trading", *Food Research Institute Studies*, February.

8

New Concepts Concerning Futures Markets and Prices

Holbrook Working[†]

Research on futures markets during the last 40 years has produced results that have required drastic revision or replacement of a great part of the previously accepted theory of futures markets and of the behaviour, not only of futures prices, but of the general class of prices that may be called "anticipatory". New light has been thrown on the behaviour of businessmen, including speculators, and on the functioning of the price system.

To say that new theory has been required poorly expresses the consequences of the research, because the main results have not been "theory" in the usual economic sense of that term. One may better follow an example set by Conant when he faced a similar problem in undertaking to explain the nature of science, and said, "... science emerges from the other progressive activities of man to the extent that new concepts arise from experiments and observations, and the new concepts in turn lead to further experiments and observations" (Conant, 1947, p. 24). The chief result of the research to be considered here has been the emergence of a series of new concepts arising from observation and statistical analysis. They are listed in Table 1, with a parallel statement of the concept that they partially or wholly displace, and their emergence is traced in succeeding sections of the paper, where the main evidence on which the new concepts rest is summarised.

Empirical research has played a leading role in the advancement of economic knowledge and understanding that is described here, but the role has been a different one than economists have ordinarily thought that such research would play in advancing their science – if economics be a

---

[*] This paper was first published in *The American Economic Review* 52(3), pp. 431–59 (1962) and is reprinted with the permission of The American Economic Association.
[†] Holbrook Working, 1895–1985.

Table 1 Concepts concerning futures markets

New concepts*	Displaced concepts
1. *Open-contract concept:* Futures markets serve primarily to facilitate contract holding (1922).	Futures markets serve primarily to facilitate buying and selling. (Disproved.)
2. *Hedging-market concept:* Futures markets depend for their existence primarily on hedging (1935; 1946).†	Futures markets depend for their existence primarily on speculation. (Disproved.)
3. *Multipurpose concept of hedging:* Hedging is done for a variety of different purposes and must be defined as the use of futures contracts as a temporary substitute for a merchandising contract, without specifying the purpose (1953).	Hedging is done solely to avoid or reduce risk. (Disproved.)
4. *Price-of-storage concept:* Storage of a commodity is a service supplied often at a price that is reflected in intertemporal price spreads, and because the holding of commodity stocks can afford also a "convenience yield", the price for storing small stocks is often negative (1933; 1949).	Storage of a commodity is a service that is supplied only in response to an assured or expected financial return equal to or greater than the cost, the latter calculable ordinarily without regard to the quantity of stocks to be stored. (Disproved.)
5. *Concept of reliably anticipatory prices:* Futures prices tend to be highly reliable estimates of what should be expected on the basis of *contemporarily available information* concerning present and probable future demand and supply; price changes are mainly appropriate market responses to changes in information on supply and demand prospects (1934; 1949).	Futures prices are highly unreliable estimates of what should be expected on the basis of existing information; their changes are largely unwarranted. (A wholly unproved inference; accumulating evidence mainly supports the new concept.)
6. *Market-balance concept:* A significant tendency for futures prices to rise during the life of each future is not uniformly present in futures markets, and when it exists is to be attributed chiefly to lack of balance in the market (1960).	Aversion to risk-taking, leading to risk premiums, produces a general tendency toward "normal backwardation" in futures markets, statistically measurable as a tendency for the price of any future to be higher in the delivery month than several months earlier.

*Dates shown in parentheses are years in which the concepts may be said to have emerged. Where two dates are shown, the earlier one is the date of publication of evidence recognised as challenging the older concept; the later date, that of first-known publication of at least the substance of the new concept.

†A statement of this concept was first prepared for publication by H.S. Irwin about 1946, but actual publication was delayed until 1954 (pp. 435–6).

science.[1] In the final section below we re-examine some prevalent views on the function of empirical research in economics, and conclude that such research can serve economics in the same way that Conant describes it as serving the natural sciences.

The practical question of economic usefulness of futures markets is only incidentally referred to in the present paper, inasmuch as our concern here is to trace the advance of economic understanding rather than to discuss the practical uses of such understanding. It will be readily observable, however, that the improved understanding contributed by each of the new concepts except the first, which bears only on the technical question of correct measurement, tends either toward refuting common criticisms of futures markets or toward indicating greater usefulness than has ordinarily been attributed to them, or both.

THE OPEN-CONTRACT CONCEPT

First of the six new concepts to emerge, and a necessary precursor of much that was to follow, was the concept that the business of futures markets should be measured primarily in terms of volume of contracts outstanding – so-called "open contracts". This was a revolutionary concept, carrying with it recognition that the traditional main function of markets, transfer of ownership, is not a significant function of a futures market. In futures markets there is little buying and selling in the usual sense, with its connotation of transfer of ownership. Instead, futures markets exist chiefly to facilitate the holding of contracts; the making and offsetting of those contracts, misleadingly called buying and selling,[2] is only incidental to the main function of such markets.

The revolutionary character of this concept was so little recognised at the outset that the concept was accepted in its original form without controversy,[3] and without explicit record of who deserves credit for the innovation. The concept emerged effectively about 1922 during consideration of what data should be collected by the newly-created Grain Futures Administration. Credit for it belongs chiefly, I think, to J. W. T. Duvel, who was shortly to become the first administrator of that federal agency, and during his long occupancy of that position gave research a prominent place in the work of that regulatory body.[4]

Directing attention to open contracts in the study of futures markets had an effect similar to that produced on the study of medicine and related science by Pasteur's discovery of bacteria; it led to study of the *causes* of phenomena, and thus toward true understanding, where previously only symptoms had been considered and understanding had been frustrated. And, incidentally, terms new to economics were required, such as "long position" and "short position" and "long hedging" and "short hedging".

The main fruits thus far of adoption of the open-contract concept have flowed from the fact that it led to quantitative study of hedging, which was obviously not significantly measurable in terms of trading statistics. It thus led directly to a new understanding of the parts played by hedging and speculation, respectively, in the origin and functioning of futures markets – a new understanding reflected in the hedging-market concept to be considered next; and it contributed significantly to emergence of the multipurpose concept of hedging. Studies of speculation have not benefited so much from the open-contract concept, partly at least because persistence of the mistaken idea that speculation is well measured by trading has considerably retarded study of speculation in terms of open contracts; but studies of the statistics of open contracts, as we shall see in the next section, have recently raised questions that bear especially on speculative behaviour.

THE HEDGING-MARKET CONCEPT

Futures markets have usually been regarded, in the past, as essentially *speculative* markets. Although they rather early won recognition as useful for hedging, their hedging use was treated as a fortunate by-product, neither necessary to the existence of such a market, nor very closely related quantitatively to the amount of speculation on the market. At Chicago, where dealings in forward contracts first took on the essential characteristics of a modern futures market, dealing in futures was initially regarded in the grain trade itself as a disreputable speculative business; for more than a decade the Chicago Board of Trade refused to allow such transactions in its quarters.[5] The opinions of economists are reflected in familiar treatments of the theory of futures markets that begin by supposing the only participants in the market to be speculators.

The first step toward a radically altered view of futures markets came with the discovery of what might justly be called Irwin's Law. After statistics of open contracts had been collected by the Grain Futures Administration for only a few years, a staff member of that agency, H. S. Irwin, noted that the volume of open contracts varied seasonally in accordance with seasonal changes in the volume of commercial stocks subject to hedging (Irwin, 1935). Irwin was not then willing to conclude that speculation tended to enter and leave a market in direct response to variation in the amount of hedging, and sought explanation otherwise.[6] Hoffman (1941, pp. 33–41) later studied additional evidence on the correspondence between commercial stocks and open futures contracts, and found it observable in year-to-year variations as well as within the year. With the aid of statistics of hedging, available for the last three years of the period that he studied, he brought to light the fact that the net amount of hedging (short minus long), which he supposed was the hedging variable logically to be considered, did not always vary in close correspondence with commercial stocks.

Presently Irwin undertook a study of the origins of futures markets in butter and eggs, taking advantage of the fact that there were then people still alive who had witnessed the early stages of emergence of futures markets in those commodities. Verbal accounts by such people, supplemented by extensive study of contemporary published information, convinced Irwin that the prime incentive to emergence of those futures markets came from hedgers rather than from speculators. This conclusion, though consistent with the statistical evidence noted above, ran contrary to an established belief, and Irwin was unable to get publication for his original paper. By the time he resorted to private publication in 1954, in expanded form (Irwin, 1954), the present writer had profited from a reading of Irwin's original manuscript in 1946, had found that his conclusion conformed with a great and varied mass of other information, and had argued that futures markets ought not to be regarded as primarily speculative, but as primarily hedging markets (Working, 1953a, pp. 318–39[7] and 1954a).

The statistical data on hedging and speculation, as they accumulated, produced evidence that the relationship was not a simple one connecting speculation with net hedging, as it had seemed natural to assume. Statistical analysis presently revealed that long hedging serves only in part to offset short hedging, and in part creates a need for short speculation. With this circumstance taken into account, the amount of speculation was shown to vary in much closer correspondence with the amount of hedging than had been evident previously. Figure 1 illustrates the correspondence found for variations through time in one market.[7a] A similarly close correspondence was found for variations between markets, using five-year averages (Working, 1960b, p. 198).

THE MULTIPURPOSE CONCEPT OF HEDGING

According to the traditional concept, hedging consists in matching one risk with an opposing risk, and hedging in futures is effective because changes in spot prices of a commodity tend to be accompanied by like changes in the futures price. The fact that hedging usually involves more than risk avoidance has long been known to hedgers themselves and to some economists. Wiese, a businessman, seems to have been the first person to criticise the traditional concept as seriously misleading (Wiese, 1952, p. 113). An attempt by Graf (1953) to test the efficacy of hedging produced the first evidence which was recognised as showing that the simple risk-avoidance concept was seriously misleading to economists (Working, 1953b, p. 544–45).

Quantitative studies, perhaps surprisingly, have contributed more to understanding of hedging than have verbal inquiries concerning business motivation. Particularly noteworthy among these were studies of the quantitative relation between hedged stockholding and market "carrying

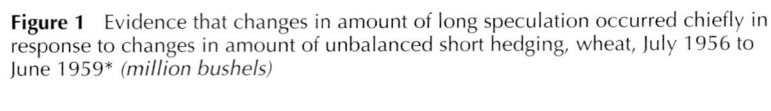

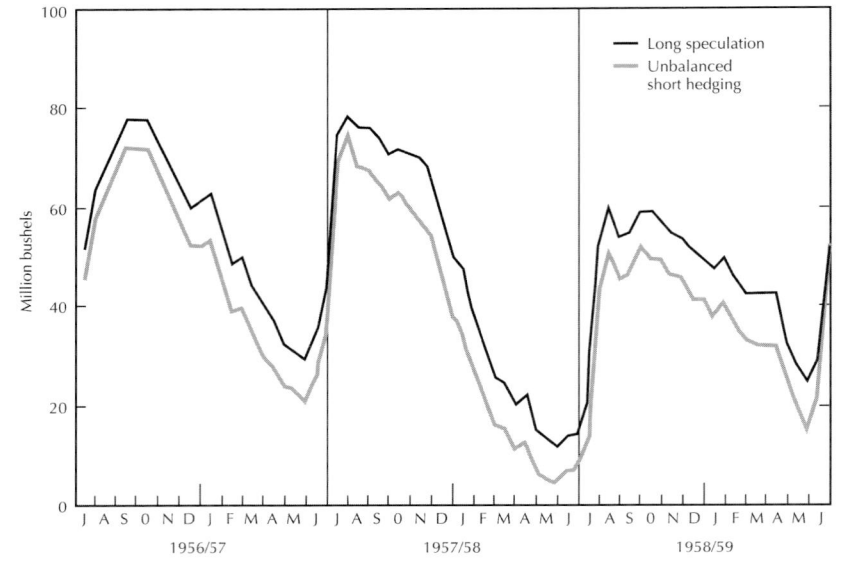

Figure 1 Evidence that changes in amount of long speculation occurred chiefly in response to changes in amount of unbalanced short hedging, wheat, July 1956 to June 1959* *(million bushels)*

*Data from Working (May 1960 p. 208) and like data for month-ends. Unbalanced short hedging is the total of open short-hedging contracts, less an appropriate fraction of open long-hedging contracts.

charges", which brought recognition that much hedging was done to assure profits, not merely to avoid risk (Working, 1953b, pp. 556–57); studies of flour-mill practice in maintaining a rough inventory balance involving the three items, physical stocks, forward merchandising contracts, and futures contracts (Working, 1953b, p. 549), which produced evidence that risk avoidance was inadequate as an explanation of flour-mill hedging;[8] and statistical inquiries intended to test the efficacy of hedging, which as noted above, produced demonstrably false conclusions explicable only on the ground that acceptance of the pure risk-avoidance concept of hedging had led to the making of a test that grossly misrepresented the effects of hedging as actually practiced. And even non-quantitative inquiry has been sufficient to establish that some hedging is practised with neither the intent nor the effect of reducing risk.

Business hedging is done for a variety of reasons, which differ according to circumstances. Consequently it is necessary to recognise the existence of several different categories of hedging and to consider them separately with respect to business purposes and economic effects.

Carrying-charge hedging is done in connection with the holding of commodity stocks for direct profit from storage (rather than merely to facilitate the operation of a producing or merchandising business). Whereas the traditional concept implies that hedging is merely a collateral operation that, in this application, would influence the stockholding only through making it a less risky business, the main effect of carrying-charge hedging is to transform the operation from one that seeks profit by anticipating changes in price level to one that seeks profit from anticipating changes in price relations. Whether a businessman regards such a transformation of the operation as desirable or not depends less on differences between inherent riskiness of the two sorts of operation than on whether he personally feels better able to predict changes in price levels or changes in price relationships. In general, producers of a commodity and those who use it as a raw material, when they undertake storage for profit, prefer to seek the profit by anticipating changes in price level, and therefore either do not hedge, or hedge selectively (see below): it is chiefly merchants, whose merchandising business requires close attention to price differences according to grade, quality, and location, who choose to seek storage profits by anticipating changes in price relations, thus using to best advantage their special knowledge concerning price relations.

Whereas the traditional hedging concept represents the hedger as thinking in terms of possible loss from his stockholding being offset by gain on the futures contracts held as a hedge, the carrying-charge hedger thinks rather in terms of change in "basis" – that is, change in the spot-future price relation. And the decision that he makes is not primarily whether to hedge or not, but whether to store or not.[9]

Operational hedging is done chiefly to facilitate operations involved in a merchandising or processing business. It normally entails the placing and "lifting" of hedges in such quick succession that expectable changes in the spot-future price relation over the interval can be largely ignored; and it is this fact which chiefly distinguishes operational hedging from carrying-charge hedging. Because the intervals over which individual operational hedges are carried tend to be short, the amount of risk reduction accomplished tends to be small – quite insufficient to explain the observed prevalence of operational hedging. Besides reducing risk, to an extent that the hedger may or may not consider significantly advantageous, it leads to economies through simplifying business decisions and allowing operations to proceed more steadily than otherwise.

Illustrations of the principal advantages of operational hedging may well be drawn from flour-milling because it was study of flour-mill hedging that first led to recognition of the special characteristics of operational hedging. In this industry buying and selling decisions are facilitated because it is easier for the mill buyer to judge prices on particular lots of wheat in terms of their relation to other wheat prices than in terms of absolute level. And similarly it is easier to judge the price offered by a potential flour buyer in relation to a present wheat price than in relation to the price that may have to be paid on a later wheat purchase. When hedging is practice, it becomes logical to make these buying and selling judgements on the easier basis. And wheat buying in particular can be carried on more steadily than otherwise because the basis is subject to less fluctuation than is the price.

These business advantages of operational hedging, however, depend on the existence of a high correlation between changes in spot prices and changes in futures prices over short intervals – day to day and even within the day. Such correlation of short-interval price changes is not always present, and in its absence there tends to be little operational hedging even though the broader correspondence between changes in spot and futures prices be close enough to permit effective risk reduction through hedging. Flour mills west of the Rocky Mountains in the United States usually do not hedge, because spot wheat prices in that area do not move in sufficiently close day-to-day correspondence with futures prices in mid-western markets to permit satisfactory operational hedging.[10]

Selective hedging is the hedging of commodity stocks under a practice of hedging or not hedging according to price expectations. Because the stocks are hedged when a price decline is expected, the purpose of the hedging is not risk avoidance, in the strict sense, but avoidance of loss. Published studies of hedging in the grain trade of the United States have indicated the presence of selectivity in hedging, through reports that hedging was done "to some extent" or "occasionally", chiefly among country elevators (US Federal Trade Commission, 1920–1926, Vol. 1, pp. 213–14 and Mehl, 1931). Efforts on the part of such small firms to gain profits from appraisal of price prospects, as is implied by selectivity in hedging, appear ordinarily ill-advised. But personal inquiry among large and well-managed firms in the grain trade has revealed that, though hedging is their standard practice in most parts of the country, they sometimes hedge incompletely. To the extent that they allow circumstances in individual instances to influence the decision whether to hedge unsold stocks or not, they hedge selectively. Outside the grain and cotton trades, which appear to be the principal ones in which routine hedging is accepted as standard practice in most parts of the country, selectivity hedging is so common as to suggest that in a considerable number of futures markets the greater part of the short hedging done may be selective.[11]

When hedging is done selectively, the advantage of the hedging to the individual firm may often (perhaps usually) be measured approximately by the amount of loss avoided directly by the hedging. Though curtailment of the amount of unsold stocks is an alternative means of restricting loss at a time of expected price decline, it is a means that few firms are able to use freely, owing to operating needs for carrying stocks. Selective hedging almost inevitably yields large advantages to any merchandising or processing firm that is able to anticipate price changes reasonably well. From an economic standpoint, selective hedging deserves appraisal as simply one aspect of the use of futures markets as a means by which handlers of a commodity increase the efficiency of their participation in the price-forming process, instead of largely withdrawing from such participation, as they do when they practice routine carrying-charge or operational hedging. Futures markets that receive a large amount of speculative hedging tend also to have a considerable amount of "speculation" by handlers of the commodity. Inasmuch as selective hedging must come chiefly from those handlers of the commodity who commonly hold substantial stocks, and who therefore take a long position by merely refraining from hedging, it is to be presumed that the "speculative" use of futures by handlers of the commodity comes mainly from dealers whose business requires relatively little holding of physical stocks.

Anticipatory hedging, which also is ordinarily guided by price expectations, differs from selective hedging in that the hedging contract is not matched by either an equivalent stock of goods or a formal merchandising commitment that it may be said to offset. It takes either of two principal forms: (a) purchase contracts in futures acquired by processors (or manufacturers) to cover raw material "requirements";[12] or (b) sales contracts in futures by producers, made in advance of the completion of production. In either of these forms the anticipatory hedge serves as a temporary substitute for a merchandising contract that will be made later. In the one case it serves as a substitute for immediate purchase of the raw material on a merchandising contract; in the other case it serves as a substitute for a forward sale of the specific goods that are in course of production. The purpose of the hedge may be said to be to take advantage of the current price; or, bearing in mind the usual availability of an alternative means of doing that, merely to gain some advantage of convenience or economy through the choice made between alternatives.

The best presently available statistics that give some indication of the prevalence of anticipatory hedging are those on long hedging, which reflect also the similarly motivated forward merchandising contracts, to the extent that they are hedged, that many processors use as a means of anticipating requirements. Nearly all long hedging consists either of anticipatory hedging or of hedging of forward merchandising contracts (often called unfilled orders), that are themselves anticipatory in the same sense as the anticipatory hedging. On the latest date (September 28, 1956), for which there are statistics of open contracts in cotton that allow accurately segregating "matching" contracts from those that were purely speculative, long hedging contracts totalled 1,001 thousand bales, as compared with 268 thousand bales of long speculative contracts (Working, 1960b, p. 192). This was at a time when the amount of short hedges to be carried was near its peak.

Pure *risk-avoidance hedging*, though unimportant or virtually non-existent in modern business practice, may have played a significant part in the early history of futures markets. In the absence of records concerning the uniformity of hedging by firms using the early futures markets, however, it is impossible to know to what extent early hedging, described as done "to reduce risk", actually had that purpose in the strict sense, or was selective hedging, done to avoid incurring loss from an expected price decline.

Recognition of the fact that hedging is done for a variety of purposes requires defining hedging otherwise than has been customary. The verb "to hedge", in the sense of avoidance or shelter, has no general connotation of avoidance especially of risk; the gambling practice of hedging bets is perhaps as often used to offset an expected loss as to reduce risk in the strict sense; and there is in any case no good reason why the word "hedging" as applied to business practices should be restricted to operations such as gamblers call hedging. For present purposes we need to define only hedging in futures. All the uses of futures that are commonly called "hedging" will be comprised, and all other uses excluded, if we characterise hedging as the use of futures contracts as a temporary substitute for a merchandising contract that is to be made later.[13]

Inclusion as hedging of the practices characterised above as selective hedging and anticipatory hedging requires either regarding hedging as sometimes closely akin to speculation,[14] or defining speculation otherwise than has been usual in economics texts. In ordinary usage and in much economic discussion the word "speculation" refers to buying and selling (or, more accurately, holding) property purely for the sake of gain from price change, and not merely as an incident to the normal conduct of a producing or merchandising business or of investment. So it is usual to distinguish between speculation in securities and the holding of securities for capital gain; many people tend to think it a strained use of the term when speculation in commodities is defined to include the holding by a farmer of part of his crop for a few weeks after harvest; and business purchasing agents object to being said to speculate when they seek to time their buying, within reasonable limits, in accordance with their judgement of price prospects. If speculation is defined in accordance with ordinary usage of the term, hedging and speculation in futures are always distinguishable.[15]

THE PRICE-OF-STORAGE CONCEPT

Though much uncertainty has existed about the relations of future prices to spot prices, one proposition long stood unquestioned: there seemed to be no generally necessary relationship between prices that depend on the abundance or scarcity of currently available supplies – spot prices for example – and prices on futures contracts applying to supplies from a subsequent harvest. To be more specific, it was regarded as clear that, in the presence of a current relative scarcity of wheat, prospects for an abundant harvest in the following summer would have no significant bearing on prices paid for existing supplies.

Statistical studies published in 1933, supplemented by further evidence later, showed this belief to be untrue.[16] What happens in fact is that any change in price of a distant, new-crop, wheat future tends to be accompanied by an equal change in prices paid for wheat from currently available supplies.[17] A comparative shortage of currently available supplies of wheat has its direct effect on the spread between the spot price and the futures price, causing wheat from current supplies to sell at a premium over the expected future price of new-crop wheat. In effect, then, the spot price is determined as the sum of the futures price, dependent primarily on expectations, plus a premium dependent on the shortage of currently available supplies.

These observed facts of price behaviour were contrary to what was logically deducible from the prevalent concept. Introduction of the concept that intertemporal price differences constitute prices of storage (Working, 1949c) provided a key to understanding the observed facts and served in effect as a summary of the facts. The most significant of these facts from the standpoint of economic theory is that spot prices of storable commodities tend to respond sensitively to changes in distant expectations. Figure 2 illustrates this tendency. The great price advance that it shows in the new-crop December corn future, owing to crop deterioration in July and August 1947, was accompanied throughout by a like advance in the price of the old-crop September future. If the traditional concept of price dependence on currently available supplies were valid, the price of the September future, comparatively high in June because of shortage of old-crop corn supplies, would have held fairly steady during June and early July, until deterioration in new-crop prospects caused the December future to rise to and above the price of the September future, and thereafter would have followed the course of the December future, but at a slightly lower level.

The main reason for listing the price-of-storage concept as marking a significant step in the advance of economic science is that it seems capable of displacing the belief that spot prices are commonly little affected by changes in distant expectations. That opinion is not founded on any factual observation, but has its basis in an assumption embedded in the

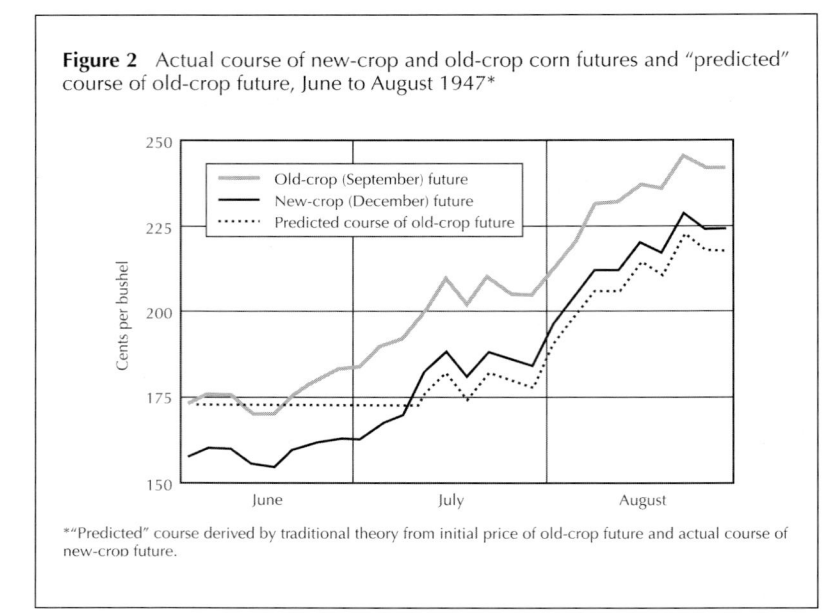

Figure 2 Actual course of new-crop and old-crop corn futures and "predicted" course of old-crop future, June to August 1947*

*"Predicted" course derived by traditional theory from initial price of old-crop future and actual course of new-crop future.

conventional exposition of price formation, an exposition that is ordinarily taken to mean that "supply", as a determinant of spot price, means currently available physical supplies. A belief so derived will not be corrected merely by citing observational evidence to the contrary,[18] but only through providing an evidently valid conceptual scheme that requires bringing expectations into account as determinants of spot prices.[19]

THE CONCEPT OF RELIABLY ANTICIPATORY PRICES

Prices on futures markets (and also spot prices of many commodities and prices on the stock market) are observed to change frequently and in an apparently erratic manner. These price changes have traditionally been conceived to be in large part random fluctuations, with some degree of cyclicality impressed upon them. Evidence of the substantial falsity of that concept has long been widely known, but the conflict between concept and evidence could be visible only to a person who was well aware of the distinction between random variation and random walk.[20] The so-called "fluctuations" of futures prices, of some other commodity prices, and of stock prices, exhibit close approximations to pure random walk.[21]

When attention was drawn to that characteristic of certain prices, about a quarter-century ago, the discovery aroused no comment and no apparent lasting interest on the part of either statisticians or economists.[22] It was a factual observation that carried no economic meaning, even to people who understood the technical distinction between random variation and random walk. But presently the economic meaning of the discovery was found: pure random walk in a futures price is the price behaviour that would result from perfect functioning of a futures market, the perfect futures market being defined as one in which the market price would constitute at all times the best estimate that could be made, from currently available information, of what the price would be at the delivery date of the futures contracts (Working, 1949b, p. 160).

The observations that the behaviour of futures prices corresponded closely to random walk thus led to the economic concept that futures prices are *reliably* anticipatory; that is, they represent close approximations to the best possible current appraisals of prospects for the future (Working, 1958). Conceiving the "fluctuations"[23] of futures prices to be mainly appropriate responses to valid changes in expectations produces a great change in thought regarding them, as compared with regarding the price movements as mainly lacking economic justification.

Further discussion of the evidence that futures prices are reliably anticipatory requires dealing with a problem of the meaning of the "reliability" in this context. Custom has established the idea that reliability of uncertain expectations is to be tested by correspondence between the expectations in the sense of correspondence between the actual expectation and what ought to be expected in the light of available information. At times, in order to avoid confusion, I shall speak of the latter as *anticipatory reliability*.

The inference, from evidence of approximate randomness of change in futures prices, that those prices have a high degree of anticipatory reliability was initially open to serious question on the ground of suspicion that the price changes were in fact much less nearly random than early statistical tests seemed to indicate. Randomness of walk in a statistical series is not a specific characteristic detectable by any one specific test, but a term that designates absence of any systematic characteristic – absence of any sort of "structure", as Cowles and Jones aptly expressed it (Cowles and Jones, 1937). Because the kinds of structure commonly believed present in futures prices and in stock prices are trends and cycles (in a somewhat loose usage of those terms), those kinds of structure were the first looked for as evidence of nonrandomness; and the statistical tests used were familiar ones that had been proved appropriate for revealing structure of those sorts in other data, namely, tests for the presence of simple autocorrelation.[24] Failure of those tests to reveal an appreciable degree of structure in futures prices, however, left open the question whether other sorts of structure might not be present.

Two principal lines of inquiry have been followed in the search for kinds of structure that would not be revealed by simple autocorrelation analysis, or equivalent methods. One line of inquiry has reasoned that the price movements to be correlated should be selected according to magnitude of the movement, without regard to the time taken to accomplish a movement of that amount. By proceeding thus, Alexander (1961) has recently shown the presence of an appreciable tendency for stock-price movements of 5% or more (in an index of industrial stocks) to be followed by further price movement in the same direction.[25] Houthakker (1961, p. 166) has published results that, though calculated and presented from a point of view different from that of Alexander, can be interpreted as indicating presence of the same sort of tendency in prices of wheat and corn futures that Alexander found in the stock-price index.

Another line of attack involved the design of a new type of statistical test that appeared capable of revealing any such tendency toward continuity of movement in futures prices, if it were present, and also capable of revealing a tendency for a price movement in one direction to be followed by an opposite movement (reaction), but without uniformity in timing of the reaction, which might sometimes be prompt and sometimes considerably delayed. A test devised for this purpose promptly yielded statistically significant evidence of structure in futures prices (Brinegar, 1954) but further research was required to determine the kind of structure that was indicated.[26] The principal component of the structure was presently identified as such as would be produced by a tendency for the price effects of new market information to be partially dispersed over a considerable time interval rather than all concentrated within the day on which the new information became available (Working, 1956 and 1958). Later it was found possible to determine approximately the fraction of the average price effect that was so dispersed and the average time span of dispersal, and also to derive similar approximations for a secondary sort of structure, involving excessive price movement on individual days, with reaction dispersed on the average over a period of a few adjacent days (Larson, 1960). The latter, rather weak, tendency was found to be associated with relatively small price movements, and the former tendency, mainly with the larger price movements.

The results produced by the two lines of inquiry summarised above appear at present to be entirely consistent. Those from the first approach are the more readily interpretable, up to a certain point. Those from the second approach lend themselves more readily to appraisal of the degree of anticipatory reliability that can be attributed to futures prices, or to such other prices as deserve similar appraisal. The most notable feature of the evidence from these two lines of inquiry, however, is not quantitative but qualitative. Each has indicated that the main kind of structure present in anticipatory (speculative) prices is roughly the opposite of that which has been chiefly charged against such prices. Whereas these prices commonly have been thought to "fluctuate" too much, showing large movements up or down that tended to be followed by reaction, in a roughly wavelike pattern, the evidence is that reaction is common only from very small movements, of short duration, which may be regarded as caused by minor accidental disturbances; that in the main the imperfection of behaviour in futures markets takes the form of retardation of price response to information that warrants price change. This retardation, being an effect similar to that of shock-absorbers on an automobile, tends toward the avoidance of excessive initial movement and subsequent reaction.

The concept that futures prices are reliably anticipatory, in the full sense of that term, evidently does not correspond wholly to the facts. In that respect it is like the concept that gases are perfectly elastic, as represented by Boyle's Law, $VP = C$. If the economic concept of anticipatory reliability of futures prices is subject to a good deal more supplementary qualification than is the physical concept of perfect elasticity of gases, it appears nevertheless to correspond more closely to full reality than does any other concept that might be used for fairly simple representation of the main facts. It avoids the gross misrepresentation involved in the concept that the variations in futures prices are in large part unwarranted, wave-like, fluctuations; and its own misrepresentation is of a sort that is inevitable in a concept that, for the sake of simplicity, must fall short of representing the full facts.

RISK PREMIUMS vs. THE MARKET-BALANCE CONCEPT

The concept that risk-bearing commands a reward has been applied at several related points in the theory of futures markets, but has recently attracted attention principally in connection with discussion of J. M. Keynes' "theory of normal backwardation". The market-balance concept, recently proposed by Gray (1960 and 1961) as a substitute for the risk-premium concept in that context, has been advanced on the basis of evidence that may be thought inconclusive,[27] and has been applied only to explaining differences in seasonal trends[28] of futures prices in different markets – not obviously a matter of great importance. The reasons for nevertheless discussing the market-balance concept in the company of the better-proved and more demonstrably fruitful concepts discussed previously are, first, that the grounds for considering the merits of the market-balance concept are stronger than appears from the statistical evidence alone; and second, that there is evident need for taking other considerations into account, along with risk, at several points in the theory of futures markets where the tendency has been to consider only the risk aspect.

The earlier studies that sought to test Keynes' theory of normal backwardation took seasonal trend in futures prices to be a measure of risk premium (as Keynes apparently had done), and focused on the question whether the statistics gave evidence of a tendency to such seasonal trend or not. Telser at first denied that they did (Telser, 1958), whereas Houthakker (1957 and 1961) and Cootner (1960) found evidence of seasonal trends consistent with the Keynesian hypothesis. Gray, meanwhile, had been making studies of some of the smaller futures markets, which had previously received little attention from economists, and was thus led to put together the pertinent evidence for a considerable number of different markets, large and small. The diversity in magnitude, and even in direction, of apparent seasonal trend that he found among the various markets appeared to him not explainable on the basis of differences between risk premiums in different markets (Gray, 1961, p. 258), whereas the differences did appear explainable on other grounds that could be comprised by the term "market balance". The term was suggested by the evidence that the differences in seasonal trend were largely associated with differences between markets in amount of speculation relative to hedging – or, more strictly, relative to the potential amounts of hedging, inasmuch as a shortage of speculation in a futures market must restrict the amount of hedging actually done.

The market-balance concept thus emerged because examination of a wider range of observational data than had been considered before posed a new question for answer. Viewed as bearing only on the original question – whether futures prices tend to show an upward seasonal trend – the new data could be interpreted as supporting an affirmative answer, in agreement with the general tenor of previous evidence; but the new data posed also the question of how to account for the occurrence of seasonal trends that in some markets and circumstances appeared to be significantly positive, and in others, significantly negative. The answer that was suggested, in the form of the market-balance concept, does not require rejecting the risk-premium concept, but regarding it as providing at best only a partial explanation of the tendencies to seasonal trend observable in futures prices.

Speculators' earnings have received significant direct study only by Stewart (1949), and his inquiry, based on detailed records for nearly 9,000 speculators in grain futures, and covering a time period of nine years, was singularly unproductive of positive conclusions. Such unproductiveness in a large-scale investigation by an able economist skilled in statistical analysis and with a wealth of data at his command must indicate need for reconsidering prevalent economic ideas about the bases of speculative profits, which Stewart sought to explore. The "sample" of speculators that he dealt with showed a strong over-all tendency for them to incur losses, and though many among them gained profits, Stewart was unable to find an objective characteristic by which to classify speculators into groups with appreciably different profit experience.

If speculation be regarded as a skilled occupation (not mere risk-taking), in which the making of profits depends both on special skills and on continuously maintained knowledge of current economic conditions and prospects relevant to prices of particular commodities, Stewart's results become understandable. He had no information by which to classify speculators according to special speculative skills, or according to the amount and quality of their pertinent knowledge; hence his study could throw no direct light on the relations of speculative rewards either to special speculative skills or to amounts of pertinent economic information that they used.[29]

In order that speculation should contribute to economically desirable price formation, speculators must keep well informed concerning the pertinent economic facts, and must be able to appraise those facts properly. The evidence considered earlier that futures prices are reliably anticipatory, to a fairly high degree, indicates that a large proportion of speculation in futures markets has been based on sound information, properly appraised. Such speculation deserves to command a wage, and that fact is ignored when the returns from speculation are viewed solely as a reward for risk-bearing.

Data published by the Commodity Exchange Authority in recent years concerning some of the smaller futures markets have revealed that much of what is classed by the CEA as speculation in them comes from persons and firms dealing in the commodity, such as are ordinarily expected to be hedgers.[30] If one accepts the argument that speculation should be defined as suggested earlier (endnote 15) this raises the question whether a considerable part of the contract holdings that have been classed as "speculative" in those markets, might not more properly be treated as anticipatory hedging. In any case the evidence reveals that, for those commodities, the futures markets have served in important degree to allow dealers in the commodity to exercise their price-forming function more freely than they otherwise could. In the absence of a futures market, price formation is almost wholly in the hands of dealers (using that term to comprise all handlers of the commodity); and many of them must exercise that function under the handicap that their need to accumulate stocks in order to maintain their position as merchandisers may force them to act in support of a price that they regard as too high, or obstacles to stock accumulation at a given time may bar effective market expression of an opinion that the current price is unduly low.[31] The presence of a futures market permits dealers to exercise price judgement without such restraints. The dealer who must accumulate stocks at a certain time for merchandising reasons may sell futures in equivalent amount and thus avoid exerting a price-supporting influence when he expects a price decline; and the dealer who expects a price advance, but cannot well accumulate physical stocks of the commodity at the time, can buy and hold futures instead.

Figure 3

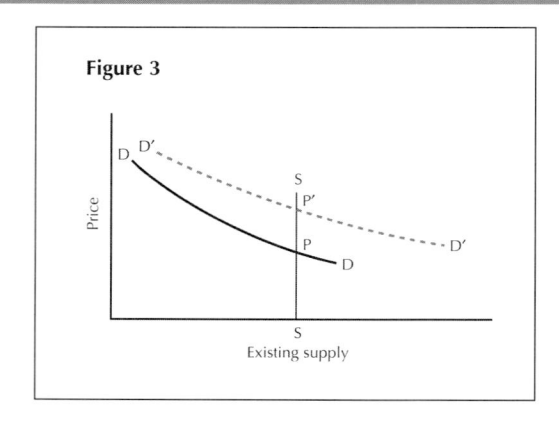

[Figure 3: Price vs. Existing supply. Demand curves DD and D'D' intersecting vertical supply line SS at points P and P' respectively.]

Thus the introduction of a futures market could result in substantially improved price behaviour even though the futures market were used only by dealers in the commodity, the better price behaviour being attributable to the opportunity given dealers to exert price influence strictly in accordance with price judgement instead of exerting price influence, or refraining from doing so, largely in response to immediate merchandising or processing needs. Existence of a futures market may tend toward better support of prices during rapid marketing of a crop shortly after harvest, not primarily because speculators demand a lower risk premium than do dealers (if that be true), but chiefly because the futures market allows stocks to be "carried" either as physical stocks held unhedged or through the holding of long futures contracts that offset short hedging. This tends to change the balance of the market in the sense of the market-balance concept.

Because potential carriers of stocks have differing opinions regarding the price at which they can afford to buy for carrying, there exists at any time a demand curve for stock-carrying such as is represented by *DD* in Figure 3. Introducing the opportunity for stocks to be carried through the holding of futures contracts tends to move this curve rightward to a position such as *D'D'*, giving an intersection at *P'* with the vertical, representing total stocks to be carried, at a higher level than before. Such a rightward shift clearly tends to occur if the futures market attracts many speculators who would not otherwise deal in the commodity, and it might occur also if the futures market were patronised only by dealers in the commodity.[32]

The conclusion, considered in an earlier section, that hedging is done for a variety of reasons, not merely to reduce risk, carried no necessary implication that speculation should be regarded as other than pure risk-bearing; but here we have seen further evidence that futures markets are viewed much too narrowly when they are regarded as serving merely to allow transfer of risk from hedgers or speculators: the risk-premium concept has appeared inadequate, by itself, to account for observed differences in seasonal trends of futures prices. Direct consideration of returns to speculators has produced evidence that seems to require regarding the returns as mainly a wage, often negative, that varies greatly according to ability and knowledge of the speculator. And information that has become available concerning futures markets in which much of the hedging is selective, and much of the "speculation" comes from persons who deal in the commodity otherwise, suggests that they serve in large part as means by which dealers take risks discriminatingly, and make their price judgements bear more effectively on price formation.

THE PLACE OF EMPIRICAL RESEARCH IN ECONOMICS

Around 1920 and subsequently, when economics was turning strongly toward doing more empirical research, it was common for economic theorists to say that the function of empirical research was to verify theory. One may still sometimes hear that opinion expressed. It is an opinion that has its foundations in the philosophy of science that was elaborated by John Stuart Mill (1900, esp. Bk. 3, Ch. 11; Bk. 6, Ch. 9) or else in reasoning similar to that of Mill. The feature of Mill's philosophy that gave grounds for that view had been widely recognised as fallacious before the end of the 19th century, and Marshall, in his *Principles*, recorded dissent from the view that economics is peculiarly a deductive science (1922, p. 29). Economic theorists have in fact rarely shown a strong feeling of need for empirical verification of theory. More typical of their attitude, at least prior to recent years, has been the aphorism, "If statistics conflict with theory, so much the worse for the statistics".

Empirical research has long been recognised as serving to clothe theory in a body of realistic circumstance that is necessary, or at least advantageous, for the practical application of theory to specific economic problems. But that use of empirical research is concerned with applied economics, not with advancing economics as a science. Though Marshall rejected Mill's classification of the sciences according to method, which had permitted regarding economics as peculiarly a deductive science with correspondingly limited opportunity for effective use of empirical research, Marshall accepted the 19th century view that the accomplishments of a science are to be measured by the laws of nature that it has discovered (see below); and he failed to discriminate between practical art and science. Predicting the tides "at London Bridge or at Gloucester", to which Marshall likened the tasks of economics (1922, p. 32), is not an undertaking that advances science.

A third view, which has had much influence in guiding empirical research on the part of economists aiming to advance economics as a science, stems from the opinion that the chief shortcoming of economics has been a failure to quantify its laws. One of the grounds for that view has been the recognised value in natural science of quantitative investigation. The research that we have been considering here offers many examples of the values of quantitative work in economics also. But the argument for quantifying economic theory depends more particularly on acceptance of the view, prevalent in natural science as well as in economics until about the beginning of the 20th century, that the substance of science is the body of "laws of nature" that it has discovered. Marshall's statement that, "A science progresses by increasing the number and exactness of its laws ..." (Marshall, 1922, p. 30) is a representative expression of the view. The turn of thinking in the natural sciences toward the conception of science that Conant has expressed particularly well may have been started by observation that the Darwinian concept of evolution continued to stand and to mark a great advance in biological science even though Darwin's "laws" of evolution became in considerable part discredited. In physical science, the results of the Michelson–Morley experiment (1887) posed an evidently severe conceptual problem, and those results were followed by a series of other discoveries that focused attention similarly on the role of concepts in science (Jeans, 1948, Ch. 8).

Quantifying economic theory can produce clear advantages for applications of theory to practical economic and business problems, but it is not so clear that quantifying received theory leads often to advancing understanding of economic phenomena. In the research considered here, the main useful results of attempts to quantify received theory have come through demonstration that the theory was fallacious – fallacious because of having started from wrong assumptions. The extreme example of such demonstration arose from an attempt to quantify the theory that price relations between old-crop and new-crop futures depended on expectations regarding the size of the new crop, which proved wholly untrue. After the price-of-storage concept had provided the basis for a new theory, quantification of the price-of-storage function involved in it produced some significant new information in the initial case, most fully revealed in (Working, 1953b, p. 556), where changes in the function from period to period are shown. Thereafter, though numerous additional quantifications of the function in specific instances could be useful for purposes of practical application, additional quantifications seemed not to offer promise of yielding further advances in understanding.[33]

Economics has been forced to go through a difficult process of learning how to make effective use of observation as a continuing means of advance, after having become accustomed to rely principally on deduction from a body of observational information that remained comparatively static because it was limited largely to what could be learned from "common observation" and from introspection. The difficulty of the learning process has been enhanced by a tradition in economics of accepting problems of evident practical importance as the ones at which research should be directed. Economists have been slow to recognise the error in Mill's attempt to explain the relatively primitive state of economics as a science on the ground that, "In every department of human affairs, practice long precedes Science: systematic inquiry into the modes of action of the powers of nature, is the tardy product of a long course of efforts to use those powers for practical ends" (Mill, 1898, p. 1).

Natural science did not develop out of practice. That route led to the empirical accomplishments of Watt, Burbank and Edison, not to the accomplishments of Newton, Darwin, Mendel or Heisenberg. Science has advanced by attacking manageable problems of acquiring understanding of observed phenomena with little or no immediate regard to evident practical importance of the problems. Seeking support for their efforts, scientists have often laid stress on the practical usefulness of scientific advance. Not infrequently they have chosen to push a particular line of inquiry because of hope that it would serve such a practical end as unlocking the mystery of cancer. But the route of science toward such an end is always that of acquiring information that promises to yield greater understanding. For example, the research economist with a scientific viewpoint who may be considering the study of futures markets does not ask himself whether futures markets are economically important institutions, concerning which conclusions are needed for practical purposes; he asks instead whether study of futures markets offers favourable opportunity for gaining understanding of the behaviour of businessmen, including speculators, and of the behaviour of prices of the class that we have called anticipatory. He makes his decision on principles like those that govern the decision of a geneticist considering the study of fruit flies.

The place of empirical inquiry in economic science, it appears to me, is that of making observations, statistical or otherwise, and conducting analysis such that one may be able to say, paraphrasing Conant, that in economics new concepts arise from observation and analysis, and the new concepts in turn lead to further observation and analysis.

1 Most natural scientists including Conant (1947, pp. 26–7), have felt during at least the past half-century that economics did not fully qualify as a science. Conant (1947, pp. 118–19) reserved judgement on the possibilities of economics becoming a science.

2 Futures contracts, used chiefly as purely financial rather than as merchandising contracts, are executed under terms that provide explicitly for settlement by financial transfer rather than by any transfer of commodity ownership (though allowing the latter as an alternative means of settlement). If the execution of such contracts went under an appropriately distinctive name instead of being called "buying and selling", people would not so often think it a perversion of sound business practice that transfer of commodity ownership

occurs only infrequently under those contracts. In fact, it is excessive forcing of settlement by transfer of commodity ownership, in connection with corners and squeezes, that is a perversion of the use of futures contracts.

3 This was probably because of the evident significance of open contracts for the theory of hedging, and because initial acceptance of the importance of open contracts did not require granting open contracts a central position in consideration of futures markets, such as they were eventually to command, but only granting them attention. Collection of the data on open contracts was indeed opposed by those who sought to avoid the expense entailed in supplying data, and even more vigorously opposed on the ground that the statistics of open contracts revealed information that, it was alleged, should not be made public. The latter contention, it should be noted, implied recognition of the pertinence of the data in study of market functioning.

4 The decision to collect statistics of open contracts is recorded in (US Department of Agriculture, 1923, p. 691). Preparation of the Federal Trade Commission's monumental *Report on the Grain Trade* was then in progress and economists of the Commission, promptly adopting the idea that open contracts should be studied, undertook compilation of open-contract data for some earlier years (US Federal Trade Commission, 1920-26, Vol. 7, pp. 8–89, 124ff). W. H. S. Stevens, who, with Francis Walker, was in general charge of the Federal Trade Commission inquiry, has told me that G. P. Watkins, of the Commission Staff, may have been responsible for the idea that statistics of open contracts should be collected; but the fact that Watkins, instead of presenting an argument of his own for use of open-contract statistics, quotes Duvel on the subject (US Federal Trade Commission, 1920–26, Vol. 7, p. 134) leads me to think that principal credit belongs to Duvel.

5 Irwin (1954, pp. 69ff.) dates futures trading in Chicago from near the beginning of the 1850s, and my reading of Taylor's *History* (1917) leads me to a similar conclusion. Dealings in futures were first admitted to the Chicago Board of Trade in 1865 (Taylor, 1917, Vol. 1, p. 331).

6 His suggestion was that the seasonal increase in amount of speculative buying of wheat at harvest time "arises chiefly from the wheat-growing sections tributary to each market, and that it is done largely when the wheat is sold", and he was able to cite some evidence in support of the hypothesis (Irwin, 1935, p. 45); but the hypothesis subsequently proved inadequate to account for the observed facts.

7 Inadvertent omission from that paper of due credit to Irwin was corrected in (Working, 1955).

7a Further research seeking interpretation of the evidence in Figure 1 has led the author to two conclusions that deserve to be noted, though they cannot be elaborated, here. The estimation procedure used to arrive at the data for the figure assumes fairly complete reporting of hedging, both short and long, and in the absence of that condition tends to produce a spurious degree of stability in the ratio between estimated amounts of long speculation and of short hedging, hence to exaggerate the closeness of response of speculation to hedging. The condition for reliability of the estimates in this respect is probably better met for wheat than for any other commodity, and is met best in periods when the amount of short hedging is large. Secondly, theoretical study of the mechanism by which one may suppose that the adjustment between amounts of speculation and of hedging is accomplished seems to make the concept of incomplete balancing between short and long hedging appropriate theoretically only in connection with comparatively long-time adjustments; its use in the accompanying figure to exhibit evidence on short-time adjustment is then to be regarded as only a graphic device. Statistical analysis of the data on lines that now seem theoretically appropriate have shown a degree of correspondence between estimated amounts of speculation and of hedging broadly similar to that indicated in the figure above.

8 This led presently to at least a partial explanation of the ineffectiveness of verbal inquiry

into the motivation of hedging: after conclusions concerning the reasons for mill hedging had been published, a mill manager, asked why millers say that they hedge to avoid risk instead of giving the real reasons, promptly replied, "Because it is hard to explain the real reasons to a person who doesn't understand the milling business".

9 In consequence, the term "carrying-charge hedging" tends to be used sometimes to designate the combined operation of storage and hedging for profit. In trade usage the combined operation is sometimes called "earning the carrying charge". It is an operation not properly divisible into two parts, for without the hedging the storage becomes a quite different operation, from the business point of view, requiring use of different information to conduct it successfully, and leading often to a different choice of the intervals over which storage is undertaken.

10 Mills in the area tributary to Seattle make little use of the alternative of hedging in the Seattle futures market because that is a very "thin" market, with consequent high costs of hedging.

11 The prevalence of selectivity in hedging on the smaller futures markets has at least two possible explanations. One, attributing smallness of the futures market to lack of speculative interest in the commodity, would explain the prevalence of selective hedging on the ground that the market lacks enough speculation to support routine hedging by any large proportion of potential hedgers, thus discouraging the development of much operational hedging and making storage for profit from expected price increases, with hedging when price decrease is expected, seem more profitable than carrying-charge hedging.

Another possibility is that the smaller futures markets have as little business as they do because of circumstances apart from the futures market that lead potential hedgers to prefer to rely on price judgements rather than on judgements concerning spot-future price relationships. Recognition of the multiple purposes of hedging should open the way to research that would give better evidence than we now have to explain why some futures markets are as little used as they are.

12 Such use of futures contracts was first given legal status as hedging under the Commodity Exchange Act by an amendment enacted in 1956 (70 Stat. 630).

13 This definition was originally given with slightly different wording in Working (1953b, p. 560).

14 For administrative purposes of distinguishing between anticipatory hedging of raw material requirements and speculation, the Commodity Exchange Act, as amended in 1956, restricts such anticipatory hedging to "... an amount of such commodity the purchase of which for future delivery shall not exceed such person's unfilled anticipated requirements for processing or manufacturing during a specified operating period not in excess of one year: *provided*, that such purchase is made and liquidated in an orderly manner and in accordance with sound commercial practice ..." In effect this provision draws the same line between hedging and speculation as does the definition of speculation given in the next footnote.

15 For this purpose, speculation in commodities may be defined as the holding of a net long or a net short position, for gain, and not as a normal incident to operating a producing, merchandising, or processing business. The reference to net position is this definition, proposed in (Working, 1960b, p. 187), serves to exclude arbitrage from speculation. Though the two have something in common, they need to be distinguished, and that is done more conveniently by defining speculation to exclude arbitrage than by adopting a new term to mean "speculation other than arbitrage".

16 See especially Working (1933, pp. 188–203 and 1934a, pp. 199–205). These studies, dealing with prices in a single exporting country, were later supplemented by a similar inquiry into price behaviour in the major import market for the same commodity (Hoos and Working, 1940). Comparisons of behaviour between these two widely differing sorts of markets helped to indicate which behaviour characteristics were special to a particular sort of

market and which were more generally observable. In both markets prices of the near future responded sensitively to distant price expectations, but the influences that bore particularly on the price of the near future, causing its price movement to diverge more or less from that of the distant future, were appreciably different in the two markets (Hoos and Working, 1940, esp. pp. 102–5, 110–26, 136–38).

17 This statement is true also for most other commodities that have futures markets, but is inapplicable to prices of a commodity, like potatoes or onions, that is incapable of economical interseasonal storage. Gray (1960, pp. 305ff.) has noted that it seems not to apply to coffee in the United States. Interpretation of his evidence requires recognition that the "price of storage", reflected in the price difference between futures for different dates, is dependent on the *expected* level of stocks during the interval between those dates, and that such expectations regarding future domestic stocks of an imported commodity may be affected by some of the same circumstances that influence the price of a distant future.

18 One evidence of this fact appears in the short-lived controversy that developed between the present author and Roland P. Vaile (Working, 1948; Vaile, 1948; Working, 1949a).

19 Two recent expositions by leading economists (Samuelson, 1957 and Houthakker, 1959) that may seem to ignore or reject the price-of-storage concept were not so intended. They sought to present a theory of price formation under conditions of certainty, leaving the effects of uncertainty to be treated in later elaboration of the theory, and assumed that under conditions of certainty, stocks would be carried only in the expectation of a direct return for storage equal to the "costs" of storage, the latter treated as being invariant with respect to the quantity to be stored. In fact, however, the observed willingness of dealers and processors to carry stocks of wheat, for example, at the end of a crop year for a negative price of storage arises only in small part from the presence of uncertainty. Each of three other considerations has more influence, namely: costs of abnormal acceleration of the movement of new-crop wheat from harvest field to mill rollers; temporal dispersion of harvest times; and geographical dispersion of flour mills among areas with different harvest times.

If convenience of exposition is well served by proceeding in two stages, the first of which assumes that stocks will fall to zero in the absence of an expected return from storage equal to the full "costs" of storage, the simplifying assumptions involved at the first stage should include instantaneous harvest of the entire crop, and costs of movement from producer to consumer that are invariant with respect to rate of movement. The supposed advantage of the two-stage exposition, with price of storage ignored at the first stage, will be found illusory, I think, when economists have fully assimilated the price-of-storage concept.

20 A familiar illustration of random variation is the variation in number of spots that turn up when a pair of "true" dice is thrown after being thoroughly shaken each time. By proceeding from that illustration, one may illustrate random walk by drawing on a figure a line that moves forward one space for each throw of the pair of dice, and up or down by a number of units equal to the number of spots minus seven (seven being the "expected number of spots on each throw). Randomness, in the statistical sense, however, does not necessarily involve absence of known causes for the events designated as random, but only that individual events are unpredictable from knowledge of previous events in the series. Thus prices may change from known causes, but if the changes are unpredictable from any knowledge of previous prices and price changes, as the changes in the curve drawn on the basis of dice-throwing are unpredictable from the previous course of the curve, the price series exhibits random walk.

21 This appears to be true of commodity prices only within time spans not longer than about one year, but true of stock prices over much longer time spans.

22 The two early papers on the subject were complementary in nature. Mine (1934b) stressed the difference between random variation and random walk, and presented a 2400-term random-walk series as a device both for illustrating the close conformity of actual price

series with the random-walk model and for testing such evidences of departure from that model as might be observed. Cowles and Jones (1937) gave a great amount of statistical evidence in the course of exploring the question whether appreciable non-randomness of stock-price changes might be found at some difference-intervals though not at others. See also (Cowles, 1960), in which he gave revised results that eliminated the effects of inadvertent use in (Cowles and Jones, 1937) of one index number series that had employed undesirable averaging. Kendall (1953), much later and in apparent ignorance of earlier work, provided a test of the hypothesis that nonrandomness of change might be revealed by serial correlations of orders higher than the first, using a constant difference-interval. The only substantial indication of significant serial correlation that Kendall found was in a first-order coefficient for cotton prices; and that, as shown in (Working, 1960c), was a correlation attributable to averaging in the construction of the cotton price series used, hence invalid as evidence that cotton prices showed any departure from random walk.

23 The quotation marks are repeated here as a reminder that the price changes referred to do not have a dominantly wave-like character such as tends to be implied by calling them fluctuations.

24 By "simple" autocorrelation I mean autocorrelation in the sense in which that term has been commonly understood, namely correlation between items uniformly spaced in time, or in the case of correlation of first differences, differences taken over uniform time intervals, uniformly spaced. It would be reasonable to regard any form of structure in a time series as involving some form of autocorrelation.

25 Price moves of 5% were the smallest ones considered by Alexander. In an earlier work of mine, I applied the same principle as did Alexander to study of very small price movements of futures, and found in the smallest movements, of which there may be several hundred within a day, a substantial tendency to negative (rather than positive) correlation (Working, 1954b, p. 11). Like Alexander, I found that the evidence of structure that was revealed by considering movements of given size tended to become invisible under consideration of movements over given time intervals.

Analysis of correlation by movement size is laborious (my work along that line was done before electronic computers became widely available), and partly for that reason I turned to another approach, described below, that seemed to hold as good, or perhaps better, possibilities of uncovering the presence of obscure sorts of structure in price "fluctuations".

26 The difficulty in identifying the kind of structure indicated arose partly from inherent difficulty in predicting how a new sort of test, designed to be responsive to a variety of different sorts of structure, would respond to particular sorts of structure; but that inherent difficulty was complicated by the fact that the kind of structure indicated was of an unexpected sort.

27 The purely statistical evidence in (Gray, 1961, p. 257), for example, did not include a specific test of the hypothesis that the observed variation among apparent seasonal trends might be explainable as consistent with existence of a uniform tendency toward positive seasonal trend in all the markets and periods considered.

28 I speak of "seasonal trends" rather than "bias," which has been the term more commonly used, because it designates what is actually measured, and then interpreted as evidence of bias; and I mean the term to comprise trends measured with respect to some pertinent seasonal variable such as Cootner (1960, pp. 400–1) has suggested should be used, as well as trends measured with respect to calendar dates.

29 The over-all tendency toward losses by the speculators in Stewart's sample must be attributed chiefly to lack of full representativeness of the sample; it included relatively few, if any, professional speculators, and may have been unrepresentative otherwise.

30 Such data appear in (Stewart, 1941, p. 22), and subsequently in a number of CEA "surveys" for several markets and dates.

31 Particularly clear examples are afforded by the onion market, which is one of those in which a large proportion of the futures contracts classed as speculative was found to be held by dealers in the commodity (Working, 1960a, p. 7). Onions from the later summer crop, which are stored for use over a period extending into April, are sold by growers to "country shippers", who store them. In order to maintain his competitive position as a dealer, the country shipper must buy when growers want to sell, and he is thus severely limited in opportunity to adjust his stock-holding in accordance with expectations of price change. Country shippers sell in turn to "dealers" in the large cities, who cannot economically accumulate large stocks because refrigerated storage, on which they would ordinarily need to rely, is much more expensive than "common storage" in the cold climates where most of the onions are produced. Consequently the city dealer, who may be better informed about the nationwide supply and demand situation for onions than most country shippers, and financially better able to carry risks of price change, is handicapped in expressing his price opinions through adjustment of stock-holdings.

32 The effect in the case of a futures market used only by dealers is not necessarily to move the curve to the right because DD, representing amounts of stock that dealers are willing to hold unhedged in the absence of opportunity for hedging, cannot be taken to represent also the amounts that they would be willing to hold unhedged, at the same prices, in the presence of opportunity for hedging. The possibility that such a rightward shift might occur even in the case of a futures market used only by dealers was suggested to me by the great apparent effect of the onion futures market in supporting the post-harvest price level for that commodity, which seemed difficult to explain solely on the ground of price support given by pure speculators (persons not connected with the industry) in view of the fact that, according to the available information, such speculators accounted for only about half of the long "speculative" holdings of onion futures at the times when such support was principally needed (Working, 1960a, p. 7).

There is a general tendency for performance of the storage function for any seasonally produced commodity to concentrate largely at some one stage of the marketing process. The concentration is most commonly at terminal markets, but it may be at country points, as for onions, or in the hands of processors of the commodity, as appears to be the case with rubber; but in any case, such concentration tends to restrict opportunity for exerting price influence through the holding of physical stocks mainly to those handlers of the commodity who operate at that stage. Existence of a futures market, allowing the many other handlers of the commodity to readily "carry" stocks by holding futures contracts against hedged stocks, would tend therefore to produce such a rightward shift as is illustrated in Figure 3, even though only handlers of the commodity used the futures market.

33 The further quantifications provided in (Brennan, 1958) and (Telser, 1958) may nevertheless be regarded as useful for corroboration of the original findings.

BIBLIOGRAPHY

Alexander, S. S., 1961, "Price Movements in Speculative Markets: Trends or Random Walks", *Industrial Management Review*, May, pp. 7–26.

Brennan, M. J., 1958, "The Supply of Storage", *American Economic Review* 47, March, pp. 50–72.

Brinegar, C. S., 1954, "A Statistical Analysis of Speculative Price Behavior", Phd dissertation (Stanford University).

Conant, J. B., 1947, *On Understanding Science* (New Haven).

Cootner, P. H., 1960, "Returns on Speculators: Telser versus Keynes", *Journal of Political Economy* 48, August, pp. 396–404 & 415–18.

Cowles, A. and H. E. Jones, 1937, "Some *a Posteriori* Probabilities in Stock Market Action", *Econometrica* 5, July, pp. 280–94.

Cowles, A., 1960, "A Revision of Previous Conclusions Regarding Stock Price Behavior", *Econometrica* 28, October, pp. 909–15.

Graf, T. F., 1953, "Hedging – How Effective Is It?" *Journal of Farm Economics* 35, August, pp. 398–413.

Gray, R. W., November 1960, "The Characteristic Bias in Some Thin Futures Markets", *Food Research Institute Studies* 1, November, pp. 296–312.

Gray, R. W., 1961, "The Search for a Risk Premium", *Journal of Political Economy* 69, June, pp. 250–60.

Hoffman, G. W., 1941, Grain Prices and the Futures Market: USDA. Technical Bulletin, 747, January.

Hoos, S. and H. Working, 1940, "Price Relations of Liverpool Wheat Futures", *Wheat Studies* 17, (Food Research Institute, Stanford), November, pp. 101–43.

Houthakker, H. S., 1957, "Can Speculators Forecast Prices?' *Review of Economics and Statistics* 39, pp. 143–51.

Houthakker, H. S., 1959, "The Scope and Limits of Futures Trading", in Moses Abramovitz *et al.*, *The Allocation of Economic Resources*, Stanford.

Houthakker, H. S., 1961, "Systematic and Random Elements in Short-Term Price Movements", *American Economic Review* 51, May, pp. 164–72.

Irwin, H. S., 1935, "Seasonal Cycles in Aggregates of Wheat-Futures Contracts", *Journal of Political Economy* 43, pp. 34–49.

Irwin, H. S., 1954, *Evolution of Futures Trading* (Madison, Wisconsin).

Jeans, J., 1948, *The Growth of Physical Science* (New York).

Kendall, M. G., 1953, "The Analysis of Economic Time Series", *Journal of Royal Statistical Society* 116, pp. 11–34.

Larson, A. B., 1960, "Measurement of a Random Process in Futures Prices", *Food Research Institute*, November, pp. 313–24.

Marshall, A., 1922, *Principles of Economics*, Eighth edition (London).

Mehl, J. M., 1931, Hedging in Grain Futures: USDA, Circular 151, June.

Mill, J. S., 1900, *Principles of Political Economy*, Revised edition (New York).

Mill, J. S., 1898, *A System of Logic*, People's edition (London).

Samuelson, P. A., 1957, "Intertemporal Price Equilibrium: A Prologue to the Theory of Speculation", *Weltwirtschaft. Archiv.* 79, December, pp. 181–221.

Stewart, B., 1941, Trading in Wool Top Futures: USDA. Circular 604, August.

Stewart, B., 1949, An Analysis of Speculative Trading in Grain Futures: USDA. Technical Bulletin 1001, October.

Taylor, C. H., 1917, *History of the Board of Trade of the City of Chicago* (Chicago).

Telser, L. G., 1958, "Futures Trading and the Storage of Cotton and Wheat", *Journal of Political Economy* 66, June, pp. 233–55; reprinted as Chapter 6 of the present volume.

Telser, L. G., 1960, "Rejoinder", *Journal of Political Economy* 68, pp. 404–15.

Vaile, R. S., 1948, "Inverse Carrying Charges in Futures Markets", *Journal of Farm Economics* 30, August, pp. 574–75.

Wiese, V. A., 1952, "Use of Commodity Exchanges by Local Grain Marketing Organizations," *Proceedings, Chicago Board of Trade Annual Symposium*, September, pp. 108–16.

Working, H., 1933, "Price Relations Between July and September Wheat Futures at Chicago Since 1885", *Wheat Studies*, (Food Research Institute, Stanford) 9, March, pp. 187–238.

Working, H., 1934a, "Price Relations Between May and New-Crop Wheat Futures at Chicago Since 1885", *Wheat Studies* (Food Research Institute, Stanford) 10, February, pp. 183–228.

Working, H., 1934b, "A Random-Difference Series for Use in the Analysis of Time Series", *Journal of American Statistical Association* 29, March, pp. 11–24.

Working, H., 1948, "Theory of the Inverse Carrying Charge in Futures Markets", *Journal of Farm Economics* 30, February, pp. 1–28; reprinted as Chapter 4 of the present volume.

Working, H., 1949a, "Professor Vaile and the Theory of Inverse Carrying Charges", *Journal of Farm Economics* 31, February, pp. 168–72; reprinted as Chapter 5 of the present volume.

Working, H., 1949b, "The Investigation of Economic Expectations", *American Economic Review, Proc.*, 39, pp. 150–66.

Working, H., 1949c, "The Theory of Price of Storage", *American Economic Review* 39, December, pp. 1254–62.

Working, H., 1953a, "Futures Trading and Hedging", *American Economic Review* 43, pp. 314–43.

Working, H., 1953b, "Hedging Reconsidered", *Journal of Farm Economics* 35, November, pp. 544–61.

Working, H., 1954a, "Whose Markets? Evidence on Some Aspects of Futures Trading", *Journal of Marketing* 29, July, pp. 1–11.

Working, H., 1954b, "Price Effects of Scalping and Day Trading", *Proceedings of Chicago Board of Trade Annual Symposium* September, pp. 114–39.

Working, H., 1955, "Review of Irwin, Evolution of Futures Trading", *Journal of Farm Economics* 37, May, pp. 377–80.

Working, H., 1956, "New Ideas and Methods for Price Research", *Journal of Farm Economics* 38, December, pp. 1427–36.

Working, H., 1958, "A Theory of Anticipatory Prices", *American Economic Review Proc.* 48, May, pp. 188–99.

Working, H., 1960a, "Price Effects of Futures Trading", *Food Research Institute Studies* 1, May, pp. 3–31.

Working, H., 1960b, "Speculation on Hedging Markets", *Food Research Institute Studies* 1, February, pp. 185–220; reprinted as Chapter 7 of the present volume.

Working, H., 1960c, "Note on the Correlation of First Differences of Averages in a Random Chain", *Econometrica* 28, October, pp. 916–18.

US Department of Agriculture, 1923, *Annual Report Department of Agriculture* (Washington DC).

US Federal Trade Commission, 1920–1926, *Report on the Grain Trade*, 7 volumes (Washington DC).

9

*The Pricing of Commodity Contracts**
Fischer Black†

The contract price on a forward contract stays fixed for the life of the contract, while a futures contract is rewritten every day. The value of a futures contract is zero at the start of each day. The expected change in the futures price satisfies a formula like the capital asset pricing model. If changes in the futures price are independent of the return on the market, the futures price is the expected spot price. The futures market is unique in the guidance it provides for producers, distributors, and users of commodities. Using assumptions like those used in deriving the original option formula, we find formulas for the values of forward contracts and commodity options in terms of the futures price and other variables.

The market for contracts related to commodities is not widely understood. Futures contracts and forward contracts are often thought to be identical, and many people don't know about the existence of commodity options. One of the aims of this paper is to clarify the meaning of each of these contracts.[1]

The spot price of a commodity is the price at which it can be bought or sold for immediate delivery. We will write p for the spot price, or $p(t)$ for the spot price at time t.

The spot price of an agricultural commodity tends to have a seasonal pattern: it is high just before a harvest and low just after a harvest. The spot price of a commodity such as gold, however, fluctuates more randomly.

Predictable patterns in the movement of the spot price do not generally imply profit opportunities. The spot price can rise steadily at any rate

---

*This paper was first published in the *Journal of Financial Economics* 3, pp. 167–179 (1976) and is reprinted with the permission of Elsevier Science.
†Fischer Black, 1938–1995.

lower than the storage cost for the commodity (including interest) without giving rise to a profit opportunity for those with empty storage facilities. The spot price can fall during a harvest period without giving rise to a profit opportunity for growers, so long as it is costly to accelerate the harvest.

The futures price of a commodity is the price at which one can agree to buy or sell it at a given time in the future without putting up any money now. We will write x for the futures price, or $x(t, t^*)$ for the futures price at time t for a transaction that will occur at time t^*.

For example, suppose that it is possible today to enter into a contract to buy gold six months from now at US$160 an ounce, without either party to the contract being compensated by the other. Both parties may put up collateral to guarantee their ability to fulfil the contract, but if the futures price remains at US$160 an ounce for the next six months, the collateral will not be touched. If the contract is left unchanged for six months, then the gold and the money will change hands at that time. In this situation, we say that the six month futures price of gold is US$160 an ounce.

The futures price is very much like the odds on a sports bet. If the odds on a particular baseball game between Boston and Chicago are 2:1 in favour of Boston, and if we ignore the bookie's profit, then a person who bets on Chicago wins US$2 or loses US$1. No money changes hands until after the game. The odds adjust to balance the demand for bets on Chicago and the demand for bets on Boston. At 2:1, balance occurs if twice as many bets are placed on Boston as on Chicago.

Similarly, the futures price adjusts to balance demand to buy the commodity in the future with demand to sell the commodity in the future. Whenever a contract is opened, there is someone on each side. The person who agrees to buy is long the commodity, and the person who agrees to sell is short. This means that when we add up all positions in contracts of this kind, and count short positions as negative, we always come out with zero. The total long interest in commodity contracts of any type must equal the total short interest.

When the two times that specify a futures price are equal, the futures price must equal the spot price.

$$x(t, t) \equiv p(t) \qquad (1)$$

Expression (1) holds for all times t. For example, it says that the May futures price will be equal to the May spot price in May, and the September futures price will be equal to the September spot price in September.

Now let us define the three kinds of commodity contracts: forward contracts, futures contract, and option contracts. Roughly speaking, a forward contract is a contract to buy or sell at a price that stays fixed for the life of the contract; a futures contract is settled every day and rewritten at the new futures price; and an option contract can be exercised by the holder when it matures, if it has not been closed out earlier.

We will write v for the value of a forward contract, u for the value of a futures contract, and w for the value of an option contract. Each of these values will depend on the current futures price $x(t, t^*)$ with the same transaction time t^* as the contract, and on the current time t, as well as on other variables. So we will write $v(x, t)$, $u(x, t)$, and $w(x, t)$. The value of the short side of any contract will be just the negative of the value of the long side. So we will treat v, u, and w as the values of a forward contract to buy, a long futures contract and an option to buy.

The value of a forward contract depends also on the price c at which the commodity will be bought, and the time t^* at which the transaction will take place. We will sometimes write $v(x, t, c, t^*)$ for the value of a long forward contract. From a discussion above, we know that the futures price is that price at which a forward contract has a current value of zero. We can write this condition as

$$v(x, t, x, t^*) \equiv 0 \qquad (2)$$

In effect, (2) says that the value of a forward contract when it is initiated is always zero. When it is initiated, the contract price c is always equal to the current futures price $x(t, t^*)$.

Increasing the futures price increases the value of a long forward contract, and decreasing the futures price decreases the value of the contract. Thus we have

$$v(x, t, c, t^*) > 0, \quad x > c$$
$$v(x, t, c, t^*) < 0, \quad x < c \qquad (3)$$

The value of a forward contract may be either positive or negative.

When the time comes for the transaction to take place, the value of the forward contract will be equal to the spot price minus the contract price. But by (1), the futures price $x(t, t^*)$ will be equal to the spot price at that time. Thus the value of the forward contract will be the forward price minus the spot price.

$$v(x, t^*, c, t^*) = x - c \qquad (4)$$

Later we will use (4) as the main boundary condition for a differential equation describing the value of a forward contract.

The difference between a futures contract and a forward contract is that the futures contract is rewritten every day with a new contract practice equal to the corresponding futures price. A futures contract is like a series of forward contracts. Each day, yesterday's contract is settled, and today's contract is written with a contract price equal to the futures price with the same maturity as the futures contract.

Equation (2) shows that the value of a forward contract with a contract price equal to the futures price is zero. Thus the value of a futures contract is reset to zero every day. If the investor has made money, he will be given his gains immediately. If he has lost money, he will have to pay his losses immediately. Thus we have

$$u(x, t) \equiv 0 \qquad (5)$$

Technically, (5) applies only to the end of the day, after the futures contract has been rewritten. During the day, the futures contract may have a positive or negative value, and its value will be equal to the value of the corresponding forward contract.

Note that the futures price and the value of a futures contract are not at all the same thing. The futures price refers to a transaction at times t^* and is never zero. The value of a futures contract refers to time t and is always zero (at the end of the day).

In the organised US futures markets, both parties to a futures contract must post collateral with a broker. This helps to ensure that the losing party each day will have funds available to pay the winning party. The amount of collateral required varies from broker to broker.

The form in which the collateral can be posted also varies from broker to broker. Most brokers allow the collateral to take the form of Treasury Bills or marginable securities if the amount exceeds a certain minimum. The brokers encourage cash collateral, however, because they earn the interest on customers' cash balances.

The value of a futures customer's account with a broker is entirely the value of his collateral (at the end of the day). The value of his futures contracts is zero. The value of the collateral posted to ensure performance of a futures contract is not the value of the contract.

As futures contracts are settled each day, the value of each customer's collateral is adjusted. When the futures price goes up, those with long positions have money added to their collateral, and those with short positions have money taken away from their collateral. If a customer at any time has more collateral than his broker requires, he may withdraw the excess. If he has less than his broker requires, he will have to put up additional collateral immediately.

Commodity options have a bad image in the US, because they were recently used to defraud investors of many millions of dollars. There are no organised commodity options markets in this country. In the UK, however, commodity options have a long and relatively respectable history.

A commodity option is an option to buy a fixed quantity of a specified commodity at a fixed time in the future and at a specified price. It differs from a security option in that it can't be exercised before the fixed future date. Thus it is a "European option" rather than an "American option".

A commodity option differs from a forward contract because the holder of the option can choose whether or not he wants to buy the commodity at the specified price. With a forward contract, he has no choice: he must buy it, even if the spot price at the time of the transaction is lower than the price he pays.

At maturity, the value of a commodity option is the spot price minus the contract price, if that is positive, or zero. Writing c^* for the exercise price of the option, and noting that the futures price equals the spot price at maturity, we have

$$w(x, t^*) = x - c^* \quad x \geqq c^*$$
$$= 0 \quad x < c^* \quad (6)$$

Expression (6) looks like the expression for the value of a security option at maturity as a function of the security price.

THE BEHAVIOUR OF THE FUTURES PRICE

Changes in the futures price for a given commodity at a given maturity give rise to gains and losses for investors with long or short positions in the corresponding futures contracts. An investor with a position in the futures market is bearing risk even though the value of his position at the end of each day is zero. His position may also have a positive or negative expected dollar return, even though his investment in the position is zero.

Since his investment is zero, it is not possible to talk about the percentage or fractional return on the investor's position in the futures market. Both his risk and his expected return must be defined in dollar terms.

In deriving expressions for the behaviour of the futures price, we will assume that taxes are zero. However, tax factors will generally affect the behaviour of the futures price. There are two peculiarities in the tax laws that make them important.

First, the IRS assumes that a gain or loss on a futures contract is realised only when the contract is closed out. The IRS does not recognise, for tax purposes, the fact that a futures contract is effectively settled and rewritten every day. This makes possible strategies for deferring the taxation of capital gains. For example, the investor can open a number of different contracts, both long and short. The contracts that develop losses are closed out early, and are replaced with different contracts so that the long and short positions stay balanced. The contracts that develop gains are allowed to run unrealised into the next tax year. In the next year, the process can be repeated. Whether this process is likely to be profitable depends on the special factors affecting each investor, including the size of the transaction costs he pays.

Second, the IRS treats a gain or loss on a long futures position that is closed out more than six months after it is opened as a long-term capital gain or loss, while it treats a gain or loss on a short futures position as a short-term capital gain or loss no matter how long the position is left open. Thus if the investor opens both long and short contracts, and if he realises losses on the short contracts and gains on the long contracts, he can convert short-term gains (from other transactions) into long-term gains. Again, whether this makes sense for a particular investor will depend on his transaction costs and other factors.

However, we will assume that the capital asset pricing model applies at each instant of time.[2] This means that investors will be compensated only for bearing risk that cannot be diversified away. If the risk in a futures contract is independent of the risk of changes in value of all assets taken together, then investors will not have to be paid for taking that risk. In effect, they don't have to take the risk because they can diversify it away.

The usual capital asset pricing formula is

$$E(\tilde{R}_i) - R = \beta_i[E(\tilde{R}_m) - R] \tag{7}$$

In this expression, $\tilde{R}_i$ is the return on asset i, expressed as a fraction of its initial value; R is the return on short-term interest-bearing securities; and $\tilde{R}_m$ is the return on the market portfolio of all assets taken together. The coefficient β_i is a measure of the extent to which the risk of asset i cannot be diversified away. It is defined by

$$\beta_i = \text{cov}(\tilde{R}_i, \tilde{R}_m)/\text{var}(\tilde{R}_m) \tag{8}$$

The market portfolio referred to above includes corporate securities, personal assets such as real estate, and assets held by non-corporate businesses. To the extent that stocks of commodities are held by corporations, they are implicitly included in the market portfolio. To the extent that they are held by individuals and non-corporate businesses, they are explicitly included in the market portfolio. This market portfolio cannot be observed, of course. It is a theoretical construct.

Commodity contracts, however, are not included in the market portfolio. Commodity contracts are pure bets, in that there is a short position for every long position. So when we are taking all assets together, futures contracts, forward contracts, and commodity options all net out to zero.

Equation (7) cannot be applied directly to a futures contract, because the initial value of the contract is zero. So we will rewrite the equation so that it applies to dollar returns rather than percentage returns.

Let us assume that asset i has no dividends or other distributions over the period. Then its fractional return is its end-of-period price minus its start-of-period price, divided by its start-of-period price. Writing P_{i0} for the start-of-period price of asset i, writing $\tilde{P}_{i1}$ for its end-of-period price, and substituting from (8), we can rewrite (7) as

$$E\{(\tilde{P}_{i1} - P_{i0})/P_{i0}\} - R = [\text{cov}\,\{(\tilde{P}_{i1} - P_{i0})/P_{i0}, \tilde{R}_m\}/\text{var}\,(\tilde{R}_m)][E(\tilde{R}_m) - R] \quad (9)$$

Multiplying through by P_{i0}, we get an expression for the expected dollar return on an asset,

$$E(\tilde{P}_{i1} - P_{i0}) - RP_{i0} = [\text{cov}\,(\tilde{P}_{i1} - P_{i0}, \tilde{R}_m)/\text{var}\,(\tilde{R}_m)][E(\tilde{R}_m) - R] \quad (10)$$

The start-of-period value of a futures contract is zero, so we get P_{i0} equal to zero. The end-of-period value of a futures contract, before the contract is rewritten and its value set to zero, is the change in the futures price over the period. In practice, commodity exchanges set daily limits which constrain the reported change in the futures price and the daily gains and losses of traders. We will assume that these limits do not exist. So we set $\tilde{P}_{i1}$ equal to $\Delta \tilde{P}$, the change in the futures price over the period,

$$E(\Delta \tilde{P}) = [\text{cov}\,(\Delta \tilde{P}, \tilde{R}_m)/\text{var}\,(\tilde{R}_m)][E(\tilde{R}_m) - R] \quad (11)$$

In effect, we have applied expression (10) to a futures contract, and have come up with expression (11), which refers to the change in the futures price. For the rest of this section, we can forget about futures contracts and work only with the futures price.

Writing β^* for the first factor on the right-hand side of (11), we have

$$E(\Delta \tilde{P}) = \beta^*[E(\tilde{R}_m) - R] \quad (12)$$

Expression (12) says that the expected change in the futures price is proportional to the "dollar beta" of the futures price. If the covariance of the change in the futures price with the return on the market portfolio is zero, then the expected change in the futures price will be zero,[3]

$$E(\Delta \tilde{P}) = 0 \quad \text{when} \quad \text{cov}\,(\Delta \tilde{P}, \tilde{R}_m) = 0 \quad (13)$$

Expressions (12) and (13) say that the expected change in the futures price can be positive, zero, or negative. It would be very surprising if the β^* of a futures price were exactly zero, but it may be approximately zero for many commodities. For these commodities, neither those with long futures positions nor those with short futures have significantly positive expected dollar returns.

FUTURES PRICES AND SPOT PRICES

When (13) holds at all points in time, the expected change in the futures price will always be zero. This means that the expected futures price at any time t' in the future, where t' is between the current time t and the transaction time t^*, will be equal to the current futures price. The mean of the distribution of possible futures prices at time t' will be the current futures price.[4]

But the futures price at time t^* is the spot price at time t^*, from expression (1). So the mean of the distribution of possible spot prices at time t^* will be the current futures price, when (13) always holds.

Even when (13) doesn't hold, we may still be able to use (12) to estimate the mean of the distribution of possible spot prices at time t^*. To use (12), though, we need to know β^* at each point in time between t and t^*, and we need to know $E(\tilde{R}_m) - R$.

A farmer may not want to know the mean of the distribution of possible spot prices at time t^*. He may be interested in the discounted value of the distribution of possible spot prices. In fact, it seems plausible that he can make his investment decisions as if β^* were zero, even if it is not zero. He can assume that the β^* is zero, and that the futures price is the expected spot price.

To see why this is so, note that he can hedge his investments by taking a short position in the futures market. By taking the right position in the futures market, he can make the β of his overall position zero. Assuming that the farmer is not concerned about risk that can be diversified away, he should make the same investment decisions whether or not he actually takes offsetting positions in the futures market.

In fact, futures prices provide a wealth of valuable information for those who produce, store, and use commodities. Looking at futures prices for various transaction months, participants in this market can decide on the best times to plant, harvest, buy for storage, sell from storage, or process the commodity. A change in a futures price at time t is related to changes in the anticipated distribution of spot prices at time t^*. It is not directly related to changes in the spot price at time t. In practice, however, changes in spot prices and changes in futures prices will often be highly correlated.

Both spot prices and futures prices are affected by general shifts in the cost of producing the commodity, and by general shifts in the demand for the commodity. These are probably the most important factors affecting commodity prices. But an event like the arrival of a prime producing season for the commodity will cause the spot price to fall, without having any predictable effect on the futures price.

Change in commodity prices are also affected by such factors as the interest rate, the cost of storing the commodity, and the β of the commodity itself.[5] These factors may affect both the spot price and the futures price, but in different ways.

Commodity holdings are assets that form part of investors' portfolios, either directly or indirectly. The returns on such assets must be defined to include such things as the saving to a user of commodities from not running out in the middle of a production run, or the benefit to anyone storing the commodity of having stocks on hand when there is an unusual surge in demand. The returns on commodity holdings must be defined net of all storage costs, including deterioration, theft and insurance premiums. When the returns on commodity holdings are defined in this way, they should obey the capital asset pricing model, as expressed by (7), like any other asset. If the β of the commodity is zero, as given in (7), then we would expect the β* of a futures contract to be approximately zero too, as given in (12). And vice versa.

The notion that commodity holdings are priced like other assets means that investors who own commodities are able to diversify away that part of the risk that can be diversified away. One way this can happen is through futures markets: those who own commodities can take short positions and those who hold diversified portfolios of assets can include long positions in commodity contracts.

But there are other ways that the risk in commodity holdings can be largely diversified away. The most common way for risk to be spread is through a corporation. The risk of a corporation's business or assets is passed on to the holders of the corporation's liabilities, especially its stockholders. The stockholders have, or could have, well diversified portfolios of which this stock is only a small part.

Thus if stocks of a commodity are held by a corporation, there will normally be no need for the risk to be spread through the futures market. (There are special cases, however, such as where the corporation has lots of debt outstanding and the lenders insist that the commodity risk be hedged through the futures market). There are corporations at every stage in a commodity's life cycle: production, distribution and processing. Even agricultural commodities are generally produced by corporations these days, though the stock may be closely held. Any of these corporate entities can take title to the stocks of commodities, no matter where they are located, and thus spread the risk to those who are in the best position to bear it. For example, canners of tomatoes often buy a farmer's crop before the vines are planted. They may even supply the vines.

This means that a futures market does not have a unique role in the allocation of risk. Corporations in the commodity business play the same role. Which kind of market is best for this role depends on the specifics of such things as transaction costs and taxes in each individual case. It seems clear that corporations do a better job for most commodities, because organised futures markets don't even exist for most commodities. Where they do exist, most of the risk is still transferred through corporations rather than through futures markets.

Thus there is no reason to believe that the existence of a futures market has any predictable effect on the path of the spot price over time. It is primarily the storage of a commodity that reduces fluctuations in its price over time. Storage will occur whether or not there is any way of transferring risk. If there were no way to transfer risk, the price of a seasonal commodity might be somewhat higher before the prime production periods than it is now. But since there are good ways to transfer risk without using the futures market, even this benefit of futures markets is minimal.

I believe that futures markets exist because in some situations they provide an inexpensive way to transfer risk, and because many people both in the business and out like to gamble on commodity prices. Neither of these counts as a major benefit to society. The big benefit from futures markets is the side effect: the fact that participants in the futures markets can make production, storage, and processing decisions by looking at the pattern of futures prices, even if they don't take positions in that market.

This, of course, assumes that futures markets are efficient. It assumes that futures prices incorporate all available information about the future spot price of a commodity. It assumes that investors act quickly on any information they receive, so that the price reacts quickly to the arrival of the information. So quickly that individual traders find it very difficult to make money consistently by trading on information.

THE PRICING OF FORWARD CONTRACTS AND COMMODITY OPTIONS

We have already discussed the pricing of futures contracts and the behaviour of futures prices. In order to derive formulas for the other kinds of commodity contracts, we must make a few more assumptions.

First, let us assume that the fractional change in the futures price over any interval is distributed log-normally, with a known variance rate s^2. The derivations would go through with little change if we assumed that the variance rate is a known function of the time between t and t^*, but we will assume that the variance rate is constant.

Second, let us assume that all of the parameters of the capital asset pricing model, including the expected return on the market, the variance of the return on the market, and the short-term interest rate, are constant through time.

Third, let us continue to assume that taxes and transaction costs are zero.

Under these assumptions, it makes sense to write the value of a commodity contract only as a function of the corresponding futures price and time. If we did not assume the parameters of the capital asset pricing model were constant, then the value of a commodity contract might also depend on those parameters. Implicitly, of course, the value of the contract still depends on the transaction price and the transaction time.

Now let us use the same procedure that led to the formula for an option on a security.[6] We can create a riskless hedge by taking a long position in the option and a short position in the futures contract with the same transaction date. Since the value of a futures contract is always zero, the equity in this position is just the value of the option.

The size of the short position in the futures contract that makes the combined position riskless is the derivative of $w(x, t)$ with respect to x, which we will write w_1. Thus the change in the value of the hedged position over the time interval Δt is

$$\Delta w - w_1 \Delta x \tag{14}$$

Expanding Δw, and noting that the return on the hedge must be at the instantaneous riskless rate r, we have the differential equation[7]

$$w_2 = rw - \tfrac{1}{2} s^2 x^2 w_{11} \tag{15}$$

Note that this is like the differential equation for an option on a security, but with one term missing. The term is missing because the value of a futures contract is zero, while the value of a security is positive.

The main boundary condition for this equation is expression (6).[8] Using standard methods to solve (15) and (16), we obtain the following formula for the value of a commodity option:

$$w(x, t) = e^{r(t-t^*)} [x N(d_1) - c^* N(d_2)]$$

$$d_1 = \left[\ln \frac{x}{c^*} + \frac{s^2}{2} (t^* - t) \right] \bigg/ s\sqrt{(t^* - t)}$$

$$d_2 = \left[\ln \frac{x}{c^*} - \frac{s^2}{2} (t^* - t) \right] \bigg/ s\sqrt{(t^* - t)} \tag{16}$$

This formula can be obtained from the original option formula by substituting $xe^{r(t-t^*)}$ for x everywhere in the original formula.[9] It is the same as the value of an option on a security that pays a continuous dividend at a rate equal to the stock price times the interest rate, when the option can only be exercised at maturity.[10] Again, this happens because the investment in a futures contract is zero, so an interest rate factor drops out of the formula.

Equation (16) applies to a "European" commodity option, that can only be exercised at maturity. If the commodity option can be exercised before maturity, the problem of finding its value becomes much more complex.[11] Among other things, its value will depend on the spot price and on futures prices with various transaction dates before the option expires.

Equation (16) also assumes that taxes are zero. But if commodity options are taxed like security options, then there will be substantial tax benefits for high tax bracket investors who write commodity options.[12] These benefits may be passed on in part or in full to buyers of commodity options in the form of lower prices. So taxes may reduce the values of commodity options.

Compared with the formula for a commodity option, the formula for the value of a forward contract is very simple. The differential equation it must satisfy is the same. Substituting $v(x, t)$ for $w(x, t)$ in (15), we have

$$v_2 = rv - \tfrac{1}{2}s^2 x^2 v_{11} \tag{17}$$

The main boundary condition is (4), which we can rewrite as

$$v(x, t^*) = x - c \tag{18}$$

The solution to (17) and (18) plus the implicit boundary conditions is

$$v(x, t) = (x - c)\,e^{r(t - t^*)} \tag{19}$$

Expression (19) says that the value of a forward contract is the difference between the futures price and the forward contract price, discounted to the present at the short-term interest rate. It is independent of any measure of risk. It does not depend on the variance rate of the fractional change in the futures price or on the covariance rate between the change in the futures price and the return on the market.

1 For an introduction to commodity markets, see Chicago Board of Trade (1973).
2 For an introduction to the capital asset pricing model, see Jensen (1972). The behaviour of futures prices in a model of capital market equilibrium was first discussed by Dusak (1973).
3 In the data she analysed on wheat, corn, and soybean futures, Dusak (1973) found covariances that were close to zero.
4 The question of the relation between the futures price and the expected spot price is discussed under somewhat different assumptions by Cootner (1960a, 1960b) and Telser (1960).

5 Some of the factors affecting changes in the spot price are discussed by Brennan (1958) and Telser (1958).
6 The original option formula was derived by Black and Scholes (1973). Further results were obtained by Merton (1973).
7 For the details of this expansion, see Black and Scholes (1973, p. 642 or p. 646).
8 Another boundary condition and a regularity condition are needed to make the solution to (15) and (16) unique. The boundary condition is $w(0, t) = 0$. The need for these additional conditions was not noted in Black and Scholes (1973).
9 Thorp (1973) obtains the same formula for a similar problem, related to the value of a security option when an investor who sells the underlying stock short does not receive interest on the proceeds of the short sale.
10 Merton (1973) discusses the valuation of options on dividend-paying securities. The formula he obtains (f. 62) should be (16), but he forgets to substitute $xe^{r(t-t^*)}$ for x in d_1 and d_2.
11 See Merton (1973) for a discussion of some of the complexities in finding a value for an option that can be exercised early.
12 For a discussion of tax factors in the pricing of options, see Black (1975).

BIBLIOGRAPHY

Black, F., 1975, "Fact and fantasy in the use of options", *Financial Analysts Journal* 31, July/August.

Black, F. and M. Scholes, 1973, "The pricing of options and corporate liabilities", *Journal of Political Economy* 81, May/June, pp. 637–54.

Brennan, M. J., 1958, "The supply of storage", *American Economic Review* 48, March, pp. 50–72.

Chicago Board of Trade, 1973, "Commodity trading manual" (Chicago: Board of Trade of the City of Chicago).

Cootner, P. H., 1960a, "Returns to speculators: Telser versus Keynes", *Journal of Political Economy* 68, August, pp. 396–404.

Cootner, P. H., 1960b, "Rejoinder", *Journal of Political Economy* 68, August, pp. 415–18.

Dusak, K., 1973, "Futures trading and investor returns: An investigation of commodity market risk premiums", *Journal of Political Economy* 81, November/December, pp. 1387–406; reprinted as Chapter 24 of the present volume.

Jensen, M. C., 1972, "Capital markets: Theory and evidence", *Bell Journal of Economics and Management Science* 3, Autumn, pp. 357–98.

Merton, R. C., 1973, "The theory of rational option pricing", *Bell Journal of Economics and Management Science* 4, Spring, pp. 141–83.

Telser, L., 1958, "Futures trading and the storage of cotton and wheat", *Journal of Political Economy* 66, June, pp. 233–55; reprinted as Chapter 6 of the present volume.

Telser, L., 1960, "Returns to speculators: Telser versus Keynes, Reply", *Journal of Political Economy* 67, August, pp. 404–15.

Thorp, E., 1973, "Extensions of the Black–Scholes options model", *Bulletin of the International Statistical Institute, Proceedings of the 39th Session*, pp. 522–29.

3

THE RATIONALE, STRUCTURE AND PERFORMANCE OF FUTURES MARKETS

Introduction
Lester G. Telser

The first chapter in this section by Abbott Payson Usher, best known for his famous book *The History of Mechanical Invention* (1929), is an undeservedly overlooked contribution to the history of organised markets. Not only does it give a concise account of the evolution of important financial instruments from bills of lading to warehouse receipts, but it also explains why these are important for the development of organised markets. The standard description of hedging focuses on the short hedge; Usher is refreshingly different – he explains the long hedge. He also explains the distinction between a squeeze and a corner, and correctly describes the circumstances conducive to each. I especially wish to emphasise his contention that corners are far more likely to occur within an unorganised rather than organised exchange.

In the second chapter, I show that the formal theory of futures markets corresponds one-to-one with the formal theory of money. Because the abstract of this article explains the purpose and results, a summary here is unnecessary. This chapter develops certain themes more briefly introduced in the fourth chapter on the costs and benefits of organised futures markets by Higinbotham and Telser.

A clearinghouse is to an organised futures market as a central bank is to commercial banks. This is one of the main correspondences between the theory of futures markets and monetary theory. Therefore, the third chapter by James T. Moser, a hitherto unpublished comprehensive history of clearinghouses for organised futures markets, describes evidence essential to support this claim of correspondence between the two theories. The clearinghouse rule, marking futures contracts to the market on a nearly continuous basis, permits Black to obtain his formula for pricing options on commodities. Futures contracts acquire more fungibility and, therefore, liquidity if the clearinghouse stands behind each transaction. A futures contract is to a forward contract as cash is to a bank cheque. Payment by cash does not depend on the credit standing of the buyer, but a seller of something will not accept a bank cheque in payment without assurances that there are enough funds in the bank account to cover the cheque. Similarly, a futures contract in an organised market with complete clearing, in Moser's usage, is a fungible instrument backed by the clearinghouse. In contrast, the validity of a forward contract hinges on the reliability of the parties to it. A third person would not accept a forward contract unless they believe it will almost certainly be honoured. Anybody with a serious interest in futures trading should read this history of the evolution of the clearinghouse.

The fourth chapter by Harlow N. Higinbotham and Telser is a theoretical and empirical study that explains the conditions conducive to the establishment of an organised futures market in a commodity or financial instrument. This topic is important both to students of futures markets and to those who seek promising new candidates for futures trading. The basic theory has its roots in monetary theory. There are close analogies between moving from barter to exchange of commodities for money and moving from trade among a small circle of merchants, who know each other, to the development of organised futures markets. This article furnishes important evidence about the relationship between the size of a margin and the size of a limit price move set by a futures exchange. Note 2 states there is a positive correlation of 0.812 between these two variables for the 17 most actively traded commodities. The margin figures were furnished by a large brokerage firm, they are not the minimal margins given by the rules of the exchanges. Indeed the correlation, while still positive, is smaller between the limit price and the exchange set minimal margin. This result explains daily price limits as a way of giving agents, the brokers, time to contact their principals for new instructions when circumstances have led to a surprisingly big price change. The bigger the daily price limit, the more margin a broker requires as protection against loss. In a market in which all trade is among principals, neither margin nor a daily price limit would be needed.

The fifth chapter on the history of futures markets by Dennis W. Carlton brings Usher's narrative up to date. It is also an avowed exercise in the structure, conduct and performance of organised futures markets. It fills in details that complement the preceding selection on the costs and benefits of organised futures markets. Carlton claims that government price support programmes inhibit futures trading by reducing price variability. It would be more accurate to say that when the market price falls to the Government's support level, sellers withdraw from the futures market because the price will not fall below the support level, although buyers would be willing to remain who are betting the chance that circumstances may change and prices rise. Silver is a good example. Trading in silver futures did not begin until after the US Government demonetised silver, thereby abandoning conversion of its silver certificates into silver coins at the price that had been fixed by the first US Secretary of the Treasury, Alexander Hamilton. The market price of silver had risen so much that the Government could not satisfy the claims given by the outstanding silver certificates. Carlton claims that increased volatility of the pertinent prices of financial instruments explains why futures trading has arisen for them. However, he shows that this explanation works for the T-Bills futures contract then traded on the Chicago Board of Trade but not for the S&P 500 futures contract, traded on the Chicago Mercantile Exchange.

Carlton mentions an important change leading to the introduction of organised markets in financial futures – cash settlement as a substitute for delivery in a futures contract. Indeed this is very important. If the cash settlement price comes from a competitive market, then it is not subject to the usual objections raised against it. Cash settlement for a futures contract traded in Illinois was allowed for the first time in the Eurodollar contract traded on the Chicago Mercantile Exchange. Once this precedent was established by US regulators it superseded decisions by Illinois courts that futures contracts without a provision for delivery were gambles, hence not enforceable in an Illinois court. The Eurodollar futures contract became such a great success that a host of other financial instruments, as well as others for which delivery would be very costly, became attractive candidates for organised futures trading. Carlton discusses competition among exchanges as analogous to competition among firms. There are worries that an organised futures market may become a monopoly and thereby harm the public interest. However, one must recognise that competition within an exchange among its members protects the public interest even if the commodity is traded in only one exchange. The nature of the rules within an exchange is paramount. Thus, an exchange rule establishing a minimum commission is against the public interest, but one setting a minimal margin may advance the public interest. It is because costs are lower when trade concentrates in one market that only one market survives. A single market with appropriate rules and competing members best serves the public interest.

The last chapter in this section by Frank H. Easterbrook is an essay on competition within a futures market. It emphasises that fraud is a necessary but not a sufficient condition for a successful corner in a futures contract. A thorough analysis of the various forms fraud might take, together with a survey of the relatively few cases in which corners were alleged to have occurred and the even fewer when they may have occurred, leads Easterbrook to the conclusion that organised futures markets come as close to the economist's ideal of a competitive market as we are likely to find in this world. He carefully weighs some sources of possible conflict between the private and the public interest and finds that organised futures markets resolve these better than can outside regulators.

10

The Technique of Mediaeval and Modern Produce Markets*

Abbott Payson Usher[†]

Mediaeval ordinances prohibited speculative transactions and were particularly severe against resale without displacement of the goods. It was supposed that gains made by conveying goods from one place to another were legitimate and that gains entirely attributable to changes in value were not. The function of the middleman was supposed to consist entirely in the movement of commodities from one place to another. According to the letter of the law, speculation was illegal, but the prohibitions could not be enforced and the arbitrage transactions between different places were not free from speculation as was supposed. Under the prevailing conditions of trade, changes in value in a period of time could not be separated from the differences in value in different markets. The purchase and sale in the distant markets were not simultaneous. Purchase in the low markets of a producing region preceded by a considerable period the eventual sale in the consuming centre. The interval of time that must needs elapse introduced a definitely speculative element into a transaction that was officially tolerated because it was supposed to be free from the taint of speculative gain. There were some communities where life was so distinctly self-centred that trade with distant markets was relatively unimportant, but such extreme localism was not characteristic of the late mediaeval period. For the most part, trading relations were elaborately developed.

The changes in the technique of market organisation in the 18th and 19th centuries have made it possible to distinguish sharply between the truly speculative time transactions and the essentially non-speculative

---

*This paper was first published in *The Journal of Political Economy* 23, pp. 365–88 (1915) and is reprinted with the permission of the University of Chicago Press.
[†]Abbott Payson Usher, 1883–1965.

transactions between different places. The accomplishment of this result turns upon the full recognition of the essential interdependence of the markets that constitute a market system, and upon the development of contracts for future delivery. Grudgingly the community has come to recognise that speculation is inevitable and necessary, but speculative gains are still associated in the mind of many citizens with dishonesty, gambling, and predatory activity. Because the sale of commodities without displacement seems to involve no effort, but merely chance, the profits are deemed to be tainted. The modern market system is thus misunderstood because of a firmly rooted prejudice, and the great improvement in the technique of trade almost unrecognised.

Speculation is to be distinguished from gambling by the nature of the contingency. Gambling is concerned with pure contingency apart from any other consideration. The outcome of any uncertain event can become the basis of a wagering contract. The results of games, races, political contests and the like are the characteristic field of the wager. Attention is concentrated wholly upon the occurrence or non-occurrence of the event. In an election bet, for instance, there is no implication that either party will be directly concerned in the outcome; so far as wagering is concerned, they might as well bet upon the turn of dice. A speculative transaction involves an element of contingency. It assumes that something is going to happen of which no one knows precisely what the outcome will be, but the speculator is interested in the consequences of the event. To bet on the outcome of a horse race is in itself pure gambling. The same event may contribute an essential fact to a speculative transaction. Suppose a person has bought a relatively unknown horse thinking the animal seriously underrated because of poor training and driving. The horse is taken in hand with a view to ultimate sale when its true powers have been revealed. The value of the horse can be demonstrated only by a series of successful performances on the race track, so that the owner is taking a chance, as it were, upon the outcome of the races. It will be readily seen, however, that the place of these races in the owner's interest is very different from the importance attached to the same events by persons who have given money to a bookmaker on the same horse. To the owner the race is merely a way of proving to others the accuracy of opinions long held by him. It is part of a larger situation. His gain is to be derived from establishing a different opinion as to the value of the horse. The gambler is interested merely in winning or losing. To him the race is a bare fact without consequences. Speculation is thus an attempt to gain by anticipating changes in the values of commodities. Gambling is a seeking of gain and excitement from the occurrence or non-occurrence of any uncertain event. Speculation is concerned with the content and significance of events affecting the valuation of commodities, gambling with the bare fact that something has occurred.

Mediaeval speculation is not to be distinguished from modern speculation by the antithesis between time differences and place differences. All speculation involves the element of time. But essential differences may arise in the mode of handling the goods during the time interval. In the Middle Ages, the speculator in produce was practically limited in his operations by the amount of his personal wealth. Today, goods held for speculation are largely carried on credit. In abstract terms, the difference may seem slight, but in reality it involves a complete transformation of the technique of trade, and the organisation which today makes possible the extension of credit in this field also brought to an end the confusion between the speculative and non-speculative elements of dealing in produce.

The necessity of speculating upon personal capital in the Middle Ages greatly restricted the scope of professional operations. All owners of property were obliged to speculate more or less, and the owners of large estates became involved in considerable ventures. It was illegal to purchase grain for speculative hoards, and there is reason to believe that the prohibitions were enforced in a measure. We may feel some assurance that large hoards were not formed by direct purchase in the markets, but the laws could not oblige an owner to sell except in times of extreme dearth, so that the owners and landed proprietors could legally store the rents in kind received from the estate. In regions which yielded a substantial surplus above ordinary local needs the hoards of the tithe barns and manor houses were considerable. These stores were the basis of much wholesale buying at all times, and were the main source of reliance in the years of dearth. Persons of small means were obliged by necessity to sell their grain in the local market more or less promptly. Unless the small cultivator was peculiarly needy his grain was sold off little by little according to the possibilities of using the straw for the cattle. The drying and curing of the grain was thus provided for automatically by leaving it unthreshed until it could be sold and consumed. This practice also insured a fairly steady supply for the local market throughout the season. Force of circumstances thus made the small cultivator the dominant resource of the market from week to week. In regions having a surplus those who could postpone sale for an indeterminate period found it to their advantage to do so, and they became by force of circumstances a class of unprofessional speculators who held grain for six, seven or eight years at times. Grain in store was usually kept in hermetically sealed pits. There was considerable risk of deterioration, but such methods of storage are excelled only by the most elaborate elevator construction of modern times. The existence of these hoards was of moment to the professional trader. Where such supplies existed the merchant from the large town found it advantageous to deal directly with the wealthy proprietors. Purchases could thus be made in bulk and without regard to the market regulations that were so frequently designed to discourage the wholesale trader. The professional trader was more likely to confine his speculation to the current season; the proprietor took the risks of loss through deterioration and of protracted waiting for a year of dearth.

The nearest approach to an application of credit to produce speculation in the Middle Ages was the purchase of a standing crop. This was definitely prohibited, but it is certain that the ordinances were not enforced. This transaction was a sale of the crop sealed by the payment of a small sum of earnest money. A merchant was thus enabled to secure a considerable supply of grain at harvest without immediate outlay.

Speculation in produce is primarily founded upon the exact determination of the relation between the visible and the total supply. In modern times, statistical information is available which confines individual opinion within fairly narrow limits. In the Middle Ages, the visible supply constituted a smaller portion of the total supply, and the total supply was hardly more than a matter of pure conjecture. The margin of possible gain for the professional trader was thus considerably increased. The great manoeuvres of the modern markets are founded upon superiority of knowledge of conditions affecting both demand and supply. General sources of information are so considerable that the trader's gain is based upon acquisition of more precise details and upon skill in drawing deductions from his facts. The history of the famous Patten wheat deal and of the Bull deal in cotton are interesting illustrations of modern successes. In the late Middle Ages the ignorance of the total supply was so complete that the speculator gains of the merchants were made by refraining from giving the public any enlightenment as to total supplies and studiously creating misapprehensions. Some of the most systematic manoeuvres of this type occurred in the vicinity of Paris in the latter half of the seventeenth century, just before the passing of the old order. A large portion of the grain supply of Paris came by water from the upper Marne and Seine. These merchants shipped their grain from the more distant sources of supply, and then, instead of allowing the boats to come through to Paris, they stopped them 15 or 20 miles outside and unloaded there. Sometimes there was a pretence of holding the grain for conversion into flour; most frequently it was merely stored in secret. The arrivals at Paris could be considerably diminished. Rumours would then spread of relative dearth in the Seine and Marne valleys. Prices would rise. The supplies in the vicinity could then be sold at the advanced prices if the quantities released from store at any one time were not considerable. Such a falsification of the market was made possible by the almost complete ignorance of the amount of the hoardings held in store by the wealthy proprietors of the country districts. The absence of information was of course a natural outcome of the conditions which created such hoards. The possibility of carrying produce on credit has resulted in more immediate sale of the crop and the storage of the greater portion of the actual stock in warehouses that are more or less public. The portion of the supply visible at any one time is much greater than in the past, and as nearly all the crop is sold in the course of the season, accurate seasonal crop statistics become possible. This brings the total supply within the range of certain knowledge.

The organisation of the modern markets has extended the functions of the middleman. In the old days his only recognised function was the transportation of the commodity from place to place. Now, apart from the speculative function that we now recognise, there are also a number of non-speculative functions.

The future contract makes it essential that some means be found of trading in the particular commodity as freely as would be possible if the entire supply were actually of uniform quality. The contractor can only agree to deliver certain quantities, and as the specific lot of goods to be delivered is not designated, the quality must be described. Such future transactions imply that each portion of the supply is substantially as good as any. In fact, the most even-running commodities present differences of quality. Organised speculation thus involves a grading system. Judgements of quality are standardised, rendered independent of the individual caprice of the parties trading, carefully defined and described so that the adjustments with reference to quality can be impartially and certainly made.

Financing the storage of the commodity during sale is inevitably associated with speculation. The possible changes in value during storage make the transaction speculative in part at least, and the amount of capital value that must lie idle pending sale constitutes a specially serious problem in these days of concentrated trade. When general farming was the rule, sale of a portion of the crop was a necessary means of securing money to pay taxes and other special obligations. The means of subsistence were raised on the farm. Today, the farm is devoted to more highly specialised agriculture. In some places, the agricultural community is actually dependent on central markets for some means of subsistence and for most general articles of consumption. There is more need of ready money. Postponement of sale by the farmer is less feasible than in the past. He desires to sell his crop immediately after the harvest. Professional traders must thus provide means during the harvest period for purchasing the great staple crops almost entire, and with their bankers they must carry the financial burden until the stocks can be sold. This function of the middlemen is essentially new because of changed conditions in the marketing of staples. The significance of this function has not been adequately appreciated.

The modern methods of marketing the cereal crops create technical problems of conditioning. If the grain is cured before it is threshed there is little danger of trouble from overheating and deterioration. The older methods thus made it possible to dispense with much elaborate curing that is indispensable when the crop is marketed rapidly and massed in elevators. The value of all these products is profoundly affected by the care with which they are handled in the elevators, so that the middleman finds a new source of gain in the manipulation of the product during storage.

Increased freedom to speculate has in fact narrowed the range of speculation. The activities of speculative traders are more evident, of course; much that was concealed is now given wide publicity, concentration has brought together in specific exchanges activities that were formerly spread at large through the storehouses of producing regions or receiving ports. The increased visibility of speculation disposes us to think of our age as characteristically speculative, and the change in law lends support to such a view. But such a generalisation is superficial. The change in the technique of trade cannot be described in such terms. It is an error to say that mediaeval trade was largely non-speculative and modern trade highly speculative. The speculative elements in mediaeval trade were not very frankly recognised, but they were present. The achievement of modern commercial organisation lies in the separation of the speculative and non-speculative elements involved. During the Middle Ages all transactions involved speculation; today some transactions are purely speculative and others wholly devoid of speculation. Today a trader may choose to speculate or to avoid speculation and seek gain in a purely industrial or commercial operation. The organisation of speculative trade has restricted the field of speculative gains and losses, actually reducing the proportionate importance of such transactions.

The organisation of produce speculation has obscured in a measure the distinction between speculation and gambling. It is not so clearly evident today that the speculator actually owns produce, and this has presented a real problem. In the modern markets many transactions are settled by ring settlement or set-off. The business of the different traders on an exchange during the day is naturally settled in the simplest way. If A has bought wheat and later sold a similar amount there is really nothing to be done but pay the differences in cash. It is likewise possible to bring together a group of transactions which involve several parties who have dealt in similar lots. The whole series of purchases and sales can perhaps be liquidated by a single transfer of warehouse certificates for money, so that the parties eliminated do not actually go through the form of buying and selling produce. Critics of the exchanges have endeavoured to discredit these operations by declaring that they are in fact mere wagers upon the rise and fall of prices. It cannot be denied that it is possible to make wagers upon the movement of prices. There may be some wagering in the exchanges, but it is certainly not characteristic. The operations on the floor of the exchange are wagers neither in form nor in intent. Transactions are based upon actual rights to acquire property or upon obligations to deliver property. The goods are represented only by documents of title that are symbols of property, but this does not make the transaction less real. The US Supreme Court has upheld the exchange, and the doctrine of intent that is involved is one of the most fundamental legal principles.[1]

TECHNIQUE OF MODERN MARKETING

Organised speculation is based upon contracts for future delivery which make it possible to sell for specified prices goods which are to be delivered in the future. These contracts may assume a variety of forms. They may be divided in general into "to arrive" contracts and term contracts, and term contracts may be of two kinds, specific grade contracts or basic grade contracts. The sale of goods in transit with agreement to deliver at the stated price immediately upon arrival is a form of contract that is naturally adapted to the conditions of trading in receiving ports or consuming markets. This mode of doing business grew up in connection with the maritime trade of London and Amsterdam. Cargoes were sold while still at sea, and time of arrival was naturally made the time of delivery. Such contracts are also applied to goods in transit by rail, though the shorter interval of time likely to elapse makes such a contract slightly less speculative than the marine contracts. These contracts are usually made upon the basis of samples sent in advance of the general cargo or upon the understanding that the goods must be of fair average quality (the so-called f.a.q. basis). Disputes as to quality would in such cases be adjudicated by a committee of the trading association and deductions from the price allowed if the stuff were below grade. Such contracts can therefore be used without any system of grading. The strict term contracts, however, require a formal and systematic grading. The precise nature of the grades established can vary within wide limits, but some system is presupposed by the character of the contract. The obligation of such a contract is not to deliver a specific lot, but merely to deliver, within specific time limits, a certain quantity of stuff, so that there must be some definition of the quality of the goods to be delivered. Two modes of defining the qualities are open: the seller may be required to deliver a specific grade of goods at the price stated, with permission perhaps to deliver higher grades without compensation, or he may be allowed to deliver stuff of several grades at prices to be computed by additions to the price of a basic grade if the goods are above the base chosen and by subtraction from the basic price if below the grade in terms of which the price is quoted.

These varieties of form are the outcome of different trading conditions.

The relative advantages of the term contract and the "to arrive" contract are related to the slightly different problems of marketing in producing and consuming centres. The producing market will have little occasion for the "to arrive" contract; the consuming market will find it possible to use both forms, though in many cases the "to arrive" contract seems to be better adapted to the needs of trade than the other form. For this reason, it would be a serious error to regard the "to arrive" form as a rudimentary term contract, and in tracing origins and studying tendencies it is essential to remember the complexity of the problem.[2] The two forms of term contracts are likewise an outcome of differences in conditions. Some commodities can be handled most readily in a particular market upon a specific contract; others can be handled only upon a basic contract. When the number of grades is small and proportions fairly certain, the specific contract has become the characteristic form, as it is in the wheat pit of the Chicago Board of Trade. With a great multiplicity of grades and much uncertainty as to the proportion of each grade from year to year, as in the cotton trade, a basis contract is probably essential.

In view of all these commodities of form it might seem that historical treatment of the growth of the modern system would be impracticable, but the course of development is not complicated. The Dutch in the seventeenth century used all forms of speculative contracts, and their speculation tended to degenerate into pure gambling entirely detached from actual buying and selling of goods. In England, in the 18th century, the "to arrive" contract was elaborately developed and placed on a secure basis by reason of the development of the bill of lading into a negotiable symbol of property. In the East India trade at London and in the iron trade at Glasgow, the dock warrant was developed and at Glasgow became a purely general certificate of ownership of a particular quantity of a specified grade of goods. This development of negotiable symbols of property was a fundamental step as it afforded the possibility of using the various future contracts without the dangers that had been fully revealed by Dutch experience. Finally, in the grain trade of the western United States, the term contract was developed into an elaborately developed instrument that seems to represent the final form.

In the Middle Ages the law of the market insisted upon the physical presence of the goods to be bought and sold. The market could deal only in such supplies as were physically visible. The inconvenience and dangers of such limitations became serious with the rise of wholesale marketing. The essential interdependence of producing and consuming markets could not be recognised adequately until each market was made competent to trade in terms of the whole supply to be found in the entire group of related markets. The stability of the large markets was greatly increased by making it possible to buy and sell not merely the goods physically present, but goods in transit and goods actually in the hands of traders on another market. The significance of this interdependence of markets has become doubly clear since the great improvements in communication have made it possible for dealers to engage in operations simultaneously in widely separated markets. The full development of this system of trading has been confined to the period subsequent to the opening of the Atlantic cable, but the origin of the system reaches farther back into the past. This modern system of trading rests upon two types of instrument: the future contracts already described, and symbols of property, such as bills of lading, dock warrants, and warehouse certificates. The early forms of future trading have been discussed already and the necessity of other instruments can be clearly perceived in the tendency of Dutch speculation to degenerate into gambling on differences.

The new legal doctrines which were to complete the technical foundation of the modern speculative system appear first in the law merchant and the English decisions associated with it. Neither the bill of lading nor the dock warrant was itself new, but both instruments acquired new legal attributes in the course of the 18th century. Originally mere receipts of goods and contracts for carriage or storage, they became negotiable instruments whose delivery when properly endorsed constituted delivery of title.

The formative periods in the legal history of the bill of lading in England are the 16th and 18th centuries. The bill became common and acquired its general form in the course of the sixteenth century; the legal doctrine of negotiability was not fully developed until the latter part of the 18th century. From these general facts one is tempted to lay down the general proposition that the bill as a receipt for goods and contract of affreightment became definitely settled in the early period, but did not become a symbol of title negotiable by endorsement until the 18th century. This conception of the development of the bill should probably be qualified, as the sale of floating cargoes and transfer of title by endorsement of bills certainly occurred in fact long before it was solidly established in legal doctrine. A number of bills of lading are published in the *Select Pleas in the Court of Admiralty*.[3] The form of the instrument is evidently unsettled in a number of respects, and a real development is evident. The documents suggest in every aspect the origin of the instrument and seem to be merely receipts for goods and contracts of affreightment,[4] but this narrower view of the bill is invalidated by the editor's heading with reference to the bill of November 7, 1539. The bill was drawn for a consignment of iron from Bilbao to London. The iron was sold while afloat, the bill of lading was endorsed to the buyer, and the goods were delivered to him.[5] A decision of Savary, the noted French authority on commercial law in the late 17th century, would also suggest that actual use of bills of lading was not limited by acknowledged doctrine. Savary says: "It is asked if a bill of lading should be deemed valid if it merely states what merchandise has been received by the master of the vessel without mention of the consignee. It is absolutely essential that the bill contain the name of the consignee, otherwise it is a fraud".[6] With the rise of speculative trade it was the practice to draw bills in blank with the intention of filling in the name of the consignee when the goods had been sold, thus it becomes interesting to speculate as to the inferences that may be properly drawn from Savary's statement and from the passage in the Marine Code of 1681 to which he refers. It is difficult to avoid the conclusion that merchants made frequent use of bills of lading in ways that were not recognised by the courts, so that one must avoid the narrow view of the matter. However, there are plenty of reasons for supposing that such deliveries by endorsement must have been rare. The practice of endorsement of bills of exchange was only just beginning in the north of Europe in this period and was not generally adopted until the middle of the 17th century. Furthermore, the fact that the full recognition of the negotiability of the bill of lading was postponed till the eighteenth century is presumptive evidence that the practice was not widespread. Had there been many cases, the problems would have come to the notice of the courts earlier. The number of significant cases between 1750 and 1790 is eloquent evidence of the close relation of case law to the needs of the community.

The modern law takes form in the 18th century. The more important cases are: *Fearon v. Bowers*, March 28, 1753; *Wright, assignee of Scott v. Campbell*, 1767; *Caldwell et al. v. Ball*, May 17, 1786; *Lickbarrow v. Mason*, 1787; and a second trial in 1794. The principle of negotiability is definitely stated in the earliest of these cases. Justice Lee said in summing up, "To be sure, nakedly considered, a bill of lading transfers property and a right to assign that property by endorsement".[7] The legal problems centred in no small measure around the nature of negotiability. There was disposition on the part of some to assume that the degree of negotiability was precisely similar to that of a bill of exchange. This doctrine was not accepted by the courts, and in the course of the period the difference between this aspect of the two bills was clearly brought out. *Wright v. Campbell* involved the right of a factor to sell goods consigned to him by his principal while they were in transit.[8] *Caldwell v. Ball* involved the problem of precedence of different copies of the bill of lading when the endorsements were different, though constructively the same.[9] The case of *Lickbarrow v. Mason* involved two problems: stoppage *in transitu* in case of the insolvency of the original consignee, and the validity of bills endorsed in blank.[10] The complexity of the case, its prominence, and long judicial history made it the controlling case on the legal doctrines involved. It may be regarded as practically completing the legal doctrine of negotiability.

The instruments of title which grew out of the warehousing system are closely analogous to the bill of lading, but the economic and legal history is absolutely distinct. These warrants or warehouse receipts arose much later than the bill of lading, and despite their economic significance they have not yet acquired a legal standing comparable to the bill of lading. Furthermore, the law of the different countries is quite distinct. There was apparently a parallel growth of such instruments in Holland, England, and France. In France and England the forms of the instrument were different, in Holland the tendencies were at the outset essentially similar to the English tendencies, but the movement seems to have lost its force in the latter part of the 18th century, so that the history of the instrument in Holland was without notable consequences. The actual history of the warrant is still hopelessly obscure, and the disproportionate emphasis placed upon the English system and its history has tended to create additional misapprehensions in a subject already fertile in difficulties.[11] Hecht declares that the economic importance of the warrant and its legal development were "a product of English trade and customary mercantile law",[12] but he does not support his contention and the history of the warrant in France and Holland[13] would seem to lead to different historical conclusions. England may have been quicker to adopt a new device with beneficial results to her commerce, or the greater volume of her trade may have given a greater significance to a commercial system whose technical details were well understood in both France and Holland. It is not very satisfactory to ascribe the increase in English trade to the development of the warrant system. The general decline of trade in both France and Holland toward the close of the 18th century affords a more natural explanation of the relative importance of the progress of the technique of trading at this time.

The general similarity of warrants and bills of lading and the frequent association of both types of instrument under the general term "document of title" has led some German writers to suppose that the legal properties of the instruments are the same. The neglect of case law is unfortunate. Both warrants and delivery orders are to be distinguished from bills of lading with respect to the legal meaning of "negotiability", and the warrants and delivery orders differ from each other.[14]

> Goods in stores, free or bonded, can be made the subjects of security, or transfer on sales, by means of delivery orders ... A delivery order, like a cheque, assumes three parties ... The usual terms of the order are simple enough. It is – "Deliver to A. B, or his order, so many goods, identified by marks and numbers, or so many bushels of grain from a particular lot lying in your store". It is signed by the owner, and is in favour of the particular party therein named. That order is not of the least use to the grantee until he has gone with it to the storekeeper, and has got the storekeeper to transfer the goods to the grantee's name[15] ...
>
> A delivery order very often is transferred from hand to hand. The original grantee endorses it "Deliver to so and so", and it may be endorsed twice or thrice over. It would be a mistake, however, to imagine that the delivery order, though

251

capable of endorsation, is a negotiable instrument[16] ... If you are the endorsee of a delivery order you are not in the position of the holder of a negotiable instrument like a bill; because, in the case of the delivery order, you are subject to all the exceptions arising out of the real contract between the original grantor and the original grantee. One important consequence is that the original grantor of the delivery order can hold the goods for the unpaid price against any endorsee whatever, even against a bona fide endorsee for full value given.[17]

The Scotch iron warrants are issued by iron masters and couched in approximately these terms, "I will deliver so many tons of iron of a specified brand, to any person who shall lodge this document with me after such and such a date". The warrants pass from hand without endorsation. They

> are treated in practice as if they were negotiable instruments. Now, the position of these warrants in law, according to the older authorities, is that they are not negotiable instruments: the law does not, or did not, accept or adopt them as such ... It is attempted to make these iron warrants negotiable by agreeing that anybody who holds them for value shall be entitled absolutely to delivery, and that he shall have no concern with the state of accounts between the iron master and the original purchaser of the warrant. The law says, or said, that it is not to be allowed, and therefore these warrants stand, or stood, in no better position in law than proper delivery orders. Indeed, it is doubtful if they are not in a worse position, because a proper delivery order is expressed in favour of a certain named person, while the warrants are blank or to bearer.[18]

It is needless to cite the cases upon which these statements are based. The law thus distinguishes between delivery orders, the Scotch warrants, and the dock warrants of the law of England as typified in the East and West India dock warrants of London. Evidently, too, the economic significance of these instruments has not been limited to the field within which they can safely be used under a strict interpretation of the law. Agreement among businessmen and regard for such commercial usages have tended to give these instruments in substance the flexibility possessed in fact by the bill of lading. The peculiar circumstances of the rise of the warehousing system at London was doubtless of material importance in the establishment of these practices.

TRANSACTIONS OF THE MODERN MARKETS

The transactions of modern commerce which contain no element of speculation fall into two general classes that are distinct both in form and in purpose. There are various forms of arbitrage dealings which are designed to secure a certain gain by reason of excessive differences in the prices current on different markets. There are various forms of hedging designed to free the manufacturer or middleman from the risk of a change in price during the process of manufacture or sale. Arbitrage dealings thus result in small but certain gains; hedge transactions are properly neutral, involving neither a net gain nor a net loss. When the manufacturer or middleman hedges, it is his purpose to confine the change of profit to his mercantile transaction. He avoids all risk of gain or loss by reason of changes in the price of the raw material in order to confine his attention to the technical problems of the process of manufacture and sale. These types of non-speculative transactions are dependent upon the mechanism that is usually thought of in connection with speculation. The various forms of future contracts are essential, and the practice of buying or selling in a particular place when the goods are physically located elsewhere is also characteristic. These transactions are not possible unless there are speculative and spot markets drawn together in a closely organised market system. The non-speculative transactions involve the same technical elements as the speculative transactions; the different results are due to the different combinations of the basic transactions. A future contract may be speculative or non-speculative, or speculative for one party and non-speculative to the other. A short sale may constitute part of a hedge or part of a daring speculative coup. The meaning of a particular purchase or sale cannot be deduced from its form; all its connections must be known. The much-discussed future contracts and short sales are indeed mere incidents of larger transactions, parts of a larger whole to which they are inseparably related. The larger aspects of marketing, too, are so closely associated that the non-speculative aspects cannot exist independently of the speculative aspects. It is this complex web of interdependent elements that constitutes the difference between the loosely related markets of the Middle Ages and the integrated market system of today. The arbitrage transactions and the hedge are of fundamental importance in maintaining the close correspondence between price on different markets that is characteristic of our organised market system.

The most typical form of arbitrage brings together a spot purchase and a sale under a term contract. Such a mode of dealing is characteristic of exporting regions where there is a keen competition for the product so that exportation is not a matter of course. In regions that seek a vent for a large surplus, the transaction is somewhat altered, though the underlying features are the same. Australian and Indian wheat are consigned to London agents to be sold on commission. Sale in some English or European port is assumed. Notice of the departure of the vessel is forwarded; samples and documents of title will also be sent and will presumably arrive considerably in advance of the ship. The vessel may be sent out with directions to call at Gibraltar for orders as to final destination. The London commission agent proceeds to sell the cargo while it is still afloat, on a "to arrive" contract. Purchase and sale are not simultaneous, and in that sense the actual character of the deal is for a time indeterminate. The shipment of the wheat may involve a real speculation or it may be sold quickly and become in essence an arbitrage transaction.

Among the various primary markets in the producing regions of the United States another form of transaction is not uncommon. It is not a true arbitrage deal because it does not contemplate actual shipment of goods by the operator. The transaction is affected by a simultaneous purchase and sale of term contracts in the high and low markets. A term contract is bought in the low market, and a contract for an equal quantity sold short in the high market. It is assumed that the operations of other parties will bring the markets closer together and afford a small but certain gain.

> Thus let us assume that on a given day in June the price of September wheat on the Minneapolis Chamber of Commerce is [US]$1 per bushel, and the price on the Chicago Board of Trade for the same wheat is $1.04 and that an arbitrageur considers this difference too large and anticipates a coming together of the two prices. Accordingly, he buys on a future contract in Minneapolis and sells short in Chicago at the prices indicated. Let us now suppose that in the course of a week the Minneapolis price rises to $1.04 and the Chicago price to 1.07\frac{1}{2}$ and that the arbitrageur closes out his transactions at these prices. By closing out his purchase in Minneapolis by a sale at $1.04 he makes four cents; and by covering his short sales at Chicago by a purchase at 1.07\frac{1}{2}$ he loses 3$\frac{1}{2}$ cents, thus clearing a gross profit of $\frac{1}{2}$ cent.[19]

This mode of trading is also applied to other types of price differences, and the practice is doubtless significant, but it would seem that its importance to the market is somewhat different from that of the other forms of arbitrage. In many respects this type of transaction seems to be particularly adapted to maintain relations between different primary markets that are receiving supplies from the producing regions. Actual shipments from market to market are in such circumstances a less convenient means of keeping markets "in line" than changes in the flow of the crop from those districts which can reach both of the markets concerned.

The hedge has been closely associated with a number of significant industrial changes. The transaction is widely used today in connection with flour-milling, meat-packing, cotton-spinning, and to some extent in the coffee trade. All these industries have been transformed or have grown up since the rise of the modern methods of marketing. The development of large-scale production in milling and packing would scarcely have been possible were it not for the hedges, and cotton-spinning could not be conducted upon such a narrow margin of profit if the risk of changes in value in the raw product were not eliminated. The nature of the change will perhaps be most readily appreciated with reference to flour-milling, as there has been a less general alteration of the place of the occupation in social life. The risk of loss to the miller through fluctuations in the price of grain was eliminated in the old days by transferring the risk to the consumer. The well-to-do and middle-class people were largely accustomed to buy grain and have it ground for their own use according to needs. The sale of raw wheat thus played a more prominent part in retail marketing a century and a half ago than it does today. The miller charged a small fee or took a portion of the meal as his toll. In the smaller towns only the poorer people bought finished flour or bread. In the larger towns the trade in flour and bread was rather more considerable, but even in towns like Paris and London much wheat was bought by townspeople for their needs and ground at their expense in mills near the city. The milling business was non-speculative, but it was necessarily conducted on a small scale with moderate equipment. Dependence upon local slaughter-houses was an equally prominent feature in the life of the past. Absence of refrigeration and of means of rapid transportation rendered the preparation of all meat products a distinctly local affair. Furthermore, there could be no question of risk from change in values of the raw product in the interval between the purchase of the creature and the disposition of the prepared meat. There was no appreciable interval. The butchers' trade was thus non-speculative, though the consumers did not buy the live creatures.

In the course of the last half-century, milling and packing have become capitalistic enterprises in no small portion of the western world. Flour consumed throughout the United States and in parts of Europe is milled in Minneapolis. Beef products consumed in the United States and in Europe are prepared in St. Louis and in Chicago. The raw material must be purchased months before the finished product can be sold. A change in market conditions might destroy entirely the mercantile profits of a highly efficient plant. Such enterprises can be conducted only if it is possible to reduce them to a non-speculative basis comparable to the conditions of the old craft organisation of days gone by. The future market affords a means of avoiding the speculation on the raw product.

The essential feature of the hedge is the combination of sale and purchase at both moments of contact with the market, both at the beginning of the process of manufacture and at the time of sale. Raw materials must be purchased for production in the spot market. With this transaction is coupled a term contract calling for the delivery of the same quantity of goods during the month in which the finished product will be ready for sale. The sale of the future is at the same price as the spot purchase. The miller is thus on both sides of the market. When the time comes for delivery of raw product under the future contract, the miller must go into the spot market to buy wheat. He is thus under an obligation to buy at the time he enters the flour market as a seller of finished product. At both moments he is buying and selling. Gains on one transaction will clearly balance losses on the other. The manufacturer is consequently independent of changes in the values of the raw materials.

The manufacturer can manage more readily if he has a large contract with the government for the supply of the army or some such service. In this case, at the time of bidding on the contract, he knows the prices of all the future options for several months in advance, and he can thus calculate pretty exactly what his raw materials will cost. If his bid is accepted, he can buy on future contracts for the entire period that he will be working on that order. The cost of his raw material will thus be settled at the outset. He is clear of risk and can make his money on the process of manufacture.

In the old days all speculation was for rising prices. Goods were bought and held back from the market in expectation of a rise. If the market was ill informed, the holding back of goods might cause a considerable increase in prices, and, if the goods were carefully unloaded without at any time revealing the extent of the supply concealed, the operators might realise considerable profits. Such operations were a serious problem in the Parisian grain trade in the late seventeenth century, and probably this was a characteristic form of bull speculation in the older markets. The essence of the transaction was to curtail the visible supply, to have large supplies concealed in close proximity to the consuming market, and to dole out these invisible supplies with scrupulous care. The relative isolation of different markets made such transactions relatively easy. The supply did not really come into sight until it arrived in the market place where it was to be sold to the consumer. In modern market systems such transactions are impossible because the supply comes into full public notice in the wholesale markets of the producing regions. The great consuming markets are today so well informed of possible supplies that they are in some cases distinctly non-speculative in tone. The wheat markets of London, for instance, are essentially non-speculative. The seaboard cities of the United States are also essentially non-speculative wheat markets. The tone of a modern market, however, is the product of many complex circumstances, so that it is impossible to generalise.

Speculative transactions of the modern markets assume that the market is informed of the general circumstances of trade. The gains of the operators are not secured by deceiving the public, but are based upon the accuracy of their inferences from the facts available. The facts are more or less generally known. The general body of public information is supplemented to a certain extent by private effort, but it is safe to say that the known facts are practically accessible to anyone who really wishes to get them. The great traders of the modern markets owe their success to shrewd inferences, wide experience, and command of credit in the commercial community.

Two types of speculation are now possible: speculation for a rise and speculation for a fall. The method of speculating for a rise is entirely different from the older transactions of the Middle Ages and the early modern periods. The speculation for a fall is entirely new.

The nature of modern speculation, however, is not to be understood from a mere designation of the contracts made on each side. Speculation is a continuous process based upon differing interpretations placed upon market conditions. In the large wholesale markets it has become a sort of party contest between "bulls" and "bears".

The notable speculative operations of the modern exchange centre around the general situation known as the "squeeze". The name is derived from the uncomfortable position the bears are in toward the close of a month when they have undertaken to deliver larger amounts of stuff than are readily to be had in the market. The competition of the bears for stuff to deliver on "short" contracts forces prices up, so that there is a double significance in the metaphor: it represents in part the notion that the bears are subjected to pressure by the bulls, in part the idea that the forces in the market push prices up to figures that are not actually representative of existing conditions. The squeeze is frequently confused with the corner, particularly by outsiders and by academic writers. The market operators are not likely to use the term "corner", and though their attempts to deny the occurrence of corners have not been well received, their intention of drawing a sharp distinction between the corner and the squeeze would seem to be well founded. The corner is the characteristic speculative transaction of the unorganised markets. The operator buys actual stuff with the intent to store it for a while. When he has secured substantial control of the supply, he begins to sell at such prices as he chooses because none can compete with him. The only limit is the ability and disposition of the consumer to pay the price asked rather than do without. The transaction, it will be observed, rests entirely with the individual operator. If his means are sufficient he can make a corner at any time. The squeeze is different in every essential particular. The bull operator buys both spot stuff and futures, but at the same time he must sell to the trade. It is his object to induce the bears to sell more stuff for delivery in some month in the future than they will then be able to secure except at greatly enhanced prices. Consequently, he has an interest in depleting the stocks on the primary market by sales to the trade.

This continuous selling to the trade in the interval before the squeeze is the most essential difference between the squeeze and the old corner. It is also worthy of note that a squeeze operation cannot be worked up at will by either bulls or bears. There will be no squeeze unless the bears sell excessive amounts on short contracts; even if the bears are really too optimistic about the future there can be no squeeze unless the bulls are willing to accept the challenge. The squeeze will arise only in those circumstances which produce a marked difference of opinion. In such an operation the party whose judgement of the conditions was the more accurate will gain. The famous Leiter deal in wheat of 1898 was disastrous to Mr. Leiter; the Patten wheat deal in 1909 was as conspicuously successful. These operations arise when real scarcity occurs for reasons that were not anticipated by the bears – either an unexpected shortage of crop or more likely an inadequate estimate of demand. The European demand for American wheat is variable because it depends in no small measure upon the harvests in other parts of the world. In many sections crops are not so well reported, so that wide differences of opinion may well exist. Both the Leiter and Patten operations were based upon inferences with reference to European demand, but Leiter failed to realise the significance of crop prospects. There had been several short crops and there was an unusual European demand which others did not foresee. The growing crop, however, was promising and ultimately proved to be large. The final offerings of the bears were based on certain knowledge of the abundance of the harvests in the great wheat-producing regions of the world. Patten was careful both in his wheat deal and in the cotton deal in which he was associated in 1909–1910. The last stages in the series of operations were in both cases dominated by the crop reports.

These episodes are without exception the most spectacular of modern speculative transactions. They exhibit the working of the modern markets when subjected to most unusual and extreme conditions. Under similar circumstances the mediaeval markets would have failed utterly. These modern markets revealed in each instance a remarkably prompt understanding of the situation.

1 L. A. Kinsey & Co. v. Board of Trade of Chicago, October, 1904, 198 US 236.
2 Emery, *Speculation on the Stock and Produce Exchanges of the United States*. In a general description of the rise of speculation (pp. 32–8), Mr. Emery says of these "to arrive" contracts (p. 35), "Their old importance as insurance against fluctuating prices has disappeared with the advent of the improved methods of the speculative market". Inasmuch as the speculative grain trade of London is still almost entirely based on such contracts and as similar contracts are common in German trade, it would seem that Mr. Emery is inclined to generalise from American conditions. There is a real danger in forgetting that such a subject cannot be adequately treated within the limits of the history of any single country. Diversities of form attributable to differences in essential conditions are one of the serious problems in the description of modern methods of speculative organisation. It is as yet too soon to announce the undoubted superiority of any particular form of doing speculative business.
3 *Seldon Society Publications*, Vols. VI and XI. For the Mediterranean history of the bill of lading, see Goldschmidt, *Universalgeschichte des Handelsrechts*, I, Part I, pp. 341–42. In general: W. P. Bennet, 1914, *The History and Present Position of the Bill of Lading*, Cambridge.
4 *Select Pleas in the Court of Admiralty*, I, 61, 89, 93; II, 59, 61, 63; bills dated, respectively, October 22, 1538; November 7, 1539; June 29, 1541; November 28, 1549; May 6, 1554; February 19, 1557; December 15, 1570.

5 *Ibid.*, I, 88–9.
6 Savary, 1721, *Parfait négociant*, 8th edition, II, 656, Parère XC.
7 I. H. Blackstone 364. *Fearon* v. *Bowers*, March 1753, cited in a note under report of *Lickbarrow v. Mason*.
8 Burrows Report, 2050, and in various other reports.
9 Term Report 205.
10 Term Report 63, 1787: 1 H. Blackstone 357, February 1790; reversal in Exchequer Chamber; 2 H. Blackstone 211; new trial ordered by the House of Lords; 5 Term Report 683, July 1794.
11 F. Hecht, 1884, *Die Warrants*, Stuttgart; O. C. Fischer, 1908, *Die wirtschaftliche Entwickelung des Warrantverkehrs in Europa und Amerika*, Berlin.
12 Hecht, *op. cit.*, p. 4.
13 Fischer, *op. cit.*, pp. 71–2, 86; Hecht, *op. cit.*, p. 3. Unfortunately neither cites evidence for these significant historical statements.
14 R. V. Campbell, 1890, *Principles of Mercantile Law*, Edinburgh, p. 115: "These bills of lading are clearly distinguishable from delivery orders. Such orders, as well as iron scrip warrants and debenture bonds issued by companies without authority of an Act of Parliament, may not be negotiable instruments: but a bill of lading is in many ways like a bill of exchange . . .".
15 Campbell, *op. cit.*, p. 105.
16 *Ibid.*, p. 106.
17 *Ibid.*, p. 107.
18 *Ibid.*, p. 111.
19 S. S. Huebner, 1911, "The Functions of Produce Exchanges", in *Annals of American Academy of Political and Social Science* 38(2), p. 22.

11

*Futures and Actual Markets: How They Are Related**

Lester G. Telser
University of Chicago

> *The underlying theme of the argument is the isomorphism between the theory of money and the theory of futures markets. This is established by showing the presence of a number of correspondences between the two. Among these are borrowing and lending in monetary theory and short and long hedging in futures theory. There is a real and nominal interest rate implicit in futures like those in monetary theory. Futures contracts themselves correspond to bank notes. Inventory firms are like banks. Open contracts are like the total stock of money, and hedging commitments are like reserves. Developing the full implications of the isomorphism will enrich both subjects.*

In order to explore how futures markets affect the market for the commodity or asset underlying the futures contract, this study takes as its central theme the analogy between futures and money. The relation between monetary and real phenomena is isomorphic to the relation between a futures contract and the underlying asset. Success in establishing the validity of this isomorphism will have several advantages. First, it will throw new light on some important aspects of futures trading by pointing out the analogies with the corresponding monetary phenomena, which are better understood. This enables students of futures to learn from the lessons of the monetary branch of economics. Second, students of money also can learn from that knowledge about futures, which

---

*This paper was first published in the *Journal of Business* 59(2), part 2, pp. S5–S20 (1986) and is reprinted with the permission of the University of Chicago Press. The author wishes to thank Meyer Burstein, Frank Easterbrook, Milton Friedman, Victor Goldberg, Harlow N. Higinbotham, Robert Townsend and Jeffrey C. Williams for their helpful comments on an earlier draft of this paper. I wish to acknowledge support from a grant of the Chicago Board of Trade that enabled me to write this paper. The views expressed here are my own and do not necessarily reflect those of the Chicago Board of Trade. The theme of this paper is an extension of the theory of Telser and Higinbotham (1977) and Telser (1981).

has been more thoroughly analysed than the corresponding monetary phenomena. By demonstrating that there is an isomorphism between money and futures, these two branches of economics can illuminate each other. This approach is especially useful for the analysis of hedging. Though it may seem a long way around to approach the subject in this way, I believe that in the end a better understanding of how futures trading affects the underlying actual commodities justifies the means.

I begin with a brief review of how the modern banking system began and of how the use of money and credit developed. Because my primary interest lies in futures, this survey focuses on those events most closely related to the corresponding ones in the history of futures trading.

Certain commodities became widely used during the evolution of human society. Those who had these commodities found they could readily trade them for other goods. As a consequence, the widely used commodities acquired an additional attribute – they became mediums of exchange. The more a commodity is used as a means of payment, the more widely acceptable it becomes. The earliest examples of such commodities are believed to be copper, silver and gold. Once a commodity becomes widely acceptable as a means of payment, it obtains yet another important attribute – it becomes used as a store of value. Some people use such commodities as the asset in which they will hold their savings, savings being the excess of their receipts over their current expenditures. They are willing to have their savings in this form because they believe they can exchange these commodities later for the particular goods they will want then. In this way, the commodity used as a means of payment also becomes useful as a temporary abode of purchasing power.

Using certain commodities as a store of value is an important stage in the history. The owner of such a commodity may store it himself in a safe place under his own control, or he can seek some trustworthy person who can store it for him. Silversmiths and goldsmiths are common examples of the latter. They had to store silver or gold for their own trade. We may well imagine that it would soon occur to others and to the smiths themselves that they could offer their facilities as a safe place to store precious metals. Moreover, they could do so at a lower cost than the owners of gold or silver would incur by storing the goods themselves.

The next step in this development was the discovery by these storage firms that they could lend some of the gold or silver left on deposit with them to borrowers who were willing to pay them interest for the loans. The owners of the gold or silver who had deposits at the inventory firms regarded their assets as a store of value available for later use or to trade for other goods later on. The inventory firm learned that it could lend some of these deposits of the precious metal, without risk that it would be unable to return the precious metals to their owners when the time came for doing so. Instead of allowing these stocks to lie idle, the inventory firm could lend them to those who could make better use of these stocks and were willing to pay for this. Not only would this allow the inventory firm to obtain some additional revenue, but it would also have some other important consequences. Instead of being a burden to store these assets so that the inventory firm would demand payment from the owners for doing so, the storage of the precious metals would become a source of gain to the inventory firm. Consequently, such firms would actively seek those who would be willing and able to deposit such commodities with them. As evidence of the deposit, the inventory firm would issue a warehouse receipt to the owner. Nor is this all. The goldsmith would not need to lend actual gold to a borrower. Instead he could give the borrower a warehouse receipt for the gold that the borrower could use in exchange for the commodities he wished to acquire with the proceeds of his loan. The evidence of ownership, the warehouse receipt, or, stated differently, the certificate of deposit would be just as useful as a store of value or a medium of exchange as the actual gold or silver, and it would be cheaper to make than the actual precious metals. Of course, the warehouse receipt is not useful for making jewellery or coins. For this, one needs the physical commodity. But for the commercial purposes of the inventory firm that borrows and lends, the certificate is just as good as the actual commodity. In these ways the physical stock of gold and silver could support a mass of bills of deposit greater than the underlying stock of the physical inventories themselves.

Up to this point in this idealised history there is no mention of money or of a monetary unit. The warehouse receipt for the gold or silver states the quantity, the name of the firm issuing the receipt, and the name of the person entitled to claim the actual commodity. As an interesting aside, recall that the term "pound sterling" refers to a physical quantity, a pound of silver. The value of the certificate of deposit depends on the rate of exchange between 1 unit of the commodity to which it is a claim and other goods. There would not be a single price for this physical commodity. Instead there would be a separate price for each commodity for which it could be exchanged. We may also imagine that it would soon occur to everyone that, by stating all prices in terms of a physical quantity for one of these precious metals, it would become easier to calculate exchange rates between any pair of goods. Hence the physical unit of the precious metal would become the numeraire for all commodities. In this way the value of the certificate of deposit would depend on how much people are willing to pay in exchange for 1 unit of the precious metal. This certificate of deposit is a private money that depends on the faith and credit of the firms holding the physical stocks of the precious metals. Neither the king nor the state need enter the scene in order to determine the validity of the money or to its value in exchange.

The warehouse receipt would be useful as a medium of exchange primarily to the larger merchants. For most ordinary transactions the medium of exchange was the coin of the realm. These small coins could also serve as a store of value for those people who wished to accumulate savings. Even among the larger merchants, use of the certificates of deposit would pose some difficulties. There must be trust in the validity of the certificate itself, meaning confidence both in the genuineness of the certificates themselves and in the reliability of the firm that issues the certificate, so that potential depositors will believe that the issuer will convert the certificates into the underlying asset according to the conditions it promised. The asset itself that underlies the certificate must be well defined and of an ascertainable quality. Assuming these conditions are satisfied, these warehouse receipts become a private money with general purchasing power. The stock of this money results from the deposits of the owners of the previous metals and from the loans made by the inventory firms to those who borrow from them. Although each inventory firm holds enough reserves to prevent insolvency, the total stock of money is a multiple of the stock of the underlying physical asset. This reminds us of a similar fact about futures where the stock of open commitments is a multiple of the stock of the asset underlying the futures contracts. What causes the multiple expansion of the certificates of deposit?

Let anyone with a valid warehouse receipt for the physical commodity come to the inventory firm. The latter would be willing to accept this receipt for deposit and to open an account (or to credit the existing account if this person has one) for the total quantity of physical units represented by this warehouse receipt. The inventory firm is willing to do this on the basis of a slip of paper presented to it and does not require the physical good itself as evidence before doing these things. The claim to the asset is just as good for this purpose as is the asset itself. The depositor entering this "bank" (for brevity it is now convenient to refer to the inventory firm as a bank) neither needs nor wishes to know whether the warehouse receipt that he wants to deposit in the bank represents a previous direct deposit of the physical asset itself by someone else in some other bank or whether it originated as a loan from the issuer of the certificate of deposit. In either case the circulation of these warehouse receipts among the merchants and the transfer back and forth from the banks to the merchants can and does result in a stock of claims to the physical assets that is a multiple of the actual quantity of the physical assets underlying these claims. The ratio of the total claims to the physical inventory depends on the reserve ratios of the individual banks.

It is well known in monetary theory that the excess of the claims to the assets over the actual physical quantity of these assets is not the result of imprudent behaviour by any individual bank. On the contrary, as far as each individual bank knows, it has enough reserves in the form of its own physical assets to meet the claims of its depositors. One who owns a warehouse receipt for the physical asset regards this receipt as equivalent to the actual commodity itself for the purpose of using it as a store of value or as a medium of exchange. Of course, it is not equivalent to the physical good for the purpose of making jewellery or coins. Each individual bank that receives a deposit in the form of a warehouse receipt regards this slip of paper as equivalent to the physical commodity for its purposes of establishing its own reserves. That is, a bank receiving a warehouse receipt drawn on another bank can regard this as a claim to the physical commodity underlying the warehouse receipt. Hence it can treat this deposit as equivalent to a deposit of the physical good itself. Moreover, it can use this warehouse receipt to obtain the actual commodity from the bank that issued this receipt if it so desires. It is this perception of the individuals that the claims to the underlying physical asset are equivalent to the possession of the physical good itself that lies at the root of this argument explaining why the stock of claims can be a multiple of the stock of the physical good. Remember, it is less costly to use the paper claims as a store of value and a medium of exchange than it is to use the actual underlying physical commodities.

In order to see more clearly the correspondence with futures, it is necessary to examine more closely the nature of the transaction between the lender, who is the bank, and the borrower, who is a merchant. The parties to the loan agree on the terms, which include when the loan will be repaid and how much the borrower will pay the lender for the loan. Although the latter is usually thought of as an interest rate, for the purpose of showing the isomorphism between money and futures it is now convenient to express the cost of the loan to the borrower in a different way. The interest rate is implicit from a comparison between the initial quantity that is lent and the quantity that the borrower agrees to repay the lender. Thus if the borrower obtains one unit for a period of one year and agrees to repay 1.1 units, then the interest rate is 10% per year. This transaction is equivalent to the following. Let the borrower buy one unit of the commodity and simultaneously sell 1/1.1 units of the commodity for delivery in one year. Hence borrowing is equivalent to a pair of transactions composed of a purchase of the spot commodity and the simultaneous sale of the future commodity. An annual interest rate of 10% is equivalent to a futures price of 0.9090. This alternative way of describing a loan implies that a loan is equivalent to a short hedge.

A short hedge is a pair of transactions in which a purchase of an actual commodity accompanies a simultaneous sale of an equal quantity of futures contracts. However, according to the description above of a loan, the borrower repays the loan. In a short hedge this would be equivalent to delivering the actual commodity in fulfilment of the obligation incurred

265

by the sale of futures contracts. It may, therefore, be objected that, since actual delivery on a futures contract is uncommon, the analogy between a short hedge and borrowing is invalid. In the usual short hedge, the buyer of the actual commodity lifts his hedge well before the maturity of the futures contract. He does so in this way. When he sells the actual commodity, he buys back the futures contract at the then-prevailing price. However, there is a loan transaction that does correspond to the usual form of a short hedge. It is one in which the borrower can repay the loan according to prescribed terms at any time up to the final date when the loan becomes due. This is known as a call loan. Before describing this in more detail, it is useful to consider some other similarities between money and futures.

In this idealised history the banks originate as inventory firms that store precious metals for others. The warehouse receipts issued by these banks to the owners of the metal may not specify the name of the owner. Instead it may be payable to the bearer. If so, such warehouse receipts have a special name – they are called "bank notes". However, the bank notes issued by the different banks are not necessarily perfect substitutes for each other. If bank notes issued by different banks were perfect substitutes, then they would have the same price. In this case a note issued by one bank would be as acceptable as one issued by another. But since the bank notes are private money secured by the faith and credit of the individual bank, it does not necessarily follow that all banks are equally acceptable sources of these notes. A bank that is less likely to redeem its obligations would find that its notes command a lower price than would another bank regarded as being more reliable.

The banks also make loans on terms mutually acceptable to themselves and to the borrowers, who receive bank notes from them. The price for a loan implies an interest rate. Equivalently, the terms of a loan determine the quantity of bank notes that a borrower must return to the bank in fulfilment of his debt. These terms imply the future prices of bank notes. That is, if a person who borrows 1 unit today must return 1.1 units in 1 year, then the futures price of a bank note of the particular bank is 1/1.1. Although these futures prices for loans are not determined in one location on an organised exchange, they are the result of competition between borrowers and lenders. Different banks issue their own notes, which are not necessarily acceptable at par for notes issued by other banks. Each bank in this history constitutes its own market, in which it deals with potential customers for loans. Nevertheless, it would seem, therefore, that there is a difference between the credit market, which is composed of many local markets, and the futures market, which is a central market.

In the development of futures markets, the storage firms – the warehouses – correspond to the private banks. The warehouse receipts of these storage firms resemble the certificates of deposit of the banks.

Warehouse receipts payable to the bearer are like bank notes. Just as notes issued by different banks are not necessarily equally acceptable, so too the warehouse receipts of different warehouses are not equally acceptable. Therefore, warehouse receipts are not a fungible, liquid instrument, whether they are notes of different banks or warehouse receipts for wheat of different storage firms. Moreover, it is interesting to note that, at an early stage in the history of the grain trade, grain terminals in Chicago made transactions like those described above by the banks in, say, Renaissance Italy.[1]

The history of money and banking shows that the emergence of a common medium of exchange resulted from actions of the state. The decisive event at the end of the seventeenth century was the establishment of the Bank of England. Though legally a private bank, it had close ties to the government. It was one of the first central banks to appear on the scene. The Bank of England, by virtue of the functions it performed for the government and by becoming the lender of last resort, had a major effect on money, credit, and commerce. The liabilities of the Bank of England became the reserves of the private merchant banks. The role of gold bullion, while becoming increasingly less important, did not finally disappear as a major factor until 1930. The important point is this. The bank notes issued by the Bank of England were universally acceptable throughout the realm and became the standard monetary unit. This means that a holder of a Bank of England note could use it as a means of payment, and it would be acceptable to the person to whom it was offered without regard to the credit standing of the payor. The importance of a universally acceptable medium of exchange is very great. A note issued by a private bank was not necessarily universally acceptable. It would depend on the credit standing of that bank. This was true in England throughout the nineteenth century. It was also true in the United States until the National Bank Act, which became law in February 1863 and allowed the creation of National Bank Notes. Even to this day there is an important vestige of individual bank notes – a certified cheque. A cheque, certified or not, is not legal tender. Your creditor need not accept your personal cheque or a certified cheque. He can require payment in cash. A traveller's cheque is another example of a financial instrument similar to a bank note because, like a bank note, a traveller's cheque is a liability of the issuer.

For futures markets the appearance of an instrument equivalent to currency is not the result of governmental intervention. The futures contract, perfectly fungible and liquid, valid without regard to the identity of the buyer or the seller, was the invention of a private enterprise, namely, those who established the first organised futures market. Had the banks formed an exchange in which they could trade their bank notes among themselves in terms of a standardised contract, there may well have arisen a private common monetary unit similar to a futures contract. The organised futures market, entering the scene about a century and a half after the Bank of England, got no inspiration from banking history. This market evolved in order to satisfy the needs of the traders for an instrument that would serve as a trading vehicle in a particular commodity (the first was, in fact, wheat but would not be actual wheat). Each contract is a perfect substitute for another of the same maturity. Such substitutability is not true of different warehouse receipts for wheat. By restricting trade in these contracts to the members of the exchange who must satisfy prescribed requirements of financial integrity and moral character and by defining the terms of the contract appropriately, there came into being a medium of exchange and a store of value in terms of the agricultural commodities first traded on these exchanges similar in its essential traits to currency. However, the particular futures contracts are denominated in units of wheat, corn or oats instead of units of gold or silver.

A futures contract in wheat, for instance, is both a temporary abode of purchasing power and a store of value in wheat. For both these purposes it is superior to actual wheat because the transaction costs for wheat futures contracts are lower than for actual wheat. There are several reasons for this. A trader in wheat futures can avoid the cost of inspection or grading. The faith and credit of the organised market stands behind each futures contract in this sense. Each transaction between a buyer and a seller, who must both be members of the exchange, is binding on them. Though they may act as agents on behalf of their customers, who are the principal parties to the contract, it is the agents, the buyer and the seller, the exchange members, who are responsible for fulfilling the contract's terms. Even if their customers default, it still remains the responsibility of the members, the agents of their customers, to carry out their contractual obligations. Subsequently, the members have the right to require customers in default to make good. Because the futures contracts of the same maturity are perfect substitutes for each other and are valid without regard to the identity of the buyer and seller, futures contracts correspond to currency with respect to the asset underlying them. Thus a futures contract in wheat is to a warehouse receipt for wheat as currency is to a cheque of an equal amount. Admittedly, for large transactions it is inconvenient to use currency – a certified cheque is better. The point is that the cheque must be certified before it can become nearly as acceptable as currency.

To verify the claim that hedging in a futures market corresponds to a loan, we must carefully examine what occurs in a hedge. It is convenient to begin by confining attention to short hedging, which corresponds to borrowing, and then to discuss long hedging, which corresponds to lending. At the outset it is necessary to recognise that hedging is not a literal loan in the sense that a short hedger borrows a commodity and subsequently returns exactly the same thing with interest to the lender. A short hedger buys the actual commodity in the spot market and at about the same time sells an equivalent quantity of futures contracts. This pair of transactions does have the essential features of a loan since the short hedger frees himself from much of the risk of a change in the price of the commodity. An owner of an asset bears the risk of a change in its price because, when he buys it, he does not know at what price he can later sell it. In contrast, the borrower of a good does not bear a risk of a change in its price during the period of the loan. This is because the terms of the loan specify how much the borrower must repay, which is equivalent to determining the futures price of the borrowed good.

A numerical examples helps to illustrate the argument. I shall follow Keynes, who, in turn, attributes the concept of the "own rate of interest" to Sraffa.[2] Assume that the spot price of wheat is US$4.00 per bushel and the futures price for delivery in one year is US$4.30 per bushel. The money rate of interest is 12%. The sum of US$4.00 will obtain US$4.48 in one year that can be used to buy 4.48/4.30 bushels of wheat for delivery in one year. So one bushel of spot wheat becomes 1.042 bushels of wheat in one year, and the own rate of interest in wheat is 4.2%. Hicks gives another example, which, for the sake of variety, shows how to lend a commodity – say, coffee – for one year. This corresponds to a long hedge. In Hicks's example, you can lend the commodity for one year by making three transactions as follows: (a) sell coffee spot; (b) lend money proceeds for 1 year; and (c) buy coffee futures. If a long hedger begins as the owner of an asset that he lends by selling spot and buying futures, it follows from this argument that, insofar as he is an owner, a long hedger does bear the risk of an adverse change in the price of the asset when he is repaid the loan with interest. The repayment occurs when he lifts the long hedge by buying spot and selling futures. Therefore, in contrast to short hedging, which corresponds to borrowing and removes price risk from the short hedger, long hedging, which is lending, does not remove price risk from the long hedger. The motive for long hedging is the same as the motive for lending. People lend when their actual receipts exceed their desired expenditures at the prevailing interest rate. The same considerations explain why there is long hedging. The higher is the spot price relative to the futures price, the greater is the interest rate obtained from a loan of the commodity. Hence the incentive for long hedging increases with the excess of the spot price over the futures price. This explanation of long hedging suggests the importance of seasonal futures. For agricultural commodities one would expect long hedging to be greatest toward the end of the crop year. For commodities for which the demand is subject to seasonal variation while the supply is not, one would expect long hedging to peak when consumption is at its peak relative to the output rate. For example, long hedging of gasoline would be highest during the summer according to this argument since this is when gasoline consumption reaches its seasonal maximum. By the same argument, long hedging of fuel oil would be highest during the winter, which is when consumption of this commodity is at its maximum.

Like money interest rates, there are two own-interest rates – the nominal and the real. The real own-interest rate equals the nominal own-interest rate minus the expected percentage rate of change of the price of the underlying asset. Therefore, this corresponds to the real money rate of interest, which is the nominal interest rate less the expected percentage rate of change in the price level. Hence, in this respect as well, there is a correspondence between monetary theory and the theory of futures.

The examples above seem to require actual delivery on the futures contract. In fact, delivery on futures contracts is uncommon even when it is feasible, and it is unnecessary in any case. A hedge lifted before the maturity date of the futures contracts is still equivalent to a loan, albeit not an ordinary one but instead a call loan. The latter is one for which the interest rate adjusts continuously and always equals the current shortest term rate. Therefore, the loan is continuously renewable at each instant in time at the currently prevailing instantaneous interest rate. Such loans are equivalent to hedges that can be lifted at any time up to the maturity date of the futures contract.

On the basis of an early Illinois court decision (*Kerting v. Sturtevant* (Ill.) 181 A 517), futures contracts without provision for delivery were not legally enforceable contracts because the court found them to be gambles. Consequently, in order that a futures contract be legally enforceable, provision for delivery became a standard feature of futures contracts. However, delivery is not a standard feature of futures contracts. In principle, a futures contract settled on the basis of the prevailing spot price determined in a competitive spot market has the same desirable economic attributes as one that permits delivery. Futures contracts without a delivery provision are said to be settled on a cash basis. Put differently, whether or not there is actual delivery on a futures contract, the price of the contract at the maturity date equals the spot price of the underlying asset. We can easily convince ourselves of the equivalence between futures contracts with and without provision for delivery by considering what would happen in case delivery is contemplated. The seller of the futures contract who does not have the actual asset could buy it or a claim to it in the form of a warehouse receipt in the spot market and transfer title of these assets to the buyer of the futures. (The clearinghouse facilitates the necessary details.) The buyer of the futures contract may or may not actually want delivery. Suppose not. Then he could sell the title to the asset he has acquired as delivered on his futures contract at the prevailing spot price in the spot market. Plainly, these transactions are equivalent to settling the futures contract at maturity by using the prevailing cash price. Provided there is a competitive spot market, cash settlement works at least as well as delivery. Moreover, settlement on the basis of actual delivery works well only as long as the spot (and futures) market is competitive. Actual delivery is unnecessary. A competitive spot market is sufficient to ensure smooth expiration at maturity of the futures contract whether or not it has an explicit provision for delivery.

Yet another alternative to an ordinary hedge is now available, namely, options. A put, an option to sell a futures contract of a given maturity at a preassigned price (called the strike price), is another way to accomplish the same purpose as a short hedge. A person may buy the spot commodity and buy a put with a suitable strike price. This pair of transactions corresponds to an option to borrow at a stipulated interest rate as an alternative to an actual loan. An option to sell affords price protection to an inventory holder similar to a sale of the futures contract under fairly general and plausible conditions. For a long hedge, the call, an option to buy futures at a stipulated strike price, supplies the alternative to buying futures. A forward purchase of the actual commodity combined with the purchase of a call gives a substitute for a long hedge. Just as a put combined with a purchase of the actual commodity gives an option to borrow, a call combined with a purchase of the actual commodity gives an option to lend. An active, competitive, liquid option market is one that facilitates trade in a close substitute for futures contracts as an alternative to hedging. Despite these advantages of options, full-scale options trading became a reality only within the past decade as a result of new federal regulations. An 1874 Illinois statute prohibited trade in options. Nevertheless, the compelling advantages of options led to active trade on an informal basis in the corridors of the Chicago Board of Trade. The options then went under an alias – they were called privileges. These privileges usually expired in, at most, 1 week and did not constitute a financial instrument like modern options.[3]

In view of the argument asserting that option on futures furnish a substitute for hedgers, it is fair to ask in what respects, if any, options differ from futures. Consider a put, for example, that gives the buyer the option to sell a specified quantity of futures contracts at the strike price. A short hedger using a put as an alternative to a short position in futures contracts pays a certain, known amount for the put option. If the spot price is higher at the time he sells his stock than it was when he bought it, then his return on the spot position is reduced by the cost of the put option. If the spot and futures prices are both lower, then the loss on his spot position is offset by the gain on the futures position less the cost of the option. Had he placed the hedge using a futures contract directly instead of a put option, he would have obtained the same protection while avoiding the cost of the put. However, hedging by means of a put gives the hedger a guaranteed price on this short position. To get the same effect using a short futures commitment instead of a put the hedger would have to place orders to sell and buy futures at stipulated prices, equivalent to a stop-loss order. There is no guarantee that such orders can be executed precisely at the specified price. For this reason the hedger would have to place his price order above the equivalent strike price for a put option. He would then have a good chance of getting execution close to the desired price but no certainty of this. The cost of a put option is equivalent to an insurance premium paid for the certainty of execution at the specified price. The more liquid and competitive the futures market, the more likely is execution of an order from the hedger at the specified futures price and the lower must be the cost of a put option that would be as attractive an alternative as a traditional short hedge using futures contracts. It remains to be seen whether options and futures can both survive side by side.

There is still another interesting parallel between futures and money. An owner of a particular asset can often obtain good hedging protection from a futures contract in a different but related asset. For instance, if the correlation between the price of soy oil and the price of corn oil is high enough, then one can hedge a stock of corn oil by selling soy oil futures. As a result, the soy oil futures market attracts a broader range of transactions from all closely related food oils so that the soy oil market becomes more liquid (no pun intended) than if there were separate markets for each of the food oils. This makes the soy oil futures contract more fungible. As a result, a futures market based on one underlying asset becomes a useful trading instrument with a broader domain over a whole set of related assets. Each of these futures contracts becomes the currency of trade over a wider domain.[4]

The buyer of the spot commodity with a customer in hand needs no short hedge. Short hedging is useful only if the inventory holder uses the futures contract as a temporary substitute for the actual sale of the inventory. He can use the stock of the spot commodity for his business purposes in the same way that one who borrows money can use the proceeds in his business. A long hedger, the one in the example given by Hicks, is a lender of the commodity. He can use the purchase of futures as a temporary substitute for buying the actual commodity. For hedgers, short or long, the futures contract substitutes for a transaction in the actual commodity or asset. Similarly, options – puts and calls – substitute for short and long positions in futures contracts and thereby furnish an alternative to short and long hedges, respectively, using futures contracts.

By accepting the argument establishing an isomorphism between money and credit on the one hand and futures and hedging on the other, an important conclusion follows, showing how futures trading affects the underlying asset. Recall that a major purpose of borrowing and lending is to change a given stream of receipts into one that can better approximate a preferred stream of expenditures. Hedging does this. Consider this for the example of wheat. This commodity has two plantings, late fall and early spring, and two harvests, one for winter and the other for spring wheat. Therefore, there is a given stream of wheat "receipts" determined by the effects of the laws of nature on the cost of producing wheat. People want a steady stream of consumption of wheat products over time, given constant prices of these products over time. So there is the problem of changing the given stream of wheat production into the preferred stream of wheat consumption. By hedging the wheat inventory, merchants can accommodate these preferences. They sell wheat futures as a temporary substitute for the actual sale of their stocks of wheat. They are in effect borrowing wheat that they repay as they sell their actual stocks of wheat in the interval between harvests. All agricultural products subject to a seasonal pattern of production fit this argument. Non-agricultural commodities with seasonal production, such as metals, also fit this argument. Products that have a seasonal consumption pattern and can be produced at a lower total cost if the rate of production is steady also satisfy this argument. Petroleum products are examples. Apparent exceptions to this explanation are financial futures. What theory applies to them?

Consider the stock index futures. Here is a case in which the argument sketching the history of money and banking is relevant and that shows how the theory of futures illuminates monetary theory. A contract in a stock index futures traded on an organised exchange is equivalent to a special kind of money, one that has purchasing power over a bundle of stocks. We are accustomed to thinking of money as something with purchasing power over all goods. In fact, different components of the money supply have their own advantages for specific purposes. For instance, at current prices a 20-dollar bill is not useful for buying the daily newspaper; neither is a 100-dollar bill useful to pay a taxi fare. Each of these transactions has a better component of money – a 1-dollar bill for the daily newspaper and a 20-dollar bill for a taxi fare. Similarly, the stock index futures contract is a type of specialised money useful for trading a bundle of stocks. Nor is this all. An owner of a portfolio of stocks can use stock index futures as a hedge. The higher the correlation between the value of his portfolio and the price of the stock index futures contract, the greater is the hedge protection.

If this argument about stock index futures is correct, then what explains the long delay before their appearance on an organised futures market? A large part of the answer to this question is the legal obstacles to stock index futures. For such contracts it is more convenient, if not imperative, to use cash settlement, not actual delivery at the maturity date of the contract. (Actual delivery would require a spot contract in the stock index, which is a feasible, though perhaps more costly, alternative than cash settlement.) This could not be without changes in the law, and the law is slow to change. In fact, the first futures contract permitted by the Commodity Futures Trading Commission (CFTC) to use cash settlement instead of delivery was the Eurodollar contract traded on the Chicago Mercantile Exchange.

Until recently, discussion of futures was usually confined to the agricultural commodities where futures trading originated – wheat, corn, oats, rye, cotton and so on. These are all storable goods. (The storability of a good varies inversely with the cost of storing it.) Important futures markets sprang up for other storable agricultural commodities as well, such as butter, eggs, lard, coffee and sugar. Some metals were actively traded outside the United States, notably, on the London Metal Exchange. Owing to this, for many years theories of futures trading took it for granted that these were the only kinds of commodities suitable for futures trading. Two important developments in the past 20 years have changed these views. First, trading in live cattle and feeder cattle on the Chicago Mercantile Exchange shows that storability of the commodity underlying the futures contract is not necessary. The success of these cattle contracts owes much to the presence of highly competitive spot markets in these commodities. Given a wide interest in a commodity capable of supporting the expense of trade on an organised exchange and given competitive spot markets, it is now clear that successful trade can develop in a non-storable commodity. Second, trade in financial futures, beginning with US Government National Mortgage Association (GNMA) futures contracts in 1975 and going on to the recent Standard and Poors 500 futures contract, has become very successful. In fact, there were important interludes of active trade in financial futures, such as foreign-exchange futures in the 1920s, when central banks allowed these rates to be set by free market forces. I have given my reasons for the success of trading in stock index futures. Now consider financial futures where the underlying asset is a debt instrument, such as the GNMAs, Treasury bills, Treasury bonds, and so on. Here I believe the argument is straightforward. A necessary condition for successful trade of a futures contract is price variability of the underlying asset. This is a source of risk to owners of these assets and it creates a demand for hedging. It is a direct implication of the theory linking hedges to loans so that short hedgers are borrowers of the asset hedged. Borrowers can avoid the risk of changes in the price of the asset they borrow. The greater is the price variability, the greater is this source of risk. Starting in the early 1970s, interest rates in the United States and elsewhere became more variable and less predictable. Hence one of the necessary conditions for successful futures trading came into being. The other necessary condition had long been present, namely, a very large number of persons with positions in these financial instruments. Combining these two factors – greater interest rate volatility and the large number of potential traders in these assets – made conditions ripe for successful futures trading in financial instruments.

The argument claiming that there is an isomorphism between the theory of money and the theory of futures implies that there is an equivalence between the stock of money and the stock of open commitments of futures contracts. Much of the stock of money is the result of loans from lenders, the banks, to their customers on terms that include the interest rate, the duration of the loan, the size of the loan, and so on. The borrower gets a deposit in the bank that lent to him. He can write cheques on this deposit. Similarly, commitments in futures contracts result from transactions between buyers and sellers on mutually agreeable terms, including the price and quantity of the futures contact. To the clearing-house of an organised futures market, the short commitments are its assets and the long commitments its liabilities. The asset underlying the futures contracts corresponds to the reserves underlying the stock of money. One of the main differences between the stock of money and the stock of open commitments lies in this. In the case of money, the central bank can control the stock of high-powered money, while in the case of futures no governmental authority controls the size of the open commitments (save perhaps indirectly in the case of futures contracts in governmental assets such as Treasury bills, Treasury bonds and so on). As I have argued above, one of the most important effects of the existence of futures markets on the underlying assets they trade is analogous to money. Just as money represents a means of payment and a temporary abode of purchasing power, so too do futures contracts. The futures contracts conserve real resources because they are financial instruments that are less costly to create and to use than are the underlying assets they represent. Such uses of futures contracts and options obtain the benefits of specialisation. The assets that the futures contracts can be said to represent become available for those purposes where they have a comparative advantage. For example, an inventory of wheat can be used to make flour instead of being used as a medium of exchange. For the latter purpose it is cheaper to use paper claims to actual wheat instead of the actual wheat itself. This benefit of futures trading depends on its existence but not on its actual magnitude. The actual size of the open commitments results from the trades of the individuals each pursuing his own interests as well as he can. By virtue of the saving in real resources made possible by trading futures contracts, organised futures markets have a real effect on the economy – they increase real income. Similarly, the saving in real resources made possible by the use of money also has a real effect on the economy and also increases real income.

1 There is an excellent history of the development of futures trading in the United States in US Federal Trade Commission (1920). See also the articles by Williams (1982, 1984). Carlton (1984) contains a history of futures trading.
2 The mature views of Keynes on the analogy between monetary theory and futures trade is in Keynes (1936, chapter 17). The analysis of Hicks is in Hicks (1946, chapter 11).
3 Mehl (1934) describes the trade in privileges on the Chicago Board of Trade.
4 See Carlton (1983) for a thorough discussion of this argument.

BIBLIOGRAPHY

Carlton, Dennis W., 1983, "Futures trading, market interrelationships, and industry structure", *American Journal of Agricultural Economics* 65, May, pp. 380–87.

Carlton, Dennis W., 1984, "Futures markets: Their purpose, their history, their growth, their successes and failures", *Journal of Futures Markets* 4, Fall, pp. 237–71; reprinted as Chapter 14 of the present volume.

Hicks, J. R., 1946, *Value and Capital*, Second edition (Oxford University Press).

Keynes, J. M., 1936, *The General Theory of Employment, Interest and Money* (New York: Harcourt, Brace & Co).

Mehl, Paul, 1934, *Trading in Privileges on the Chicago Board of Trade*, Circular 323 (Washington DC: US Department of Agriculture).

Telser, L. G. and H. N. Higinbotham, 1977, "Organised futures markets: Costs and benefits", *Journal of Political Economy* 85, October, pp. 969–1000; reprinted as Chapter 13 of the present volume.

Telser, Lester G., 1981, "Why there are organised futures markets", *Journal of Law and Economics* 24, April, pp. 1–22.

US Federal Trade Commission, 1920, *Report on the Grain Trade*, (Washington DC: US Government Printing Office).

Williams, Jeffrey C., 1982, "The origin of futures markets", *Agricultural History* 56, January, pp. 306–16.

Williams, Jeffrey C., 1984, "Fractional reserve banking in grain", *Journal of Money, Credit and Banking* 16, November, pp. 488–96.

12

*Origins of the Modern Exchange Clearinghouse: A History of Early Clearing and Settlement Methods at Futures Exchanges**

James T. Moser
Federal Reserve Bank of Chicago

This chapter studies innovations in futures contracting before 1926. In that year the Chicago Board of Trade Clearing Corporation (BOTCC) began intermediating futures contracts. As more than 80% of US futures contracts were traded at the Chicago Board of Trade, this step established complete clearing as the standard for clearing and settling derivatives contracts. In the hundred years preceding the BOTCC, clearing and settling methods moved from bilaterally negotiated arrangements to practices, though now automated, which were very similar to those used today. In this chapter we will explore how contract terms evolved to conform to the ways they were cleared. Defining futures contracts as substitutes for directly-related cash transactions enables a discussion of the evolution of controls over the non-performance risk associated with contracts, and how these controls are incorporated into exchange methods for clearing contracts. Three clearing methods will be discussed: direct, ringing and complete, and the incidence and operation of each will be described in detail.

The development of the Chicago Board of Trade Clearing Corporation is discussed in relation to two principal objections to adoption of complete clearing at the CBOT; namely, anti-gambling provisions and

---

*The author would like to thank the Chicago Board of Trade who made their archives available for this research and to Owen Gregory who assisted in my access of these archives. The chapter has benefited from the comments of Herb Baer, Bob Clair, Jernie France, Geoff Miller, Lester Telser, Jeffrey Williams and participants at seminars held at the Department of Finance University of Illinois-Urbanna/Champaign and at the 1993 meetings of the Federal Reserve System Committee on Financial Structure and Regulation. Luis F. Vilarin provided valuable research assistance. Bernie Flores tracked down many obscure references. The analysis and conclusions of this chapter are those of the author and do not suggest concurrence by the author's previous or present employers.

privacy concerns. As we shall see, the Christie case overcame the gambling concern. The privacy concerns were mitigated by the Grain Futures Administration by giving the exchange members a choice between reporting their trades to the Department of Agriculture or to an exchange-controlled clearinghouse. Once these objections were overcome, CBOT members adopted a complete clearing system.

The plan of this chapter is as follows. The second section describes clearinghouse operations, the third offers a history of clearing operations preceding the CBOTCC. The fourth section describes the development of the CBOTCC and the fifth section summarises the paper.

A definition of futures contracts

The chapter uses a broad definition of futures contracting. The definition recognises the force of contractual obligations in two distinct regions of the state space: contract performance and contract non-performance. In performance states, one counterparty (the short position) delivers the underlying asset to its counterparty, the long position. At delivery, the long position pays the short position according to the contract terms. This portion of the definition conforms to the standard definition of a futures contract as an obligation between counterparties to make a future-dated exchange at a price determined at the contract's inception. However, the standard definition is insufficient in two senses. First, it omits the counterparty's choice not to perform as stipulated by the terms of the contract. This choice will be optimal to one side of the contract in non-performance states. Rights to exercise such choices are usefully construed as non-performance options and these options have value. Contract counterparties recognise that the cost of absolute performance assurances can exceed the value of trading benefits and act as barriers to trade. The mutual provision of non-performance options substitutes for absolute performance assurances to overcome these barriers and enables the realisation of trading benefits.

Second, the standard futures definition obscures institutional incentives to innovate contract design. These incentives stem from needs to reduce credit risk exposures by reducing probabilities for non-performance states and mitigating loss amounts when non-performance occurs. Further, the chapter develops connections between loss-sharing arrangements and innovations in contract terms. Exchanges adopting complete clearinghouses internalise non-performance losses increasing their incentives to innovative in ways that reduce these exposures.

Definitions of futures contracts that omit the non-performance option are common. Emery (1896, p. 46) defines a futures contract as a

> contract for the future delivery of some commodity, without reference to specific lots, made under rules of some commercial body, in a set form, by which the

conditions as to the unit of amount, the quality, and the time of delivery are stereotyped, and only the determination of the total amount and the price is left open to the contracting parties

This definition is typical in that its definition of futures contracting is based on contractual details.[1] I define futures contracts as enforceable substitutes for transactions in cash, commodities or assets.

The definition serves two purposes. First, it broadens the category of contracts called futures. Williams (1982), for example, argues that contracts traded at the Buffalo Board of Trade during the 1840s might be classed as futures because their terms were similar to those later adopted by the Chicago exchanges. The success of the Buffalo contracts developed from shared commercial interest in lessening non-performance costs. Chicago merchants, having similar interests, adopted similar contract terms. Both the New York merchants and the Chicago exchange members faced the potential for non-performance loss and responded by adapting their contract terms to control loss exposures. For the purposes of this chapter, the relevant commonality is the economic interest to limit losses.

The second purpose served by this definition follows from the first. Many futures-contract terms are best understood as efforts to minimise non-performance costs subject to available loss-sharing arrangements. The specific measures adopted to control losses are determined by the extant legal environment. Contracts traded at the Buffalo Board of Trade rapidly developed the use of performance bonds (margins) and delivery standards.[2] However, enforcing contract performance beyond this point proved costly. Subsequent changes in commercial law enable the Chicago exchanges to surpass the Buffalo precedent, ultimately offering performance guarantees, in this sense, the contracts traded at the Buffalo exchange served the same commercial purposes as the futures contracts exchanged in Chicago. Differences in contract details stem more from differences in legal environment than to differences of economic purpose.

Non-performance issues are often ignored because failures are infrequent. An understanding of the economic principles determining success is useful. This chapter follows Coase (1937) in arguing that the record of successfully managed non-performance risk is largely due to the internalisation of information and incentives obtained when exchange-affiliated clearinghouses guarantee performance.

The contention of this chapter is that the evolution of clearing arrangements was importantly influenced by the needs of members to control their risk of losses from contract non-performance. Thus, exchange policy on contract details like margin and marking contracts to market stems from its interest in the clearing mechanism.

A MODERN CONTEXT FOR UNDERSTANDING CLEARINGHOUSE OPERATIONS

A general description of modern clearing and settlement operations puts the development of the CBOTCC into a useful context. Clearing is the process of reconciling and resolving obligations between counterparties. Settlement, the last step of the clearing process, extinguishes the current liabilities between counterparties. This section develops clearing and settlement in two subsections. Clearing is first examined in the absence of non-performance, then the problem of non-performance contracts is developed.

The clearing of futures contracts is accomplished in three steps: registration; offsetting; and settlement.

Clearinghouses initiate the clearing process with the registration of traded contracts. This registration identifies the contractual counterparties and records their respective liabilities. Contract standardisation simplifies registration. In banking, contract standardisation is obtained by restricting payment on cheques to a single medium of exchange. Clearinghouse acceptance of a draft presented by a clearing member registers a claim against another member of the clearinghouse. Banks that are not members of the clearinghouse gain access to clearing facilities by opening correspondent accounts at member banks. Cheques presented for payment at correspondent banks are registered in the name of the clearing bank and processed through the clearinghouse. Payments made to the clearing bank are then credited to the accounts of its correspondents.

Standardisation of futures contracts is more complex. Items exchanged vary not only according to the commodity underlying the contract, but also according to the month of delivery. Thus, the registration of contracts is by type of contract and by delivery month. At futures exchanges, registration occurs as the buy and sell sides of traded contracts are matched.[3] Futures clearinghouses require non-member futures commission merchants (FCMs) to "give up" their trades to member FCMs.[4] A "give up" occurs when the non-member FCM relinquishes the trade to a member FCM. Like the correspondent services offered in banking, the non-member "gives up" their claim to the clearing member who clears the contract through the clearing organisation and then adjusts the accounts of the non-member.

Registration at a central clearinghouse simplifies the second stage in the clearing process; the offsetting of claims. Aggregation of the related transactions of each member of the clearinghouse enables identification of offsetting commitments. Offset occurs when the aggregated claims against any member are netted against the aggregate of the member's claims against all other members. In banking, the due-from and due-to claims of each clearing member are netted one against the other resulting in payment obligations into or from the member's clearinghouse account.

In futures markets, the clearinghouse nets the buy and sell orders registered to each clearing member. The current liabilities of the clearinghouse and its members are based on the net of these obligations. Thus, clearing reduces the number of liabilities by relying on the fungibility of individual contracts.

In the third step, contract settlement extinguishes the current payment liabilities of the counterparties to the contract. Banking settlements occur when the accounts of the clearing members are adjusted to reflect amounts paid. On payment the obligations of all parties are satisfied.[5] In futures markets, outstanding contracts are settled periodically by marking them to market. Generally, marks are either the most recent market-determined price for each contract or, at the contract's delivery date, the cash-market price of the underlying asset.[6] All outstanding contracts are marked to the settlement price. As contracts are marked to market, payments are determined by the netted obligations. Increases in settlement prices result in gains to long positions and losses to short positions. Increases in settlement prices result in gains to long positions and losses to short positions. Conversely, decreases in settlement prices result in losses to long positions and gains to short positions. Settlement occurs when payments due to the net positions of clearing members are made.

Unlike drafts, the futures contract generally remains outstanding following settlement. However, like drafts, settlement sets the current payment amount between counterparties to zero. Credit risk, that is the risk that one counterparty may fail to meet his obligations, does remain; periodic settlement reduces this risk to a proportion of the amounts of price change realised at the end of the next settlement period.

Non-performance problems

The non-performance problem is well illustrated by the experience of several Peoria, Illinois gains elevators offering forward contracts to local farmers. Quoting from the Federal Trade Commission's *Report on the Grain Trade*:

> Contracting for grain at a fixed price has proven an unsatisfactory practice with many elevators. The principle objection thereto is that if prices are in advance of those stipulated in the contract when the time of delivery arrives the farmer becomes dissatisfied and often refuses to fulfill the contract. If the elevator then attempts to enforce it the usual result is that the farmer transfers his business to another elevator. His dissatisfaction easily spreads to other farmers, especially if the elevator in question is an independent or one of a line company and many result in serious loss of business.[7]

Forward contracting is motivated by the expectation of benefits; in the above case, these beneficiaries are farmers and grain elevators. However, despite the motivation to enter into forward contracts, subsequent performance is conditional on the realisation of prices at the delivery date.

In some states, a counterparty will find failure to perform contract terms preferable to realising losses due to his performance of the contract. Recognising this, counterparties have incentives to restrict their non-performance opportunities. Doing so improves their access to the benefits of forward contracting. Along similar lines, Smith and Warner (1979) show how restrictions imposed by bond covenants lower debt costs by lessening the default risk of corporation-issued bonds. Similarly, counterparties are motivated to adopt contract terms that restrict their non-performance opportunities. These restrictions lessen both the likelihood of non-performance and the extent of losses should non-performance become unavoidable. Despite these incentives, contract modifications that assure contract performance are costly. When these costs exceed the benefits derived from further assurances of performance, it becomes optimal for counterparties to exchange non-performance options.[8]

Edwards (1984) distinguishes between bank and present-day futures clearinghouses. Bank clearinghouses settle by netting payments due between members and issuing credits and debits on the clearinghouse accounts of its members. The CBCH is obligated only to the extent of a member's account balance. Thus, it does not provide a full guarantee of member performance. Futures clearing houses guarantee performance of cleared contracts. At the CBOT where, by one estimate, six sevenths of all US contracts traded, clearing of futures contracts before 1925 was similar to that provide by CBCHs.

THE EVOLUTION OF CLEARINGHOUSES AND CONTRACT TERMS

Three clearing methods were developed before the formation of the CBOTCC. These are: clearing by direct settlement; clearing through rings; and complete clearing.

Direct settlement

Direct settlement is the bilateral reconciliation of contractual commitments. It is obtained through delivery or by offset between the original counterparties. For example, A contracts with B to sell 5,000 bushels of wheat in May at US$1.00 per bushel. There are three categories of possible outcomes in a system of direct settlements.

First, the specified terms of the contract can be performed. Thus, the contract is settled when A delivers 5,000 bushels of wheat to B in May and B pays US$5,000 to A. Settlement, then, is by direct delivery.

Second, the parties can settle the contract prior to May by establishing a mutually agreed price at which both are willing to extinguish the pending liabilities of the other. This method is called direct offset. I extend the previous example to illustrate. Suppose A and B agree to a second contract (in addition to the one above) in March as follows: B commits to deliver 5,000 bushels in May to A and A commits to pay B US$0.95 per bushel, or

US$4,750. The two contracts could be settled in May as follows: A delivers 5,000 bushels to B and B pays A US$5,000. Then B delivers 5,000 bushels to A and A pays B US$4,750. The net from settling both contracts is a US$250 payment by B to A; the wheat deliveries between them cancel. Alternately, both parties benefit by recognising in March that the earlier contract has been offset on payment by B of US$250. This payment is called a payment of the difference. The benefit from paying the difference is shared by A and B. Specifically, both avoid the expense of transferring title and both enjoy the benefit of reduced recording-keeping expenses

It might be objected that the present value of US$250 paid in March is greater than US$250 paid in May, thus B would refuse to settle on these terms. However, recall that the price for the March settlement is mutually agreed upon. B will agree to settle early provided the difference amount paid in March is less than the price change expected to be realised in May. A is willing to take an amount smaller than his expected price change because the payment amount can be invested. Thus, a mutually agreeable settlement price in March would be based on the present value of a settlement occurring in May.

Alternately, B could be compensated by a payment of interest from A on profit realised by A in March. Some exchanges required payment of interest on the profit.[9] Note, however, these approaches are equivalent provided the interest rate used by the exchange to determine interest on profits is equal to the market rate over the same term for an equal-risk investment.[10]

As a last possible outcome, one party could fail to perform and the contract is then settled based on standing enforcement procedures.[11] As this contract is described, recognition of non-performance can only occur in May. Before May, neither party has a performance obligation. Inferences might be drawn about the ability of a counterparty to perform and these inferences might be recognised as completely accurate; nevertheless, until a counterparty fails to perform a contract term, the contract stands. This aspect of the contract elevates risk in two ways. First, the possibility of accumulating substantial losses increases as the time remaining in the contract increases. Second, the failing counterparty has incentives to gamble in the hope of resurrecting his net worth and further increasing credit risk. Recognition of these additional sources of risk motivates adoption of contract terms that impose periodic demonstrations of continued performance capability. Inclusion of these provisions curtails the buildup of additional losses and restricts incentives to gamble on resurrection.

A HISTORY OF DIRECT SETTLEMENTS

Direct settlement is the oldest clearing arrangement. Emery (1896, pp. 35–6) describes trading in the warrants of the East India Company in 1733.

Here, the warrants were bearer instruments that gave title on a future date to a warehouse receipt for a quantity of metal (iron being frequently mentioned). Endorsement signified the sale of the warrant. Thus, transferred warrants were directly settled at the time of sale. The early form of these warrants was for specific lots of a metal, later "general warrants" were adopted which gave title to a quantity of the specified grade of metal.[12]

These warrants did not trade on exchanges organised to facilitate trading in these contracts. Thus, the resolution of legal disputes arising from trading in warrant contracts was obtained through the court system which proved costly. Obtaining a less costly route to handle trade disputes served as an impetus for the formation of exchanges and trading associations.[13] Section 7 of the Acts incorporating the CBOT in 1859 providing for exchange arbitration decisions is described by Andreas (1894, volume 3, p. 326) as having "the force and effect" of the judgements of the Circuit Court. An 1884 article in the Chicago Tribune (January 1, 1884, p. 9) illustrates the importance of this provision in lessening the cost of resolving trade disputes. The article compared the timeliness and cost of resolving the 1883 McGeoch contract dispute to the alternative of a court settlement. J. R. Bensley, receiver in the dispute, estimated that, had the issue been resolved in the courts: "... it probably would have taken ten years to settle the estate, and probably the creditors would have realised about 15 cents on the dollar ...". Instead the dispute, involving contracts having a total value of US$6 million, was resolved in six weeks at a total cost, including legal fees and damages, of US$20,000.

Ellison (1905, p. 15) records that trading in the Liverpool Cotton Association began shortly after trading in "to arrive" contracts for cotton in London.[14] By 1802 the practice of warehouse inspections before offering a bid had become too time-consuming and bids were based on samples brought from the warehouse. The account Dumbell (1927, p. 199) gives of the cotton market says that in 1826 the Liverpool market for "to arrive" contracts had developed to the point of attracting legal intervention: the case of *Bryan v. Lewis* determined that uncovered short selling amounted to wagering. By 1832, a market report had become necessary to keep current with the Liverpool market. Ellison (1905, p. 292) quoting the *General Circular*, says the report "gave an account of the imports, sales, deliveries, stocks, and prices current ...". During the 1840s, buyers retained a right of inspection before completion of the transaction. This became cumbersome as the buyer could keep the cotton off the market until completing his inspection. The practice was replaced and the system of samples was augmented with credible assurances that lot quality matched that of the samples.[15]

Indications are that contracts were directly settled in the early years of the Liverpool market. Ellison (1905, p. 244), describes the market before 1860 as follows:

> The merchant sold his cotton through a selling broker; the spinner purchased it through a buying broker. There were brokers who bought and sold, but they were an exception to the rule, and comparatively few in number.

Preparation of market reports also suggests direct settlement. To obtain the amount of cotton sold, Ellison (1905, p. 292) states:

> The sellers furnished an account of all cotton sold, but the buyers returns only the purchasers for export speculation, the balance, after deducting these two items, being put down as deliveries to customers.

The evidence implies that measures were taken by members of the Liverpool Association to assure contract performance. The early measures were left to negotiation between counterparties to fit existing circumstances. Ellison (1905) describes an 1825 transaction involving the placement of "two letters of credit of US$50,000 each" on a contract for 6,000 bales of cotton estimated to be worth US$500,000. By 1871, the system of voluntary performance bonding was replaced by a rule stipulating performance bonding of all contracts The necessity of an association rule to replace individual arrangements suggests that voluntary bonding had proved inadequate.

The effectiveness of Liverpool clearing methods is indicated by rule changes adopted at that exchange. Ellison (1905, pp. 292–95) describes these clearing problems as "numerous disputes arising out of the gigantic speculative transactions developed by the occurrences incidental to the American [Civil] War". Dumbell (1927, p. 196) says that "the confusion of the war years forced the [Liverpool] Association to prescribe for itself a constitution and a gradually increasing number of rules and bye-laws". Williams (1982, p. 306, fn. 2) describes a vote on the adoption of Association rules on June 17, 1864. Ellison (1905, pp. 325–26) states that the lack of grade standards created bargaining situations that contributed to contract-settlement problems:

> At other times the importer would discover that his property had been sold "short", in which case he would refuse to part with it except at a smart premium on current prices.

These comments suggest that the direct settlement of individually arranged contract terms used in earlier years had proved unwieldy in the volatile markets of the 1860s. The members of the Association responded by standardising contract terms, particularly grade standards as grade standards increase contract fungibility. As for transfers of the underlying commodity, contracts became close substitutes for one another, increasing the ability to offset positions.

Contracts traded at the Buffalo Board of Trade appear to be the first instances of extensive futures contracting in the United States which arose as traffic in grain via the Great Lakes increased. The early market was limited to cash transactions financed by bank advances. The Buffalo Board of Trade was organised in 1844. Williams (1982, p. 310) documents an extensive futures market by 1847 stating that: "most often the forward sales considerably outnumbered sales of flour on the spot". Settlements were direct, Williams (1982, p. 314) describes them as "between pairs of parties".

The Buffalo Association adopted measures to ensure contract performance. Contracts were most often settled by payment of differences rather than delivery. Williams (1982, p. 315 and 1986, p. 125) says that contract offset was obtained because "individual traders were themselves adept at enforcing the terms of contracts and keeping them comparable to others ...". Contract offset was assisted by standardising deliverable grain with stipulations on the source of deliverable grain or flour. Market participants understood the quality implied by these locations, enabling them to substitute contracts for one another. Later the Chicago markets would adopt grain-quality standards. Both approaches increased the fungibility of contracts.

The measures proved effective. Williams (1982, p. 313) quotes Buffalo newspaper descriptions of that market's response to late-spring ice on the Great Lakes which prevented delivery on May contracts. Prices were high "... as contracts are interested in having high prices maintained that the damages for non-fulfillment of contracts may be corresponding". In other words, long positions expected full compensation. Apparently, this expectation was justified as a subsequent news account stated:

> All the houses which contracted to deliver breadstuffs have so far settled without litigation or delay, except one, though the balances in many cases have been very heavy ...

Futures contracts for oil traded at several exchanges before the formation of Standard Oil.[16] The rules of these exchanges suggest that settlements were exclusively direct. Rules of the Titusville Oil Exchange do not provide for anything other than direct settlement of its contracts.[17] The rules of the nearby Oil City Exchange likewise provide only for direct settlements.[18] A *Petroleum Age* article further describes contracting at the Oil City Exchange, stating "futures were regulated to suit the convenience of either party".[19] This lack of uniform contract terms would have impeded contract offsets except those between original counterparties. Proceedings from the 1883 annual meeting of the New York Petroleum Exchange suggest the extent of trading on that exchange. Cash transactions and futures contracts for 1.5 billion barrels were exchanged during 1882. That exchange stipulated standard contracts to be 1,000 barrels, suggesting the number of contracts exchanged. Like the Pennsylvania exchanges, these appear to have been directly settled, as the rules of the New York exchange omit provisions to enable more advanced clearing methods.

Rules for each of the petroleum exchanges contained provisions to control non-performance risk. The rules of the Oil City exchange allowed counterparties to "call for mutual deposit on margin of not more than 10% of the contract price". The Titusville and New York Petroleum Exchange had similar provisions. Similarly, contract non-performance could result in the suspension of trading privileges or expulsion from the exchange.

To aid the clearing of its contracts, the New York Petroleum Exchange engaged the Seaboard Bank as its clearinghouse agent. Its agreement with Seaboard required each member of the exchange to complete a daily statement of "all Oil coming in and going out". This statement would be sent with a certified check "if the difference is against the sheet, or with the Oil, if the sheet shows more going out than coming in". Therefore, payments were made when deliveries were accepted and oil certificates were presented when deliveries were made. Members not fulfilling these requirements were in default of their contracts and faced fines or expulsion from the exchange.

Exchanges using more advanced clearing mechanisms generally permitted direct offset. The New York Produce Exchange required clearing through designated trust companies acting as its clearinghouse only for contracts held overnight.[20] Emery (1896, p. 66) describes the right for traders to offset any contracts made during the day directly. Chicago Board of Trade rules permitted direct settlements even after incorporation of its clearinghouse. Morris Townley, CBOT counsel in an option rendered to the exchange's board of directors, states:

> ... it is entirely lawful and proper for members to make trades, stipulating as between themselves that such trades are not subject to clearance through the Board of Trade Clearing Corporation. Such trades would be ex-clearinghouse and would be settled by direct delivery between buyer and seller.[21]

Similarly, the Italian Bourse offered complete clearing as an option. Counterparties submitted contracts to the clearinghouse within three days of the trade to gain a clearinghouse guarantee. Otherwise, settlement was left to the original counterparties.[22]

SETTLEMENT BY RINGING

Direct settlement limits settlement to the original counterparties. This is particularly important for contract offset because the ability to obtain direct offset depends on the joint interests of both counterparties. Contract fungibility simplifies these joint interests. This simplification means that settlements for many contracts can be accomplished simultaneously.

Table 1 Summary of contracts and counterparties

Counterparty	Buy price	Sell price	Profit (loss)
A		1.00	
	0.95		0.05
B	1.00		
		0.95	(0.05)
	0.97		
C		0.97	
	0.93		0.04
D		0.93	

Ring settlements however, are relatively informal arrangements between three or more counterparties having an interest in settling their contracts. Incentives to enter a ring stemmed from shared interests in reducing exposure to counterparty risk and reducing the cost of maintaining open positions. To achieve these benefits, participants in a ring were required to accept substitutes for their original counterparties. Exchanges adopted rules and practices that facilitated ringing settlements thereby enabling access to those benefits.

To illustrate settlement by ringing, consider four parties with positions in a contract requiring delivery of 5,000 bushels of wheat in May. A sold to B at US$1.00; B sold to A at US$0.95; C sold to B at US$0.97; and D sold to C at US$0.93. Table 1 enumerates these positions.

Long positions are listed as entries under buy price. Entries under sell price denote short positions. Profits (losses) based on direct settlements are listed for each transaction reversing a previously listed transaction.

The three possibilities previously examined are available here; that is, contracts can be settled between original counterparties through deliveries, offset or by standing rules for contract non-performance. Thus, the two contracts between A and B can be directly settled through offset. A fourth possibility, settlement by ringing, becomes available provided each party regards the contracts as fungible. Clearly, the contracts are fungible with respect to the commodity delivered and the date of delivery. Differences in the prices of the contract can be settled as before at mutually agreed prices. However, complete fungibility requires the acceptance of substitutes for original counterparties. Thus, B must regard the exposure stemming from a substitute contract with D as no worse than the exposure from their original contract with C. Given the equivalence of nominal contract terms and the substitutability of credit risk, the contracts are fungible and settlement by ringing becomes possible.

The equivalence of nominal contract terms is achieved by standardising contract terms. The literature on contract standardisation emphasises the early development of standard deliverable grades. Standardisation was in the interest of participants in these markets because it created benchmark contracts. Absent non-performance concerns, individual commercial interests can be restated in terms of the benchmark contract.[23] Trades in the benchmark contract can then substitute for trades in specific-but-similar commodities. Transactions in these separate and usually illiquid markets for specific commodities are replaced by transactions in the liquid market for the benchmark commodity.[24] The literature that recognises the importance of contract standardisation omits consideration of credit risk concerns; like the standard definition of futures contracting, this literature ignores the non-performance option. Nevertheless, contract fungibility depends on the substitutability of counterparty credit risk.

Working through a ringing settlement helps to establish the importance of counterparty credit risk. A ring formed by the counterparties A, B, C and D would need to arrive at a mutually agreed price at which all contracts will be settled.[25] This price must produce pay-offs identical to those obtained using the present value of the expected settlement in May. The most recent futures price provides this.[26] Taking the contract between C and D as the most recent price, the settlement price for the ring is 0.93. Reconstructing the previous table based on this price gives Table 2.

The bold cells in Table 2 denote settlement prices. An odd number of rows for a counterparty denote an open position after ringing. Therefore, after ringing, A and C have offset their contracts and B and D become counterparties to one open contract with B long and D short.

From Table 2, A records a net profit of 0.05 and B records a net loss of (0.05). Both amounts are the same as what would have been determined had their contracts been directly settled.[27] Thus, as for the profits and losses from their two contracts, A and B will be indifferent between

Table 2 Illustration of clearing through a ring

Counterparty	Buy price	Sell price	Profit (loss)	Net profit
A	0.93	1.00	0.07	
	0.95	0.93	(0.02)	0.05
B	1.00	0.93	(0.07)	
	0.93	0.95	0.02	
	0.97	0.93	(0.04)	(0.09)
C	0.93	0.97	0.04	
	0.93	0.93	0.00	0.04
D	0.93	0.93	0.00	0.00

settling directly and settling through the ring. C has opted to offset his contract made with B entirely, while B has opted to remain long in the contract. Recall that a system of direct settlements requires that both counterparties are willing to offset. To obtain a direct settlement, C might find it necessary to give up part of his 0.04 profit per unit to obtain an offset of the contract with B. A ring settlement enables C to avoid this bargaining problem, establishing a weak preference for settling through the ring compared with a direct settlement.[28]

The interests of counterparty B in the contract with C deserve special attention. Note that B loses his bargaining position with C upon agreeing to the ring's settlement price. In addition, B has a concern that his counterparty risk will increase. B's decision regarding counterparty risk has several dimensions. First, since C has an unrealised gain of 0.04, prices must rise 0.04 before B has any loss exposure. Replacing C with a substitute counterparty of equivalent creditworthiness at the current settlement price increases B's loss exposure. B will prefer a ringing settlement only if the new counterparty poses less credit risk than presently posed by C and his unrealised gain. This first dimension of B's decision implies that direct settlement is preferred.

The second dimension of B's decision compares his alternative counterparty risk without C's unrealised gain. B's present counterparty risk is subject to the contract performance of D in that D's failure to perform can lead to C being unable to complete his obligations with B.[29] Thus, the loss that B may suffer is conditional on how much loss C incurs from his remaining open contracts. An informed decision by B requires information on the status of his positions with C. In addition, B will be concerned with the dependency of those positions on all other contracts affecting C's performance. The exposure from directly contracting with D may be less than the exposure from a contract with C whose performance is dependent on the contract between C and D. To illustrate this, let us consider two alternative contracts each having no performance dependencies. In a contract between B and C, C fails if state one occurs but performs in all other states. In a separate contract between B and D, D fails if state two occurs but performs in all other states. To obtain an upper limit on B's current risk, add the condition that C's performance on the contract with B depends on the performance of D. With these conditions, B suffers a loss on the occurrence of states one or two. In contrast, directly contracting with D reduces B's exposure to the occurrence of state two only. Thus, B's upper limit is decreased from its present level. This second dimension of B's decision is determined by the extent and importance of these dependencies. As they increase, B's preference for a ringing settlement increases.[30]

Besides affecting counterparty risk, ringing can affect the cost of maintaining open positions.[31] Maintaining margin deposits is a significant portion of this cost. To illustrate, assume each counterparty maintains a margin deposit of 0.05 per bushel; that is, US$250 per contract. From Table 1: A and C have two contracts each, B has three and D has one. Margin deposits while these contracts remain open are: US$500 for A and C, US$750 for B and US$250 for D. With direct settlements, A and B would recognise that their cross exposures are nil, so under direct settlement rules A deposits no margin and B deposits US$250 for his open position with C. Direct settlement does not reduce the deposits of C because his two contracts remain open. D maintains a US$250 deposit for his one open position with C.

In a ring settlement, D is substituted for C so that margin deposits are US$250 each for B and D and zero for A and C. Thus, without ringing, C incurs the opportunity cost of maintaining deposits of US$500. With respect to their costs of maintaining margin deposits, A, B and D are indifferent between settling directly or through rings. C, on the other hand, prefers ringing.

Summarising, A is indifferent between direct and ringing settlements. B's preference is determined by comparing the value of his bargaining power over C with his assessment of the effect that entry into the ring will have on his credit risk. C strongly prefers a ringing settlement as he avoids a weak bargaining position and reduces his cost of maintaining open positions. Finally, D shares counterparty substitution risk with B. Therefore, the interests of B and C can be in conflict. An optimal ringing rule should enable C to improve his position without imposing costs on B or D. The exchanges obtained this result by recognising ringing settlements while making ring entry voluntary. However, the rules stipulated that once a ring was entered, its results were binding on its participants. From the above, B's minimum condition for entry into the ring is reduced loss exposure after a ring settlement. On satisfying this condition, B will enter the ring and allow himself to be bound by its settlements.

Rules enabling settlement through rings must provide finality for all offsets arranged through rings. Referring to the above example, finality is obtained when neither B nor D can enforce a claim against C should their substituted counterparty fail to perform. Exchange practice clearly intended finality.[32] The courts upheld the principle of offset finality on *Oldershaw v. Knoles*. Referring to the 1879 decision of this case in which a commission merchant arranged for a substitute counterparty who later failed, Bisbee and Simonds (1886, p. 158) state:

> The customer was held bound by this similar transaction on the part of his commission merchant; because, in employing the merchant, the customer was taken as intending that the business should be done according to the custom or usage of that market, whether or not he knew of such custom or usage.

Thus, counterparty C could obtain offset in his original contract with B and be assured that termination of this liability was final. The finality established by this decision did not depend on either specific commodities or apply to the original counterparties to the contract. The practice of the trade did not require delivery of specific lots, nor were commission merchants prevented from obtaining contract offsets with substitute counterparties. As these practices facilitated the trading of contracts for the benefit of customers, these customers could not choose to invalidate their contracts based on these practices on realising losses unrelated to these practices.[33]

To aid ring settlements, exchanges adopted centralised mechanisms for payments and deliveries. These arrangements performed like bank clearinghouses.[34] Counterparties realising a loss following a ring settlement submitted a record of their offset contracts, along with a suitable draft, to the clearing facility. These offset contracts were matched with the offset contracts submitted by counterparties realising gains. The clearing facility credited a clearing account in the amount of drafts received and debited this account when it disbursed payments to counterparties realising gains. Deliveries were made by passing warehouse receipts to the clearing facility. The receipts were then passed to parties taking delivery. Clearing fees were charged for each contract settled through the clearing facility.[35]

Although the earliest futures clearinghouses operated like CBCHs, the timing of this innovation suggests it originated with innovations in the clearing of stock transactions. The London Stock Exchange adopted a clearinghouse in 1874. Referring to that innovation, Stanley Jevons (1903, pp. 281–82) writes:

> An important extension of the clearing principle was effected by the establishment, in 1874, of the London Stock Exchange, which undertakes to clear, not sums of money, but quantities of stock.

Ellison (1905, p. 341) refers to an operating clearinghouse for the Liverpool Cotton Association in 1879. Emery (1896, p. 69, fn. 1) says the Liverpool clearinghouse was adopted in 1876, stating that this was the first clearinghouse for a produce market.

Nevertheless, clearinghouse operations originated in banking and the operations of bank clearinghouses probably were an important influence in the development of futures clearinghouses. Spahr (1926, pp. 70–1) dates the London CBCH to 1773. The London CBCH formalised clearing operations that previously took place between bank messengers frequenting bank-direct coffee houses. Gorton (1985) and Gorton and Mullineaux (1987) study the economic forces motivating formation of the New York clearinghouse in 1853 as drafts on individual bank deposits replaced bank-issued claims on *specie*.

Like direct settlement, ringing leaves the resolution of non-performance to individual counterparties. A letter by T. P. Newcomer to the Secretary of the Chicago Board of Trade illustrates this. The letter complains that margins were collected for his positions by his broker. The brokerage firm did not meet its margin calls and subsequently failed. Newcomer accurately describes his position with the CBOT stating: "We understand your Board is not a collecting agency and do not expect you to get us our money ...".[36]

A 1923 letter from the CBOT Rules Committee (in BOD October 2, 1923) further describes the position of a trader with respect to the CBOT, stating:

> That part of the regulation referring to the financial standing of a correspondent should be understood to mean that the principal should keep himself well informed, as business transactions between the two would warrant, as to the financial condition of his correspondent, so as to protect himself and the trade in general against any losses which might occur through the correspondent becoming insolvent.

Thus, each counterparty retained responsibility for monitoring the financial condition of his counterparties and to collect from them any payments due. The exchange did not take on this responsibility.

The exchanges gave their members several routes to control non-performance risk. The first of these was margin. Like direct-settlement clearing systems, exchange rules enabled counterparties to call for margin. Two forms of margin deposit could be required of contract counterparties. The first, original margin, was generally limited to no more than 10% of the value of the contract at its most recent futures price. This established an upper limit on the liquid assets that an exchange member would need to maintain. The limit curtailed the ability of members to call for excessive margin to force a solvent but illiquid counterparty into default. Margin amounts were set by mutual assent; members calling for margin were required to post amounts equal to those called for on their positions.[37]

The second form of margin, often called variation margin, was based on the amount of the difference between the contract price and the current settlement price. This amount applied only to the counterparty with an unrealised loss from the contract. Amounts paid as variation margin were also kept on deposit, they were not paid out until the contract was offset.

The right to assess margin was well recognised by the courts. The Illinois Supreme Court, in *Denton et al. v. Jackson*, held that without an agreement between counterparties on margin, contract margining was governed by the rules of the exchange. If the rules of the exchange enabled calls for margin, failure to meet a margin call put the counterparty into default.[38]

Exchange rules generally provided that margin amounts called from members had to be placed in accounts agreed to by the counterparties or with a bank approved by the exchange. A 1915 amendment to CBOT rules permitted members to fulfil margin requirements with cash or securities.[39] The rules stipulated the timing of these deposits with an expectation of rapid compliance. Rules often stipulated that margin amounts called for in the morning had to be deposited by early afternoon.[40] During this period the form used for customer-trade confirmation by Edwards, Wood & Co. Stock Brokers and Commission Merchants included a preprinted notice that customer positions can be closed out "when margins are running out without giving further notice". Like most, this firm was leaving no doubt that it could close out positions as it deemed necessary.

Failure to make a required margin deposit was a default on the contract. This provision enabled members questioning the ability of their counterparties to perform the terms of a contract to prove financial ability by posting a suitable margin amount. Failure to post the margin was non-performance, enabling members to curtail the accumulation of further losses and prevent gambles to resurrect net worth.

Rules requiring margin deposits were also facilitated by the clearinghouse. The ability to offset contracts and, by that, substitute counterparties required notification rules. The clearinghouse kept track of contract offset, enabling identification of counterparties. However, because not revealing the names of their principals was customary for commission merchants, the commission merchants obliged themselves to fulfil the terms of contracts.[41] Thus, exchange rules generally regulated calls for margin between commission merchants and not their customers, the actual principals. The commission merchants, in turn, arranged for margin deposits from the actual principals. The calls for margin from actual principals who were not members of the exchange were not subject to exchange limits on margin requirements.

Periodic contract settlement, now called marking to market, was not generally adopted by the exchanges. An exception was the Liverpool Cotton Association. Ellison (1905, pp. 354–56) says that periodic settlements were adopted by that exchange in 1883 because of heavy broker losses incurred during a corner. Forrester (1931, pp. 196–207) states: "Liverpool has weekly settlements; all outstanding contracts are reduced to weekly settlement price and all differences must be cleared". Liverpool's adoption of periodic settlement followed its adoption at the London Stock Exchange. Forrester (1931, pp. 196–207) says the motivation for both organisations was the same: "to prevent plungers without capital and unduly optimistic speculators from proceeding so far as to hurt the market before a check is applied". Periodic settlement curtailed non-performance losses in two ways. First, it lessened the probability of incurring a loss by imposing repeated demonstrations of financial ability to perform. Second, the accumulation of losses was curtailed.

Another route to control non-performance risk was to have control over the financial integrity of exchange members. The CBOT took early steps to lessen credit risk by regulating membership based on their financial ability. On March 27, 1863, its membership adopted a rule stating:

> Any member of the association making contracts either written or verbal, and failing to comply with the terms of such contract, shall, upon representation of an aggrieved member of the Board of Directors, accompanied by satisfactory evidence of the facts, be suspended by them from all privileges of membership in the association until such contract is equitably or satisfactorily arranged and settled.[42]

Thus, failure to comply with the terms of a contract could result in the loss of membership in the association, the value of this membership stemming from the right to trade contracts "on 'Change'". The principle of controlling non-performance through controlling membership was illustrated a year later in a debate over initiation fees. The FTC (1920, 2, p. 72) quotes from that debate:

> The amount of initiation fee is not one of the questions taken into account when a man is proposed for membership. The character and standing of the applicant is the only matter for consideration.

In 1873, the CBOT extended its efforts by making non-performance of any contract, on or off the exchange, grounds for requiring a demonstration of financial ability. Andreas (1894, volume 3, p. 299) gives the text of the rule:

> Any member of this Association who fails to comply with and meet any business obligation or contract, may, on complaint of any member of this Association, be required to make an exhibit of his financial condition on oath to the Directory of this Board, which shall be open to any aggrieved member; and should such member, failing as aforesaid, refuse to make such statement, he shall be expelled from the Association.

Thus, exchange rules gave the member methods to obtain protection from non-performance losses. However, the usefulness of these measures depended on the ability to detect the risk that a counterparty will fail. This dependency weakened the effectiveness that could be offered by these measures. In particular, failure of one's counterparty could result from failure on a contract made by that counterparty with yet another member. Thus, like dominoes, contract failures could cause a string of seemingly unrelated counterparties to fail. The ringing system, because it left assessment of counterparty risk to individual members who lacked the ability to detect the exposures of their counterparties, was susceptible to these systemic failures. The 1902 bankruptcy of George Phillips serves to illustrate this problem. Losses from the Phillips bankruptcy reached the accounts of 748 members, more than 42% of the CBOT membership.

The US exchanges generally permitted settlement of contracts by ringing. Arrangements for payments from counterparties varied between arrangements with banks to handle payments to adoption of a clearinghouse within the association. The CBOT developed a clearinghouse to handle difference payments in 1883. A contemporary description of its operations appear in the *Chicago Tribune* (January 1, 1884, p. 9) which states

> It takes no cognizance of the transactions on the board, but simply plays the part of a common fund, to which each member pays the excess of his daily debit over his daily credits, or receives the excess in case the later aggregate be greater than the former.

This clearinghouse began operating on September 24, 1883. Summarising its operations during its first 14 weeks, Chicago's *Tribune* reported that 26,986 cheques had been processed. That newspaper reported that under the previous system 260,000 cheques would have been required.[43] The *Annual Report* for that year stated that the clearinghouse "meets a want which has long been felt by the trade". The *Annual Report* of the following year suggests that the initial role of clearing operations was limited, stating: "we heartily indorse the proposed system of checking trades through the medium of the same Clearing House force which was adopted last Saturday". The comment suggests that contract registration initially occurred only at contract settlement. The introduction of contract registration in 1884 extended clearing operations beyond the handling of difference payments and beyond the clearing operations conducted by banks. After this date, traders intending to clear contract settlements through the clearinghouse were required to turn in cheque slips at the inception of these contracts which gave the clearinghouse a record of all outstanding contractual obligations.

The extent of clearing operations in the years before CBOTCC cannot be directly ascertained. Indications of activity by the CBOT clearinghouse are reported in a series of monthly reports to the Board of Directors. These provide the dollar amount of clearings occurring during the month for all contracts. Since the exchange did not report trading volume, the proportion of contracts processed through the clearinghouse cannot be determined.

Nevertheless, indications of the importance of the CBOT clearinghouse do come through in various communications. A 1906 letter (in BOD, October, 17 1906) describes the consequences realised by a member on being denied access to the clearinghouse: "It started a run on my firm. Those who had credits with us wanted their money at once, while those who owed us refused to pay, waiting, as some of them said, till they could settle for 10%". In 1910 the exchange found it had to rely on its clearinghouse so that it could offer option contracts. Trading in "privileges", the early term for what are today called options on futures contracts, was prohibited by an Illinois anti-gambling statute. This put the exchange at a disadvantage as other states permitted trading of these contracts.[44] The exchange determined in 1910 that the law did not restrict option trading if the investor held the underlying futures contract. To assure its compliance with anti-gambling statutes, the CBOT renamed these options contracts "indemnities" and required all indemnities to be cleared through the clearinghouse. This enabled the clearinghouse to verify that counterparties to indemnity contracts had previously registered futures contracts consistent with their indemnity positions.

Other evidence points to an increasing role for the CBOT clearinghouse, thereby confirming its importance. A 1903 change to the rules required that loss or damages from defaulted contracts be paid through the clearinghouse. While this left the determination of these amounts to the affected parties, their processing through the clearinghouse gave the exchange a record of the performance of each of its members.

In 1917 the exchange altered its rules by requiring notification from the clearinghouse when a member was called for margin. The member making the deposit was required to notify the clearinghouse of his compliance. On failure to make the required margin deposit, the calling party had the right to offset the contract on the exchange floor. The defaulted contract then became "due and shall be payable through the Clearing House ... the same as though the said contract had fully matured".

A federal stamp tax imposed during the First World War was charged on futures transactions. Members were required to report their transactions to Internal Revenue, the federal tax-collecting agency, so that tax assessments could be made. To verify their reported numbers of transactions, the clearinghouse was required to file reports with the agency giving total numbers of transactions. This implies that most contracts were being handled by the clearinghouse.

Ringing, generally with a facilitating clearinghouse or bank, was the predominate method of clearing at the US exchanges. Emery (1896, p. 66) quotes from a copy of the clearing sheet used at the New York Produce Exchange: the clearing bank is intended "to facilitate the payment of differences on the deliveries, direct settlements and rings of the previous day".[45] Rules for Lard and Provisions contracts traded at the New York Produce Exchange stipulated that on agreement to form a ring, the parties were compelled "to settle their differences on said contract with each other, on the basis of the settlement price". This rule bound exchange members to perform on the basis of the settlements reached by the ring. Emery (1896, p. 68) says the New York Cotton Exchange, established on

April 8, 1871, used direct and ringing settlements with no clearinghouse until 1896. Boyle (1931) describes clearing at the New Orleans Cotton Exchange. The exchange, organised on September 19, 1871, allowed contracts to "be settled through the Clearing House of the Cotton Exchange, or by offset between members, or by ringing out and paying only the balance due". Forrester (1931) describes the London Corn Exchange as relying on ringing settlements, Huebner (1911, p. 301) reports that the New York Coffee Exchange allowed ring settlement and that the London Metal Exchange continued to use the ringing method until recent times. Forrester (1931, p. 196) describes the London Metal Exchange as follows:

> There are two classes of members: (1) the ring members, who are principals and deal direct; and (2) the brokers and dealers, who, although members, are without those privileges. There are 40 seats in the ring, but the membership is 155, including 94 firms.

Settlement with complete clearing

Complete clearing interposes the clearinghouse as a counterparty to each side of every exchange-traded contract. Contracts agreed to on the floor of the exchange and accepted for clearing require the clearinghouse to take the buy side of every contract to sell and the sell side of every contract to buy. This role substitutes the credit risk of the clearinghouse for the credit risk of individual counterparties. Thus, contracts exchanged in a complete clearing system are completely fungible: grade standards imply that commodities underlying contracts are the same and completed clearing implies that all contracts have equivalent credit risks.

Table 3 adds the clearinghouse as counterparty to the contracts used to illustrate the ringing system.

As before, the figures in bold are the settlement prices obtained from D's trade. The first five columns repeat Table 2. The four columns on the right give the position of the clearinghouse in each of these contracts. It can be seen to have the opposite side of each contract. Naturally, these contracts offset one another in performance states. Thus, in states of the world where non-performance options expire unexercised, the outcome from a complete clearing system is identical to that of a ringing system.

The routes taken by payments do differ in a complete clearing system. In complete clearing, the cash payment made by A goes to the clearinghouse. In turn, the clearinghouse makes a cash payment to B. However, since the cash-flows of counterparties A and B remain unchanged, A and B are indifferent between a ringing settlement and complete clearing. Thus, from the perspective of A and B, the complete clearing system operates like a ringing system augmented by a clearinghouse to simplify payments between counterparties.

Table 3 Illustration of complete clearing

Counterparty	Trading partners				Clearinghouse			
	Buy price	Sell price	Profit (loss)	Net profit	Buy price	Sell price	Profit (loss)	Net profit
A	0.93	1.00	0.07		1.00	0.93	(0.07)	
	0.95	0.93	(0.02)	0.05	0.93	0.95	0.02	(0.05)
B	1.00	0.93	(0.07)		0.93	1.00	0.07	
	0.93	0.95	0.02		0.95	0.93	(0.02)	
	0.97	0.93	(0.04)	(0.09)	0.93	0.97	0.04	0.09
C	0.93	0.97	0.04		0.97	0.93	(0.04)	
	0.93	0.93	0.00	0.04	0.93	0.93	0.00	(0.04)
D	0.93	0.93	0.00		0.93	0.93	0.00	

Contrasting these offsets with those made through the ringing settlement, B's original and substituted counterparty is the clearinghouse. Recall that B's minimum condition for entering a ring settlement was a reduction in counterparty loss exposure. If provided with a choice between ringing settlements and settlements through a complete clearing house, B again requires lower loss exposures as his minimum condition for accepting complete clearing. With complete clearing, B is assured of no contract dependencies – a result that, in itself, will frequently obtain reduced loss exposure. Given B's participation in a ringing settlement, counterparty C's ability to offset his contract with B is unchanged by the adoption of complete clearing. C prefers complete clearing when it reduces his dependence on B's decision to participate. Otherwise, C is indifferent between ringing and complete clearing. Finally like B, D's preferences are determined after comparing the loss exposures obtained from each settlement arrangement.

Treating the adoption of a clearing system as a permanent choice, B will prefer a complete clearing system if he expects his loss exposure from the clearinghouse to be generally less than typically obtained from ringing arrangements. While some counterparties may pose less exposure than the clearinghouse, the avoidance of contract dependencies obtained by complete clearing will frequently imply lower loss exposures than those realised with ringing arrangements. Electing not to adopt a complete clearing system implies that B regards the proposed complete clearinghouse as a greater risk than his typical counterparty.

C's consideration of these alternative clearing systems focuses on his cost concerns. C can anticipate the increased costs he incurs each time that B elects not to participate in ringing settlements. C is indifferent between clearing arrangements provided B's participation is assured. Complete clearing makes C's offset automatic rather than an option for B, thus C's vote is entirely based on his higher expected costs under a ringing system.

As all traders can expect to find themselves on occasion in B's or C's position, ballots for the adoption of complete clearing express a consensus of the members' assessments of the net values of these future loss exposures and cost savings. Traders favouring the adoption of complete clearing will be those expecting either substantial savings on their margin deposits or less counterparty loss exposure. Traders with modest required-margin deposits will expect little savings from the adoption of complete clearing. If these traders can also sufficiently manage their counterparty loss exposures through a ringing system, they will lack incentives to vote for the adoption of a complete clearinghouse.

Once adopted, complete clearing shifts the realisation of non-performance losses to the clearinghouse. The loss-sharing arrangements made between members of the clearinghouse serve to allocate any losses among these members. Their need to control loss exposures from non-performance motivates the clearinghouse to adopt measures that reduce its exposure to non-performance risk.

The adoption of rules by the clearinghouse is constrained in two ways. First, the additional cost of adhering to clearinghouse rules cannot exceed the value added from the adoption of a complete clearing system. Therefore, members exert an external influence on clearinghouse decisions. Second, clearinghouse loss-sharing arrangements imply loss exposures from all open contracts. Members will avoid any loss-sharing arrangement that does not adequately compensate for this risk. As a lack of participation in the loss-sharing arrangement implies the clearinghouse is non-viable, this results in internal pressures on clearinghouse decisions.

Internal pressure motivates the pricing of clearing services to compensate participants in the loss-sharing arrangement for their exposure to losses from contract non-performance. However, compensatory payments reduce the value of benefits obtained from complete clearing, creating external pressure to limit rates of compensation. The compensation rate for the complete clearinghouse is its contract-clearing charge per unit of loss exposure. By pricing its services at a level which just satisfies its external pressures, its rate of compensation can be controlled by adopting trading rules that limit loss exposures. However, as trading rules impose explicit and implicit costs, these costs, when added to the clearing fees, would invoke external pressure against the clearinghouse. So, the adoption of contract rules requires clearing fees at or below indifference point of a majority of members. The clearinghouse satisfying this externally imposed boundary price will find its marginal condition is met when reducing clearing-fee revenues just equals the incremental cost stemming from a rule change.[46] Parties to the clearinghouse arrangement will be those members who find this rate of compensation adequate for their resulting levels of exposure. Risk-neutral clearinghouse members will demand compensation just equal to expected losses.

The analysis here suggests that the adoption of complete clearing systems would stem from a need to reduce counterparty exposure and to lower the cost of maintaining open positions. The following comment provides direct evidence of the importance attached to reductions in margin costs. After the Kansas City exchange adopted complete clearing, an observer contrasted its complete clearing with its former arrangements, stating:

> Under the system the tying up of large sums of money in margins, in event that a long or short on the other end refuses to "ring out", is avoided. Thus, an evil which tends to concentrate future trading into the hands of the stronger firms is eliminated.[47]

At times, exchange members avoided the cost of carrying these balances by simply not conforming to rules requiring quick responses to calls for margin. The following 1920 complaint from the Rogers Grain Co. to the Board of Directors (BOD October 26, 1920) of the CBOT illustrates:

> This rule has practically become a dead letter with many. Very few members put up margins until after the close even though they are called at Nine O'clock in the morning, while the rules provide that margins shall be put up and evidence is submitted of same within one hour ...

This response reduced the cost of carrying margin balances, but increased non-performance risk and, notably, further elevated margin requirements to safeguard against this risk.

Complete clearing systems originated in Europe. Emery (1896, pp. 71–2) indicates that European coffee exchanges featured complete systems. In France, a complete clearinghouse was called a *Caisse de Liquidation* and in Germany, a *Liquidationskasse*. Hirschstein (1931, p. 213) says that after 1924 the produce exchanges of Germany adopted complete clearing systems, modelling them on a system used for many years by a futures exchange operating in Hamburg. De Lavergne's (1931, p. 218) description of French clearing is consistent with the system described by Emery in the late 19th century:

> To replace the collateral security which might theoretically be required, but in fact is not, there has been set up at some of the exchanges a bureau of settlement (Caisse de Liquidation) in the form of an independent corporation. Such bureaus are attached to the exchanges at Havre, Lille, and Roubaix, as well as the sugar exchange at Paris.
>
> When a contract is entered into, the function of the bureau of settlement is to substitute for the original contract between buyer and seller, two new and

distinct contracts – one between the bureau as buyer and the original seller, and the other between the buyer and the bureau as seller. As a result of this operation, the individual buyers and sellers have no direct relations with one another, but each has a contract with the bureau.

In the United States, officials of the New York coffee exchange proposed to copy the clearing system used in the European coffee exchanges. Emery (1896, p. 72) says this proposal was rejected – apparently more than once. The FTC (1920, volume 2, p. 17) study says that complete clearing was first adopted by the Minneapolis Chamber of Commerce in 1891.[48] That exchange, later renamed the Minneapolis Grain Exchange, was first organised in October 1881. The FTC (1920, volume 2, p. 146) describes the adoption of Rule VI in 1891 establishing its complete clearinghouse:

> Section 1. All transactions made in grain during the day shall be cleared through the clearing association, unless otherwise agreed upon by the parties to the transaction. Upon acceptance by the manager of such transactions, the clearing association assumes the position of buyer to the seller and seller to the buyer in respect to such transactions and the last settling price shall be considered as the contract price.

The operation of the Association's clearinghouse is described in Annals (1911, p. 232) as follows:

> Most of the larger firms own memberships in it and it has been found to be almost a vital necessity to the trade. Certainly it insures less friction than the old way of trading and also facilitates business generally. When the trades are checked at the close of the session the member gives a check to the clearing house for margins or in case the market has fluctuated in favor of their customers they receive a check. It does away with a great deal of trouble. To settle with the clearing house at a certain time every day is a far different matter than calling each other for margins.

Rule VII provided the Chamber of Commerce with the right to control its exposure to non-performance risk by requiring margins. Quoting from the FTC (1920, volume 2, p. 146) report, the rule states:

> Section 1. The manager of this Association may call from purchasers below the market and from sellers above the market such reasonable margins as in his judgement may be necessary for the protection of the association. Such margins to be placed to the credit of the party paying the same and to be retained by the manager, in whole or in part, as he may deem necessary until the trades for which such margins have been paid have been settled.

The Board of Trade of Kansas City (Missouri) was organised in 1869, but by the mid-1890s was regarded as unsuccessful. In March 1899, the Kansas City exchange organised a Board of Trade Clearing Company modelled on the clearing corporation of the Minneapolis Chamber of Commerce. Referring to control over its exposure to non-performance risk, George G. Lee, clearinghouse manager, is quoted in Annals (1911, p. 227) as saying: "As the clearinghouse is responsible for all trades put through it, close tab must be kept on the position of each member". A clearing manager of the Kansas City exchange has wide latitude in setting margin requirements. The Duluth Board of Trade, first organised in 1881, incorporated its clearing association in 1909 and according to the FTC (1920, volume 2, p. 158) its rules were based on those adopted by the Minneapolis Association.

The complete clearing systems established in the United States before 1925 all followed the Minneapolis model for loss-sharing. Each was incorporated with share purchases limited to members of the exchange. Shares entitled clearing members to clear the trades of the exchange and to charge for this service. No other members were permitted to clear trades. Incorporation of the clearinghouse limited its owners' losses to the value of their shares and the sum of their deposits held by the clearinghouse.

Evolution of clearing systems

This section summarises the three methods of clearing. Markets employing these methods have certain commonalities: each market sought to increase the ability of members to obtain contract offset. This ability increased the benefits that could be derived from transacting in benchmark commodities rather than making similar but more costly transactions in the actual commodities.

Participants in direct settlement markets were able to offset contracts, provided they could be assured of the performance of their counterparties. The rules developed by markets relying on direct settlement systems were in response to the problem of contract non-performance. The intent of these responses was to lessen the frequency and extent of non-performance problems. Successful avoidance of these problems moved the members of these exchanges closer to obtaining full-contract offset. The two most common rules adopted covered exchange arbitrations of disputes and the right to collect margin deposits. Arbitration of disputes through the exchange avoided the slower and more costly resolutions available through the civil courts. The provision of the rights of a member to collect margin moved this source of protection from aright under common law where adequate time had to be allowed for the posting of the required margin. Entitling members to require margin enabled quicker access to this protection and established margin non-performance as a contract default. This later feature enabled members to protect themselves against build-ups of non-performance risk.

Ringing systems took advantage of the increased contract fungibility obtained from the adoption of standardised contract terms. Participants in a ring could be assured that the nominal terms of all contracts entering the ring were identical. Thus, offsets could be obtained by ring participants who accepted an externally imposed settlement price and were willing to accept the loss exposures implied by substituted counterparties. The benefits obtained from ringing were the reduction in contract dependencies and reductions in the cost of maintaining open positions; this later reduction coming principally through lower margining requirements as contracts were offset. Thus, access to these features enabled members to be more cost-efficient in their control over non-performance risks. Exchange rules were adapted by binding the participants to every ringing settlement to the agreements made by the ring.

Complete clearing established contracts as completely fungible. As every contract accepted by the clearinghouse involved the clearinghouse as a counterparty, the credit risk of every contract was identical. Exchanges created clearinghouses as synthetic members of the exchange and members of the clearinghouse became participants in a loss-sharing arrangement. Faced with non-performance risk, clearinghouses adopted rules to limit their exposures to non-performance losses and sought compensation for exposures remaining.

Therefore, clearing evolved from arrangements negotiated between individuals to become rules imposed by the exchanges. Flexibility lost during this evolution resulted in lower risks of contract non-performance and lower costs of maintaining open positions. The actions of these memberships were consistent with predictions offered by Smith and Warner (1979); by binding themselves to externally determined rules they achieved lower rates of contract default and lower costs of operation.

THE ADOPTION OF COMPLETE CLEARING AT THE CBOT

By 1900, precedents for operation of the CBOTCC were established. Complete clearing was well established at several exchanges in Europe and the United States and procedures enabling clearinghouse control over their exposure to non-performance risk had been developed. In 1903, some members of the CBOT saw a need for establishing a system of complete clearing. Two impediments to its adoption were generally recognised. First, anti-gambling statues prohibited trades that lacked an intent to deliver. Since the clearinghouse became a counterparty in all contracts, a clearinghouse delivery intent was necessary to avoid the charge that its contract positions were wagers. As this intent was clearly absent, the operations of the clearinghouse could be construed as gambling and, therefore, illegal. Second, Emery (1896, p. 89) says that complete clearing required regular disclosure of the positions taken by members of the exchange. Once these impediments were removed, the exchange moved to a system of complete clearing. Meanwhile, the members approved a series of measures that increased the importance of the existing clearinghouse.

The US Supreme Court resolved the first of the objections to complete clearing in May of 1905. The case was the *Board of Trade v. The Christie Grain and Stock Company*. Writing for the majority, Justice Holmes, comparing direct settlement with settlement through a clearinghouse, wrote:

> The contracts made in the pits are contracts between the members. We must suppose that from the beginning as now, if a member had a contract with another member to buy a certain amount of wheat at a certain time and another to sell the same amount at the same time, it would be necessary to exchange warehouse receipts. We must suppose that then as now, a settlement would be made by the payment of differences, after the analogy of the clearing house. This naturally would take place no less that the contracts were made in good faith for actual delivery, since the actual delivery would be to leave the parties just where they were before. Set-off has all the effects of delivery ...
>
> The fact that contracts are satisfied in this way by set-off and the payment of differences detracts in no degree from the good faith of the parties, and if the parties know when they make such contracts that they are very likely to have a chance to satisfy them in that way and intend to make use of it, that fact is perfectly consistent with a serious business purpose and an intent that the contract shall mean what it says.

The decision implied that the stated intent of a contract to deliver sufficiently distinguished the contract from a wager. Thus, even though a clearing process might involve transactions lacking a delivery intent, the original intent of the contract predominated. Therefore, methods used by the exchanges to clear contracts could not detract from the legitimacy of their business purpose.

Exchange records of 1903 refer to a petition listing more than 20 signatures proposing the adoption of a new system of clearing. The system proposed was called the Horsford System.[49] Action on this proposal was slow. In 1907 the committee assigned with formalising the Horsford proposal reported on their programme. Their plan stipulated that

> [t]he Clearing-House at the end of each day will (unless the contracting members otherwise agree) be substituted as the buyer from the seller and the seller to the buyer in every contract, and so remain until such contract is performed or settled.

The proposed initial capitalisation was US$1,000,000, to be obtained from the sale of shares to members of the exchange. As protection from non-performance losses, the plan enabled the proposed clearinghouse to collect margins according to standing rules between members. There is no record that this plan ever came to a vote.[50]

Subsequent proposals for changes to the system of clearing were voted on by the CBOT membership. The dates of these votes and their vote tallies are shown in Table 4.

Table 4 CBOT ballots for changes to complete clearing

Date of vote	Number voting for adoption	Number voting against adoption
September 25, 1911	75	656
January 1, 1915	272	528
October 4, 1920	228	502
January 27, 1922	418	423
September 3, 1925	601	281

The table shows that interest in changing the CBOT system of clearing was increasing. Rule changes proposed in the 1911 ballot would have reformulated clearinghouse procedures, but did not incorporate the clearinghouse.[51] Contracts would be offset through the clearinghouse at the daily settlement price and irrespective of counterparty. Thus, the identity of counterparties would not be known. The proposal included daily cash settlements with collections by the clearinghouse. In addition, the clearinghouse would collect margins on all open positions. Thus, with respect to measures to control non-performance losses, the clearinghouse would have functioned as a complete clearinghouse. However, the proposed clearinghouse would not offer guarantees of contract performance. The proposed rules were clear on this:

> in no case shall the said clearinghouse, its officers, or the Association incur any liability thereby except for the prompt payment and accounting of the moneys so received.

In case of default, the clearinghouse proposed to the BOD (July 20, 1911) would take control of all the deposits of the defaulting member and distribute them on a *pro rata* basis to all members involved in contracts with the defaulting member. Thus, the 1911 proposal arranged for loss-sharing among the counterparties of a non-performing member. Additional rules stipulating margin and daily marking of contracts to market were intended to limit the extent of losses that might be incurred.

No description of the 1915 proposal is available. At the annual meeting on January 11 of that year, the BOD (January 11, 1915) noted that the petitions had hung for many weeks and that most members were against it.

In 1917, a proposal was under consideration to amend CBOT Rule XXII, adding a Section 8. It was described to the BOD (August 27, 1917) as a change that would incorporate the clearinghouse with the intention "that said Clearing House Corporation shall become substituted as the seller to each buyer, and the buyer to each such seller". Proposed initial capitalisation was US$2,000,000. Should capitalisation fall below US$750,000, the exchange would revert to its previous method of clearing trades. Rule changes included provisions for margin assessment by the clearinghouse and daily marking of contracts to market. Before the proposal could come to a vote, the United States entered the First World War. Herbert Hoover, appointed US Food Administrator, wrote to the Board of Directors on August 14, 1917 requesting that the exchange halt trading in wheat futures. Interruptions in trading futures during the war years appear to have led to tabling of the clearinghouse proposal.

The 1920 ballot considered adding a section to CBOT Rule XXII to incorporate a clearinghouse.[52] The proposal required the clearinghouse to "assume the obligations" of contracts made on the exchange. Stockholders of the clearinghouse were entitled to clear trades in proportion to the number of shares held in the corporation. The CBOTCC would hold a first lien on the shares of its stockholders. In the event of failure of a clearing member, this lien enabled the CBOTCC to sell the member's shares in the clearinghouse using the funds obtained to cover the loss. If these funds were inadequate to cover the loss, the remaining members could choose between supplying additional funds or liquidating their shares. Thus, the loss-sharing arrangement under this proposal was limited to clearing members; their losses would be in proportion to the number of trades cleared, but liabilities were limited to the amount of capital. Rules of the proposed clearinghouse, particularly its divided restrictions, suggest the clearinghouse would have a fund of US$500,000 available to cover its liabilities. The proposal included provisions to limit the extent of losses. Margins would be required on all open contracts and all contracts were to be marked to market daily.

The Grain Futures Act, more often called the Capper–Tincher bill, was signed by President Harding on September 21, 1922. The Act gave the Department of Agriculture regulatory authority over the futures exchanges.[53] To carry out its authority, the Grain Futures Administration published Miscellaneous Circular 10 on June 22, 1923. The circular specified the general rules and regulations for the oversight role of the GFA.[54] The circular documents two primary activities of the GFA. First, it was to collect and record information on trading activity occurring at the exchanges. Second, it was to monitor the policies of the exchanges.

To implement the first of these roles, the circular stipulated that clearing members provide daily reports of their aggregate futures positions and their cash transactions stemming from these positions. The exception to aggregate of trading activity came when the net position of an individual trader exceeded GFA-specified levels. In these cases, the clearing member was required to report the aggregated trades of these individual traders. If trades were not made through a clearing member, individual traders were required to provide this information. Thus, whether reported by clearing members or the individual traders, the reporting arrangement would alert the GFA to a developing corner or squeeze.

The reporting requirements had two aspects that favoured the clearinghouse proposal. First, the aggregate cost of reporting information to the GFA was much less if reports were completed by clearing members only. The contract registration requirements of the clearinghouse provided all needed information. Second, because most often clearing members were required to report only their aggregate positions, the GFA would have no record of the trades of individual traders. This aspect of the regulation removed the remaining objection to adoption of a complete clearinghouse. Traders could choose between revealing their trades to a clearinghouse operating under control of the exchange or to the Department of Agriculture.

Initially, the CBOT continued to resist federal oversight. After the Grain Futures Act was signed into law, the exchange filed a case testing its constitutionality. On April 16, 1923, the US Supreme Court sustained the legislation.

The 1922 complete-clearing proposal was being considered as the Capper–Tincher bill was debated. This one-paragraph proposal to the BOD (January 19, 1922) would have inserted Section $9\frac{1}{2}$ into rule XXII. That section treated the clearinghouse as an artificial exchange member stating:

> If such clearing corporation shall be substituted in trades for members of this Association, such clearing corporation shall have all of the rights, and be subject to all the liabilities, of such members ...

The absence of further specifics in this proposal suggests that they did not differ substantially from that voted on in 1920. A newspaper report on the day of the balloting summarised the strong points of the proposal, stating: "the new plan could be used as the basis of taxes and would also answer requirements of the Capper–Tincher bill that certain records be kept". The tax referred to in the newspaper account is the stamp tax collected on futures transactions. The reporting requirements of the Capper–Tincher bill were the reporting requirements that would later be spelled out by the previously described Department of Agriculture circular. The newspaper report also summarised the position against the clearing house proposal. The Chicago *Daily Tribune* (January 27, 1922) quotes John Hill Jr. as saying: "the average 'pit trader' might be technically guilty of violating the 'bucket shop' law of Illinois". This was because substitution of the clearinghouse for original counterparties might be construed as a futures transaction. As this transaction would not take place in a trading pit, it might be regarded as a bucket trade. Illinois law prohibited these. However, the 1905 *Christie* decision would appear to invalidate this assumption: if transactions made in the course of clearinghouse operations could not be construed as wagers, conceiving of these same transactions as bucket trades is difficult

Over the next two years, the Board of Directors' attitude toward adopting a complete clearinghouse was mixed. A petition dated July 1, 1925 that came before the Board requested balloting on an amendment to Rule XXII to enable establishment of a general clearing corporation. The Board denied the petition with the explanation: "The subject has been before the membership on a number of occasions and defeated in every instance". An *American Elevator and Grain Trade* editorial entitled "Rubbing it in" (August 15, 1925, p. 94) remarked that by making the comment, the Board risked losing the support of Secretary of Agriculture Jardine who was, they reported, "staving off tinkering legislation".

Despite the attitude of its Board of Directors, member interest in complete clearing was evident. CBOT members voted on September 8, 1924 in favour of a proposal to establish a cotton contract on that exchange. Specifications for this contract, which began trading on December 12, 1924, included a requirement that all trades in that contract be cleared through a complete clearinghouse established solely to clear cotton contracts. This clearinghouse, referred to as the Chicago Cotton Clearing Corporation, began its operations on January 10, 1925. Bylaws for the corporation required a guaranty fund, formed from contributions of the stockholders of the corporation. Members could clear contracts in proportion to their contributions to the fund. Section 19 of the Bylaws provided that the liabilities and rights of buyer and seller were to be assumed by the Clearing Corporation once the trade has been accepted. To control its risk, contracts were marked to market daily. Price limits of two cents per bale were specified, which limited the payment amounts that could be assessed. Clearing members were required to post margin as determined by the Board of Directors, but no less than US$3.00 per bale nor more than US$25.00 per bale. Margin amounts were to be deposited in approved banks. In addition, the clearing manager could call for margin "to meet the variations in the market at any time during the day".[55] The attitude of members toward complete clearing is expressed in a comment of A. T. Martin, Cotton Committee member, reporting to the BOD (December 14, 1925), who wrote: "The committee feels that a broader interest will develop in cotton as soon as the general clearinghouse plan becomes operative and speculative conditions in cotton become more attractive".

A corner in early 1925 prompted Secretary Jardine of the Department of Agriculture to step up the campaign for exchange reforms. Among these reforms was a complete clearinghouse. At the Toledo meetings of the Ohio Grain Dealer Association (June 23 and 24, 1925) CBOT president Frank Carey reported formation of a special committee to consider the Jardine proposals.[56] This committee sent questionnaires to the 1600 exchange members. The preface to the questionnaire stated that the concern to be addressed was the wide price swings encountered when the "heavy trading of a limited number of professional speculators" becomes exaggerated by "large participation on the part of the public". The questionnaire then asked for comments on measures to limit these price fluctuations. The committee asked how positions might be disclosed, whether positions should be limited, whether option trading should be prohibited and whether price changes should be limited.[57] Responses to the questionnaire were positive. The Directors of the CBOT again put the clearinghouse issue before the membership for a vote.

On September 3, 1925, the CBOT membership approved changes to the Bylaws to provide for a complete clearinghouse. The clearinghouse would be organised as a corporation with 1,000 shares authorised at a par value of US$2,500 each. Clearing members were authorised to clear trades in quantities proportionate to the number of shares held in the corporation. Once the clearing corporation accepted a trade "the buyer shall be deemed to have bought such commodity from the clearinghouse, and the seller shall be deemed to have sold such commodity to the clearing house". Section 54 of the document includes a restriction on the liability of the CBOTCC, specifically:

> If any check given by any member to the clearing house under section 53, fails to certify on the following morning, all trades made by such member on the previous day shall be deemed rejected by the clearinghouse, and the clearing house shall be not be substituted on such trades.

Thus, the clearinghouse guarantee would not generally begin until the following business day. To control its non-performance risks, contracts would be marked to market daily and margins would be set by the clearinghouse. General margin amounts were limited to 10% of contract value, but individual members could be assessed larger margins where "the market position or financial standing of a particular member is such as to render his trades as unduly hazardous ...". These extra margin provisions required a three-quarter vote of the clearinghouse governors. In addition, the rule changes included price limits. Because these limits restricted the price change that could be marked to market, they limited the settlement amounts that could be called for by the clearinghouse.

A status report to the Board of Directors dated November 2, 1925 states that 680 shares in the clearinghouse had been sold, giving a total capitalisation of US$1,600,000. By that time three or four members of the previous clearinghouse said they did not intend to subscribe, while "about a dozen new names were added". A meeting of the Board on December 29, 1925 handled the staffing needs of the new clearinghouse.

The CBOTCC began operating on January 4, 1926. CBOT president Frank L. Carey, commenting on the beginning of operations, said: "For ten years this proposal had from time to time been the subject of lively discussion". Clearing of wheat contracts at the CBOTCC began Saturday, January 30, 1926.

Thus, the CBOT adopted complete clearing once the anti-gambling and privacy objections were met but the GFA did not impose complete clearing. The clearinghouse proposed in 1911 would have satisfied the contract-reporting requirements of the GFA without performance guarantees. The exchange had pursued improvements to its clearing facilities for years. The GFA circular, like the earlier *Christie* decision, removed an objection to the adoption of a complete clearinghouse. It did this by nullifying the privacy objection. Once the privacy objection was overcome, exchange members adopted complete clearing.

Before the adoption of complete clearing provisions, exchange rules left control over non-performance risk to individual counterparties. Members were required to monitor their counterparties and control their exposure through the protections offered by the rules of the exchange. The pre-1926 clearing arrangement enabled protection against contract non-performance, provided all contracts made with a counterparty were known but the inability of members to monitor the net exposures of a counterparty to all other contracts proved insurmountable when contracts were cleared through rings. Adoption of complete clearing shifted control over non-performance from individual members of the exchange to the clearinghouse. This shift was accomplished by creating a synthetic counterparty endowed with the ability to monitor the net exposure of all its counterparties. Combining this information advantage with previously developed mechanisms to control non-performance risk substantially reduced both the incidence of non-performance and the extent of losses realised from non-performance.

The motivation to develop complete clearing is similar to that of the banking industry. Gorton and Mullineaux (1987) argue that bank clearinghouses organise available information more efficiently than separate markets for the drafts on individual accounts. These authors argue this development is consistent with equilibria described by Coase (1937). Evidence of this consistency is drawn from two banking environments: non-panic and panic. During non-panic times, clearinghouses collected and disseminated information on the ability of member banks to meet the payment obligations imposed by drafts drawn on the accounts of these banks. This organisation provided a low-cost substitute for information reflected through market-determined prices for drafts. During panics, bank clearinghouses adopted a supervisory and regulatory role. In effect, they internalised the costs of industry failure and acted to minimise these costs. Specifically, they provided liquidity through the provision of collateralised loans and disciplined members with expulsion from the clearinghouse.

Complete clearing organises credit-risk information at the level of the futures clearinghouse. During non-panic times, complete clearinghouses collect information on the ability to exchange members to meet their potential payment obligations. The performance guarantee offered by complete clearinghouses motivates the incorporation of this information into contract specifications to obtain reductions in the risk of contract non-performance. Incorporation of this information substantially lessens the risk of non-performance loss, increasing the value of futures contracting. During financial crises, complete clearinghouses are empowered to reduce their non-performance risk exposures further by imposing additional margin requirements on individual members. This action decreases the risk of market failure. Thus, complete clearinghouses, acting to protect their own interests, increase the value of futures contracting by reducing the realisation of losses from contract non-performance. The realisation of these benefits owes much to the centralisation of information. Thus, like Gorton and Mullineaux, the clearinghouse arrangements of futures exchanges can be interpreted along the lines offered by Coase (1937). They replace the information flows offered by market-determined contract prices with a centrally coordinated information flow. Unlike the CBCH however, this information is not disseminated, but incorporated into contract specifications to lessen losses from contract non-performance.

CONCLUSION

Future contracts are defined as substitutes for associated cash transactions. This definition enables a discussion of the evolution of controls over non-performance risk. Three clearing methods were discussed: direct, ringing and complete. Direct-clearing systems feature bilateral contracts with terms specified by the counterparties to the contracts. Exchanges relying on direct clearing systems serve chiefly as mediators in trade disputes. Ringing is shown to help contract offset by increasing the number of potential counterparties. Increased ability to obtain contract offset is valuable because counterparties can reduce the number of dependencies of their outstanding contracts and can reduce the costs incurred by maintaining open positions. Entry into a ring settlement was voluntary; but on joining the ring, exchange rules bound the participants of the ring to its settlements. Exchanges that cleared through ringing methods generally adopted a clearinghouse to handle payments.

Complete clearing interposes the clearinghouse as a counterparty to every contract. This measure ensures that contracts are fungible with respect to both the underlying commodity and counterparty risk. Exchange members benefit from complete clearing because contract offset is automatic rather than dependent on counterparty interest in offset. The loss-sharing arrangements of the complete clearinghouse produce exposure to loss from every open contract. Members of the clearinghouse respond by requiring compensation for this risk. As the amount of this compensation reduces the value of complete clearing, the amount of this compensation is limited. Participants of the loss-sharing arrangement will substitute rules for pricing up to the point where the marginal value of risk reduction obtained from rules equals the marginal benefit from compensation.

Development of the CBOTCC was examined. Two principal objections to the adoption of complete clearing with anti-gambling provisions and privacy. The *Christie* case overcame the gambling concern and the GFA overcame privacy concerns by giving the exchanges a choice between reporting their trades to the Department of Agriculture or to an exchange-controlled clearinghouse. Once these objections were overcome, the members of the CBOT adopted a complete clearing system.

The success of the CBOTCC may have obscured its contribution to the development of present-day financial institutions. Because contract failure is generally infrequent and losses from these failures seldom disrupt markets, little attention is paid to the clearinghouses of the future market. Despite this lack of attention, the development of clearing systems that provide contract surety to a diverse group of counterparties is striking. Recent regulatory expressions of concern involving credit risk in the rapidly evolving OTC, or "swap", markets show the importance of contract surety. Efforts by participants of the OTC market to obtain greater contract surety are similar to the efforts made toward this end in the early years of the futures industry. OTC operators face the additional obstacle of mimicking too closely the features of futures contracts aimed at lessening non-performance risk. Should OTC contracts look too much like futures contracts, present legislation dictates they will be subjected to the regulatory burdens faced by the futures exchanges. Thus, like the early developments in futures markets, attempts to minimise counter-party risk in today's OTC markets are constrained by the existing legal structure.

1 For a recent example of the standard definition see Kolb (1991, p. 4). In contrast see Edwards (1984, p. 225) who takes a position consistent with that taken here stating that the clearinghouse "transforms what would otherwise be forward contracts into highly liquid futures contracts".
2 See Williams (1980, pp. 140–45) for a discussion of early margin rules. Hill (1990) reviews the literature on grade standards.
3 Unmatched buy and sell sides are referred to as "out trades".
4 I use the term "give up" to demonstrate the functional equivalence between present usage of the term "give up" and the relationship between clearing members and their associated non-member firms. In present usage, a "give up" occurs when a thinly capitalised FCM noted for good order execution is engaged to handle a large order. The FCM's thin capitalisation prevents him from taking the order. Under a give-up arrangement, he executes the order, then gives it up to a better capitalised member. Similarly, FCMs who are not members of the clearinghouse execute orders. Because they lack clearinghouse capital, these orders must be given up to a member of the clearinghouse.

5 The practice of using offsets rather than cash settlements dates to banking practice in early Florence. See Lane and Mueller (1985, p. 81).
6 Marks are determined by the settlement committee which generally follows the rules outlined in the text. In exceptional circumstances, these committees can determine marks by substituting their assessed valuations for market-determined prices.
7 Hereafter references to this six volume report will be given as FTC followed by the date, volume and page. This quote is from FTC (1920, 1, p. 113).
8 This is the incompleteness referred to in Kane (1980).
9 Forrester (1931) indicates the rules of the Liverpool Cotton Association required payment of interest on profits. As of 1882, the General Rules of the New York Produce Exchange also specified payment of interest on profits. It is doubtful, however, that these rules were intended to solve this present-value problem. More likely, the rules' substitution of one commercial transaction for another avoided the appearance of futures trading without intending delivery, an appearance which might lead to a charge of gambling.
10 Since there is some possibility of default on these interest payments, the rate is likely to exceed the default-free rate of interest.
11 One motivation for the formation of trading associations is to move arbitration of contract disputes out of the courts. See Milgrom, North and Weingast (1990) for an early example.
12 Nevin and Davis (1970, pp. 17–9) suggest that assignability of contracts developed much earlier. Thus, it would not be surprising to find similar contracts trading well before 1733.
13 For similar instances, see Powell's (1984) description of the development of arbitration in England.
14 Ellison (1905, p. 15) reports that listed prices for "to arrive" contracts were recorded in London in 1781.
15 Ellison (1905, pp. 126–33). Williams (1982, p. 311, fn. 28) cites an 1864 market summary in the Manchester Guardian indicating that the Liverpool Association replaced contracting for lots with contracting for grades by 1864.
16 Weiner (1991) attributes the decline of petroleum futures to the development of the Standard Oil Trust in the mid-1890s. I am indebted to Rob Weiner who provided copies of most of the petroleum-futures references which are used here.
17 General and local rules of the Titusville Oil Exchange, 1878.
18 Constitution and Bylaws of the Oil City Oil Exchange, 1876.
19 "Speculative Halls – The Oil City Oil Exchange", *The Petroleum Age* 4(4), May 1885.
20 Exchange rules were developed to the point of considering the problem of failure of banks holding margin deposits. These rules stated that "the loss shall be borne by the party or parties to whom it may be found said margins are due, taking the average price of like deliveries on the day such Bank or Trust Company failed as a basis of settlement". The rule made ownership of the balance contingent on realised prices rather than providing the original depositor with a claim on the balance.
21 Chicago Board of Trade, Board of Directors meeting January 1, 1926. (Hereafter cited BOD and the date.) Williams (1986, p. 307) also concludes clearing through the clearinghouse was not mandatory until the 1920s.
22 I am indebted to Giorgio Szegö who described these aspects of early practices at the Italian Bourse to me.
23 This practice is currently referred to as "pricing off the futures".
24 The significance of the ability to transact in futures markets is demonstrated by a 1923 letter from a J. C. Wood. Wood compares transacting in the cash markets and the futures markets: "In other words, the service performed by the broker in the execution of orders for 'seller month' or 'future month delivery' in the pit is an entirely different service than the service performed by the cash broker, whose work is largely specialized and carries with it technical knowledge and represents an expensive activity". BOD November 23, 1923.

25 Counterparty D has no incentive to enter this ring as he has no contracts to offset. D is included to demonstrate that his interests are not damaged by the settlements arrived at within the ring.
26 This is because arbitrage restricts futures prices to the cost of acquiring and financing a position in the underlying asset. Thus, the most recent futures price provides the current "cost of carry" for the underlying commodity. Therefore, this price gives the present value of a May settlement.
27 This is true despite settling the contracts at 0.93 rather than at 0.95 as in the direct-settlement example. As long as all gains and losses are computed from the same price and payments are netted, the payoffs are the same for all closed contracts regardless of settlement price. Use of the most recent price eliminates loss and gain carryovers for contracts remaining open. Cox, Ingersoll and Ross (1981) show that the investment implication of non-zero carryover amounts for these open contracts can affect the cost of carrying a futures contract.
28 The preference is weak because at a profit of 0.04 from direct settlement C is indifferent.
29 Counterparty D will also be concerned with counterparty substitution. As his concerns are the same as B's, the discussion which follows focuses on B's concern only.
30 Although not explicitly stated, the example implies that these risks do not diversify away.
31 These arguments for the cost of maintaining open futures positions closely follow Baer, France, Moser (1994).
32 An FTC report observed in 1920 that while counterparty substitution was implicitly "recognised in the rules of various exchanges, only in Chicago and St. Louis is it set up as a ring of traders executing contracts". FTC volume 2, p. 284
33 *Oldershaw v. Knoles* (4 Bradwell 63–73 and 6 Bradwell 325–333) built on two prior cases: *Horton v. Morgan*, 19 NY 170, an 1859 case in New York involving transfers of stock and *Bailey v. Bensley* 87 Ill 556, an 1877 case involving CBOT futures.
34 Nevin and Davis (1970, p. 6) indicate that similar clearing systems were employed by the French in the 13th century. Merchandise was bought and sold at fairs with transactions debited or credited accordingly by an on-the-spot banker. At the close of the fair, all transactions were cleared with settlement made in a single payment as needed between the banker and each merchant. This substantially reduced the number of necessary cash transactions.
35 A case described by Parker (1911) suggests that demand for clearing services can be substantial. The membership of a German exchange in 1903 sought to avoid government regulation by moving to another building and dropping its clearing house arrangements. Soon after, a private firm offered to clear settlements for those who chose to patronise it.
36 Minutes of the September 18, 1900 meeting of the CBOT Board of Directors as found in the archives of that exchange.
37 The CBOT attitude toward margin determination is expressed in a letter from a special committee which considered exchange-determined margin: "mandatory rules are impossible and that anything else would operate simply in the nature of a suggestion and would not only be unenforceable but ill-advised, because of the fact that each member of this exchange governs his transactions with his customers by his own ideas of credit". BOD September 17, 1912.
38 Bisbee and Simonds, (1884) (p. 150). The court went on to add that absent a contractual stipulation for margin and absent an exchange, common law enabled a call for margin. Under common law, a reasonable period had to be provided to meet the margin requirement before the contract could be regarded in default.
39 The rule does not stipulate the types of allowable securities. This determination was left to the counterparties.
40 The CBOT adopted a one-hour rule in 1887. It required members to meet calls for margin

within one banking hour. Prior to that date three banking hours were allowed.
41 See Bisbee and Simonds (1884, pp. 182–83). In 1894, the CBOT sought to control bucketshop operators by requiring brokers to name their principals. This would enable customers to determine if their orders had been filled on the exchange floor or simply bucketed. Failure to give the name was punishable by suspension or expulsion. See Lurie (1979, pp. 227–28). However, the name given would be the principal's counterparty. Most often this was the commission merchant acting for the other counterparty, not the actual counterparty.
42 Quoted from Andreas (1894, volume 3, p. 351). The following communication illustrates a typical settlement between counterparties. "We beg to advice you that a private settlement has been arranged on the Sept. Barley on which we yesterday reported default. This settlement is satisfactory to all parties concerned; consequently we ask that our request for the appointment of a Committee to determine a settlement price be withdrawn". BOD meeting October 7, 1919.
43 A newspaper report of cheque clearing by the Chicago Clearing Association indicates that growth in check clearings had declined. This was explained as due to the operations of the clearinghouse. (*Chicago Tribune*, December 3, 1883, p. 6.)
44 In March, 1901 a group of members organised an exchange in Milwaukee partially to avoid the Illinois restriction on option contracts. See FTC volume 2, p. 138. Subsequently, the CBOT felt that it was losing volume to this exchange and to the Winnipeg exchange which also permitted trading of option contracts.
45 It is not clear when the New York Produce Exchange adopted a clearing bank. Trading at that exchange began in 1877. A *Chicago Tribune* article summarising business developments in 1883 reported that the New York exchange formed an in-house clearinghouse on October 29, 1883 on a three-week trial basis. The operation halted after the trial period.
46 This marginal condition presumes that most members are indifferent to an addition or loss in the number of members. If additional members are valued, the sum of clearing fees and the cost of rules may need further reduction to offset the value lost when members dissatisfied with the cost structure leave the exchange.
47 "The Exchanges of Minneapolis, Duluth, Kansas City, Mo., Omaha, Buffalo, Philadelphia, Milwaukee and Toledo", *Annals of the American Academy of Social Science*, 1911, pp. 227–52. (Hereafter, citations from this source will be listed in the text as Annals and the page number.)
48 I am indebted to Lester Telser who provided this reference.
49 Extensive efforts were made to identify the source of the term "Horsford" which in one reference is referred to as "Hosford". Henry E. Hosford & Co., a commission brokerage firm, is listed as holding a membership in 1878 which was suspended in 1880. I was unable to locate a document associating the proposal with this individual.
50 See BOD October 20, 1907. The committee report includes a revenue projection for the proposed clearing house based on a yearly average for the previous six years for the standing clearing house. Their projection of US$73,680.45 in revenue implies the standing clearing house had averaged 1,473,609 grain contract clearings per year.
51 A letter to the Board of Directors dated October 30, 1923 contradicts this, listing the vote on this date as considering incorporation of the clearinghouse. The proposal voted on does not indicate that incorporation was under consideration.
52 This left intact rules providing for the existing methods of clearing. Should the clearinghouse fail, these rules would enable the clearing and settlement of trades to continue under the previous system.
53 See Lurie (1979) for a description of the events leading to enactment.
54 The FTC *Report on the Grain Trade* had noted the difficulty in determining trading activity giving the Minneapolis Chamber of Commerce as the exception. This exception was attributed to the clearing system used by that exchange. (FTC volume 2, p.7.)
55 A letter from the president of the clearing corporation in August 1925 argued that minimum

margins were not required during "dull markets" and suggesting that contributions to the guaranty fund be counted toward margin requirements between clearing members. The proposal was not adopted.

56 See *American Elevator and Grain Trade*, July 15, 1925.

57 "Acts to Prevent Wide Price Swings", *American Elevator and Grain Trade*, July 15, 1925, p. 28.

BIBLIOGRAPHY

Andreas, A. T., 1894, *History of Chicago from the Earliest Time to the Present Time.*

Baer, Herbert, Virginia Grace France and James T. Moser, 1994, "Opportunity Cost and Prudentiality: A Model of Futures Clearinghouse Behavior", Working paper (Federal Reserve Bank of Chicago).

Bisbee, Lewis H. and John C. Simonds, 1884, *The Board of Trade and the Produce Exchange: Their History, Methods and Law* (Chicago: Callaghan).

Boyle, James E., 1931, "Cotton Seed Oil Exchanges", *Annals of the American Academy of Social Sciences*, p. 171.

Brennan, Michael J., 1986, "A Theory of Price Limits", *Journal of Financial Economics* 16, pp. 213-33.

Chicago Board of Trade, *Annual Report*, various issues.

Coase, Ronald H., 1937, "The Nature of the Firm", *Economics* 4, pp. 386-405.

Cox, J., J. Ingersoll and S. Ross, 1981, "The Relation Between Forward Prices and Futures Prices", *Journal of Financial Economics* 9, pp. 321-46.

De Lavergne, A., 1931, "Commodity Exchanges in France", *Annals of the American Academy of Social Sciences*, pp. 218-22.

Dumbell, Stanley, 1927, "The Origin of Cotton Futures", *Economic History* 1, pp. 193-201.

Edwards, Franklin, 1984, "The Clearing Association in Futures Markets: Guarantor and Regulator", in *The Industrial Organization of Futures Markets*, Ronald W. Anderson (ed.), pp. 225-54 (Lexington Books).

Ellison, Thomas, 1886, *The Cotton Trade of Great Britain* (London: Effingham Wilson, Royal Exchange).

Ellison, Thomas, 1905, *Gleenings and Reminiscences* (Liverpool: Henry Young).

Emery, Henry Crosby, 1896, *Speculation on the Stock and Produce Exchanges of the United States* (New York: Greenwood Press).

Federal Trade Commission, 1920, *Report on the Grain Trade* (six volumes), (Washington DC).

Forrester, R. B., 1931, "Commodity Exchanges in England", *Annals of the American Academy of Social Sciences* 151, pp. 196-207.

Gorton, Gary, 1985, "Clearinghouses and the Origins of Central Banking in the U.S.", *Journal of Economic History* 42, pp. 277-84.

Gorton, Gary and Donald J. Mullineaux, 1987, "The Joint Production of Confidence: Endogenous Regulation and Nineteenth Century Commercial-Bank Clearinghouses", *Journal of Money, Credit, and Banking* 19, pp. 457-68.

Hill, Lowell, D., 1990, *Grain Grades and Standards: Historical Issues Shaping the Future*, (Champaign: University of Illinois Press).

Hirschstein, Hans, 1931, "Commodity Exchanges in Germany", *Annals of the American Academy of Social Sciences* 151, pp. 208-17.

Hoffman, G. Wright, 1931, "Governmental Regulation of Exchanges", *Annals of the American Academy of Social Sciences* 151, pp. 39-55.

Huebner, G. G., 1911, "The Coffee Market", *Annals of the American Academy of Social Sciences*, pp. 292-302.

Jevons, W. Stanley, 1903, *Money and the Mechanism of Exchange* (New York: D. Appleton).

Kane, Edward J., 1980, "Markets Incompleteness and Divergences Between Forward and Futures Interest Rates", *Journal of Finance* 35, pp. 221-34.

Kolb, Robert W., 1991, *Understanding Futures Markets* (Miami: Kolb Publishing Co.).

Kulp, C. A., 1931, "Possibilities of Organized Markets in Various Commodities", in *Annals of the American Academy of Social Sciences*, pp. 190-91.

Lane, Frederick C. and Reinhold C. Mueller, 1985, *Money and Banking in Medieval and Renaissance Venice*, volume 1 (Johns Hopkins University Press).

Loman, H. J., 1931, "Commodity Exchange Clearing System", *Annals of the American Academy of Social Sciences* 151, pp. 100-09.

Lurie, Jonathan, 1979, *The Chicago Board of Trade 1859–1904: The Dynamics of Self-Regulation* (Urbana: University of Illinois Press).

Milgrom, Paul R., Douglass C. North and Barry R. Weingast, 1990, "The Role of Institutions in the Revival of Trade: The Law Merchant Private Judges and the Champagne Fairs", *Economics and Politics* 2, pp. 1–23.

Nevin, Edward and E. W. Davis, 1970, *The London Clearing Banks* (London: Elek Books).

Odle, Thomas, 1964, "Entrepreneurial Cooperation on the Great Lakes: The Origin of the Methods of American Grain Marketing", *Business History Review* 38, pp. 439-55.

Parker, Carl, 1911, "Governmental Regulation of Speculation" in *Annals of the American Academy of Social Sciences*, pp. 126-54.

Powell, Edward, 1984, "Settlement of Disputes by Arbitration in Fifteenth-Century England", *Law and History Review* 2, pp. 21-43.

Smith Jr., Clifford W. and Jerold B. Warner, 1979, "On Financial Contracting: An Analysis of Bond Covenants", *Journal of Financial Economics* 7, pp. 117-61.

Spahr, Walter Earl, 1926, *The Clearing and Collection of Cheques* (New York: The Banker's Publishing Corporation).

Taylor, Charles H., 1917, *History of the Board of Trade of the City of Chicago* (Chicago).

Telser, Lester, 1986, "Futures and Actual Markets: How They Are Related", *Journal of Business* S5-S20; reprinted as Chapter 11 of the present volume.

Weiner, Robert J., 1991, "Origins of Futures Trading: The Oil Exchanges in the 19th Century", Working paper 301 (Brandeis University).

Williams, Jeffrey C., 1980, "The Economic Function of Futures Markets", PhD dissertation (Yale University).

Williams, Jeffrey C., 1982, "The Origin of Futures Markets", *Agricultural History* 56, pp. 306-16.

Williams, Jeffrey C., 1986, *The Economic Function of Futures Markets* (Cambridge University Press).

13

*Organised Futures Markets: Costs and Benefits**

Lester G. Telser and Harlow N. Higinbotham
University of Chicago

> *A futures contract is to a forward contract as payment in currency is to payment by cheque. An organised market facilitates trade among strangers. Such a market trades a standardised contract under appropriate rules. The equilibrium distribution of market clearing prices is asymptotically normal with a standard deviation that varies inversely with the volume of trade, given underlying supply and demand conditions. Empirical relations giving the commission and margin per contract as a function of the volume of trade and outstanding commitments for 23 commodities support the theory. Also, comparisons of pertinent aspects of 51 commodities divided into active, less active, and dormant groups are consistent with the theory.*

Of all of the hundreds of thousands of commodities in the economy only a few have ever been traded on an organised futures market. This is a puzzle. The basic idea we use to give an answer to this puzzle draws on the theory of money. An organised futures market creates a medium of exchange, a futures contract, with many of the attributes of money. A futures contract facilitates trade in the commodity in the same way that the use of money in exchange has advantages over trade by barter under normal conditions. Nor is this all. A futures contract is a temporary abode of purchasing power in terms of the commodity. It is these aspects of a futures contract and not the more common view that futures contracts

---

*This paper was first published in the *Journal of Political Economy* 85(5), pp. 969–1000 (1977) and is reprinted with the permission of the University of Chicago Press. The authors wish to thank the financial support of the National Science Foundation. Members of the Industrial Organization Workshop and of the Seminar in Applied Price Theory of the University of Chicago gave helpful comments on earlier drafts. Charles Cox, Edward Lazear, George Stigler and an anonymous referee deserve our thanks. All remaining errors and defects are our sole responsibility.

enable hedgers to avoid risks that explain the benefits of an organised futures market.

A transaction on a physical commodity in the real world has many unique characteristics stemming partly from the identities of the parties to the transaction, their reliability, credit worthiness, promptness, honesty, and flexibility; the qualities of the good; and the circumstances of the trade. These particulars make a transaction in the physical commodity resemble a barter trade. The two parties in a mutually acceptable trade have costs like those that arise in the double coincidence of trade by barter. The introduction of a standard futures contract by an organised futures market creates a financial instrument that can be traded without knowing the actual identity of the two parties in the transaction. The seller of the futures contract incurs a liability to the organised futures market, and the buyer acquires an asset from this market. Neither need have concern about the integrity of the other in the same sense that one who accepts a US$10 banknote in payment for something need not worry about the credit rating of the buyer. This argument implies that any commodity not made to order can benefit from the introduction of an organised futures market. The latter has a cost. Therefore, the use of this commercial invention, the futures contract, appears only for those commodities where the benefits outweigh the costs.

The development of a formal theory of these costs and benefits begins with a stochastic model of market clearing prices which has its roots in the economics of information and search. This model shows that the distribution of market clearing prices is asymptotically normal. The standard deviation of the limiting normal distribution depends on certain properties of the underlying schedules of bids and offers. Also, the realised market clearing price approaches the asymptotic market clearing price in probability as the number of traders increases. This theory is a marriage of economics and probability.

The standard deviation of the distribution of market clearing prices measures the liquidity of the market. There is a cost of lowering the standard deviation and there are benefits. The optimal amount of liquidity takes both the costs and benefits into account and results in some positive value of the standard deviation.

Although the standard deviation is an important element in our theory and seems to be a roundabout way of introducing risk, our emphasis is different and leads to different implications from a theory of futures markets based on the concept of reducing risk. An organised futures market is not necessary in order to obtain the advantages of hedging. These can result from a system of long-term contracts. However, an organised futures market can reduce the costs of moral hazards by the introduction of a standard contract. A long-term contract between two parties depends on their integrity. In a situation with many participants, a standard contract with the backing of the organised market has less moral hazard. Ours is not a theory of why there may be advantages in long-term contracts; it is a theory of the net benefit of an organised exchange. There is some common ground since a long-term contract and a futures contract both refer to dates in the future. Hence an organised futures market is more likely to occur in commodities where the timing of transactions is more important. This only gives a set of necessary conditions, and it does not furnish sufficient conditions for the emergence of an organised futures market.

More generally, this theory asserts that a necessary feature of an organised market is a standard contract traded in that market. This standard contract need not be a futures contract. It may be a certificate that allows the bearer to obtain on demand a specified quantity of a good. Certain forms of certified warehouse receipts, gold and silver certificates, and similar financial instruments, such as shares of stock, have these properties. If the standard contract represents a physical good, then some legal entity is liable for fulfilling the terms of the contract on demand. This person would hold sufficient reserves to maintain the probability of default at an optimal level. Reserves equal to a fraction of the outstanding commitments of the standard contracts may suffice. These standard contracts may circulate among the traders at market-determined prices, and the participants may never wish to convert their standard contracts into the physical good. This is plain in the case of shares of stock since rarely do the shareholders wish to liquidate the corporation.

This argument implies that the number and identity of the participants in a potential organised market is important. If this number is small, if the potential participants know each other, and if there is little turnover of these participants, then they may not need a standard contract to deal with each other. If there is a large number of potential participants and a rapid turnover of these traders, then this raises the benefit of a standard contract as well as the benefit of having an organised exchange. The members of the exchange meet certain requirements so that those who deal with them can rely on their integrity. The members of the organised exchange also face the problem that their clients may lack integrity. That is, as the number of potential participants in an organised market increases, the cost of having them all trade on their own account in the organised market may rise more than proportionally. This is not only because of the increasing congestion this would cause but also because there may be increasing costs of having one organisation certify all of the potential traders.

Even without an organised market in a standard contract the traders may have confidence in each other if each is a legal representative of well-known principals. This is an important consideration when the trade is in financial instruments that represent actual goods or assets instead of trade in the physical goods or assets themselves. The latter does occur in, say, a flea market. Even if the buyer is a good judge of the commodity before him in the flea market, there is still the problem that he must know that the seller is the legal owner of the good or is the agent of the legal owner. These are important problems in the real world that are often overlooked in some formal theories of market exchange.

The theory herein owes a debt to some important contributions by others. Most notable is Holbrook Working, who always put the emphasis on the functions of a futures market aside from its advantages to hedgers. In some of his work he came close to regarding a futures contract as a temporary abode of purchasing power in the physical commodity (1953, 1967). H. S. Houthakker, also points out the similarities between money and futures contracts in his 1959 article. In contrast, most of the economists who have written on the subject, beginning with Marshall's *Industry and Trade* (1920), follow the convention of focusing their attention on hedging and speculation, which loses sight of the more fundamental properties of an organised exchange.

There is another important aspect of the subject that is peripheral to our main interest, and we do not discuss it. This refers to the actual rules of the exchange and the operation of the market. This topic deserves the close attention of economists because it deals with the practical problem of creating a set of conditions that can make a real market function according to the theoretical model of a perfectly competitive market. Informed observers of actual futures markets know that this is a difficult task. An organised futures market cannot survive unless it does approximate a perfectly competitive market.

PROPERTIES OF AN ORGANISED EXCHANGE

Before presenting a theory of the costs and benefits of an organised futures market, it is helpful to describe some of its important properties. Trade occurs in one physical place, the floor of the exchange, during specified hours, called the trading session. The traders cry out bids and offers, making a bilateral auction market. Only members of the exchange may trade on the floor, and no member of the exchange may make transactions off the floor of the exchange. Members of the exchange either trade for themselves or under the instructions of their clients who are typically not members of the exchange. Many organised exchanges are responsible for ensuring that the terms of a transaction are fulfilled. The two parties in a transaction agree on price and quantity of the given futures contract. All other terms of the contract are specific to the rules of the exchange.

The exchange determines the number of members who may trade on the floor. A membership is often called a "seat" on the exchange. A member may sell his seat to a non-member. However, this is subject to some control by the exchange. The exchange investigates the character of potential members, and it may refuse permission to a potential buyer of a seat. The price of the seat is mutually agreed upon by the seller and buyer and is not subject to control by the exchange.

The exchange can discipline its members by imposing fines, suspending their trading privileges, or by expelling them. The oldest organised futures exchanges, such as the Chicago Board of Trade, have evolved elaborate rules as a result of their long experience. These rules intend to give those who trade on the exchange confidence in the reliability of the transaction executed on the exchange. Members of the exchange must execute the orders of the public, that is, non-members, before they execute their own trades. Exchange members who make fictitious trades are subject to penalties. Such fictitious trades intend to record prices that will mislead others. The exchange defines the terms of a contract such that all contracts of a given class are perfect substitutes and such that the validity of a transaction in that contract does not depend on the identity of the principals. Thus, a standard contract for the delivery of 5,000 bushels of wheat in July 1976 executed on the floor of the Chicago Board of Trade in January 1976 at a price mutually agreeable to the parties is as well defined as currency. A futures contract is to a forward contract as currency is to a cheque drawn against a demand deposit in a commercial bank. The validity of a genuine US$10 bill, one printed in the US Bureau of Engraving, does not depend on who offers it in payment, while the validity of a US$10 cheque depends on the identity of the person who writes or presents it and on the identity of the bank. Similarly, the validity of a July futures contract on the Chicago Board of Trade is as good as the faith and credit of the Chicago Board of Trade, while the validity of a forward contract for July delivery depends on the integrity of the buyer and seller in the transaction.

A forward contract shares with a futures contract the important property that the buyer and the seller agree in the present on the terms of a transaction that will be completed at a specified time in the future. The important distinction between the two kinds of contracts lies in this. In a forward contract the actual identity of the buyer and seller is important. Neither has recourse in case of dispute to a third party other than a court of law. The validity of the forward contract depends on the good faith of the two parties themselves. A futures contract has a third party, the organised exchange or its designated representative, that guarantees the validity of the contract and will enforce the terms. When A sells forward to B, the consummation of the transaction depends on their honesty. When A sells a futures contract to B, A incurs a liability to the organised exchange and B acquires an asset from the organised exchange. The exchange, or its clearing-house, enters the transaction as a third party. It records the sale by A as an asset on its books and the purchase by B as a liability on its books. The outstanding commitments in a futures contract constitute the open interest in that contract and correspond to the stock of money. This analogy with money is important.

It follows from these arguments that the introduction of an organised futures market has consequences resembling those that occur when money is introduced into an economy to facilitate trade. Before the introduction of money, we may assume there is trade by barter with all of the disadvantages of trade by barter. Money provides a means of making trades at a lower cost and it is a temporary abode of purchasing power. Similarly, a futures contract facilitates trade in the commodity and it is a temporary abode of purchasing power in that commodity. A futures contract in wheat is to actual wheat as a US$10 banknote is to a market basket of US$10 worth of actual commodities.

Price quotations on an organised exchange convey reliable information about mutually agreed-upon terms of genuine transactions. Some individuals may contrive to gain from deception by violations of the rules or by exploiting defects in their wording, thereby violating the spirit of the rules. The result of long experience leads the exchange to make more elaborate rules in order to reduce the expected return to potential violations. In an organised market with bona fide transactions among honest men, the transaction prices convey such information as would approximate the equilibrium of a competitive market. Departures from a competitive equilibrium, called corners, cannot occur unless there is a violation of the rules of exchange. It appears that a necessary condition for a corner is the ability of one or more traders to deceive others by having the record show false transactions and false prices.

The conditions for the emergence and survival of an organised exchange are costly to bring about. Without these costs every commodity would have an organised market since surely such markets are beneficial. Therefore, to explain the presence or the absence of an organised futures market in a commodity requires a theory of the costs and benefits of such a market.

THE DISTRIBUTION OF MARKET CLEARING PRICES

A futures contract enables trade to occur in the present with reference to dates in the future. It is equivalent in its effects to a means of classifying different grades or qualities of a commodity. It sorts trades with respect to time. Without futures trading a given set of spot transactions would be more heterogeneous with respect to the preferred timing of the traders. The set includes trades that may have taken place earlier or would take place later and which do occur now because the traders lack the alternative of trading in futures contracts. The use of futures contracts sorts the transactions with respect to time so that transactions in contracts of a given maturity date are more nearly alike with respect to those attributes that are correlated with time.

This sorting out of transactions with respect to time that results from the possibility of trading futures contracts has several effects. It can lower the dispersion of the distribution of market clearing prices without necessarily changing the mean price of this distribution. Therefore, a given number of transactions can occur at more nearly equal prices. Also, the traders incur less delay in making their transactions at mutually acceptable prices. For these reasons futures trading increases the liquidity of the market.

An organised market has another important property. It tends to attract trade from a wide geographical territory. Since characteristics and qualities of a commodity are correlated with their location in space for reasons similar to those giving a correlation between timing and relevant aspects of the good, the pooling of trades from many locations into a central market increases the heterogeneity of the potential transactions in the market. This in turn increases the dispersion of the distribution of market clearing prices. An offsetting force is the larger volume of trade attracted to the central market that reduces the dispersion of the distribution of market clearing prices.

Underlying these arguments is a basic theory about market clearing prices. The traders present in a market at a given moment can be regarded as if they were a random sample from the underlying population of traders. The latter has a distribution of minimal acceptable prices for offers and maximal acceptable prices for bids. The distribution of offers is given as follows:

$$U(p) = \int_0^p u(r)\,dr \qquad (1)$$

where $U(p)$ is the cumulative distribution of the offers that would accept a price not less than p. Similarly, the pertinent distribution of bids is

$$1 - V(p) = \int_p^\infty v(r)\,dr \qquad (2)$$

giving the cumulative proportion of the bids willing to pay a price equal to or greater than p. Observe that in this theory the asymptotic supply curve has a slope given by $dU(p)/dp = u(p) \geq 0$, and the asymptotic demand curve has a slope given by $(d/dp)[1 - V(p)] = -v(p) \leq 0$. The equilibrium price in the population, p_e, is the solution of the equation

$$U(p) = 1 - V(p) \qquad (3)$$

As the number of traders in the market increases indefinitely, the market clearing price approaches the solution given by (3).

Under mild assumptions about the cumulative densities, $U(\cdot)$ and $V(\cdot)$, this theory implies that the distribution of the market clearing prices is asymptotically normal. This result also holds even if individual traders each make bids or offers for more than one unit of the good, provided each trader has a negatively sloping excess demand and is small, in an appropriate sense, relative to the whole market. Since the details are available elsewhere (Telser, 1978), the proof here is brief. The population quantile of offers that can expect acceptance at the asymptotic market clearing price p_e is $U(p_e)$. Given a random sample of m offers, calculate the sample fractile, which is some price, corresponding to the population fractile $U(p_e)$. If (3) has a unique solution and if $U(p)$ has a derivative in a neighbourhood of p_e and at p_e, then the random variable $\sqrt{m}(p_m - p_e)$ converges in distribution to normality with mean zero and standard deviation $\sigma(p_m)$, where

$$\sigma(p_m) = \{1/u(p_e)\}\{\sqrt{[U(p_e)(1 - U(p_e))/m]}\} \quad (4)$$

Similarly, for a random sample of n bids, the random variable $\sqrt{n}(p_n - p_e)$ converges in distribution to normality with mean zero and standard deviation $\sigma(p_n)$, where the expression for $\sigma(p_n)$ is the same as for $\sigma(p_m)$, after making the appropriate substitutions: $1 - V$ for U, V for $1 - U$, v for u, and n for m. By hypothesis, the samples of bids and offers are independent. Therefore, the distribution of market clearing prices is also asymptotically normal with mean zero and standard deviation $\sigma(p_{m,n})$, where

$$\sigma^2(p_{m,n}) = \sigma^2(p_m) + \sigma^2(p_n) \quad (5)$$

Several important propositions follow from this analysis. First, the standard deviation of the distribution of market clearing prices is a decreasing function of the square root of the number of transactions. Second, the standard deviations are lower the larger the slopes of the cumulative densities. The slopes measure the homogeneity of the offers or bids in the population. If all offers were the same in the population then $dU(p_e)/dp = \infty$, which corresponds to an infinitely elastic supply. Consequently, (4) asserts, $\sigma(p_m) = 0$. If all offers in the population were different so that dU/dp approaches zero, then $U(p)$ would approach an improper uniform distribution on the halfline $[0, \infty]$ and $\sigma(p_m)$ would increase indefinitely. Therefore, the homogeneity of offers varies inversely with dU/dp. Third, the standard deviation has a maximum if $V(p_e) = U(p_e) = \frac{1}{2}$.[1]

According to this theory the standard deviation of the distribution of market clearing prices approaches zero as the number of potential transactions per unit time increases. A given batch of transactions has one market clearing price at which mutually acceptable trades can occur. It is the size of this batch that determines $\sigma(p_{m,n})$. The cost is an increasing function of m and n and there is also a benefit. The optimal volume of trade depends on both. The analysis of costs and benefits in the next section determines the optimal value of σ and shows that it is positive. In this way we obtain a theory of the equilibrium price distribution with both the mean and σ explained by the theory.

A THEORY OF THE NET BENEFIT OF AN ORGANISED FUTURES MARKET

The preceding section shows that the distribution of market clearing prices is asymptotically normal with a standard deviation, σ, given by

$$\sigma = \alpha K x^{-1/2} \qquad (6)$$

where x is the volume of trade corresponding to m and n in the preceding section, K a parameter that depends on the underlying supply and demand for the commodity, and α represents certain exogenous factors that we consider in more detail below. For present purposes it is convenient to write (6) in a more general form:

$$\sigma = g(x, \alpha) \qquad (7)$$

and, in conformity with (6), $g_x < 0$, $g_\alpha > 0$, and $g_{\alpha x} < 0$. We assume that the same function $g(\cdot)$ applies to all commodities and that differences among commodities express themselves via different values of α.

The total cost function of having and operating an organised market is

$$c = h(x, X, \sigma) \qquad (8)$$

where X is the size of the outstanding commitment of futures contracts, called the open interest. The presence of x, the volume of trade, as an argument of this total cost function needs no discussion. Assume $h_x > 0$. The open commitment, X, enters for two reasons. First, there is a cost of record keeping. A buyer and seller in the market agree on the price and quantity. This transaction creates a financial instrument on the balance sheet of the futures market, or, more exactly, its clearinghouse. The seller of a futures contract has a liability to the clearinghouse and the buyer has an asset of the clearinghouse. The clearinghouse must keep track of the identity of its debtors, called the shorts, and its creditors, the longs. This imposes a cost that is an increasing function of X. Some of the trade on an organised market comes from outsiders who are not members of the exchange. These trades go through the exchange members who carry out the orders of their customers. Hence the exchange members incur costs as a result of the actions they take on behalf of their customers. These costs depend on x, X, and σ and are included in the function $h(\cdot)$. Accordingly, $h_X > 0$.

The total cost also depends on both X and σ because an exchange member is liable to the clearinghouse for the commitments of his customers. The commodity price may change adversely to the customer of the exchange member. This can impose a loss on the exchange member although he can liquidate the customer's commitment in order to limit the loss. The exchange member normally requires a security deposit, called the margin, from his customer in order to protect himself against loss. This gives a cost per futures contract to the customer that is proportional to the margin per contract. If the only traders on the floor of the exchange were principals then it would not be necessary to have margins and this component of the total cost could be avoided. The higher is σ, the larger is the margin per unit of X and the higher is the total cost. Therefore, $h_X > 0$, $h_\sigma > 0$, and $h_{X\sigma} > 0$.

The net marginal cost with respect to volume is dh/dx given as follows:

$$dh/dx = h_x + h_\sigma g_x \qquad (9)$$

Since the theory stipulates $h_x > 0$, $h_\sigma > 0$, and $g_x < 0$, the net marginal cost can be negative. If the net marginal cost is positive, then, since $g_x < 0$, dh/dx can be a U-shaped function of x. It is more convenient to return to the analysis of this function in conjunction with the discussion of the equilibrium conditions given below.

The total benefit, denoted by b, is an increasing function of x and X. Since σ is the standard deviation of the distribution of market clearing prices, if it enters the benefit function at all, it should do so as a "bad". That is, an increase in σ lowers the total benefit. There would be little point in making σ an argument in the benefit function because this would simply complicate the algebra without yielding new results. The theory does assume that the total cost is an increasing function of σ. Consequently, as it stands, the net benefit, which is the total benefit minus the total cost, is a decreasing function of σ. This means that σ now enters the net benefit function as a bad.

There is another separate issue related to price variability. One reason often given for having an organised futures market is to enable hedging by dealers in the actual commodity. Someone who has a commitment in the actual good, according to this argument, can take an offsetting position in commodity futures contracts and thereby reduce the effects of a change in the price of the commodity. The price risk per unit of the commodity is an increasing function of the unpredictable price variability, denoted by β. Therefore, the benefit of an organised futures market is an increasing function of β. Write the benefit function as follows:

$$b = f(x, X, N, \beta) \tag{10}$$

Assume that f_x, f_X, and f_β are all positive. Now if β represents the unpredictable price variability, it should also affect σ, which is the standard deviation of the distribution of market clearing prices. Hence we modify (7) and write

$$\sigma = g(x, \alpha, \beta) \qquad g_\alpha > 0 \text{ and } g_\beta > 0 \tag{7'}$$

This formulation distinguishes between those factors as given by the parameter α that affect σ and those which affect price variability over longer periods of time as represented by β. For instance, seasonal variation can raise σ and does not necessarily affect β (see n. 1). This formulation shows that the effect of a rise of β on the net benefit is indeterminate because both the cost and the benefit are increasing functions of β.

The parameter N in the benefit function represents the number of traders who can benefit from the organised market. Recall that we regard a futures contract as analogous to a form of money that facilitates trade in the commodity. A futures contract corresponds to trade with a common medium of exchange. Since a common medium of exchange is more useful the larger the number of participants, the total benefit increases with N, $f_N > 0$. Also, let $f_{xN} > 0$ and $f_{XN} > 0$.

Assume that the equilibrium values of x and X give the maximal net benefit

$$J(x, X) = b - c \tag{11}$$

and assume that the observed values of x and X are the equilibrium values. A sufficient condition for the existence of an equilibrium is that the net benefit function, $J(x, X)$, is strongly concave in x and X. Hence the equilibrium is given by the necessary conditions for a maximum of the net benefit subject to the non-negativity constraints, x and $X \geq 0$. We obtain

$$df/dx - dh/dx \leq 0 \qquad \text{and} \qquad df/dX - dh/dX \leq 0 \tag{12}$$

The terms df/dx and df/dX giving the marginal benefit with respect to the volume of trade and open interest correspond to the demand for the services of an organised exchange. The terms dh/dx and dh/dX giving the marginal cost with respect to the volume of trade and open commitments correspond to the supply of services of an organised exchange. If the equilibrium values of x and X are positive, there is equality in both equilibrium conditions of (12). The two conditions are independent although one refers to x and the other to X, and under suitable conditions x is the rate of change of X. One can have $X = 0$ and $x > 0$, giving the implication of an upward jump of X. One may have $X > 0$ and $x = 0$ so that although there are positive outstanding commitments, there happens to be no trade. Finally, if the equilibrium values of x and X are both zero there is no organised market in the good. This theory states the absence of an organised market as a corner solution of the equilibrium conditions.

There is the complication that the net marginal cost, dh/dx, is not necessarily monotonic. Under the hypothesis that the net benefit function is strongly concave and that the benefit function $f(x, X, N, \beta)$ is concave in the first two arguments, there are definite restrictions on the properties of dh/dx. Thus, dh/dx must be increasing wherever it is negative. If dh/dx is positive and decreasing, then it must decrease less rapidly than df/dx at corresponding values of x.

Let t denote the commission per transaction. The theory asserts that in equilibrium with $x > 0$ the commission per transaction satisfies the equation

$$t = dh/dx = df/dx \qquad (13)$$

In practice the situation is more complicated. First, the members of the exchange determine the smallest unit that may be traded and the minimum allowable price change per contract. For example, a wheat futures contract is 5,000 bushels and the minimum allowable price change per contract is US$12.50 on the Chicago Board of Trade. There are floor traders on the exchange who do not carry an open position overnight and who make many trades during a trading session. Their activities may increase x and lower σ. The return to these traders depends on their cost and is a function of the minimum quantities that may be traded and the minimum allowable price change. The marginal cost also depends on the rate of arrival of information, the level of trading activity, and the distribution of the sizes of the transactions.

There is another important aspect of an organised market. There are traders and brokers who specialise in the commodity and who furnish an inventory of their services during a trading session almost independently of the volume of trade during the session. The number of these specialists in the long run varies directly with the expected volume of trade. For an inactive commodity the commission per unit transaction may include a fixed component that remunerates the specialists for the services they provide by their continued presence on the floor even when trade is light. As the expected volume of trade rises relative to the stock of outstanding contracts, this phenomenon diminishes in importance.

Let m denote the margin per contract. Assume that the cost of the margin is proportional to m. In equilibrium with $X > 0$, the margin per contract, up to the factor of proportionality to represent the cost, satisfies the equation

$$m = df/dX = dh/dX \qquad (14)$$

In practice the size of the margin depends on the nature of the commitment, whether it is speculative, a straddle, a spread, or a hedge. It also depends on the exchange rule that determines the maximum allowable change in the price per contract during a single trading session. Trade during a session cannot continue unless price changes remain within prescribed limits. As a result there is a relation between the margin per contract and the maximum allowable price charge per contract. Recall that the members of the exchange often act on the instructions of their customers and that many of the principals do not directly participate in the trade on the floor of the exchange. This relation between the margin per contract and the maximum allowable price change per contract gives all of the brokers time to consult their customers to determine whether the customers wish to continue their commitments in the face of adverse price change by furnishing additional margin. Otherwise, a broker would close out the commitment of his customer. The larger is the maximal allowable price change during a trading session, the larger is the margin that the broker demands from his customer. The margin is between one and two times the daily maximal allowable price change per contract.[2] This argument implies that in those organised markets where the principals trade directly it would be unnecessary to stop trading so that there would be no maximum allowable price change. In such markets margins would also not be needed.

We study the properties of this theory in two different ways. First, we see how changes in the values of the exogenous parameters, α, β, and N affect the equilibrium values of x and X. Second, we study how the margin and the commission depend on these exogenous parameters. We maintain the hypothesis that the functional forms of the benefit and cost functions are the same for all commodities and that differences among the commodities express themselves via the values of the exogenous parameters.

Assume positive values of the initial equilibrium of x and X so that the equilibrium conditions given by (12) are equalities. Write the first-order conditions of the maximum of $J = b - c$ in the form $J_x = J_X = 0$. Use the symbol z to denote one of the exogenous variables, α, β, or N.

Differentiate the equilibrium conditions with respect to z and obtain the equations as follows:

$$\begin{bmatrix} J_{xx} & J_{xX} \\ J_{Xx} & J_{XX} \end{bmatrix} \begin{bmatrix} dx/dz \\ dX/dz \end{bmatrix} + \begin{bmatrix} J_{zx} \\ J_{zX} \end{bmatrix} = 0 \quad (15)$$

Call the matrix on the left-hand side of (15) M. The hypothesis that J is strongly concave implies that M is a negative definite matrix. Hence multiplication of the expression in (15) on the left by the row vector $[dx/dz\ dX/dz]$ gives

$$(dx/dz)J_{zx} + (dX/dz)J_{zX} > 0 \quad (16)$$

If $z = \beta$, then $J_{zx} = J_{\beta x}$ and $J_{xX} = J_{\beta X}$. Consequently,

$$\begin{cases} J_{\beta x} = f_{x\beta} - h_{x\sigma}g_{\beta} - h_{\sigma}g_{\beta x} - h_{\sigma\sigma}g_{\beta}g_{x} \\ J_{\beta X} = f_{X\beta} - h_{X\sigma}g_{\beta} \end{cases} \quad (17)$$

Both of these expressions are of an indeterminate sign. Therefore, (16) does not fix the signs of $dx/d\beta$ and $dX/d\beta$. This is an important result because it means that a rise in exogenous price variability represented by the parameter β has an effect on x and X that depends on the actual numerical values of the terms in (15) and that knowledge of the signs of the components is not sufficient to predict the effects. The same conclusion applies to α since

$$J_{\alpha x} = -h_{x\sigma}g_{\alpha} - h_{\sigma\sigma}g_{\alpha}g_{x} \quad (18)$$

$$J_{\alpha X} = -h_{X\sigma}g_{\alpha} < 0 \quad (19)$$

The indeterminacy of the sign of $J_{\alpha x}$ is due to the negative value of g_x. As we shall see from the empirical results, the futures give strong support to the view that $g_x < 0$.

Take $z = N$. Then

$$J_{Nx} = f_{xN} > 0 \quad \text{and} \quad J_{NX} = f_{XN} > 0 \quad (20)$$

It is now convenient to write out the solution of (15) for $z = N$. We obtain

$$dx/dN = (-J_{XX}f_{xN} + J_{Xx}f_{XN})/\det M \quad (21)$$

$$dX/dN = (J_{Xx}f_{xN} - J_{xx}f_{XN})/\det M \quad (22)$$

The only terms with an open sign in (21) and (22) involve J_{Xx}. If J_{Xx} is positive, then both dx/dN and dX/dN must be positive. If so, the larger the number of potential participants in the market the larger is the equilibrium volume of trade and equilibrium open interest:

$$J_{Xx} = f_{Xx} - h_{Xx} - h_{\sigma X}g_x \tag{23}$$

Assume X and x are complements in the benefit function so that $f_{Xx} > 0$. By the preceding argument, $h_{\sigma X} > 0$ and $g_x < 0$. Hence two of the three terms in J_{Xx} would be positive. Even if $J_{Xx} > 0$, one may have h_{xX} of either sign. To settle some of these questions we now consider the additional evidence that commissions and margins can furnish.

Equations (13) and (14) give the equilibrium relating the commission t to the marginal benefit and marginal cost and the margin m to the marginal benefit and marginal cost. That is,

$$t = f_x = h_x + g_x h_\sigma \quad \text{and} \quad m = f_X = h_X \tag{24}$$

Hence t and m are functions of α, β, and N. If one can identify the cost or the benefit function using econometric techniques, then this would contribute to our knowledge of the relevant parameters and it would remove some of the indeterminacies above. To this end, we proceed as follows:

$$t_x = f_{xx} = h_{xx} + g_x(2h_{\sigma x} + g_x h_{\sigma\sigma}) + h_\sigma g_{xx} \tag{25}$$

$$t_X = f_{Xx} = h_{Xx} + g_x h_{\sigma X} \tag{26}$$

$$t_\alpha = (h_{x\sigma} + h_{\sigma\sigma}g_x)g_\alpha + h_\sigma g_{x\alpha}$$
$$\text{(recalling that } \alpha \text{ does not appear in } f) \tag{27}$$

$$t_\beta = f_{x\beta} = (h_{x\sigma} + h_{\sigma\sigma}g_x)g_\beta + h_\sigma g_{x\beta} \tag{28}$$

$$t_N = f_{xN}$$
$$\text{(recalling that } N \text{ does not appear in } h \text{ or } g \text{ by hypothesis)} \tag{29}$$

Now one may write

$$dt = t_x dx + t_X dX + t_\alpha da + t_\beta d\beta - t_N dN \tag{30}$$

in two different ways by using either the partials from the cost side or the partials from the demand side. Taking a convex combination of the partials from the cost and the demand side gives a continuum of possible ways of writing dt in terms of the underlying structure. That is,

$$dt = [\theta(\delta h/\delta x) + (1-\theta)f_{xx}]dx + [\theta(\delta h/\delta X) + (1-\theta)f_{Xx}]dX$$
$$+ [\theta(\delta h/\delta \alpha) + (1-\theta)0]d\alpha + [\theta(\delta h/d\beta) + (1-\theta)f_{x\beta}]d\beta$$
$$+ [0 + (1-\theta)f_{xN}]dN \tag{31}$$

where $0 \leq \theta \leq 1$, and $\delta h/\delta x$, $\delta h/\delta X$, $\delta h/\delta \alpha$, $\delta h/\delta \beta$, and $\delta h/\delta n$ denote the right-hand sides of (25)–(29) where this is applicable. Since the cost function excludes N while the benefit function includes N, a regression of t on the variables on the right-hand side of (31) that excludes dN requires $\theta = 1$, so that the cost function is identifiable.

A similar argument applies to the margin. We have

$$m_x = f_{Xx} = h_{Xx} + h_{X\sigma}g_x \tag{32}$$

$$m_X = f_{XX} = h_{XX} \tag{33}$$

$$m_\alpha = h_{X\sigma}g_\alpha \quad [\text{cf. (27)}] \tag{34}$$

$$m_\beta = f_{X\beta} = h_{X\sigma}g_\beta \tag{35}$$

$$m_N = f_{XN} \quad [\text{cf. (29)}] \tag{36}$$

Therefore, a regression of m on the explanatory variables excluding dN would identify the pertinent parameters of the cost function.

This theory determines the equilibrium values of σ together with the equilibrium values of x and X. Given x and X, the distribution of market clearing prices is asymptotically normal with a standard deviation depending on x. The values of x and X give the maximum net benefit. Hence the theory determines the optimal value of σ which takes into account the marginal benefit and the marginal cost of trade on an organised exchange. Consequently, σ is an endogenous variable in this theory.

The section, 'Effect of Price Variability' continues this analysis empirically.

THE SOURCES OF PRICE VARIABILITY

Since the empirical evidence refers to physical commodities, not financial instruments such as foreign exchange or bonds which sometimes also have organised futures markets, we parcel out the sources of price variability among the explanatory variables regarded as random variables in the following function:

$$\log p_{it} = a_{0i} + a_{1i} \log q_{it} + a_{2i} \log y_t + u_{it} \tag{37}$$

where

p_{it} = US price of commodity i in year t during the month of peak stocks divided by the US wholesale price index,[3]

q_{it} = stocks of commodity i in year t during the month of peak stocks in the United States or the world, depending on which is more suitable,

y_t = US gross national income divided by the US consumer price index,
u_{it} = residual.

The coefficient a_{1i} gives the elasticity of the price with respect to the stock, called EPQ; the coefficient a_{2i} gives the elasticity of the price with respect to deflated national income, called EPY; VPAV is the coefficient of variation of the annual average price of the commodity; STDERR is the standard deviation of the residual. We also measure the variability of monthly prices in terms of a parameter called VPM2 (see the Appendix for the details). All of these measures are in percentage terms and are dimensionless numbers comparable across commodities.[4]

In terms of the theory in the preceding section, these statistics are among the determinants of β which enter the benefit function. They are also among the determinants of α which affect σ in (7'). We claim that α and β are increasing functions of EPQ, EPY, VPAV, STDERR, and VPM2.

We take the explanatory variables in (37) as exogenous. First, the more elastic the underlying supply and demand curves for the commodity the closer a_{1i} is to zero. Hence annual variability of stocks has less effect on annual price variability as a_{1i} approaches zero. For example, the better the substitutes for a given commodity in either demand or supply the closer a_{1i} is to zero. Second, price is less responsive to income as a_{2i} approaches zero. The level of income affects the demand for the commodity. As a result, a_{2i} is the compound effect of the size of the income elasticity of demand and the size of the elasticity of supply. The larger is the income elasticity of demand, the larger is a_{2i}. The smaller is the elasticity of supply, the larger is a_{2i}. Therefore, a high income elasticity of demand and a low elasticity of supply combine to give a large value of a_{2i}. The standard deviation of the residuals measures the effects of exogenous factors on price variability that are uncorrelated with the variables explicitly included. Since VPM2 gives the mean value of the standard deviation of prices around the monthly means, it measure the effects of short-term price variability in contrast to the preceding statistics, which give measures of different aspects of annual price variability.

The sample has 51 commodities for the period 1959–1971. We divide the sample into three subsamples of equal size such that the first has 17 actively traded goods, the second has 17 less actively traded goods, and the third has 17 with little or no trade as far as we know. Each commodity in a subsample has two companions, one in each of the other two subsamples, with roughly similar gross characteristics – metal, meat, fowl, wood, etc. We use three criteria to place a commodity into a subsample – the size of the open interest, the volume of trade, and the duration and timing of trading activity. Each subsample has commodities difficult to place. For example, we put plywood and broilers in the active group and wool and oats in the inactive group, although the latter two have a larger open interest than the former two. We took as decisive the fact that plywood and broilers have a larger volume of trade. If trade in a commodity came to a halt at some point in the sample period and did not revive during the remainder of the sample period, we put that commodity into the inactive group. This explains our placement of cottonseed meal, coffee, and zinc. A few commodities in the inactive group are actively traded in London, such as tin and zinc. Cocoa is active in both London and New York. We put cocoa in the active group. Copper, in the active group, is more active in London than in New York. There have been attempts to start futures markets for some commodities in the dormant group that so far have failed. These commodities are nickel, apples, hams, and rice. We have been unable to discover any attempts to organise futures trading for some commodities in the dormant group, including newsprint, lamb, veal, and frozen strawberries. Except for platinum, our active group includes every commodity actively traded on an organised futures market in the United States during the sample period.

In some cases the absence of an organised futures market probably results from highly stable prices due to governmental policies or actions of private concerns. Foreign exchange, though not in our sample, best illustrates this point. Market determined foreign exchange rates began in 1971. Before the United States closed its gold window, the fiscal authorities of various nations had a coordinated price-support program for foreign exchange that maintained a web of fixed exchange rates. Prior to 1971 futures markets in foreign exchange were unimportant. Similarly, the US government price-support program in peanuts and tobacco stifles interest in futures trading in these commodities. Trading in cotton, a commodity with an organised spot market in New York since the eighteenth century, virtually expired when government stocks were large and growing. Trading did not revive until after the disappearance of the government inventory. Published prices of petroleum crude oil and newsprint are very stable. This seems to have the same effect as a price-support program, but we know of no government policy as the explanation. Perhaps the Texas Railroad Commission is to blame for stable crude-oil prices. The end of our sample period precedes the Arabian oil embargo to the United States by 2 years.

Accepting this classification of the commodities into the three groups, we now examine the pertinent evidence. Table 1 gives the means and standard deviations of the variables related to price variability.

According to Table 1, VPAV is highest for the active group, next highest for the inactive group, and lowest for the dormant group. This pattern could occur by chance with a probability of 1%. Both VPM2 and STDERR

Table 1 Means and standard deviations of selected characteristics related to price variability of 17 active, 17 inactive, and 17 dormant commodities

Parameter	Active 1–17	Inactive 18–34	Dormant 35–51	Total 1–51	Significance level*
VPAV	16.4610	12.7476	11.5544	13.5877	0.010
	(6.6498)	(4.7874)	(6.0370)	(6.1319)	
VPM2	5.5363	4.7726	2.9326	4.4139	0.016†
	(3.0607)	(3.9698)	(2.3594)	(3.3277)	
EPQ	−0.6007	−0.3628	−0.2322	−0.3986	0.071
	(0.8914)	(0.9100)	(0.4596)	(0.7814)	
QRSK	11.6102	8.2079	4.8470	8.2217	0.006
	(8.9936)	(7.1992)	(3.5411)	(7.3660)	
EPY	0.1557	0.1118	0.06297	0.1102	0.170
	(0.9810)	(0.4470)	(0.5439)	(0.6841)	
RSQ	0.5264	0.4962	0.4863	0.5029	0.176
	(0.2706)	(0.2660)	(0.2628)	(0.2617)	
STDERR	11.3271	8.7645	6.6858	8.9258	0.004
	(4.6238)	(4.3014)	(4.0692)	(4.6622)	
DW	1.6900	1.6929	1.6412	1.6747	0.199
	(0.4803)	(0.4734)	(0.6224)	(0.5197)	
Seasonal of prices	29.4229	22.6903	14.2013	21.4421	0.018†
	(35.1045)	(19.33)	(12.3177)	(24.71)	
Seasonal of stocks	49.1009	50.2590	60.1442	52.8282	0.090
	(33.5402)	(30.5552)	(33.6604)	(31.9250)	

*The numbers in parentheses are the estimated standard deviations of the population. The statistic as follows:

$$[(m_1 - m_2)/\sqrt{(n_1 s_1^2 + n_2 s_2^2)}]\sqrt{[n_1 n_2(n_1 + n_2 - 2)/(n_1 + n_2)]}$$

has a t-distribution where m_i denotes the sample mean and s_i^2 the sample variance. Let $i = 1$ denote the active, $i = 2$ the inactive, and $i = 3$ the dormant. The theory predicts the signs of $m_2 - m_1$ and $m_3 - m_2$. The significance level gives the probability of observing the two independent t-statistics on the hypothesis that the true differences are zero (Cramèr 1946, sec. 31.2.1).

†There is a slight complication because we lack monthly price series for two commodities: tomato paste, which is in the inactive group, and strawberries, which is in the dormant group. Therefore, we cannot calculate VPM2 and the price seasonal for these two commodities. Consequently, with respect to these two parameters, the sample size is 16 for both the inactive and the dormant group. The significance levels take this into account.

show the same pattern. Therefore, prices are most variable for the actively traded group, next most variable for the inactive group, and least variable for the dormant group.

Consider now the pattern with respect to the sources of price variability. The EPQ approaches zero as we move from the active to the dormant group. This is clearer if, instead of calculating the percentage price change with respect to a 1% change in stocks, we compute the percentage price change with respect to a 1% change of the coefficient of variation of stocks, a parameter denoted QRSK (quantity risk). The inverse relation between trading activity and QRSK is significant at the 99% level. Also, EPY decreases with trading activity but the pattern has a low level of significance, 83%. Finally, observe that the R^2 (RSQ) and DW statistics do not differ significantly among the three subsamples.

Table 1 shows two other aspects of the difference among commodities that are relevant for our argument. First, the seasonal index of prices is positively correlated with the marginal cost of storage and with the holding of stocks. Observe that the price seasonal varies inversely with trading activity. That is, the actively traded commodities show a more pronounced seasonal price pattern than do the less actively traded ones. The pattern is significant at the 98% level. It is also true that there is a more highly seasonal pattern of stocks for the actively traded goods than for the others but with a significance level of only 91%. (The stock seasonal is the trough stock level divided by the peak level, where the peak and trough refer to an appropriate 12-month period.) Nevertheless, consistent with our theory that the introduction of futures trading permits a sorting of transactions over time, we find more seasonality for the actively traded goods, which implies a greater benefit from organised futures trading in these goods.

EFFECTS OF PRICE VARIABILITY ON THE OPEN INTEREST AND THE VOLUME OF TRADE

The preceding material shows the presence of an inverse association between trading activity and price variability across the three subsamples. According to the theory of the costs and benefits of futures trading given above, the relation between the volume of trade, open interest, and various measures of price variability does not necessarily imply that more price variability raises the volume of trade and open interest. Although greater price variability implies a greater benefit from organised futures trade, it also implies a greater cost. The greater cost is from two sources. First, the higher the price variability, the higher is the cost of holding a given open commitment. Second, the higher the price variability, the larger is the dispersion of the distribution of market clearing prices for a given volume of trade, which is a cost-increasing phenomenon. Therefore, estimates of how the volume of trade and open interest depend on various measures of price variability can give useful information about the nature of the costs and benefits. The purpose of this section is to furnish some empirical evidence in order to narrow the range of theoretical alternatives.

We now use a sample of 25 commodities, which adds data for eight commodities in the inactive sample to the 17 actively traded commodities.[5] The eight additional observations include all commodities in the less active group for which there was some non-negligible volume of trade and open interest during the sample period. Table 2 lists the commodities and the pertinent data. The velocity of the open interest, the average annual volume of trade divided by the average end-of-quarter

Table 2 Data by commodity

No. and commodity	Volume of trade (10^9 US$)	Open interest (10^6 US$)	Stocks (10^6 US$)	Seasonal of stocks	VPAV (%)	STDERR (%)	VPM2 (%)
1. Cotton	1.116	52.6	1374	32.559	16.928	11.82	2.85
2. Soybeans	27.510	552.8	1453	14.127	10.661	5.53	4.42
3. Wheat	9.173	195.8	1418	48.441	15.824	8.10	4.15
4. Cocoa	2.096	124.0	82	20.987	26.902	7.58	7.90
5. Sugar	2.646	110.5	264	42.614	10.875	6.57	5.21
6. Copper	1.073	31.1	244	100.000	23.794	7.72	2.14
7. Silver	5.935	376.3	450	100.000	27.701	11.81	3.88
8. Corn	7.522	216.8	3045	30.354	8.149	5.67	3.49
9. Soymeal	1.808	94.1	497	14.127	12.509	9.58	5.34
10. Soy oil	3.705	95.5	351	14.127	15.304	13.22	6.31
11. Plywood	0.835	16.9	503	96.053	8.614	14.95	6.96
12. Broilers	0.508	10.0	299	75.868	8.348	19.75	2.71
13. Cattle	4.984	170.6	5951	79.178	14.227	6.68	3.28
14. Eggs	1.649	23.8	982	92.738	12.967	19.70	8.24
15. Frozen orange juice	0.715	24.6	181	32.478	25.687	14.21	4.79
16. Pork bellies	16.573	192.5	620	75.448	21.582	14.44	7.54
17. Potatoes	0.659	19.3	416	25.619	19.766	15.24	14.90
18. Wool	0.231	22.1	205	41.190	17.823	16.82	3.38
19. Cottonseed meal	0.002	0.1	149	7.470	11.884	10.85	5.02
20. Rye	0.610	14.2	25	40.079	6.681	3.97	3.37
21. Coffee	0.246	17.8	173	37.735	15.107	13.05	3.63
22. Propane	0.018	3.2	318	48.859	7.390	4.04	1.23
23. Tin	0.003	0.4	72	38.279	20.070	7.19	4.05
24. Zinc	0.011	1.5	45	94.934	11.201	5.17	1.74
25. Oats	0.457	17.2	539	34.598	4.672	4.49	3.06
26. Milo	0.023	2.4	445	41.020	8.184	7.22	3.56
27. Cottonseed oil	0.475	16.0	170	32.426	16.078	8.29	5.95
28. Lumber	0.390	10.1	625	96.053	15.303	8.47	4.07
29. Turkeys	0.162	4.6	384	27.326	9.160	5.88	3.98
30. Hogs	0.374	8.2	1947	83.226	17.770	10.50	5.94
31. Butter	0.056	0.0	114	48.426	6.908	3.10	1.88
32. Tomato paste	0.0	0.0	102	12.114	17.149	10.27	4.77
33. Lard	0.051	71.7	10	65.426	16.321	16.76	6.41
34. Onions	0.0	0.0	14	5.242	15.008	12.91	19.10
35. Tobacco	0.0	0.0	2755	65.026	9.345	3.36	0.83
36. Peanuts	0.0	0.0	282	15.705	9.088	4.82	2.79
37. Rice	0.0	0.0	389	12.629	4.446	4.90	1.74
38. Tea	0.0	0.0	34	60.000	7.549	4.33	1.59
39. Coal	0.0	0.0	394	86.170	25.416	13.42	1.77
40. Aluminium	0.0	0.1	728	100.000	7.620	3.39	1.23
41. Nickel	0.0	0.0	129	100.000	22.544	8.27	2.54
42. Barley	0.0	0.0	412	34.088	8.243	6.58	3.28
43. Petroleum crude oil	0.0	0.0	779	93.637	4.901	1.25	0.66
44. Peanut oil	0.0	0.0	14	15.705	12.173	13.55	4.36
45. Newsprint	0.0	0.0	88	100.000	5.459	1.35	0.56
46. Lamb	0.0	0.0	249	75.000	13.363	8.32	3.86
47. Veal	0.0	0.0	278	79.178	12.571	6.39	5.34
48. Cheese	0.0	0.0	158	74.445	19.852	4.27	1.93
49. Strawberries	0.0	0.0	19	41.806	11.850	10.10	2.93
50. Hams	0.0	0.0	1057	67.833	8.701	5.22	4.08
51. Apples	0.0	0.0	197	1.230	13.304	14.13	10.35

Note — A detailed description of the sources and of the methods for estimating these figures is available to readers on request.

Table 3 Simple correlations between log volume, log open, and selected measures of price variability, means, and standard deviations for 25 commodities

Variables	Simple correlations					Mean	SD
	Log open	Log volume	VPM2	STDERR	VPAV		
Log open	1	0.93267	−0.02780	−0.21903	0.25128	3.75921	1.29777
Log volume		1	0.05703	−0.17166	0.13909	0.31176	1.36342
VPM2			1	0.35044	0.24315	5.09960	2.60072
STDERR				1	0.24568	10.56160	4.51223
VPAV					1	15.29728	6.22404

open interest, ranges from a low of 14 for coffee to a high of 86 for pork bellies. These are large numbers by comparison with, say, the velocity of the money supply. Soybean futures contracts have a daily average volume of trade of US$110 millions, and for pork bellies this figure is US$66 millions! Table 3 gives the simple correlations, means, and standard deviations of the variables in the regression analysis. Observe that the simple correlations between the log of open interest and price variability are all small, and the same is true of the corresponding correlations for the log of volume of trade. However, since the various measures of price variability are dimensionless pure numbers, while the volume of trade and open interest are in dollars and depend on the size of the demand for the commodities, the two sets of variables are not comparable. Also, the theory implies the presence of a scale variable as a separate factor to explain the size of the open interest and volume of trading in the commodity. The level of stocks suggests itself as one way of measuring scale. This has the disadvantage of introducing serious measurement errors for some commodities. As an alternative, one can use the volume of trade as a scale variable to explain the open interest and the open interest as a scale variable to explain the volume of trade. In effect this treats the system of equations (15) for candidates for direct estimation by least squares.

Substitute β for z in (15) and write out the result as follows:

$$J_{xx}(dx/d\beta) + J_{Xx}(dX/d\beta) = -J_{x\beta} \tag{38}$$

$$J_{xX}(dx/d\beta) + J_{XX}(dX/d\beta) = -J_{X\beta} \tag{39}$$

Take $dx/d\beta$ as the dependent variable in the first and $dX/d\beta$ as the dependent variable in the second to give the following representation:

$$dx = -(J_{Xx}/J_{xx})\,dX - (J_{x\beta}/J_{xx})\,d\beta \tag{40}$$

$$dX = -(J_{xX}/J_{XX})\,dx - (J_{X\beta}/J_{XX})\,d\beta \tag{41}$$

Table 4 Selected statistics from regressions relating futures trading activity to price variability for a sample of 25 commodity futures

	Dependent variables			
	Log open interest		Log volume of trade	
Explanatory variables	Elasticity	t-ratio of coefficient	Elasticity	t-ratio of coefficient
Log volume	0.8591	12.412	...	...
Log open	...	...	1.0302	12.412
VPM2	−0.2406	−1.240	0.2793	1.321
STDERR	−0.2182	−0.920	0.1329	0.504
VPAV	0.1769	2.235	−0.4790	−1.773
R^2	0.9020	...	0.8953	...

Take a linear approximation to these equations and use the three measures of price variability for β. We estimate by least squares the following equations:

$$\log x = a_0 + a_1 \log X + a_2(\text{VPM2}) + a_3(\text{STDERR}) + a_4(\text{VPAV}) + \varepsilon \quad (42)$$

$$\log X = c_0 + c_1 \log x + c_2(\text{VPM2}) + c_3(\text{STDERR}) + c_4(\text{VPAV}) + \zeta \quad (43)$$

There is the complication that x and X are endogenous variables. It seems plausible that the random disturbances are more likely to shift the benefit functions than the cost functions, because changes in weather affect production of agricultural goods and cyclical factors affect the demand for all goods, which shifts the hedging demand and, therefore, the benefit function. Therefore, ε will be positively correlated with $\log X$ and ζ will be positively correlated with $\log x$. Hence least-squares estimates of the coefficients a_1 and c_1 are both upwardly biased. We deal directly with one aspect of this problem below by studying how commissions and margins depend on the pertinent variables as given in (24).

Table 4 gives the estimates of the elasticities derived from (42) and (43) together with the t-ratios of the coefficients. The coefficient of $\log X$ in (42) corresponds to $-(J_{Xx}/J_{xx})$ in (40), while the coefficient of $\log x$ in (43) corresponds to $-(J_{Xx}/J_{XX})$ in (41). The other coefficients in (42) and (43) relate similarly to their counterparts in (40) and (41). Consider the regression with the log of the open interest as the dependent variable. Of the three measures of price variability, the only one that enters with a large t-ratio is VPAV, the coefficient of variation of annual prices. It has a t-ratio of 2.2 and an elasticity of 0.18. The other two measures of price variability enter with t-ratios around 1 and have negative elasticities. Also

Table 5 Selected statistics from regressions to explain log of open interest and log of volume of trade for 25 commodity futures

	Dependent variables			
	Log open interest		Log volume	
Explanatory variables	Elasticity	t-ratio of coefficient	Elasticity	t-ratio of coefficient
Log stock	0.4451	2.13	0.5594	2.47
VPM2	0.0167	0.03	0.3010	0.55
STDERR	−0.9155	−1.49	−0.8122	−1.24
VPAV	1.2109	1.93	0.8031	1.20
R^2	0.3048	...	0.2893	...

observe that in the regression equation with $\log x$ as the dependent variable each coefficient of price variability enters with a sign opposite to that which it has in the regression with $\log X$ as the dependent variable. One is immediately tempted to conclude from these results that a rise in some aspect of price variability tends to move the open interest and the volume of trade in opposite directions. This may be true, but the high correlation between $\log x$ and $\log X$ is consistent with an alternative and a simpler explanation of these results. The simple correlation between $\log X$ and $\log x$ is 0.933 according to Table 3. Also, the multiple R^2 in the log open regression in Table 4 is 0.902 ($R = 0.95$). One may then say that these regressions are nearly identities and it would follow by simple arithmetic that the coefficients of price variability in one equation are nearly perfectly predictable by the coefficients in the other. For instance, solve for $\log x$ in the regression equation with $\log X$ as the dependent variable and you will come close to the estimated regression with $\log x$ as the dependent variable.

For these reasons it is of some interest to try an alternative to these equations which uses the log of stocks as a scale variable. These results appear in Table 5. Here we see that each measure of price variability has a coefficient of the same sign in the regressions with $\log X$ and $\log x$ as the dependent variables. This is hardly surprising given the high positive correlation between $\log X$ and $\log x$. Nevertheless it is worth noting that an increase in stocks of 1% will lead to a larger percentage increase in the volume of trade than in the open interest.[6]

We now study how the commission per transaction and the margin per contract depends on the pertinent variables according to the theory in the fourth section.[7] We assume that the cost and benefit functions are of the same form for all commodities and that differences among commodities express themselves via different values of the exogenous variables, α, β, and N. In equilibrium the commission, t, and the margin, m, satisfy (24). The argument in the fourth section shows that the cost functions are identified in a regression that excludes the scale variable N and includes the other exogenous variables related to α and β. We have two regressions that we estimate by ordinary least squares. In the first the commission is the dependent variable and in the second the margin is the dependent variable. In addition to the proxies for α and β, we include among the explanatory variables $\log x$, $\log X$, and the average price per futures contract over the sample period, a variable we label CONPR (contract price). The relevance of CONPR as a determinant of m hardly needs discussion. Plainly, the potential cost to a broker of a customer default is an increasing function of the size of the customer's commitment or open interest. Hence m should be an increasing function of CONPR. It is not equally plain why CONPR should affect the commission. Empirically, one can hardly doubt that t is an increasing function of CONPR. This result is consistent with the belief that it is more costly to arrange a mutually satisfactory trade for a contract of larger value, such as coffee, than for one of lesser value, such as potatoes, even holding the volume of trade constant. That is, a broker's execution according to his instructions is more costly for a contract of a larger value. It may not be superfluous to add that this argument would be consistent with the theory of optimal search.

If a good measure of scale were available, this would enable the application of simultaneous-equations techniques in an attempt to estimate the structural relations of the model. The only candidate is an estimate of stocks by commodity, and this is subject to considerable measurement error. Therefore, we focus our attention on the cost side which is identifiable and we accept a possible simultaneous-equations bias insofar as $\log x$ and $\log X$ are both positively correlated with the variable we omit, the scale factor. This gives an upward bias to the coefficients of both $\log x$ and $\log X$ if t is positively related to N and m is positively related to N, where N denotes the scale factor.

We claim that the regressions given in Table 6 using data in Table 7 furnish estimates of the slopes of the commission and margin with respect to the various variables for the cost equation. Thus, the slope of t with respect to x in Table 6 gives an estimate of t_x on the cost side, which is the right-hand side of (25), and so on. Similarly, the slope of m with respect to x gives an estimate of $h_{Xx} + h_{X\sigma}g_x$ according to the right-hand side of (32), and so on. With the theoretical interpretations in hand as given by (25)–(29) and (32)–(35), consider the evidence in Table 6.

Both the commission and the margin regressions display a similar pattern of results.[8] With only one exception, all of the explanatory

Table 6 Selected statistics from regressions to explain commissions and margins for a sample of 23 commodity futures

	Dependent variables			
	Commission		Margin	
Explanatory variables	Elasticity	t-Ratio of Coefficient	Elasticity	t-Ratio of Coefficient
Log volume	−0.218	−2.72	−0.419	−2.80
Log open interest	0.141	1.58	0.449	2.69
CONPR	0.536	4.62	0.729	3.36
STDERR	−0.189	−1.75	−0.425	−2.10
VPAV	0.076	0.72	−0.265	−1.33
VPM2	0.130	1.48	0.415	2.53
Constant	−0.1997	−0.01	−1,227.000	−1.59
R^2	0.688	...	0.662	...
SE	8.678	...	397.100	...

Table 7 Average contract price (CONPR), commission (COMM), and margin (MARG) for 23 commodity futures contracts in US dollars per contract

Commodity	CONPR	COMM	MARG
Cotton	14,368	70	1,300
Soybeans	13,715	50	2,500
Wheat	8,858	45	1,250
Cocoa	8,511	70	2,400
Sugar	8,034	75	2,000
Copper	9,949	50	750
Silver	7,926	50	1,200
Corn	6,100	40	1,000
Soy meal	7,527	40	1,500
Soy oil	6,005	40	1,000
Plywood	5,947	40	700
Broilers	6,250	35	500
Cattle	10,925	50	900
Eggs	8,071	40	700
Frozen concentrated orange juice	6,712	45	450
Pork bellies	14,425	60	1,500
Potatoes	1,892	35	750
Wool	6,947	50	1,500
Rye	5,941	40	1,500
Coffee	12,764	80	2,000
Oats	3,357	35	750
Lumber	4,198	55	700
Hogs	4,087	40	900

variables have coefficients of the same sign in both regression equations. The higher is the volume of trade the lower is the commission and the margin. The higher is the open interest, the higher is the commission and the margin. The higher is the average price per contract, the higher is the commission and margin. Although the t-ratios indicate a high level of significance by the usual standards, one should regard this with some scepticism because, for one thing, simultaneous-equations bias is present due to the omission of a scale variable that may be positively correlated with $\log x$ and $\log X$. The coefficients of $\log x$ and $\log X$ both give strong support to the belief that $g_x < 0$ (cf. [25] and [32]). The negative coefficient of $\log x$ in the margin regression is also consistent with the view that a rise in the volume of trade for a given open interest lowers the marginal cost with respect to the open interest ($h_{Xx} < 0$).

Now consider the regression coefficients of the various measures of price variability in Table 6. In the commission regressions these coefficients refer to t_α or t_β in (27) and (28). If $h_{x\sigma} + h_{\sigma\sigma}g_x$ is negative because g_x is negative, then the coefficient of the measure of price variability can also be negative. Nor is this all. Given the hypothesis about the signs of the various partials in (27) and (28), a negative value of g_x is *necessary* but is not sufficient for an observed negative value of the coefficient of a price variability parameter; STDERR does have a negative coefficient in the commission regression. Both VPAV and VPM2 have positive coefficients in the commission regression.

Next consider the margin regression. Equations (34) and (35) predict positive coefficients of the variables that are positively related to price variability if $h_{X\sigma}$ is positive and if X is not an argument in $g(\cdot)$. According to Table 6, VPM2 does have a positive coefficient with an elasticity of 0.415 and a t-ratio of 2.5. Recall that VPM2 measures the coefficient of variation around the monthly price seasonal so that it reflects shorter-term sources of price variability than either VPAV or STDERR. Both STDERR and VPAV have negative coefficients. The elasticities are -0.425 and -0.265, and the t-ratios are -2.1 and -1.3. These results are not consistent with the present model. There are two possible explanations of this discrepancy of which the first is econometric and the second is economic. The regression equation omits the scale variation, N. According to the theory, N is positively related to x and X and is also positively related to m (and to t) as is shown in (36) (and [29]). This imparts a positive bias to the estimates of the coefficients of $\log x$ and $\log X$. It is also possible that N is correlated with the exogenous variables α and β. If N is negatively related to α and β, then this would impart a downward bias to the estimates of the coefficients of price variability. We do have a weak proxy for the scale variable given by the log of stocks, a variable which we regard with suspicion because of measurement error for some commodities. Nevertheless, we do find negative, albeit small, correlations between the three measures of price variability and the log of stocks. The simple correlations, denoted by $R(\cdot)$, are as follows:

$$R(\log \text{stock}, \text{VPAV}) = -0.149, \quad R(\log \text{stock}, \text{STDERR}) = -0.064$$

$$R(\log \text{stock}, \text{VPM2}) = -0.045$$

Observe that it is the variable with the correlation closest to zero that has a positive effect on the margin.

This econometric argument agrees with an economic argument. The size of the elasticity of the excess demand function increases with the stock of outstanding commitments. According to (4), this tends to lower σ. We believe that a closer investigation of the function $\sigma = g(x, \alpha, \beta)$ is desirable since these arguments suggest that the variable X should enter this function. Such research would directly study the determinants of liquidity in the market.

SUMMARY AND CONCLUSIONS

In an organised market the participants trade a standardised contract such that each unit of the contract is a perfect substitute for any other unit. The identities of the parties in any mutually agreeable transaction do not affect the terms of exchange. The organised market itself or some other institution deliberately creates a homogenous good that can be traded anonymously by the participants or their agents.

Although the discussion centres on organised futures markets, the basic theory applies equally well to any organised market. The benefit of an organised market is an increasing function of the number of potential participants. It is also an increasing function of the turnover of the potential participants in that market. It would not be necessary for a small group of traders who know each other well and who have had and will continue to have contacts with each other to bear the expense of organising a formal market. In such markets the terms of sale often do depend on the identity of the parties in addition to the characteristics of the goods. An organised market deals in a highly fungible good that is readily traded among strangers. In an organised market the transactions prices alone convey a considerable amount of useful information to those who are not currently trading in the market. In those markets where heterogeneous goods are traded and where the identity of the buyer and seller affects the terms of trade, the transaction price alone conveys only partial information to outsiders.

In addition to scale, price variability affects the benefit of having an organised market. It also affects the cost. There is more price variability for those goods that have an organised futures market than for the goods that lack such markets. It does not follow that futures trading causes greater price variability. The organisation of a futures market is the response to an increase of price variability. For example, when a government allows the price of a good to fluctuate and abandons its attempt to control the price, this may create an incentive to organise a futures market if the potential scale of operation is large enough, as witness the creation of organised futures markets in some foreign exchange.

We find that the volume of trade increases relative to the open interest, the higher is the level of the open interest. As the open interest in a commodity declines, the volume of trade declines even more rapidly. According to the empirical results relating commissions to $\log x$ and $\log X$, commissions are higher, the less active the trade in the commodity. Let price variability be the driving force as represented by the parameter β. Let the price of some good become more stable. The open interest declines and the volumes of trade falls relative to the open interest. This raises the commission and the margin. A sequence of events now begins that may well end in a corner equilibrium with no trade and no open interest in the commodity. In this way the theory predicts the disappearance of an organised market. Similarly, a rise in price variability may lead to the appearance of an organised market.

The best concise summary of the theory is as follows: an organised market facilitates trade among strangers.

APPENDIX: DESCRIPTION OF CROSS-SECTION DATA AND SOURCES

I. Trading activity and market structure variables

Seasonal of stocks. Average ratio of trough-month stocks to peak-month stocks, 1959–1971, expressed as a percentage.

Open. Average number of open contracts, end-of-quarter, 1959–1971 or subperiod of active futures trading. *Journal of Commerce and Commercial*.

Opnint. Open × average dollar value of one contract, 1959–1971 or subperiod of active futures trading, in millions of nominal dollars.

Stocks. Value of average privately held stocks in millions of nominal dollars, 1959–1971 or subperiod of active futures trading. Computations are described in Appendix II, Average US Stocks Conceptually, stocks are computed as pipeline stocks plus average intraseasonal inventories. They give the peak holdings during the year.

Volume. Average annual volume of futures transactions on all US contract markets, 1959–1971 or subperiod of active futures trading, × average dollar value of one contract, expressed in billions of nominal dollars. Association of Commodity Exchange Firms, Inc.

II. Variables derived from monthly price series P_{im}

VPAV. Coefficient of variation across years of $\Sigma_{m=1}^{12} P_{im}$, 1959–71. When weighted average yearly prices are available (ie, weighted by quantities sold), such data are used.

VPM2. Let P^*_{im} equal P_{im} deflated by a 12-month moving average, centred at the sixth month. For each month m, let σ_m equal 100 × standard deviation of P^*_{im}; VPM2 equals the mean of σ_m across months.

Seasonal of prices. Maximum own rate of interest, at annual rate, displayed by averages across years of P_{im} (one for each month), 1959–71. Example: if $P \cdot_{DEC}$ and $P \cdot_{SEP}$ represent average December and September prices, and if $P \cdot_{DEC} > P \cdot_{NOV} > P \cdot_{OCT} > P \cdot_{SEP}$, then season equals 400 log $(P \cdot_{DEC}/P \cdot_{SEP})$, if $P \cdot_{DEC}$ and $P \cdot_{SEP}$ are the maximising pair. Months are indicated in Table 1.

III. Variables from regressions of price on stocks and income

For each commodity, regress via ordinary least squares:

$$\log\left(\frac{\text{PSK}}{\text{WPI}}\right)_t = a_0 + a_1 \log(\text{QSK})_t + a_2 \log\left(\frac{\text{GNI}}{\text{CPI}}\right)_t + u_t, \; 1959\text{–}71$$

where

QSK = peak-month stocks
PSK = US nominal spot (monthly) price
GNI = US gross national income. *Statistical Abstract*
WPI = US wholesale price index. *Statistical Abstract*
CPI = US consumer price index. *Statistical Abstract*.

DW. Durbin–Watson coefficient.
EPQ. $\hat{a}_1$.
EPY. $\hat{a}_2$.
ERR. Standard error of regression × 100 (ie, expressed as a percentage coefficient of variation).
QRSK. $|\hat{a}_1|$ × coefficient of variation across years of QSK.
RSQ. Coefficient of determination (R^2).
STDERR.[9] ERR × $[DW(1 - 0.25DW)]^{1/2}$.

---

[1] The effect of heterogeneity on the standard deviation is given as follows. Assume there are n normal distributions, denoted by N_i, $i = 1, \ldots, n$. Let α_i be the probability of a random drawing from N_i. Let μ_i denote the mean of N_i and σ_i the standard deviation of N_i. Let X denote a random variable. Then

$$\mu = EX = \sum_i E(X:N_i)\Pr(N_i) = \sum_i \alpha_i \mu_i \quad \text{(i)}$$

Also,

$$EX^2 = \sum_i E(X^2:N_i)\Pr(N_i) = \sum_i \alpha_i(\sigma_i^2 + \mu_i^2) \quad \text{(ii)}$$

$E(X^2:N_i) = \sigma_i^2 + \mu_i^2$. Since var $X = EX^2 - (EX)^2$, it follows that

$$\text{var } X = \Sigma \alpha_i(\sigma_i^2 + \mu_i^2) - (\Sigma \alpha_i \mu_i)^2 \quad \text{(iii)}$$

However, $\Sigma \alpha_i(\mu_i - \mu)^2 = \Sigma \alpha_i \mu_i^2 - \mu^2$, so that (iii) becomes

$$\text{var } X = \Sigma \alpha_i \sigma_i^2 + \Sigma \alpha_i (\mu_i - \mu)^2 \tag{iv}$$

If $\sigma_i^2 = \sigma^2$ for all i then (iv) reduces to

$$\text{var } X = \sigma^2 + \Sigma \alpha_i (\mu_i - \mu)^2 \tag{v}$$

Therefore,

$$\text{var } X - \text{var } X_i = \Sigma \alpha_i (\mu_i - \mu)^2 \geq 0 \tag{vi}$$

with equality if and only if $\mu_i = \mu$ for all i. This shows that heterogeneity raises the standard deviation of the distribution of the random variable X. For the application to the distribution of market clearing prices,

$$\sigma_i^2 = F_i(1 - F_i)/[m_i(\partial F_i/\partial p)^2] \tag{vii}$$

where m_i denotes the size of the random sample from N_i. Let p_e denote the asymptotic market clearing price in market i. If $F_i(p_e^i) \neq F_j(p_e^j)$ $i \neq j$, then there are differences both with respect to μ_i and with respect to σ_i.

2 For the 17 commodities in the sample of actively traded goods, the simple correlation between the maximal allowable daily price change per contract and the margin per contract is 0.812. The margin increases by US$1.972 per contract for each US$1 increase in the maximal allowable daily price change per contract. For the 23-commodity sample, the simple correlation is 0.796 and the slope is different. For each US$1.00 increase in the maximal daily allowable price change per contract, the margin rises by US$1.744. The 23-commodity sample includes six less actively traded ones. These results show that the increment in margin per dollar increment in daily price limit is lower for the less actively traded commodities.

3 Under a system of market-determined foreign exchange rates, the nominal price in US dollars of internationally traded goods, the goods in our sample for the most part, would depend on the factors that determine the US aggregate price level. Under fixed exchange rates this is no longer true. The prices in US dollars of internationally traded goods respond to the forces determining the US aggregate price level only insofar as these forces also influence the world market for these goods. The US economy is a substantial fraction of the world economy. Therefore, even with fixed foreign exchange rates, the prices of internationally traded goods denominated in US dollars depends on the factors influencing the aggregate price level in the United States. Our analysis treats the US relative price as an endogenous variable. It is, therefore, consistent with floating exchange rates determined by free market forces or with fixed exchange rates and the US economy as a substantial fraction of the world economy. To the extent that the US dollar was a reserve currency during the time of fixed exchange rates, the case for treating the US price level as a function of US monetary and fiscal policy is even stronger. Our empirical experiments with various alternatives led us to the formulation we use.

4 Even if the R^2 in (37) were one for one or even for all commodities, this would not affect the logic of the decomposition of the sources of price variability in terms of the explanatory variables of (37). One may regard (37) as giving an analysis of variance of the price in terms of the variability of the explanatory variables, $\log q_{it}$ and $\log y_t$, regarded as random variables. To the extent that $\log y_t$ and $\log q_{it}$ are predictable one or more years in advance, it would be desirable to represent this in the analysis of variance. For present purposes this is outside the scope of our analysis.

5 The 25 commodities in this sample are as follows: 1–18, 20, 21, 25, 27, 28, 29, and 30 (see Table 2 for the names of these commodities).

6 It has not escaped our attention that one may use the regression equations in Table 5 to predict the log of the open interest and the log of the volume of trade for the 26 (=51 − 25)

commodities which now lack an active organised futures market. In our earlier work we did precisely this. We have doubts about the reliability of these results for several reasons. The most important caveat is with respect to the measure of stocks to represent the scale of potential interest in the commodity. Empirically, it turns out to be a very important variable for explaining lack of interest in a commodity. That is, commodities with a small potential market are bad candidates for an organised futures market. In some cases there are large government holdings of the commodity (eg, peanuts and tobacco), and it is difficult to determine how to measure the relevant stocks. (For empirical work on this problem, see Telser 1958.) There may also be a price-stabilising effect of organised futures trading so that the relevant measure of price variability between traded and untraded goods is larger than would appear from the observed differences in price variability. (For a theoretical analysis of the stabilising effects on prices of profitable speculation, see Telser 1959.) A third important factor, the tax incentive for trading in futures contracts in the United States, does not appear in our theoretical or empirical analysis. Such trading may allow a shift of capital gains and losses from various speculations between years and thereby lowers the income tax of the speculator, who can take advantage of the lower tax rate on capital gains. The trade believes this is important. (For a careful analysis of portfolio aspects of futures trading, see Dusak 1973.)

7 We use a sample of 23 commodities. The two missing commodities are cottonseed oil and turkeys. Both have been inactive since 1965. We lack data giving the commissions and margins for these two, and this explains their absence from our sample. The commission per transaction and the margin per contract were obtained from brokerage firms and refer to the first half of 1976. The sample period gives averages of the explanatory variables for an earlier period. This should cause no problem. One may argue that the commissions and margins do in fact depend on long-term averages. In any case, since these variables appear as the dependent variables in the regression equations, the presence of measurement error does not bias the estimates of the regression coefficients. However, it does lower the R^2 and increase the standard errors of the estimated regression coefficients.

8 The simple correlation between m and t is 0.643. This seems too low to explain the similarity in the pattern of results.

9 The indicated adjustment is given as follows:

Let
$$u_t = v_t + \rho u_{t-1}, \; Eu_t v_\tau = 0, \; t \neq \tau, \; Ev_t^2 = \sigma^2$$

Then
$$\operatorname{var} u_t = E(v_t^2 + 2\rho v_t u_{t-1} + \rho^2 u_{t-1}^2) - \rho^2 (Eu_{t-1})^2$$
$$\operatorname{var} u_t = \sigma^2/(1 - \rho^2) \tag{A1}$$

Since
$$DW \equiv \Sigma (u_{t+1} - u_t)^2 / \Sigma u_t^2$$
$$E(DW) = 2[1 - E(\Sigma u_{t+1} u_t / \Sigma u_t^2)]$$

Hence
$$E(DW) = 2(1 - \rho) \tag{A2}$$

Substitution of (A2) into (A1) implies the indicated formula.

BIBLIOGRAPHY

Arrow, Kenneth J., 1971, *Essays in the Theory of Risk Bearing* (Chicago: Markham).

Arrow, Kenneth J., 1975, "Vertical Integration and Communication", *Bell Journal of Economics* 6, Spring, pp. 173–83.

Bresciani–Turroni, C., 1931, "L'Influence de la speculation sur les fluctuations des prix du coton", *L'Egypte contemporaine*, 22, March, pp. 308–42.

Cramèr, Harald, 1946, *Mathematical Methods of Statistics* (Princeton University Press).

Demsetz, Harold, 1968, "The Cost of Transacting", *Q.J.E.*, 82, February, pp. 33–53.

Dusak, Katherine, 1973, "Futures Trading and Investor Returns: An Investigation of Commodity Market Risk Premiums", *J.P.E.*, 81(6), pp. 1387–406; reprinted as Chapter 24 of the present volume.

Higinbotham, Harlow N., 1976, "The Demand for Hedging in Grain Futures Markets", PhD dissertation (University of Chicago).

Houthakker, H. S., 1959, "The Scope and Limits of Futures Trading", In *The Allocation of Economic Resources*, edited by Moses Abromovitz *et al.* (Stanford University Press).

Marshall, Alfred, 1920, *Industry and Trade*, Third edition (London: Macmillan).

Telser, Lester G., 1955, "Safety First and Hedging", *Review of Economic Studies*, 23, December, pp. 1–16.

Telser, L. G., 1958, "Futures Trading and the Storage of Cotton and Wheat", *Journal of Political Economics*, 66(3), pp. 233–55; reprinted as Chapter 6 of the present volume.

Telser, L. G., 1959, "A Theory of Speculation Relating Profitability and Stability", *Review of Economics and Statistics* 41, pp. 295–301; reprinted as Chapter 19 of the present volume.

Telser, L. G., 1967, "The Supply of Speculative Services in Wheat, Corn, and Soybeans", *Food Research Institute Studies* 7, supplement, pp. 131–76.

Telser, L. G., 1978, *Economic Theory and the Core* (University of Chicago Press).

Working, Holbrook, 1948, "Theory of the Inverse Carrying Charge in Futures Markets", *Journal of Farm Economics* 30, pp. 1–28; reprinted as Chapter 4 of the present volume.

Working, H., 1953, "Futures Trading and Hedging", *A.E.R.* 43, pp. 314–43.

Working, H., 1967, "Toward a Theory Concerning Floor Trading on Commodity Exchanges", *Food Research Institute Studies* 7, supplement, pp. 5–48.

14

*Futures Markets: Their Purpose, Their History, Their Growth, Their Successes and Failures**

Dennis W. Carlton
University of Chicago

Futures trading has exploded since 1970. As the number of futures markets has grown and the number of participants increased, numerous policy questions regarding futures markets and their regulation have arisen. Many of the most important of these policy questions are discussed by the articles in this volume. In assessing regulations that will influence how futures markets will evolve, it is useful to know how futures markets have behaved in the past. This article provides that historical perspective.

The first question is whether futures markets should be regulated at all. If a firm has an idea to produce a better mousetrap, the firm is allowed to produce it and see whether it sells. The firm is not told how to produce it, what payment terms to use, what maximum amount to sell to any one person, or how to change the mousetrap over time. Yet, current and proposed regulations of futures markets would constrain futures exchanges in all these ways. Is a futures contract so different from a mousetrap that close regulation of one, but not the other, is required?

To answer this question or any question related to the regulation of futures markets, one must first understand why futures markets develop, what functions they perform, and what characteristics commodities should have if they are to be successfully traded on a futures market.

---

*This paper was first published in *The Journal of Futures Markets* 4(3), pp. 237–71 (1984) and is reprinted with the permission of John Wiley and Sons. The author wishes to thank the Columbia Center for the Study of Futures Markets, NSF, and the Law and Economics Program of the University of Chicago for support. He also thanks Virginia France and Kenneth McLaughlin for research assistance, Mary Frances Kirchner for providing data, and Walter Blum, Peter Berck, Frank Easterbrook, Stephen Figlewski, Daniel Fischel, Kenneth French, Anita Garten, Joseph Isenberg, Geoffrey Miller, Frank Rose, Roger Rutz, Stephen Selig, Jeffrey Williams and participants at the Arden House Conference on Futures Markets, February 16, 1984 for helpful comments.

Once one knows what futures markets do (ie, what "product" they provide), it is necessary to understand how the futures "industry" works. That is, one can look at how and why the industry has grown, how long the products the industry produces (ie, a particular futures markets) last, and how vigorous is the competition between the industry members (ie, the exchanges).

In this analysis, exchanges are treated as a type of firm. True, they are nonprofit and have many distinctive institutional features. Still, their objective is to succeed by generating volume and it is useful to view exchanges as firms in competition (or potential competition) with each other. It is important to treat futures exchanges as a type of firm in competition with other firms because competition between firms can often be a substitute for regulation. The greater the degree of competition, the more likely it is that exchanges themselves promulgate rules and regulations that benefit and protect consumers in much the same way that competition in other markets protects consumers. The importance of competition and the incentives of exchanges seem to have received insufficient attention in assessments of the effects of regulations on futures markets.

This article provides an industrial organisation study of the futures industry. It looks at the product, the producers, the customers, the recent growth, the rate of product innovation and turnover, the concentration in the industry, and the degree of competition. In the first section the basic groundwork is laid for understanding what futures markets do. The second section builds on the first to identify the important characteristics of commodities traded on futures markets. The third section documents the growth in the number of futures markets and the shift in the mix of the commodities traded. The fourth section pays special attention to the recent growth in futures market trading, and, using the results of the third section, discusses some reasons for this recent growth. The fifth section analyses the length of time futures markets typically last (ie, the rate of product turnover in the futures industry), and the last section examines the important issue of competition between exchanges.

FUTURES MARKETS – AN OVERVIEW

Futures versus forward contracts

When an individual buys a December futures contract in wheat on an organised exchange, he agrees to pay the current futures price to receive wheat in December. An individual can satisfy his obligation either by physically receiving the wheat and paying for it, or alternatively by cancelling his obligation. He can cancel his obligation by entering into an offsetting transaction in which he sells a December wheat futures contract. The individual receives a gain or loss depending on whether the price at which he sells is above or below the one at which he bought. For every buyer, there is a corresponding seller. An exchange typically has a clearing organisation which interpositions itself so that it is a seller to each buyer and a buyer to each seller. In this way, each buyer or seller is dealing with a known entity, the clearing organisation.[1]

Futures contracts on organised exchanges are standardised.[2] If Joe buys a December wheat future on a certain exchange, it is the same as the December wheat future that Harry sells on that exchange. If Joe wants to offset his position (equivalent to selling his futures contract), he needs to find a buyer. Because the exchange has a standard wheat futures contract, it is easier to find a willing buyer than if each wheat contract were specialised to the particular buyer (or seller). The exchange provides a market (a trading floor) at which Joe and Harry (and all other prospective buyers and sellers) can trade and, through its clearing organisation, guarantees both sides of the trade.[3] Further, the exchange market is impersonal. A buyer and seller typically will act through intermediaries (futures commission merchants and floor brokers) and thus will not know the identity of each other. However, even if a buyer knows and distrusts the seller, he would still trade since buyers and sellers are trading (ultimately) with the clearing organisation of the exchange and not with each other.

A futures market attempts to lower transaction costs and generate liquidity. An exchange facilitates trade by removing uncertainty about the reliability of the other side of the trade, and by developing a standardised contract, so that a large enough group of interested traders will exist so that positions can be offset without large price disruptions. Most futures contracts are not entered into with the intention of physically taking or making delivery. Fewer than 1% of futures contracts were settled by delivery in 1983.[4]

It is important to distinguish a futures contract from a forward contract (see Black, 1976). A forward contract, like a futures contract, is an agreement between two parties in which the parties incur an obligation to each other to make or take delivery of a particular commodity at a certain date for a given price. A forward contract, unlike a futures contract, is typically a nonstandardised agreement between the two parties that is difficult to transfer to other parties.

For example, Joe agrees on Monday to deliver one bushel of a specified grade of wheat to Harry for US$2 per bushel on Thursday. Each party relies on the other to fulfil his side of the contract. Harry may later want to assign his contract to a third party. But a third party may be reluctant to take the forward contract. The third party may not know Joe, and, in any case, the particular quality of wheat, specified in the contract, may be a bit different from what the third party wants. That is, if products (and trading partners) are not perfectly homogeneous (and no product or trade ever is), the ability to resell forward contracts may be quite limited. Unlike futures contracts, forward contracts for a commodity typically are made with the intention of making or taking physical delivery and usually differ in their details (eg, grade, delivery point, time of delivery) from transaction to transaction.

It is, of course, possible to conceive of a standardised forward contract that all buyers and sellers use and that can be resold. That is, a standardised forward contract that allows a person who is long to satisfy his obligation by selling his long contract to someone else (or paying someone to take it). If such a market developed, it would differ from an organised futures market in only one basic respect: the transaction would be between individuals rather than between an individual and (ultimately) the clearing organisation of an exchange. The transaction costs of having to guarantee the performance of the other side of the trade could be quite costly. For example, suppose Joe makes a forward contract to sell one bushel of wheat to Harry for US$2. Harry later decides he does not want the wheat, and sells the forward contract to Tom. Tom is a crook and now Joe is worried that his forward contract is useless, unless he pursues costly legal remedies.

Although the trading in forward contracts may sound a bit cumbersome, there have been several times when active markets in forward contracts have arisen. For example, in the 1850s trading in forward contracts was common in corn and wheat. A bit later, trading in forward contracts arose in cotton. These active forward markets set the stage for the development of futures markets in these commodities in the 1860s and 1870s.[5] Trade in forward contracts also set the stage for the futures markets in soybean meal and most recently in GNMAs. For example, in 1950 trading in forward contracts of soybean meal was common. One soybean meal forward contract was sold and resold 30 times.[6] Recently an active forward market in GNMAs arose (Shriver, 1978). The market grew so large that the government commissioned a study to investigate whether the market should be regulated (Shriver, 1978). In several of these active forward markets, problems arose with guaranteeing trades. "Some of the endorsers of contracts were difficult to locate and collect from when the contracts matured."[7] The establishment of a futures market solved these problems in a more efficient manner than could individual efforts on the part of buyers and sellers.[8]

Value of futures markets
The key to understanding the purpose of a futures market (or its close cousin, an active market in forward contracts) is to understand why someone would want to be able to retain the flexibility to undo a commitment. That is, it is necessary to understand why forward contracting with physical delivery (and no trading in forward contracts) is not a perfect substitute for a futures market.

First, let us examine one of the most common reasons for futures markets, the ability to hedge or reduce one's risk. For example, suppose that in September a grain elevator has 1000 bushels of wheat which it intends to sell on December 16. The grain elevator can eliminate the uncertainty over what the price will be in December by selling (shorting) in September a December futures contract (which comes due December 16) for the 1000 bushels at price P_F. If the grain elevator sells in the cash market on December 16 for P_S, it will receive $1000 P_S$. But on December 16, the futures price will equal the spot price, P_S; hence the gain or loss on the futures contracts will be $1000(P_F - P_S)$, and total receipts will be $1000 P_F$. That is, before December 16, the grain elevator can use the futures market to eliminate price risk and guarantee revenue of $1000 P_F$.

One important real-world complication is that the variety or location of the wheat in the grain elevator may differ from that specified in the futures contract. This means that on December 16 the futures price of wheat will not necessarily equal the cash price of wheat in the grain elevator. This is just another way of saying that the price of wheat in Chicago can differ from that in a town in Iowa. The more heterogeneous are the price movements of wheat in Chicago and wheat in Iowa, the less able is the Iowa farmer to use the futures market to hedge. (The risk caused by the differences between the futures price and the spot price is called "basis risk".) There is now clearly a tradeoff. To make it easy to hedge, one wants the specification of the commodity in the futures contract to closely resemble one's own product. In the extreme, however, that would lead to an idiosyncratic contract for every participant and no liquidity. As a result, it would be difficult to easily undo one's futures position.

But clearly the price risk (including the basis risk) would be completely eliminated if the 1000 bushels could be sold in September to a buyer on a forward contract stipulating delivery in December. What then are the advantages of using a futures market instead of a forward market? The main advantage is liquidity, which affects the terms on which risk can be transferred. A buyer may not know his future demand and, especially if he cannot resell the forward contract, may be reluctant to make a commitment to the grain elevator (ie, the buyer will discount the price heavily because of his inability to resell). If the set of traders who must bear the risk of price fluctuations is restricted to those taking or making physical delivery of the good, a greater risk premium will be paid than if a futures market (with many more participants) is used. We might expect then that, as markets in forward contracting develop and become more liquid, the price of the forward contract should rise.[9] Hieronymus (1977, p. 83) states that this did in fact occur in soybean oil around 1950.

A related point is that at any instant it may be difficult to find a buyer quickly in the cash market. Search among buyers takes time. In contrast, the liquid futures market provides an almost instantaneous transaction. Suppose a farmer suddenly delivers 1000 bushels to a warehouse; the warehouse buys it, immediately hedges it on the futures market, and starts searching for a buyer. Later in the day (after much price movement) a flour miller calls to purchase 1000 bushels next week. The warehouse makes the forward contract at a fixed price (related to but not identical to the futures market) and then offsets its short position on the futures market. In this way, the warehouse has used the futures market to minimise its exposure to the risk of holding unsold cash wheat in inventory. In most cash markets, a quick sale requires a combination of search costs plus a price discount. Without a futures market, the optimal time to remain exposed to risk (time spent holding unsold and unhedged inventory) would be higher. It would be too costly to eliminate risk immediately by a quick sale in the cash market.[10] Futures markets therefore confer a benefit by lowering transaction costs. These lower transaction costs arise from having a smaller amount of unsold and unhedged inventory, more accurate price setting, and less search.

Another reason, in addition to hedging, for using a futures market is to speculate on one's beliefs about prices. If Joe thinks prices will fall and Harry thinks they will rise, then each may be willing to use a futures market. The futures market will produce a price that reflects some average of the beliefs of the market participants. That is, the futures market can be viewed as forecasting prices based on many people's different sets of information. Unlike the price of a forward contract (with no trading), a futures price does not weigh beliefs in proportion to planned physical purchases of the commodity, but instead in proportion to the size of one's futures position; which reflects, in part, the strength of one's beliefs.

By providing a forecast of price, a futures market can better enable firms to efficiently plan their future production. Of course, any improvement in efficiency must come from improved forecasts of futures prices (or from lowered costs of forecasting). If a futures market attracts uninformed traders, it is conceivable that price predictions could become less accurate. However, this seems unlikely because any deterioration in the accuracy of price predictions would attract informed investors who would have an incentive to use their knowledge to earn high profits and thereby drive the poorly informed from the market by inflicting losses on them.

CHARACTERISTICS OF COMMODITIES TRADED ON FUTURES MARKETS

With this very brief overview of futures trading, we can now identify the most important features that a commodity traded on a futures exchange should possess to be successful. By identifying the necessary characteristics, we can better understand why trading has grown in some markets and not in others. The most important features we now discuss are (a) uncertainty, (b) price correlations across slightly different products, (c) large potential number of interested participants and industrial structure, (d) large value of transactions, and (e) price freely determined and absence of regulation.

Uncertainty

Obviously, the key reason futures markets arise is uncertainty. It is useful to spell out exactly what uncertainty is relevant. If there is uncertainty about price, then buyers or sellers with fixed quantity commitments could use a futures market to hedge their risks. People with different beliefs about price could use a futures market to speculate on their beliefs.

If there were little or no uncertainty about price, current holders of the product with intentions to sell a fixed amount in the future (or buyers with intentions to buy a fixed amount in the future) would face no price risk and would not need a futures market to hedge. On the other hand, producers of the product may face considerable *revenue* uncertainty even if price is not highly uncertain, as long as output is stochastic.[11] (Clearly, overall supply uncertainty translates into price uncertainty in a manner determined by the elasticities of supply and demand.) The revenue uncertainty could induce producers to use futures markets.[12]

Price correlations across slightly different products

A futures contract is for a commodity standardised in grade and location, while physical products traded can differ in specification and location. A futures market is most valuable when prices across different specifications and locations are highly correlated.[13] The ability to establish grades with well-known price differentials between them (as would occur, for example, if transport costs were stable or very low) will broaden the appeal of a futures market.

The benefit from an additional futures market to shift risk and provide price predictions depends upon the pre-existing ability to shift risk and predict price using other futures markets in other commodities. For example, those who store (feed) barley can use corn futures to hedge barley since the prices of barley and corn are correlated. If the uncertainty present in the price of one commodity is very similar to that in another, we are less likely to expect two separate futures markets.

An elaboration of this point is to recognise that a substitute for a futures market to shift risk is for other activities to shift risk. A risk-averse grain elevator operator who stores wheat could (a) store a portfolio of grain, (b) sell shares in the equity market and let the stockholders diversify away the risk, or (c) diversify into other lines of business. This suggests that there may well be a close link (and a relatively unexplored one) between the industrial organisation of a market and whether futures trading exists. We will return to this point below.

Large potential number of interested participants and industrial structure
The larger is the *number* of different firms involved in producing and distributing the goods, the larger is the potential number of users of a futures market, and hence the greater the likelihood that a futures market will develop. For example, if a flour maker might buy wheat in the future from a grain elevator, there are two firms *each* of whom may want to hedge in a futures market. If instead the flour producer also owns the grain elevator, it is as if the integrated firm internally offsets the desired futures trades of the flour division and wheat division. Vertical integration (or systematic use of long-term forward contracts) can be viewed as a substitute for futures trading. If Joe is always long on a certain contract and Harry always short, a combined firm of Joe and Harry will net out to a zero futures position. The less vertically integrated are firms, the more likely is a futures market. A futures market makes it easy to find a trading partner. Vertical integration (or long-term forward contracting) does the same thing.

Even the demand for speculation will depend on industrial structure. If a flour firm fulfils most of its wheat requirements internally, its need to predict future wheat prices is different from if it buys cash wheat. A vertically integrated firm supplants some of the price function (ie, does not exclusively use prices to allocate), and therefore relies less on price predictions than a nonintegrated firm. Hence, an integrated firm is less likely than a nonintegrated firm to have developed a price prediction on which it would be willing to speculate in a futures market.

Large value of transactions
Clearly, if all else is constant, the greater is the value of the total product sold, the larger is the value of reducing risk. Presumably, when more is at stake, risk-averse people will want to hedge more. People will want to speculate more as well. The reason is that people have a greater incentive to invest in prediction when transaction values are large. When many people are each making their own predictions, they are more likely to feel they have knowledge that can be profitably used in futures markets.

Price freely determined and absence of regulation
If equilibrium price is heavily influenced by government regulations or controlled by one firm, the likelihood of finding a futures market decreases. In the first case, price may not be able to vary. In the second, there is the obvious danger that the price settler could manipulate cash prices to take advantage of his futures positions.[14]

NUMBER, VOLUME, AND TYPES OF COMMODITIES TRADED IN FUTURES MARKETS

Although it is easy to identify some important characteristics that make a commodity suitable for a futures market, it is extremely difficult to predict which futures markets will succeed. Exchanges still spend considerable effort to introduce a new contract, only to find little interest in the contract. In this and the next section, we analyse the number of futures markets, their changing composition, their volume of trading, and explanations for the recent growth of trading. As we will see in the last section, these factors are relevant in assessing the competitiveness of the futures industry.

In tabulating the number of futures markets, one is immediately faced with the difficulty that, while a contract may still technically be in existence, little or no trading may be occurring. Tabulating almost extinct contracts does not seem particularly informative. Therefore, the analysis has relied on the financial section of the *Chicago Journal of Commerce* and its successor the Midwest Edition of the *Wall Street Journal* (reporting prices for July 15 or the closest date to July 15) to tabulate all futures markets in existence since 1921. (Henceforth, I refer to these data sources as the *Wall Street Journal*.) The advantage of this data source is that it automatically screens inactive or very low volume markets since those markets are not listed. The disadvantage is that changes in reporting standards or emphasis on Chicago markets or simply lapses in reporting could skew the results. However, based on some simple checks, the data appear sufficiently accurate to show the general historical trends.

To get an idea of the comprehensiveness of the *Wall Street Journal* listings we can compare their listings to the number of existing futures markets. In 1983, the *Wall Street Journal* listed 58 markets, while, as of December 1983, the Futures Industry Association reported 85 markets with positive trading.[15] In 1975, the corresponding figures were 38 for the *Wall Street Journal* and 79 for the Futures Industry Association. For 1960, the *Wall Street Journal* listed 38 markets, while the Futures Industry Association reported 55 markets with trading. These numbers suggest that the absolute discrepancy between the number of markets listed and the number with some (perhaps very little) trading is greatest in the most recent past. We can also compare the volume of trading in the markets listed in the *Wall Street Journal* to the total volume of trading. The markets listed in the *Wall Street Journal* account for the bulk of all futures trading. For example, in 1983 the markets listed in the *Wall Street Journal* accounted for about 98% of the total volume of contracts traded.[16]

Futures trading began at the Chicago Board of Trade in the 1860s. Between then and now, numerous different commodities have, at one time or another, been traded on futures markets. Since 1921, 79 different types of commodities have been listed in the *Wall Street Journal*. (By

Figure 1 Number of futures markets (1921–83)

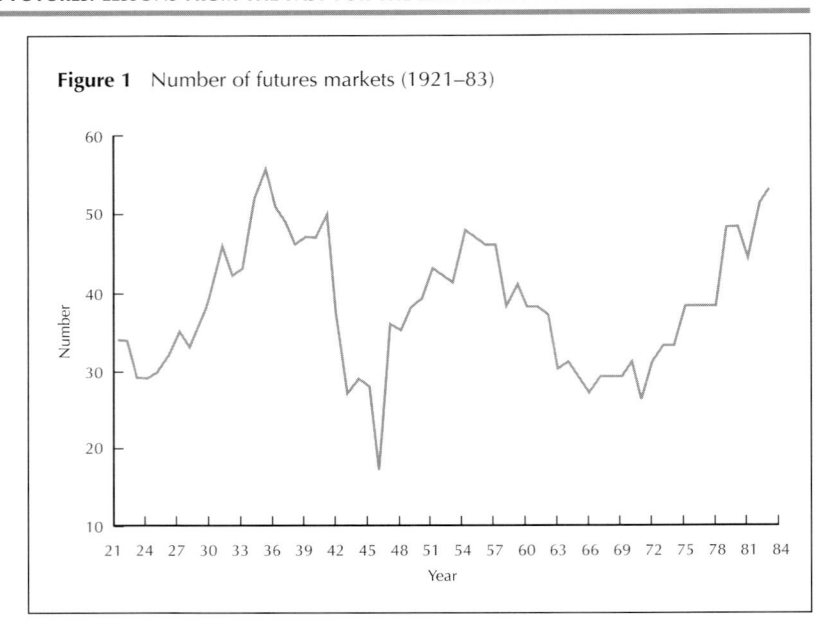

different type, I mean commodities that are not different specifications of each other. For example the wheat traded at the Chicago Board of Trade is considered here to be the same type of commodity as the wheat traded in Minneapolis.) The different commodities, together with the year in which they first appeared in the *Wall Street Journal* are presented in Table 1.

The number of different types of commodities understates the number of markets. For example, in 1921 there were 15 different types of commodities traded, but 34 markets. Many grains simultaneously traded on several different exchanges. In 1921, various varieties of wheat were traded on five different exchanges: the Chicago Board of Trade, the Kansas City Board of Trade, the St. Louis Merchant's Exchange, Minneapolis Chamber of Commerce Exchange, and the Duluth Board of Trade. In 1983, there were 42 different types of commodities and 53 listed markets.

Figure 1 shows the number of futures markets over time. Figure 1 shows that two of the largest increases in the number of futures markets occurred between 1928–1935 when the total number of futures markets increased by 23, and between 1978 and 1983 when the increase was 15.[17] Figure 1 also shows that during the relatively stable period of the mid-1950s through the early 1960s, the number of futures markets dropped from 46 in 1957 to 30 in 1963. These facts suggest that turbulent times, caused (or characterised) by rapid deflation or inflation, lead to a

Table 1 Different commodities traded since 1921 by group with year of first appearance in the *Wall Street Journal*

Financial		Foods		Grains		Industrial materials		Live stock		Metal		Oilseed	
Year	Commodity	Year	Commodity	Year	Commodity	Year	Commodity	Year	Commodity	Year	Commodity	Year	Commodity
1982	Bank CD	1949	Apples	1921	Barley	1954	Burlap	1979	Beef	1935	Copper	1929	Cottonseed Meal
1972	British Pound	1921	Butter	1930	Bran	1921	Cotton	1924	Bellies	1975	Gold	1921	Cottonseed Oil
1972	Canadian Dollar	1934	Canned Goods	1923	Cloverseed	1935	Gasoline	1969	Broilers	1955	Scrap Iron	1929	Cottonseed
1979	Commercial Paper	1935	Cheese	1921	Corn	1979	Heating Oil	1965	Cattle	1934	Lead	1921	Flaxseed
1972	Deutschmark	1927	Cocoa	1930	Middlings	1971	Lumber	1929	Hides	1968	Palladium	1937	Soybeans
1982	Eurodollar	1921	Coffee	1921	Oats	1935	Oil	1930	Hogs	1959	Platinum	1940	Soybean Meal
1976	GNMA	1921	Eggs	1982	Rice	1970	Plywood	1921	Lard	1931	Silver	1941	Soybean Oil
1973	Japanese Yen	1967	Frozen Concentrate Orange Juice	1921	Rye	1926	Rubber	1921	Pork	1929	Tin	1981	Sunflower Seeds
1974	Mexican Peso	1931	Molasses	1930	Shorts	1929	Silk	1921	Ribs	1934	Zinc		
1982	NYSF 500	1944	Onions	1945	Sorghums	1931	Wool	1935	Tallow				
1982	SP 500	1937	Pepper	1921	Wheat			1949	Turkeys				
1972	Swiss Franc	1931	Potatoes										
1976	US T-Bill	1921	Sugar										
1978	US T-Bond	1935	Tobacco										
1982	US T-Note												
1982	Value Line												

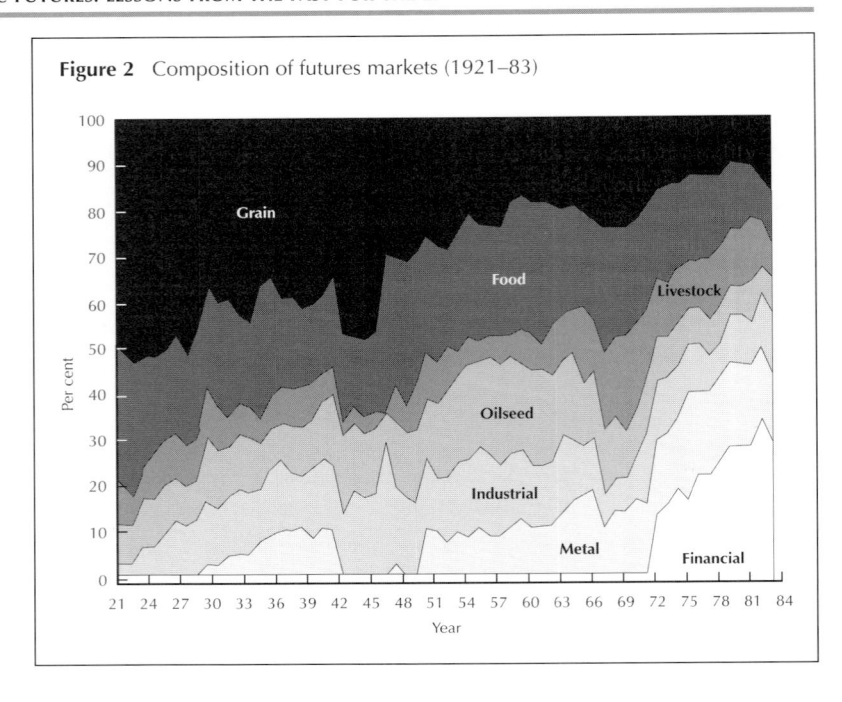

Figure 2 Composition of futures markets (1921–83)

greater number of futures markets. The simple correlation of the change in the number of markets with the absolute value of the annual inflation rate over this period is 0.26 and is statistically significant at the 10% level.[18] One interpretation of this result is that inflation is associated with increases in relative price dispersion (see Fisher, 1981, and the sources cited therein for empirical verification of this) and that the resulting increased uncertainty leads to the establishment of additional futures markets. (See Carlton, 1982, 1983a, 1983b, for a more detailed analysis of the effect of inflation on the organisation of markets.)

Government regulations, especially those affecting agricultural prices, have often had enormous impacts on futures trading. During World War II, government regulations rendered several futures markets useless and by 1946 there were only 17. By 1947, the number had more than doubled to 36.

Figure 2 illustrates the changing composition of futures markets based on the number of futures markets listed in the *Wall Street Journal*. As Figure 2 shows, in 1921 grains alone accounted for 50% of the futures markets and grains and foodstuffs together accounted for about 80% of them. Although their importance did diminish, grains and foodstuffs continued to dominate futures markets until recently. As late as 1967, they still accounted for over half of all futures markets, even though in absolute terms the number of food plus grain markets has generally declined since the mid-1930s. By 1983, the picture had changed dramatically. Grains and goods together accounted for 28% of all futures markets, financial futures accounted for 28% and metals for 15%.

It is also possible to use volume data to measure the growth of futures trading and to measure the shift in the composition of commodities traded. One difficulty with volume data is that contracts for different commodities are not worth the same amount of money so that the economic significance of total volume across different contracts is a bit unclear.[19] For example, a 1000-bushel wheat contract is worth about 20% of a 5000 bushel contract. For grains, it is possible to convert all contracts in one grain to equivalent bushels to overcome this problem. But it is not clear what adding one silver to one wheat contract means, since different values (and more importantly, different risks) are involved in the two contracts. (No one has ever suggested that it is meaningful to construct an aggregate output measure by adding units, ie, number of cars + number of houses.) Still, volume data does give one a general idea about how well-used futures markets are and the trend in that use.[20]

The story told by volume data is similar in its broad implications to that told by the data on number of markets. The Chicago Board of Trade (CBOT) publishes data on annual volume of futures trading for all grains (which is defined to include soybeans) traded on the CBOT, Kansas City Board of Trade, the Minneapolis Grain Exchange and the MidAmerica Commodity Exchange (formerly the Chicago Open Board of Trade) since 1921. Those data are presented graphically in Figure 3, which shows that trading in grains dropped off beginning in the 1930s (at the same time the number of grain futures markets expanded) and did not attain its 1930 level of trading until about 30 years later.[21] During the 1950s futures trading in grains was fairly constant. But during the 1960s, it began to rise. By 1970, it had more than doubled from its 1960 value. By 1980 it had risen from its 1970 value by a factor of about six! This rise in trading over a 10-year period was clearly unprecedented in the previous 50 years of trading. The next section will discuss some reasons for this growth.

The Futures Industry Association, Inc. publishes annual volume information on all contracts traded since 1954.[22] That information is presented in Figure 4. Like Figure 3, it shows the explosive growth that has occurred since 1970. Between 1960 and 1970, futures trading roughly tripled. Between 1970 and 1980, futures trading rose by a factor of about 10.

The data on the composition of futures markets by volume are presented in Table 2. They confirm the impressions of Figure 2 and show the shift away from grains, oilseeds and foods and toward financial futures. The combined share of foodstuffs, grain, and oilseeds fell from 94% in 1960 to 63% in 1970 to 32% in 1983. In just a three-year period, the importance of financial futures, including currencies, more than doubled from 16% in 1980 to 38% in 1983. Moreover, the growth of financial

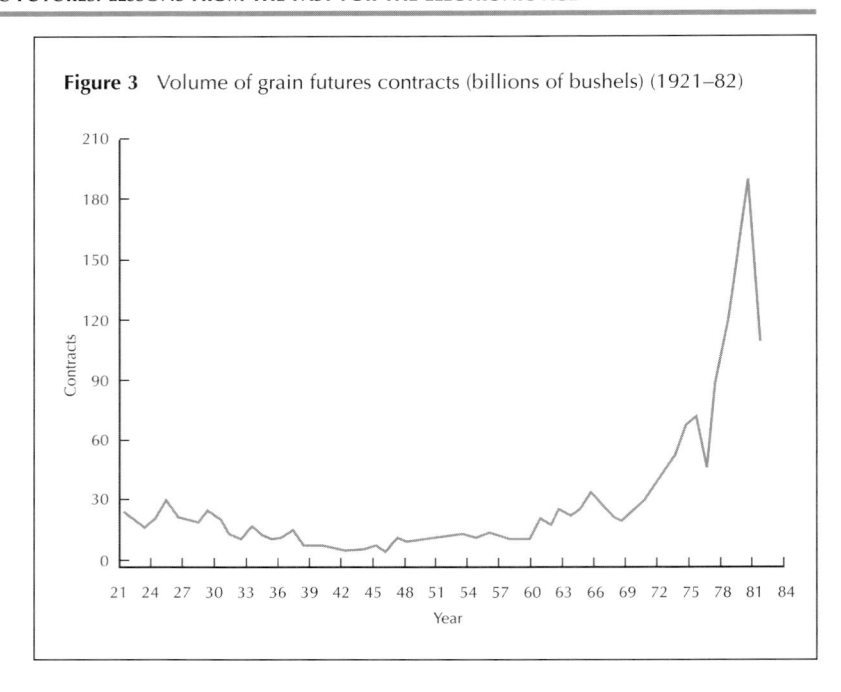

Figure 3 Volume of grain futures contracts (billions of bushels) (1921–82)

futures seems likely to continue. Of the 54 applications, pending as of June 1983, for a new futures market, 39 were for financial futures.[23]

REASONS FOR THE RECENT GROWTH IN THE NUMBER AND USE OF FUTURES MARKETS

In the previous section we discussed the rapid increase in the number of futures markets and the unprecedented growth in trading volume. We saw that the introduction of a new type of market – financial futures – was a major event. In this section we draw on the second section and explore some of the reasons for this growth in the number and trading volume of futures markets. Because of their importance, we discuss financial futures separately.

Nonfinancial futures

(i) Uncertainty and price correlations. The 1970s and early 1980s were years filled with turbulent change. The oil shocks of 1973 and 1979, the first serious recession since the 1950s, and the unprecedented inflation coincided with uncertain nominal and real prices.[24] As explained in the second section, when price uncertainty increases, the use of futures markets should increase.[25] Conversely, if the remainder of the 1980s prove to be more stable than the 1970s, we would expect a decrease in futures trading.

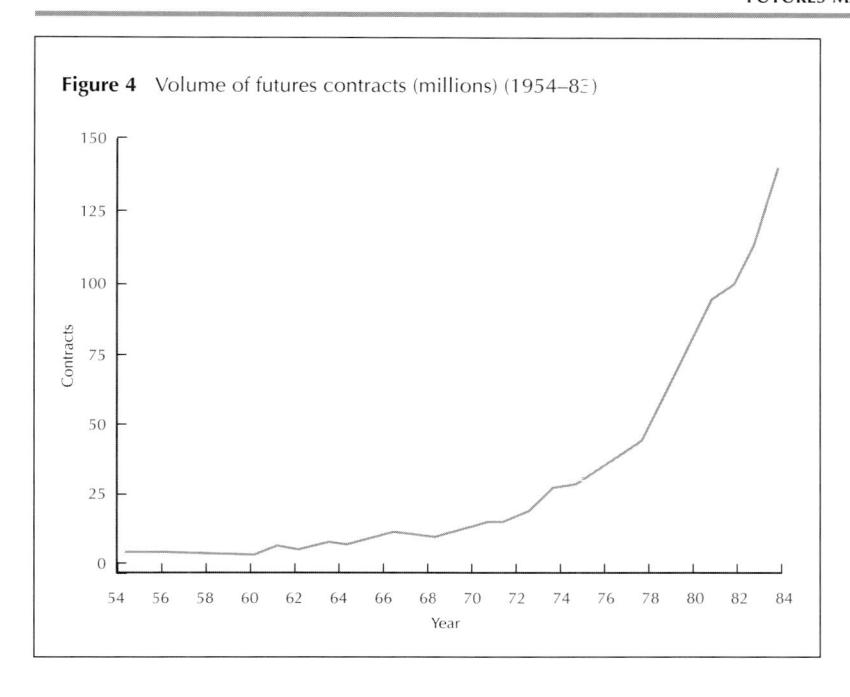

Figure 4 Volume of futures contracts (millions) (1954–83)

Table 2 Composition of trade by volume by product category (as a percentage of total trades)[a]

Product category	1960 (%)	1970 (%)	1980 (%)	1983 (%)
Grain	31[b]	17[b]	22	13
Oilseeds/products	39	30	19	15
Livestock/products	—	28	14	8
Foodstuffs	24	16	6	4
Industrial materials	4	0	4	3
Metals	3	9	17	19
Financial instruments	—	—	12	30
Currencies	—	—	4	8

[a]Source: CFTC 1979, 1983. Before 1977, fiscal year ends in June. Beginning with 1977, fiscal year ends in September.
[b]Based on a standardised 5000-bushel contract.

Another effect of increased price uncertainty is to increase basis risk by lowering the price correlations of different products.[26] As explained in the second section, as price correlations fall, there will be a need for more futures markets.

It is easy to peg contracts for physical delivery to the futures price.[27] This is a benefit during highly uncertain times, when setting price is difficult. This means that futures markets save transaction costs in setting price and that such savings will be greatest when inflation makes it difficult to use the price mechanism. We expect that during uncertain times (see Carlton, 1982), futures prices will be used more often to index prices, and therefore that the hedging use of futures would increase. Even if hedging use does not increase, the value of the price-discovery function of futures markets does increase.

Based on the preceding discussion, one would expect volume and inflation to be related. There does seem to be a strong link between futures volume and inflation for some, but not all, grains. Carlton (1983a), indicates that, other things constant, each 1% increase in inflation is associated with a 9% increase in trading in oats futures, a 6% increase in wheat and corn futures, and no statistically significant effect on soybeans futures. I am not aware of analyses of volume and inflation for other commodities, but would expect significant relationships to emerge, especially for metals and financial futures. This suggests that trading volume could fall if inflation (and the accompanying real price uncertainty) diminishes.

(ii) Number of participants, industrial structure, and value of transactions. As explained in the second section, futures trading will be greater: the greater the number of participants in the cash market, the greater the vertical disintegration of the market, and the greater the value of transactions. Undoubtedly, each of these factors contributes to the explanation for increased futures trading. But none of these factors is probably as important as the increased price uncertainty to explaining the recent dramatic increases in the volume of trading in nonfinancial futures.

(iii) Price freely determined and absence of regulation. As explained in the second section, if price is not determined by market forces, the need for a futures market may disappear. For example, government agricultural programmes have often reduced the need for futures trading. If the government is effectively influencing the price, there is no need for a futures market. Since the early 1970s, the government has generally held much lower stocks of grains through the Commodity Credit Corporation (CCC) than in previous years. For example, since 1973 the CCC has not held any stocks of soybeans. In the late 1960s it was not unusual for the CCC to hold in excess of 10% of annual soybean production in storage.[28] Thus, the widespread increase in futures trading in agricultural products in the 1970s is due, at least in part, to the absence of major agricultural programmes that remove substantial uncertainty from grain prices.[29]

Financial futures

The recent introduction of financial futures and the large trading volume they have generated have raised several policy concerns. Therefore, the

Table 3 Coefficient of variation on price of 3-month T-bill rates (based on quarterly data)[a]

Period	Coefficient of variation ($\times 10^2$)
1932–39	0.13
1940–49	0.08
1950–59	0.21
1960–69	0.32
1970–79	0.42

[a] The author thanks F. Mishkin for providing the data on prices.

reasons for the growth of financial futures deserve special discussion. As the discussion makes clear, the economic forces explaining the popularity of many existing and proposed financial futures were probably present before the 1970s and 1980s. Therefore, it seems correct to regard financial futures as a clever product innovation. We now discuss some of the most important reasons for the popularity of financial futures.

(i) Economic uncertainty, value of transactions, and number of participants. Financial futures were introduced in the 1970s, a decade that we have already characterised as one of change. Yet, the increased uncertainty explains only part of the reason for the success of this new type of futures. The large value of the relevant spot market and the large number of potential participants are key reasons for the success of these new futures markets.

There have been numerous financial futures that deal with interest rates. Obviously, interest rates have become much more volatile recently. Table 3 reports the coefficient of variation of the price of a three-month T-bill. The 1970s stand out. However, the coefficient of variation is small in comparison with the coefficient of variation for some other commodities traded on futures markets. For example, over the recent period, a typical coefficient of variation for the nominal price of wheat would be several times those listed in Table 3. But, interest rates directly affect every long-lived transaction; hence, the value of transactions and the number of individuals affected by fluctuations in interest rates are enormous when compared with similar measures for commodities in other futures markets. It is this large value of transactions and number of potential participants, together with the increased volatility of interest rates, that explains the recent success of futures markets involving interest rates.

The new futures markets in stock indices and the proposed markets in such things as number of housing starts and new car sales do not represent a response to increased uncertainty in the underlying cash market. For example, the coefficient of variation of the S&P 500 has not risen over time. In 1930–1931, it was 0.27, in 1940–1941 it was 0.11, in 1950–1951 it was 0.15, in 1960–1961 it was 0.12, in 1970–1971 it was 0.17, and in 1980–1981 it was 0.10.[30] Similarly, the percentage fluctuations in housing starts or in car sales are unlikely to be much higher in the future than they have been in the past.[31] As in the case of interest rate futures, it is probably best to explain this type of futures market as resulting from the fact that these markets (ie, stock, housing, and car markets) represent a large cash value, and that there are a large number of potential participants in such futures markets. For example, in 1982 the value of all shares listed on the New York Stock Exchange exceeded one trillion dollars, a decrease of 17% in real terms since 1970. The value of bonds traded on the New York Stock Exchange was about three quarters of a trillion dollars, a real increase of 174% since 1970. The value of government debt in 1982 was about 1.2 trillion dollars, a real increase of 24% since 1970.[32]

Another type of financial futures, the currency futures, also owe part of their success to the size of the relevant cash market; ie, foreign trade. For example, in 1972, 6.5% of GNP was for exports. In 1982, that number was 10%.[33] The continued growth of foreign trade should lead to continued increases in futures trading in foreign currencies. (The effect on currency futures of floating exchange rates is discussed below.)

(ii) Regulation and industrial structure. Government regulation impinges on the operation of futures markets in several ways. First, since 1974 the CFTC has had jurisdiction over all futures markets and must approve each new one.[34] This means that there is a regulatory delay (or a barrier) in introducing new contracts. Undoubtedly, the contract-approval process at the CFTC raises the costs and therefore inhibits the introduction of new contracts. Still, since 1974 the CFTC has approved many new contracts and has approved some innovative proposals. For example, allowing cash settlement has had an important effect on the types of feasible futures markets, especially financial futures.

Second, government regulations can control price so closely that futures markets are not needed. Scrapping these regulations can then create the need for futures markets. For example, the abolition of the Bretton Woods agreement on foreign exchange led directly to the establishment of futures in foreign currencies. Before 1972, foreign exchange rates were fixed from time to time and did not fluctuate continuously based on market forces. Once exchange rates were allowed to float freely in the marketplace, the International Monetary Market at the Chicago Mercantile Exchange began trading futures successfully in foreign currencies.

Third, government regulations can affect industry structure, and, as explained in the second section, thereby influence the need for futures markets. The deregulation of the [US] banking sector produced major changes in the competitiveness of that industry and in the risks of being in that industry. Regulation had enabled many small firms to survive. Small firms in a deregulated environment may be less able to spread risk through a diversified portfolio of investments than large firms. Futures markets in financial instruments, for the reasons explained earlier, enable all firms directly (or indirectly, through intermediaries who rely on futures markets) to spread risk. Without this ability to spread risk, we might expect the financial industry to become more concentrated as firms grow in order to spread their risk.[35]

(iii) Other reasons. A future in a financial asset like a bond does not really add a market that is based on new "states of the world". If there is a liquid cash market in bonds, the market already provides a vehicle to infer the term structure of interest rates. Futures markets provide an alternative vehicle from which to infer the same term structure. If one believes that interest rates in six months will be lower than today, one can either hold bonds and enjoy the capital gain in price, or alternatively be long a future and enjoy the capital gain.[36]

These arguments apply with only slightly less force to stock index futures. For example, the S&P 500 already is predicting the discounted value of corporate profits. A future on the S&P 500 is predicting the discounted value of profits from a future day forward. Therefore, the new information from the futures market is the separate prediction of the S&P 500's discounted profits from today until the expiration date of the futures contract. If one ignores this small additional bit of information, there is an equivalence between trading in the "cash" S&P and trading in the futures.

This means that for financial instruments futures markets can sometimes provide an almost perfect substitute to a cash purchase of a security. In fact, futures markets lower transaction costs so that it might be cheaper to carry out certain transactions in the futures market rather than the cash market.

One concern with financial futures is that different government laws and regulations apply to the cash market (ie, IRS tax laws, Federal Reserve restrictions on security margins, the SEC rules on insider trading) than apply to futures markets. Therefore, if a futures market can serve as a good substitute for the cash market, one of the reasons for the growth of futures markets may be their ability to get around some of the government laws which apply only to the cash market. It seems undeniable that there is truth to this argument, but it is extremely difficult to estimate its importance. We now examine the effect of some specific differences in the laws applicable to cash and futures markets. The three areas we discuss are (a) tax, (b) margins, and (c) insider trading.

(a) Tax treatment. The tax treatment of futures is an extremely complicated subject (see Hamada and Scholes, 1983; and Rudnick and Carlisle, 1983).

The basic message is that the various IRS rules can create incentives for some investors to use futures positions instead of positions in the cash securities in order to minimise their taxes. To the extent that futures markets lower transaction costs and make it cheaper for certain groups (eg, brokers) to trade, a futures market might even change the identity of the relevant marginal investor in various types of securities.

One example of a tax effect is that gains from futures are taxed at a maximum of 32%.[37] One might argue that individuals are encouraged to earn a living by investing in futures rather than by working for a high salary, or by investing in a cash security for a short-term capital gain, both of which can be taxed at 50% (at the margin).

It is difficult to determine how important these tax effects are in contributing to the interest in financial futures. The possibility of inducing people to become futures traders rather than high-salaried executives is probably not all that important in explaining growth in futures. The recent [US] income tax provision requiring futures traders to "mark to market" all their year-end positions for tax purposes has raised the effective tax on futures traders.[38] Prior to this change, futures could be used to roll gains over from year to year, thus postponing taxes. This recent increase in the effective tax on futures traders, together with the recent cuts in marginal tax rates on salary, coincided with some of the largest volume increases in the history of futures trading.

(b) Margins. Another commonly given reason for growth in financial futures has to do with margins.[39] There are government restrictions on margins in securities but none in futures.[40] If government regulations of margins have the effect of restraining the credit available to investors in securities from what would be available in an unregulated environment, then futures markets can be used to avoid these restrictions on credit whenever futures and cash markets are substitutes. Moreover, if one believes that government restrictions on short selling are effective and more onerous than on buying, one would expect the futures market to be relatively attractive to those who would like to go short. Again, as in the tax case, it is difficult to assess the quantitative importance of these effects on futures volume. However, the likely effect of margin requirements on buying behaviour is small, since presumably these restrictions most affect broker loans, and only a tiny fraction of stock is owned through margin loans from brokers.[41]

There is the concern with some financial futures that they are not in the public interest precisely because they allow an investor to circumvent Federal margin regulation. One reaction to this concern is that it may be a benefit to bypass margin requirements since the economic justification for these margin requirements has always been a bit unclear.

(c) Insider trading. Another possible reason for the predicted popularity of several proposed financial futures markets is that they will generate trading because they provide a means to escape SEC restrictions on insider trading which do not apply to futures markets. If, for example, one had inside information on IBM, one is currently prohibited from trading in IBM stock. If, however, the insider were able to trade in a futures contract on a high-technology stock subindex of which IBM comprised a significant fraction, he could legally profit from his information. There are two responses to this line of reasoning.

First, as discussed elsewhere (Carlton and Fischel, 1983b), the harms of insider trading may be exaggerated and, in any case, the effectiveness of the SEC in restricting insider trading in securities has been quite limited.[42] Second, if insider trading becomes widespread in futures markets but not in securities markets, it would make the futures markets riskier for noninsiders and could lead to a decline in the liquidity of those futures markets. In that case, exchanges themselves would be expected to promulgate rules to curb the practice, or alternatively to design subindexes broad enough to avoid the problem.

If one feels that current federal laws against insider trading are necessary, ie, that banning insider trading through federal regulation is efficient (see Carlton and Fischel (1983b), for a discussion of the issue) and that neither firms nor exchanges alone can adequately deal with insider trading either through contract design or formal rules, then concern about insider trading in futures to bypass security laws is relevant. However, in light of the evidence on the ineffectiveness of SEC restrictions to limit insider trading, it seems unlikely that the ability to avoid insider trading restrictions on securities could provide a primary reason for the success of a futures market.

THE LONGEVITY OF A FUTURES MARKET

In studying an industry, it is usual to ask whether it is easy for new products to be introduced or whether products, once established, are hard to displace. The answers will have an important bearing on the competitiveness of the industry, the topic of the sixth section. In this section, we examine the typical lifetime of a futures market, the product produced by futures exchanges.

Between 1921 and 1983, more than 180 different futures markets existed.[43] The majority of futures contracts fail within 10 years of their introduction. If we use futures markets in existence (based on listings in the *Wall Street Journal*) between 1921 and 1983, the average lifetime (starting in 1921) is a little under 12 years while the median is around seven years. If we exclude markets in existence in 1921, the average drops to about nine years and the median to five years. The implication is that

Table 4 Distribution of lifetimes for futures contracts 1921–83[a]

Age (years)	Frequency (%)	Cumulative frequency (%)
2	31	31
5	15	46
9	11	57
19	22	78
29	12	90
39	5	96
49	2	97
63	3	100

[a]Note: frequencies do not add to 100 because of rounding.

those contracts that existed in 1921 were above-average in terms of their success. (The only contracts which have been continuously listed in the *Wall Street Journal* over the entire period are wheat on the Minneapolis Grain Exchange, oats on the Chicago Board of Trade, and cotton on the New York Cotton Exchange.) If instead one excludes all contracts currently trading in 1983, the average life falls to about ten years and the median to about five. This is, of course, expected. The "winners" (ie, contracts not yet extinct) have a longer life than the "losers".

The histogram in Table 4 reveals just how skewed the distribution of lifetimes is for the 1921–1983 sample. A large fraction of contracts die within two years of their initial listing. One difficulty with Table 4 is that it mixes together contracts that are still in existence with those that have died. This could create a bias in the estimates of a distribution of lifetimes, since existing contracts will clearly have a life longer than their current age. One way to avoid this problem is to calculate survival rates by age. That is, calculate the probability that a contract of age X will live to age $X + 1$ for various values of X. This calculation can allow one to calculate cumulative death rates by age without any biases. The calculations (based on a "hazard" function) are presented in Table 5 for 20 years.

Table 5 shows that the survival rates increase fairly steadily from age 1 to age 7. Most of the deaths occur in the early years of a contract. About 16% of all contracts die in their first year and over 40% of all contracts die by age 6. But if a contract makes it to age 6, it can expect to remain around for a good while. For example, by age 20, 69% of contracts have died. That means that of the surviving 57% in year 6, about half of those last more than an additional 14 years.

Table 6 reports the same type of information as Table 4 by type of commodity. Table 6 suffers from the limitation that the financial contracts

Table 5 Survival and death rates of futures markets (hazard function method)

Age (years)	Probability of surviving from age $X-1$ to age X	Probability of dying at age X or less
1	0.84	0.16
2	0.89	0.25
3	0.91	0.31
4	0.92	0.37
5	0.95	0.40
6	0.95	0.43
7	0.97	0.45
8	0.96	0.47
9	0.96	0.49
10	0.99	0.50
11	0.92	0.54
12	1.00	0.54
13	0.95	0.56
14	0.93	0.59
15	0.91	0.62
16	0.90	0.66
17	0.93	0.68
18	1.00	0.68
19	1.00	0.68
20	0.97	0.69

Table 6 Cumulative frequency of contract life by group 1921–83

Age (years)	Financial (%)	Food (%)	Grain (%)	Industrial materials (%)	Livestock (%)	Metals (%)	Oilseed products (%)
2	55	32	30	26	24	21	32
5	55	46	46	42	52	42	36
9	70	54	54	47	57	67	50
19	100	71	68	79	86	88	73
29	—	86	88	84	100	96	82
39	—	93	94	89	—	100	95
49	—	93	96	95	—	—	100
63	—	100	100	100	—	—	—
Average (years)	5.5	13.6	13.7	14.7	9.4	9.1	13.8
Median (years)	2.0	7.5	7.5	13.0	5.0	8.0	10.5
Total number of markets	20	28	50	19	21	24	22

Table 7 Cumulative death rates by commodity group (hazard function method)

Age (years)	Financial products (%)	Foodstuffs (%)	Grain (%)	Industrial materials (%)	Livestock (%)	Metals (%)	Oilseed/ products (%)
2	33	32	25	13	24	21	29
5	33	46	42	25	52	38	33
10	33	54	51	32	57	62	48
20	—	65	68	65	76	80	71
40	—	83	87	83	100	—	81
60	—	100	91	91	—	—	100

are so new that the current histogram is not likely to be representative of their lives. For the other groups, with the exception of metals, there is a striking consistency – roughly 50–55% of all contracts have had lives of less than nine years. Of these other groups, industrial materials had the longest average life of 14.7 years while metals had the shortest of 9.1 years. Of these other groups, industrial materials had the longest median lifetime of 13.0 years, while livestock had the shortest median lifetime of 5.0 years.

Table 6, like Table 4, reflects the experience of failed plus still existing contracts. An alternative is to calculate cumulative death rates, based on annual survival rates (hazard function method) as was done in Table 5. Table 7 presents those results and shows that for all groups other than industrial materials, roughly 20–30% of contracts fail within two years and that (with the further exception of financial products) about 50–60% of contracts fail within 10 years. Futures contracts in industrial materials seem to have higher success rates than other contract types during their first 10 years.

The consistent story that emerges from all these tables on death rates is that the rate of product failure is high. Those contracts that do succeed last for many years. It is interesting to note that of the top five markets in 1983 (comprising 46% of trading volume), three were less than 10 years old.[44] One can conclude that the demand for the individual futures contracts is highly variable. Old contracts fail at the same time that new contracts are being introduced. Trading volume over the life of an individual contract can be highly variable and, not infrequently, can vanish.

EXCHANGES AND COMPETITION

One can view a futures exchange as a type of firm. Although exchanges are nonprofit and have many distinctive institutional features, their objective is to succeed by generating volume. It is useful to view exchanges as competing (or potentially competing) with each other.[45] As in other markets, competition is a substitute for regulation. The more competition there is, the more likely it is that exchanges themselves will promulgate rules and regulations that benefit and protect consumers in much the same way as competition in other markets protects consumers.

To assess the level of competition between exchanges, it is necessary to focus not just on similar futures contracts (eg, silver futures of the Chicago Board of Trade and the Commodity Exchange, Inc.) but also on potential competition. Because volume and liquidity are important to any futures contract, each futures contract has elements of natural monopoly. Because liquidity would be increased, it *may* be more efficient to have all trading in a given contract concentrated on one exchange, instead of split between two exchanges. But as long as there is competition to become the one successful exchange, competition can protect consumers.

There may also be economies of scope associated with trading related products. For example, it may reduce transaction costs to have soybeans and soybean meal or wheat and corn traded on the same exchange rather than on two different exchanges. Floor traders who know farm products might trade all these different products each day.

The issues of natural monopoly and economics of scope raise the question whether exchanges can develop a dominant position in an area and then use that position to exert market power. Theoretically they could exert their market power by raising the cost of using their exchanges either by raising commissions or by adopting rules and regulations unfavourable to the users of the exchange.[46] The concern is not necessarily with actual competition between two exchanges in an identical product but, rather, whether any one exchange, by virtue of its liquidity in other markets, has a substantial advantage over other exchanges in markets of a particular type. If so, potential competition for the right to become the one successful futures market for a particular commodity may not protect consumers.[47] Therefore, understanding how the various exchanges have fared and using the results of the third, fourth and fifth sections to assess the likely level of competition is important in an analysis of regulation, since competition between exchanges can be viewed as a substitute for regulation.

Between 1921 and 1983, there were over 20 different US exchanges. In 1983, there were 11 futures exchanges authorised by the CFTC. Table 8 presents information on three of the largest exchanges (largest by volume in 1983). The largest exchange during the period 1921–1983 was the Chicago Board of Trade. In 1921, the CBOT had contracts in both grains (5) and livestock (3).[48] From 1921 to 1947, the CBOT always had grain contracts, and usually a smaller number (1 or 2) of livestock and industrial materials. Between 1947 and 1970, it added several oilseed product contracts (soybean oil and soybean meal). From 1970 on, it had contracts in grains, industrial materials, livestock, metals, and oilseed products. Finally, in the mid-1970s it added financial futures. It currently has no

Table 8 Importance of the Chicago Board of Trade (CBOT), Chicago Mercantile Exchange (CME), and the Commodity Exchange, Inc. (COMEX)[a]

	Market share by number of contracts (%)			Market share by volume (%)		
	CBOT	CME	COMEX	CBOT	CME	COMEX
1921	24	21	—			
1930	18	10	10			
1940	17	4	19			
1950	18	15	15			
1960	24	11	13	64	15	3
1970	35	16	10	58	24	6
1980	27	35	6	49	24	12
1983	23	30	6	45	27	14

[a]Source: number of contracts based on listings in *Wall Street Journal*. Volume figures from Futures Industry Association, Inc.

futures markets in livestock and has never had a successful futures market in food. The history of the CBOT is one of gradual diversification into a variety of very different products.

The number of contracts on the CBOT is presented in Figure 5 (based on listings in the *Wall Street Journal*). The interesting feature of Figure 5 is their stability. From 1921 to 1975, the number of markets usually varied between 6 and 10. From 1970 to 1983 the number of markets averaged 11, with a high of 14 in 1979. The importance of the CBOT by number of contracts and by volume is illustrated in Table 8. The CBOT's importance measured by volume has diminished since 1960 from 64% to 45% in 1983. Its share by number of markets, however, has, if anything, increased slightly since 1947. In 1983, it accounted for about 45% of all futures contracts traded and 23% of all futures markets (listed in the *Wall Street Journal*).

Another important exchange is the Chicago Mercantile Exchange (CME). From 1921 until 1947, the CME specialised almost exclusively in foodstuffs, the one category in which the CBOT has never achieved success. After 1947, the CME quite often had livestock futures which gradually became more important. So, for example, in 1969 the *Wall Street Journal* reported two food and three livestock futures for the CME. According to data supplied by the Futures Industry Association, livestock futures accounted for about 86% of the trading volume on the CME in 1969. Since 1970, the CME (including the International Monetary Market) has instituted contracts in industrial materials, financial futures, and metals. By 1983, the CME no longer offered contracts in the food group. In 1983, the CME accounted for about 27% of all futures contracts traded and 30% of all markets (listed in the *Wall Street Journal*).

Figure 5 Number of futures markets on the Chicago Board of Trade (1921–83)

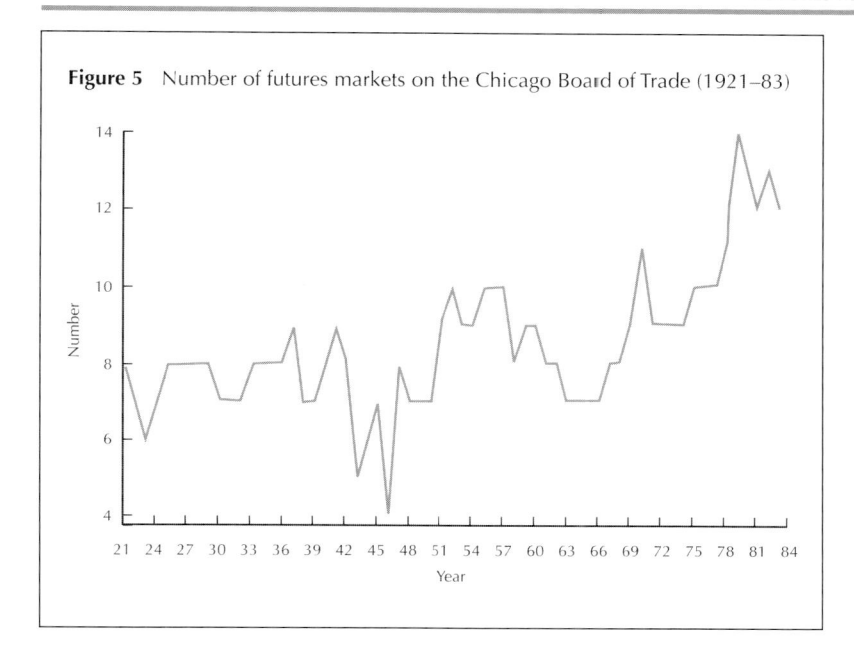

The Commodity Exchange, Inc. (COMEX) presents a contrasting picture to both the CBOT and CME. Until the 1970s the COMEX typically had contracts in industrial materials, livestock, and metals. Since 1971, COMEX has specialised almost exclusively in metals. In 1983, it accounted for about 14% of all futures trading and for 6% of all markets (listed in the *Wall Street Journal*).

The other exchanges (existing as of 1983) are much smaller and have typically offered contracts in only one or perhaps two groups of products for most of their history. Recently some branching out has occurred. For example, the Kansas City Board of Trade, which until recently had specialised only in grains, now offers financial futures, which in 1983 accounted for 44% of its volume.

Do some exchanges consistently do better than others when a contract is introduced? Table 9 shows for selected exchanges the median life of contracts for all contracts in existence between 1921 and 1983 and the death rates by various ages.[49]

Several interesting facts emerge from Table 9. In terms of death rates, the two largest exchanges, the CBOT and CME, are quite similar. (See note to Table 9 for abbreviations.) Some of the other exchanges, like the CSCE, the KCBT, and COMEX do about as well or better than either the CBOT or SME, while some other exchanges, like the NYPE, do worse. There is no indication from Table 9 that the two dominant exchanges in 1983 (the CME and the CBOT) have a decided advantage over all other exchanges.

Table 9 Success of futures markets for selected exchanges[a]

Exchange	Cumulative probability of contact death by selected exchange (hazard function method) years				Median life (years) for contracts 1921–83
	2	5	10	20	
CBOT[b]	0.25	0.40	0.46	0.56	8.5
CME (including IMM)[c]	0.22	0.41	0.44	0.61	7.0
COMEX[d]	0.21	0.33	0.55	0.86	8.5
CSCE[e]	0.14	0.14	0.14	0.14	22.0
KCBT[f]	0.28	0.49	0.49	0.78	4.0
NYME[g]	0.09	0.49	0.66	0.66	3.5
NYPE[h]	0.60	0.70	0.80	0.90	1.5

[a]Note: some exchanges – eg, the CSCE – were formed as a result of a consolidation of independent exchanges. We treat contracts traded on the independent exchanges as part of the data set for the exchange formed by consolidation.
[b]CBOT = Chicago Board of Trade.
[c]CME = Chicago Mercantile Exchange (includes IMM).
[d]COMEX = Commodity Exchange, Inc.
[e]CSCE = Coffee Sugar Cocoa Exchange.
[f]KCBT = Kansas City Board of Trade.
[g]NYME = New York Mercantile Exchange.
[h]NYPE = New York Produce Exchange.

The data on median life, though less reliable for the reasons explained earlier, confirm this impression.

The death rates in Table 9 are based only on contracts listed in the *Wall Street Journal*. The analysis is thus limited to contracts that have already achieved some initial degree of success. Some exchanges may be much worse than others at introducing contracts that succeed even initially, and that fact would not be reflected in Table 9. Using and updating the data published in Silber (1981), and using his definition of success as the trading of 10,000 contracts three years after introduction (inclusive of the year of introduction), I have calculated success rates by selected exchanges for contracts introduced between 1960 and 1981. Those results appear in Table 10.

In contrast to Table 9, Table 10 shows that the CBOT has a much higher success rate than its rivals. The four largest exchanges (largest by volume in 1983) (CBOT, CME, COMEX, CSCE) usually do much better than the smaller exchanges. The one exception is the MACE, whose success derives primarily from introducing successful contracts of other exchanges in smaller contract denominations.

It is a bit tricky to draw inferences about competition from Table 10. For example, the CBOT has introduced fewer new contracts than the CME.

Table 10 Success rates based on all contracts introduced by selected exchanges 1960–81[a]

Exchange	No. of new contracts introduced	Probability of success
CBOT	19	0.58
CME	43	0.35
COMEX	8	0.25
CSCE	9	0.33
KCBT	3	0
MACE[b]	7	0.57
MPLS[c]	3	0
NYCE[d]	6	0.17
NYFE[e]	8	0
NYME	28	0.11
NYPE	4	0

[a] Data source: Silber (1981), Futures Industry Association, Inc.
[b] MACE = MidAmerica Commodity Exchange.
[c] MPLS = Minneapolis Grain Exchange.
[d] NYCE = New York Cotton Exchange.
[e] NYFE = New York Futures Exchange. (See notes to Table 9 for explanations of other abbreviations.)

Even though the CME's failure rate is higher, it introduced 15 successful contracts compared with the CBOT's 11. The disparity between the CBOT and the CME may reflect the aggressiveness of the CME to try less-certain contracts rather than any decided advantage of the CBOT. Table 10 does suggest though that effective competition from the very smallest exchanges may be less important than competition among the top exchanges.

Table 11 reports which exchanges have the largest number of contracts for each product category and the number of exchanges offering futures contracts in each category. Table 11 shows that with the exception of oilseed products and livestock, there are currently at least three exchanges trading in each of the five remaining product categories.

If exchanges are specialised and have advantages over other exchanges in their specific areas, they should do better when they introduce a contract in their specialised field than when any competing exchanges introduce contracts. For example, when a grain contract is introduced, it may have the greatest probability of succeeding if it is introduced on the CBOT. The less is the built-in advantage of any one exchange, the stronger is the case that competition protects consumers.

Table 12 presents the median lifetime and the death rates for certain years for selected exchanges by product categories. Table 12 shows that

Table 11 Fraction of contracts accounted for by the largest exchange – name of exchange – total number of exchanges offering at least one product in group · name of exchange offering greatest number of contracts –

	Financial	Food	Grain	Industrial materials	Livestock	Metals	Oilseed products
1921	—	7/10 CME 2	5/17 CBOT 5	1/1 NYCE[a] 1	3/3 CBOT 1	—	1/3 DULU[b] 3 MPLS[c] NYPE
1950	—	6/10 CSCE 2	4/10 CBOT 3	3/6 NYCE 4	2/4 CME 3	4/4 COMEX 1	2/5 MPHS 3 NYPE
1960	—	5/11 CSCE 3	4/7 CBOT 3	3/5 NYCE 3	2/3 CBOT 3	3/4 COMEX 2	3/8 CBOT 4
1970	—	3/7 CSCE 4	4/7 CBOT 3	2/3 NYCE 2	3/6 CME 3	2/5 COMEX 3 NYME	3/3 CBOT 1
1980	8/13 CME 3	4/7 CSCE 4	3/5 CBOT 3	2/5 CME 4	5/6 CME 2	3/9 COMEX 4 NYME	3/3 CBOT 1
1983	10/15 CME 4	4/6 CSCE 3	3/9 CBOT 5	3/7 NYME 5	4/4 CME 1	3/8 COMEX 4	3/4 CBOT 2

[a] NYCE = New York Cotton Exchange.
[b] DULU = Duluth Board of Trade.
[c] MPLS = Minneapolis Grain Exchange. (See notes to Tables 9 and 10 for explanations of other abbreviations.)

Table 12 Success by type of good by exchange

Product category	Exchanges	Probability of failure within five years (hazard function method)	Median life for contracts 1921–83
Financials	CBOT	0.47	2.0
	CME[a]	0.43	10.5
Grain	CBOT[a]	0.42	22.5
	KCBT	0.50	7.5
	MPLS	0.29	24.0
Livestock	CBOT	0.38	10.0
	CME[a]	0.63	4.0
Metal	COMEX[a]	0.23	9.0
	NYME	0.50	4.5
Oilseed products	CBOT[a]	0.33	19.0
	MPHS	0.17	11.5
Food	CME	0.54	5.0
	CSCE[a]	0.14	22.0
Industrial	COMEX	0.50	5.0
	NYCE[a]	0.00	54.5

[a]Indicates 1983 volume leader in the indicated category. See notes to preceding tables for explanations of abbreviations.

for financials, grains, livestock, and oilseeds, the 1983 volume leader is either less successful or about as successful as a listed rival. In industrial materials, foods, and metals, the 1983 volume leader seems significantly more successful than its listed rival. Table 12 suggests that, despite the fact that there are only a handful of competitors in many product categories, the volume leader is not so dominant as to render competition unimportant in several product categories. On the other hand, Table XII is based exclusively on *Wall Street Journal* listings and this suffers from the same shortcomings as Table 9. Therefore, caution is advised when drawing inferences. (Because of the small number of innovations by product category by exchange, a reliable analogue to Table 10 cannot be presented.)

The numbers in Table 8 showed that regardless of whether one uses shares by number or by volume, futures exchanges are part of a highly concentrated industry.[50] Using the four exchanges with the greatest trading volume, the four-firm concentration ratio in 1970 was 71% by number of markets and 92% by volume.[51] In 1983, those two percentages were 66 and 90, respectively. Table 8 also showed a large reshuffling of market shares, especially by volume, among the three largest volume exchanges (in 1983) in the previous 25 years.

Some economists would characterise a market with only three large firms as noncompetitive. However, economists have long recognised that fewness of firms alone does not necessarily indicate a lack of competition. Other industry characteristics must be analysed before conclusions about competition can be reached (see, eg, Scherer, 1980).

The theory of oligopoly (Stigler, 1964) predicts that competitions is more likely to emerge the greater is the difference in products among firms and the greater is the instability of demand in the industry. The products sold in the futures industry (ie, different futures markets) are very heterogeneous. As the third through to the fifth sections documented, the demand for individual products (ie, one particular futures market) is highly variable and difficult to predict. Moreover, as the previous sections have shown, overall demand is growing and product innovation is occurring rapidly. In such circumstances, the theory of oligopoly teaches that it is not unusual to expect competition rather than collusion to emerge. In other words, a straightforward application of the theory of oligopoly to the industry of futures markets suggests that a handful of potential competitors may be all that is needed to protect consumers.[52]

In addition to competition between domestic exchanges, there are other sources of competition that exchanges face. For example, foreign futures exchanges can be viewed as competing with US exchanges. Moreover, especially for financial futures markets, it is important to realise that cash and forward markets can provide a low-cost substitute to the futures market (see the fourth section). In several financial instruments such as government bonds and foreign exchange, there are already very liquid cash and forward markets. This means that for the participants in these markets[53] it would be relatively easy to take delivery of forward contracts and resell in the cash market since the financial "commodities" are quite standardised.[54] Thus, any discrepancy between the forward or cash market and the futures market could be easily arbitraged by the active participants in the existing cash and forward markets. The presence of the well-established forward and cash market protects the futures market. If the futures market were temporarily or systematically manipulated, arbitrageurs would rush in to eliminate the abnormality. This is analogous to the situation in industrial organisations in which the presence of vertically integrated firms (who have the option of self-supply) as participants in an input market can ensure the competitive performance of that market. The presence of such firms protects other, non-vertically integrated firms.

The important point from a policy viewpoint is that the forces of competition may protect consumers most in those futures markets in commodities with well-established forward and cash markets. Paradoxically, the need for regulatory scrutiny to protect consumers may therefore be least in some of the financial futures markets, precisely the markets whose growth has brought out the call for increased scrutiny of futures markets.

CONCLUSION

This article has attempted to provide a study of futures markets as a background for addressing current policy concerns. One must know what futures markets do before one can sensibly analyse some of these concerns. Organised futures markets provide a low-cost way of allowing risks to be transferred. They also allow people with different beliefs to speculate on those beliefs and thereby affect futures prices. If futures markets were either outlawed or made more costly to use, there would be greater reliance on forward contracting.[55] As we saw earlier, forward contracting is not a perfect substitute for futures markets and can often be more costly to use than futures markets.

Aside from regulation, participants in futures markets can look to competition to protect them. In assessing proposals affecting futures markets, one must understand that the futures "industry", like any other industry, is composed of competing firms (exchanges). One must also understand the environment within which firms compete. That environment has changed drastically within the past 15 years. There has been an explosion in the output of the futures industry. There has been rapid product innovation and a reshuffling of industry shares especially among the large exchanges. Products now, as in the past, have high failure rates and often contracts die soon after introduction. Being a firm in the futures industry is risky. In addition to competition between exchanges, there is also competition between brokers on the same futures exchange, and between futures exchanges and the less-well-organised cash and forward contracting markets.

The private gain to the organisers of a futures market and the social gain need not necessarily coincide. Although this could form a rational foundation to regulate futures markets, in fact the same could be said of any product. For example, the developer of a new consumer product does not necessarily capture the social gain of his product. Whether this causes too much or too little product variety is unclear (see eg, Spence, 1976a,b). Yet, no one has seriously considered explicitly regulating the variety of consumer products or the way in which they can be sold. In other words, a case for restrictions on, for example, the type of products (eg, new markets) that can be offered by futures exchanges or the way in which contracts can be sold (eg, margins) must rest not upon theoretical foundations alone, but upon an empirical judgement that unregulated futures markets will create a harm that could be avoided at low cost through regulation. The existing evidence suggests that competition in futures contracts may not be all that different from competition in other products. In such a case, we need not worry about special regulations for futures markets for the same reason that we do not worry about special regulations for those other products.

APPENDIX

Table A1[a]	
Year	Number of markets with trading
1983	85
1982	81
1981	86
1980	83
1979	81
1978	71
1977	65
1976	75
1975	79
1974	88
1973	75
1972	68
1971	68
1970	69
1969	57
1968	54
1967	53
1966	52
1965	54
1964	57
1963	57
1962	52
1961	51
1960	54
1959	58
1958	58
1957	60
1956	63
1955	59
1954	56

[a]Source: Futures Industry Association, Inc., Annual Series.

1 Typically a buyer or seller does not deal directly with a clearing organisation, but instead deals first with a futures commission merchant who in turn deals with the clearing organisation. The mechanics of futures trading is explained in detail in Selig (1981). The role of the clearing association is described in Edwards (1983).
2 See Telser and Higginbotham (1977).
3 Technically, the clearing organisation only guarantees its trade with the entity dealing with it, eg, a firm (ie, a futures commission merchant) through whom an individual is placing a trade. See Selig (1981).
4 See CFTC (1983, p. 114).
5 See Hieronymus (1977, pp. 76–83). See also, J. Williams (1982).
6 See Hieronymus (1977, p. 83).
7 See Hieronymus (1977, p. 83).

8 Several provisions of futures markets such as "mark to market", margins, and clearinghouses can be interpreted as a means to better guarantee performance of trades. The reader should note that the CFTC has taken the position that, under appropriate conditions, forward contracts can constitute illegal off-exchange futures contracts. See, eg, *CFTC v. National Coal Exchange*, paragraph 21, 424 in *Commodity Futures Law Reporter*.

9 This prediction must be qualified by the results of Anderson and Danthine (1983) and Hirschleiffer (1983), both of whom show that the elasticities of supply and demand are relevant in determining the risk premium. In the case of fairly inelastic demand, the statement in the text is correct.

10 If a futures market does not develop, an alternative is for corporations, owned by many shareholders, to develop as buyers or sellers and thereby help spread the risk. This and related points are developed further in the second section.

11 See Anderson and Danthine (1983) and Hirschleiffer (1983).

12 It may make sense to set up a futures market not in prices but in quantities. Using an approach similar to Weitzman (1974), one could ascertain whether a futures market based on price or one based on quantity is better for minimising risk. The Coffee Sugar and Cocoa Exchange has recently proposed such nonprice futures based on number of housing starts, new car sales, and corporate profits.

13 That is, there should be little basis risk. The basis is defined as the difference between the futures price and the spot price. If the basis is unchanged over time, a futures market can provide a perfect hedge.

14 See, however, Anderson and Sundaresan (1983) for an analysis of futures markets when the cash price is not competitively determined.

15 The Futures Industry Association, Inc. is a trade association and is the source for volume futures reported by the CFTC.

16 Henceforth, unless indicated otherwise, the *Wall Street Journal* listings are used to indicate a market's existence. In Table A1 in the Appendix, the data available from the Futures Industry Association are used to tabulate the annual number of markets (with trading) in existence since 1954.

17 During the 1928–1935 period, the increase in the number of futures markets came through foodstuffs which increased from 8 to 15, grains which increased from 15 to 19, industrial materials which increased from 4 to 8, livestock which increased from 3 to 4 and metals which increased from 0 to 5. During the 1978–1983 period, the main increase in futures contracts has come from financial products which increased from 9 to 15, grains which increased from 5 to 9, and industrial materials which increased from 3 to 7.

18 Based on observations for 1930–1983 excluding 1941–1946 using the GNP implicit price deflator as a measure of the price level. Going back to 1921, and using the CPI as a measure of the price level, the correlation is about the same, but is no longer statistically significant. The correlation of the number of markets with the absolute value of inflation was not statistically significant in either case.

19 Of course, the economic significance of the number of futures markets is also unclear when the markets are very different.

20 A good measure of the output of futures markets would be total fees paid to brokers and clearinghouses.

21 Interestingly, the annual volume of trade between 1881 and 1920 was about at the same level as in the 1920s. See Hieronymous (1977, p. 23).

22 Data provided by Mary Frances Kirchener.

23 Tabulation received from Susan Philips.

24 Numerous studies (see, eg, Fischer, 1981) have linked increased inflation and increased real price uncertainty.

25 If inflation is associated with increased uncertainty, the social loss of eliminating futures markets will be greatest during inflationary times (see Carlton, 1983c).

26 An offsetting effect is to alter the product variety produced so that it more clearly conforms with a commodity traded on the futures market – thereby lowering basis risk (see Carlton, 1982).
27 Surveys have shown that many agricultural and livestock products are sold on forward contracts with price indexed to a futures price. For example, in 1976, about 50% of surveyed farmers said their forward contracts were "based" on the futures price (Helmuth, 1977).
28 Source: *CBOT Annual* 1970, 1981, 1982.
29 Government influence on price through agricultural programmes may be reappearing. For example, in 1981–1982, the US government, through the CCC, held its largest stock of corn since 1972–1973. Source: *CBOT Annual*, 1982, 1981, 1978. A reappearance of government influence on price should lead to a decline in future volume. In fact, trading volume in grains in 1982 was well below that in 1981.
30 Calculated from the end of month S&P index for 24 months.
31 To the extent that fluctuations in housing starts and car sales are closely correlated with general movements in the stock market, there may not be a need for futures markets in these two areas if a futures market in the stock market exists.
32 See *1983 Fact Book*, New York Stock Exchange, *Economic Report of the President* (1983), p. 77, Tables B-79 and B-54.
33 *Economic Report of the President* (1983), Table B-2.
34 Futures markets in grains were first federally regulated in the United States under the Grain Futures Act of 1922. (This act replaced the Futures Trading Act of 1921 which was declared unconstitutional.) The Commodity Exchange Act of 1936 extended federal regulation to additional products and extended the scope of regulation.
35 Small firms could diversify by selling shares to stockholders who diversify. But another way to diversify is to hold a portfolio of projects. There is some evidence that this second form of diversification is a common one when managers cannot be monitored perfectly (Amihud and Lev, 1981).
36 This does not mean that the accuracy of the price prediction is unchanged by the existence of a futures market since more people might trade with a futures market. The point is that in financial futures, a cash transaction and a futures transaction can be very good substitutes.
37 This tax effect applies to all futures markets. For tax effects specific to financial markets, see Hamada and Scholes (1983).
38 As of June 23, 1981, futures positions at year end are marked to market for tax purposes. Sec. 1256 of Internal Revenue Code.
39 For a discussion of margins in securities and futures, see Telser (1981) and Figlewski (1984).
40 Regulations G, T, and U of the Federal Reserve Act set forth margin requirements on securities. The basic requirements on stock are that a 50% margin is required for purchase of stock (ie, your broker can lend you only 50% of your security) and 150% margin is required for a short sale (ie, the broker must hold the full cash value plus 50% more).
41 Less than 2% of stock is owned on margin. Source: *Federal Reserve Bulletin*, and *Fact Book* 1983. New York Stock Exchange.
42 The evidence on the effectiveness of SEC restriction on insider trading is reviewed in Carlton and Fischel (1983).
43 In this and the subsequent calculations, a single continuous contract is defined as either a single contract continuously listed or two contracts on the same exchange for the same commodity of slightly different grades that overlapped or were sequential. For example, if a grade of eggs is changed from one year to the next, the market is considered to be one, not two. Two markets are considered to be separate if they are nonoverlapping or nonsequential in time, or on different exchanges. The analysis in this section is based on listings in the *Wall Street Journal*.

44 Those three markets were T-bonds on the Chicago Board of Trade, gold on the Commodity Exchange, Inc., and the S&P 500 on the Chicago Mercantile Exchange.
45 For an examination of recent competition between exchanges, see Silber (1981). It is also important to note that even within an exchange, there are competitive issues to examine. For example, brokers on the same exchange compete with each other, and different contracts on the same exchange compete with each other. Analysing these types of competition is also necessary in order to accurately assess the need for regulation.
46 As the result of a settlement in the early 1970s of an antitrust suit, the CBOT does not allow its traders to fix commission rates among themselves. Presumably, other exchanges are aware of this anti-trust issue. Since the rules of the exchange are akin to a choice of product quality (see Fischel and Grossman, 1984), choosing rules can be analysed as quality choice as in Stigler (1965).
47 Of course, competition to become the one dominant exchange in an area could protect consumers, but the protection is more direct if competition is present for each new market that arises.
48 Based on listings in the *Wall Street Journal*.
49 Recall that there is a bias in reporting statistics based on contracts still in existence. The failure rates (based on hazard functions) avoid this bias. The reader is also warned that failure rates, especially for the small exchanges and for older ages, can be subject to large sampling error.
50 Relying on share of contracts by number will often understate one exchange's dominance of current trading. For example, the CBOT accounted for 87% of all grain contracts traded and 92% of contracts traded in grain plus oilseed products in 1982. It accounted for less than 50% of the markets in grains and in grains plus oilseed products. Source: *CFTC Annual Report 1982*, pp. 108–9.

On the other hand, it is the potential to introduce new contracts – ie, competition for the market – that is relevant to determining whether market power exists. That is why presence in the market, rather than percentage of volume, may be a better indicator of the level of competition.

It is also worth pointing out that trading volume is heavily concentrated in a handful of markets. For example, in 1983, the top five markets accounted for 45% of all trading. The top five markets, in order, were T-bonds (CBOT), soybeans (CBOT), corn (CBOT), gold (COMEX), and S&P 500 (CME). Source: Futures Industry Association, Inc.
51 The four exchanges with the largest trading volume in 1970 were the CBOT, CME, COMEX and NYME and in 1983 were the CBOT, CME, COMEX and CSCE.
52 An additional complication in analysing competition between exchanges has to do with cross-memberships. That is, it is possible to be a member of several exchanges at once. A complete analysis of this issue requires identifying how each exchange responds to its members and which members control decision making. Indications are that independent decision making among exchanges is a reasonable working hypothesis.

Conditions of entry are also an important consideration in assessing a market. Although there do not seem to be any entry barriers (aside from regulation), each of the largest exchanges entered long ago. Within the past five years, one futures exchange has failed, and two have begun. One (the NYFE) currently has a very successful stock index contract.
53 These markets have participants different from futures market. For example, a dealer in government securities will have to have an excellent reputation and credit backing before anyone would sign a forward contract with him. There apparently are no margins in these forward markets. Reputation of dealers is what guarantees performance.
54 This is less true of a commodity like wheat. Taking delivery involves storage costs, and finding a buyer at the delivery location may not be easy.
55 Alternatively, as discussed in the second section, industrial structure could change.

BIBLIOGRAPHY

Amihud, Y. and B. Lev. 1981, "Risk Reduction as a Managerial Motive for Conglomerate Mergers", *Bell Journal of Economics* 12, Autumn, pp. 605–17.

Anderson, R. and J. Danthine, 1983, "Hedger Diversity in Futures Markets", *Economic Journal* 93, pp. 370–89.

Anderson R. and M. Sundaresan, 1983, "Futures Markets and Monopoly", Working paper 63 (Columbia Center for Futures Markets).

Black, F., 1976, "The Pricing of Commodity Contracts", *Journal of Financial Economics* 3, pp. 167–79.

Carlton, D., 1982, "The Disruptive Effect of Inflation on the Organization of Markets", in *Inflation*, R. Hall, (ed.), pp. 139–52, (University of Chicago Press).

Carlton, D., 1983a, "Futures Volume, Market Interrelationships, and Industry Structure", *American Journal of Agricultural Economics* 65, May, pp. 380–87.

Carlton, D., 1983b, "The Cost of Eliminating a Futures Market and the Effect of Inflation of Market Interrelationships", Working paper 69 (Columbia Center for Futures Markets, Chicago Board of Trade, *Statistical Annual*), various issues.

Carlton, D., and D. Fischel, 1983, "The Regulation of Insider Trading", *Stanford Law Review* 35, May, pp. 857–95.

CFTC Annual Reports, various issues.

Economic Report of the President, various issues.

Edwards, F., 1983, "The Clearing Association in Futures Markets: Guarantor and Regulator", *The Journal of Futures Markets* 3(4), pp. 369–92.

Fact Book 1983, New York Stock Exchange.

Federal Reserve Bulletin, various issues.

Figlewski, S., 1984, "Margins and Market Integrity: Margin Setting for Stock Index Futures and Options", *The Journal of Futures Markets* 4(3), pp. 385–416.

Fischel, D. and S. Grossman, 1984, "Customer Protection in Futures and Securities Markets", *Journal of Futures Markets* 4(3), pp. 271–93.

Fischer, S., 1981, "Relative Shocks, Relative Price Variability, and Inflation", *Brookings Papers on Economic Activity* 2, pp. 381–431.

Hamada, R., and M. Scholes, 1983, "Taxes and Corporate Financial Management", October, mimeo.

Helmuth, J., 1977, *Grain Pricing*, September, CFTC.

Hieronymus, T., 1977, *Economics of Futures Trading*, Commodity Research Bureau, Inc.

Hirschleiffer, D., 1983, "Essays in Futures Markets", Unpublished PhD dissertation (University of Chicago).

Rudnick, R. and L. Carlisle, 1983, "Commodities and Financial Futures", *Journal of Taxation of Investments*, Autumn, pp. 45–55.

Selig, S., 1981, Statement of Stephen F. Selig Before the Subcommittee on Monopolies and Commercial Law of the House Committee on the Judiciary Regarding the Operation of the Bankruptcy Code with Respect to the Commodities Industry, 23 July.

Scherer, F. M., 1980, *Industrial Market Structure and Economic Performance*, 2nd C. P., Rand McNally.

Shriver, R. and Associates, 1978, "Analysis and Report on Alternative Approaches to Regulating the Trading of GNMA Securities", prepared for HUD, November 7.

Silber, W., 1981, "Innovation, Competition, and New Contract Design in Futures Markets", *The Journal of Futures Markets* 1(2), pp. 123–55.

Spence, M., 1976a, "Product Differentiation and Welfare", *American Economic Review* 66, May, pp. 407–14.

Spence, M., 1976b, "Product Selection, Fixed Costs, and Monopolistic Competition", *Review of Economic Studies* 43, June, pp. 217–35.

Stigler, G., 1964, "The Theory of Oligopoly", in *The Organisation of Industry*, Irwin, pp. 39–63.

Stigler, G., 1965, "Price and Non-Price Competition", in *The Organisation of Industry*, Irwin, pp. 23–8.

Telser, L., 1981, "Margins and Futures Contracts", *Journal of Futures Markets* 1, pp. 225–53.

Telser, L. and H. Higinbotham, 1977, "Organized Futures Markets: Costs and Benefits", *Journal of Political Economy* 85, October, pp. 969–1000; reprinted as Chapter 13 of the present volume.

Weitzman, M., 1974, "Prices Versus Quantities", *Review of Economic Studies* 41, October, pp. 477–91.

Williams, J., 1982, "The Origin of Futures Markets", *Agricultural History* 56, January, pp. 306–16.

15

Monopoly, Manipulation, and the Regulation of Futures Markets*

Frank H. Easterbrook
University of Chicago

> *Future markets are hard to manipulate or monopolize because of actual and potential competition from the cash commodity. Either manipulation or monopoly depends on secrecy. The futures exchanges have the appropriate incentive to establish the optimal degree of secrecy, which promotes trading even as it facilitates manipulation. The loss from monopoly and manipulation cannot exceed the gains from trading any contract, so it is never right to limit entry.*

In the textbook model of competition all buyers and sellers are tiny compared with the market as a whole. These people trade fungible commodities in an ongoing auction. Every seller's product is the same as every other's. All participants in the market are sophisticated. Every buyer continues to purchase so long as the price is less than the value he places on the commodity; every seller continues to sell so long as the price is greater than the marginal cost of production. New buyers and sellers can jump in at an instant's notice, and if they find the going rough they can jump out again. No one can take advantage of a trading partner because someone else will offer the prospective victim a better deal. Information about bids and offers spreads instantly to all participants and would-be entrants. Every beneficial trade takes place. The trading leads to an outcome that is best for each buyer, each seller, and society as a whole.

Of course textbook economies occur only in textbooks. It is commonplace to lament the fact that traders are not microscopic compared

---

*This paper was first published in the *Journal of Business* 59(2), pp. S103–27 (1986), and is reprinted with the permission of the University of Chicago Press. The author wishes to thank Dennis W. Carlton, Tom Coleman, Daniel R. Fischel, John H. Stassen, Lester Telser, and the participants in the Workshop in Applied Price Theory at the University of Chicago for helpful comments on an earlier draft. This paper was substantially completed before the author joined the court of appeals and was revised in minor ways thereafter.

with the market, traders are gullible, entry and exit not quick and free, information not cheap and ubiquitous, products not identical. These departures from the utopian vision of perfect competition give rise to calls for intervention by the government to "perfect" the operation of our regrettably imperfect markets. If the intervention of the government is afflicted with the same costs and failures as the markets, we shall have to sigh in resignation.

Yet some markets come close to the model. Agricultural and financial markets have standard products handled by many sophisticated traders. Products are comparable. Information spreads quickly. True, all of us can't be farmers or bankers, and even farmers cannot produce more wheat instantly or move it costlessly to where it is demanded, so the free entry and exit condition does not hold. Farmers and bankers do not always honour their contracts, so their promises are not really interchangeable. And of course all of us can't be glued to video monitors displaying price information. So perfection cannot be achieved. But a market can come close.

Futures markets come closer than cash markets. A futures contract is a claim to the underlying commodity but because of its standard terms is more easily traded than the commodity itself. Every contract (for a given commodity and time) is identical to every other, so contracts are fungible. The contracts are made with or guaranteed by a clearing house, so the performance of a contract does not depend on the identity, wealth, or inclinations of the person holding it. Anyone can buy a contract: entry and exit are quick and cheap; contracts can be sold nationwide without transportation cost; the supply of contracts can grow or shrink quickly. Traders in futures markets are largely professionals – both those who use futures contracts to hedge and those who trade for a living. They have up-to-the-minute information on price. Futures markets accordingly are all but indistinguishable from the textbook models of perfect competition.

In most product markets, from razor blades to iron ore to computers, firms may introduce products as they wish. If they fail, both the firms and the customers suffer. (Customers may be left without service or performance of warranties.) Few people think that the departure of the computer market from the model of perfect competition is a reason to restrict entry, which would force the market even farther away from the model.

Yet both entry into and the daily conduct of futures markets are heavily regulated. The usual reason given is that futures markets suffer from monopoly and manipulation. Traders can "corner the market" to their profit, or they can manipulate prices. So they can. The question I investigate is whether the threat of monopoly and manipulation is so serious (compared with the threat in other markets) that entry should be restricted. The first section shows how monopolisation could occur and assesses the costs of monopoly in futures markets. The second section takes up the way futures markets respond to these costs. It looks at the kinds of precautions exchanges take and whether they are apt to take the appropriate level of precaution. The third section looks at some cases of actual or asserted monopolisation to see whether there were substantial losses and whether the private mechanisms worked. The final section evaluates the case for regulation of entry and compares the costs of monopoly against the costs of preventing the operation of markets altogether.

THE NATURE AND EFFECTS OF MONOPOLY IN FUTURES MARKETS

In markets for commodities such as wheat and automobiles, monopoly means controlling the supply. A merger or effective cartel of all automobile manufacturers would enable the sellers to reduce output. Consumers bid more for the remaining supply. Producers get monopoly profits. Consumers who would buy at the competitive price but are unwilling to pay the higher price lose the difference between the value they place on a car and the real cost of producing one. The foregone consumers' surplus is the allocative welfare loss of monopoly. The prospect of obtaining the monopoly profit also induces would-be monopolists to spend resources getting and keeping their position, which increases the total cost of monopoly. (See Posner 1976, pp. 8–22, 237–55.) The higher price attracts new entry, however, and the monopolist cannot hold onto its position without barricading this entry. The cartel must not only prevent entry but also prevent "cheating" by members – that is, the practice of increasing one's own output hoping to sell more at the enhanced price. The allocative loss is limited by cheating and new entry; the sooner and the more the cheating, and the quicker the entry, the less the loss.[1]

A futures market is a slice of the market in the underlying commodity. A contract may entitle the buyer to delivery of wheat in Chicago in March. Accordingly it seems simpler to monopolise the supply. The would-be monopolist need not capture the entire supply of wheat; it is enough to get hold of wheat deliverable in Chicago in March. But it turns out not to be so easy. The monopolist of cars, wheat, or ore can extract an overcharge because people need the commodity. Someone who needs wheat does not need a futures contract for wheat; he can purchase the wheat directly. Thus a person with a monopoly of futures contracts to deliver wheat in Chicago in March cannot easily extract an overcharge. People who want wheat will not transact with him.

The general point is that for many purposes transactions in the commodity are substitutes for transactions in futures contracts.[2] Because buyers can use whichever is cheapest, they cannot easily be exploited in futures markets. The "monopolist" of a futures contract must find someone to stand on the other side of the transaction, and no one voluntarily subjects himself to monopoly when a competitive price is available. If it becomes known that one person has a monopolistic position on one side of the market, the contract dies. The putative monopolist cannot find anyone to exploit.

"Monopoly" in a futures market therefore turns out to be a species of fraud. The putative monopolist must conceal his position and intentions from his would-be trading partners, for otherwise he won't have partners. People do not knowingly submit to exploitation. In fact, a reasonable definition of monopolisation of a futures market is any explicit or implicit misrepresentation of the size of one's position in relation to the rest of the open interest. The misrepresentation may take two forms. (Each can occur on either the buying or the selling side, and for ease of exposition I do not distinguish them.)

The first, which I call a Position Fraud, is the unexpected assembly or maintenance of a large position on one side of the market. (I forswear the usual terminology of "squeezes", "corners", and the like, not only because people do not agree on the meaning of these terms, but also because this jargon makes the argument hard to follow.) The party may acquire a large portion of the existing contracts, thus undercutting the usual assumption that every trader is "small" in relation to the market. Or the party may simply decline to liquidate his position, so that at the very close of trading a formerly small holding becomes large in relation to the open contracts. The holder of these contracts then demands or tenders delivery (depending on whether he is long or short). Holders of opposite positions, surprised by the sudden demand or tender, unable either to make or take delivery without incurring large costs, and unable to find other parties with whom to close out their positions, must pay a premium to negotiate around the demand. If the adverse parties had been aware of the large position, they could have stayed out of the market, liquidated their positions earlier, or prepared to take or make delivery (such preparations could greatly reduce the costs or compliance and therefore the premium available to the holder of the large position).

The second manoeuvre, the Ownership Fraud, entails owning both futures contracts and the underlying commodity. For example, the perpetrator buys up the supplies of wheat available for delivery in Chicago in March. Then he enters "long" contracts giving him a right to delivery, and he insists on delivery. The "shorts" (the people who have promised to deliver wheat) are embarrassed by their inability to obtain supplies to fulfil their obligations. They are shocked to discover that the person entitled to delivery already owns the deliverable supply. They must either spend large sums to put more supplies into a deliverable position or pay a premium to the holder of the long interests. Again secrecy is important. People would not knowingly sell wheat short to someone who already owned the wheat; and if they did so by accident and found out in time, they could bring new supplies to market to reduce the premium they must pay to cancel their obligations. Of course the defrauder must pay a premium to get control of the deliverable supply; when the fraud is over he must unload the commodity at a loss.[3] He seeks to make more by jacking up the price of the futures contract than he loses on the cash commodity.

Both types of fraud leave a characteristic trail. There will be substantial concentration of ownership on one side of the market and sudden changes of price around the end of the contract's term. As soon as the contracts have been settled prices return to the level established by the terms of supply and demand in the underlying commodity. The sudden price fluctuation produces a transfer of wealth among participants in the market and also enables us to identify suspicious events. The manipulation may be secret at the start but cannot be secret at the end. An undisclosed manipulation is an unsuccessful manipulation. (A flurry of wash sales might lengthen the period during which the price change occurs but cannot eliminate the change, at least not without eliminating the profit from the transaction.)

What are the losses from these frauds? There are transfers of wealth from victim to perpetrator, but these shifts of wealth are not real economic loss. No resources are used up. The transfers are not even unambiguous losses to the victims. The risk of manipulation in futures markets is known. People transact in these markets voluntarily, and they will not do so if the anticipated costs (including the loss on becoming a victim) are less than the anticipated returns. The opportunities outside futures markets (including investing in the cash commodity) will determine the returns required to justify participation in futures. When manipulation is possible, the terms of trade adjust: there is no reason to think that participants in futures markets make smaller net profits on account of manipulation.

The real costs of manipulation *come from the adjustments* of terms that prevent the occurrence of anticipated net losses. The adjustment for the risk of manipulation drives a wedge between the futures price and the anticipated price of the cash commodity. This gap – needed to compensate the participants for risk – also makes the futures contract less valuable as a tool for hedging. Someone who owns the underlying commodity and wants to reduce the risk of a change in its price cannot obtain an exactly offsetting position in the futures contract but must continue to bear some of the risk. The futures contract becomes a little less liquid and therefore a little less useful as a money substitute. The price signals sent by the futures market also become a little less accurate. A farmer looking at futures prices in order to decide how much corn to plant will miscalculate by a little on account of the change.[4] And if a futures exchange alters the contract to reduce the possibility of manipulation – say, by permitting delivery in three places instead of one, thus enlarging the deliverable supply and making it harder to commit the Ownership Fraud – the change creates the same sorts of loss. A contract with a defined supply spanning three cities is less useful to someone who wants to hedge against wheat in Chicago, and so forth. The larger area means that there is no single index of value; a greater supply produces greater uncertainty in price.

Manipulations also create some direct welfare losses. People who are surprised by the state of the market may expend extra resources moving new supplies to a deliverable condition. The costs of shipping commodities at the last minute are wasted. People make take precautions such as moving additional supplies earlier; again these are wasteful. Finally, people may spend resources trying to execute these manipulations or to shift the costs of manipulations to others. These resources are lost to society.

It is not possible to quantify these costs, but they are not likely to be large. Manipulations are rare – since they must be disclosed eventually, it is hard to commit more than one. Each manipulation affects only a single contract among the many that cover each commodity. The number of recorded manipulations is a minuscule fraction of the total number of futures contracts that have been traded. Thus the change in price needed to compensate people for the risk of manipulation will be small, and the losses entailed in creating and taking precautions against manipulation also must be small. Hedgers can obtain almost perfect hedges; farmers can obtain almost perfect price signals to guide future conduct.

Although we cannot know the social costs of monopoly and manipulation in futures markets, we can know the *maximum* costs. They will be less than the social benefits of the futures contract itself. No one enters into a futures transaction unless he expects to obtain benefits exceeding expected costs – including the costs of being the victim of manipulation. When private costs become too high, people shift to using forward contracts and other transactions in the underlying commodity rather than futures markets. For reasons I have sketched, the private costs of being a victim, which include the transfer of wealth to the perpetrator, exceed the social costs. Similarly the social gains from futures markets exceed the expected private gains. The social gains include benefits derived from more accurate planting by farmers, more accurate investing by owners of Treasury bills, and so on, yet the futures exchanges cannot charge the farmers or investors for all these benefits. The benefits flow from the price signals sent by the futures markets, not from the actual purchase and sale of contracts. Thus those who never trade get some benefits from futures markets. (A process of linkage discussed in the second section ensures, however, that the futures exchanges take into account some of the value these people place on information.)

Because participation in futures markets is voluntary, people will cease dealing in a given contract when expected private gains fall to zero. Expected private gains fall to zero while social gains are still positive (because social gains exceed private ones while social costs of monopoly are less than private costs). The futures market in a given commodity will close down – with or without regulation – before monopoly and manipulation create net social costs. Thus whenever we observe a futures market in operation, we may be confident that there are net social benefits; however large the costs of monopoly and manipulation may be, they are not large enough to extinguish the gains from having the market.

MARKET RESPONSES TO THE COSTS OF MONOPOLY

Traders can protect themselves against monopoly and manipulation by withdrawing. As soon as they learn of the existence of a manipulative strategy, they close their positions. They can go on to transact in another contract free from the attempted manipulation. Entry and exit are so easy that monopoly cannot thrive.

This is not an interesting response, though, when the would-be monopolist can keep his position secret. Then the response must be organised rather than individual. The exchange on which the futures contract trades will handle the collective response.

Some futures contracts hardly require an organised response to the possibility of monopoly. Both of the frauds discussed above depend on the ability of the manipulator to obtain a "large" position in relation to the open interest and the deliverable supply. For some futures contracts this is all but impossible. Consider, for example, a futures contract covering Treasury securities of a particular maturity. The securities are readily available. They can be carried from one end of the country to the other quickly and for trivial cost, and so all of them are potentially available for delivery. No one could corner the supply of them. Futures in precious metals or other easily shipped, fungible commodities also are almost impossible to manipulate because of the large deliverable supply.[5]

Or consider a contract based on the price of some other security, such as a contract on the Standard & Poors 500 stock index. The value of this index depends on the price of a large basket of stocks. The futures contract on the index is not settled by delivery or even by transactions to set a price; the cash settlement price of the futures contract is derived mechanically by inspecting the value of the index. No one can manipulate this contract even by having 100% of the futures interest. To manipulate the futures price you must manipulate the price in the other market that determines the futures price.

Because the reference market is much larger than the futures market, manipulation becomes almost impossible. Suppose someone wanted to manipulate the price of an index in stocks of the electronic industry. He would have to jack up the closing price of IBM, Xerox, and other electronics stocks, on the day specified in the futures contract. This is no mean feat; if he makes a higher bid price, he must be prepared to buy. Securities markets are very liquid, and even a small increase in the bid price may bring a flood of offers. The manipulator must accept these offers through the closing, else the price will return to where it was. Few people or institutions have the assets to manipulate the price of IBM stock even for an instant.[6] And things get worse. Having jacked up the price of the securities in the index, and thus having realised more on the futures contract, the manipulator is stuck with a lot of IBM and Xerox stock that cannot be sold for what the manipulator paid. The manipulator supported the price; with the manipulator selling rather than buying, the price will fall. The manipulator's gain in the futures market must exceed his loss in the securities market, and because the securities market is so much larger, this is unlikely.[7] Manipulation of this sort is not logically impossible, but it should be sufficiently rare that this risk may be discounted in the absence of evidence that such manipulations occur. None has ever been established. (See Federal Reserve System 1985, pp. VII-5–VII-13).

I return, then, to commodities such as grains in which manipulation is more likely. Here a few simple precautions will go a long way to reducing the amount of manipulation. Because monopolisation of futures markets is a form of fraud, the best antidote is information. A futures exchange can monitor the positions the traders hold; if a trader appears to have too large a share of the open interest, the exchange can instruct him to reduce his holdings. (Position limits perform some of this task automatically, and an exchange can monitor holdings to ensure that even amounts less than the limit do not become excessive in relation to the rest of the market.) If the trader will not sell, the exchange can publicise the situation. Most traders believe that publicity would prevent the realisation of profit and will act to avoid it.[8] Publicity enables other traders to make appropriate responses; it also means that in the future other people will be less likely to trade with the potential offender. The exchange might formally bar a person from trading (sanctions can include bans for life). "Shunning" of professional traders by other traders also could be effective. The value of a trader's seat is open to forfeiture if the trader misbehaves. Traders who commit an ownership fraud may find people unwilling to sell them the cash commodity in the future for the market price, and this too is an effective sanction. (Recall that the Ownership Fraud involves buying the cash commodity for less than the perpetrator will demand to settle the contract, which implies a loss for its trading partners.)

A futures exchange has a large arsenal of other weapons against monopolisation. By establishing daily price-change limits the exchange reduces the ability of a manipulator to capitalise on a position. The exchange may elect to require liquidation of a contract at a fixed price if the probability of monopolisation becomes too high. It could change the terms of delivery, enlarging the deliverable supply to frustrate a manipulative manoeuvre. Indeed, the exchange could build any of these precautions into the terms of the contract.

Every futures exchange uses some or all of these methods – and every exchange used at least some of them before [US] federal regulation began with the Grain Futures Act of 1922 (42 Stat. 998). The exchanges today frequently impose position limits more restrictive than those required by regulations, and they frequently intervene quietly to require traders to reduce their positions. (See Federal Reserve System 1985, pp. VII-17–VII-35.) Although interventions that alter the terms of the contract or require immediate liquidation are much less frequent, these too occur.[9]

It is plainly in the interest of exchanges to define the terms of contracts and establish rules that reduce the amount of monopoly and manipulation. Each exchange must attract business, which will not be forthcoming (in the desired volume) if the contracts are needlessly risky. Traders control the terms by their ability not to trade. They can buy forward contracts on the underlying commodity or simply stay out of the market. Each exchange also faces competition from contracts designed by other exchanges. Traders can take their business elsewhere in a twinkling of an eye. This competition, together with the availability of substitutes for futures markets, ensures that the exchanges will do a great deal to police transactions. See Fischel (1986). An exchange that neglects to take precautions – and to find ways to certify that it will make these precautions effective – cannot long survive.

The question remains whether an exchange will take the right precautions. To say it will take some is not to say it will take enough. There is an optimal amount of precaution in any market, an amount that stops short of eliminating fraud. The only way to have no manipulation is to have no market. Traders are willing to suffer the risk of manipulation in exchange for the benefits of the futures market, and they want to conserve on enforcement costs. The exchange should continue taking additional precautions only until the social gains at the margin just equal the social costs. But of course an exchange looks only at the private costs and benefits – those it can appropriate for itself and its members. The question, then, is whether the private costs and benefits facing a futures exchange closely approximate the social costs and benefits.

The answer is almost certainly yes. The traders incur both the private and the social gains and suffer the losses from futures trading and manipulation. These traders constantly transact with the exchanges and among themselves. So long as no significant effects are felt only by third parties, strangers to these contracts, the exchanges have the proper incentives to design optimal contracts and take optimal precautions.[10] And here almost every affected party is in on the contract.

Only "almost" because those who plan transactions in reliance on the price information produced by a futures exchange may stand outside the circle of contracting parties. For example, people deciding whether to plant wheat may rely on the information about future prices of wheat contained in the quotations from a futures exchange. These producers (and therefore the rest of us) gain when the prices permit more astute decisions about production, yet the futures exchange has no obvious way to charge for this service. Accordingly, the exchanges will take too few precautions designed to preserve the accuracy of price.

This is not likely to be a serious shortcoming, however. Many who deal in wheat are likely to hedge by transacting in the futures markets, and if they do so the exchanges will be induced to take the appropriate precautions. Even if the occasional farmer disdains the futures markets, the farmer expects to sell to someone else who does. If the person to whom the farmer sells hedges by trading in the futures market, then again this transmits the appropriate incentives to the exchange. The social benefits of accurate prices that are not reflected in futures transactions of one sort or another are likely to be small. Certainly there is no reason to think that futures markets take less care than other markets to make prices accurate for the benefit of such parties.[11] (This is not to say that they take the optimal degree of care; it is only to say that they are likely to be driven in the appropriate direction to a greater degree than many markets that are unregulated.)

It has been argued, however, that futures markets do not respond to these (and other) costs and benefits completely because of the way they operate. The futures exchanges sell trading services. They charge by the trade, and they cannot engage in price discrimination to collect extra from those who gain the most from less manipulation and more accurate prices. A reduction in the number of episodes of monopolisation does not necessarily lead to more trading. Consequently, the argument concludes, the exchange will not take optimal precautions (Edwards and Edwards 1984, pp. 354–55). This argument is fallacious. Although a futures exchange sells transactional services, and although a reduction in manipulation does not necessarily lead to an increase in the number of transactions precisely reflecting the gains from less manipulation, it does not follow that the futures exchange cannot charge for the value of its precautions. The exchange can increase the price it collects per trade. Traders will pay higher prices per trade so long as the change is less than the value of the reduction in the costs of manipulation. The futures exchange can charge for its precautions even if, by increasing the accuracy of the price signals, the precautions lead to fewer trades.

In sum, there is no reason to think that futures exchanges will systematically take suboptimal precautions against monopoly and manipulation. Edwards and Edwards (1984) eventually come to this conclusion. The traders (and the traders' trading partners) reap the gains and suffer the losses of manipulation, so they will compensate the exchanges for taking the appropriate steps. Each exchange must do so, or it will lose business to another exchange that selects the appropriate mix of precautions. No exchange will stamp out manipulation because that is not worth the cost. The cost of catching each episode rises as they become more rare, and the value of stamping out that last offence is small. The optimal rate of crime is always greater than zero, and so too no manufacturer tries to eliminate defects from its products.

If it is necessary to choose between the level of precaution set by the exchanges and the level set by a regulatory agency, several considerations suggest that the exchange will do better than the agency. One consideration favours the regulator: the regulator can consider 100% of the social gains and losses from monopoly and manipulations; a futures exchange, while approaching that, will always fall short. In principle, then, the regulator could promulgate the perfect rule. But practice falls far short of principle. Consider these three difficulties.

First, a regulatory agency must act on the basis of estimates of costs and benefits. The exchange, too, acts on the basis of estimates. A regulator who estimates wrongly suffers no loss; loss falls rather on private parties for whom beneficial transactions have been foreclosed. The regulator may suffer a political loss, but the political "market" does not transmit price signals very well. A futures exchange that estimates wrongly suffers real losses. The exchange can reward employees who analyse correctly and penalise those who do not. The regulatory agency pays its employees at a fixed civil-service scale, quite impervious to profit and loss. You should trust those who wager with their own money to do the calculations correctly. They may be wrong, but they are less likely to err than are the regulators, who are wagering with other people's money.

Second, the futures exchanges compete with one another. We may suppose (despite the first consideration above) that both exchanges and regulators make haphazard estimates of the appropriate level of precautions to take. When an exchange estimates wrongly people do less trading there and move their business to other exchanges that are more nearly right. When a regulatory agency guesses wrong, it is hard for "customers" to take their business elsewhere. Armen Alchian and Gary Becker have shown that, even if participants in markets are stupid and respond poorly to incentives, the reactions of their customers lead the market in the direction of efficiency. (See Alchian 1950; Becker 1976 and Sowell 1980.) A form of natural selection is at work. Businesses that miscalculate are sifted out. Certainly a process of selection is at work in the futures business. Carlton (1984) illustrates the rapid birth and death of both contracts and exchanges. No similar process eliminates regulatory agencies.

Third, regulatory agencies can cause futures exchanges to diverge from efficient practices in ways the exchanges cannot achieve by themselves. Regulators can impose costly rules that make it hard for new futures exchanges to compete with existing ones. They can prevent the entry of new exchanges altogether or delay existing exchanges from offering new contracts. They can award new contracts to the least successful exchanges in order to buoy them up, protecting them from competition by others. Traders lose as the penalty for failure goes down and as contracts migrate toward less efficient forums. The exchanges cannot do these things; only the government can. Yet as I emphasised at the outset, persistent monopoly is possible only if new competition may be nipped in the bud. The ability of government to reduce competition and produce monopoly profits for the exchanges makes it an attractive nuisance. (See Stigler 1975; and Salop, Scheffman, and Schwartz, 1986.) Consider the airline business. For a long time the [US] Civil Aeronautics Board justified its existence as necessary to ensure safe, low-cost air transport. To that end it excluded new entry and tightly regulated existing firms. It awarded "plum" routes to failing airlines. Since deregulation in 1978 fares have fallen, air travel is safer, and there is more frequent service (provided by new commuter airlines) to small cities. The experiences showed that the principal function of regulation had been to prop up a cartel of the major carriers. We should incur the risk that regulation will be put to anticompetitive ends only if there is strong evidence that government has a comparative advantage over private conduct in solving important failures in markets. There is neither a logical argument nor evidence showing such a comparative advantage in futures markets.[12]

REGULATION AND THE HISTORY OF MANIPULATION

No one seeking to demonstrate the existence of monopoly and fraud in futures markets would have much difficulty doing so. The Hutchinson wheat fraud of 1867, the Harper Deal of 1887, the Cargill wheat corner of 1963, and many others are part of the lore of the markets.[13] But of course no one wanting to demonstrate the existence of monopoly in the underlying commodities or fraud in the issuance of securities of firms that deal in commodities would want for examples either. The question is not whether there is monopoly and manipulation but how much there is – and whether the amount has declined with the recent increase in the amount of regulation by enough to cover the costs of that regulation.

These questions are not easily answered. One problem is the lack of a benchmark. The number of futures contracts offered has increased through the years. This naturally leads to an increase in the opportunities for manipulation. But we cannot count the total number of attempted manipulations. Perhaps exchanges handle attempted manipulations with greater secrecy now than they used to. I have been unable to persuade the exchanges to furnish me with any data about the number or circumstances of private interventions.

Then there is a problem of determining whether any change in the amount of manipulation was produced by the exchanges or the government. Both the exchanges and the government have altered their rules over time. Even if we could determine whether the exchanges' rules or governmental regulations were responsible for a change in the frequency of manipulation – which is unlikely – we could not determine who was responsible for the rules and regulations. Do futures exchanges adopt their rules freely, or are some of the rules responses to or anticipations of regulatory pressure? Do regulators adopt their regulations on the basis of independent judgements about the benefits of regulation, or are some of these regulations merely copies of the exchanges' private rules (or, worse, efforts to reduce rather than enhance competition)? These difficulties make it all but impossible to determine whether the amount of monopoly and manipulation has decreased and, if so, why.

It may be helpful to recall that monopoly and manipulation (which I have used almost as synonyms) are not the same as exploiting an advantageous position in the market. Markets offer people the opportunity to assess anticipated supply and demand. Someone who takes a long position (buying the right to delivery or to settlement) does so because he believes or fears the price will rise. The stronger his belief, the more long contracts he will buy. Both hedgers and speculators, on both sides of the market, serve valuable functions in bringing to light new information, both about supply and about demand, that makes the price more accurate and the futures market more useful. The reward for this effort (and the associated risk) is the ability to obtain a profit from coming to the right conclusion. If a shortage of supply develops and the price rises, we cannot logically turn around and accuse the holder of a large long position of "manipulation"; that would remove the lure of profit that makes the market work.

Any effort to find and denounce "manipulation" therefore must differentiate that conduct from astute anticipation of changes in the conditions of supply and demand. The usual legal definitions of manipulation invoke notions such as "intentional wrongdoing" or creation of "artificial" conditions. The most-cited definition is that behaviour is manipulative when "conduct has been intentionally engaged in which has resulted in a price which does not reflect basic forces of supply and demand".[14] This gets us nowhere.

An effort to isolate which "forces of supply and demand" are "basic" and which are not is doomed to failure. What is a "basic" demand? Economists think of supply and demand as givens. People demand what they demand, and never mind the reasons why. If people want to purchase wheat to admire its beauty rather than to mill it into flour, they may be weird, but their demand is real. Is a demand for gold to be put into a safe "basic"? How about a demand for wheat to be held in silos in anticipation of famine? People may want warehouse receipts for wheat in order to bake bread, but they may also want them (as they want Treasury bills) for their close equivalence to money. There is no way to say what demand is real and what is artificial. The addition of "intent" does not help. No one accumulates futures contracts – for reasons good or ill – unaware of what he is doing. Everyone in the futures market intends to make as much money as he can. Scrutiny of intent therefore is not likely to assist in the search for manipulation.

Yet the legal tests nonetheless point in the right direction. For reasons I have explained, manipulation is a form of fraud. A person who accumulates a substantial long position could be counting on one of two things to produce profit: either a storage of the commodity or the ignorance of other traders about what the first is doing. Someone who buys long positions because he understands the supply of the commodity better than other traders is engaged in normal economic behaviour; his actions drive the price in the direction it should move. Someone who is betting on his ability to conceal his own position from others, and to profit solely from that concealment, is engaged in fraud; his actions likely drive the price in the wrong direction. The person who seeks profit solely from concealment makes today's price less, not more, accurate as a predictor of future prices. The decrease in accuracy is a source of economic loss. Moreover, the inaccurate price is likely to cause people to ship the commodity toward the place where the price suddenly rises, only to ship it back again later. This, too, is a source of economic loss.

Concealment by itself may be beneficial. Concealing one's position may be useful in obtaining the profit from one's ability to predict actual conditions of supply and demand, and hedgers need secrecy to avoid revealing their positions in the cash commodity. Secrecy is a valuable commercial strategy, which increases the value of searching for new information. The essential distinction is between secret strategies necessary to capture the value of new information about underlying conditions and secrecy designed to cause prices to diverge from those that reflect the underlying conditions.

So a reasonable economic definition of manipulation is conduct in which the profit flows solely from the trader's ability to conceal his position from other traders and the trades do not move price more quickly in the direction that reflects long-run conditions of supply and demand. Someone searching for manipulation might look for asymmetric information. He also might look for the telltale sign of sudden price fluctuations. When the closing price on a futures contract significantly diverges from the price of the cash commodity immediately before and after, this is strong evidence that someone has reduced the accuracy of the market price and inflicted real economic loss on participants in the market. Courts usually look for both concealment and sudden swings in price.

A search for examples of manipulation in light of this test turns up remarkably few. You can find one in 1931, another in 1957, and a few more before and since.[15] While thousands of suits are filed each year alleging violations of the [US] anti-trust statutes, and goodly numbers of plaintiffs win antitrust judgments every year, most years go by without a single judgment of liability for a futures manipulation. Even some of the adjudicated incidents of manipulation demonstrate its rarity.

Take, for example, the Ownership Fraud involving the May 1963 wheat contract on the Chicago Board of Trade. Cargill secured 2,471,000 of the 2,804,000 bushels of wheat available for delivery in Chicago. Then it took a long position of almost 2,000,000 bushels and demanded delivery. On the last day of trading Cargill owned about 62% of the long interest. The shorts discovered to their chagrin that Cargill owned the wheat available to be delivered and would not sell to them for less than US2.28\frac{1}{4}$ per bushel – the same price it wanted to settle delivery obligations under the contract. Cargill got its US2.28\frac{1}{4}$. Yet just before the last day of trading on the May contract, Cargill had sold 600,000 bushels of wheat for US$2.09, and right after the settlement cash wheat in Chicago sold for between US$2.03 and US$2.15. The court found Cargill liable for manipulation, pointing out that this situation could be called "artificial" because such a price fluctuation had never been observed before: the court made the rarity of the price changes the tip-off of the offence.[16] It also concluded that Cargill had used its special knowledge to advantage – it profited not because it knew more about the demand and supply of wheat in the cash market but because it alone knew who owned the deliverable wheat in Chicago.[17] (The court also might have observed that Cargill alone knew that it would demand delivery, which most longs do not.)

When there is no fraud, there also is no manipulation. The May 1979 Maine potato futures contract, like the May 1963 Chicago wheat contract, was marked by a sudden price rise. But this time evidence showed that the rise was caused by an unexpectedly poor crop. One trader accumulated almost all the long interest in the contract, but others knew what this trader was doing, and, according to the opinion of the court, they also anticipated that he would stand for delivery.[18] The exchange ultimately stopped trading in the contract and required an orderly liquidation – a demonstration of the ability and willingness of a futures exchange to prevent even suspicious conduct.[19] The point: there is often a large price change and a large profit without manipulation. The number of cases of real manipulation seems to be small indeed.

THE METHODS AND LIMITS OF REGULATION

Regulation of futures markets may take three principal forms. The first is to punish monopolistic or manipulative behaviour. The second is to establish rules for the administration of futures contracts. The third is to prohibit the writing of contracts unless a regulator concludes that the contract is not subject to manipulation.

The first approach allows the exchanges to offer such products and take such precautions as they choose. Then an agency or the courts may punish those who engage in monopolistic or manipulative conduct. This form of regulation by deterring wrongful conduct is the prevailing one in most markets. People make their own arrangements, and the [US] anti-trust laws (administered by the courts, the Department of Justice, and the Federal Trade Commission) punish wrongdoers. It has also been the method of regulating futures markets for most of US history.

The second approach allows exchanges freely to offer futures contracts and requires them to establish precautions, such as reporting requirements and position limits, that reduce the probability of manipulation. An agency reviews the requirements the exchanges establish and, if it concludes that these are inadequate, requires changes. This approach has been used (though to varying degrees) since the Grain Futures Act of 1922.[20] It relies heavily on the self-interest of the futures exchanges to identify and implement effective precautions. If the markets function well without rigorous constraints, then the agency does not intervene. This approach is also the one in use today for the securities markets. Anyone may offer a new security without obtaining the SEC's approval. The regulatory agency requires some disclosure by the issuer of the characteristics of the security, and it also requires the stock exchanges to adopt and enforce some simple rules of honest conduct. The system of regulation assumes that informed traders should be left to their own judgement, to decide well or poorly as they choose, both in buying stocks and in designing rules. The SEC largely defers to the exchanges' rules – especially those that have been successful through the years. (Loss (1983) describes the rules.)

The third approach, entry regulation, requires an exchange to demonstrate both that futures contracts are not subject to manipulation and that they are otherwise in the public interest. Entry regulation was applied to futures markets for the first time in the United States by the Commodity Futures Trading Commission Act of 1974 (7 USC §§2–22). Under this approach the exchange must justify each contract. The burden is on the exchange to support the contract rather than on a court or an agency to expose flaws in the contract. It is obviously the most effective in preventing monopolisation – you cannot monopolise a market that does not exist – but also has the highest costs.

The costs of deterrence, the first method, are least. The cost of regulation of the exchanges, the second method, are higher because the regulator may require the exchanges to take excessive precautions. The regulator may set the position limit too low, for example, thus reducing the value of the contract in hedging and also reducing the liquidity of the market. This form of regulation is unlikely to be beneficial unless regulators systematically do better than the exchanges at finding the optimal precautions. For the reasons discussed in the second section, it is unlikely that the government has a comparative advantage here.

Entry controls almost certainly are not cost-justified. For the reasons discussed in the first section, the costs of monopoly and manipulation in a futures market cannot exceed the other benefits of the futures contract in question. The participants in futures markets bear the costs of monopoly and manipulation; if these costs exceed the benefits from the futures contract, they will stop trading.

Even if the costs of monopoly are high, and futures exchanges do not take appropriate precautions, the costs of having no contract at all exceed the costs of having a manipulable contract. Recall from the first section that the cost of monopoly is the forgone value that the buyers of a product could receive. Monopoly means a lesser output at a higher price; people do less trading in monopoly-prone contracts because the implicit price (including the cost of being victimised) is higher. The value to the buyers is lost altogether, however, when the contract is abandoned or not introduced. Consider: which is worse, a world in which videocassette recorders (VCRs) are subject to monopolisation, or one in which VCRs do not exist at all because of fear that, if they existed, someone might monopolise them? In the former world the buyers priced out of the market lose the value they place on having a VCR; in the latter world everyone who would buy a VCR loses. We do not prohibit the manufacture of products that might become monopolised; we do not even prohibit the manufacture of products such as power lawn mowers that might prove dangerous. Instead we punish misconduct when it occurs. What is true of the market in VCRs and lawn mowers is true of the market in futures contracts.

To make matters worse, the absence of a futures contract makes it easier to monopolise the market in the underlying commodity. Recall again how monopoly works. The sellers reduce their output, so customers pay more for what remains. So long as they can keep output down, and keep new entrants out, they stand to profit. New entry, and cheating by members of the cartel, ultimately undercuts the monopoly, and prices fall again. The existence of a futures market makes cartels harder to form and maintain. Producers participate in futures markets to hedge their risks. They will sell futures contracts, locking in a price for some time ahead. The existence of these price commitments on the part of producers is a form of prearranged "cheating" on the cartel. It is very hard to organise to drive down output when sellers have commitments to deliver. The futures market commits sellers to deliver and enables buyers to obtain the competitive price for some time to come. By the time all these prearranged commitments have expired, new entry from other producers is likely, so the cartel never really gets off the ground.

There is another perspective from which to assess the costs and benefits of both entry control and regulation of the exchanges. That is to ask how often the costs are incurred. The costs of prohibiting the introduction of a contract apply to every transaction that would have been consummated. The costs of regulation of the operation of the futures exchanges apply to every transaction, too. If a stringent position limit makes a contract less useful for hedging or makes the market less liquid, that cost is reflected in every transaction because the price of the futures contract becomes less accurate. The gains of regulation, on the other hand, are realised only to the extent that the regulation prevents particular episodes of monopoly or manipulation. As the first section showed, the beneficiaries from a reduction in monopoly are the marginal traders – those who are priced out of the market if monopoly raises the effective cost of participating. Regulation therefore trades a certain loss on all trades under existing contracts for a possible gain on some trades on some contracts. This is worthwhile only if monopoly and manipulation, when it occurs, is quite costly indeed. Yet we have no reason to conclude that monopoly in futures markets is particularly costly. There are so many substitutes for futures markets (the purchase of the underlying commodity on forward contracts and the purchase of options on securities are just two among many) that the allocative welfare loss from monopoly and manipulation of futures markets is bound to be small. The swap of certain losses on many trades for possible gains on a few trades is unlikely to be worthwhile.[21]

Whether this trade-off is worth making depends in part on how often manipulation occurs and whether deterrence of violations is likely to be effective. If monopoly and manipulation are rare, it is cheaper to penalise the wrongdoer in that rare case than to penalise all participants in every case by establishing costly precautions.[22] Why spend resources scrutinising each contract and each trade in advance when very few are undesirable? Regulation by deterrence comes at lower cost than regulation by prior scrutiny.

Of course whether manipulation is "rare" depends in turn on the effectiveness of the sanctions meted out on offenders. The discussion in the third section suggests that offences are indeed rare. It is easy to see why. Sanctions deter conduct by increasing the loss associated with that undesired conduct. The effective sanction depends on the probability that the offence will be detected and on the size of the penalty imposed when detection occurs. The penalty is the sum of any fine actually collected by the government and the effective private sanction.

It is very hard to commit an undetected monopoly or manipulation in futures markets. All futures exchanges require traders to report substantial holdings. These reports are confidential, but they provide evidence with which holdings can be traced. (In litigation arising out of asserted monopolisation and manipulation, courts customarily can compute to the hundredth of a percent every trader's share of the open interest throughout the life of the contract.) When one trader owns a large portion of the open interest at the end of a contract, others find out – and fast. The trader with the big stake can't capitalise on his position except by demanding a higher price to settle the contract. The traders quickly learn with whom they must deal to close their positions. The holder of a large position may say that the position was not acquired with the purpose or effect of manipulation; large positions and manipulation certainly are not synonymous. The point, rather, is that no one is likely to obtain gains from monopoly or manipulation without being identified, so penalties can be imposed. This is not like robbery, where the principal task is finding the culprit.

Offences may be harder to detect when they involve more than one market. For example, an effort to influence the price of a futures contract that reflects the value of a stock index will take the form of an effort to change the price of the stocks in the index. The offender will attempt to disguise these efforts as normal trading, perhaps the trading of a few large blocs at the last minute. A regulatory agency may have a comparative advantage at monitoring substantial trades in order to determine whether they have cross-market effects. Even here, however, it is possible to use computers to detect abnormal price movements, and the markets can investigate appropriately. If the bloc traders also hold substantial positions in the futures markets, there are grounds for investigation. It should be rare indeed for someone to engage in manipulation without attracting notice. (A manipulation too small to notice also does not impose substantial costs. Because we do not want to deter all offences, however trivial, the existence of such small manipulations is not a compelling reason for regulation.)

Once a person has been certified as a monopolist or manipulator, it is relatively easy to impose a substantial penalty. Traders are not judgement-proof. Many have substantial wealth. People holding seats on the exchange may forfeit the value of the seats; traders who do not hold seats may forfeit the value of their reputations (shunning by other traders can impose enormous losses). So deterrence is likely to be an effective, low-cost method of dealing with wrongdoing in futures markets.

Notwithstanding the utility of deterrence, many people believe that futures markets impose substantial costs on unsuspecting third parties and that futures exchanges systematically underestimate the size of these costs. For example, they may contend that futures markets are especially risky and that because the professional traders can protect themselves the bulk of the cost of manipulation will fall on "outsiders" – less sophisticated, less wealthy investors. Certainly most traders in futures markets are sophisticated, and many are wealthy. It is hard to evaluate arguments that these traders arrange for losses to fall elsewhere. But let us suppose for the moment that they do. What is the appropriate response? One possibility is that the government should prevent sophisticated, wealthy people from trading a futures contract with their eyes open just on the off chance that widows and orphans might gain from the ban. The other possibility is that the government should prevent widows and orphans from doing themselves injury. This is the less costly alternative. In securities markets certain privately sold, risky securities may be sold only to "accredited investors" – people of substantial income, wealth, or knowledge.[23] Similar restrictions could protect less sophisticated investors in futures markets without injuring people who use futures markets for hedging. I do not recommend such restrictions; there is no solid evidence that exchanges fail to take adequate precautions against the exploitation of the unsophisticated. (Some exchanges and dealers establish minimum account sizes designed to keep unsophisticated traders from the markets.) But if there were such evidence, the response should be limited.

What remains is the persistent belief that the existence of futures contracts increases the likelihood of manipulation in other markets. This is the stated ground of the legislative and regulatory antipathy to contracts on "narrow-based" indices. (See Federal Reserve System (1985), making an argument of a sort characterised in Fischel (1986) and Fischel and Grossman (1984).) I have indicated in earlier discussion that such cross-market manipulation is unlikely to be profitable. Now assume that this is wrong and that futures contracts create opportunities for profitable manipulation. It still does not follow that prohibition of the futures contract is the optimal solution. Regulation by deterrence remains the least-cost option, just as in markets for other products. It is possible to detect manipulatory conduct and impose appropriate sanctions.

No one believes that the right response to corporate officers' trading on material inside information is to prohibit corporate officers from owning or trading stock. The officers' ownership of stock is highly valuable to other participants in the corporate venture because it aligns the officers' interests with those of other investors. It is foolish to sacrifice these gains in order to reduce the amount of inappropriate trading. We therefore permit trading but penalise wrongful trades. The same principles apply to futures markets. Futures contracts are highly valuable in hedging or distributing risk, in creating liquidity, and so on. It is foolish to sacrifice these benefits when a system of penalties imposed on wrongful conduct can preserve them.

CONCLUSION

The exchanges that establish futures contracts are led by the demands of traders to establish the optimal precautions against monopoly and manipulation. These offences are not nearly as costly in futures markets as they are in the underlying product markets, so "optimal" precautions do not seek to eradicate the misconduct. Instead a futures exchange seeks some mix of care and deterrence that reduces the frequency of these offences so long as it is beneficial to do so. The exiting evidence suggests that there are some episodes of monopoly and manipulation, but not very many. This is what we should expect to see.

Because the exchanges seek to select the right mix of precaution and penalty, there is no strong justification for either imposing regulations on trading or restricting the creation of new contracts. Because the loss from manipulation cannot exceed the value of the contract itself, it is never right to limit new entry.

1 If entry is immediate, even the producer of 100% of the supply must charge the competitive price (Baumol, Panzar, and Willig 1982).
2 The cash commodity is not a perfect substitute, for otherwise we would not see futures markets. Futures are useful in hedging risks, and they also create additional liquidity akin to money. (See Telser and Higinbotham 1977, and Telser 1981.) I return to these features below.
3 Some traders have control of the deliverable supply because they monopolise it. Governments monopolise the production of their currencies, and some nations participate in cartels monopolising physical commodities (such as tin and cocoa) traded in futures markets. These markets are therefore more open to manipulation by the monopolist, and we should expect to see less trading in them, relative to the trade in the underlying commodity. I put these unusual markets to one side in this paper. For treatment of their special properties, see Anderson and Sundaresan (1983).
4 A farmer who looked at the prices during a manipulation would of course be sent wildly astray. But no one makes planting decisions on the basis of the price in the last few days of the current contract, the only price seriously affected by a manipulation. People make decisions on the basis of prices predicted for some time hence by open contracts. (See Marcus and Modest (1984), explaining how futures markets influence production decisions. See also Forsyth, Palfrey, and Plott (1984), confirming predictions that futures markets lead to more accurate prices in spot markets.)
5 Every once in a while someone tries. Between 1978 and 1980 the price of silver futures rose rapidly and then fell. The result: a welter of charges and countercharges of manipulation. The principal targets of lawsuits are members of the Hunt family who, it is alleged, took substantial long positions and demanded delivery, driving up the price. (For example, *Friedman v. Bache Halsey Stuart Shields, Inc.*, 738 F. 2d (D.C. Cir. 1984), one of the many suits.) I do not try to sort out these charges. But if it is true that the Hunts were trying to monopolise silver and silver futures, they were apparently the largest victims of their scam.

Their desire to take delivery month after month became known, and people scoured the world to find deliverable supplies. The price rose, but the Hunts discovered that they could not sell the silver for what they paid; the price was driven by their demand, not by the demands of others. Apparently they lost more than a billion dollars. Not the stuff that encourages future manipulations! When a tactic generates only losses for the perpetrators, it suffocates itself. Neither the exchange nor the government need be concerned.

6 It would be easier to affect the price of IBM or any other stock by spreading lies about the firm's prospects. Fraud could affect the value of any commodity, from stock to oil, underlying a futures contract. The defrauder would need to make the false news believable, however, and then escape detection. No set of precautions can prevent futures frauds that involve lies about the commodity – but then this sort of fraud also does not appear to be common. I have not seen any claim that a futures fraud has been committed in this way.

7 In late April 1984 the press expressed concern that the Chicago Board Options Exchange's (CBOE) options contract on the Standard & Poors 100 stock index had been manipulated through coordinated last-minute block transactions that jacked up the price of the securities. A quick check revealed that the implied losses on reselling the securities would be approximately five times the apparent gains on the futures contract. (See *16 Securities Regulation and Law Report* 723, 750 (April 27, May 4, 1984).) The CBOE established that no manipulation had occurred. (*Report of the Special Index Study Committee*, CBOE (August 22, 1984).) So far as I can tell, this is the only time a manipulation in a financial options contract has been alleged, and no one has ever seriously alleged, let alone documented, a manipulation of a financial futures contract (Federal Reserve System 1985, pp. VII-5–VII-13). In principle a defrauder could sell the underlying securities short to "recover the corpse". This, too, seems to be significant only in principle. Short sales require the posting of 100% margin and thus are very expensive; they also are so closely monitored that a strategy involving large-scale buying and simultaneous short selling would be detected.

8 Secrecy is the order of the day in futures markets, and for good reasons. Hedgers do not want others to learn their positions because that would convey information valuable to their competitors in the underlying commodity. Traders seek secrecy even knowing that the secrecy possessed by others facilitate frauds. That traders subject themselves to occasional frauds as the price of general secrecy speaks volumes about both the value of secrecy and the infrequency of fraud.

9 For example, *Sam Wong & Son, Inc. v. New York Mercantile Exchange*, 735 F.2d 653 (2d Cir. 1984), recounting in some detail the New York Mercantile Exchange's handling of Maine potato futures in the spring of 1979.

10 Coase (1960) points out that the private solution will diverge from the optimal one when transactions costs are high and that they are likely to be so in multiparty cases. But the exchanges themselves are in a position to coordinate the actions of the many traders. The exchanges set the terms of the contracts and then see whether people want to trade. The action of the exchanges replaces the need for an express series of negotiations among the many potential traders.

11 The difficulty of appropriating the value of accurate prices extends beyond futures markets, and is only casually related to the problem of monopoly and manipulation. For example, the interest rates at auctions of Treasury bills may be used by many people to plan transactions, and it is not possible to charge those who refer to the information. Because of the webs of contracting among producers, hedgers, and other traders, futures markets probably come closer than other markets to taking optimal precautions.

12 The Federal Reserve System's 1985 study does not perform any comparative analysis. It also finds few problems and recommends little additional regulation. It never considers whether the current level of regulation is excessive.

13 These and others discussed below are discussed in Hieronymus (1977), Johnson (1981), and

McDermott (1979). See also *Cargill, Inc. v. Hardin*, 452 F.2d 1154 (8th Cir. 1971), cert. denied, 406 US 932 (1972) (finding liability for the conduct in 1963).

14 *Cargill, Inc. v. Hardin* (n. 13 above), 452 F.2d at 1163. See also, eg, *In re Cox*, CCH Commodity Futures Law Reports ¶21,767 at 27,076 (CFTC 1983).

15 *Peto v. Howard*, 101 F.2d 353 (7th Cir. 1939) (July 1931 corn in Chicago); *Volkart Bros., Inc.*, 20 A.D. 306 (1961) (October 1957 cotton in New Orleans), reverse *sub nom. Volkart Bros., Inc. v. Freeman*, 311 F.2d 52 (5th Cir. 1962). See also the cases collected in the sources cited in n. 13 above.

16 Rarity of the fluctuation is a common element of the claim of manipulation. For example, *In re Cox* (n. 14 above) (comparative tables showing rapid price fluctuations, a dramatic change in the relation among Chicago, Kansas City, and Minneapolis prices that could not be accounted for by transportation costs, and the lack of similar fluctuations and price relations in other years.)

17 *Cargill, Inc. v. Hardin* (n. 13 above), 452 F. 2d at 1159: "It is important to note here that Cargill was in possession of a valuable piece of information that other traders and grain dealers in Chicago did not have access to, namely that it owned the bulk of deliverable wheat in Chicago." (See also 452 F. 2d at 1170.)

18 *Sam Wong & Son, Inc. v. New York Mercantile Exchange* (n. 9 above).

19 The court held that the exchange was not liable for failure to stop trading sooner or to redefine delivery obligations to increase deliverable supply. It concluded, however, that Anthony Spinale, the disappointed holder of the long contracts, might be entitled to recover if he could prove that the exchange stopped trading out of malice or in order to give other traders an advantage. As the court observed, there is nothing wrong with guessing the deliverable supply and insisting on maximum profit. "The Exchange is not a social club. We see no reason to believe that if a surplus of potatoes had developed, the shorts would have exhibited a higher quality of mercy to Spinale than he did to them" (*Sam Wong & Son, Inc. v. New York Mercantile Exchange* (n. 9 above), 735 F.2d at 678).

20 See Grain Futures Act, 42 Stat. 998 (1922); Commodity Exchange Act, 49 Stat. 1491 (1936), both amended and substantially superseded by the Commodity Futures Trading Commission Act of 1974, 7 USC §§2–22.

21 This discussion is closely related to the trade-off between allocative losses and productive efficiency gains in mergers. A merger may enable the surviving firm to reduce its real costs of production even while it gets the monopoly power to increase price. The lower costs of production increase efficiency, while the monopolistic reduction in output decreases efficiency. The lower cost of production apply to all units of output, however, while the efficiency loss from monopoly applies only to the forgone units (the reduction in output). Unless the loss from each unit is forgone, the savings from lowering the actual costs of production on each unit will exceed the losses from the cutback in output. (For discussions, see, eg, Williamson (1968) and Fisher, Lande, and Vandaele (1983).) This is one reason why US anti-trust law does not bar all mergers but concentrates instead on the very large mergers that could cause substantial reductions in output. The parallel to futures contracts is plain. A futures contract with fewer precautions is "more efficient" because it can be "produced" at lower cost, and these savings apply to all transactions. Unless the contract creates a substantial prospect of monopoly and consequent cutbacks in output, the savings from the lower-cost production of transactions exceed the potential losses from the monopoly, which apply only to the marginal units.

22 Several scholars discuss the trade-offs between regulation and the use of penalties and show why, as the number of violations falls, the use of penalties becomes relatively more attractive (Schavell 1984a, 1984b; Wittman 1977, 1984).

23 See Section 4(6) of the Securities Exchange Act of 1934 and SEC Rules 215, 501–5 (defining "accredited investor" and specifying certain securities for which status as an accredited investor is a necessary condition of purchase).

BIBLIOGRAPHY

Alchian, A. A., 1950, "Uncertainty, evolution, and economic theory", *Journal of Political Economy* 58, June, pp. 211–21.

Anderson, R. and M. Sundaresan, 1983, "Futures markets and monopoly", Working Paper CSFM-63 (New York: Columbia University), Business School, July.

Baumol, W. J., J. C. Panzar and R. D. Willig, 1982, *Contestable Markets and Industry Structure*, (New York: Harcourt Brace Jovanovich).

Becker, G. S., 1976, "Irrational behavior and economic theory", in G. S. Becker, *The Economic Approach to Human Behavior* (University of Chicago Press).

Carlton, D. W., 1984, "Futures markets: Their purpose, their history, their growth, their successes and failures", *Journal of Futures Markets* 4, Fall, pp. 237–71; reprinted as Chapter 14 of the present volume.

Coase, R. H., 1960, "The problem of social cost", *Journal of Law and Economics* 3, October, pp. 1–44.

Edwards, L. N. and F. R. Edwards, 1984, "A legal and economic analysis of manipulation in futures markets", *Journal of Futures Markets* 4, Fall, pp. 333–66.

Federal Reserve System, 1985, *A Study of the Effects on the Economy of Trading in Futures Markets* (Washington, DC: US Government Printing Office).

Fischel, D. R., 1986, "Regulatory conflict and entry regulation of new futures contracts", *Journal of Business* 59(2).

Fischel, D. R. and S. J. Grossman, 1984, "Customer protection in futures and security markets", *Journal of Futures Markets* 4, Fall, pp. 273–95.

Fisher, A. A., R. H. Lande and W. Vandaele, 1983, "Could a merger lead to both a monopoly and a lower price?" *California Law Review* 71, December, pp. 1697–706.

Forsyth, R., T. R. Palfrey and C. R. Plott, 1984, "Futures markets and informational efficiency: A laboratory examination", *Journal of Finance* 39, September, pp. 955–81.

Hieronymus, T. A., 1977, *The Economics of Futures Trading for Commercial and Personal Profit*, second ed. (New York: Commodity Research Bureau).

Johnson, P. McB., 1981, "Commodity market manipulation", *Washington and Lee Law Review*, 38 Summer, pp. 725–79.

Loss, L., 1983, *Fundamentals of Securities Regulation* (Boston: Little Brown).

McDermott, E. T., 1979, "Defining manipulation in commodity future trading: The futures 'squeeze'", *Northwestern University Law Review* 74, Spring, pp. 202–25.

Marcus, A. J. and D. M. Modest, 1984, "Futures markets and production decisions", *Journal of Political Economy* 92, June, pp. 409–26.

Posner, R. A., 1976, *Antitrust Law: An Economic Perspective* (University of Chicago Press).

Salop, S., D. Scheffman and W. A. Schwartz, 1986, "A bidding analysis of special interest regulation: Raising rivals' costs in a rent-seeking society", In R. Rogowski and B. Yandle (eds.), *Regulation and Competitive Strategy* (Washington DC: Federal Trade Commission).

Schavell, S., 1984a, "Liability for harm versus regulation of safety", *Journal of Legal Studies* 13, June, pp. 357–74.

Schavell, S., 1984b, "A model of the optimal use of liability and safety regulation", *Rand Journal of Economics* 15, Summer, pp. 271–80.

Sowell, T., 1980, *Knowledge and Decisions* (New York: Basic).

Stigler, G. J., 1975, *The Citizen and the State* (University of Chicago Press).

Telser, L. G., 1981, "Margins and futures contracts", *Journal of Futures Markets* 4, Summer, pp. 225–53.

Telser, L. G. and H. N. Higinbotham, 1977, "Organised futures markets: Costs and benefits", *Journal of Political Economy* 85, October, pp. 969–1000; reprinted as Chapter 13 of the present volume.

Williamson, O. E., 1968, "Economics as an antitrust defense: The welfare tradeoffs", *American Economic Review* 58, March, pp. 18–36.

Wittman, D., 1977, "Prior regulation versus post liability: The choice between input and output monitoring", *Journal of Legal Studies* 6, January, pp. 193–244.

Wittman, D., 1984, "Liability for harm or restitution for benefit", *Journal of Legal Studies* 13, January, pp. 57–80.

4

EFFECTS OF SPECULATION

Introduction

Lester G. Telser

The first chapter in this section by R. H. Hooker is one of the first statistical studies of the effects of speculation. It was presented to the Royal Statistical Society and published in its journal in 1901. This article, as the author states, draws upon his 1896 study done jointly with A. W. Flux. Historically it has significance, a modern reader must be greatly impressed by the sophisticated analysis of so difficult a subject undertaken more than a century ago. It was not only in Germany that one could find enemies of futures markets. I wish to add to Hooker's description of the successful attempt in Germany to stop trading in futures, an earlier effort in the United States that came to naught. Congressman Reagan, Democrat from Texas, during the debate in the House on the Sherman Antitrust bill, spring 1890, introduced an amendment to prohibit futures trading in the United States. This amendment was defeated.

Hooker divides his analysis into two parts, the first shows that the suspension had no effect on the level of prices of wheat or rye in Berlin, and the second shows that wheat prices were more variable in Berlin as a result of the suspension. Both parts require careful examination of the price data to make comparisons as accurately as possible. The study of the effects on the price levels is less complicated than on price variability largely because of many alternative ways to measure variability. Hooker describes three of these, of which only the first two are pertinent. The first is the standard deviation of daily prices and the second is the Mean Daily Price Movement, which he calls MDM. However, he does not define the latter clearly, at least by modern standards of notation. It is the mean of the absolute value of daily price differences for which the formula is

$$MDM = \sum_{t=1}^{n+1} |p_{t+1} - p_t|/n$$

Hence, the MDM must be non-negative. Hooker explains why he

removes data for 1898 from his analysis of price stability. It is owing to the Leiter corner in the May wheat future on the Chicago Board of Trade that would distort the comparison by raising the estimate of the MDM for Chicago prices. He summarises his main conclusions on price variability in Table 16. It shows that the MDM in Berlin was higher during the suspension than before, omitting the 1898 data from the benchmark figures for the Chicago Board of Trade. (Note that there seems to be typos in the original table corrected herein; all the MDMs should be positive so there should be no minus signs. In the appendix Hooker presents simple correlations by year of daily prices for Berlin and Chicago, Berlin and Liverpool, and Liverpool and Chicago; laborious hand calculations in those days. He also comments on some surprising values for these correlations and believes there would be fewer anomalous results had he used day-to-day price differences for the correlations instead of the price levels. In the light of current technology, this is a simple task whereas previously it would have been a major undertaking. The integrity of the investigator and the quality of his work are exemplary.

The next selection, Chapter 17 from Keynes' *General Theory*, is one of the most famous essays on speculation in economics. Many of its ideas are now so familiar that few are aware of their source. Investment depends on animal spirits. The fetish of liquidity, comforting but fallacious, encourages people to embark on risky ventures. Each person believes he can rescue his capital if things go awry but collectively this is impossible. Here one finds the Keynesian beauty contest. Speculators try to predict the behaviour of other speculators not fundamental factors. As a writer of economics, Keynes has no peer.

The following four chapters do not attain the heights of Keynes' prose on speculation. Their purpose is to construct formal models of speculation to see if it can be destabilising. William J. Baumol's 1957 article is the first to present a mathematical model of speculation. It takes up this challenge by using a differential equation to represent the behaviour of speculators. Baumol claims that even profitable speculation can increase the variability of the price. As you will note in the following chapter, I found this approach to be wanting. This article is more than a criticism of Baumol. It has a model of speculation that calculates its effect on prices over an interval in which the beginning and the ending net positions of the speculators are the same so that the speculators can realise their profits. Over such an interval, the model shows that profitable speculation is stabilising in the sense that it reduces the variance of the price.

The third chapter in this quartet by Michael J. Farrell considers whether profitable speculation can be destabilising. If the non-speculative demand depends only on the current price, that he labels the independence assumption, and if stability is measured by the variance of prices, then Farrell proves that profitable speculation is stabilising, that is, reduces the variance, if and only if the non-speculative demand is linear. In his model the effect of speculation is measured over a period such that speculators begin and end with the same quantity of the commodities in which they are speculating. Hence, a necessary but not a sufficient condition for profitable speculation to be destabilising is non-linearity of the non-speculative demand function. In addition to the complications Farrell raises, there is another he does not consider, the measure of stability appropriate for the form of a non-linear non-speculative demand function. To measure stability by the variance of prices is reasonable for linear demand but not necessarily for non-linear demand. Presumably there is a suitable measure of stability induced by the particular non-linearity of the non-speculative demand function. Farrell goes on to show that in the presence of transaction costs, a linear demand is no longer necessary for the proposition that profitable speculation reduces the variance. Farrell's analysis definitively settles a number of theoretical issues. No future work on this problem can afford to ignore Farrell's results.

The fourth chapter by Oliver D. Hart and David M. Kreps introduces a new feature into the model for studying whether profitable speculation can be destabilising. It explicitly distinguishes between what speculators know when they open their speculative positions and what has happened when they close their positions. In their model speculators store the commodity and consumers buy more, the lower is the current price. Hence, their consumers obey the law of demand, but the consumers do not allow their current consumption to depend on their views about prices later. Random shocks enter via demand. *Ex ante* speculators respond rationally to a noisy signal about next period's demand. Most of the time they are wrong *ex post* and sometimes they are right *ex post*. When they are wrong, they incur losses, and when they are right, they get profits. Whether speculators are right or wrong, their activities raise the price initially and lower it subsequently. However, when they are right, one may say they have stabilised the price, and when they are wrong, one may say they have destabilised it. If the speculators' forecasts do come to pass, then, although they have raised prices while acquiring stocks, they diminish the rise that would have occurred in their absence. If the speculators' forecasts do not come to pass, then they unduly raise the price while acquiring inventory, and unduly lower it when they subsequently dispose of their inventories. Their profits on average can be positive although on average they have destabilised the price.

In the Hart–Kreps model and in reality stocks can move in only one direction, from the present to the future. If a glut in the future is anticipated, then there is no way to move the excess supplies from the future back to the present. If there were short sales, speculators could depress the current price and stimulate a higher level of current consumption but the laws of nature are in force placing limits on how much additional consumption could occur. However, the absence of short sales does not affect the validity of the Hart–Kreps model. To produce counter-examples to the proposition that positive speculative profits imply price stabilisation, it suffices to confine attention to the effects of storage.

Hart and Kreps conclude, correctly in my view, that no economist they cite in their opening paragraph from Adam Smith to Paul Samuelson would dispute their analysis or deny their conclusions. Moreover, their statements about the relevance of price stabilisation for consumer welfare are very important. The reader ought to ponder these with great care. Nevertheless, in their model it remains true that *ex post* profitable speculation does stabilise the price while the realisation of positive profits on average does not.

16

The Suspension of the Berlin Produce Exchange and its Effect upon Corn Prices*

R. H. Hooker

The matter which I propose to bring before you today cannot claim to be entirely novel. Much of the work has been done in the course of the preparation of the report of a Committee appointed by the Economic Science and Statistics Section of the British Association, in 1896, to investigate the subject of "Future Dealings in Raw Produce". This report, ultimately drawn up by Professor A. W. Flux (Secretary to the Committee) and myself, was submitted to the British Association in September, 1900, at Bradford and embodies, amongst other results, the main conclusions of the investigation which I now submit to your notice. This particular investigation was by far the most laborious piece of work in connection with the report in question; but our deductions are there stated in the briefest possible manner, and I venture to think that an examination of the material on which these deductions rest, and more especially of the methods in which we dealt with the numerous little statistical problems encountered at every turn, is worthy of being set forth in some detail. Very considerable difficulties were met with in answering the two apparently simple questions involved in the title, and it is a discussion of methods, quite as much as of results, that I hope to elicit today.

I must admit that in preparing this paper I have to some extent had in view an object which I trust will not be considered presumptuous. It is nowadays far too common a practice to draw conclusions from sets of figures which have hardly been submitted to even the most perfunctory of examinations, and theoretical edifices are constantly being built upon the flimsiest of statistical foundations. I propose accordingly to try to take my readers with me along the paths – some of them "blind" – which we have ourselves followed, and to point out the various tests which had to

---

*Read before the Royal Statistical Society, December 17, 1901. C. S. Loch, Esq., B.A., Vice-President, in the Chair.

be applied at every step before we could feel assured of the solidity of the ground we were treading upon. With this object in my mind I nearly yielded to the temptation to entitle the paper, "An exercise in handling statistical material", and I lay the work before the Society in the hope that friendly masters will correct the exercise.

The question of the influence of future dealings upon the price of commodities is one that was brought into considerable prominence during the period of great depression of prices about 1894–1895, and numerous writers in England, on the Continent, and in America, have been emphatic in asserting that the system of contracting for future delivery[1] has "ruined agriculture". More especially, as I understand it, are the operations of speculators – who desire neither to receive nor to deliver the goods sold, but merely to make a profit out of the difference in price at different dates – supposed to have forced down the price of corn and commodities in general. Into the theory of the subject I do not propose to enter: there is ample literature, and it is not necessary to do more here than refer to the publications of C. W. Smith, Professor Rühland, W. Mancke, C. Wood Davis, among many who hold that futures are pernicious, and of H. C. Emery, F. Goldenbaum, as well as to the above-mentioned British Association Report, on the other side.

We have, however, of late years had the advantage of an actual experiment in this connection. So impressed was the Agrarian party in the German Parliament with the arguments of those who urged the abolition of futures, that, during the discussion on the "Exchanges Bill" in 1896, they induced the Government to accept an amendment prohibiting this class of transaction in grain and mill products upon the Exchange. This law came into force on the January 1, 1897.

There has been considerable misapprehension in Britain and elsewhere as to the effect of this law, or rather of this particular section, for the law also deals with other matters and applies to stock exchanges. A brief statement of the circumstances is necessary in order to understand the position of affairs.[2] The law does not prohibit contracts for future delivery; the exact words are: "Futures transactions according to Exchange procedure in corn and mill products are prohibited".[3] And sec. 48 of the Act defines transactions "according to Exchange procedure" to be those which are, *inter alia*, made in accordance with business conditions laid down by the Committee of the Exchange. Now the Berlin Exchange has laid down no conditions in this connection, and the transactions are therefore not "according to Exchange procedure". Contracts for future delivery are recognised by the Commercial Code; hence future dealings are not in themselves illegal, and some modifications in the form of contract in use at Berlin were sufficient to allow of the operations being recognised as ordinary commercial (*handelsrechtlich*), and not Exchange (*börsenmässig*), transactions.[4]

So far as regards the contract, therefore, there is nothing in the procedure of the Berlin Produce Exchange to cause its suspension. The conflict turned on quite a different matter. The Act provides for Exchange regulations (*Börsen-Ordnung*) to be drawn up for the conduct of each Exchange, which must have the approval of the Provincial Government; and by sec. 4 of the Act the latter can require the inclusion, among the Committee of the Exchange, of representatives of agriculture. This power was duly exercised, and the Berlin Produce Exchange was required to accept certain agriculturists on their Committee. The Exchange, however, objected to these "outsiders", and by declining to accept the persons nominated, acted in violation of the law, with the result that the Produce Exchange was absolutely suspended on January 1, 1897. The members at first met for the conduct of business in a building known as the Fairy Palace, from which they were in a few months ejected, the place being held to be an Exchange within the meaning of the Act, as the result of police proceedings. The merchants afterwards transacted their business "in their own offices", each taking a room in a large building. A settlement between the Government and the dealers was finally reached in March, 1900, the obnoxious clauses as to the nomination of agriculturists being modified. Under the revised regulations the Exchange is now entitled to select five from among ten persons nominated by the *Landes Oekonomie Kollegium*. Certain other minor alterations were also introduced, particularly in the form of contract, but these are immaterial here.

Under these circumstances it will be asked, if contracts for future delivery are not illegal, and have indeed been made throughout the period, how can Berlin be regarded as an illustration of a market with no business in futures? For the following two reasons. First, because there being no Exchange, there could by the terms of the Act (sec. 29) be no official quotations. Hence from 1897 till 1900 there are no official prices, either of spot or futures, at Berlin. Any influence which futures may be supposed to exert upon spot prices is thus eliminated. Secondly, there being no Exchange, and there being consequently considerable uncertainty as to the verdict in the event of any disputed transaction being brought before a court of law, the genuine grain dealers took care to do all their business with those whom they could trust. By this means brokers who merely speculated in differences found their occupation gone, and turned to other branches of business – possibly went on to the Stock Exchange. To what extent the gambling element may return now that the Produce Exchange has been reopened is a difficult question, and one that time alone can answer. One result of the suspension has certainly been the removal of quite nine-tenths of the purely speculative transactions which took place on the Produce Exchange.[5] This is almost as important as the non-quotation of futures; since it is sometimes the purely gambling transactions which are alone adduced as the cause of the depression of prices. Berlin therefore offers us, during these three years, the example of a market with no gambling – as it is usually understood – and with no quotations of prices of grain for future delivery.

We can now state the problem to be investigated. What has been the effect on the prices of grain? The problem resolves itself into two main questions: Has the price of grain at Berlin been (1) raised or lowered, (2) more steady, during the three years 1897–1899 than previously?

Before actually attacking these questions, it may be remarked that opponents of the existing system urged that the effect of the abolition of futures would be to raise the price of grain.[6] Of late, however, this position, at least in Germany, appears to have been abandoned, and partisans of the Act are now satisfied with the contention that the price of grain in Germany has been steadier since 1897 than before that date. Economic theory, on the other hand, indicates that an active futures market should tend to impart greater steadiness to the prices of the commodities dealt in.

It will be well also to consider first what effect the closure of the Produce Exchange might be expected to produce upon prices. For it must not be overlooked, in examining whether a particular cause will produce a certain event, that if *a priori* considerations would lead us to expect that event, its ascertained existence is no evidence for or against the influence of the particular cause. In the present instance, if the operations on the Berlin Produce Exchange are not of sufficient importance to materially influence the world's price of grain, the abolition of futures or gambling there would have very little effect on the average price, unless Berlin were absolutely cut off from the rest of the world. And thus, were we to find that the level of price has not been raised there, we should by no means have disproved the contention of the opponents of gambling that the system lowered prices. The Berlin market is certainly an important one, more especially as regards rye, for which grain it probably is (or was), as claimed by the merchants there, the most important in Europe or out of it. But as regards wheat, the business at Chicago, New York, and Liverpool is on a far larger scale, and the Paris market is also quite probably more important. In addition to this, Germany is a grain-importing country, but at the same time is not as a rule a sufficiently big customer to interfere seriously with the supply of the outside world, and the price should accordingly approximate to that of other countries (due regard being had to customs duties, etc.). I do not therefore think it possible that the abolition of futures could have had any effect in raising German wheat prices. And even if it could, the rest of the world would have followed its lead, unless the increase were very small, so that the effect could not be gauged. This consideration renders vain any expectation of a rise in the price of corn, supposing that it could be brought about by such a cause. Consequently, we must expect no increase in German grain prices (as compared with other markets), and, finding none, we should not disprove the contention of those who allege that future dealings depress prices; although we should illustrate practically the fallacy of the prognostication that the abolition of futures would raise prices.

As regards steadiness of the price, this seems to be on a different footing; I see no *prima facie* reason why any increased steadiness, in the absence of other causes affecting our comparison, should not be attributed to the one cause known to have arisen. Greater stability (or instability) would point to good (or evil) effects of the abolition of gambling, while no change in this respect would imply that the existence or non-existence of the system was indifferent.

COMPARISON OF THE AVERAGE PRICE AT BERLIN AND OTHER MARKETS

To ascertain whether the price of grain was raised or lowered after the suspension of the Exchange is not a very difficult matter, and a comparison of the average prices of wheat and rye at Berlin with those at other markets will yield an answer to this question.

The data at our disposal for this comparison, so far as regards wheat, consisted mainly of the daily records of spot prices for 1892–1900 at Berlin, given in the *Vierteljahrshefte*, and the daily records of cash prices at Chicago, given in the Annual Reports of the Chicago Board of Trade for 1892–1900.[7] The Chicago quotations presented little difficulty, but the Berlin figures required to be more carefully scrutinised; for we had of course to satisfy ourselves in the first place as to what these prices were during the period when there were no official quotations.

To remedy the absence of quotations (for this was naturally regarded by both parties as a great obstacle to business), the Prussian Chambers of Agriculture published, through the medium of their "Central Quotation Office", the prices of actual spot transactions. These are supplied to the Central Office "from private information", ie, through the various local chambers of agriculture; and in default of other statistics the Government has reproduced the figures for Berlin in its annual summaries,[8] but with the warning that they are not official, and should only be used with caution. It may be noted that the accuracy of these figures has been challenged, chiefly as being too high, by the Berlin Society of Corn and Produce Dealers. On the other hand, the Agrarians accuse the former Berlin Exchange quotations of being below the prices at which transactions were actually concluded. However this may be, these accusations will not, as will be seen, interfere with the validity of our calculations. The point we took particular care to verify is, that the Central Office quotations apply strictly to the same grade as the prices quoted up to 1896 on the Exchange, viz. (in the case of wheat) grain of "good sound quality,

Table 1 Average price of wheat per cental (100 lbs)

Year	Berlin		Chicago*		Difference		Average difference	
	s	d	s	d	s	d	s	d
1892	8	−1/8	5	6 1/8	2	6		
1893	6	10 5/8	4	8 1/2	2	2 1/8		
1894	6	2	3	10 5/8	2	3 3/8	2	4 1/2
1895	6	5 5/8	4	2 3/8	2	3 1/4		
1896	7	1	4	5 3/8	2	7 5/8		
1897	7	11 1/4	5	6 7/8	2	4 3/8		
1898	8	9 1/4	6	3 1/4	2	6	2	4
1899	7	1 7/8	5	1 1/4	2	1 5/8		
1900	6	11 1/2	4	11 1/8	2	−1/8	—	

*The Chicago figures have undergone revision since our Report was presented to the British Association, and some errors have been corrected.

weighing not less than 755 grammes per litre". I will add that to me the comparison of these prices appears quite legitimate, and, having thus exercised the caution recommended by the German Government, I am content to use the figures.

As already noticed, the Berlin Produce Exchange was reopened towards the end of March, 1900. The daily prices for the latter year quoted in the *Vierteljahrsheft*, I, 1901, are still from the same source as in 1897–1899, as official Exchange figures are only available from April 1. It is stated that these official figures in many cases differ materially from the Central Quotation Office prices. Under the circumstances I do not see how we can form an opinion as to the effect of the reopening of the Produce Exchange on these outside prices, and I have not attempted to draw any conclusions from the statistics relating to 1900.

We may now, I think, safely proceed to draw up our first table, viz., a comparison of the average price of wheat at Berlin and Chicago during the years 1892–1900. I exclude Liverpool and New York from this comparison, mainly because we had daily records for only five years, ie, for only one year prior to the suspension of the Berlin Produce Exchange. It may be mentioned that these averages in Table 1 are the arithmetic means of the daily prices at these towns; and that, in converting the prices into English measures, the mark has been taken as $1s.$, the doppelzentner as 220 lbs., the cent as $\frac{1}{2}d.$, and the bushel of wheat as 60 lbs.

From Table 1 it would seem that, as compared with Chicago, the Berlin price has scarcely been affected.

But a more reliable test may be deduced from a consideration of the excess of price at Berlin over the average of the prices at the chief corn markets of the world. To calculate the average from the daily prices would have entailed enormous labour, apart from the difficulty of obtaining

Table 2 Average annual spot prices of wheat and rye at the chief markets of Europe and America, 1892–1900 *(in marks per tonne)*

Market	1892	1893	1894	1895	1896	1897	1898	1899	1900
Wheat									
Vienna	166	141	125	125	133	184	210	170	147
Budapest	156	131	115	115	124	176	199	155	127
St. Petersburg	162	138	122	109	112	133	155	140	124
Antwerp (a)	143	117	96	103	118	150	152	135	135
Antwerp (b)	151	122	100	106	124	162	164	133	134
Amsterdam	124	116	91	98	112	137	146	126	127
London (a)	153	131	115	113	129	145	165	126	133
London (b)	142	124	108	108	124	142	161	123	130
London (c)	157	130	110	115	132	157	167	137	137
England, Gazette	142	123	107	108	123	142	159	121	127
Liverpool (a)	163	134	113	119	134	162	172	143	143
Liverpool (b)	157	135	116	123	129	159	174	139	143
New York (a)	140	117	105	110	114	140	151	125	127
New York (b)	134	112	92	101	110	135	142	118	119
Average	149.3	126.5	108.2	110.9	122.0	151.7	165.5	135.1	132.2
Berlin	176.4	151.5	136.1	142.5	156.2	174.8	193.2	157.5	153.2
Difference	27.1	25.0	27.9	31.6	34.2	23.1	27.7	22.4	21.0
Rye									
Vienna	148	115	98	109	119	137	153	131	126
Budapest	136	104	88	97	107	126	142	119	112
St. Petersburg	150	108	85	80	73	85	105	109	95
Amsterdam (a)	133	106	83	83	85	97	119	121	117
Amsterdam (b)	143	114	93	87	84	93	117	124	118
Average	142.0	109.4	89.4	91.2	93.6	107.6	127.2	120.8	113.6
Berlin	176.3	133.7	117.8	119.8	118.1	130.1	146.3	146.0	142.6
Difference	34.3	24.3	28.4	28.6	24.5	22.5	19.1	25.2	29.0

access to the daily quotations at each market for a series of years. Data for such a comparison are, however, given in the *Vierteljahrshefte*, 1901, part 1. In this publication there is a table showing the average annual prices of wheat and rye at all the principal markets. From this I have extracted all the spot prices and reproduced them for the years 1892–1900 in Table 2. There are 14 such sets of prices for wheat, and five for rye, more than one set of figures, representing different quantities of wheat, being available for some of the more important markets. The omission of Paris may be remarked. For this there are two reasons first, *Vierteljahrshefte* only quotes the price of futures for this city; and secondly, even if spot prices had been available, the quotations would have vitiated the results, because the French Government in 1898 suspended the customs duty on wheat for a

Table 3 Average excess of corn prices at Berlin over the "World's Price"

	1892–96		1897–99	
	Per cental		Per cental	
	s	d	s	d
Wheat	1	3⁷/₈	1	1¹/₄
Rye	1	3¹/₄	1	−¹/₈

few months, thereby moderating the price of wheat at that market.

In Table 2, the grades of wheat are described in the *Vierteljahrshefte* as follows: Vienna, *Theiss (früh. Banat)*: Budapest, *Average quality*; St. Petersburg, *Saxouka*; Antwerp, (a) *Danube, average*, (b) *La Plata, average*; Amsterdam, *Odessa*; London, (a) *English white*, (b) *English red*, (c) *Californian*; Liverpool, (a) *Californian*, (b) *Northern Duluth*; New York, (a) *Northern Spring No. 1*, (b) *Red Winter No. 2*. The grades of rye are: Vienna, *Pester Boden*; Budapest, *Average quality*; St. Petersburg (not stated); Amsterdam, (a) *Azov*, (b) *Petersburg*.

Averaging the figures in Table 2, and subtracting this average from the Berlin price, we find that the excess of the latter was less during 1897–1899 that during 1892–1896, as appears from the summary in Table 3.

The comparison is, as regards rye, I think, not quite correct. For the Berlin wheat prices for 1897–1899 I have used the averages of the "Central Office" quotations; but the *Vierteljahrshefte* does not reproduce the corresponding daily prices of rye. For the latter grain I have therefore taken the annual averages quoted in that publication; but these are derived from prices published by the Municipal Statistical Office of Berlin, and relate to rye of "good average quality" – not apparently of a particular grade, as is the case in earlier years. In 1898 and 1899, however, the monthly averages of the "Central Office" prices of rye, which relate to corn of the same grade, are printed. I find that in these two years they average 2 marks per tonne (1⅛d. per cental) higher than the Municipality's prices, so that the reduction in the price of rye, as shown in Table 3, instead of being 3⅛d., is only 2d. per cental. Similarly if the Municipality's average for wheat be taken, the apparent reduction in the price of wheat is increased to 4⅝d. per cental. But the "Central Office" prices appear the more comparable; thus, abiding by them, and assuming the mean difference between Berlin and the rest of the world to have been normal during the five years 1892–1896, we find that wheat and rye prices were reduced during the suspension of the Exchange by 2⅝d. and 2d. respectively.

If the New York prices be omitted from Table 2, the Berlin price of wheat shows a reduction, compared with Europe, of about 3d. instead of 2⅝d.

Table 4 Grain freight rates

Year	From Chicago	From Odessa	
	Per cental	Per ton	
	US$	s	d
1892	0.3287	10	–
1893	0.3410	10	6
1894	0.3250	11	6
1895	0.3200	9	–
1896	0.3350	10	–
1897	0.3360	10	–
1898	0.3435	9	6
1899	0.2972	11	6
1900	0.2948	8	6

A factor which must not be overlooked in comparing prices of articles of international commerce at different periods is the question of freights, and a parenthesis will not be out of place here to ascertain whether the above-noted changes in price can be accounted for by reductions in ocean freights. I have accordingly reproduced in Table 4 the average freights from Chicago and Odessa. (The Chicago figures are taken from the *United States Statistical Abstract*, and represent the annual average freight rates on grain, through from Chicago to Liverpool by all rail to seaboard and thence by steamer; the Odessa figures, from the *Shipping World Year-Book* for 1901, are the grain freights in January of each year to the United Kingdom and Continent (steamer).

There has been practically no change. The freights from Chicago show an average reduction of $1/400d.$ per cental in 1897–1899 as compared with 1892–1896; while from Odessa there is an average increase of $1/12d.$ Freights may therefore certainly be ignored among possible causes of the reduction in price during the past decade.

The duty on wheat and rye in Germany is, and has been throughout the period, 1s. 7d. per 100 lbs. This statement, by the way, simple as it looks, is not conclusive as to the effect of the duties, and affords an instance of the ramifications which have to be followed up in statistical investigations.

The duties quoted above are those applicable to imports from Russia, which is among the "most-favoured nations" in this respect, and the country whence the great mass of foreign grain is usually derived, so that it practically determines the price. That is to say, the import duty practically determines the price, provided the country does not produce the whole of its requirements; in the latter event the price tends to sink to that prevailing in other countries. I had occasion to look into this question some time ago, and came to the conclusion, if we may judge from the

Table 5* Production and total supply of wheat and rye grain in Germany *(in doppelzentners, 000's omitted)*

Cereal years (September to August)	Production	Net imports	Total supply	Ratio of production to total supply
Wheat				Per cent
1893–94	34,050	8,896	42,946	79.3
1894–95	33,364	12,731	46,095	72.4
1895–96	31,718	14,550	46,268	68.6
1896–97	34,199	12,439	46,638	73.3
1897–98	32,632	10,554	43,186	75.6
1898–99	36,076	13,585	49,661	72.6
1899–1900	38,474	10,199	48,673	79.0
Rye				
1893–94	89.419	4,438	93,857	95.3
1894–95	83,430	6,666	90,096	92.6
1895–96	77,249	9,416	86,665	89.1
1896–97	85,330	8,785	94,115	90.7
1897–98	81,705	8,699	90,404	90.4
1898–99	90,322	4,315	94,637	95.4
1899–1900	87,758	6,588	94,346	93.0

*Trade figures from German monthly trade returns (special trade), and harvests from *Vierteljahrshefte*, extra number, 1900.

experience of France during the past thirty years, that the whole effect of the import duty is felt so long as the home production does not exceed something like 95% of the consumption. This proportion may very possibly vary in different countries. It therefore became necessary to ascertain whether we had here any such disturbing element, ie, whether the home production had materially increased in 1897–1899. An inspection of Table 5 shows that there is no such material increase in the ratio of production to total supply. This consideration may therefore be eliminated. (It should be mentioned that the German harvest returns prior to 1893 are officially stated to be too low; probably, I think, by 10–20%, but there are no data by which to judge. This is my reason for not going back beyond 1893–1894 in Table 5.)

As already pointed out, the absence of a rise in price proves little; but we have here some evidence of a decline. I do not see that this can be due to any other cause than to the absence of a well organised market. Sellers, being uncertain as to the real position of the market and as to the prices that should be ruling, were fain to accept something less than they might have obtained if current quotations were freely accessible to all; while buyers intending to resell, uncertain of what profits they were likely to make, had to lower their offers in order to minimise their risk.

The harvests are given, from the German official returns, in the last table. I dare not venture an opinion as to whether there is any approach to truth in the statement which will doubtless be made, that – at $2\frac{5}{8}d.$ per cental of wheat and $2d.$ per cental of rye – this legislation has deprived the German farmers of £5–6,000,000 in the three years 1897–1899 on account of these two cereals alone. Should there be any truth in *either* of the counter accusations of the two parties in Berlin, viz., that the former Exchange quotations were too low, or that the "Central Office" prices are too high, this sum would be proportionately increased.

During our visit to Berlin [in 1900], we found that the theory there current of the evil effects of future dealings differed materially from that usually advanced in this country. I confess I find it difficult to follow the arguments of those who allege the system to be the cause of the fall in prices. So far as I can understand their contention, it is supposed to depress prices because, the commodities being sold some months ahead, the sellers desire to purchase the goods, in order to deliver, at a lower price than that at which they sold: these "bears" are supposed to be in the majority, and are hence able to beat the "bulls" down to their own price. Further, the selling of "wind wheat", ie, stuff which is non-existent, to an enormous extent, creates a huge fictitious supply, which would have the effect of depressing prices. The theory of the German Agrarians, on the other hand, appears to rest on a less shadowy foundation. It is, if I understand it aright, that the system of selling for future delivery allows the seller time to get his supplies from abroad, and thus, by favouring importations and increasing the supply available, to depress the price. This hypothesis offers a much more tangible point of attack, and being apparently the recognised theory in Germany, we find that a great controversy rages around the question of the magnitude of the imports. It does not appear to me that this hypothesis is any more correct than the others: the country must import what it requires; and, in fact, Table 5 shows that there was no diminution of imports after 1896.

In this connection our attention was called at Berlin to certain features of the grain trade of the cereal year 1897–1898. Stress was laid, by advocates of the futures system, upon the great losses incurred by Germany in that year, through traders having exported wheat immediately after harvest; with the result that Germany, being dependent upon external sources for part of her supply, was obliged to import a correspondingly increased quantity during the spring of 1898, of course at the enhanced price then prevailing owing to the Leiter corner.

The theoretical reason given by advocates of futures business for this movement of grain was, that the system of contracting for future delivery relieves the market from the abundant supplies thrown upon it at harvest, and thus prevents a slump during the autumn of the year, to the great advantage of the producer. Owing to the non-existence of an organised futures market, this safeguard was lacking, and, the home price being disproportionately low, it became more profitable to sell abroad than at home. We had already in England been confronted with the opposite hypothesis that gambling depressed the price at harvest time (owing to the supposed domination of the "bears"), and had accordingly given some little attention to the question. Our examination was into the annual course of prices in this country before and after 1870 (this being the approximate date when the system began to assume prominence), and we came to the conclusion that this latter hypothesis certainly had no foundation. At the same time the figures did not yield very much evidence that the annual curve has been smoother since 1870. The possibility of obtaining confirmation of our conclusion from the modern German experience, and perhaps of demonstrating the theory statistically, appeared desirable, and I have therefore reproduced the figures of the monthly imports and exports of wheat during the nine years 1892–1900 (Table 6) and compared the prices with Chicago month by month (Table 7).

To a certain extent the figures bear out the contention of unusual shipments abroad in the autumn of 1897, and the price at Berlin was abnormally low in September of that year. A similar phenomenon occurred in 1899; but in the latter instance the German harvest had proved abundant, and heavier exports are accordingly comprehensible. In 1897, on the other hand, the German harvest was deficient, having been about 150,000 tons below the yield of 1896. But the imports during the spring of 1898 were not larger than were warranted by the previous short harvest – not so large, indeed, as in 1896. It may, however, probably be with justice maintained that, in view of the price, Germany did not import a single ton more than was sufficient to meet bare necessities (compare the total supply in 1897–1898 in Table 5), so that, but for the shipments in the previous autumn, the quantity would have been still smaller. On the whole, I am disposed to agree that there is considerable evidence in favour of the contention that Germany suffered from injudicious exports favoured by relatively low prices at home.

The question of German importations appears to be complicated by the system of "import-permits", which allows an importer, under certain conditions, to introduce into Germany wheat free of duty against an equivalent export of wheat and flour. The regulations concerning these import-permits have been subjected to modifications during the past decade.[9] Our excursion into this question is thus leading us into difficulties, and as the point does not appear likely to prove of material service in connection with the main subject – the level of price – I have not pursued the inquiry further.

As regards the general theoretical point that the futures system should

Table 6 German imports and exports of wheat *(in Doppelzentners. 000's omitted)*

Month	1892	1893	1894	1895	1896	1897	1898	1899	1900
Imports									
January	1582	567	729	845	1717	1227	1331	1384	1537
February	952	679	372	578	817	490	540	764	556
March	1209	638	796	479	740	655	528	766	582
April	1060	294	513	1324	1943	1129	1331	1608	1455
May	1176	404	683	1332	1060	884	1090	1038	958
June	1060	667	1000	1368	1243	878	1175	1174	1169
July	3267	730	1709	1867	1984	1618	1969	1898	1685
August	984	1156	1196	1212	1281	886	1468	1107	954
September	483	523	984	908	1042	654	1159	814	826
October	421	467	1175	1533	1813	1725	1957	1666	1531
November	432	474	1332	888	1927	874	1267	753	807
December	335	435	1049	1049	962	777	959	738	878
Year	12,962	7035	11,538	13,382	16,527	11,795	14,775	13,709	12,939
Exports									
January	—	1	1	95	52	37	159	113	242
February	—	—	—	21	26	31	193	70	222
March	—	—	—	43	33	48	182	165	309
April	1	—	—	94	67	10	68	168	293
May	—	—	58	38	24	126	60	137	257
June	—	—	85	15	19	77	92	107	116
July	—	—	113	14	10	64	17	118	126
August	—	—	55	16	34	91	7	49	56
September	—	—	42	43	82	155	76	212	279
October	—	—	103	138	128	211	133	333	439
November	—	—	146	113	171	371	161	275	320
December	—	—	187	69	106	392	198	227	291
Year	2	3	792	699	752	1714	1348	1974	2951
Wheat flour									
Imports	{	1892–96, wheat and rye flour				{ 364	282	420	337
Exports	{	not separated				{ 449	384	332	320

tend to mitigate a slump in the autumn, Table 7 shows that, in 1897 and 1899, there was a decided autumn slump at Berlin (relatively to Chicago). The four earlier years, however, show a similar tendency, though not so marked. 1898, like 1892, is of no service in this connection, because prices were falling from a very high level, and as they fell from the maximum much more slowly at Berlin than at Chicago, the excess at the European market is very large for a time. Examining the actual monthly prices (not reproduced here) in the four cereal years 1893–1894 to 1896–1897, I find that prices were low in the autumn of 1894–1895 and 1895–1896, but high

Table 7 Excess of Berlin wheat price over Chicago *(in pence per cental)*

Month	1892	1893	1894	1895	1896	1897	1898	1899	1900
	d	d	d	d	d	d	d	d	d
January	43$\frac{1}{4}$	20$\frac{7}{8}$	27$\frac{7}{8}$	30$\frac{1}{8}$	32$\frac{1}{4}$	30	24$\frac{3}{4}$	32$\frac{1}{2}$	27$\frac{3}{8}$
February	37$\frac{1}{8}$	21$\frac{1}{4}$	29	31$\frac{1}{4}$	31	31$\frac{1}{4}$	22$\frac{1}{8}$	28$\frac{5}{8}$	27$\frac{5}{8}$
March	36$\frac{3}{8}$	19$\frac{3}{4}$	29$\frac{1}{4}$	32$\frac{1}{4}$	32$\frac{1}{4}$	28$\frac{1}{4}$	21	27	28$\frac{7}{8}$
April	35$\frac{5}{8}$	21$\frac{1}{2}$	27	30$\frac{1}{8}$	32$\frac{1}{8}$	27$\frac{7}{8}$	24$\frac{1}{4}$	24$\frac{3}{4}$	27$\frac{5}{8}$
May	34$\frac{1}{8}$	27$\frac{5}{8}$	28$\frac{1}{8}$	26$\frac{1}{8}$	34$\frac{3}{8}$	27$\frac{1}{4}$	2$\frac{1}{2}$	25$\frac{3}{8}$	28$\frac{1}{8}$
June	31$\frac{5}{8}$	31$\frac{5}{8}$	26$\frac{3}{4}$	22$\frac{3}{8}$	32$\frac{1}{2}$	28$\frac{3}{4}$	35$\frac{1}{4}$	25$\frac{1}{8}$	22$\frac{1}{4}$
July	29$\frac{3}{8}$	33$\frac{3}{8}$	30	21$\frac{1}{2}$	29$\frac{3}{4}$	27$\frac{3}{8}$	40$\frac{7}{8}$	25$\frac{7}{8}$	21$\frac{5}{8}$
August	22$\frac{5}{8}$	33$\frac{7}{8}$	28$\frac{7}{8}$	22$\frac{3}{8}$	31$\frac{5}{8}$	26$\frac{7}{8}$	40$\frac{1}{8}$	24$\frac{3}{4}$	22$\frac{7}{8}$
September	22	25$\frac{3}{4}$	27$\frac{1}{2}$	24$\frac{1}{2}$	33$\frac{1}{4}$	23$\frac{7}{8}$	39$\frac{5}{8}$	22$\frac{3}{8}$	21$\frac{5}{8}$
October	22$\frac{7}{8}$	25$\frac{1}{8}$	25	26$\frac{1}{8}$	31$\frac{1}{4}$	28$\frac{5}{8}$	39$\frac{1}{2}$	23$\frac{1}{8}$	21$\frac{1}{4}$
November	23$\frac{1}{2}$	26$\frac{5}{8}$	25	30$\frac{1}{4}$	30	29$\frac{1}{2}$	37$\frac{5}{8}$	23$\frac{3}{4}$	22$\frac{3}{8}$
December	21$\frac{3}{8}$	26$\frac{5}{8}$	27$\frac{1}{4}$	31$\frac{5}{8}$	28$\frac{7}{8}$	29$\frac{1}{2}$	35$\frac{3}{4}$	24$\frac{3}{8}$	23

in the autumn of 1893–1894 and 1896–1897: these four years therefore show no evidence of an autumn depression. But the experience of 1897 and 1899 seems to point to the steadying effect of an active futures market; although it must be confessed that six years are hardly sufficient to form a very trustworthy guide on such a question.

STEADINESS OF THE PRICE

Our attention was directed to the desirability of investigating the effect of futures upon the steadiness of the price of commodities by the writings of W. Mancke, in 1898. The principal book issued by him in this connection is entitled, *What and Who determined Wheat Prices in the Harvest Year 1897–1898?* with two others in continuation.[10] These mainly consist of diagrams illustrating the daily price of wheat at various markets, and the author draws the conclusion therefrom that the price of wheat at Berlin has been far more steady than in other countries since 1897. To us, who had up till then found nothing to invalidate the view suggested by theory that future dealings should tend rather to steady prices than otherwise, these conclusions were unexpected; and we felt that our inquiry could hardly be satisfactorily terminated so long as they remained unrefuted.

The diagrams published by Herr Mancke wear a plausible appearance of conclusiveness. They are unfortunately not for the most part accompanied by tables showing the actual prices on which they are based; and, although the author has taken very considerable pains with his statistics, yet a little further research is really required. The diagrams in fact afford an excellent illustration of the taunt often hurled at us by scoffers, that "statistics will prove anything". If in this statement the word *prove* is replaced by *show*, there is little to cavil at. Statistics, when their validity is thoroughly examined, and when all outside influences affecting them have been taken into consideration, will only prove one thing. The difficulty is that these other influences can rarely be gauged. I propose therefore to submit these data to a somewhat more exhaustive examination, and in the end I hope that you will concur in my "proof" of conclusions directly opposed to those of Herr Mancke.

In attacking these diagrams we (or rather I) made more than one failure before locating the flaws which we felt sure must exist somewhere; but after some trouble we found two points which seemed to require careful investigation.

In the first place the Berlin prices represent spot quotations of a particular grade, whereas the prices for foreign markets appeared to be the futures price of the current or nearest month. Moreover, these foreign prices rest "upon private telegrams" from the city concerned, and represent the quotation of the day converted into German equivalents, inclusive of charges for freight, duty, etc. Now it appeared to us that in an investigation which had as its object the comparison of Berlin with cities where a different system was supposed to influence the price, the one thing to be carefully avoided was a comparison of two different kinds of prices, at least in an inquiry confined to the period when Berlin had only the one quotation. The spot price at Berlin ought as far as possible to be compared with spot prices elsewhere. Futures and spot *might* be so closely bound up as to allow of the use of either, but then again they might not; and, until this point was determined, results derived from such a comparison could not be held to be conclusive. Again, we considered that it was eminently undesirable to include freight charges, which are variable; and to add these variable amounts to, say, the Chicago price would tend to increase fictitiously the unsteadiness of Chicago.[11] Besides which the amount of wheat sent to Germany from America in certain years, and still more from Liverpool, is very small, and such freights must be purely nominal. For these reasons we decided to collect our own data, except as regards Berlin, for which, as already noted, the figures given by Herr Mancke are accepted as being the best.

In the second place, although the author shows much greater stability at Berlin than at the other markets, he contents himself with the mere assertion that prior to 1897 Berlin followed these other markets closely; but he adduces no figures for earlier years in support of this statement.

Our first task was of course to get the records of prices in a handy form, and some time was spent in copying out lists of daily prices at New York, Chicago, Liverpool, and Berlin. The main source of the statistics was Broomhall's *Corn Trade News*, from which have been taken all the quotations for Liverpool and New York, as well as futures (closing price) at Chicago. We found this paper better suited to our purpose than

Dornbusch's *Evening Corn Trade List*, although the latter has been useful as a check in cases where the prices given in the *Corn Trade News* appeared sufficiently doubtful as to suggest misprints. The cash prices at Chicago (not given in the *Corn Trade News*) have been taken from the Annual Reports of the Chicago Board of Trade, and apply to the standard grade there, No. 2 Spring; the Berlin prices, as before, from the *Vierteljahrshefte*.

The task of copying out the figures was far from being as simple as it appears, and two or three nice points necessitating settlement arose in the course of the work.

For New York we were confronted with the primal difficulty of selecting winter or spring wheat for the spot price: we copied out both, and, as it happened, owing to the absence of quotations for lengthened periods, only the one or the other was at times available. This would naturally affect an average from one year to another, the more so as the relative movements of the two sometimes appear puzzling, but I doubt whether it is of material import to the average steadiness. An endeavour was made to ascertain the cause of the relative movements of spring and winter wheat at New York, and a certain amount of statistical material was collected with this object. We did not, however, succeed in eliciting any facts of moment; and as we subsequently found that the New York figures were not indispensable, we abandoned the pursuit. Still, the point at first occasioned us some trouble, and may be mentioned as one of the blind alleys we had to explore.

Liverpool was more difficult. The *Corn Trade News* quotes the daily spot prices of a whole series of kinds of wheat. We naturally desired to utilise the prices of No. 2 Red Winter – the standard grade at that market; but we found that it was frequently non-existent, or at least not quoted. After making inquiries, we decided to record the prices of No. 2 Red Winter, No. 2 Hard Kansas, and No. 2 Western Winter, whenever such prices were available, and to use what we could. The result was that for 1897–1900 we obtained an almost complete set of quotations for No. 2 Hard Kansas, but that for 1896 we had no one of these kinds quoted throughout the year. As will be seen later, therefore, neither the average price nor the "standard deviation" can be given for 1896 at Liverpool; but we could, as will be explained later, give a "mean daily movement". The file of newspapers at our disposal did not permit us to carry the Liverpool figures back beyond 1896.

As regards futures at Liverpool, we had the choice of working on the "closing price" – the only one quoted by Broomhall for Chicago and New York – or on the "basis for calls"; i.e., the "official" price, which rules the settlement of contracts previously entered into. The latter appeared the more desirable, but for the reasons already alluded to it would (or might) not be justifiable to compare basis for calls at Liverpool with closing price at Chicago. We therefore extracted both sets of figures. This consideration need not, as matters turned out, really have affected us, for we instituted no comparisons between futures in different localities. Our object being to investigate the influence of the modern trading system on the prices paid for real wheat, the only use we made of these futures was to see how they moved relatively to spot prices, in order to test the validity of Herr Mancke's employment of futures for one place and spot for another. Another example of apparently unnecessary labour performed in order to know what to select!

The usual measure of variability indicated by $\bar{x}$ statistical theory is the standard deviation; ie, $\sigma = \sqrt{\Sigma(x - \bar{x})^2/n}$, where $\bar{x}$ is the average price, x the price on any given day, and n the number of days. It is obvious, however, that the price at one market follows that at another very closely, and it was not to be expected that the standard deviation of the one set of prices would differ materially from that of the other in any given year. What we required was a measure of the steadiness (freedom from rapid changes) from day to day, and we finally decided to adopt the "mean daily movement". This M.D.M. (as I shall denote it) $= \Sigma(x_1 \sim x_2)/n$, where x_1 is the price on any one day, and x_2 that on the following day. This is the measure adopted by Herr Mancke, except that he uses the total daily movement, ie, $\Sigma(x_1 \sim x_2)$, instead of the mean, and it certainly seems the most satisfactory.

These considerations will be best illustrated by Figure 1, on which I have plotted two imaginary sets of prices, one of which may be said to follow the other fairly closely. The σ is clearly no measure of their relative variability, because it is the same in both cases (3.4); but they have a very different M.D.M., that of the dotted line being double the other (1.7 against 0.8). It is with "curves" resembling these that we have to deal, and although the σ has been worked out, but little use has been made of it, and our deductions are based chiefly upon comparisons of the M.D.M.

Another advantage of the M.D.M. is that it may be determined practically in cases where one particular grade of wheat is not quoted throughout the year. If two grades are quoted for different periods which overlap, we are able, without serious error, to shift from one grade to another and give the M.D.M. for the whole period. For instance, in August, 1898, we have at Liverpool the price of No. 2 Kansas quoted from the 1st to 11th, and 16th to 18th; on the other days there was apparently none. No. 2 Red Winter is quoted on and after 9th August. The difference between the two on the 9th and 10th is $2d$.; we can thus add the daily changes that occurred in Red Winter after the 10th to those which occurred in Hard Kansas before that date; for experience shows that, although the two grades might in a considerable period diverge from each other by a few pence, the total of the daily changes will not differ practically by more than those few pence at the end of that period, and

Figure 1

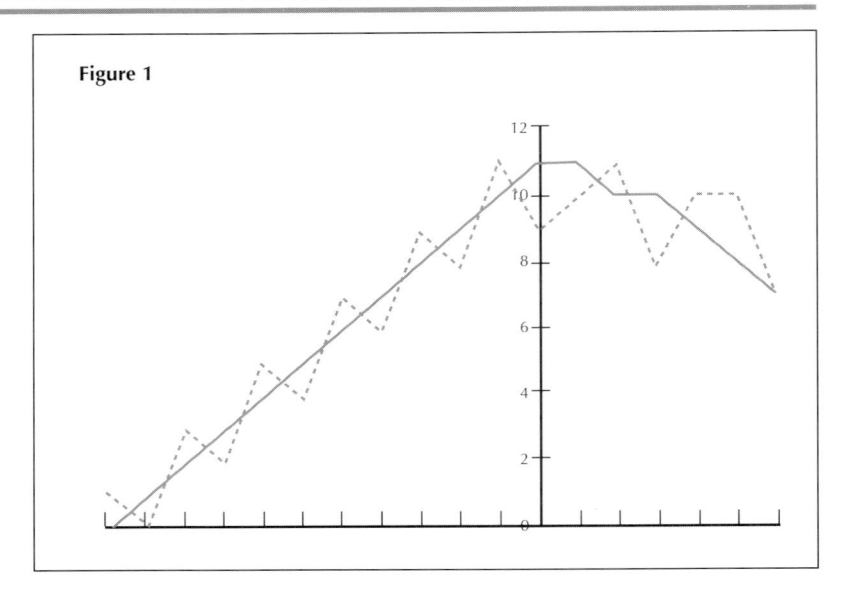

this slight difference is lost in averaging over a year. Similarly, we can go back to Hard Kansas (new) when it comes into market early in September. (This might not be necessary were it not that, later on, Red Winter drops out of the quotations.) Clearly, however, there is greater danger of error in getting an average or a σ for August, because we have shifted onto a different level: the mean of one portion is $2d.$ above the other at the date of overlapping; while any subsequent divergence, however slight, would affect all the subsequent figures, and the mean of that subsequent period would be increased by that amount. Expressed mathematically, interpolation is in our case much simpler and safer, if the M.D.M. is used, than if the σ is to be obtained.

It will be observed, however, that I have nevertheless given a mean and σ for 1898 at Liverpool in the following tables. For these two calculations, August has been omitted, and the year consists of eleven months only. This must be borne in mind in comparing the Liverpool price for 1898 with other averages. The average price of Hard Kansas during August would, supposing corresponding supplies to be in existence, probably be about $6s.\ 6d.$; so that, could it have been included, the average for the year would have been reduced by $1d.$, while the σ would not differ from that given by as much as this.

The M.D.M. for 1896 at Liverpool has been obtained in a similar manner, but in this case no figure representing the mean or σ can be given.

I may mention that the average price, σ, and M.D.M., were all as a matter of fact worked out for each month, and the σ for the year (σ_y) then calculated from the monthly $\sigma(\sigma_m)$ by the use of the formula

$$\sigma_y = \sqrt{\frac{\Sigma n_m (x_y - x_m)^2 + \Sigma n_m \sigma_m^2}{n_y}}$$

(in which x_y = average of year, x_m = average of each month, n_y and n_m number of days in the year and each month respectively).

Before making the actual comparisons, we had yet another point to consider. The prices of futures quoted are given as a single price, while the spot price at Liverpool and the cash at Chicago is given as a range, and the Berlin price as an average, of the day (but each of course applying always to the particular grade). In considering variations which require to be measured in quantities as small as $\frac{1}{8}d.$, we thought it necessary to find an answer to the question: Is the average of several daily prices liable to exhibit sensibly less variation than a single closing price or than the extreme quotation of a range? (Theoretically, Mr. Yule points out that we might expect it to be somewhat less.) The latter quotation we found we were compelled to use for Liverpool; for the reason that, while No. 2 Hard Kansas was usually quoted, there were occasions when the price was simply denominated Hard Kansas, and a greatly increased range – the lower limit dropping – rendered it probable that the price covered No. 3 (an inferior quality) as well as No. 2, so that we dared not average.

To test this point we worked out the average price and M.D.M. for each month (and year) at Chicago in 1892 and 1893, and Berlin (futures) in 1896, first working out the average daily price (midway between the two extremes), and then taking only the highest quotation. Of course, in the first place, the average price is somewhat lower. The M.D.M. also tends to be very slightly lower, but so slightly that the difference in taking either is evidently quite immaterial. The annual results are given in Table 8.

From this we concluded that it was permissible to use the highest quotation; and wherever we have been confronted with such a range this has been done. The lowest of the range could of course not be taken for the reason mentioned above, viz., that we suspected that, at Liverpool, it sometimes referred to a lower grade of wheat.

Having now settled these preliminary considerations, and ascertained that our data are comparable, we may proceed to the actual comparisons. The first point to be determined is whether futures, even for the current month, are more or less steady than the spot price. It was sufficiently clear, from inspection, that futures for distant months, generally speaking, fluctuate less than those for near months. Theory leads us to expect this. The futures price should forecast the value at the time of delivery, and should depend upon anticipations of the probable supply (relative to requirements) at that future date; account being also taken of the cost of

Table 8 Comparison of average with extreme of a range *(in cents per bushel and marks per tonne)*

Market, year, etc.		Average price	MDM
Chicago, 1892	Cash, using average of range	79	⁵/₈ (0.57)
	Cash, using highest of range	79³/₈	⁵/₈ (0.63)
Chicago, 1893	Cash, using average of range	67⁵/₈	⁵/₈ (0.66)
	Cash, using highest of range	67⁷/₈	⁵/₈ (0.69)
Chicago, 1896	Futures, using closing price	63¹/₂	⁷/₈ (0.84)
	Cash, using highest of range	64	³/₄ (0.77)
Berlin, 1896	Futures, using average of range	155.7	0.87
	Futures, using highest of range	156.1	0.90
	Futures, using closing price	156.3	0.85

holding the wheat pending the maturation of the contract, etc., etc. In such calculations influences of a momentary character affecting the immediate supply and demand should naturally count for little. The effect of a storm for instance, which interferes with traffic, would probably not be felt at all some months later, unless it were of widespread severity and caused serious damage. In practice, however, weather appears to play an important part in price movements, since it affects the growing crops. And temporary causes do also, as a matter of experience, affect the price of futures; for otherwise the difference between spot and futures would be greater than the normal (cost of holding for the period, etc.), and would offer chances of profit in one direction, inducing a rush of buyers or sellers on one side until the difference regained normal dimensions. A rise or fall in the price of near futures, unless extremely trifling, is thus accompanied by a corresponding rise or fall, which tends to be less in amount, in more distant futures; so that the result is, as already stated, somewhat less fluctuation in distant futures than in near futures or spot.

It may not be out of place to refer here to the statement frequently made, that spot prices are ruled by futures. The connection between the two is extremely close, and it seems impossible to say that the one regulates the other. But in view of what is said above, I think there is quite as much ground for saying that spot prices regulate futures – though I do not mean to exclude the influence of the latter upon the former. In this connection it may be noticed that in 1897 there was no rise in the price of wheat in the world's markets until July: in fact, the minimum price of the year occurred in that month, the price having fallen gradually, but very slightly, since January, and evidently depending upon the supply of the current cereal year. And the prices for September–December delivery were quoted no higher: no account was taken of a forthcoming deficiency. Yet two months afterwards the price, spot and futures alike, had risen by more than 2s. per cental. It was not until the new wheat had practically come in that the price rose. In 1896 no rise in price took place until September. Again, during May, 1897, there were reports of great damage to crops in France. This could not be expected to materially affect current prices,[12] and I can find no reflection of the storm in the price of futures. It would be idle to deny that false rumours are frequently – shall I say, constantly? – set afloat on the Exchanges; but in this instance the French crop did prove short, and the shortness certainly seems to have been in part attributable to the damage during May.

The Leiter corner is another instance of distant futures being governed by spot (current month futures in this case probably ruled spot). It was perfectly well known that the corner applied to May wheat, yet the price for September and December (1898) rose when spot rose (though, of course, not to the same extent), and fell with it in June. There was at that time nothing to indicate that the harvest of 1898 would be short, and there was no reason, apart from the rise in May wheat, why prices for September delivery should have risen between March and May. I have thought it of interest to reproduce in Figure 2 the daily prices of wheat at Chicago during the months of April, May, and June, 1898. This illustrates the movements of futures and spot just explained, and is also an extreme instance of the greater relative stability of distant futures.

These examples, I think, afford some ground for the opinion that current prices govern futures as much as vice versa; if indeed the influence of one can be said to be greater than that of the other.

In comparing futures with spot, we confined ourselves to the years 1896 and 1897 (we had not statistics at our disposal for earlier years). Since we have seen that distant vary less than near futures, it is clear that we must use the data for as near a month as possible; preferably the current month. In 1896 and 1897 quotations for futures are made for several months, including the current and post-current. About 1898, however, we find a change, and the tendency lately has been to concentrate trading onto a few important months of the year. Consequently we find that, except for these particular months, no futures are quoted less than two months ahead; and to take the nearest month would involve us in the use of an average of prices for varying dates ahead, which might vitiate our conclusions.[13]

The first thing we did, inasmuch as we have two sets of quotations for Liverpool, was to compare the relative variability of the closing price with the basis for calls. The latter, being an official price determined by the Exchange authorities, would, we should expect, tend to represent better the tone of the market at a given hour than would the price at a given moment, and might accordingly be somewhat steadier than the closing price. There appears to be very little difference, but it is in the direction indicated, as appears from Table 9.

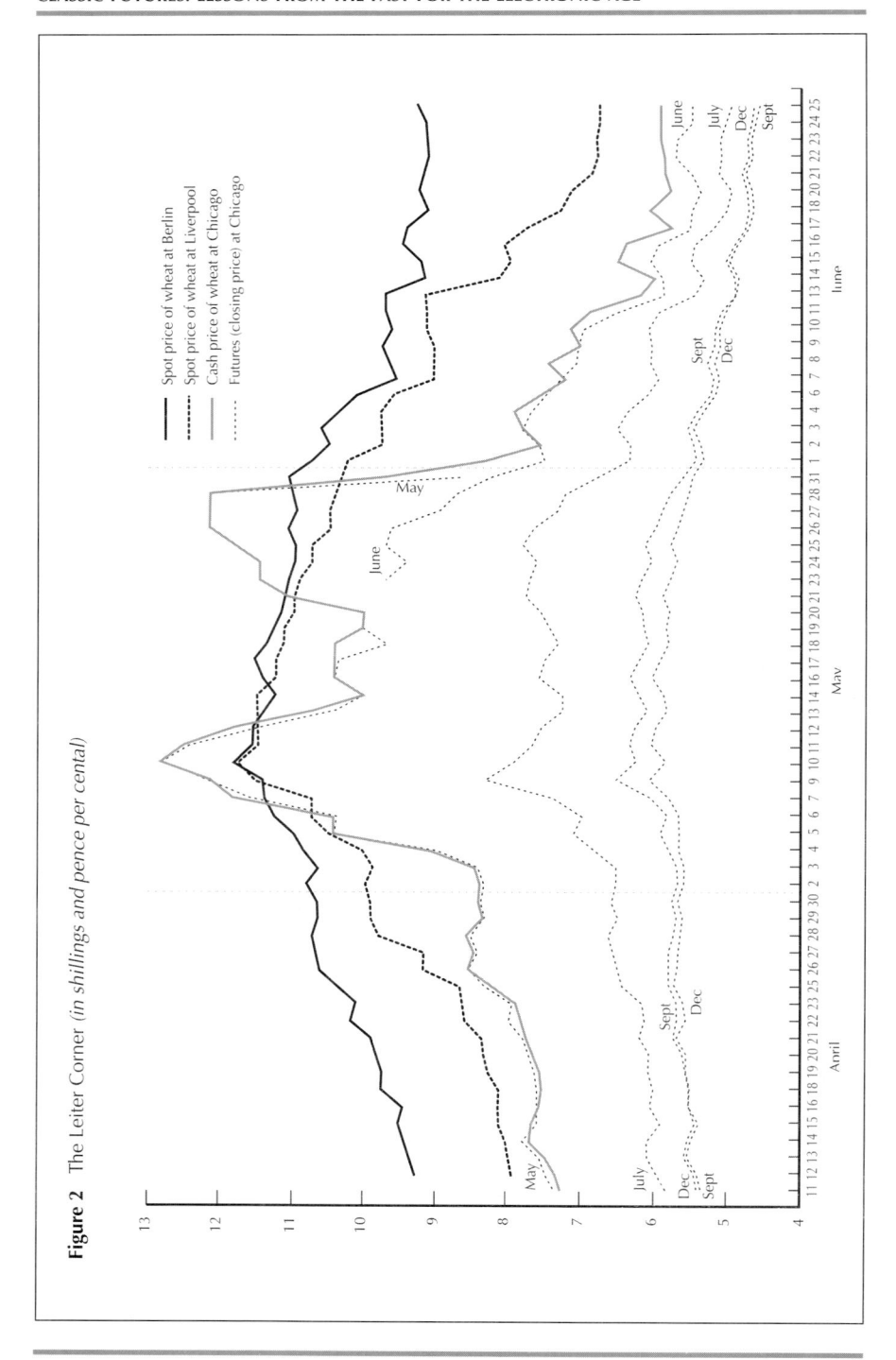

Figure 2 The Leiter Corner *(in shillings and pence per cental)*

Table 9 Liverpool. Basis for calls and closing price of futures. *Per cental* (1896)

	Average	σ	MDM
	s d	d	d
Basis for calls	5 7 3/8	7 3/4	0.52
Closing price	5 7 1/4	7 3/4	0.54

We remarked that in the basis for calls there was never any fraction of a penny smaller than a halfpenny, whereas the closing price is quoted to the nearest eighth of a penny. This raised the question as to whether a price taking account of very small fractions would show more variability than one ignoring them. In the first case there would certainly be a greater number of changes; in the second, the changes, although not so numerous, would never be very small, and the total change over any period might be expected to be the same in both cases. It would seem possible that during a period of great steadiness the sum of the very small changes might exceed the sum of the few larger ones, but that where the fluctuations were violent the result would be the same. As regards our other comparisons, however, the point is immaterial, because $\frac{1}{8}$ cent per bushel, $\frac{1}{8}d.$ per cental, and 25 pfennigs per tonne[14] (which are the smallest changes recorded in each case) are nearly equivalent, and thus the four cities are on the same footing.

In Table 10 are given certain comparisons of spot prices with futures (current month closing price). Although we have daily spot prices at Liverpool, yet there are in the latter city two special market days in the week (Tuesday and Friday), and a careful examination of the Liverpool spot quotations shows that during periods when price changes are not violent, the majority of the changes occur on these market days; that is to say, that there is probably comparatively little trade on other days; and to compare the sum (or average) of the daily changes on the futures market with the sum of the bi-weekly changes would be wrong, and might lead to untrustworthy results. When fluctuations are at all rapid, however, changes are equally to be noted at Liverpool on all days of the week. And this consideration will of course not affect the validity of the comparisons which I shall presently make between Berlin and Liverpool in different years, since the same conditions hold throughout the period.

Apart from New York, therefore, we may draw the deduction that spot prices, to judge by the M.D.M., tend to be more stable than futures. The New York figures, as I have already remarked, I do not readily understand; and the large excess in the price of spot over futures at this market is also curious.

One other explanation yet remains to be given. In all comparisons with

Table 10 Spot and futures prices at certain markets

Locality and year			Average price	σ	MDM
			cents	cents	cents
New York, 1896	Futures (close)	per bushel	71 3/4	8	3/4
	Spot (winter)	per bushel	78 1/8	9 7/8	7/8
	Spot (spring)	per bushel	74 1/4	8 1/4	3/4
New York, 1897	Futures (close)	per bushel	88 1/8	8 7/8	1
	Spot (spring)	per bushel	91	9 1/8	1
			s d	d	d
Liverpool, 1896	Futures (close)	per cental	5 7 1/4	7 3/4	1/2
	Spot	per cental	—	—	3/8
Liverpool, 1897	Futures (basis for calls)	per cental	6 8 5/8	9 1/4	3/4
	Spot	per cental	6 9 5/8	9 5/8	1/2
			marks	marks	marks
Berlin, 1896	Futures (various months)	per tonne	156.1	—	0.90
	Spot	per tonne	156.0	10.2	0.64

Chicago I have been obliged to use the "cash" quotations for that city. These "cash" prices are, I understand, not spot prices, but the same as what I have in this paper called "current month futures"; and I have been unable to obtain any spot prices for Chicago. This procedure is, however, not open to the objections indicated above as applying to the method adopted by Herr Mancke. All the conclusions I draw from the differences in price at Berlin and Chicago are based upon comparisons of the periods 1892–1896 and 1897–1899. We have at Berlin a set of spot prices obtained under different conditions in each of these periods, and the object is to ascertain whether one set is lower than the other. For *this* purpose it is quite immaterial what standard is used for comparison, provided that it remain the same throughout the eight years; and by using "cash" prices we secure a standard of stability, which, although not the same as spot prices, is nevertheless equally reliable for measuring the variation in stability at Berlin during the two periods. Herr Mancke merely shows that spot prices at Berlin are less stable than, for instance, cash at Chicago during a few months in one of the periods. Similar considerations apply to the comparisons of the level of price.

We may now take the figures given by Herr Mancke[15] for the three months April, May, June, 1898; with which he shows, by comparing Berlin spot prices with futures at Liverpool, New York, Chicago (and including freights, duty, etc.), and other towns, that Berlin was much steadier than the other cities during the Leiter corner. The figures[16] are given in Table 11. By taking spot prices at these same towns we get the figures given in Table 12.

Thus it appears that, while Berlin is still the steadier, we have

Table 11 Total daily movement during three months, Herr Mancke's figures *(in shillings and pence per cental)*

Month	Berlin		Liverpool		New York		Chicago (cash?)	
	s	d	s	d	s	d	s	d
1st to 30th April, 1898	1	10³/₈	3	2	2	9⁵/₈	2	8³/₄
1st to 31st May, 1898	2	9³/₄	5	7¹/₈	13	—	10	1³/₈
1st to 30th June, 1898	3	—	7	−⁵/₈	4	3⁵/₈	5	5³/₈
Total	7	8¹/₈	15	9³/₄	20	1¹/₄	18	3¹/₂

Table 12 Total daily movement during three months; spot prices *(in shillings and pence per cental)*

Month	Berlin		Liverpool		New York		Chicago (cash?)	
	s	d	s	d	s	d	s	d
1st to 30th April, 1898	1	10³/₈	2	2	2	7³/₈	2	3³/₄
1st to 31st May, 1898	2	9³/₄	3	2	13	2¹/₂	11	9⁵/₈
1st to 30th June, 1898	3	—	4	−¹/₂	3	10⁷/₈	5	3³/₈
Total	7	8¹/₈	9	4¹/₂	19	8³/₄	19	4³/₄

nevertheless reduced the apparent greater instability of Liverpool by three-quarters, and our criticism of the comparison made by Herr Mancke is, I think, justified. Spot prices at one town should in such a comparison be collated with spot prices at the others.

The other point which Herr Mancke did not prove was that Berlin prices corresponded more closely to those outside prior to 1897. To test this the σ and M.D.M. have been worked out for Berlin and Chicago for the years 1892–1900, together with the corresponding figures for New York for 1896–1899 and Liverpool for 1896–1900. Spot prices are of course taken in all cases (except Chicago). The standard deviations are shown in Table 13, and the mean daily movements in Table 14.

From Table 14 we may draw up a summary showing the difference between Berlin and Chicago during the periods when the former did and did not possess a Produce Exchange.

From this we see that the excess of instability at Chicago was greater during 1897–1899 than during 1892–1896. At first sight, it might therefore appear that the suspension of the Berlin Exchange had contributed to render the German price more steady. But this is not so; for it is entirely due to the year 1898, and Table 14 shows that Liverpool in 1898 was, relatively to Berlin, as steady as in other years, and much steadier than Chicago, showing clearly that the greater unsteadiness due to the Leiter corner was a phenomenon affecting America to a far greater extent than Europe. Comparing the average of 1897–1899 at Berlin with that at

Table 13 Standard deviations (*in pence per cental*)

Locality	1892	1893	1894	1895	1896	1897	1898	1899	1900
	d.	d.	d.	d.	d.	d.	d.	d.	d.
Berlin	12	$3\frac{3}{8}$	$3\frac{1}{4}$	$3\frac{3}{4}$	$5\frac{3}{8}$	$6\frac{3}{8}$	$11\frac{1}{4}$	3	$1\frac{1}{2}$
Liverpool	—	—	—	—	—	$9\frac{5}{8}$	$17\frac{1}{2}$	$1\frac{1}{2}$	$1\frac{3}{4}$
New York*	—	—	—	—	$8\frac{1}{4}$	$7\frac{5}{8}$	$19\frac{1}{8}$	$3\frac{1}{2}$	—
Chicago	5	$4\frac{7}{8}$	$2\frac{1}{2}$	$6\frac{3}{8}$	$6\frac{3}{8}$	$7\frac{1}{8}$	$22\frac{1}{8}$	$2\frac{1}{8}$	$4\frac{1}{2}$

*Spring wheat in 1897; winter wheat in other years.

Table 14 Mean daily movements (*in pence per cental*)

Locality	1892	1893	1894	1895	1896	1897	1898	1899	1900
	d.	d.	d.	d.	d.	d.	d.	d.	d.
Berlin	$\frac{1}{2}$	$\frac{1}{8}$	$\frac{1}{4}$	$\frac{1}{4}$	$\frac{3}{8}$	$\frac{1}{2}$	$\frac{5}{8}$	$\frac{3}{8}$	$\frac{1}{4}$
Liverpool	—	—	—	—	$\frac{3}{8}$	$\frac{1}{2}$	$\frac{3}{4}$	$\frac{1}{4}$	$\frac{3}{8}$
New York*	—	—	—	—	$\frac{3}{4}$	$\frac{7}{8}$	$1\frac{1}{2}$	$\frac{1}{2}$	—
Chicago	$\frac{1}{2}$	$\frac{5}{8}$	$\frac{1}{2}$	$\frac{5}{8}$	$\frac{5}{8}$	$\frac{3}{4}$	$1\frac{1}{2}$	$\frac{1}{2}$	$\frac{1}{2}$

	Berlin	Chicago	Excess of Chicago over Berlin
Average M.D.M. 1892–96	$\frac{5}{16}$	$\frac{9}{16}$	$\frac{1}{4}$
Average M.D.M. 1897–99	$\frac{1}{2}$	1	$\frac{1}{2}$

Liverpool, we find that it is identically the same ($\frac{1}{2}d.$), both cities being equally more stable than Chicago. And as Liverpool has an active futures market, we clearly cannot ascribe the effect to the absence of such a market. If we exclude the demonstrably exceptional year 1898, we find that the mean M.D.M. of 1897 and 1899 was $\frac{7}{16}d.$ at Berlin, and $\frac{5}{8}d.$ at Chicago; so that on the average of these two years Berlin was, relatively to Chicago, a trifle less steady than previously. The theory therefore that the absence of futures has steadied the market is untenable; while the hypothesis that Berlin followed the markets abroad more accurately prior to 1897, is equally shown to have no foundation in fact (see *Appendix*).

The year 1892 is an exception: the M.D.M. of that year was as great at Berlin as at Chicago. This phenomenon is due to the fact that Chicago recovered very rapidly from the high prices during 1891, while at Berlin the fall was very much slower, and took place mostly during 1892 (see Table 15). Thus at Chicago prices fell comparatively little during 1892,

Table 15* Monthly prices of wheat in Berlin and Chicago in 1891–92 (*in marks per tonne*)

Month	Berlin		Chicago		Difference	
	1891	1892	1891	1892	1891	1892
January	189.6	214.3	140.5	134.7	49.1	79.6
February	196.0	203.8	146.0	135.0	50.0	68.8
March	209.2	195.6	151.4	128.2	57.8	67.4
April	226.4	190.0	164.4	124.6	62.0	65.4
May	241.0	189.8	159.6	127.3	81.4	61.5
June	232.5	182.7	148.4	127.1	84.1	55.6
July	237.3	174.1	137.5	119.6	99.8	54.5
August	236.2	159.4	153.8	118.3	82.4	41.1
September	234.1	152.8	147.0	112.3	87.1	40.5
October	226.6	153.3	147.7	110.7	78.9	42.6
November	233.6	152.5	144.8	109.6	88.8	42.9
December	228.3	148.6	140.6	108.6	87.7	40.0
Year	224.2	176.4	148.5	121.3	75.7	55.1

*In this table (taken from the *Vierteljahrshefte* for 1895) the Chicago prices are for No. 2 Red Winter. Not having daily prices for Berlin in 1891, I have been unable to calculate the M.D.M. The standard deviations of the monthly averages from the yearly mean are as follows:
Berlin, 15.9 in 1891; 21.8 in 1892.
Chicago, 7.5 in 1891; 9.1 in 1892.

while at Berlin the fall was much greater, so that at the latter city the M.D.M. includes movements which had to some extent already been discounted at Chicago in the previous year (see Table 15). It will be noted also that during this period the average price at Berlin was also abnormally high, reaching nearly 100 marks per tonne more than the Chicago price (Red Winter), although the duty was only 35 marks (compare also the large standard deviation in both years at Berlin). The cause of this state of affairs is to be found in the "famine" of 1891, which was essentially a Russian famine. A large portion of Germany's imports are usually derived from her eastern neighbour, who prohibited the export of grain in 1891–1892, and we should under such circumstances expect to find German prices more affected than American. In short, Berlin prices were higher and subjected to more violent movements in 1891–1892 than Chicago, because the scarcity was in Europe; whereas in 1898 Chicago prices were subjected to more violent movements because the shortage was in America. If we omit these years 1892 and 1898, and compare only 1893–1896 with 1897 and 1899, we have the summary given in Table 16, and I see no reason why it should not be entitled to as much credence as the others.

The result of our investigation may now be summed up in a single sentence. The conditions existent at Berlin during the suspension of the Produce Exchange, while causing the greatest hindrance to the trade, have

Table 16 Average prices and mean daily movements at Berlin and Chicago (*in shillings and pence per cental*)

	Berlin	Chicago	Difference between Berlin and Chicago	Change during period
	s. d.	s. d.	s. d.	d.
Average price, 1893–96	6 $7\frac{1}{2}$	4 $3\frac{3}{4}$	2 $3\frac{3}{4}$ }	$1\frac{1}{4}$
Average price 1897 and 1899	7 $6\frac{1}{2}$	5 4	2 $2\frac{1}{2}$	
Average MDM 1893–96	$-\frac{5}{8}$	$-\frac{5}{8}$	$-\frac{3}{8}$ }	$-\frac{3}{16}$
Average MDM 1897 and 1899	$-\frac{5}{8}$	$-\frac{5}{8}$	$-\frac{3}{16}$	

not induced a rise in the prices of grain, and they have not imparted greater stability to those prices; if, indeed, they have not exercised a deleterious effect in both these directions.

Into the question of the dislocation of the grain trade, more especially the international trade, of Germany, I have not entered, but have confined myself to a discussion of the effect on prices. We heard much, when in Germany, of the difficulties under which merchants laboured during the period; but this is a subject that scarcely admits of statistical treatment, and hence it falls outside my province, and outside the province of the Society.

APPENDIX

I have thought it advisable to correlate the daily prices for Berlin, Chicago, and Liverpool, in order to demonstrate mathematically whether the Berlin market was as intimately connected with the rest of the world during the period 1897–1899 as previously. The results are shown in the following Table A, in which r (the coefficient of correlation[17]) $= \Sigma xy/n\sigma_1\sigma_2$, and p (the probable error) $= 0.67(1 - r^2)/\sqrt{n}$.

There is thus no evidence that Berlin was more closely connected with Chicago prior to 1897 than subsequently. But some of the coefficients obtained are rather remarkable, and need some explanation. The very high coefficient in 1892, in particular, seems curious in view of the statement made earlier that Chicago fell more rapidly than Berlin from the high prices of 1891–1892. I think it is due to the rather exceptional circumstance that the Chicago price, like the Berlin, was falling steadily throughout the year, although not by the same amount (*cf.* the σ). Correlation thus appears, in this instance, not to measure the causal connection between the fall at the two markets. The very low coefficients of 1893 and 1899 are also unexpected. It seems probable that in years

Table A Coefficients of correlation and probable errors

	Correlation coefficients			Probable errors		
Year	Berlin and Chicago	Berlin and Liverpool	Liverpool and Chicago	Berlin and Chicago	Berlin and Liverpool	Liverpool and Chicago
1892	+0.966	—	—	0.002	—	—
1893	+0.287	—	—	0.036	—	—
1894	+0.836	—	—	0.012	—	—
1895	+0.775	—	—	0.016	—	—
1896	+0.865	—	—	0.010	—	—
1897	+0.929	+0.929	+0.958	0.005	0.005	0.003
1898*	+0.903	+0.909	+0.953	0.008	0.007	0.004
1899	+0.430	+0.397	+0.634	0.032	0.033	0.024
1900	+0.823	+0.670	+0.651	0.013	0.022	0.023

*Eleven months only: no quotation for August at Liverpool. The coefficient for the whole twelve months of 1898 between Berlin and Chicago is 0.908, probable error 0.007.

when prices remain at about the same level (σ very small), changes due to local conditions may be, relatively, more important, and there may thus be less apparent dependence of one market upon another. The high coefficients of 1897 and 1898 are only to be expected, owing to the great influence of the Leiter corner, and should not be interpreted as necessarily pointing to greater dependence of Berlin upon the outside world during the suspension of the Exchange.

The results of this calculation have thus proved somewhat unsatisfactory. But the application of correlation to economics may almost be said to be still in its infancy, and an instance where the conclusions which may be legitimately drawn from its use are very restricted is therefore, I hope, not out of place, and may serve as a warning that care is as necessary to the proper comprehension of such calculations as in the interpretation of results deduced by more elementary treatment. Although it is thus clear that ordinary correlation does not yield an answer to the particular question I was anxious to elucidate, I have nevertheless let the work stand – a final example of apparently useless labour – as an object-lesson of how not to do it is often of educational value. With some hesitation I submit that, in a comparison of prices at different markets, correlation may be a measure of the difference between the effect of world-wide influences (common to both markets) and local influences upon the price. But, bearing in mind what has been said earlier as to the greater suitability of the M.D.M., rather than the σ, as a measure of stability in such inquiries, I would suggest that the problem – to what extent do the fluctuations at one market follow those at another? – requires to be attacked by the use of some formula which should correlate the differences between the prices on consecutive days, instead of the differences from the average price.

1 I may as well say at once that I shall use the word "futures" as an abbreviation for "contracts for future delivery". By such transactions are understood contracts to deliver goods during a certain month, the precise day of the month being at seller's option. There are thus prices of "current month futures", sometimes called "cash", which usually differ slightly from "spot" quotations, but may be practically identical towards the end of the month.
2 The Exchanges Act is dealt with by Prof. Lexis in the *Economic Journal*, vii, p. 368; while Prof. Flux in a later number (x, p. 245) has given the history of the controversy in the Berlin Produce Exchange. Readers who wish for further details may be referred to F. Goldenbaum's exhaustive article in Schmoller's *Jahrbücher*, August, 1900, and February, 1901.
3 "Der börsenmässige Terminhandel in Getreide und Mühlenfabrikaten ist untersagt" (sec. 50).
4 For particulars of these see Prof. Flux's and F. Goldenbaum's articles, *loc. cit.*
5 On this point I found, during a visit to Berlin in April [1900], that opinion was unanimous. I may take this opportunity of acknowledging the great courtesy with which Prof. Flux and myself were received during this visit, and the extreme willingness with which information was afforded us, alike by friends and opponents of the futures system. It is not too much to say that we could hardly have arrived at a clear perception of what is, after all, a somewhat tangled position, without the assistance so freely rendered.
6 Eg, Mr. Hatch's Report of Committee on Agriculture on Dealing in Fictitious Farm Products (United States House of Representatives, Fifty-third Congress, first session, Report No. 969).
7 I have to thank Mr. G. J. S. Broomhall and Mr. S. Woods (Editor of "Dornbusch's Lists") for kindly filling some gaps in my data.
8 *Vierteljahrshefte des Deutschen Reichs*. Part 1 of each year.
9 I do not know why the exports should have practically started suddenly in May, 1894 (this applies equally in the case of rye). Provision is made in the Customs Act of 1879 for import-permits, but effect may possibly not have been immediately given to this. Nor do I know whether the whole of the export under this privilege is in the form of flour or not.
10 (1) *Was und Wer bestimmte die Weizenpreise im Ernte-Jahr 1897–98?*; (2) *Die Bewerthung des Weizens auf den Weltmärkten seit Inkrafttreten des Börsengesetzes*; (3) *Was und Wer bestimmte die Weizen und Roggenpreise im Ernte-Jahr 1898–99?* Published by the author, Berlin.
11 Because freights tend to rise when the price is enhanced.
12 Except to the extent that a probable short harvest would involve greater reliance on stocks left over from the previous season, and so tempt owners to hold the wheat in hand.
13 The Chicago Board of Trade recently took steps towards restricting this concentration of the trade onto a few special months, which probably facilitated the business of those who dealt only in futures, but possibly tended to minimise the cash trading. By a rule passed in 1900, no wheat price may be quoted on the Exchange for more than two months ahead, except that May wheat may be quoted from the October 1. The change appears commendable.
14 Though the Berlin price is far more frequently quoted to the nearest mark only.
15 *Was und Wer bestimmte die Weizenpreise im Erntejahr 1897–98?* p. 8.
16 The changes recorded are those between the first and last day of each month: the change between the last day of one month and the first of the next is thus not recorded (I have of course followed this in Table 12). I have no New York spot quotation on May 31, or June 1/2. My figures for Chicago would have come out lower were it not for a drop of 35 cents per bushel (cash) on the last day of May, which does not appear in Herr Mancke's tables, so that he would appear to be dealing with some other futures at least on that day.

17 See Yule, 1897, *Journal*, 60, p. 812, etc.; or Bowley, *Elements of Statistics*. For much help in the lengthy calculations involved on this piece of work, I have to thank Mr. R. J. Thompson and Mr. J. A. P. Mackenzie.

DISCUSSION ON MR. HOOKER'S PAPER

Sir Robert Giffen desired to express the pleasure and satisfaction which he was sure they had all experienced in listening to the reading of the paper. They must recognise the extreme industry and skill with which Mr. Hooker had treated the subject and compiled his tables. The impression made upon his mind was that Mr. Hooker had completely proved his case, and that taking the experience of Berlin, those who agitated for stopping speculation and suppressing dealings in futures on the ground that they would thereby attain a more stable and a higher price, must be held entirely in the wrong. A great market on which dealings in futures took place had been stopped for three or four years, and it was found that the stoppage of speculation had not tended in any way to raise the price of wheat, and that the farmers throughout the world were not one whit better off than they were before. Of course it was quite true that it was not enough to stop one market if the remaining markets were still open. But that, he believed, was not the view of the people who agitated: they were satisfied that if they could stop speculation in their own markets, they would get better prices for their wheat. The great wonder was that the people who agitated did not take a more general view of the question. Speculation they knew took place on the stock exchanges throughout the world, and not only with regard to wheat, but with regard to other produce and other commodities; and he believed that the conclusion of all the great leaders in these markets, and of economists who had considered the subject, was that on the whole the tendency of such speculation was to equalise prices over a given period, and to prevent an extreme fall on the one side or an extreme rise upon the other. The reason was that there were so many people with capital and with acute brains interested in the subject, that no man could take advantage of another, and if there was a fall it was checked as soon as possible by the wise people who perceived that there would be a reaction, and if there was a rise, it was checked when it came near the top by the action of those who saw that the rise could not last, and who therefore sold in good time the stocks of which they were possessed as dealers in the market. Those who agitated in the particular case of wheat were bound either to produce illustrations from other departments of the subject in support of their case, or to acknowledge that the arguments which they used were unsound. This they had failed to do, and their case failed absolutely. Thanks to Mr. Hooker, the members of the Society and of the public were now able to deal with the question from experience as well as theory, and Mr. Hooker had done great service by this work.

Sir John Glover joined with the last speaker in the recognition of the industry and skill which the paper disclosed. He did not think that the experience of Berlin in this matter could be held to be conclusive, because, that being the only market closed, the German merchant who was disposed to speculate had only to send his order to Paris, or Amsterdam, or London, or to the New York or Chicago markets, which the German Government had not the power to close. Again, the influences on markets generally of dealings in futures had not so far shown that the law of supply and demand had thereby lost any of its force. That law in the matter of wheat and rye was governed by a hundred things, of which speculation was only one. The crops, seasons, adjacent famines, the action of other countries (as in the case of Russia), the stoppage of exports – these were all material things to be considered. Another thing which struck him was that, if an experiment in this direction was to be made, Berlin was not the best place in which to make it, because in Germany there were very heavy import duties in operation – duties which had been repeatedly increased, and the pressure of which must interfere very materially with the natural course of supply and demand on that market. Then Germany was not to any extent an exporting country for grain. That would affect the question, since the system of import permits opened the door to very peculiar dealings. For his own part he should certainly agree with the conclusion of Mr. Hooker, that the experiment in Berlin had proved nothing. Even in America, he thought the influence of the Leiter corner ought to be disregarded. It was a phenomenon with remarkable circumstances, and most happily for mankind it was a dead failure, and he did not think that for a long time anybody would be likely to repeat it.

Sir W. Thiselton-Dyer said that the great value of the paper lay in the fact that it supplied the critical apparatus upon which the British Association report was founded, and therefore very much enhanced the convincing character of that admirable document. *A priori* one had little difficulty in arriving at the same conclusions as Mr. Hooker. Dealings in futures were of the nature of a bet as to what would happen at a distant date. Even speculators were not wholly destitute of intelligence, and it was obvious that consciously or unconsciously they would work upon what was really a statistical basis. It was not surprising, then, that futures were more constant than immediate prices, because the minor fluctuations to which the latter were subject were smoothed out in working on an average. Again, though local gambles might conceivably have a temporarily disturbing effect on a particular market, futures were dealt in over so wide an area and so large a scale that they must add to the steadiness of the market.

The prospects of the probable wheat crop of the world in any year were watched with the most minute attention at all the great centres at which it was dealt with. Very close estimates of the probable crop were formed in advance, and upon these the future prices would be calculated. The world price of wheat was a function of two variables – climatic conditions, and the appetite of man for bread. These were physical facts which no gambling could affect. One might gamble about the death-rate, but it would be powerless to affect actual mortality.

Mr. J. C. Pillman, of the London Corn Exchange, complained of the want of practical knowledge shown by the author, who had drawn deductions in purely academic fashion from certain figures in justification of option trading. No one could perhaps rightly appreciate the effect of option dealings upon legitimate trade except those who were themselves engaged in legitimate trade, and who had been engaged in it prior to the introduction of this system. The *raison d'être* of an option market was to enable people to sell what they did not possess. It was not necessary for the distribution of the produce of the world. When it was said that it did not in any way interfere with the law of supply and demand, they might take as a typical instance the years that had been already referred to in the paper, viz., of 1897–1898. In January, 1897, they knew of the failure of the Argentine wheat crop, and of the drought in Australia, which prevented any shipments coming from Australia; in fact, in that year Australia had to import wheat from America. They knew also in the month of May of the failure of the Indian crop. Yet in spite of these three important factors, the price of wheat fell from January to June from about 33s. to 27s. – a fall of 6s. a quarter. The value of wheat was entirely influenced by the dealings in options in the international markets. Mr. Hooker referred to the fact that there had been damage to the crops in France in May or June of that year, but in the trade they did not know of any damage to the French crops of any serious importance till the month of August. Then came also the failure of the Italian crops, which caused the bread riots in Italy in the spring of 1898. As he had pointed out before the Commission on National Granaries, in November, 1897, the statistical position of wheat consequent upon these several failures, according to the law of supply and demand, justified a price of 50s. a quarter. But the price at which English farmers were then selling the harvest of 1897 did not average 35s. a quarter – in a year of exceptional shortage – or about the cost of production. The price of 50s. was not reached until the farmers had parted with every bushel of their wheat, so that the rise in the months of April, May, and June, 1898, mentioned by Mr. Hooker, was no benefit to the producer at all. The power of the "bear" speculation in futures kept down the price, and at that price the farmer had to sell his produce, and it was only in that year of actual shortage that farmers in this country were enabled to get anything near the cost of production, which he took to be something like 33s. a quarter. These dealings were worse than bets, because they influenced the business of those who refused to participate in them. They undermined values, and had been the main cause of agricultural depression not only in this country but throughout the world. America at the present time was the dominating influence in the world so far as wheat values were concerned, and it was in the United States that these futures markets prevailed. The price of wheat today was made by the speculator, who held no stock and never intended to have any. These markets in futures were characterised as being similar to the Stock Exchange, but they were very dissimilar. A man could sell the whole crop of the United States for next May, and by throwing at any moment this fictitious quantity of stuff upon the market, he naturally reduced the value, and had all the intervening days, weeks, and months in order to cover those sales. The Stock Exchange, on the other hand, by its limited accounts, protected itself against "bear" operations of an unlimited character by enforcing settlements within thirty days. Bankers had brought about the passage of an enactment called Leeman's Act in 1866, which prevented anyone selling a single share of any bank in the United Kingdom unless at the same time he gave the number of the scrip. That was in order to prevent a "bear" raiding bank stocks and undermining the financial institutions of the country. The protection of the producers was surely as important as that of our financial institutions. The existence of futures markets formed one of the reasons why such short stocks were kept in the United Kingdom. This fact constituted a national danger, as this process of fictitious trading has paralysed importers, who have found that stock carrying is ruinous under the play of the option system. A system which was designed to enable people to make money by depreciating the value of an article upon which agriculture was built, should be abolished.

Dr. B. W. Ginsburg desired to refer to the remarks of the last speaker only, and to speak himself only from his commercial experience. In the first place he would insist that whatever might be the theory of the Stock Exchange, the common practice there was to allow dealings to be extended indefinitely. But looking at the main point of the paper, he supposed that the ultimate business of every market was to pass the existing supplies of the commodity dealt in to those who prepared it for the consumer. The gambling, or future dealing, was merely an excrescence. Thus his own dealings in sugar were to that extent comparable with those in wheat. As a refiner of sugar he had found that he needed so many tons of sugar each week to keep the refinery going, and it was necessary to make arrangements for supplies for some time ahead, even when the close attention which the refiner had to give to the market showed him that there was a probability that prices would go lower. If when he had to buy to keep his works going – though he knew he was buying at a loss – he was to be debarred from hedging against the loss he was making, there would be practically an end of business. It was impossible to prevent a man making use of his knowledge and opportunities, and if a man in the trade could speculate – if it were called speculation – it could not be possible to say that others who thought they had the knowledge, even if they had not the demand for the stuff, should not deal too. The suggestion would bring about a gross interference with liberty. He was sorry that Mr. Pillman had introduced his point as to the restriction of dealings in futures from the idea that their abolition would tend to increase the stocks of grain in the country. No one felt more strongly than did he (the speaker) as to the importance of keeping up those stocks. But that was for national purposes that the safety of the country might be ensured in time of war. The decline in stocks of most commodities in this country was due to the fact that modern improvement in transport made the arrival of goods from abroad much more certain and punctual than formerly, and thus encouraged buyers to live more and more from hand to mouth. It would be outrageous to force upon a small class of traders the obligation to keep up those stocks for purposes which affected the security of the nation as a whole. If the nation needed larger stocks of grain, the Government must see to that provision.

Mr. Sydney Young pointed out that Mr. Hooker laboured under a disadvantage owing to his not having a practical knowledge of his subject. The closing of the Berlin Produce Exchange was London's opportunity. The London Produce Exchange, which had only a short time previously opened its corn department, was at once placed on a firm basis by the large dealings of the Berlin operators.

Mr. Hooker seemed to have confused in his paper[1] cash prices with spot prices. If he meant by cash prices the price of the day of the current month in the futures market, he (the speaker) was afraid that the results given in the tables would be misleading, for the following reason, that Mr. Hooker had taken cash prices as a steady figure, whereas they showed the greatest fluctuations owing to the "bears" having at times to cover at the end of each month. Again, cash prices were not regulated by the future prices, but by the trade of the day. Whereas if he meant the spot price of wheat, that was the price paid for parcels of wheat on the spot, he would be again at fault in his tables, as the two classes of prices could not be compared or spoken of in the same breath.

Mr. R. H. Rew protested against the doctrine that "he who drives fat cattle must himself be fat". He looked upon it as a great advantage to themselves as a Statistical Society that they should have the presence of gentlemen like Mr. Pillman and Mr. Sydney Young, who were qualified to express expert opinions, but at the same time he demurred to the suggestion that this subject was one that could not be dealt with from a purely scientific and statistical standpoint. While accepting any statement from Mr. Pillman upon matters within his knowledge with the greatest deference, he would suggest that he would greatly strengthen the case he put forward if, in addition to his personal opinion, he would give them a little of the dry light of statistics, such as Mr. Hooker had so admirably thrown upon the subject. He had read and heard much on this subject, but he thought the advocates of the theory that options or futures had a depressing effect upon prices had suffered from the absence of that close tackling of the subject from a statistical standpoint that on the other side was represented by this paper. He suggested that Mr. Pillman and those who thought with him should bring forward for consideration a paper on somewhat similar lines to that which was under discussion. Thereby certain points, at present somewhat confused, might be elucidated. To instance only one: Mr. Pillman had expressed the opinion that the price of wheat at a certain period ought to have been 50s. He would ask on what statistical data that opinion was founded? In conclusion, he expressed his view that this was an admirable instance of the sort of paper which this Society wished to encourage.

Mr. G. Udny Yule wished to add his tribute of praise to the author for the extremely careful and dispassionate way in which the whole paper had been worked out.

Major P. G. Craigie thought they ought to look at the question of the effect of these quotations of futures on current prices not only from the point of view of their effect upon the wheat markets, but also as regards other commodities which were the subject of speculation. In spite of the effect on the mind of any statistician of a useful and carefully argued paper like the present, they could not shut their eyes to the fact that throughout the world there was a considerable class of politicians who still maintained that the policy of the German Act was a right policy, and were striving hard to induce other Governments to take similar action. The Austrian Government had lately had before their Parliament a measure for the regulation of their local exchanges, and the vote of that Parliament had pointed to the adoption of even more drastic measures than the Austrian Ministry had contemplated. As regarded wool at least, he believed that some action, though possibly not on entirely the same lines, had been attempted in Italy. It behoved this Society to welcome and encourage Mr. Hooker, and those who had discussed the matter in the way it was handled here, to look a little further into its bearings and to see if they could corroborate these conclusions by the same laborious processes and the same admirable analysis of the conditions of one or two other industries, so that they might have before them a still larger basis of discussion. With regard to the paper itself, he would ask Mr. Hooker in replying, to tell them what the (a) and (b) were in his prices on a particular market in Table 2; did they represent particular grades of wheat, or different markets differently organised? He would also ask why Chicago was not among the list of markets in that table, in which the average was taken to compare with Berlin.

Mr. T. A. Welton said that in his view, having had a good deal to do with sales of sugar, a great deal of the demoralisation which had been complained of was due to the fact that the speculators operating had been strong men, and had more consistency and ability than the ordinary importer or consumer. He had seen importers absolutely demoralised and panic-stricken, and the ordinary consumer – the man who bought from week to week – afraid to buy more than a week's supply, because he was impressed with the probability that there would be a further fall. When the "bears" so far succeeded in frightening everybody, it produced a very calamitous influence on the market; but that was not due to any law, but only to the fact that human nature is weak, and that the individual elements on one side might often be a good deal weaker than those on the other. These circumstances might bring about conditions which tended to keep consumers' stocks low. He agreed with previous speakers that the Berlin market was rather an unfortunate one in which to have tried such an experiment. It seemed from the tables that the consumption of grain for the food of man in that part of the world was two-thirds rye and one-third wheat, so that wheat was not the food of the mass of the population in the same way that it was in this country or in France. He thought the article of superior quality would be considerably affected by the fluctuations of the inferior article, and so be taken out of the category of ordinary dealings where wheat was universally employed. The results of experiments in the markets of Paris or London would be far more edifying.

Mr. R. W. Dunham asked whether Mr. Hooker in his experience of Berlin had come across what was called "straddling" in the corn market here, where the corn merchant and the future operator, having a cargo of wheat coming from a long distance, speculated, and made his profit on his future, so that he, as a corn merchant, was able to sell the cargo merely to get rid of it, and at a price lower than he could otherwise afford to accept in order to pay him. Again, he believed that in America the people there who operated in the future market owned or controlled the warehouses all along the line to Minneapolis and right away to Dakota. Dealings in wheat futures had depressed the price of wheat in Dakota very considerably, and he contended that this was cause and effect.

Mr. C. S. Loch (Chairman) said that the net result of the debate appeared to be, that Mr. Hooker had analysed the effect of futures on the corn market with the greatest care, showing that, so far as the evidence went, they did not cause increased prices or increased unsteadiness. On the other hand, these conclusions were challenged, and it was desired to impose on the market what amounted to a system of protection by rendering quotations of futures illegal. It was a pity that those who took this trend, had not submitted evidence proving the connection between the dealing in futures and the low prices of corn in England which they had quoted. As it was proposed to introduce into Austria the system adopted in Berlin, it was important that the subject should be dealt with on a broader basis, so as to include other markets. Considering the bearing of the question, such an investigation could hardly be deemed other than "practical", in the sense in which members of the Statistical Society desired to be practical, though they might not themselves be engaged in the transactions of the markets.

Mr. Hooker, in thanking the meeting for the kind way in which they had received his paper, said that he agreed so much with Sir R. Giffen and Sir W. Thiselton-Dyer, that no reply was required in their case. Sir John Glover seemed to think that his results might be vitiated by the German import duties, but these had not changed during the period, and so the comparison held good. He was in agreement with the speaker as to the law of supply and demand and had in fact already pointed out that no increase in German prices was to be expected. Grain prices were now "world prices", and could not be materially dissociated from them in any particular locality.

Mr. Pillman appeared to believe in the theory that future dealings depressed prices. He (Mr. Hooker) thought that the main argument for this theory was the fact that the fall in prices and the growth of the futures system were practically coincident in time. But this was no proof; and there were plenty of people who would say that the sole cause of the fall in wheat prices was the fall in silver, which was equally coincident. Undoubtedly temporary depressions could be and were manipulated under the futures system, in the same way as corners; but he did not see that the system could be answerable for a persistent fall during several years. In his view the main causes of the decline in grain prices were increased transport facilities and reduction of freights. In this connection he was willing to admit that future dealings in *real* wheat had contributed something towards the reduction of price in this country, because a knowledge of the exact quantities required at future dates was an important factor in facilitating regular shipments and smoothing the whole traffic organisation. But this applied only to genuine, and not to "wind", wheat. He further pointed out that the price of maize – a commodity subjected to as much gambling as any – had not declined on the farm at the principal centre of production, namely, in Iowa.[2]

With regard to Mr. Young, it would appear that there was some confusion between cash and spot prices, and this would be corrected before the paper finally appeared in the *Journal*. Major Craigie's suggestion that Table 2 would be clearer if the grades of wheat were specified would also be carried out. Chicago was omitted from that table because, as explained in the text, the table included only the spot prices recorded in the *Vierteljahrshefte*, which gave only futures quotations for that market.

He was aware, as Mr. Welton said, that rye was a commoner article of diet than wheat in Germany; but he had shown that the former grain had also suffered a diminution of nearly the same amount during the suspension of the exchange. As regarded Mr. Dunham's point, the purchase of grain afloat, and its re-sale prior to arrival if opportunity of profit occurred, was a class of business which had been specifically mentioned to him when in Berlin as having been entirely lost to Germany. Owing to the uncertainty prevailing abroad as to the exact position in that country, and the common belief that futures had been abolished, merchants in other countries were extremely shy of selling their goods to Germans, not feeling sure that the transaction might not be afterwards repudiated.

Finally, he thought that the experience of Berlin was an example of what this country might expect if future dealings were abolished here. Were it possible to have an international prohibition of the system, it was conceivable that prices might be raised to a slight extent in England, by an amount probably representing the increased difficulties in getting grain from abroad; but American farmers would not gain anything. If the futures system had any permanent effect on prices, it was probably in the direction of reducing the price to the consumer, but not at the expense of the favourably situated producer.

1 This point was later dealt with by the author.
2 See *Journal*, December, 1900, 63, p. 675.

17

*The State of Long-Term Expectation**

John Maynard Keynes†

The scale of investment depends on the relation between the rate of interest and the schedule of the marginal efficiency of capital corresponding to different scales of current investment, whilst the marginal efficiency of capital depends on the relation between the supply price of a capital-asset and its prospective yield. In this chapter we shall consider in more detail some of the factors which determine the prospective yield of an asset.

The considerations upon which expectations of prospective yields are based are partly existing facts which we can assume to be known more or less for certain, and partly future events which can only be forecasted with more or less confidence. Amongst the first may be mentioned the existing stock of various types of capital-assets and of capital-assets in general and the strength of the existing consumers' demand for goods which require for their efficient production a relatively larger assistance from capital. Amongst the latter are future changes in the type and quantity of the stock of capital-assets and in the tastes of the consumer, the strength of effective demand from time to time during the life of the investment under consideration, and the changes in the wage-unit in terms of money which may occur during its life. We may sum up the state of psychological expectation which covers the latter as being the *state of long-term expectation* – as distinguished from the short-term expectation upon the basis of which a producer estimates what he will get for a product when it is finished if he decides to begin producing it today with the existing plant.

It would be foolish, in forming our expectations, to attach great weight to matters which are very uncertain.[1] It is reasonable, therefore, to be guided to a considerable degree by the facts about which we feel

---

*This paper was first published as chapter 12 in *The General Theory of Employment Interest and Money*, 1936 (London: Harcourt Brace).
†John Maynard Keynes, 1883–1946.

somewhat confident, even though they may be less decisively relevant to the issue than other facts about which our knowledge is vague and scanty. For this reason the facts of the existing situation enter, in a sense disproportionately, into the formation of our long-term expectations; our usual practice being to take the existing situation and to project it into the future, modified only to the extent that we have more or less definite reasons for expecting a change.

The state of long-term expectation, upon which our decisions are based, does not solely depend, therefore, on the most probable forecast we can make. It also depends on the *confidence* with which we make this forecast – on how likely we rate the likelihood of our best forecast turning out quite wrong. If we expect large changes but are very uncertain as to what precise form these changes will take, then our confidence will be weak.

The *state of confidence*, as they term it, is a matter to which practical men always pay the closest and most anxious attention. But economists have not analysed it carefully and have been content, as a rule, to discuss it in general terms. In particular it has not been made clear that its relevance to economic problems comes in through its important influence on the schedule of the marginal efficiency of capital. There are not two separate factors affecting the rate of investment, namely, the schedule of the marginal efficiency of capital and the state of confidence. The state of confidence is relevant because it is one of the major factors determining the former, which is the same thing as the investment demand schedule.

There is, however, not much to be said about the state of confidence *a priori*. Our conclusions must mainly depend upon the actual observation of markets and business psychology.

For convenience of exposition we shall assume in the following discussion of the state of confidence that there are no changes in the rate of interest; and we shall write, throughout the following sections, as if changes in the values of investments were solely due to changes in the expectation of their prospective yields and not at all to changes in the rate of interest at which these prospective yields are capitalised. The effect of changes in the rate of interest is, however, easily superimposed on the effect of changes in the state of confidence.

The outstanding fact is the extreme precariousness of the basis of knowledge on which our estimates of prospective yield have to be made. Our knowledge of the factors which will govern the yield of an investment some years hence is usually very slight and often negligible. If we speak frankly, we have to admit that our basis of knowledge for estimating the yield ten years hence of a railway, a copper mine, a textile factory, the goodwill of a patent medicine, an Atlantic liner, a building in the City of London amounts to little and sometimes to nothing; or even five years hence. In fact, those who seriously attempt to make any such estimate are often so much in the minority that their behaviour does not govern the market.

In former times, when enterprises were mainly owned by those who undertook them or by their friends and associates, investment depended on a sufficient supply of individuals of sanguine temperament and constructive impulses who embarked on business as a way of life, not really relying on a precise calculation of prospective profit. The affair was partly a lottery, though with the ultimate result largely governed by whether the abilities and character of the managers were above or below the average. Some would fail and some would succeed. But even after the event no one would know whether the average results in terms of the sums invested had exceeded, equalled or fallen short of the prevailing rate of interest; though, if we exclude the exploitation of natural resources and monopolies, it is probable that the actual average results of investments, even during periods of progress and prosperity, have disappointed the hopes which prompted them. Business men play a mixed game of skill and chance, the average results of which to the players are not known by those who take a hand. If human nature felt no temptation to take a chance, no satisfaction (profit apart) in constructing a factory, a railway, a mine or a farm, there might not be much investment merely as a result of cold calculation.

Decisions to invest in private business of the old-fashioned type were, however, decisions largely irrevocable, not only for the community as a whole, but also for the individual. With the separation between ownership and management which prevails today and with the development of organised investment markets, a new factor of great importance has entered in, which sometimes facilitates investment but sometimes adds greatly to the stability of the system. In the absence of security markets, there is no object in frequently attempting to revalue an investment to which we are committed. But the Stock Exchange revalues many investments every day and the revaluations give a frequent opportunity to the individual (though not to the community as a whole) to revise his commitments. It is as though a farmer, having tapped his barometer after breakfast, could decide to remove his capital from the farming business between 10 and 11 in the morning and reconsider whether he should return to it later in the week. But the daily revaluations of the Stock Exchange, though they are primarily made to facilitate transfers of old investments between one individual and another, inevitably exert a decisive influence on the rate of current investment. For there is no sense in building up a new enterprise at a cost greater than that at which a similar existing enterprise can be purchased; whilst there is an inducement to spend on a new project what may seem an extravagant sum, if it can be floated off on the Stock Exchange at an immediate profit.[2] Thus certain classes of investment are governed by the average expectation of those who deal on the Stock Exchange as revealed in the price of shares, rather than by the genuine expectations of the professional entrepreneur.[3] How then are these highly significant daily, even hourly, revaluations of existing investments carried out in practice?

In practice we have tacitly agreed, as a rule, to fall back on what is, in truth, a *convention*. The essence of this convention – though it does not, of course, work out quite so simply – lies in assuming that the existing state of affairs will continue indefinitely, except insofar as we have specific reasons to expect a change. This does not mean that we really believe that the existing state of affairs will continue indefinitely. We know from extensive experience that this is most unlikely. The actual results of an investment over a long term of years very seldom agree with the initial expectation. Nor can we rationalise our behaviour by arguing that to a man in a state of ignorance errors in either direction are equally probable, so that there remains a mean actuarial expectation based on equi-probabilities. For it can easily be shown that the assumption of arithmetically equal probabilities based on a state of ignorance leads to absurdities. We are assuming, in effect, that the existing market valuation, however arrived at, is uniquely *correct* in relation to our existing knowledge of the facts which will influence the yield of the investment, and that it will only change in proportion to changes in this knowledge; though, philosophically speaking, it cannot be uniquely correct, since our existing knowledge does not provide a sufficient basis for a calculated mathematical expectation. In point of fact, all sorts of considerations enter into the market valuation which are in no way relevant to the prospective yield.

Nevertheless the above conventional method of calculation will be compatible with a considerable measure of continuity and stability in our affairs, *so long as we can rely on the maintenance of the convention*.

For if there exist organised investment markets and if we can rely on the maintenance of the convention, an investor can legitimately encourage himself with the idea that the only risk he runs is that of a genuine change in the news *over the near future*, as to the likelihood of which he can attempt to form his own judgement, and which is unlikely to be very large. For, assuming that the convention holds good, it is only these changes which can affect the value of his investment, and he need not lose his sleep merely because he has not any notion what his investment will be worth ten years hence. Thus investment becomes reasonably "safe" for the individual investor over short periods, and hence over a succession of short periods however many, if he can fairly rely on there being no breakdown in the convention and on therefore having an opportunity to revise his judgement and change his investment, before there has been time for much to happen. Investments which are "fixed" for the community are thus made "liquid" for the individual.

It has been, I am sure, on the basis of some such procedure as this that our leading investment markets have been developed. But it is not surprising that a convention, in an absolute view of things so arbitrary, should have its weak points. It is its precariousness which creates no small part of our contemporary problem of securing sufficient investment.

Some of the factors which accentuate this precariousness may be briefly mentioned.

(1) As a result of the gradual increase in the proportion of the equity in the community's aggregate capital investment which is owned by persons who do not manage and have no special knowledge of the circumstances, either actual or prospective, of the business in question, the element of real knowledge in the valuation of investments by those who own them or contemplate purchasing them has seriously declined.

(2) Day-to-day fluctuations in the profits of existing investments, which are obviously of an ephemeral and non-significant character, tend to have an altogether excessive, and even an absurd, influence on the market. It is said, for example, that the shares of American companies which manufacture ice tend to sell at a higher price in summer when their profits are seasonally high than in winter when no one wants ice. The recurrence of a bank holiday may raise the market valuation of the British railway system by several million pounds.

(3) A conventional valuation which is established as the outcome of the mass psychology of a large number of ignorant individuals is liable to change violently as the result of a sudden fluctuation of opinion due to factors which do not really make much difference to the prospective yield, since there will be no strong roots of conviction to hold it steady. In abnormal times in particular, when the hypothesis of an indefinite continuance of the existing state of affairs is less plausible than usual even though there are no express grounds to anticipate a definite change, the market will be subject to waves of optimistic and pessimistic sentiment, which are unreasoning and yet in a sense legitimate where no solid basis exists for a reasonable calculation.

(4) But there is one feature in particular which deserves our attention. It might have been supposed that competition between expert professionals, possessing judgement and knowledge beyond that of the average private investor, would correct the vagaries of the ignorant individual left to himself. It happens, however, that the energies and skill of the professional investor and speculator are mainly occupied otherwise. For most of these persons are, in fact, largely concerned, not with making superior long-term forecasts of the probable yield of an investment over its whole life, but with foreseeing changes in the conventional basis of valuation a short time ahead of the general public. They are concerned not with what an investment is really worth to a man who buys it "for keeps" but with what the market will value it at, under the influence of mass psychology, three months or a year hence. Moreover this behaviour is not

the outcome of a wrong-headed propensity. It is an inevitable result of an investment market organised along the lines described. For it is not sensible to pay 25 for an investment of which you believe the prospective yield to justify a value of 30, if you also believe that the market will value it at 20 three months hence.

Thus the professional investor is forced to concern himself with the anticipation of impending changes, in the news or in the atmosphere, of the kind by which experience shows that the mass psychology of the market is most influenced. This is the inevitable result of investment markets organised with a view to so-called "liquidity". Of the maxims of orthodox finance none, surely, is more anti-social than the fetish of liquidity, the doctrine that it is a positive virtue on the part of investment institutions to concentrate their resources upon the holding of "liquid" securities. It forgets that there is no such thing as liquidity of investment for the community as a whole. The social object of skilled investment should be to defeat the dark forces of time and ignorance which envelop our future. The actual, private object of the most skilled investment today is "to beat the gun", as the Americans so well express it – to outwit the crowd, and to pass the bad, or depreciating, half-crown to the other fellow.

This battle of wits to anticipate the basis of conventional valuation of an investment hence, rather than the prospective yield of an investment over a long term of years, does not even require gulls amongst the public to feed the maws of the professional – it can be played by professionals amongst themselves. Nor is it necessary that anyone should keep his simple faith in the conventional basis of valuation having any genuine long-term validity. For it is, so to speak, a game of Snap, of Old Maid, of Musical Chairs – a pastime in which he is victor who says "Snap" neither too soon nor too late, who passes the Old Maid to his neighbour before the game is over, who secures a chair for himself when the music stops. These games can be played with zest and enjoyment, though all the players know that it is the Old Maid which is circulating, or that when the music stops some of the players will find themselves unseated.

Or, to change the metaphor slightly, professional investment may be likened to those newspaper competitions in which the competitors have to pick out the six prettiest faces from a hundred photographs, the prize being awarded to the competitor whose choice most nearly corresponds to the average preferences of the competitors as a whole; so that each competitor has to pick, not those faces which he himself finds prettiest, but those which he thinks likeliest to catch the fancy of the other competitors, all of whom are looking at the problem from the same point of view. It is not a case of choosing those which, to the best of one's judgement, are really the prettiest, nor even those which average opinion genuinely thinks the prettiest. We have reached the third degree where we devote our intelligences to anticipating what average opinion expects the average opinion to be. And there are some, I believe, who practise the fourth, fifth and higher degrees.

If the reader interjects that there must surely be large profits to be gained from the other players in the long run by a skilled individual who, unperturbed by the prevailing pastime, continues to purchase investments on the best genuine long-term expectations he can frame, he must be answered, first of all, that there are, indeed, such serious-minded individuals and that it makes a vast difference to an investment market whether or not they predominate in their influence over the game-players. But we must also add that there are several factors which jeopardise the predominance of such individuals in modern investment markets. Investment based on genuine long-term expectation is so difficult today as to be scarcely practicable. He who attempts it must surely lead much more laborious days and run greater risks than he who tries to guess better than the crowd how the crowd will behave; and, given equal intelligence, he may make more disastrous mistakes. There is no clear evidence from experience that the investment policy which is socially advantageous coincides with that which is most profitable. It needs *more* intelligence to defeat the forces of time and our ignorance of the future than to beat the gun. Moreover, life is not long enough – human nature desires quick results: there is a peculiar zest in making money quickly, and remoter gains are discounted by the average man at a very high rate. The game of professional investment is intolerably boring and overexacting to anyone who is entirely exempt from the gambling instinct; whilst he who has it must pay to this propensity the appropriate toll. Furthermore, an investor who proposes to ignore near-term market fluctuations needs greater resources for safety and must not operate on so large a scale, if at all, with borrowed money – a further reason for the higher return from the pastime to a given stock of intelligence and resources. Finally it is the long-term investor, he who most promotes the public interest, who will in practice come in for most criticism, wherever investment funds are managed by committees or boards or banks.[4] For it is in the essence of his behaviour that he should be eccentric, unconventional and rash in the eyes of average opinion. If he is successful, that will only confirm the general belief in his rashness; and if in the short run he is unsuccessful, which is very likely, he will not receive much mercy. Worldly wisdom teaches that it is better for reputation to fail conventionally than to succeed unconventionally.

(5) So far we have had chiefly in mind the state of confidence of the speculator or speculative investor himself and may have seemed to be tacitly assuming that, if he himself is satisfied with the prospects, he has unlimited command over money at the market rate of interest. This is, of course, not the case. Thus we must also take account of the other facet of the state of confidence, namely, the confidence of the lending institutions towards those who seek to borrow from them, sometimes described as the

state of credit. A collapse in the price of equities, which has had disastrous reactions on the marginal efficiency of capital, may have been due to the weakening either of speculative confidence or of the state of credit. But whereas the weakening of either is enough to cause a collapse, recovery requires the revival of *both*. For whilst the weakening of credit is sufficient to bring about a collapse, its strengthening, though a necessary condition of recovery, is not a sufficient condition.

These considerations should not lie beyond the purview of the economist. But they must be relegated to their right perspective. If I may be allowed to appropriate the term *speculation* for the activity of forecasting the psychology of the market, and the term *enterprise* for the activity of forecasting the prospective yield of assets over their whole life, it is by no means always the case that speculation predominates over enterprise. As the organisation of investment markets improves, the risk of the predominance of speculation does, however, increase. In one of the greatest investment markets in the world, namely, New York, the influence of speculation (in the above sense) is enormous. Even outside the field of finance, Americans are apt to be unduly interested in discovering what average opinion believes average opinion to be; and this national weakness finds its nemesis in the stock market. It is rare, one is told, for an American to invest, as many Englishmen still do, "for income"; and he will not readily purchase an investment except in the hope of capital appreciation. This is only another way of saying that, when he purchases an investment, the American is attaching his hopes, not so much to its prospective yield, as to a favourable change in the conventional basis of valuation, ie, that he is, in the above sense, a speculator. Speculators may do no harm as bubbles on a steady stream of enterprise. But the position is serious when enterprise becomes the bubble on a whirlpool of speculation. When the capital development of a country becomes a by-product of the activities of a casino, the job is likely to be ill-done. The measure of success attained by Wall Street, regarded as an institution of which the proper social purpose is to direct new investment into the most profitable channels in terms of future yield, cannot be claimed as one of the outstanding triumphs of *laissez-faire* capitalism – which is not surprising, if I am right in thinking that the best brains of Wall Street have been in fact directed towards a different object.

These tendencies are a scarcely avoidable outcome of our having successfully organised "liquid" investment markets. It is usually agreed that casinos should, in the public interest, be inaccessible and expensive. And perhaps the same is true of Stock Exchanges. That the sins of the London Stock Exchange are less than those of Wall Street may be due, not so much to differences in national character, as to the fact that to the average Englishman Throgmorton Street is, compared with Wall Street to the average American, inaccessible and very expensive. The jobber's "turn", the high brokerage charges and the heavy transfer tax payable to the Exchequer, which attend dealings on the London Stock Exchange, sufficiently diminish the liquidity of the market (although the practice of fortnightly accounts operates the other way) to rule out a large proportion of the transactions characteristic of Wall Street.[5] The introduction of a substantial Government transfer tax on all transactions might prove the most serviceable reform available, with a view to mitigating the predominance of speculation over enterprise in the United States.

The spectacle of modern investment markets has sometimes moved me towards the conclusion that to make the purchase of an investment permanent and indissoluble, like marriage, except by reason of death or other grave cause might be a useful remedy for our contemporary evils. For this would force the investor to direct his mind to the long-term prospects and to those only. But a little consideration of this expedient brings us up against a dilemma, and shows us how the liquidity of investment markets often facilitates, though it sometimes impedes, the course of new investment. For the fact that each individual investor flatters himself that his commitment is "liquid" (though this cannot be true for all investors collectively) calms his nerves and makes him much more willing to run a risk. If individual purchases of investments were rendered illiquid, this might seriously impede new investment, so long as *alternative ways* in which to hold his savings are available to the individual. This is the dilemma. So long as it is open to the individual to employ his wealth in hoarding or lending *money*, the alternative of purchasing actual capital assets cannot be rendered sufficiently attractive (especially to the man who does not manage the capital assets and knows very little about them), except by organising markets wherein these assets can be easily realised for money.

The only radical cure for the crises of confidence which afflict the economic life of the modern world would be to allow the individual no choice between consuming his income and ordering the production of the specific capital-asset which, even though it be on precarious evidence, impresses him as the most promising investment available to him. It might be that, at times when he was more than usually assailed by doubts concerning the future, he would turn in his perplexity towards more consumption and less new investment. But that would avoid the disastrous, cumulative and far-reaching repercussions of its being open to him, when thus assailed by doubts, to spend his income neither on the one nor on the other.

Those who have emphasised the social dangers of the hoarding of money have, of course, had something similar to the above in mind. But they have overlooked the possibility that the phenomenon can occur without any change, or at least any commensurate change, in the hoarding of money.

Even apart from the instability due to speculation, there is the instability due to the characteristic of human nature that a large proportion of our positive activities depends on spontaneous optimism rather than on a mathematical expectation, whether moral or hedonistic or economic. Most, probably, of our decisions to do something positive, the full consequences of which will be drawn out over many days to come, can only be taken as a result of animal spirits – of a spontaneous urge to action rather than inaction, and not as the outcome of a weighted average of quantitative benefits multiplied by quantitative probabilities. Enterprise only pretends to itself to be mainly actuated by the statements in its own prospectus, however candid and sincere. Only a little more than an expedition to the South Pole, is it based on an exact calculation of benefits to come. Thus if the animal spirits are dimmed and the spontaneous optimism falters, leaving us to depend on nothing but a mathematical expectation, enterprise will fade and die – though fears of loss may have a basis no more reasonable than hopes of profit had before.

It is safe to say that enterprise which depends on hopes stretching into the future benefits the community as a whole. But individual initiative will only be adequate when reasonable calculation is supplemented and supported by animal spirits, so that the thought of ultimate loss which often overtakes pioneers, as experience undoubtedly tells us and them, is put aside as a healthy man puts aside the expectation of death.

This means, unfortunately, not only that slumps and depressions are exaggerated in degree but that economic prosperity is excessively dependent on a political and social atmosphere which is congenial to the average business man. If the fear of a Labour Government or a New Deal depresses enterprise, this need not be the result either of a reasonable calculation or of a plot with political intent – it is the mere consequence of upsetting the delicate balance of spontaneous optimism. In estimating the prospects of investment, we must have regard, therefore, to the nerves and hysteria and even the digestions and reactions to the weather of those upon whose spontaneous activity it largely depends.

We should not conclude from this that everything depends on waves of irrational psychology. On the contrary, the state of long-term expectation is often steady, and, even when it is not, the other factors exert their compensating effects. We are merely reminding ourselves that human decisions affecting the future, whether personal or political or economic, cannot depend on strict mathematical expectation, since the basis for making such calculations does not exist; and that it is our innate urge to activity which makes the wheels go round, our rational selves choosing between the alternatives as best we are able, calculating where we can, but often falling back for our motive on whim or sentiment or chance.

There are, moreover, certain important factors which somewhat mitigate in practice the effects of ignorance of the future. Owing to the operation of compound interest combined with the likelihood of obsolescence with the passage of time, there are many individual investments of which the prospective yield is legitimately dominated by the return of the comparatively near future. In the case of the most important class of very long-term investments, namely buildings, the risk can be frequently transferred from the investor to the occupier, or at least shared between them, by means of long-term contracts, the risk being outweighed in the mind of the occupier by the advantages of continuity and security of tenure. In the case of another important class of long-term investments, namely public utilities, a substantial proportion of the prospective yield is practically guaranteed by monopoly privileges coupled with the right to charge such rates as will provide a certain stipulated margin. Finally there is a growing class of investments entered upon by, or at the risk of, public authorities, which are frankly influenced in making the investment by a general presumption of there being prospective social advantages from the investments, whatever its commercial yield may prove to be within a wide range, and without seeking to be satisfied that the mathematical expectation of the yield is at least equal to the current rate of interest – though the rate which the public authority has to pay may still play a decisive part in determining the scale of investment operations which it can afford.

Thus after giving full weight to the importance of the influence of short-period changes in the state of long-term expectation as distinct from changes in the rate of interest, we are still entitled to return to the latter as exercising, at any rate, in normal circumstances, a great, though not a decisive, influence on the rate of investment. Only experience, however, can show how far management of the rate of interest is capable of continuously stimulating the appropriate volume of investment.

For my own part I am now somewhat sceptical of the success of a merely monetary policy directed towards influencing the rate of interest. I expect to see the State, which is in a position to calculate the marginal efficiency of capital-goods on long views and on the basis of the general social advantage, taking an ever greater responsibility for directly organising investment; since it seems likely that the fluctuations in the market estimation of the marginal efficiency of different types of capital, calculated on the principles I have described above, will be too great to be offset by any practicable changes in the rate of interest.

1 By "very uncertain" I do not mean the same thing as "very improbable". Cf. my *Treatise on Probability*, chapter 6, on "The Weight of Arguments".
2 In my *Treatise on Money* (vol. ii, p. 195) I pointed out that when a company's shares are quoted very high so that it can raise more capital by issuing more shares on favourable terms, this has the same effect as if it could borrow at a low rate of interest. I should now describe this by saying that a high quotation for existing equities involves an increase in the marginal

efficiency of the corresponding type of capital and therefore has the same effect (since investment depends on a comparison between the marginal efficiency of capital and the rate of interest) as a fall in the rate of interest.

3 This does not apply, of course, to classes of enterprise which are not readily marketable or to which no negotiable instrument closely corresponds. The categories falling within this exception were formerly extensive. But measured as a proportion of the total value of new investment, they are rapidly declining in importance.

4 The practice, usually considered prudent, by which an investment trust or an insurance office frequently calculates not only the income from its investment portfolio but also its capital valuation in the market may also tend to direct too much attention to short-term fluctuations in the latter.

5 It is said that, when Wall Street is active, at least a half of the purchases or sales of investments are entered upon with an intention on the part of the speculator to reverse them *the same day*. This is often true of the commodity exchanges also.

13

North Trading Room, Chicago Board of Trade, 1930
Courtesy of Chicago Board of Trade

14
Old South Trading Room, Chicago Board of Trade, 1930
Courtesy of Chicago Board of Trade

15
Art Deco Chandelier, North Trading Room Ceiling, 1930
Courtesy of Chicago Board of Trade

16
First CBOE trading floor, c. 1973
Courtesy of Chicago Board Options Exchange

17 Early days of construction to build CBOE's second trading floor, with CBOT's North Trading Room, 1974
Courtesy of Chicago Board Options Exchange

18
Chicago Mercantile Exchange trading floor, 1953
Courtesy of Chicago Mercantile Exchange Archives

19

London's Royal Exchange Building, first home to the London International Financial Futures and Options Exchange (LIFFE), from 1982 to 1991

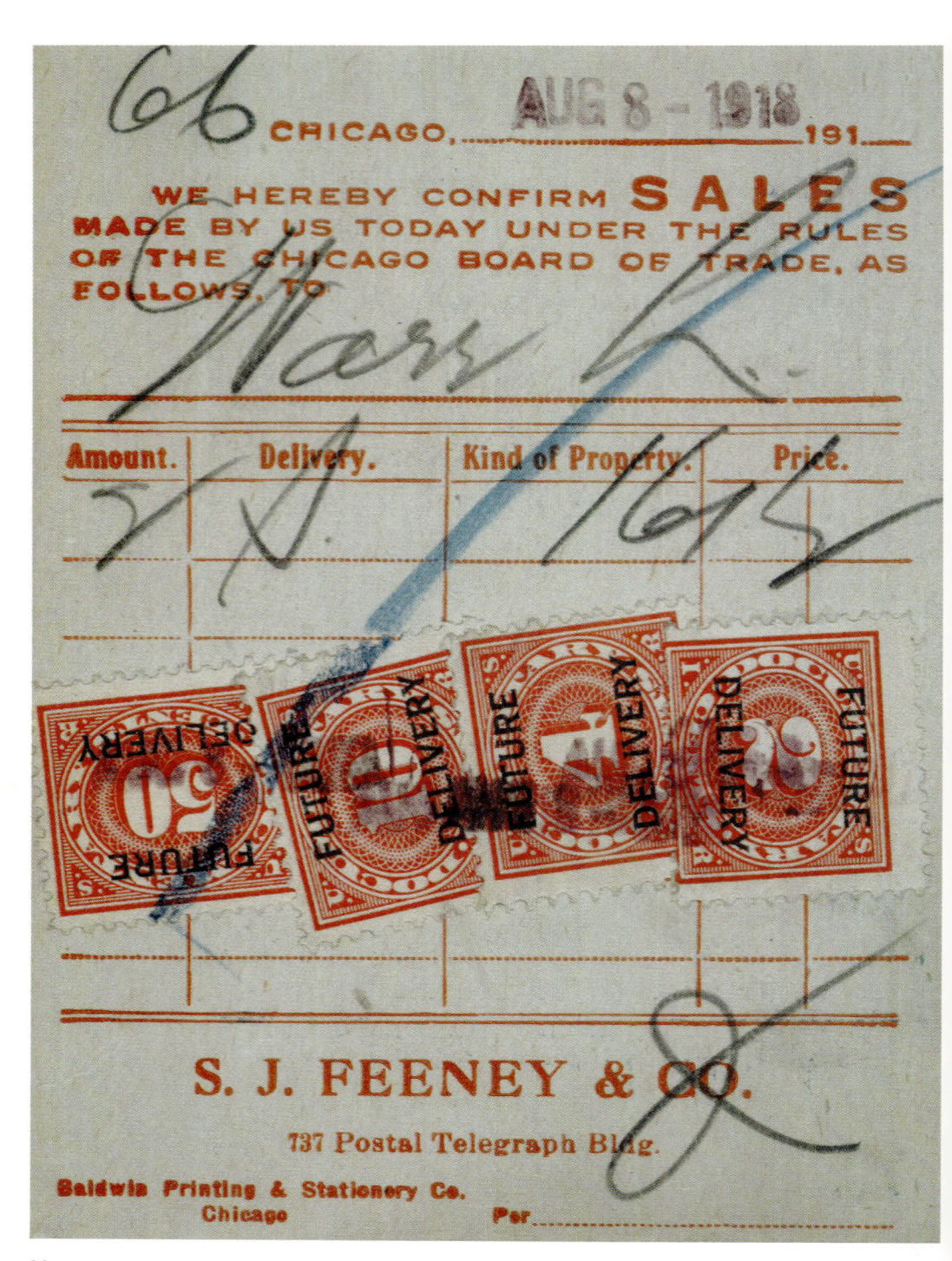

20
Old Trading Card, Chicago Board of Trade, 1918
Courtesy of Dennis A. Dutterer

21
New Electronic Trading Card, Chicago Mercantile Exchange, 1999
Courtesy of Chicago Mercantile Exchange Archives

22
Hand Signals in Open-Outcry Trading Pits of the London International Finanancial Futures and Options Exchange (LIFFE) at Cannon Bridge, 1995

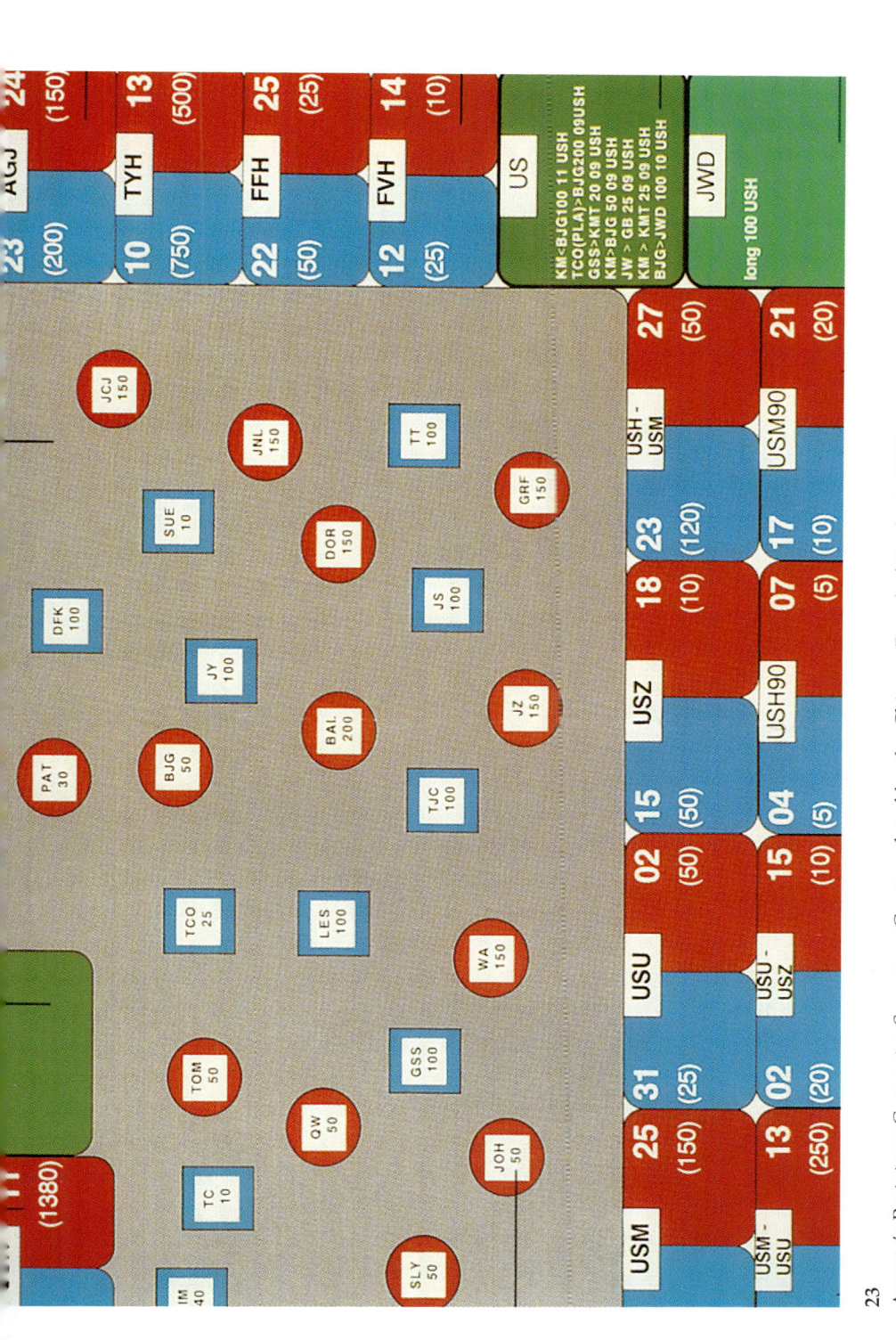

Aurora's Prototype Computer Screen, as Contemplated by the Chicago Board of Trade, in 1989
Courtesy of Chicago Board of Trade

24
The original trading ring of what was then known as the Sydney Greasy Wool Futures Exchange, during the 1960s
Courtesy of Sydney Futures Exchange

18

*Speculation, Profitability, and Stability**

William J. Baumol
New York University

Proponents of flexible exchange rates have maintained that a completely free exchange market is very likely to be stable. In particular they have argued that any profitable speculative activity in this and other markets must necessarily be stabilising. By this they appear to mean that it must, *ceteris paribus*, reduce the frequency and amplitude of price fluctuations. In this note, I dispute this allegedly universal proposition with the aid of a counter-example. Certainly this counter-example is not meant to suggest that profitable (or even unprofitable) speculation will never exert a stabilising influence. How often and to what extent speculation is stabilising remains a matter for empirical inquiry.

Perhaps a more important aim of this note is to indicate the sort of mathematical apparatus which is necessary for an analysis of the effects of speculation on stability. The techniques are precisely those which have been used in other stability analyses, and it is surprising that they do not seem to have been employed in this area. Because most of the mathematical analysis of speculation and stability has been conducted in static terms, it has failed to get to the heart of the stability question which, of course, refers to properties of the price *movements*.

NATURE OF THE ARGUMENTS

It is easy to recapitulate the basic argument which maintains that profitable speculation is necessarily stabilising. In Professor Friedman's relatively guarded words, "People who argue that speculation is generally destabilising seldom realise that this is largely equivalent to saying that speculators lose money, since speculation can be destabilising in general

---

*This paper was first published in the *Review of Economics and Statistics* 39(3), pp. 263–71 (1957) and is reprinted with the permission of the MIT Press. The author wishes to thank Professors Chandler, Dorfman, Friedman and Viner for their comments.

only if speculators on the average sell when the currency is low in price and buy when it is high".[1]

Certainly this position conflicts with what may, depending on the point of view, be described as our commonsense or our preconceived views. There is however a counter-argument.[2] This maintains that speculative profits characteristically are earned by selling *after* the price peak has been passed and by buying after the beginning of the upturn. This is because speculators know they cannot foretell the future with accuracy, and so can only hope to identify price peaks and troughs in retrospect after the price trend has been well established. By doing so they give up any chance to skim off the cream but hope in return significantly to reduce their risks. The occurrence of such speculative patterns is not implausible. Certainly, for example, this aim is inherent in the idea of buying and selling in accord with the Dow Jones indicators and in some of the other investment formulas, though there may be some question about the extent to which these schemes succeed in realising their goal.

The main point is that speculation of this variety involves purchases during the upswing and sales during the downswing. It will have some stabilising influence in that, if profitable, it involves higher priced sales than purchases, thereby forcing the higher prices down and vice versa. But it must also have a destabilising influence in accelerating both upward and downward movements because speculative sales occur when prices are falling, and purchases are made when prices have begun to rise. For this reason the speculative activity may be profitable, yet be on balance destabilising.[3]

Several questions must be answered before this argument can be accepted.

1. What precisely is the difference between a speculator and a non-speculator? Are not those to whom the profit makers sell and from whom they purchase also in some sense speculators? If so, the preceding argument breaks down, because it is not really true that speculation is on balance profitable in this situation. Rather, it just amounts to some more skilful speculators profiting at the expense of others.
2. The sort of speculative pattern just considered has both a stabilising influence, in that its sales occur at a higher price than its purchases, and a destabilising influence, in that it accelerates price movements. Can the destabilising influence ever predominate?
3. Can the acceleration of downward and upward movements, even if it is the dominating influence, ever increase the amplitude and frequency of fluctuations, or will it simply result in lengthened price plateaux at the old peak and trough levels with a reduction in the time taken in moving from peak to trough?

I shall dodge the first question in the next few sections and only deal with it indirectly in the penultimate section. Certainly there are no absolute definitions available to the positivist to settle the matter once and for all. It may be pointed out that a similar problem exists for the argument which maintains that profitable speculation must be stabilising. Indeed, in a market which trades a fixed stock of items (eg, securities) among a fixed group of traders it is a tautology that net money profit (meaning net cash withdrawal) of the group is zero. Whatever one group of traders gains another must lose, and it is necessary somehow to define the second group to consist of non-speculators. In most of the present paper this problem will be evaded by assuming that there exists a group of non-speculators on some unspecified definition and that its activities somehow result in cyclical behaviour in the price of some commodity. The effect of the entrance of speculators into the market will then be examined.

It may be remarked, however, that for practical problems the answer to the first question may not be so very difficult. For the relevant dichotomy may not be between pure speculators and pure non-speculators, but rather it may involve conscious *versus* unconscious speculators or professional *versus* amateur speculators, or even pure speculators *versus* those whose market behaviour is not primarily influenced by speculative considerations. That professional or pure or conscious speculators can profit at the expense of the hybrid residual groups, and do so in a destabilising manner is, I think, conclusively shown by the models which follow.

The purpose of the models is to provide a construction which permits measurement of the amplitude and frequency of the price movements and thereby an unambiguous answer to questions 2 and 3.

MARKET IN THE ABSENCE OF SPECULATORS

It is simply assumed here that the pattern of prices of the commodity in question is endogenously determined and that the time path of prices is perfectly cyclical, ie, sinusoidal and of constant amplitude. This time path can be represented by a second-order difference equation,[4]

$$P_t = 2aP_{t-1} - P_{t-2} + k \qquad |a| < 1 \qquad (1)$$

It is also convenient to assume $a > 0$, for reasons which will be seen presently.

The solution of this equation is

$$P_t = c \cos qt + s \sin qt + R = p \cos(qt + r) + R \qquad (2)$$

where R is a constant which represents the mean level of prices; c, s, p and

r are constants determined by initial conditions; and q is an angle given by $\cos q = a < 1$.[5]

As already mentioned, I assume that the behaviour of non-speculative traders is of such a variety that this time path will result.[6] However, it is possible to give some sort of meaningful if unconvincing economic interpretation of (1). Aside from the inherent desirability of such an interpretation it will be needed to decide how to bring the speculative elements into the model.

Suppose the non-speculative excess demand function (quantity demanded minus quantity supplied as a function of price) is

$$E_t = K - UP_t + V(P_t - P_{t-1}) + W(P_{t-1} - P_{t-2}) \tag{3}$$

where W is a positive constant; and V, U and K are constants given by

$V = W(1 - 2a)$, which is positive for $a < \frac{1}{2}$,

$U = 2W(1 - a)$ and

$K = Wk$

Direct substitution of these values into (3) together with the equilibrium condition $E_t = 0$ at once yields our basic equation (1).[7]

Equation (3) states that excess demand is dependent on current price and recent price trends. The dependence on current price is given by $A - UP_t$, where A is some positive constant. This says that an increase in price will result in a linear decrease in excess demand. The dependence on recent price trends is expressed in $K - A + V(P_t - P_{t-1}) + W(P_{t-1} - P_{t-2})$ and states that rising recent price trends make for high excess demands.[8]

SPECULATIVE BEHAVIOUR

It is now necessary to formulate a mathematical description of the speculative behaviour postulated in the first section. The idea is that speculators will do the bulk of their buying right *after* the upturn and their selling right after the downturn. The trough and the subsequent upturn are characterised by a downward movement followed by an upward movement. If the trough is at t, $(P_{t+1} - P_t)$ and $-(P_t - P_{t-1})$ will then both be positive. Similarly if P_{t+1} is the price of the first period after a downturn, both of these expressions will be negative. At all other times these expressions will be of opposite sign. This suggests that a speculative excess demand function given by

$$E_{st+1} = C[(P_{t+1} - P_t) - (P_t - P_{t-1})] = C(P_{t+1} - 2P_t + P_{t-1}) \tag{4}$$

where C is a positive constant, will have the desired properties. I shall prove that this is so. First note that (3) and (4) together with the equilibrium condition $E_t + E_{st} = 0$ give

$$0 = K - WP_t + 2WaP_{t-1} - WP_{t-2} + C(P_t - 2P_{t-1} + P_{t-2})$$
$$= K - (W - C)P_t + (2Wa - 2C)P_{t-1} - (W - C)P_{t-2}$$

From this follows

$$P_t = \frac{K}{W - C} + 2\frac{Wa - C}{W - C}P_{t-1} - P_{t-2} \qquad (5)$$

If $Wa > C$ (so that certainly $W > C$), ie, if speculative demand is not too large relative to non-speculative demand, both fractions in (5) will be positive and the second fraction will be less than unity. Thus the equation will be of precisely the same form as (1), and this too will involve a sinusoidal constant-amplitude time path for price.

We now derive two properties of our model:

Property 1: Net speculative purchases of the commodity in question are given by a sinusoidal curve of constant amplitude and the duration of each cycle is equal to that of the cycle in prices. Maximum purchases occur one period after the upturn in prices, and maximum sales occur one period after the price downturn.

Property 2: The speculative behaviour described by (4) is profitable provided the cycle is more than four periods long.

Proof of property 1: Since the form of (5) is precisely the same as that of (1), expression (2) is the solution of (5) for appropriate values of the constants in (2). That is, the time path of prices is once again given by (2). To see how this price behaviour affects speculative excess demand, substitute expression (2) for P_t into the speculative excess demand function (4). We then have

$$E_{st+1} = C[p\cos(qt + q + r) - 2p\cos(qt + r) + p\cos(qt - q + r)]$$
$$= C[p\cos(qt + r)\cos q - p\sin(qt + r)\sin q - 2p\cos(qt + r)$$
$$+ p\cos(qt + r)\cos q + p\sin(qt + r)\sin q]$$

so that, collecting terms,[9]

$$E_{st+1} = 2pC(\cos q - 1)\cos(tq + r) \qquad (6)$$

Here $2pC(\cos q - 1)$ is a negative constant since the cosine of an angle is usually less than unity, and indeed, by elementary difference equation analysis we have from (5) $\cos q = (Wa - C)/(W - C) < 1$.

Comparing this result with (2) we see that $E_{st+1} = -D(P_t - R)$, where D is a positive constant. This proves the first property,[10] since it states essentially that speculative excess demands fluctuate inversely with prices *one period earlier* (note the time subscripts of E and P) and are inversely proportionate with the deviations of the past periods' prices from the mean price level.

Proof of property 2: Speculative profit during period t is the excess supply by speculators (quantity sold minus quantity bought)[11] multiplied by the price, ie, it is, by (2) and (6)

$$-P_t E_{st} = [p\cos(qt + r) + R][D\cos(qt - q + r)]$$

where $D = -2pC(\cos q - 1) > 0$ [see (6)].

Thus profit over the cycle is the sum of these single-period revenues taken over the entire cycle beginning at period T:

$$\sum_{t=T}^{T+(360/q)-1} [pD(\cos qt + r)(\cos qt - q + r) + RD\cos(qt - q + r)]$$

I assume here that $360/q$ is an even integer. Consider first the second term in this expression. Write $G = qT + q + r$. Then this second term is the sum

$$RD[\cos G + \cos(G + q) + \cos(G + 2q) \cdots$$
$$+ \cos(G + 180 - q) + \cos(G + 180)$$
$$+ \cos(G + q + 180) + \cos(G + 2q + 180)$$
$$+ \cdots + \cos(G + 360 - q)]$$

This sum is equal to zero because for any angle Q, $\cos Q = -\cos(180 + Q)$. Turning now to the first term of the expression for profits over the cycle and taking $qt + r = Q$, the sum of any two consecutive terms may be written

$$pD[\cos(Q + q)\cos Q + \cos Q \cos(Q - q)]$$
$$= pD[\cos^2 Q \cos q - \sin Q \sin q \cos Q$$
$$+ \cos^2 Q \cos q + \cos Q \sin Q \sin q]$$
$$= 2pD\cos^2 Q \cos q$$

Total speculative profits over the cycle are then

$$pD\cos q \sum_{x=1}^{180/q} 2\cos^2(qT + 2xq + r)$$

which is positive if $0 < \cos q$, ie, if $q < 90°$, that is, if the cycle is longer than four periods.

INFLUENCE OF SPECULATORS ON THE CYCLE

It remains only to see how the postulated speculation affects the frequency and amplitude of the fluctuations. I now derive two further properties of the model.

Property 3: The frequency of the fluctuations is increased by the postulated speculative behaviour.

Property 4: The amplitude of the fluctuations is dependent on initial conditions and may or may not be increased.

Proof of property 3: There will be a complete cycle every time tq in (2) increases by 360°. That is, the cycle will be of length $t = 360/q$ and hence of frequency $q/360$. In the absence of speculation write $q = q_N$. Here $0 < \cos q_N = a < 1$. In the presence of speculation write $q = q_s$ so that $0 < \cos q_s = (Wa - C)/(W - C) < 1$. Since the fraction $(Wa - C)/(W - C)$ is smaller than a, $\cos q_s$ is less than $\cos q_N$, ie, $q_s/360 > q_N/360$. This really just confirms the intuitive notion that speculative purchases during the upswing and sales during the downswing will speed up the cyclical process.

Proof of property 4: This is obvious since, by (2), the amplitude of the fluctuations will be $2p$ whose magnitude depends on initial conditions. But a bit more can be said on this matter. Suppose that the speculators first observe the market in question to learn its behaviour and then enter that market. The time path in the non-speculative market then provides the initial conditions. I shall now show by two illustrations that even so both increases and decreases in amplitude are possible.

First some preliminaries. From (2) we may write

$$P_t = c \cos qt + s \sin qt + R$$

and

$$P_t = c' \cos q't + s' \sin q't + R$$

respectively for the non-speculative and speculative time paths, ie, for the respective solutions of (1) and (5). It is to be noted that the constant R is the same in both solutions.[12]

Set $t = 0$ in both cases and assume that the corresponding P_0s are the same (the first initial condition). We obtain $c = c' = (P_0 - R)$. If this initial price is at the mean value of the prices, ie, if $P_0 = R$ so that $c = c' = 0$, the time paths simplify to

$$P_t = s \sin qt + R \quad \text{and} \quad P_t = s' \sin q't + R$$

whose amplitudes are $2s$ and $2s'$ respectively.

Now for our two cases illustrating the effects of speculation on amplitude:

Case 1: Speculation decreases amplitude. Let the second initial condition for the speculative case be given by the P_1 of the non-speculative time path. Then since $t = 1$ we have[13]

$$s \sin q = s' \sin q'$$

or

$$s/s' = \sin q'/\sin q = \sqrt{1 - \left(\frac{Wa - C}{W - C}\right)^2} \Big/ \sqrt{1 - a^2} > 1$$

Case 2: Speculation increases amplitude: Let the second initial condition be given by the peak (or trough) of the non-speculative time path so that $qt = 90$. Then

$$s' \sin q' (90/q) = s \sin q (90/q) = s \sin 90 = s$$

or

$$s/s' = \sin q' \, 90/q < 1$$

since $0 < q'/q < 1$. This result has a simple intuitive explanation. The two sine curves which portray the time paths of prices in the speculative and non-speculative situations coincide at the mean price level, at A (Figure 1), and at the peak E of the (non-speculative) curve AC with the longer cycle. But at the latter point the other (speculative) curve, AB, since its cycle is shorter, must already be on its way down, ie, its peak must be higher than that of the non-speculative curve.

A similar restatement is possible for case 1. Here the two curves AC and AD meet two consecutive periods near the midpoint of the cycle, ie, they are roughly tangent in that vicinity. The speculative curve AD has the shorter cycle so it must reach its peak first and thereafter its peak will be at a lower level. Both curves start climbing at the same rate so the one which has the longer time in which to climb will reach the highest level.

A DIFFERENTIAL EQUATION MODEL

I outline briefly an analogous differential equation model because it permits me to illustrate another more impressive form which may be assumed by speculative destabilisation. Again I posit price behaviour in the absence of speculation to be perfectly periodic and of constant amplitude. It is then representable by the equation

$$P = R - a\ddot{P} \tag{7}$$

Figure 1

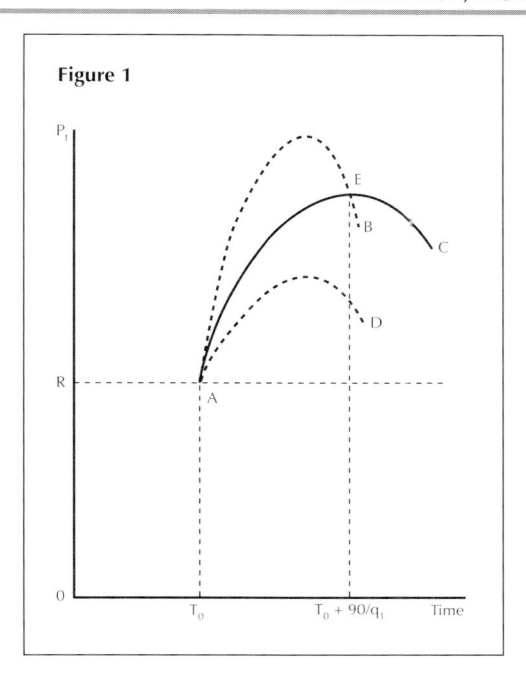

(Figure 1: axes P_t vs Time; horizontal dashed line at level R; curves emanating from point A at T_0 rising to peaks labeled E, B, D with point C to the right; vertical reference at $T_0 + 90/q_1$.)

where $\ddot{P}$ is the second derivative of P with respect to time, and a and R are positive constants.

Equation (7) can be taken to be derived from the following not entirely implausible non-speculative excess demand function

$$E = 0 = R - P - a\ddot{P} \qquad (8)$$

This says that the excess demand curve will be low near the trough when the second derivative is positive, and high in the vicinity of the peak. Such perverse behaviour seems rather characteristic in some markets!

I now assume speculative excess demand to be given by

$$E_s = U(R - P) + W\dot{P}, \quad U > 0, \quad W > 0, \quad \dot{P} \equiv dP/dt \qquad (9)$$

Roughly, this exhibits the features we desire because it has speculators concentrating their purchases when prices are simultaneously low ($P < R$) and rising ($\dot{P} > 0$). That is, they will be buying preponderantly during the beginning of the upswing, and, analogously, selling mainly during the beginning of the downswing.[14] It is also possible to show that this speculative activity can be profitable.[15]

By (8) and (9) the equilibrium condition $E + E_s = 0$ is now equivalent to

$$P(1 + U) = (1 + U)R + W\dot{P} - a\ddot{P}$$

or

$$P = R + \frac{W}{1 + U}\dot{P} - \frac{a}{1 + U}\ddot{P} \qquad (10)$$

which can be written

$$R + 2r\dot{P} - s\ddot{P}$$

This will have complex roots $r + \sqrt{r^2 - s} = r + qi$ and hence a cyclical time path provided $r^2 < s$, which will surely hold if $W^2 < a$, that is, provided the lag in speculative excess demand behind price movements is sufficiently small.

But there is an important difference between the solution of (10) and that of the non-speculative equation (7). The roots of the characteristic equation of (10) have a positive real part and the solution is

$$P = Ae^{rt}\cos(qt + B) + R, \qquad A \text{ and } B \text{ constants} \qquad (11)$$

These fluctuations are no longer of constant amplitude. Rather their amplitude will grow at a geometric rate. Thus profitable speculation has demonstrated its destabilising ability by knocking the system from its position of delicate constant amplitude balance into a time path of explosive fluctuation.[16]

A REAL NON-SPECULATIVE CYCLE

It will be recalled that both our non-speculative cycle models have involved excess demands which are influenced by price trends as well as by prices. This raises some doubts as to whether the traders in question can legitimately be classed as non-speculators. In a letter to the author, Professor Friedman has suggested that perhaps a non-speculator can only safely be defined (if this is done in terms of his demand curve) as one whose purchases are directly influenced by current prices but not by past prices or price trends. In the absence of speculation, so defined, cyclical price movements can only result from real influences. It may perhaps be conjectured that such a cycle would not be subject to any destabilising effects of profitable speculation, since the speculative influence on rates of price change would, by definition, not affect non-speculative behaviour.

I shall now argue, however, that even in such a situation speculation which is apparently profitable can be destabilising.

Suppose the supply S of some commodity varies sinusoidally, say, as the result of seasonal climatic changes, in accord with the equation

$$S = A - B\ddot{S} \quad (A, B \text{ constants}) \tag{12}$$

On Friedman's criterion the non-speculative demand function may take the form

$$D = K - VP \tag{13}$$

where K and V are constants and D and P respectively represent the quantity demanded and current price.

In the absence of speculation the equilibrium condition is $D = S$, which readily yields by substitution of $S = D = K - VP$ and $\ddot{S} = -V\ddot{P}$ into (12),

$$P = \frac{K - A}{V} - B\ddot{P}$$

This is of the same form as is (12) and involves constant amplitude sinusoidal price movements, as is to be expected.

What happens when speculation is imposed on this real cycle? The equilibrium condition (zero excess demand) is now

$$D - S + E_s = 0 \tag{14}$$

where E_s represents speculative excess demand, and is given by equation (9) in the preceding section. Substitution of this expression and (13) into (14) yields

$$S = K - VP + U(R - P) + W\dot{P}$$

which may be rewritten

$$S = C - EP + W\dot{P} \quad (C \text{ and } E \text{ constants})$$

Substituting this and the corresponding expression for $\ddot{S}$ into (12) we obtain the third-order differential equation

$$C - A = EP - W\dot{P} + BE\ddot{P} - BW\dddot{P}$$

Since complex roots come in pairs, at least one of the three roots of the characteristic equation must be real, and it cannot be negative since the terms are alternately positive and negative.[17] It follows that the time path of prices is changed by speculation from a cyclical pattern of constant amplitude into an unstable, explosive movement.

However, we must see whether this speculation can be profitable. Here the matter is not quite so clear-cut. In the short run, the time path may still be approximately cyclical, and the previous arguments apply. In the long run, the root whose real value is largest will determine the time path of P. If this is a complex root, again the analysis of the preceding section seems applicable. But if this root, r, is real, the time path of P will eventually approach ae^{rt}, where a is a constant. The speculator's excess demand as given by (9) of the previous section will then be

$$U(R - ae^{rt}) + Wrae^{rt}$$

This will be positive for appropriate values of W and U. In other words, speculators will continue to buy on a rising market and, at least in terms of the value of their assets, this would appear to be profitable. However, it is clear that difficulties can arise if speculators try to cash in these profits or if they run out of funds with which to continue their purchases.

STABILITY OF UNPEGGED FLEXIBLE EXCHANGE RATES

I digress from my central theme to recapitulate a recent discussion by Professor Viner.[18] I have just maintained that even if exchange markets would otherwise be stable, speculation may act as a destabiliser. By contrast Viner has indicated that even if exchange rate speculation were stabilising, the stability of flexible exchange rates is sometimes questionable. Together these two discussions then call into question much of the stability analysis of the proponents of flexible exchange rates.

Viner argues that the equilibrating forces which are ordinarily taken to be present in commodity markets cannot be expected to help stabilise unpegged exchange rates. In competitive markets, for example, the relative price of one consumers' good as against another is determined by costs of production and demands. If this price ratio is out of line with costs, capital will find it profitable to flow from the production of one item to the other and move prices back toward their equilibrium ratio in accord with the well-known analysis.

But the peculiarity of the exchange market is that the cost of production of the commodities traded is, for all practical purposes, nil. Governments or the central banking systems can, at nearly zero cost, expand their currencies at will, and frequently they do not resist that temptation. Moreover, a currency "... has no price ceiling or price floor derived from any direct utility as a consumers' good or any indirect utility as an ingredient, a technical coefficient, in the production of producers' or consumers' goods".[19]

If, for example, one country inflates indefinitely, the exchange rate of its currency against others can fall indefinitely unless the foreigner himself inflates with equal success. By counterbalancing inflations, a precarious and rough constancy of exchange rates might be maintained, but in the absence of any pegging mechanism, the coincidence could end at any moment.

This means also that there is little reason to expect any consistent pattern of cyclical price movement on these exchanges. Rather, the time path can plausibly be expected to be erratic, responding to the political developments in the countries involved. This may appear to limit the applicability of the arguments of the earlier part of this paper, because these arguments examined the effects of speculation on perfectly cyclical price movements; but it is possible to interpret any one of the cycles discussed as a never-to-be repeated erratic individual movement which may well approximate some actual exchange movements. These illustrations, as we have seen, are counter-examples to the assertion that profitable speculation is *necessarily* stabilising. If interpreted as single erratic movements rather than strings of regular cycles, these counter-examples may carry rather more conviction for the exchange market case.

The Viner analysis also explains my employment of linear cycle models. It is fairly easy to construct a non-linear "relaxation" cycle model in which a constant amplitude is not the produce of bizarre coincidence (see note 6 end). Characteristic of such a model is a sudden change in the basic relationships, as in the Hicks model, where attainment of full employment abruptly stops the working of the accelerator. In locating that turning point, the place where the relationships change, one has explicitly or implicitly built up a theory which explains the amplitude of the cycle. This would be very useful for our purposes, for such a theory would permit us to examine the effects of speculation on amplitude directly, and thus to grapple with an essential part of the stabilisation question.

But this I have been unable to do because I can see no way of constructing a systematic explanation of turning points of truly flexible exchange rates. If, as I believe, Viner is right, exchange rate movements will be erratic and these turning points will be erratic – or rather, they will be explainable only in terms of the political circumstances of the countries in question.

---

1 Milton Friedman, 1953, *Essays in Positive Economics*, p. 175 (Chicago). Friedman adds in a footnote: "A warning is perhaps in order that this is a simplified generalization on a complex problem. A full analysis encounters difficulties in separating 'speculative' from other transactions, defining precisely and satisfactorily 'destabilising speculation' and taking account of the effects of the mere existence of a system of flexible rates as contrasted with the effects of actual speculative transactions under such a system". For another such statement, see Friedrich A. Lutz, 1954, "The Case for Flexible Exchange Rates", *Banca*

Nazionale Del Lavoro, (31) pp. 30–2. The argument goes back much further. For further references see James A. Ross, Jr., 1938, *Speculation, Stock Prices and Industrial Fluctuations*, p. 127 footnote 1 and p. 134 footnote 19 (New York).

2 See eg, Ross, *op. cit.*, pp. 131, 134–38.

3 The main point of this counter-example then is that while the Friedman argument takes account of the levels of the variables, it neglects their time derivatives, and the time path is dependent on both. Note the analogy with Samuelson's criticism of the Hicks stability analysis, *Foundations of Economic Analysis*, 1947, pp. 269–76 (Cambridge).

4 It should be noted that no other second-order linear difference equation with constant coefficients and no lower-order equation of this variety will produce the sort of time path which is described above.

5 There are really two such angles since $\cos q = \cos -q$, but it is easy to see that this choice makes no difference to the time path indicated by the solution.

6 This involves some difficulties. It is well known that if the coefficient of P_{t-2} in (1) is ever so slightly different from -1 the price cycles will either explode or their amplitude will asymptotically approach zero. In the mathematician's language the values of the parameters of a linear difference equation which can produce cyclical behaviour of constant amplitude is a set of measure zero. As Friedman points out, it is an invalid use of inverse probability to conclude from this alone that such a constant-amplitude-equation situation is unlikely to be encountered in practice. (See Friedman, *op. cit.*, pp. 292–3.) But this excuse is rather lame here. If I were looking for a convincing construction rather than a counter-example, I would be driven to employ non-linear equations and models involving relaxation oscillations. See the final section below, for the reason I have not employed this type of model.

7 Further, it is easy to show that (3) is the only excess demand equation of this form which yields (1).

8 There is a fundamental difficulty here. Our non-speculators are influenced by price trends. Does this make speculators of the non-speculators? I know no really satisfactory answer. It can only be remarked that this influence need not be (conscious) and that, in any event, the above is only one possible interpretation of (1). We shall return to this problem below.

9 The second line in the derivation is obtained by use of the standard formula for the cosine of a sum of two angles: $\cos(A + B) = \cos A \cos B - \sin A \sin B$ and the properties $\sin(-A) = -\sin A$, $\cos(-A) = \cos A$.

10 The property is really much more general, in that it will hold for a much greater variety of time paths. It is easy to prove that if $P_t = f(t)$ and P_t attains a symmetric maximum or minimum at $t = T$, so that $f'(T - 1) = -f'(T + 1)$ then

$$dE_{st}/dt = 0 \quad \text{at} \quad t = T + 1$$

11 This is the same as quantity supplied minus quantity demanded since we are using an equilibrium model where total excess demand $E_t + E_{st} = 0$. Such a model is appropriate for most markets where there is organised speculation and the price rapidly adjusts to eliminate excess demands.

12 Proof: set $P_t = P_{t-1} = P_{t-2} = R$ in (1) to obtain $R(2 - 2a) = k$. Now make the same substitution in (5), getting $R[2(W - C) - 2(Wa - C)] = RW(2 - 2a) = K = kW$ by (3). Thus in both cases we have $R = k/2(1 - a)$.

13 The last line employs the relation $\sin B = \sqrt{1 - \cos^2 B}$ for any angle B.

14 This can be shown more rigorously. Differentiation of the solution (11) below for P yields

$$\dot{P} = Ae^{rt}[r \cos(qt + B) - q \sin(qt + B)]$$

Substituting this and expression (11) for P in (9) gives us

$$E_s = Ae^{rt}[(Wr - U)\cos(qt + B) - Wq(\sin qt + B)],$$

ie,

$$E_s = -KAe^{rt}\cos(qt + B - E)$$

where $K = \sqrt{(Wr - U)^2 + (Wq)^2}$, $\cos E = -(Wr - U)/K$ and $\sin E = Wq/K$. Set $U > rW$ in (9) so that all of these are positive and therefore $0 < E < \pi/2$. Comparing the last expression for E_s with (11) we see that speculative purchases are inversely related to price and lag behind price movements but by less than one quarter of a cycle, as desired.

15 Proof: Suppose first $W = 0$. By (9) speculative excess demand is now $E_s = U(R - P)$. In other words this is the case where speculative excess supply does not lag behind purchases. Now speculative profit over the cycle is given by

$$\int_{t_0}^{t_0 + 2\pi/q} -PE_s\, dt = U\int_{t_0}^{t_0 + 2\pi/q} P(P - R)\, dt = U\int_{P > R} P(P - R)\, dt$$

$$+ U\int_{P < R} P(P - R)\, dt = UP_1 \int_{P > R} (P - R)\, dt$$

$$+ UP_2 \int_{P < R} (P - R)\, dt \qquad P_1 > R > P_s$$

(all positive constants) by one of the mean value theorems for integrals. Since we are assuming $W = 0$, by (10) $r = 0$. Then by (11) P is sinusoidal, so the portion of P above R is a mirror image of the portion below R. Therefore the last two integrals are equal in absolute value. The first integral is positive and the second negative, so that profit over the cycle is positive.

This proves our result for $W = 0$. But (9) shows that profit over the cycle, as expressed by the first integral in this footnote, is a continuous function of W. It follows that for sufficiently small W, ie, if the lag in speculative demand is sufficiently small, the postulated speculation will be profitable.

16 It is clear that the same reasoning is applicable to a non-speculative market with damped or explosive fluctuation. The type of speculative behaviour postulated in (9) could then increase the rate of explosion or reduce the rate of dampening because the coefficient of the P term is positive.

17 No negative number can be the root of an equation like $X^3 - 3X^2 + 5X - 2 = 0$, since the substitution of a negative number for X will make every term negative, and the sum of four negative terms can clearly never be zero.

18 Jacob Viner, 1956, "Some International Aspects of Economic Stabilisation", in Leonard D. White (ed.), *The State of the Social Sciences* (Chicago).

19 Viner, *op. cit.*, p. 291.

19

*A Theory of Speculation Relating Profitability and Stability**

Lester G. Telser
University of Chicago

If a speculator makes profits consistently this implies that he possesses the ability to forecast prices or price changes with a fair degree of success. Whether he thereby stabilises price is another matter. That is the central problem of this chapter.

The proposed theory relating speculators' profits and the stability of prices is intended to be applicable to those commodities traded on organised exchanges by two kinds of people, "speculators" and "non-speculators". What distinguishes speculators from other traders in the market is that their profits depend only on the price or price change of the commodity they trade. Non-speculators' profits are determined not only by the price of the commodity traded on the organised exchange but also by the prices of other related commodities. If the non-speculators are hedgers, they can make their profits almost independent of the price level itself. What enters the excess demand curve of the non-speculators is some weighted sum of the prices of the commodities they handle. However, I shall single out only the price of the commodity traded on the organised exchange when I come to discuss their excess demand.

Some examples may make my distinction between speculators and non-speculators clearer. In the cotton market a non-speculator may be a textile manufacturer whose profit is a function of the price of raw cotton, cotton textiles, and other inputs besides raw cotton. By hedging or other means he can avoid taking a price risk and can specialise in producing cotton textiles to maximise his profits. In the foreign exchange market, one kind of non-speculator is an importer whose profit depends not only on the exchange rate but also on shipping costs, the price of the commodity in the two countries, and the like. He too by hedging can avoid bearing

---

*This paper was first published in *The Review of Economics and Statistics* 41(3), pp. 295–302 (1959) and is reprinted with the permission of the University of Chicago Press. The author wishes to thank Professor Zvi Griliches for his helpful comments and criticisms.

the risk of a change in the exchange rate and can specialise in providing a merchandising service. However, the stock market is not one for which the theory to be described seems applicable, because no reasonable distinction between speculators and other traders can be made. Perhaps the only non-speculators in that market are those corporations engaged in a new stock issue.

In what follows I show that for a fairly general model positive speculators' profits imply that they have stabilised the price. My conclusion is the opposite of that reached by Professor Baumol. In a brilliant article recently published, Baumol gives a counter-example to disprove the proposition that if speculators make profits they necessarily stabilise the price.[1] After presenting my own model I show why Professor Baumol's results are unsatisfactory.

DEFINITION OF PROFITS, STABILITY, AND EQUILIBRIUM

To prepare the ground for what follows I offer a definition of speculators' profits and a measure of price stability. Speculators' net commitments at time t are $U(t)$, a stock variable. I adopt the convention that a positive $U(t)$ means speculators are net long, and a negative $U(t)$ means speculators are net short. Their net change in commitments, a flow variable, is $U'(t)$. An excess of purchases over sales implies $U'(t)$ is positive; the converse implies that $U'(t)$ is negative. The price at time t is $p(t)$.

By "profits" I mean the realised change in the speculators' stock of money resulting from their speculative activities such that their net commitments of the commodity at the end of the speculative period at time s equals their net commitments at the beginning of the period at time 0. Thus profits, Π, are defined by

$$\Pi = -\int_0^s p(t) U'(t)\, dt \qquad U(0) = U(s) \tag{1}$$

(The minus sign before the integral is required because of the convention regarding the sign of $U'(t)$.) Integrating (1) by parts gives the relation between profits, the price change, and commitments:[2]

$$\Pi = U(0)[p(0) - p(s)] + \int_0^s U(t) p'(t)\, dt \tag{2}$$

The assumption that beginning and ending net commitments are equal implies that the speculators realise their profits. If the ending commitments did not equal the beginning commitments, one could still ask whether the speculators' profits realised till then have affected the stability of the price. At that point, however, the speculators have been caught in midstream and their effect on the price till then is temporary.

The full effect is apparent only after the speculators have realised their profits by making their beginning and ending commitments the same. Moreover, that speculative commitments are not accumulated indefinitely is one of the hallmarks of speculation.

In addition to a definition of profits, I need a measure of price stability. Prices over time exhibit turning points, various central tendencies, amplitude, and the like. Of two price series, the one with fewer turning points per time period and smaller differences between successive peaks and troughs (smaller amplitude) is often called more stable. Most economists agree that the amplitude of the price series is an important ingredient of stability, and the measure of stability I have chosen is intended primarily to reflect the amplitude of the price series.

Thus, if two price series have the same mean, I call the one with the smaller variance more stable. If the mean prices differ, I call the one with the smaller coefficient of variation more stable. By this definition, the price series having the smaller amplitude is more stable. However, it doesn't follow that the price series with fewer turning points per period of time has a smaller variance.

Suppose, for example, the "price" is given by a sine curve

$$p(t) = A \sin \beta t \tag{3}$$

The mean "price" is zero and the variance of the "price", Ep^2, is given by

$$Ep^2 = \int_0^s A^2 \sin^2 \beta t \, dt = A^2 \left[\frac{1}{2t} - \frac{1}{4\beta} \sin 2\beta t \right]_0^s \tag{4}$$

Clearly, the smaller is A, the amplitude, the smaller the variance of p. However,

$$\frac{\partial Ep^2}{\partial \beta} = -\frac{1}{4} A^2 \frac{2\beta s \cos 2\beta s - \sin 2\beta s}{\beta^2} \tag{5}$$

so that the variance decreases for a short period, s near zero, and increases for a longer period, s near 90°. Thus the effect on the variance of changes in the frequency, the number of turning points per period, depends on the length of the period.

In spite of the latter ambiguity, I prefer to use the variance of the price as a measure of its stability. Although such a measure may confound the effect of changes in frequency with changes in amplitude, still it does reflect changes in amplitude and, I believe, captures the essence of what is meant by stability.

Another preliminary step is a definition of the market's equilibrium. Denote the non-speculative commitments by V, a stock variable, and the excess demand of the non-speculators by V', a flow variable. Both U', the speculators' excess demand, and V' are functions of the price, $p(t)$, and other variables. Equilibrium in the market is determined by the condition that

$$U' + V' = 0. \qquad (6)$$

The solution of this equation with respect to the price determines its equilibrium time path.

This equilibrium condition seems to imply that, with respect to the market considered, the speculators and non-speculators are involved in a zero-sum game. Such an implication holds only if the population of traders remains the same and their total wealth is constant. However, if traders enter or leave the market or if their total wealth changes, the game is not really zero-sum. Consider the following example. Trader A sells some amount of a commodity to B and then has zero commitments (leaves the market). If the price should subsequently rise, permitting B to realise a profit when he sells the commodity to a third party, trader A has not *realised* a loss thereby, although he has *forgone* a profit. If the total wealth of the participants in the market has changed then B has not profited at the expense of the others. Thus even if speculators as a group make positive profits, and it turns out they stabilise the price, it doesn't necessarily follow that the losses of the non-speculators imply the non-speculators destabilised the price.

A FAIRLY GENERAL MODEL

The first step in my argument is the derivation of the speculators' excess demand function having the property that the speculators' realised profits are a maximum. That is, looking backward from time s, what sales and purchase strategy followed by the speculators would have permitted them to maximise their realised profits. Second, I give the relation between the stability of the price, measured by the variance, and the speculators' profits assuming the speculators' foresight is imperfect.

Taking $U(s) = U(0)$ and neglecting transactions costs, the profits that speculators realised at time s are $\Pi = -\int_0^s pU' \, dt$ from (1). Since V', the non-speculators excess demand, is a function of p and other variables, the market equilibrium equation, $U' + V' = 0$, implies that p depends on U' and those other variables. The U' that maximises profits subject to the constraint that beginning and ending commitments are equal poses a problem in the calculus of variations. It is to find the functional U' that maximises

$$\Pi = -\int_0^s pU' \, dt \quad \text{and satisfies} \quad \int_0^s U' \, dt = 0$$

Fortunately, in this case, a simple artifice permits us to solve this problem rather easily. If we first divide the time elapsed from 0 to s into discrete points, $t_1 = 0, t_2, \ldots, t_k = s$, and consider commitments at these discrete points, we may reduce the original problem involving integrals to one involving ordinary sums. Thus profits, $\tilde{\Pi} = -\Sigma_i p(t_i) U'(t_i)$ is the discrete analogue to the original problem. Finding the $U'(t_i)$'s that maximise $\tilde{\Pi}$ requires only the use of the differential calculus. However, when the discrete problem is solved, it shows the way to deal with the original one. For we can choose an arbitrary number of discrete time points in the closed interval 0 to s and by so doing find in the limit the solution to the original problem involving time as a continuous variable.

To this end let us find the $U'(t_i)$'s that maximise

$$\tilde{\Pi} = -\Sigma p(t_i) U'(t_i) \quad \text{subject to} \quad \Sigma U'(t_i) = 0 \quad i = 1, \ldots, k \quad (7)$$

In (7), the market equilibrium condition of equation (6) determines $p(t_i) = F_i[U'(t_i)]$. To maximise $\tilde{\Pi}$ requires that the $U'(t_i)$'s satisfy

$$-p(t_i) - U'(t_i) \frac{\partial F_i}{\partial U'(t_i)} + \lambda = 0 \quad i = 1, \ldots, k \quad (8)$$

Now λ is a Lagrangian multiplier and $\tilde{\Pi}$ is a maximum if the northwest principal minors of the Hessian H, given in (9), alternate in sign. Let

$$a_i = 2 \frac{\partial F_i}{\partial U'(t_i)} + U'(t_i) \frac{\partial^2 F_i}{\partial U'^2(t_i)}$$

$$H = \begin{bmatrix} 0 & 1 & 1 & \ldots & 1 \\ 1 & (-a_1) & 0 & \ldots & 0 \\ 1 & 0 & (-a_2) & \ldots & 0 \\ 1 & 0 & 0 & \ldots & (-a_k) \end{bmatrix} \quad (9)$$

The northwest principal minors are[3]

$$H_1 = \begin{bmatrix} 0 & 1 \\ 1 & (-a_1) \end{bmatrix}, \quad H_2 = \begin{bmatrix} 0 & 1 & 1 \\ 1 & (-a_1) & 0 \\ 1 & 0 & (-a_2) \end{bmatrix}, \text{ etc.}$$

$$\det H = \det H_k = (-a_k) \det H_{k-1} - (-a_1)(-a_2) \ldots (-a_{k-1}) \quad (10)$$

If $a_i > 0$ then the northwest principal minors of H will alternate in sign and the sufficient condition for $\tilde{\Pi}$ to be maximum is satisfied.

Since I shall assume below that the non-speculators' excess demand is a linear function of the price p which implies that $\partial F_i/[\partial U'(t_i)]$ is a constant and that $\partial^2 F_i/[\partial U'_2(t_i)] = 0$, the condition $a_i > 0$ holds provided $\partial F_i/[\partial U'(t_i)] = -(1/[\partial V'_i/\partial p(t_i)]) > 0$. The latter inequality is true if $(\partial V'_i/[\partial p(t_i)]) < 0$ which means that the non-speculators' excess demand varies inversely with the price.

In the limit, $k \rightarrow \infty$, and (8) is satisfied at every point in time. Hence the $U'(t)$ which maximises $\tilde{\Pi}$ and satisfies the constraint that beginning and ending commitments are equal is given by

$$p + U'\frac{\partial F}{\partial U'} - \lambda = 0 \qquad (11)$$

If $(\partial V'/\partial p) = a$, a constant, then the speculators' excess demand takes a particularly simple form,

$$p - \frac{1}{a}U' - \lambda = 0 \qquad (12)$$

since

$$\frac{\partial F}{\partial U'} = -\frac{1}{\partial V'/\partial p} = -\frac{1}{a} > 0$$

Integrating equation (12) from 0 to S,

$$\int_0^s p\,dt - \frac{1}{a}\int_0^s U'\,dt - \lambda \int_0^s dt = 0,$$

the middle term vanishes and $\lambda = \bar{p}$, the mean price from 0 to s. Replacing λ by $\bar{p}$ in (12) and rearranging terms, (12) becomes

$$U' = a(p - \bar{p}), \qquad a < 0 \qquad (13)$$

Thus if the speculators' excess demand for the period $0 < t < s$ is proportional to the difference between the actual price and the mean price, they will have realised a maximum profit during the period of their speculation from 0 to s.

When speculators are absent from the market, the equilibrium condition $V' = 0$ determines the price path $p = f(t)$. However, when speculators are present in the market, the market equilibrium condition, $U' + V' = 0$, determines a different price path, $p = g(t)$. Thus $f(t)$ denotes the equilibrium price path in the speculators' absence, and $g(t)$ denotes the equilibrium price path in the speculators' presence.

Suppose the non-speculators' excess demand function is

$$V' = h(t) + ap \qquad (14)$$

which has a constant slope with respect to p as required by the argument leading to (13). The function $h(t)$ admits the effects of seasonal factors on the non-speculators' excess demand. More generally, $h(t)$ depends on a set of variables and I require that members of a subset of these display systematic tendencies over time. If so, then that subset of systematic variables is represented by their proxy t. Moreover, $h(t)$ may also include random variables.

From (14), it follows that

$$f(t) = -(1/a)h(t) \qquad (15)$$

which gives the price path when there are only non-speculators in the market. The price path, $p = g(t)$, when speculators are present also satisfies (6), $V' + U' = 0$. Replacing V' and U' in (6) by (13) and (14) respectively, we find that $p = g(t)$ is determined by

$$a[g(t) - \bar{g}] + h(t) + ag(t) = 0 \qquad (16)$$

Using (15), $f(t)$ replaces $-(1/a)h(t)$, and (16) becomes

$$[g(t) - \bar{g}] = f(t) - g(t) \qquad (17)$$

Integrating (17) with respect to t shows that the resulting left side is zero, which implies that $\bar{g} = \bar{f}$. The mean price is the same whether speculators are present or not.

Since (17) implies $2[g(t) - \bar{g}] = f(t) - \bar{f}$, after squaring both sides of this equation and integrating with respect to t from $t = 0$ to $t = s$, we find that $4\sigma_g^2 = \sigma_f^2$. In short, the variance of prices without speculators present is four times greater than the variance of prices with speculators present. Hence in this model speculation stabilises the price. Profits are given by

$$\Pi = -a\int_0^s p(p - \bar{p})\,dt = (-a)\sigma_g^2 \qquad (18)$$

and they are positive because a is negative.

A more general model supposes the speculators change their prediction of the price during the speculative period because they have imperfect foresight and cannot realise a maximum profit. Under these circumstances their excess demand function is

$$U'(t) = \beta[p(t) - p^*(t)] \qquad (19)$$

For this excess demand function the coefficient of the difference between the actual price at time t and the predicted price $p^*(t)$ is β, a negative coefficient which no longer equals a as in (13). The function $p^*(t)$ is the speculators' prediction of the mean price over the entire period 0 to s as of time t. When the actual price exceeds their predicted price, they sell an amount proportional to the difference; when the actual price is below the predicted price, they buy an amount proportional to the difference. The speculators try to predict the mean price for the entire period 0 to s at each point of time in the interim, but they need not predict perfectly the actual price prevailing at each point in time. The better p^* predicts $\bar{p}$, the greater are the speculators' profits; and when $p^* = \bar{p}$ they realise the maximum profit. The definition of realised profits for the period 0 to s that places a restriction on beginning and ending commitments, $U(0) = U(s)$, implies that $\bar{p} = \bar{p}^*$.[4] Therefore, the speculators' predicted price is correct when averaged over their entire period of speculation. This follows from (19) since

$$\int_0^s U' \, dt = \beta \int_0^s (p - p^*) \, dt = 0 \quad \text{so that} \tag{20}$$

$$\int_0^s p \, dt = \int_0^s p^* \, dt \quad \text{and} \quad \bar{p} = \bar{p}^* \tag{21}$$

For the speculators' excess demand function, (19), profits are

$$\Pi = -\int_0^s U' p \, dt = (-\beta) \left[\int_0^s p^2 \, dt - \int_0^s p p^* \, dt \right] \tag{22}$$

Hence

$$\Pi = s(-\beta)[\sigma_p^2 - \mathrm{cov}(p, p^*)] \tag{23}$$

where

$$\mathrm{cov}(p, p^*) = \frac{1}{s} \int_0^s (p - \bar{p})(p^* - \bar{p}^*) \, dt.$$

The speculators realise the maximum profit when $p^* = \bar{p}$ and $\mathrm{cov}(p, p^*) = 0$. Otherwise, since their foresight is imperfect, their profits are reduced.

When speculators are absent, $V' = ap + h(t) = 0$ determines the same price path $p = f(t)$ as given in (15). However, with speculators present, the equilibrium price path $p = g(t)$ differs from the one given by (17). For in this case $p = g(t)$ satisfies

$$U' + V' = \beta[g(t) - p^*(t)] + h(t) + ag(t) = 0 \tag{24}$$

Since $f(t) = -(1/a)h(t)$, (24) becomes

$$\frac{\beta}{a}[g - p^*] = f - g \qquad (25)$$

By integrating both sides of (25) with respect to t, over the period 0 to s, we see that since $\bar{g} - \bar{p}^* = 0$, then $\bar{f} = \bar{g}$. That is, in this case for which speculators' foresight is imperfect and their realised profits are not a maximum, the resulting mean price is the same whether or not speculators are in the market. This surprising result derives from the equality $\bar{g} = \bar{p}^*$, which follows in turn from the condition that $U(0) = U(s)$ as required by my definition of realised profits.

With the aid of (25) we can find the relation between profits, Π, and the two measures of stability, σ_g^2 and σ_f^2. First, we may rewrite (25) using the equality $\bar{g} = \bar{f}$ and obtain,

$$f - \bar{f} = \frac{\beta}{a}(g - p^*) + g - \bar{g} \qquad (26)$$

If we square both sides of (26) and integrate from 0 to s, we find that

$$s\sigma_f^2 = s\left(\frac{\beta}{a}\right)^2 \text{var}(g - p^*) + s\sigma_g^2 + 2\frac{\beta s}{a}\text{cov}(g - p^*, g) \qquad (27)$$

Since $\text{cov}(g - p^*, g) = \sigma_g^2 - \text{cov}(p^*, g)$, the results in (23) indicate that $2(\beta s/a)\text{cov}(g - p^*, g) = 2/(-a)\Pi$. With this equation we may simplify (27) to

$$s\sigma_f^2 = s\left(\frac{\beta}{a}\right)^2 \text{var}(g - p^*) + s\sigma_g^2 + \frac{2}{(-a)}\Pi \qquad (28)$$

Equation (28) shows that if the speculators realise a positive profit, $\Pi > 0$, then every term on the right-hand side is positive and $\sigma_f^2 > \sigma_g^2$. Therefore, positive speculators' profits imply they stabilise the price. Even if they suffer losses, $\Pi < 0$, they may still stabilise the price. When they realise the maximum profit, $a = \beta$, $\bar{p}^* = \bar{p}$, and $\sigma_f^2 = 4\sigma_g^2$ as shown above.

CRITICISMS OF BAUMOL'S COUNTER-EXAMPLES

For a particular excess demand function describing the non-speculators' behaviour, Professor Baumol constructs a speculators' excess demand function that permits speculators to receive profits and yet the resulting price series exhibits less stability – that is, more frequent fluctuations and, under some conditions, greater amplitude – than in the absence of the speculators. In his counter-example, speculators concentrate their purchases just *after* the price has reached its minimum, thus their purchases occur when the price is rising; they concentrate their sales just *after* the price has reached its maximum, thus their sales occur when the price is falling. Provided speculators react quickly enough after the price exhibits a turning point and provided non-speculators' excess demand depends on price changes or recent price trends, the speculators' strategy is both profitable and destabilising.

It seems just as plausible that speculators concentrate their purchases just *before* the price reaches its minimum and their sales just *before* the price reaches its maximum. If so they likewise earn profits but stabilise the price. Which strategy speculators follow rests on their ability to forecast the price. For speculators to concentrate their purchases and sales just before the price exhibits a turning point requires that they can predict when the turning point will occur. The speculators' profits are greater if they are capable of predicting turning points than if they concentrate their purchases and sales just after turning points have occurred. The latter strategy, followed by Baumol's speculators, requires less skill but does not yield as great a remuneration. In Baumol's words, his speculators "give up any chance to skim off the cream but hope in return significantly to reduce their risks".[5]

His models are subject to several criticisms. The most important is of his unreasonable non-speculators. I am unwilling to accept his non-speculators' excess demand function on the grounds given by Professor Friedman, whose objections Baumol quotes in his article:[6]

> In a letter to the author, Professor Friedman has suggested that perhaps a non-speculator can only be safely defined (if this is done in terms of his demand curve) as one whose purchases are directly influenced by current prices but not by past prices or price trends. In the absence of speculation, so defined, cyclical price movements can only result from real influence.

Baumol's differential equation model (cf. pp. 486–7 of his chapter) imputes somewhat peculiar behaviour to the speculators. Their excess demand using my notation is

$$U' = a(p - R) - bp', \qquad a < 0, \, b < 0, \text{ and } R \text{ constant} \tag{29}$$

In the notation of the previous sections, $p^* = R + (b/a)p'$, and differentiating both sides with respect to t gives $dp^*/dt = (b/a)p''$. Thus the speculators increase their predicted price when the price path is concave up even though the price may be falling, and they decrease their predicted price when the price path is concave down even though the price may be rising.

In general, Baumol's models put no restrictions on the speculators' excess demand that would prevent the speculators from accumulating commitments indefinitely.

To meet Friedman's objections to the nature of the non-speculators' excess demand, Baumol constructs "A Real Non-Speculative Cycle" (beginning on p. 488 of his chapter). I claim, however, that this model prevents speculators from ever realising their profits and that it *necessarily* implies that the price rises indefinitely while speculators accumulate inventories indefinitely. Baumol concedes that this is a possibility but that it is not a necessary consequence of his model.

Briefly, Baumol's model is as follows. The supply of a commodity, S, is independent of the price:

$$S = A - BS'' \qquad (30)$$

In order that S vary sinusoidally, $B > 0$. The demand for the commodity, D, depends only on the current price thus

$$D = K - Vp \qquad (31)$$

in which, presumably, $V > 0$. When speculators are present whose excess demand is given by (29), the equilibrium price path satisfies

$$(a - V)p - bp' + B(a - V)p'' - Bbp''' = A - aKR \qquad (32)$$

(This equation corresponds to Baumol's equation at the bottom of page 488.) As he points out, the characteristic equation must have one real positive root and two complex conjugate roots. Making the substitution $p = e^{xt}$, we find the characteristic equation for the homogeneous differential equation derived from (32) reduces to

$$-Bbx^3 + B(a - V)x^2 - bx + (a - V) = 0 \qquad (33)$$

It is easy to verify that the positive real root of this equation is $(a - V)/b$ and the complex roots are $\pm i\sqrt{1/B}$ where $i = \sqrt{-1}$. To prove this, we can multiply $[x - (a - V/b)][x^2 + (1/B)]$ and recover equation (33). It follows that the solution to the homogeneous equation associated with (32) is given by

$$p = k_1 e^{\left(\frac{a-V}{b}\right)t} + k_2 e^{\left(i\sqrt{\frac{1}{B}}\right)t} + k_3 e^{\left(i\sqrt{\frac{1}{B}}\right)t} \qquad (34)$$

where, of course, the latter two terms are sinusoidal with constant amplitude. Thus the solution is dominated by $e^{((a-V)/b)/t}$ and, since $((a - V)/b) > 0$, the price rises indefinitely. I must emphasise again that the price necessarily rises indefinitely (cf. Baumol's page 489).[7]

Yet what is wrong with a model that implies speculators accumulate inventories indefinitely while the price rises indefinitely? Even a constant physical stock of the commodity traded would not prevent that from occurring. So long as traders may borrow from the banks using as collateral their holdings of the good, and so long as the price asked for the good rises at a rate which exceeds the interest rate, it seems that the bubble may expand indefinitely. No speculator can ever sell his commitments but surely he may borrow and reborrow without ever risking bankruptcy. Moreover, at refunding, the amount he borrows exceeds the amount he must repay on his old loans since the value of his collateral rises at the same rate at which the asked price rises. Those who are speculating become richer and richer. However, should one wish to diversify his assets by selling even a small part of his holdings of the good, the boom ends. Worse still, should the interest rate rise so much that it equals the rate of price change, then the market will collapse and the devil take the hindmost. Unless the banking system is willing to finance a hyper-inflation, eventually the boom must end. A theory of speculation which cannot explain turning points is as useless as a theory of business cycles which cannot explain why inflations and depressions come to an end.

The weakest link in Baumol's counter-examples is his specification of the non-speculators' excess demand. In my models only the current price appears in the non-speculators' excess demand. Any real cycles that may be present are included in my function $h(t)$. If I am granted that the effect of speculators on the stability of the price can be determined only if the speculators' profits are realised and that the variance is an adequate measure of the stability of the price, then the model summarised in (28) shows that when the speculators realise a positive profit they stabilise the price and even if they incur losses they may stabilise the price. If non-speculators are not affected by past prices and if speculators have to realise their profits then Baumol's counter-examples are unacceptable.

REPLY

W. J. Baumol

The last section of Dr. Telser's very interesting article sums up Professor Friedman's objections to my attempts to prove that profitable speculation need not be stabilising. The main objections boil down to two:

1. In one of my counter-examples speculators' gains consist essentially in inflated values of their holdings, which result from speculative bidding for these items. These "profits" cannot be realised since if speculators all attempted to get rid of such goods their prices would collapse. Hence I must agree that it is unreasonable to cite this as an example of destabilising *profitable* speculation.

2. In my other counter-examples it is objected that those individuals whom I have classed as non-speculators are really speculators, because they take into account rate of change of price and not just current price. It is difficult to quarrel with definitions but, as I have already pointed out, this is hardly a reasonable criterion. The practical question which has lain behind the discussion is whether the entry into a market of skilful *professional* speculators, people who have no desire to hedge their holdings, can be stabilising. Now it is clear that the remaining participants in the market, the non-speculators, the people who would like to hedge, must in their own interests consider price trends. For price changes must also affect the values of non-speculators' holdings and obligations unless in fact they have succeeded in setting up perfect hedges, which in most markets is out of the question. It is noteworthy that the firms which constitute Dr. Telser's own illustrations of non-speculators (in his first section) – textile manufacturers, importers, corporations who issue new securities – would normally be most foolish to ignore price trends in their supply–demand decisions, though they meet the author's test for non-speculators as their profits are not determined *exclusively* by the price level of the commodity in question.

In sum I still believe I have in my article shown by valid counter-examples that profitable speculation need not always be stabilising. Of course it is equally easy to demonstrate that it need not always be destabilising. This is again illustrated by Telser's model which, because it is surely far from general, can be interpreted only as a counter-example that makes this point. I believe then that we end up at my original position, that the effects of profitable speculation on stability is in part an empirical question and that attempts to settle it by *a priori* arguments must somewhere resort to fallacy.

---

1 William J. Baumol, 1957, "Speculation, Profitability and Stability", *The Review of Economics and Statistics* 39, August, pp. 263–71; reprinted as Chapter 18 of the present volume.
2 The formula for integrating by parts is $\int x\, dy = xy - \int y\, dx$. Choose $x = p$ and $dy = U'\, dt$ so that $dx = p'\, dt$ and $y = U$. Hence

$$-\int_0^s p(t)U'(t)\, dt = -[p(s)U(s) - p(0)U(0)] + \int_0^s U(t)p'(t)\, dt$$

which is equation (2) in the text when $U(s) = U(0)$.
3 By expanding the determinant of H according to the elements of the last row, we may obtain (10).
4 Note that I need only assume $U(s) = U(0)$ to get the result in the text. Thus in empirical work I would measure speculators' profits over a time period defined by the condition that their commitments are the same at the beginning and end of the period.
5 Baumol, *op. cit.*, p. 482.
6 Baumol, *op. cit.*, p. 488.

7 The conclusion in the text, that the price rises indefinitely, is strictly true only if k_1 is positive. A negative k_1 implies indefinitely falling prices. In either case the price path eventually has no turning point.

20
*Profitable Speculation**
Michael J. Farrell[†]

The behaviour of competitive markets under conditions of uncertainty is not only an immensely important problem from the point of view of economic policy, it is also a problem towards the solution of which very little has been done, either by economic theorists or by empirical economists. The analysis of the effects of speculators on the markets in which they operate is almost as large a question and almost as far from being answered. It is therefore essential to stress the strictly limited objectives of the present article.

We may perhaps begin with the remark by Friedman (1953, p. 175) on the effects of speculation in foreign exchange markets under a system of flexible exchange rates:

> People who argue that speculation is generally destabilising seldom realise that this is largely equivalent to saying that speculators lose money, since speculation can be destabilising in general only if speculators on the average sell when the currency is low in price and buy when it is high. (Footnote: A warning is perhaps in order that this is a simplified generalisation on a complex problem. A full analysis encounters difficulties in separating "speculative" from other transactions, defining precisely and satisfactorily "destabilising speculation", and taking account of the effects of the mere existence of a system of flexible rates as contrasted with the effects of actual speculative transactions under such a system.)

Although this remark is directed towards the effects of speculation in one particular market, and although arguments of this sort had undoubtedly been used earlier, this is a convenient point to start our discussion.

---
*This paper was first published in *Economica* 33, pp. 183–93 (1966) and is reprinted with the permission of Blackwell Publishers Ltd.
[†]Michael J. Farrell, 1926–1975.

The basic argument is that, to make a profit, speculators must on average sell at higher prices than those at which they bought; and that this tends to raise the lower, and to reduce the higher, prices, thus (in a sense) stabilising prices. This argument contains no assumptions about the behaviour of speculators and it is this, I suggest, that gives it its peculiar interest, for it can clearly be used as the basis of an argument (analogous to that put forward by Alchian (1950)) that, independently of any behavioural assumptions, some selection process will ensure that speculation will tend to stabilise prices. Even if the basic proposition, that profitability implies stabilisation, is valid, the conclusion that speculators tend to stabilise prices does not necessarily follow, for the argument via selection is subject to considerable difficulties, which I hope to explore elsewhere; but the present article will be restricted to investigating the validity of the basic proposition.

In contrast to previous discussions of the subject (see, for instance, Baumol (1957 and 1959), Kemp (1963), Stein (1961) and Telser (1959)), our argument will avoid making assumptions about the behaviour of speculators. A finally satisfactory analysis of the effects of speculation will doubtless involve formulating and testing hypotheses about the motives, expectations and strategies of speculators, but the very powerful argument outlined in the previous paragraph would lose its force and generality if it were made dependent on such hypotheses. We shall therefore see how far we can advance without making such assumptions. In addition, we shall avoid relying on the sort of highly particular counter-example that has been used by some writers. (Baumol (1957) used a counter-example based on unrealised paper profits, and Kemp (1963) one based on a non-speculative demand curve of perverse slope.) Rather, we shall try to derive directly necessary and sufficient conditions for the basic proposition to hold. First, of course, we must find a precise statement (or statements) of the basic proposition.

DEFINITIONS

We shall consider an abstract market for an abstract commodity – which may, of course, be a security or foreign currency – of which the only essential property is that it must be storable. We shall, however, abstract from the costs (eg, storage and interest costs) of, and returns (intangible and monetary) to, holding the commodity, so that the profitability of speculation can appropriately be defined as the difference between the proceeds of selling and the cost of purchasing. We shall use a discrete time variable $t = 1, 2, \ldots, T$, defining T periods within each of which all transactions are assumed to take place at the same price. Let q_t be net speculative sales in period t, and write the sequence of such sales in vector form $q = [q_t]$. We shall not ask who speculates or why, but merely consider what the profitability of a speculative sequence q implies about its effect on prices. We shall, however, assume that

$$q'u = 0 \qquad (1)$$

where u is a vector whose elements are unity; that is, we shall consider only *completed* speculations, where, however complex the sequence of purchases and sales may have been, the final holding of the commodity is identical to the initial holding. Such a restriction accords with the common usage of the word "speculation" – it is usually assumed that a speculator only buys in order to resell – and is moreover necessary to a satisfactory definition of profitability.

Now define P_t as that price which would obtain in period t if there were no speculative activity in any period (ie, if $q = 0$), and write the sequence of these prices in vector form as $P = [P_t]$. We shall also consider the price $P_t + p_t$ obtaining in period t in the presence of a given vector q of speculative sales, and write these prices as a vector $P + p = [P_t + p_t]$. In general, p_t is a function of P, q and t, so that we should write $p_t = f(P, q, t)$; but we shall soon have to make rather drastic simplifying assumptions.

Assuming (1) and that there are no transactions costs, the profit associated with a speculative vector q is simply defined as

$$\pi = (P + p)'q \qquad (2)$$

but the definition of stabilisation requires some discussion. The word "stabilising" is commonly used in two quite different senses when applied to the effect of some measure or other on a market. The first sense is that of rendering the equilibrium of the market more stable, and in this sense it is natural to seek definitions in terms of the behaviour of the market when displaced from equilibrium. Stabilisation in this sense is clearly important; but in this article we shall be concerned with the second sense, that of a reduction in the variation of prices. This is clearly the sense in which the word is used in the arguments quoted above.

However, granted that we wish to call speculation "stabilising" if it reduces the variation of prices, there remain a variety of ways of making this precise, corresponding to the different possible measures of variation. Of these, Telser (1959) has suggested the coefficient of variation and Kemp (1963) the variance, while a third possible measure is the mean square deviation from the mean undisturbed price. The choice among such measures seems inevitably arbitrary, and we shall choose the variance on grounds of simplicity. We therefore define the criterion

$$C = (P + p - \bar{P}u - \bar{p}u)'(P + p - \bar{P}u - \bar{p}u) - (P - \bar{P}u)'(P - \bar{P}u) \qquad (3)$$

where $\bar{P} = P'u/T$; $\bar{p} = p'u/T$, and say that the effect of a given speculative sales vector q is stabilising or destabilising as $C \leq 0$. As C is proportional to the change in the variance of prices brought about by q, it can be seen that we call q "stabilising" if it reduces the variance and "destabilising" if it increases it.

In general, it would seem quite possible that the analysis should be sensitive to the choice among the possible measures of variation, and it is therefore of interest to note that if we had chosen the mean square deviation from $\bar{P}$, we should have had a stabilisation criterion equal to $C + T\bar{p}^2$, and that the results of the third section would be unaffected by the substitution of this criterion for that of (3).

We can now formulate precisely the basic proposition, that is,

$$\pi > 0 \quad \text{implies} \quad C < 0 \tag{4}$$

That this is not true in general can be seen from the following trivial counter-example, with $T = 2$:

$$P' = [0, 0]$$
$$q' = [-1, +1]$$
$$p' = [p_1, p_2] \quad \text{where} \quad p_1 < 0 < p_2 \tag{5}$$

That this gives $\pi > 0$, $C > 0$ is hardly surprising, for we have buying lowering the price and selling raising it. If this were so, we would soon all be speculators!

Clearly, if we wish to establish the validity of the basic proposition, we must build into our assumptions the observed (hard) fact of life, that speculative buying tends to raise the price and selling to lower it. We must, in other words, make some assumption about the relationship of p and q. Bearing in mind that p_t depends in general on P, q, and t, the assumption we shall (initially) make represents a drastic simplification.

INDEPENDENCE; NO TRANSACTIONS COSTS

We shall assume that p_t depends on q_t only, and that the relationship is the same for all t, writing

$$q_t = d(p_t) \quad t = 1, 2, \ldots, T$$

We shall call this the *independence assumption*. We shall also call the function $d(p_t)$ the "non-speculative excess demand function". It describes the reaction of the remainder of the market to a given level of speculative sales, and it is natural to call this the "non-speculative" part of the market; and it is clearly an excess demand function, as it relates the quantity q_t of speculative sales the non-speculative part of the market is willing to absorb to the price, measured as a deviation from the undisturbed price P_t.

The *independence assumption* is a very strong one, and has two particularly notable implications. First, it implies that the non-speculative demand function (and the reigning price) in any period is independent of speculative activity in previous periods: we shall later consider the possibility of relaxing our assumptions in this respect. Second, it implies that the non-speculative demand function is independent of P_t; and this plays a part in the analysis of this section.

A third implication, that the non-speculative excess demand function does not vary with time, is relatively innocuous. Although in principle the *shape* of this function could change over time, this is a relatively unlikely happening, to allow for which would complicate the analysis considerably – in other words, to rule out this possibility is as justifiable as almost any simplifying assumption in economic theory. On the other hand, shifts over time in the non-speculative supply and demand functions are quite consistent with the independence assumption, for, although it follows trivially from our definitions of P and p that $d(0) = 0$, we have placed no restriction on the sequence P of undisturbed prices and have said nothing at all about the quantity traded at these prices.

Next let us add the *law of demand* to our assumptions by simply assuming that $d(p_t)$ is a monotonic decreasing function. Are our assumptions now strong enough to ensure that (4) holds? To put it another way, can we still find q and P such that π and C are both positive? Perhaps surprisingly, in general we can. To show this, let us first write C in a more convenient form:

$$C = (P + p - \bar{P}u - \bar{p}u)'(P + p - \bar{P}u - \bar{p}u) - (P - \bar{P}u)'(P - \bar{P}u) \quad (6)$$

$$= 2(P + p - \bar{P}u - \bar{p}u)'(p - \bar{p}u) - (p - \bar{p}u)'(p - \bar{p}u) \quad (7)$$

$$= 2(P + p)'p - 2T(\bar{P} + \bar{p})\bar{p} - (p - \bar{p}u)'(p - \bar{p}u) \quad (8)$$

$$= 2T(\bar{P} + \bar{p})C_1 - (p - \bar{p}u)'(p - \bar{p}u) \quad (9)$$

where

$$C_1 = (P + p)'p/T(\bar{P} + \bar{p}) - \bar{p} \quad (10)$$

If we assume that all prices are positive, then a necessary and sufficient condition for us to be able to find q and P that will make π and C both positive is that we should be able to do the same for π and C_1. From (9) it is clear that $C_1 < 0$ implies $C < 0$; while if $C_1 > 0$ we can always choose $\bar{P} + \bar{p}$ large enough (given q and therefore p) to make C positive. (C_1 determines $P + p$ only to a factor of proportionality, and by varying this factor we can give any positive value to $\bar{P} + \bar{p}$ without altering C_1.) We can also replace by

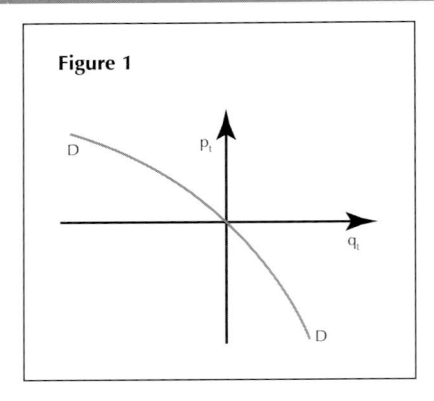

$$\pi_1 = \pi/T(\bar{P} + \bar{p}) \tag{11}$$

which clearly has the same sign.

To investigate the relationship between π_1 and C_1, it is convenient to refer to Figure 1, where p_t is plotted against q_t. On the assumptions we have made, corresponding values of q_t and p_d are represented by points on some curve DD, which represents the non-speculative demand function, has negative slope, and passes through the origin. A vector q of speculative sales and the corresponding vector p thus give a set of T points (q_t, p_t), $t = 1, 2, \ldots, T$, lying on DD. Call this set of points A, and consider the two points G_1 and G_2 given by

$$\begin{cases} G_1 = (g_1, h_1) = \left(\dfrac{1}{T}u'q, \dfrac{1}{T}u'p\right) = (0, \bar{p}) \\ G_2 = (g_2, h_2) = \left(\dfrac{(P+p)'q}{T(\bar{P}+\bar{p})}, \dfrac{(P+p)'p}{T(\bar{P}+\bar{p})}\right) \end{cases} \tag{12}$$

G_1 and G_2 are weighted means of the points of A, G_1 using equal weights $v = u/T$ and G_2 using as weights the elements of the vector $w = (P+p)/T(\bar{P}+\bar{p})$. (Note that $(P+p)/T(\bar{P}+\bar{p}) > 0$, and $(P+p)'u/T(\bar{P}+\bar{p}) = 1$.) Now

$$\pi_1 = g_2 - g_1, \quad C_1 = h_2 - h_1 \tag{13}$$

so that we can find $\pi_1 > 0$, $C_1 > 0$ if and only if we can find G_1 and G_2 with

$$g_2 > g_1, \quad h_2 > h_1 \tag{14}$$

But given G_1 we can always find G_2 satisfying (14) provided G_1 is an interior point of the convex hull of A, which is always so unless the points of A are collinear; and unless DD is a straight line, we can always find a q giving a non-collinear set of points, A. Conversely, if DD is a straight line, the points of A must be collinear and its convex hull a line segment, which, by the *law of demand*, must have negative slope; and so, since G_1 and G_2 belong to this convex hull, they cannot satisfy (14). Thus the linearity of DD is, given the *independence assumption* and the *law of demand*, a necessary and sufficient condition for the basic proposition (4) to hold.

We should perhaps note that if $T = 2$, the set A contains precisely 2 points and (4) holds whatever the form of DD, so long as it has negative slope. In other words, if we restrict ourselves to a single purchase and a single sale of the same quantity, the basic proposition is valid; but if we try to extend it (eg, by the insertion of the words "on the average") to more complicated speculative sales vectors, it ceases to be valid. This may give some insight into why such a superficially plausible proposition should turn out to be true only under very special conditions. Indeed, it is possible to regard such an extension as constituting a sort of aggregation problem, in which case the crucial importance of the linearity assumption is not surprising. (Cf. Farrell, 1954)

We may also note that sufficiency can be shown directly as follows. Suppose that $q = ap$, where a is a negative scalar. Then $\bar{p} = 0 = (\bar{P}u + \bar{p}u)'q$, and (2) becomes

$$\pi = (P + p)'q = (P + p - \bar{P}u - \bar{p}u)'q$$
$$= (P + p - \bar{P}u - \bar{p}u)'ap = a(P + p - \bar{P}u - \bar{p}u)'(p - pu) \quad (15)$$

and substituting in (7) gives

$$C = 2\pi/a - p'p < 0 \quad (16)$$

This direct proof of sufficiency is very like that given by Kemp (1963, p. 189), which in turn is somewhat similar to that given by Telser (1959, p. 298).

Let us now return to the main stream of our argument. We have shown that, on a number of restrictive assumptions, the linearity of the non-speculative excess demand function is necessary and sufficient for (4). As the necessity of linearity is the more surprising result, our next step should be to see how far it is dependent on our simplifying assumptions – the *independence assumption* and the absence of transactions costs. As we have noted, one implication of the *independence assumption* is that the non-speculative excess demand function in any period is independent of the undisturbed price in that period. This is an unrealistic assumption and one which plays an important part in the proof of necessity (although not in that of sufficiency) for it is this assumption that permits us to replace C by C_1. It would thus be of interest to see whether linearity remained necessary if the non-speculative excess demand function depended on the undisturbed price in the same period. Unfortunately, as Tolstoy might have said, although there is only one form of independence, there are infinitely many forms of dependence. To be sure, one could investigate the matter for certain specific and plausible forms of dependence, but it would be difficult to establish conclusive results in this way. In any case, the necessity of linearity depends on the absence of transactions costs, as we shall see in the next section.

TRANSACTIONS COSTS

Let us suppose that transactions costs are shared equally by buyer and seller, and that they represent a constant fraction $2d$ of the value of the sale. Then the only change we need to make in our analysis is to replace (2) by

$$\pi = (P + p)'q - d(P + p)'|q| \tag{17}$$

where we write $|q|$ for $[|q_t|]$. This gives

$$\pi_1 = \frac{(P + p)'}{T(\bar{P} + \bar{p})}(q - d|q|) = w'q - w'd|q|,$$

so that we now ask on what conditions does

$$w'q - w'd|q| > 0 \tag{18}$$

imply

$$w'p - v'p < 0 \tag{19}$$

If we write $p_t = -bq_t - e(q_t)$ (noting that $b > 0$, $e(0) = 0$), we have

$$w'p - v'p = (w - v)'(-bq - e)$$
$$= -b[(w - v)'q + (1/b)(w - v)'e]$$
$$= -b\{w'q + [(1/b)(w - v)'e]\} \tag{20}$$

so that (18) implies (19) if

$$(1/b)(w - v)'e \geq -w'd|q| \qquad \text{or} \qquad (w - v)'e \geq -w'd|bq| \tag{21}$$

It is sufficient that

$$(w_t - v_t)e_t + w_t d|bq_t| \geq 0 \text{ for all } q_t \text{ and } w_t \tag{22}$$

Suppose $e_t \leq 0$; then

$$w_t e_t - v_t e_t + w_t dbq_t \geq w_t e_t + w_t d|bq_t| \geq 0 \text{ for all } w \geq 0, \text{ if } e_t \geq -d|bq_t| \tag{23}$$

Suppose $e_t > 0$ and let there be $m > 0$ such that $w_t \geq m v_t$. Then

$$w_t e_t - v_t e_t + w_t d|bq_t| \geq v_t(me_t - e_t + md|bq_t|) \geq 0$$

$$\text{if } e_t \leq [m/(1-m)]d|bq_t| \tag{24}$$

Combining (23) and (24) we have

$$m/(1-m)d|bq_t| \geq e_t \geq -d|bq_t|, \text{ or} \tag{25}$$

$$-m/(1-m)d|bq_t| \leq p_t + bq_t \leq dbq_t, \text{ or} \tag{26}$$

$$-bq_t - m/(1-m)d|bq_t| \leq p_t \leq -bq_t + d|bq_t| \tag{27}$$

Thus (27) is a sufficient condition for (4) for all q and for $(P_t + p_t)/(\bar{P} - \bar{p}) = w_t/v_t \geq m$. Geometrically, (27) defines a pair of cones (one in the second quadrant and one in the fourth) within which DD must lie. We are free to choose any $b > 0$, and for given b the vertical width of the relevant cone at any value of q_t increases with d and with m. In other words, the magnitude of the deviations from linearity of DD that are consistent with (4) increases with the magnitude of transactions costs and decreases with the range of values of P_t that we permit.

If we replace (17) by

$$\pi = (P + p)'q - dv'|q| \tag{28}$$

so that transactions costs are proportional to the *quantity* sold rather than to its *value*, we get as sufficient conditions

$$-bq_t - d|bq_t| \leq P_t \leq -bq_t + 1/(n-1)d|bq_t| \tag{29}$$

where $(P_t + p_t)/(\bar{P} + \bar{p}) = W_t/v_t \leq n$. Equation (29) is formally similar to (27) and indeed, they are identical if $m = \frac{1}{2}$, $n = 2$. Thus the main conclusion of this section – that the greater the transactions costs, the less closely need DD approximate a straight line – does not seem very sensitive to the form of transactions costs postulated.

TEMPORAL INTERDEPENDENCE

We have so far assumed (it is one of the implications of the *independence assumption*) that p_t is independent of speculative activity in any other period. If we wish to relax this assumption we are again embarrassed by the need to consider an infinite number of possible forms of temporal interdependence; but we shall find that even the simplest form introduces great difficulties. Suppose that, instead of (6) we have

$$P_t = -aq_t - cq_{t-1} \qquad a, c > 0, \ t = 1, 2, \ldots, T \qquad (30)$$

This assumes that p_t is linear in q_t and introduces only one lagged variable. Nevertheless, consider the speculative sales vector

$$q' = [-k, 0, k] \qquad (31)$$

This gives $p' = [ak, ck, -ak]$ and a set of points $A = \{(-k, ak), (0, ck), (k, -ak)\}$, which violates not only the linearity condition of the third section but also, for any finite d, conditions (27) and (29). This result holds for any $b > 0$ and a similar one for $b < 0$; similar, though possibly more complex, arguments can be constructed if p_t depends on $q_{t-s}(s > 1)$.

The linearity condition of the third section was necessary as well as sufficient for (4) to hold for general q and P, so that we may now conclude that, in a market with temporal interdependence but no transactions costs, (4) is not true for general q and P. Moreover, since the argument of the third section required values of $(P_t + p_t)/(\bar{P} + \bar{p})$ differing only slightly from unity, we cannot make (4) true by any reasonable restriction on P.

If a market has temporal interdependence *and* transactions costs, it is difficult to say anything conclusive. The conditions (27) and (29) are sufficient but not necessary; they are already quite complicated and involve some (not necessarily severe) restrictions on the permissible values of $(P_t + p_t)/(\bar{P} + \bar{p})$. We cannot rule out the possibility of finding alternative sufficient conditions allowing for temporal interdependence, but they would presumably be more restrictive than (27) and (29) and certainly a good deal more complicated. We shall not attempt to find such conditions but merely repeat that temporal interdependence violates the sufficient conditions found in the fourth section.

Since temporal independence is crucial to our conclusions, we must consider carefully how far we are justified in assuming it. Friedman has suggested (1953, p. 269) that one should define anyone whose decisions are influenced by prices other than the current one as a speculator, so that the non-speculative excess demand function is temporally independent by definition. Baumol (1959, p. 301) objects that, since the question most interesting to the policy maker is the effect of *professional* speculators on a market which in any case contains amateur speculators, the non-speculative excess demand function should reflect the behaviour of the latter group. Given the limitations set out in the first section, we can legitimately say that the distinction between amateur and professional speculators is inappropriate to this part of the analysis, but should be considered in the selection argument.

Unfortunately, though, quite apart from amateur speculators, there are several real-world phenomena that may cause temporal interdependence. An excess demand function depends, of course, on both demand and supply, and in the case of the latter lagged responses may occur, for instance, in commodity markets (because of the technical conditions of production), or in security markets (because new issues take time to float). Similarly, demand responses may be lagged because complementary investment is needed or because adjustments (to purchasing policies or to portfolios) are made infrequently. Again, if supply and demand relate not to a flow but to a stock of the commodity (as is usually the case with security markets), p_t is a function, not of current speculative sales or purchases, but of speculative holdings, which again introduces temporal interdependence.

A way out might be sought by introducing holding costs, an aspect of the real world which we have so far ignored; and certainly, sufficiently large holding costs would invalidate the counter-example of (31). However, it is difficult to believe that holding costs of the requisite magnitude occur at all commonly, both because holding costs in well-organised speculative markets tend to be small and because the coefficients of lagged quantities in the non-speculative excess demand function may well be as large as or larger than the coefficients of current quantities.

CONCLUSION

We must now consider the conclusions at which we have arrived. So long as we maintained the *independence assumption*, our results were clear-cut and satisfactory. In the absence of transactions costs linearity of the non-speculative excess demand function is both a necessary and a sufficient condition for the validity of the basic proposition. That it should be necessary is perhaps surprisingly severe, but the introduction of transactions costs removed this necessity and led to less stringent sufficient conditions, with the permissible departures from linearity dependent on the magnitude of the transactions costs. Thus, in markets where the *independence assumption* holds and where transactions costs are large enough to cover the actual departures from linearity of the non-speculative excess demand function, the basic proposition holds and one may set about trying to answer the second part of the question: whether the selection process in the market is strong enough to ensure that speculators are (predominantly) profit makers.

However, this does not take us very far, for we have found reasons for expecting that many real-world markets will display some measure of temporal interdependence, so invalidating our sufficient conditions. Thus our search for reasonably simple and plausible sufficiency conditions for the validity of our basic proposition seems to have been in vain. But the analysis of this paper will not have been wasted if it has persuaded economists that our basic proposition is too strong to hold with any great generality and that they should therefore seek to establish weaker propositions concerning the properties of speculative markets. Such weaker propositions might be based on specific behavioural assumptions. Alternatively, since the selection argument is inevitably statistical in character, it would seem reasonable to replace our proposition about what "must" happen by another phrased in terms of probabilities.

BIBLIOGRAPHY

Alchian, A. A., 1950, "Uncertainty, Evolution and Economic Theory", *Journal of Political Economy*.

Baumol, W. J., 1957, "Speculation, Profitability and Stability", *Review of Economics and Statistics* 39 (3), August, pp. 263–71; reprinted as Chapter 18 of the present volume.

Baumol, W. J., 1959, "Reply", *Review of Economics and Statistics*.

Farrell, M. J., 1954, "Some Aggregation Problems in Demand Analysis", *Review of Economic Studies*.

Friedman, Milton, 1953, *Essays in Positive Economics*, pp. 157–203 (Chicago).

Kemp, M. C., 1963, "Speculation, Profitability and Price Stability", *Review of Economics and Statistics*.

Stein, J. L., 1961, "Destabilising Speculative Activity Can Be Profitable", *Review of Economics and Statistics*.

Telser, L. G., 1959, "A Theory of Speculation Relating Profitability and Stability", *Review of Economics and Statistics* 41, pp. 295–301; reprinted as Chapter 19 of the present volume.

21
*Price Destabilising Speculation**
Oliver D. Hart and David M. Kreps
Harvard University; Stanford University

It is sometimes asserted that rational speculative activity must result in more stable prices because speculators buy when prices are low and sell when they are high. This is incorrect. Speculators buy when the chances of price appreciation are high, selling when the chances are low. Speculative activity in an economy in which all agents are rational, have identical priors, and have access to identical information may destabilise prices, under any reasonable definition of destabilisation. It takes extremely strong conditions to ensure that speculative activity (of the commodity storage variety) "stabilises" prices, even in a very weak sense.

Do speculators stabilise prices? This old question has been the subject of a large literature, going back as far as Smith (1789/1937) and including Mill (1921), Friedman (1953), Baumol (1957), Telser (1959), Farrell (1966) and Samuelson (1971), among others. (For a comprehensive bibliography, see Goss and Yamey (1978).) The case that speculators must stabilise prices is succinctly put in the adage that (rational) speculators "buy cheap and sell dear", thereby raising low prices and lowering high prices. To quote from Friedman (1953, p. 175), "People who argue that speculation is generally destabilising seldom realise that this is largely equivalent to saying that speculators lose money, since speculation can be destabilising in general only if speculators on the average sell when the currency is low

---

*This paper was originally published in the *Journal of Political Economy* 94(5), pp. 927–52 (1986) and is reprinted with the permission of the University of Chicago Press. The authors wish to acknowledge financial assistance from the Social Science Research Council of the United Kingdom, the International Center for Economics and Related Disciplines at the London School of Economics, the National Science Foundation (grants SES-8006407 and SES-8408586), the Office of Naval Research (contract N00014-77-C-0518), Harvard University, and the Sloan Foundation. They would also like to thank the editor and two referees for helpful suggestions and comments.

in price and buy when it is high". Thus it seems that only irrational speculators could destabilise prices.

In fact, matters are more complicated than this, as a number of authors have noted. This is because there is less to the "buy cheap and sell dear" adage than meets the eye. Speculators will buy when the chances of price appreciation are high, which may or may not be when prices are low (see Kohn 1978). This observation in fact underlies a number of examples of profitable, destabilising speculation that were developed in the 1950s and 1960s (see Baumol 1957; Telser 1959; Farrell 1966). These examples are not conclusive, however, since they rely either on there being a small number of imperfectly competitive speculators or on non-speculators having irrational expectations. The purpose of the present paper is to show that speculation can be destabilising even when speculators are competitive and both speculators and on non-speculators have rational expectations.

Any theory of speculative behaviour must address the thorny question of how to define speculation. Despite the many attempts to do this in the literature, a satisfactory general definition is still not available (and probably never will be; see, eg, Johnson (1976)). We shall not attempt to give one here. Rather we will study a very specific situation in which there is a fairly natural and (we hope) non-controversial notion of speculation and speculators. We will also make no attempt to give a general definition of price destabilisation. The cases we will study are sufficiently specific that the terms stabilisation and destabilisation have simple and intuitive meanings.

Our basic model concerns the market for a commodity such as wheat (although we will show that the model can also be interpreted as one for a durable asset). The market meets at a sequence of discrete dates. There are two types of agents. One type, *non-speculators* or *consumers*, trades in the market only for purposes of immediate consumption. The second type, *speculators*, buys with the intention of holding the commodity and then selling at a higher price at a later date. (Thus in our model "speculation" is synonymous with "storage".) Exogenous uncertainty in the model comes from the fact that the non-speculative demand schedule at each date is random, drawn according to some probability distribution. (This distribution may depend on the position of demand in the previous period.) Speculators are assumed to know this probability distribution and to have rational expectations about the resulting stochastic evolution of prices. Speculators are assumed to know the current position of the demand for consumption schedule, and they may possibly possess some further information concerning future levels of this schedule. The inferences they draw from this information are always correct in the Bayesian sense, and they all have access to the same information. Consumers also know these things, but that they do so is unimportant.

They enter the market only once, buying only for immediate consumption. (Models along these lines may be found in a number of papers; see, eg, Kohn (1978) and Scheinkman and Shechtman (1983).)

Our objective is to show that such storage/speculation can lead to less stable equilibrium prices. The basic example of this runs roughly as follows. Suppose that the sequence of consumption demand schedules is independent and identically distributed but that at any date there may be portents about next period's schedule. Specifically, assume that demand in each period is either very high or very low, that the change of high demand is quite small, and that whenever next period's demand will be high, there is in this period a signal that this will happen. Suppose, however, that this signal is imperfect in the sense that, while it *always* appears prior to high-demand periods, it also *sometimes* appears prior to low-demand periods. It does the latter sufficiently infrequently that the chance of high demand next periods conditional on having observed the sign is greater than the (marginal) probability of high demand but also sufficiently frequently that it is wrong more times than not. (Numbers will be given in the body of the paper.) Now suppose that in the current period consumption demand is low and the signal is present. Speculators may then buy up some of the supplies of the commodity, anticipating the chance of a high price next period. If the signal turns out to be accurate so that the high price does occur, then this is price stabilising in the sense that prices are higher this period and lower next. But suppose the signal this period is inaccurate and moreover that there is no signal next period. Then next period speculators will dump their stored supplies onto the consumption market without providing any further demand. (The fact that the signal is absent means that high prices cannot occur in the next period, and so in this period there is no speculative motive.) This might well depress prices next period to a level lower than they would ever get without the presence of speculators. Now it is by no means clear that this leads to less stable prices in any precise sense, but (by playing with elasticities of demand for consumption) we will be able to flesh out this example and to show that destabilisation may indeed occur.

The existence of a noisy signal is not crucial to the construction of lessened stability from speculation. Destabilising speculation can occur even if speculators have perfect foresight about the future as long as this foresight is limited, in particular, if speculators at date t know (perfectly) the position of consumption demand at dates $t, t + 1, \ldots, t + k$, but nothing more. But while one can always construct an example in which speculation is destabilising for such an information structure for any finite $k > 0$, in the extreme cases $k = 0$ and $k = \infty$, speculation will always be stabilising (in a very weak sense). And for a given economy (with everything held fixed except the size of k) this will be true for large values of k.

The thrust of this paper is that speculative activity *can* destabilise prices (in reasonable circumstances) – not that it will. It should also be noted that whether or not speculation stabilises prices is in some sense the wrong question. One really ought to be interested in the welfare implications of speculation. One may feel intuitively that price stabilisation is "good", but, if so, one's intuition is faulty (see Newbery and Stiglitz (1984)). We study the impact of speculation on price stability rather than on welfare because in our model the welfare effects of speculation turn out to be complicated. The reader should, however, bear in mind throughout that welfare and price stabilisation can sometimes be related in a surprising fashion. We will indicate this in the context of our basic example and then comment on it generally in the final section.

The paper is organised as follows. In the following section, we give our "canonical" example of price destabilising speculation. In the next two sections, we present a more general model and analyse the characteristics of equilibrium. (Technical details are left for the Appendix.) The fifth section presents the limited results that we have obtained concerning when speculation might be said to be price stabilising. We sum up in the last section.

AN EXAMPLE OF DESTABILISING SPECULATION

We first present a very simple example to show how speculation can be destabilising (under any reasonable definition of stability). We begin by giving a brief description of the economy we shall be considering. Imagine a market for a storable commodity. This market meets at a sequence of dates $t = \ldots, -1, 0, 1, 2, \ldots$. Supply to this market is inelastic; we will explain the sources of this supply shortly. There are two types of demand: demand for *consumption* purposes and demand for storage or *speculative* purposes.

Consumption demand derives from consumers who are in the market just once. That is, a new generation of consumers enters at each date t and then exists forever. We suppose that consumption demand in period t depends on the equilibrium price of the commodity, p_t, and on a random taste parameter, θ_t. That is, the consumption demand schedule shifts in or out from period to period, with θ_t a description of the position of the schedule at date t. We write $D(p_t; \theta_t)$ for this demand schedule.

The second source of demand is speculative or storage demand (we use the two terms interchangeably). We suppose that there are overlapping generations of competitive speculators, each speculator living for just two periods. Speculators, who consume only the numeraire commodity, enter the market when young with a fixed endowment of the numeraire good, some of which they spend on the storable commodity. They then store this commodity for one period, possibly subject to some wastage, and supply inelastically what comes out of storage to the market when they are old.

There is also a constant exogenous inelastic supply (fresh crop) each period, which we denote by X. Speculators' storage decisions depend on the current price, p_t, and on their expectations about next period's price. These expectations will be rational.

All speculators (at date t) observe the current consumption demand parameter, θ_t. There may also be additional public information ξ_t concerning future demand. For example, ξ_t might be θ_{t+1}: the consumption demand parameter is known one period in advance. More generally, ξ_t might represent noisy information about $\theta_{t+1}, \theta_{t+2}, \ldots$.

We can now describe our example of destabilising speculation. Imagine that in each period θ_t is either $\bar{\theta}$ or $\underline{\theta}$. The sequence $\{\theta_t\}$ is an independently and identically distributed (i.i.d.) sequence, with the probability that $\theta_t = \bar{\theta}$ equal to 0.01. Consumption demand when $\theta_t = \bar{\theta}$ is perfectly elastic at a price of US$1,000. When $\theta_t = \underline{\theta}$, consumption demand is a bit more complex: from zero to 100 units, it is perfectly elastic at price US$10.00. Then the demand curve slopes downward: It is at price US$9.00 at 130 units, and it decreases thereafter, asymptoting at price US$8.00.

In each period 100 fresh units of the commodity are provided (ie, $X = 100$). Speculators can store up to 50 units, getting back three for every five stored. Speculators maximise the expected present worth of income, with a zero interest rate. Their endowment of the numeraire is assumed to be large: they can afford to purchase 50 units even if the price is US$1,000.

The key to this example is the information flow. At date t, besides θ_t there is available a signal ξ_t that is an imperfect portent of the future. Specifically, ξ_t is either $\underline{\xi}$ or $\bar{\xi}$. Whenever θ_{t+1} will equal $\bar{\theta}$, then $\xi_t = \bar{\xi}$; $\underline{\xi}$ indicates the impossibility of $\bar{\theta}$ next time. But $\xi_t = \bar{\xi}$ indicates only that $\bar{\theta}$ is possible. When $\theta_{t+1} = \underline{\theta}$, then $\xi_t = \bar{\xi}$ with probability 1/11 and $\xi_t = \underline{\xi}$ with probability 10/11. Consumption with Bayes's rule shows that when $\xi_t = \bar{\xi}$ the chance that θ_{t+1} will be $\bar{\theta}$ is 0.1. In terms of our formulation, $\{(\theta_t, \xi_t)\}$ is a four-state Markov chain, with transition matrix given in Table 1.

The idea of this example is that, very occasionally, there is huge demand for the commodity ($\theta_t = \bar{\theta}$). Such huge demand next period is foreshadowed this period by $\bar{\xi}$. But seeing $\bar{\xi}$ indicates only the possibility of huge demand next period – there remains substantial (0.9) probability that demand will be small.

What will be equilibrium price behaviour in this economy? First consider what would happen if there are no speculators in this economy at all. Then prices will be US$1,000 whenever $\theta_t = \bar{\theta}$ and US$10.00 whenever $\theta_t = \underline{\theta}$.

Now add the speculation/storage activity. Speculators will buy the commodity, up to their limit of 50 units, whenever $\theta_t = \underline{\theta}$ and $\xi_t = \bar{\xi}$. Doing so does not raise the current price above US$10.00, and there is

Table 1 Transition matrix for the Markov chain

	$\underline{\theta}\underline{\xi}$	$\underline{\theta}\bar{\xi}$	$\bar{\theta}\underline{\xi}$	$\bar{\theta}\bar{\xi}$
$\underline{\theta}\underline{\xi}$	0.9	0.1	0	0
$\underline{\theta}\bar{\xi}$	0.81	0.09	0.09	0.01
$\bar{\theta}\underline{\xi}$	0.9	0.1	0	0
$\bar{\theta}\bar{\xi}$	0.81	0.09	0.09	0.01

Note. Entries are $P(\phi_{t+1} = \text{column} | \phi_t = \text{row})$.

some prospect (a 0.1 chance) that they will be able to sell next period at US$1,000, more than enough, given their storage technology and decision criterion, to justify the gamble. When $\theta_t = \bar{\theta}$, speculators will not buy: There are no prospects for price appreciation. And when $\theta_t = \underline{\theta}$ and $\xi_t = \underline{\xi}$, speculators will not buy: the current price is no lower than US$8.00, and next period's price can be no higher than US$10.00, which yields an insufficient return on the investment.

Price behaviour with speculators, then, goes as follows. Whenever $\theta_t = \bar{\theta}$, the equilibrium price is US$1,000, just as in the case with no speculators. And when $\theta_t = \underline{\theta}$, there are two possibilities: If $\xi_t = \bar{\xi}$, then speculators will take 50 units off the market. This will ensure that supply for consumption is not more than 80 units, and the equilibrium price will be US$10.00. If $\xi_t = \underline{\xi}$ and it was the case that $\xi_{t-1} = \underline{\xi}$, then supply for consumption is 100 (speculators bought neither last period nor this), and again equilibrium price is US$10.00. The one change from the economy without speculators (in terms of equilibrium prices) comes when $\theta_t = \underline{\theta}$, $\xi_t = \underline{\xi}$, θ_{t-1} was $\underline{\theta}$, and ξ_{t-1} was $\bar{\xi}$. In this case, speculators do not buy this period, *but they did buy last period*. Supply for consumption is 130 units, and the equilibrium price is US$9.00.

In other words, without speculation, price is sometimes US$1,000 and sometimes US$10.00 (much more often the latter). With speculation, price is US$1,000 whenever it was before. But in those cases in which it used to be US$10.00, it is sometimes US$10.00 and sometimes US$9.00. (It is US$9.00 about 8% of the time). The additional variability comes when speculators buy in one period in the hope of a huge price rise and then, disappointed in the next period, dump their holdings. Note that only very rarely will their storage decision look sensible *ex post*. But it is perfectly rational ex ante because, if it is "right", it gives a huge rate of return.

The reader will doubtless note the extent to which we have cooked this example. We wished to present as stark an example as possible of price destabilisation (and one that will be an example of destabilisation under any reasonable notion of that term). Perfectly elastic consumption demand and a sharp cut-off (at 50 units) of the storage capabilities of speculators were the keys. Relaxing these would muddy the waters considerably: storage activities would help depress the very high prices if demand were less than perfectly elastic at $\bar{\theta}$. And either competition among speculators or reduced supplies for consumption (or both) would raise prices when speculators are withdrawing supplies for purposes of storage. We do not mean to suggest that speculation must destabilise prices. But this example clearly shows that, in a model that plays by all the rules of rationality and competitive agents, speculation *may* have this effect.

Moreover, we content that this example may well capture salient aspects of reality. Think of the θ_t process not as an i.i.d. sequence but as a process that very occasionally flips from one state to the other. We have in mind some very rare but very significant sea change in the structure of the economy: war, famine, or a very radical change in climate or technology. Such sea changes are often heralded in the press as being imminent. In fact, the heralds appear much more often than the changes, followed by heralds of some different and radical change. That is, these premonitions of change are wrong more often than not, but they do indicate that one is in a period in which the sea change, very unlikely (in the short run) in any case, is relatively more likely. This increase in the likelihood of the change, especially if the change is massive, may be enough to cause speculators to enter whatever markets are appropriate. Then, as is almost always the case, the premonitions turn out to have been wrong, the danger recedes, and speculators dump their holdings, depressing prices. Especially where the change is of the extremely rare and extremely catastrophic variety, one might observe a large number of episodes of price fluctuations and no observable (*ex post*) reason for these fluctuations.

This is simply a recasting of the so-called peso problem, studied by Krasker (1980), in which the peso forward rate fluctuated in anticipation of a devaluation that did not come. It provides a possible explanation for the recent empirical work on the volatility of share prices (eg, Shiller 1981a). The empirical finding is that share prices fluctuate too much given the observed fluctuations in the net present value of dividends. If some changes in dividends are of the extremely large and rare variety, however, one may require many years of data for these to show up. These changes, although rarely observed (perhaps not yet observed at all), may still cause fluctuations in share prices, given the sorts of imperfect portents that we have in mind.[1]

In the rest of the paper, we shall investigate in greater depth what drives the example above. To put it another way, we shall consider conditions that are sufficient to rule out the destabilising speculation of this example. These conditions will turn out to be surprisingly strong.

We will do everything possible to avoid discussing the welfare implications of speculation, and this simple example is useful for saying why this is. In this example, the speculators are risk-neutral expected consumption maximisers. Because of the constraint that they face in their storage capacity, even though they are competitive, they extract positive surplus from being allowed to engage in speculation. And the consumers are better off for the presence of the speculators, at least if one uses the expected consumer surplus measure. So even though prices are less stable, welfare of the consumers and speculators is Pareto-improved.

However, some party is supplying those fresh 100 units each period, and they, it might be supposed, are worse off for sometimes getting US$9.00 for their output when previously they got US$10.00. If we think of the supply as coming in the form of endowment to the speculators, then with the storage constraint we have imposed they are still better off (ex ante) if they are allowed to speculate. But if the 100 units come from some third sector of, say, consumers of the numeraire good and if members of this third sector are extraordinarily risk averse concerning their level of consumption, then we can juggle the *ex ante* probabilities of being a consumer, a speculator, or a member of this third sector so that speculation leads to an *ex ante* Pareto decline in welfare. We repeat from earlier: welfare considerations are difficult, and they may bear no particular relationship to the stability or instability of prices.

A GENERAL MODEL

The basic structure of the economy in our example is kept throughout: There are one-period consumers, whose demand is given by the schedule $D(p_t; \theta_t)$ and there are overlapping generations of speculators. We begin with two regularity assumptions about the demand schedules of the consumers.

Assumption 1. The possible values of θ_t (the possible demand schedules) are finite in number, coming from a finite set Θ. Within Θ are a "least" and a "greatest" θ, written $\underline{\theta}$ and $\bar{\theta}$, such that

$$D(p; \underline{\theta}) \leq D(p; \theta) \leq D(p; \bar{\theta}) \quad \text{for all } p \text{ and } \theta \in \Theta$$

Assumption 2. For each θ, $D(p; \theta)$ is strictly positive and continuously differentiable in p, with strictly negative derivative (for p between zero and infinity). Also, $D(0; \theta) = \infty$ and $\lim_{p \to \infty} D(p; \theta) = 0$; the usual Inada conditions hold.

Note that, by virtue of assumption 2, if we let $P(x; \theta)$ be the inverse demand function, then P has all the properties enumerated for D in assumption 2. Note also that assumption 2 is violated in the example of the second section; however, examples "close" to the one given there can be constructed that satisfy assumption 2 and that exhibit the same behaviour, although price when $\theta_t = \bar{\theta}$ will have to fall a bit from US$1,000.

Speculative demand at date t depends on the current price p_t and the distribution of next period's price, which we write F_{t+1}. This demand schedule is denoted by $D'(p_t; F_{t+1})$. The following assumptions are made.

Assumption 3. For fixed F_{t+1}, $D'(p_t; F_{t+1})$ is continuous and non-increasing in p_t.

Assumption 4. For fixed p_t, $D'(p_t; F_{t+1})$ is continuous and non-decreasing in F_{t+1}.[2]

Assumption 5. If p_t exceeds the largest value in the support of F_{t+1}, then $D'(p_t; F_{t+1}) = 0$.

In other words, no one will speculate if it is certain that they will lose money by doing so.

Assumption 6. $\mathrm{Lim}_{p_t \to \infty} D'(p_t; F_{t+1}) = 0$, uniformly in F_{t+1}.

That is, speculative demand falls to zero as p_t gets large, regardless of the promises of speculative profits.

Assumption 7. For every $\lambda > 1$, p_t and F_{t+1}, $D'(p_t, F_{t+1}) \geq D'(\lambda p_t; \lambda F_{t+1})$, where λF_{t+1} means the distribution F_{t+1} "inflated" by λ. That is, $\lambda F_{t+1}(q) = F_{t+1}(q/\lambda)$.

Of our assumptions, these last two are the least palatable. (They are made so that, subsequently, we can obtain uniqueness of equilibrium.) One way to view assumption 7 is to note that, by inflating both the current price, p_t, and the distribution of next period's price, F_{t+1}, by the factor λ, we do not change the distribution of rates of return derived from speculation. If speculators base their demand only on the rates of return from speculation, their demand will be unchanged. Indeed, one might expect speculative demand to fall, insofar as the same rates of return prevail, but at higher stakes. (And, to satisfy assumption 6, for large enough λ we will have to have strict inequality.)

The simplest story underpinning D' is that each speculator possesses a storage technology of the following sort: storage of y units at date t yields $f(y)$ units recovered at date $t + 1$. Imagine that $f(0) = 0$, $f(y) \leq y$ for all y, and f is non-decreasing, non-negative, strictly concave, and bounded above. Speculators are risk neutral with a zero discount rate in the sense that they attempt to maximise the expected revenue accrued from selling $f(y)$ units at price p_{t+1} less the cost $p_t y$ of purchasing the y units for storage. Thus each speculator will wish to store that quantity y^* that satisfies $f'(y) = p_t/E(\hat{p}_{t+1})$, where $\hat{p}_{t+1}$ represents the random variable of next period's price (the distribution of which is given by F_{t+1}), and $E(\cdot)$ denotes expectation. Of course, y^* is constrained below to be nonnegative and above by w/p_t, where w is the speculator's wealth in the numeraire good, assumed to be finite and exogenously given. Then D' is just the number of firms times the optimal (constrained) y^*. The reader can easily check that assumptions 3–7 all hold: assumption 5 because we assume that $f(y) \leq y$ and f is strictly concave, assumption 6 because of the finite wealth constraint, and assumption 7 because only rates of return matter until the constraints bind, and the constraints go in the right direction.

Many variations on this simple story can be played, but care must be taken if the assumptions are to remain. One can allow speculators to borrow in the numeraire in order to finance their purchases as long as there is some limit on credit. (Otherwise, assumption 6 is in jeopardy. Such a credit limit arises from general equilibrium considerations if there is a finite amount of wealth in the economy at each date.) A positive discount rate or risk aversion could be added. But in adding risk aversion, one must be careful to preserve the non-decreasing part of assumption 4. (If, for example, speculators had a target level for proceeds from sales, then raising F_{t+1} could lower the amount they need to store to make that target.)

Speculators base their storage decisions at date t on the publicly available information (θ_t, ξ_t), which we write ϕ_t, where ξ_t represents any information that they might possess concerning future demand (beyond what is in θ_t). We assume that all speculators concur in the following (objective) assessment.

Assumption 8. The sequence $\{\phi_t; t = \ldots, -1, 0, 1, \ldots\}$ forms a time-homogeneous Markov chain with finite state space Φ.[3]

Note that the ϕ_t may be serially correlated. This can occur even if the θ_t are uncorrelated or independent because of the additional information ξ_t. For example, if $\xi_t = \theta_{t+1}$, then $\phi_t = (\theta_t, \theta_{t+1})$ and $\phi_{t+1} = (\theta_{t+1}, \theta_{t+2})$ are dependent.

All that remains is to specify the supply side of the economy. Supplies in period t, denoted X_t, are inelastic and depend on the amount stored in the previous period. That is,

$$X_{t+1} = G(D'(p_t; F_{t+1})) \tag{1}$$

for a given function G. For example, in our simple storage story,

$$G(D'(p_t; F_{t+1})) = \underline{X} + nf\left(\frac{D'(p_t; F_{t+1})}{n}\right)$$

where $\underline{X}$ is exogenous supply each period, and n is the number of speculators. The following general assumption is made.

Assumption 9. The function G is non-decreasing and continuous, and it satisfies $G(0) > 0$ and $\lim_{y \to \infty} G(y) < \infty$. Also, $G(y) - G(0) \leq y$ for $y \geq 0$.

The last part simply says that storage is non-productive.[4]

In summary, our market model is determined by equation (1) and the supply equals demand equation,

$$X_t = D(p_t; \theta_t) + D'(p_t; F_{t+1}) \tag{2}$$

These are subject to assumptions 1–9 above.

Before giving a formal definition of equilibrium, let us briefly sketch a second interpretation of our model involving durable goods. Think of a market for houses in a given locality. In this locality is a fixed stock Y of houses, subject to no depreciation. At date t some part of this stock, say Y_t, is held by owner-occupiers (consumers). A randomly selected fraction α of these consumers are compelled by circumstances (death, a job somewhere else) to sell their houses – the stock that they own (αY_t) is inelastically supplied to the market. These departing consumers are replaced by an equal number of new consumers (new families, people taking jobs in the area) so that the number of consumers is constant. (This is not a one-house-per-consumer society; per capita holdings of the asset will change through time. Think of it this way: some of the consumers own houses, while others, finding the price of housing too high on arrival, decide to rent accommodation.) Newly arrived consumers demand houses in a fashion that depends on the current price p_t and a random taste parameter θ_t; this gives us D. Also, there are speculators who buy the houses one period and sell them the next. Imagine that these speculators live for two periods only, so their supply (at the end of their lives) is inelastic. Their demand (at the beginning of their lives) depends on the current price and the distribution of next period's price; this gives us D'. Equation (2) is immediate. As for (1), since consumers hold what speculators do not,

$$X_{t+1} = G(D'(p_t; F_{t+1})) = D'(p_t; F_{t+1}) + \alpha[Y - D'(p_t; F_{t+1})]$$

In this interpretation, we assume that owner-occupiers do not enter the housing market except when they first arrive and when they depart; they do not adjust their holdings in the intervening periods. Also, they are, on arrival, sensitive only to current price; they ignore the possible capital gains in deciding how much to demand. We are able to make some excuses for these assumptions: transaction costs for owner-occupiers are too high to make retrading profitable. Capital markets are imperfect so that utility from capital gains is nearly inconsequential, or utility from capital gains is logarithmic (so that while capital gains do affect utility levels, demand is insensitive to future prices). The fact that such imperfections must now be assumed, however, means that this interpretation of the model is not as clean as the interpretation involving a commodity.

EQUILIBRIUM

Fix the data of a model as in the third section. Let $\overline{X} = \lim_{y \to \infty} G(y)$ and $\underline{X} = G(0)$ (ie, supplies in any period lie somewhere between $\overline{X}$ and $\underline{X}$). The equilibrium price p_t will be a function of current information ϕ_t and current supplies X_t. We will look for time-homogeneous equilibria: p_t does not depend on the particular value of t. Hence, we look for prices given by some function p from $\Phi \times [\underline{X}, \overline{X}]$ into $[0, \infty]$. Imagine that some such function p is advanced as a candidate for an equilibrium. When the current state of information is $\phi = (\theta, \xi)$ and supplies are $X \in [\underline{X}, \overline{X}]$, consumption demand (at the supposed equilibrium price) will be $D(p(\phi, X); \theta)$. Storage in equilibrium will therefore be the residual of supply, or $X - D(p(\phi, X); \theta)$. This means that supply next period will be $G(X - D(p(\phi, X); \theta))$. Thus the distribution of next period's price (still assuming that the function p gives equilibrium prices) will be the distribution of

$$p(\hat{\phi}_{t+1}, G(X - D(p(\phi, X); \theta))) \qquad \text{given } \phi_t = \phi$$

where $\hat{\phi}_{t+1}$ is random. Let us write

$$F(\phi, G(X - D(p(\phi, X); \theta)), p)$$

for this distribution function. Note how the three arguments enter: $\hat{\phi}_{t+1}$ is distributed according to the Markov transition probabilities, given $\phi_t = \phi$; the second argument gives the supplies brought forward into next period; and p gives the (supposed) equilibrium price functional.

The equilibrium condition is that speculative demand does indeed absorb the residual of supply less consumption demand. This yields the following formal definition.

Definition. An *equilibrium* is a function p: $\Phi \times [\underline{X}, \overline{X}] \to [0, \infty)$ such that for every $\phi = (\theta, \xi)$ and $X \in [\underline{X}, \overline{X}]$

$$D(p(\phi, X); \theta) + D'(p(\phi, X); F(\phi, G(X - D(p(\phi, X); \theta)), p)) = X \qquad (3)$$

Note that this is a rational expectations equilibrium in which speculators act competitively. That is, the speculators correctly anticipate next period's equilibrium prices, and they take as given the total amount that will be stored this period.

Proposition 1. Under our assumptions, there exists a unique equilibrium price function, which we denote p^*. This function is continuous and strictly decreasing in its second argument. (Price is lower the greater is supply). In this equilibrium, $x - D(p^*(\phi, x); \theta)$ is increasing in x. (The more that is stored in one period, the more will be stored in the next, ceteris paribus.)

The proof of proposition 1 is left to the Appendix. But some short technical points about the proof that have economic relevance are worth making. Our method of proof is somewhat different from those that have appeared in the literature. As in Kohn (1978), we essentially compute the equilibrium by successively computing equilibria in which there are no speculators, then in which speculators will exist for only one period more, then two, and so on. In the limit, we get the equilibrium. Kohn (1978) gets convergence by a contraction mapping argument this requires strong assumptions on elastaicities. We avoid those assumptions by invoking assumption 4: speculative demand is non-decreasing in next period's prices. This allows us to use monotone mappings (instead of contraction mappings); prices computed for an economy in which there are $n + 1$ more generations of speculators are always at least as high as in an economy in which there are n more generations, with the level of current supplies held fixed.[5] Because we use iterated mappings, it is (in principle) possible to compute equilibria. Also, much of the literature assumes that the θ_t are i.i.d. and no further information is available; we are able to avoid this entirely. Typically, it is difficult to prove uniqueness in models of speculation because of the existence of Ponzi schemes. We have avoided such schemes in two steps. First, assumption 6 rules out the possibility that equilibrium prices will become unbounded. And then assumptions 4 and 5 combine to rule out bounded Ponzi schemes: speculative demand must be nearly zero when prices are close to their maximum attainable value.

Indeed, we can calculate that maximum price. Let $\bar{p}$ be the solution of $D(p; \bar{\theta}) = \underline{X}$. That is, $\bar{p}$ is the price that prevails if supply is $\underline{X}$ (the lowest possible level), the consumption demand is at its highest possible level, and there is no speculative demand.

Proposition 2. $p^* \leq \bar{p}$, and for all ξ, $p^*((\bar{\theta}, \xi), \underline{X}) = \bar{p}$.

Since we are more interested in examples than in general properties of the equilibrium, we will desist from further general development. But it is perhaps worth noting that standard results from the theory of Markov chains will (with mild regularity conditions on G) ensure that the chain $\{(\theta_t, X_t); t = 0, 1, \ldots\}$ will settle down to a long-run stationary distribution if we assume that the chain $\{\theta_t\}$ is well behaved (ie, is aperiodic and irreducible).

CASES OF (WEAKLY) STABILISING SPECULATION

Were there no speculators, the evolution of the economy would be simple. Supplies at each date would be $x = G(0)$, and the equilibrium price would be a function of θ_t, the solution to $D(p; \theta_t) = x$. Let us denote this solution by $p^c(\theta_t)$ or, for $\phi_t = (\theta_t, \xi_t)$, $p^c(\phi_t)$. Note that $\bar{p}$ is simply $p^c(\bar{\theta})$. We said above that price functions rise the longer (it is supposed) that speculators are around. The prices p^c correspond to an economy in which there will be no speculators in the future. Hence one might conclude that speculators only raise prices. But, of course, this is wrong. The monotonicity result holds for a fixed level of current supplies. When we add speculators, the distribution of supplies changes: speculators shift supplies; in periods in which they increase X_t, they may drive down prices. Indeed, we know already (proposition 2) that $p^* \leq \bar{p}$.[6]

We are interested in comparing p^* with p^c. We saw in the second section that there exist nonpathological economies for which p^* is more variable than p^c under any reasonable notion of variability. One particular property of the example is that the lowest price in the support of p^* is lower than the lowest price in the support of p^c. We now present sufficient conditions for p^* not to be more variable than p^c in this very grossest of senses. That is, we provide conditions for the lowest price in the presence of speculators to be greater than or equal to the lowest price in their absence. By virtue of proposition 2, this implies that the range of prices does not grow with the addition of speculators. In this case, we say that speculation is *weakly stabilising*, with maximal possible emphasis on "weakly".

One case in which speculation is weakly stabilising is the following. Suppose that $\phi_t = \theta_t$ and that the θ_t are i.i.d. That is, at date t agents know nothing that improves on their *ex ante* prediction of θ_{t+1}. Then given any level of X_t, the distribution of next period's prices is almost independent of the current price, and speculative demand will be larger the lower is the current price. Note well the qualifier "almost"; for fixed X_t a larger speculative demand this period means lower prices next (through the effect on X_{t+1}). The point is that the "buy-cheap-sell-dear" adage (which gives us hope for price stabilisation) does hold at the equilibrium.

Lemma. In the special case $\phi_t = \theta_t$ and $\{\theta_t\}$ an i.i.d. sequence, if $p^*(\theta, X) > p^*(\theta', X)$, then $D(p^*(\theta, X); \theta) \geq D(p^*(\theta', X); \theta')$.

We have stated the results in terms of D instead of D'. The reverse inequality holds for D'. Suppose we strengthened assumption 1 so that the possible values of θ were real numbers, and higher θ meant a consumption-demand curve that was shifted up and to the right. Then this lemma could be paraphrased: those consumers who value the good more highly get more of it. This paraphrase is precise: since X_t is a function of X_0 and $\{\theta_{t-k}; k \geq 1\}$, X_t is independent of θ_t.

To show this, suppose the converse held. Then supplies next period would be higher under θ than θ', implying that next period's price will be lower under θ. Thus, since $p^*(\theta, X) > p^*(\theta', X)$, speculative demand is lower under θ, which is a contradiction.

Does the fact that the adage holds mean that speculators stabilise prices weakly? Proposition 3 says that the answer is yes.

Proposition 3. In the special case $\phi_t = \theta_t$ and $\{\theta_t\}$ i.i.d., if we begin with $X_0 = \underline{X}$, then $p^*(\theta_t, X_t) \geq p^c(\theta)$ for all t with probability one.

To prove this, note three facts already shown. (*a*) In this special case, $p^*(\theta, x) \geq p^*(\underline{\theta}, x)$ for all x. This follows from the lemma. (*b*) In this case, $D(p^*(\theta, x); \theta) \geq D(p^*(\underline{\theta}, x); \underline{\theta})$. This is (virtually) the lemma. (*c*) The function $x - D(p^*(\theta, x); \theta)$ is increasing in x. This is part of proposition 1.

Now let $\theta_0, \theta_1, \ldots, \theta_{t-1}$ be any sequence of θ's and let $X_t(\theta_0, \ldots, \theta_{t-1})$ be defined iteratively by $X_0 = \underline{X}$ and $X_{s+1}(\theta_0, \ldots, \theta_s) = G(X_s(\theta_0, \ldots,$

$\theta_{s-1}) - D(p^*(\theta_s, X_s(\theta_0, \ldots, \theta_{s-1})); \theta_s))$. That is, X_t gives the value (in equilibrium) of supplies at date t as a function of past states of demand, where $X_0 = \underline{X}$. We claim that, for all $(\theta_0, \ldots, \theta_{t-1})$, $X_t(\theta_0, \ldots, \theta_{t-1}) \leq X_t(\underline{\theta}, \ldots, \underline{\theta})$ (where in the right-hand side there are t θ's). This follows inductively from facts b and c. Suppose that $X_s(\theta_0, \ldots, \theta_{s-1}) \leq X_s(\underline{\theta}, \ldots, \underline{\theta})$. Then

$$X_s(\theta_0, \ldots, \theta_{s-1}) - D(p^*(\theta_s, X_s(\theta_0, \ldots, \theta_{s-1})); \theta_s)$$
$$\leq X_s(\underline{\theta}, \ldots, \underline{\theta}) - D(p^*(\theta_s, X_s(\underline{\theta}, \ldots, \underline{\theta})); \theta_s)$$

by fact c

$$\leq X_s(\underline{\theta}, \ldots, \underline{\theta}) - D(p^*(\underline{\theta}, X_s(\underline{\theta}, \ldots, \underline{\theta})); \underline{\theta})$$

by fact b. Since G is nondecreasing, we obtain $X_{s+1}(\theta_0, \ldots, \theta_s) \leq X_{s+1}(\underline{\theta}, \ldots, \underline{\theta})$, completing the induction step. Thus

$$p^*(\theta_t, X_t(\theta_0, \ldots, \theta_{t-1})) \geq p^*(\theta_t, X_t(\underline{\theta}, \ldots, \underline{\theta})) \geq p^*(\underline{\theta}, X_t(\underline{\theta}, \ldots, \underline{\theta}))$$

(The last step uses fact a). Thus the lowest possible price at date t is the one at which every $\theta_s = \underline{\theta}$.

And, finally, if we start at $X_0 = \underline{X}$, then $X_1(\underline{\theta}) \geq X_0$. Applying fact c inductively shows that $X_t(\underline{\theta}, \ldots, \underline{\theta}) \geq X_{t-1}(\underline{\theta}, \ldots, \underline{\theta})$. Thus along the sequence $\theta_0 = \underline{\theta}, \theta_1 = \underline{\theta}, \ldots$, if we start at $X_0 = \underline{X}$, then prices continually decline. But they can never decline below $p^c(\underline{\theta})$, since at each stage speculative demand is at least as large as at the stage before, and so $G(D'_t) - D'_{t+1} \leq G(D'_t) - D'_t \leq G(0)$. Thus price is such that consumption demand is no more than $G(0)$; that is, price is no lower than $p^c(\underline{\theta})$.

Is there any stronger criterion by which one might judge the stability of prices? One possibility is to look at the stationary distribution of prices, using summary statistics such as the variance of the distribution or criteria of riskiness such as those in Rothschild and Stiglitz (1970) or Diamond and Stiglitz (1974). But consider: suppose that in one regime we have price equal to US$100 for 1,000 periods in a row, followed by 1,000 periods at US$200, then 1,000 at US$100, and so forth, while in a second regime the price is either US$101 or US$199, each with probability one-half, independent of previous prices. By criteria applied to the distribution of prices, the former regime is "less stable", yet this does not necessarily accord with our intuition. So perhaps we should look at the stationary distribution of price changes (in either absolute or relative terms): we could consider the variance of price changes or apply Diamond–Stiglitz or Rothschild–Stiglitz measures to them.

It is hard to find some criterion by which one might choose because greater stability is not related to anything like the welfare of consumers or any other economically meaningful quantity. We can think of circumstances in which each of these measures would be appropriate and others in which each would be inappropriate, at least intuitively. (The example of the second section, it should be noted, gives greater instability in both the distribution of prices and the distribution of price changes). And, in any event, even when $\phi_t = \theta_t$ and the θ_t are i.i.d., it seems unlikely that any stronger results can be obtained.

One problem is the following: consider what happens if some of the good put into storage is lost to wastage. In this case, average consumption with speculation will be less than average consumption without, so one expects average prices to rise. This is rigorously true if $D(p; \theta) = K(\theta) - Ap$ for constant $A > 0$ in the relevant range of prices. How does a rise in average price square with greater stability? If, moreover, demand at $\underline{\theta}$ is (nearly) perfectly elastic (and the storage capacity of speculators is insufficient to absorb $\underline{X}$), then correcting for any such shift in mean will give a (corrected) distribution of prices under speculation that falls below the distribution of prices for the economy with no speculation.[7] What about the distribution of price changes? We will not give details but simply assert that examples can be constructed that show that the answer here is ambiguous as well. If a stronger statement than the proposition is possible, we do not know what it is.

Since it is hard to strengthen the *notion* of stabilisation, we now consider whether instead it is possible to relax the *conditions* required for stabilisation. In the example of the second section, the sequence $\{\theta_t\}$ is i.i.d. Hence this alone is insufficient for the weak result in proposition 3 if agents are able to foresee something of the future. It is also insufficient to assume that $\phi_t = \theta_t$ but relax the assumption that $\{\theta_t\}$ is i.i.d. This is immediate from the example of the second section as well: we can simply redefine θ_t so there are four possible values corresponding to the four states in the Markov chain.

One might imagine that what causes the problem in the example is the imperfect nature of the signal ξ_t. Suppose, for example, that $\{\theta_t\}$ is i.i.d. and that any foreseeing that takes place is flawless: $\phi_t = (\theta_t, \theta_{t+1}, \ldots, \theta_{t+k})$ for some k. Will this suffice to obtain a result analogous to proposition 3?

The answer is no, for any finite k. Fixing k, we can construct an example in which the lowest price with speculation is lower than the lowest without. Let us sketch the example for $k = 1$ to give the basic idea.

Imagine that θ_t can take on any of three values, $\underline{\theta}$, θ^0, or $\bar{\theta}$. Consumption demand schedules are as illustrated in Figure 1. Suppose that $\bar{\theta}$ has very high probability and that $D(\cdot; \bar{\theta})$ is sufficiently high that, whenever it is known that θ_{t+1} will equal $\bar{\theta}$, speculators buy up to their capacity (less than $\underline{X}$). Moreover, the chance that $\theta_{t+2} = \bar{\theta}$ is so high that, when $\theta_{t+1} = \theta^0$ and $\theta_t = \underline{\theta}$, the anticipation of this (and the concomitant rise in p_{t+1}) causes

Figure 1

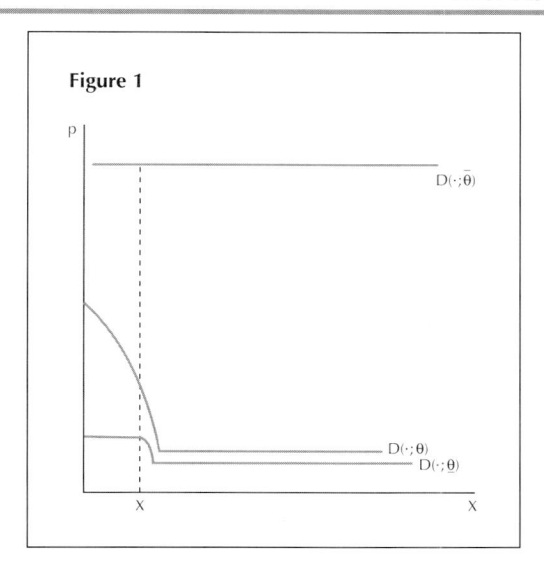

[Figure 1: plot with axis p vertical and x horizontal, showing curves labeled $D(\cdot;\bar{\theta})$ (upper) and $D(\cdot;\theta)$, $D(\cdot;\underline{\theta})$ (lower).]

speculators to buy up to their capacity. But when $\theta_{t+1} = \underline{\theta}$, the elasticity of $D(\cdot;\underline{\theta})$ implies that in period $t+1$ prices will not rise above $p^c(\underline{\theta})$; hence under no circumstances would speculators store at date t.

Start the economy with $\theta_t = \underline{\theta}$ and $\theta_{t+1} = \theta^0$. Speculators then store, anticipating that, with high probability, θ_{t+2} will equal θ and p_{t+1} will rise, given the further rise anticipated at date $t+2$. But then suppose θ_{t+2} turns out to be $\underline{\theta}$. Speculators will dump their stores, and – the way we have drawn $D(\cdot;\theta^0)$ – this will take p_{t+1} below $p^c(\underline{\theta})$. Note that prices will never fall below $p^c(\underline{\theta})$ when $\theta_t = \underline{\theta}$, but they will sometimes do so when $\theta_t = \theta^0$. That is, the lowest observed price will not correspond to the lowest θ.

This covers the case $k = 1$. For any finite k, a similar example can be constructed by having the consumption demand curves ever more wild in behaviour. On the other hand, if we fix an economy except for the specification of k (ie, fix D, D', Θ, and G) and then lengthen the foresightedness of agents by increasing k, eventually we get an asymptotic result analogous to proposition 3. Indeed for cases of very far foresightedness, we need not assume that the θ_t are i.i.d.

While the general proof (and even the statement of the result) is quite involved, there is a simple limiting case, that of perfect foresight: At date t agents know all future values of θ_{t+k}. (This violates our assumption of finite Φ; however, no technical problems arise).

Proposition 4. In the case of perfect foresight (starting with $X_0 = \underline{X}$), equilibrium prices never drop below $p^c(\underline{\theta})$.

The proof is simple. Fix any sequence $\{\theta_t\}$, and let $p_0, p_1, \ldots$ be the (deterministic) sequence of equilibrium prices. Suppose (inductively) that

$p_t \geq p^c(\theta)$. Either $p_{t+1} \geq p_t$ or $p_{t+1} < p_t$. In the former case, $p_{t+1} \geq p^c(\theta)$. And in the latter case, speculative demand is zero so that $p_{t+1} \geq p^c(\theta_{t+1}) \geq p^c(\theta)$. Since $p_0 \geq p^c(\theta)$, the result follows by induction.

More generally, we have the following result.

Proposition 5. Fix the following pieces of a model: Θ, D, D', and G. For every $\varepsilon > 0$ there exists an N such that for all models in which ϕ_t reveals (at least) $(\theta_t, \theta_{t+1}, \ldots, \theta_{t+N})$, the lowest price ever achieved, starting with $X(0) = \underline{X}$, is greater than or equal to $p^c(\underline{\theta}) - \varepsilon$.

The proof is only sketched here. (A complete proof is available from the authors on request.) Given our assumptions, one can show that for every $\delta > 0$ there is a sufficiently large N (depending on Θ, D, D', and G) that, if agents can foresee at least N periods, the support of prices next period is contained in an interval of size δ. That is, as one sees further and further into the future, surprises about that future have less and less effect on next period's prices. Now for every $\varepsilon > 0$, there is a $\beta > 0$ such that prices can be depressed to a level at or below $p^c(\underline{\theta}) - \varepsilon$ only if supplies carried forward by speculators are at least β. And for each $\beta > 0$ there is a "rate of interest" $\iota > 0$ such that speculators will not carry forward β unless there is positive probability that prices will rise by at least ι. If we take δ to be $\iota[p^c(\underline{\theta}) - \varepsilon]/2$ and N sufficiently large, this means that prices can be depressed to below $p^c(\underline{\theta}) - \varepsilon$ from above this level only if they are certain to be no lower than $p^c(\underline{\theta}) - \varepsilon$, which establishes $p^c(\underline{\theta}) - \varepsilon$ as a lower bound if we start above this level.

CONCLUSION

It is sometimes argued that speculation has the following desirable features. First, by buying cheap and selling dear, speculators stabilise prices. Second, by looking into the future and anticipating economic trends, speculators smooth the transition of the economy from one long-run equilibrium to another. We have shown, using a very simple model in which speculation is synonymous with storage, that speculation may possess neither of these features. In particular, we have presented examples in which, in spite of the fact that non-speculators and speculators alike behave rationally and speculators are competitive, speculation destabilises prices (in any meaningful sense of the word). Moreover, this is not because of a lack of foresight by speculators: in fact, making speculators more foresighted may actually worsen the problem.

As we have emphasised, the conclusion is that speculative activity *may* destabilise prices, not that it will. In fact, we have presented sufficient conditions for speculation to be stabilising (albeit in a very weak sense). These are that consumption demand is independently and identically distributed over time and that speculators have no foresight about future demand at all, or that speculators have a great deal of foresight. The extreme restrictiveness of these conditions is striking, as is the very weak criterion of stabilisation that we can show. And it is not lack of effort (at least) that causes us to report such meagre positive results.

Indeed, while we do not believe that speculative activity has no stabilising effect at all, we do find appealing the sort of short-run decrease in stability that permeates our basic example. Speculators will buy or sell according to increases and decreases in the probability of large-scale changes. They may withdraw supplies from the market, and then, when the danger recedes, they may dump those supplies back on the market. While this activity might well smooth the major transitions when they happen, it can (and, we believe, does) do so at the cost of more small-scale fluctuations for which a tangible cause will only very rarely be identifiable.

In order to proceed with our analysis, we have made a number of simplifying assumptions. One of the most serious of these is the omission of production. It is not clear, however, that including production would change our conclusions all that much. One must take care in distinguishing between instantaneous production and production with a lag. If production is instantaneous, then it can be absorbed into non-speculative demand, and our analysis is unaltered. Production with a lag raises new issues. Production with a lag (of, say, one period) is similar to storage in some respects, but there is an important difference: Storage responds (generally) to increased chances for spot price appreciation since the opportunity cost of storage is forgone consumption today of the good in question. Insofar as production has as opportunity cost forgone consumption of other goods, it responds more to increased chances of high (absolute) prices tomorrow. This can have counterintuitive effects (see, eg, Scheinkman and Schechtman 1983). Still, the sort of example we have given for destabilising storage is easily modified to give an example of destabilising production. At the same time, with lagged production, speculation may yield an important additional benefit of conveying information. Suppose, for example, that speculators (correctly) anticipate a future increase in demand for a commodity. This will increase current storage demand and hence lead to a current price rise. Producers, observing this price rise, will, given rational expectations, deduce that future demand is likely to be high and will be encouraged to invest in the future production of the commodity. As a result, future prices will be lower than otherwise (since supplies are greater), and the price path may be stabilised relative to a situation of no speculation. (This, of course, is subject to exactly the sorts of caveats around which this paper has been written.) Thus the inclusion of production may strengthen the case that (foresighted) speculative activity is stabilising, at least insofar as it gives to producers information that they would otherwise not collect.

Finally, we should emphasise that (as already noted above), while we have studied the effect of speculation on prices, the more interesting economic question concerns the effect of speculation on welfare. Conditions are known under which competitive speculation leads to a first-best in terms of total welfare (see, eg, Samuelson 1971; Scheinkman and Schechtman 1983). This is so regardless of whether speculation stabilises or destabilises prices. But these conditions involve risk neutrality, and they are results for total welfare. We noted, in the context of our basic example, that with risk-averse producers, "destabilising" speculation can lead to an *ex ante* Pareto decrease in utility. Examples in the spirit of Newbery and Stiglitz (1984) can obtain the same result for "stabilising" speculation. And if one looks at one sector only and not at overall welfare, then even assuming risk neutrality is of little help. In our basic example, consumers are better off for having prices destabilised; it is not hard to put together examples in which risk-neutral consumers are worse off with (stabilising) speculation. We have shown here that it is hard to say much about the effects of speculation on price stability. It is just as hard to say anything about the welfare implications of speculation.

APPENDIX

Technical details

Recall that $\bar{p}$ solves $D(\bar{p}, \bar{\theta}) = \underline{X}$. Let $\underline{p}$ be the solution to $D(\underline{p}, \underline{\theta}) = \overline{X}$. By assumption 2, there exists a constant γ such that, for each θ, $D(\cdot; \theta)$ is γ-Lipschitzian over the range $[\underline{p}, \bar{p}]$ and $P(\cdot; \theta)$ is γ-Lipschitzian over the range $[\underline{X}, \overline{X}]$. (Take γ to be a uniform upper bound on the first derivatives of the functions over these compact intervals.)

Lemma A1. Suppose that $q: \Phi \times [\underline{X}, \overline{X}] \to [0, \infty)$ is continuous and nonincreasing in its second argument and that $q \leq \bar{p}$. Then there exists a unique function $Tq: \Phi \times [\underline{X}, \overline{X}] \to [0, \infty)$ that solves

$$D(Tq(\phi, X); \theta) + D'(Tq(\phi, X); F(\phi, G(X - D(Tq(\phi, X); \theta)), q)) = X \quad \text{(A1)}$$

Moreover, this function Tq is strictly decreasing and 2γ-Lipschitzian in its second argument and satisfies $\underline{p} \leq Tq \leq \bar{p}$.

The operator T defined by (A1) has a simple interpretation: if prices next period are given by q and if speculators this period understand that this is so, then prices this period are given by Tq.

Proof. Refer to Figure A1. Fixing $\phi = (\theta, \xi)$ and X, we graph $D(p; \theta)$ and $D(p; \theta) + D'(p; F(\phi, G(X - D(p; \theta)), q))$. Of course, $D(p; \theta)$ is continuous and strictly decreasing in p. And $D'(p; F(\phi, G(X - D(p; \theta)), q))$ is continuous and nonincreasing in p. To see that D' is nonincreasing, note that increasing p raises the first argument and lowers (stochastically) the second. As for continuity, apply assumptions 3, 4, and 9, noting that q is, by assumption, bounded above. For $p \leq \underline{p}$, $D(p; \theta)$ and, hence, $D + D'$ exceed $\overline{X}$. For $p \geq \bar{p}$, $D(p; \theta) \leq \underline{X}$ while $D' = 0$ (since $q \leq \bar{p}$), so $D + D' \leq \underline{X}$. Thus Figure A1 is appropriate, and (for each $X \in [\underline{X}, \overline{X}]$) there exists a

Figure A1

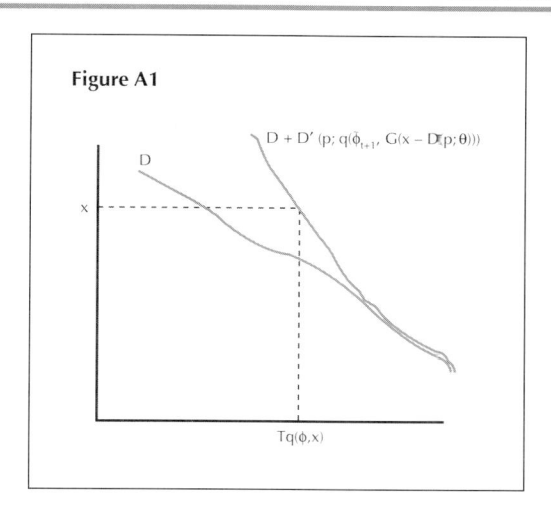

[Figure A1: Graph showing curves D and D + D'(p; q(φ_{t+1}, G(x − D(p;θ)))), with dashed lines from X on vertical axis meeting at Tq(φ,x) on horizontal axis.]

unique $Tq(\phi, X)$ that solves equation (A1) and that lies between $\underline{p}$ and $\bar{p}$.

Refer now to Figure A2. Fixing $\phi = (\theta, \xi)$, when we raise X to the level X', the level of D' falls for each p because $X - D(p; \theta)$ and, hence, $G(X - D(p; \theta))$ both increase, which causes the distribution of future prices to fall. It is immediately clear that $Tq(\phi, X') < Tq(\phi, X)$; Tq is strictly decreasing in its second argument. (The strictness comes from the strict decrease in D.) To measure how much Tq decreases, we will estimate first $p^0 - Tq(\phi, X')$ and then $Tq(\phi, X) - p^0$, where p^0 is (as shown) the solution of

$$D(p; \theta) + D'(p; F(\phi, G(X' - D(p; \theta)), q)) = X$$

Estimating $p^0 - Tq(\phi, X')$ is easy; the fact that $D + D'$ is more steeply sloped than is D together with our upper bound on the slope of the inverse demand function P imply that $p^0 - Tq(\phi, X') \leq \gamma(X' - X)$. As for $Tq(\phi, X) - p^0$, note that, if p' and p'' are such that $D(p'; \theta) - D(p''; \theta) \geq X' - X$, then $X - D(p''; \theta) \geq X' - D(p'; \theta)$, and hence $q(\phi_{t+1}, G(X - D(p''; \theta))) \leq q(\phi_{t+1}, G(X' - D(p'; \theta)))$ for each ϕ_{t+1}. Since $D(p'; \theta) - D(p''; \theta) \geq X' - X > 0$ can hold only when $p' < p''$, this condition implies that

$$D(p'; \theta) + D'(p'; F(\phi, G(X' - D(p'; \theta)), q)) > D(p''; \theta)$$
$$+ D'(p''; F(\phi, G(X - D(p''; \theta)), q))$$

Letting p^0 play the role of p' and $Tq(\phi, X)$ the role of p'', the definition of p^0 implies that $D(p^0; \theta) - D(Tq(\phi, X); \theta) < X' - X$. Thus by our bound on

Figure A2

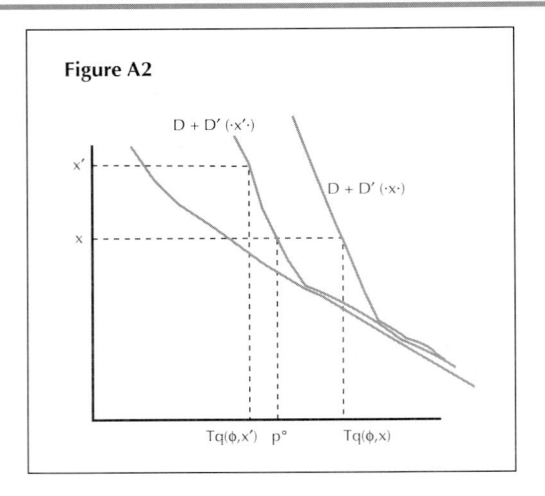

[Figure A2: Graph showing curves $D + D'(\cdot x'\cdot)$ and $D + D'(\cdot x \cdot)$ with horizontal levels x' and x, and points $Tq(\phi, x')$, p^0, $Tq(\phi, x)$ on the horizontal axis.]

the slope of the inverse demand function, $p^0 - Tq(\phi, X) \leq \gamma(X' - X)$.

With these two estimates, the proof of the lemma is complete.

The last part of this argument is a bit hard to read through, so let us rephrase it. (The reason we need this last part of the argument will become apparent shortly.) We want to know how much equilibrium prices will change as we move from supply level X to a (higher) level X'. We take the change in two pieces. First (and this corresponds to the second estimate above) imagine that the "extra" $X' - X$ is put into storage for a period — without trying to account for who is doing so. At the equilibrium price for X, speculators will react to this by lessening their demand since the extra will depress next period's price. But then equilibrium price this period will fall. The estimate simply says that the fall in equilibrium price cannot be more than the amount that causes consumers to absorb the extra since when that happens total storage will be back to its original level and speculative demand will be restored to its original level (or, owing to the lower current price, more). Second, prices must fall so that someone is really absorbing the extra units. But again the price decline cannot be more than it would take to put these units in the hands of consumers since speculative demand will only rise. Since we have (by assumption) put a uniform upper bound on the elasticity of demand of consumers, we get the estimates given above.

Lemma A2. Let q and q' be two functions from $\Phi \times [\underline{X}, \overline{X}] \to [0, \infty)$ that are both continuous and non-increasing in their second arguments and that satisfy $q \leq q' \leq \bar{p}$. Then $Tq \leq Tq'$.

Proof. Fixing $\phi = (\theta, \xi)$ and X, $q \leq q'$ implies that, for each p,

$$D'(p; F(\phi, G(X - D(p; \theta)), q)) \leq D'(p; F(\phi, G(X - D(p; \theta)), q'))$$

Thus moving from q to q' in Figure A1 amounts to a shift outward of the $D + D'$ schedule and, hence, an increase (weakly) in $Tq(\phi, X)$.

Proof of Proposition 1. We can now proceed to prove proposition 1. Define $p^0 \equiv 0$, and let $p^1 = Tp^0$, $p^2 = Tp^1$, and so on. (The interpretation is that p^n is the equilibrium price if speculators will disappear in n periods). Lemma A1 ensures that each step of the construction is feasible, that each p^n is strictly decreasing and 2γ-Lipschitzian (hence continuous) in its second argument, and that $\underline{p} \leq p^n \leq \overline{p}$ for each n. Lemma A2 ensures that $p^{n+1} \geq p^n$. Thus the limit function $p^* = \lim_{n \to \infty} p^n$ is well defined, it is non-increasing and 2γ-Lipschitzian, and it satisfies $\underline{p} \leq p^* \leq \overline{p}$. (Note that here is where we use the Lipschitzian property: simple continuity of the p^n would not necessarily imply continuity of p^*.)

Because p^* and the p^n are all equi-Lipschitzian, we can "pass to the limit" in (A1), showing that $Tp^* = p^*$ and (hence) that p^* is an equilibrium. Precisely, for every X and $\phi = (\theta, \xi)$, the continuity of G and D implies that $G(X - D(p^n(\phi, X); \theta))$ approaches $G(X - D(p^*(\phi, X); \theta))$. And thus the equicontinuity of p^* and the p^n implies that, for each ϕ_{t+1}, $p^n(\phi_{t+1}, G(X - D(p^n(\phi, X); \theta)))$ approaches $p^*(\phi_{t+1}, G(X - D(p^*(\phi, X); \theta)))$. The assumed continuity of the D' function then gives the desired result.

Application of lemma A1 once more implies that p^* (now shown equal to Tp^* for a non-increasing p^*) is strictly decreasing. To show that $X - D(p^*(\phi, X); \theta)$ is increasing in X, suppose that $X > X'$ but $X - D(p^*(\phi, X); \theta) < X' - D(p^*(\phi, X'); \theta)$. Storage is less this period under X than X', so next period's prices will be no smaller under X. And current price is lower under X. Hence speculative demand can be no lower under X, contradicting the supposition that it is strictly lower.

It remains to show that this equilibrium is unique. The first step is to show that equilibrium prices are bounded above. Since supply is never less than $\underline{X}$, we can apply assumption 6 to get this trivially: pick p large enough that $D(p; \overline{\theta}) < \underline{X}/2$ and that $D'(p; \cdot) < \underline{X}/2$. Clearly, no equilibrium price could ever exceed this p.

Next, we argue that, if $\tilde{p}$ is any other equilibrium, then $\tilde{p} \geq p^*$. To see this, we must strengthen lemma A2 to read: if q and q' are functions such that (i) q is continuous and non-increasing in its second argument and is less than or equal to $\overline{p}$, (ii) q' is such that there is some solution Tq' of (A2) and (iii) $q \leq q'$, then $Tq \leq Tq'$ (for the unique solution Tq and any solution Tq'). This strengthening is easy: simply review the argument given and/or draw the appropriate picture. With this result, proceed as follows: Since prices are non-negative, $\tilde{p} \geq p^0$. Applying the result inductively yields $T\tilde{p} = \tilde{p} \geq Tp^n = p^{n+1}$ for all n, and (hence) $\tilde{p} \geq p^*$.

Finally, suppose that we knew that there was some (ϕ^0, X^0) that achieved the maximum value in $\tilde{p}(\phi, X)/p^*(\phi, X)$. Then the distribution of "rates of return" under $\tilde{p}$ starting from (ϕ^0, X^0) will be no larger (stochastically) than the distribution of rates of return under p^* at the same point. Then assumption 7 implies that speculative demand under $\tilde{p}$ can be no larger at this point than it is under p^*. And, since $\tilde{p} \geq p^*$, consumption demand can be no higher. Thus to satisfy market clearing, $p(\phi^0, X^0) = p^*(\phi^0, X^0)$ (or else there would be a shortfall in consumption demand). Since (ϕ^0, X^0) is presumed to maximise the ratio $\tilde{p}/p^*$, this implies that $\tilde{p} \equiv p^*$.

This argument is not quite complete since it assumes that there is some (ϕ^0, X^0) that attains the maximum in the ratios. As the domain of possible (ϕ, X) is compact, if we knew that $\tilde{p}$ was continuous, we would be done. But even if $\tilde{p}$ is not necessarily continuous, the argument goes through with a bit of care: look along a sequence of (ϕ, X) that attains (in the limit) the supremum of $\tilde{p}(\phi, X)/p^*(\phi, X)$. Picking a subsequence if necessary, we can assume that $\tilde{p}(\phi', G(X - D(\overline{p}(\phi, X); \theta)))$ converges along this sequence for every ϕ'. Now apply the argument above to show that, for this limit distribution of next period's prices, there will be a non-vanishing shortfall in demand as long as the supremum of $\tilde{p}(\phi, X)/p^*(\phi, X)$ is strictly greater than one. Applying the continuity of D and D' near this limit completes the argument.

In our definition of equilibrium, we specifically assume time homogeneity. It is worth noting that time-inhomogeneous equilibria are ruled out by an argument similar to the one just given. The key is that, because of assumption 6, equilibrium prices must be uniformly bounded. So one would look for a triple (t, ϕ, X) that achieves close to the supremum of $\overline{p}_t(\phi, X)/p^*(\phi, X)$ and then proceed as above.

Proof of Proposition 2. We have already shown that $p^* \leq \overline{p}$. Suppose that for some ξ we had $p^*((\overline{\theta}, \xi), \underline{X}) < \overline{p}$. Then consumption demand at this price would exceed $\underline{X}$, a contradiction.

---

1 This would be one sort of small sample bias. Others are discussed in Kleidon (1986). Also, it should be noted that Shiller himself (1981b, p. 300) considers and rejects this explanation of his empirical findings. We do not wish to enter into any empirical controversy, but we cannot help noting that if we interpret $\theta_t = \overline{\theta}$ as Shiller's disaster and the time period as 1 year, then our numbers work out as follows. There is a 0.01 chance of a disaster in any year, the chance of observing no disasters in a period of 108 years is approximately 0.34, and the chance of observing exactly one disaster is around 0.37. Moreover, the standard deviation in the probability that a disaster will occur next period is approximately 0.03. This is not the 0.05 that is needed to explain Shiller's data, but then we have not tried to find the model that maximises the standard deviation in the probability of an incipient disaster subject to keeping the average probability of a disaster low. In any event, this explanation is not so easy to dismiss as Shiller seems to contend.
2 Since F_{t+1} is a distribution, this assumption needs clarification. The schedule D' is continuous in F_{t+1} in the sense that, if $\{F_n; n = 1, 2, \ldots\}$ have uniformly bounded supports and approach F in the weak topology, then $D'(p, F^n) \to D'(p; F)$ for all p. And D' is nondecreasing in F_{t+1} if $D'(p; F') \geq D'(p; F)$ for all p whenever F' is (first-order) stochastically larger than F.
3 Time homogeneity simply means that transition probabilities from one state to another do not depend on calendar time.

4 We note again that, since supply is inelastic and demand is independent of the amount brought forward, we are thinking of a consumption sector distinct from storers and from overlapping generations of speculators who consume only the numeraire. Random exogenous supply is easily accommodated by shifting any randomness to the demand side, ie, by absorbing it in D. Randomness in the storage technology can also be accommodated, although the model becomes a bit more complicated.
5 The force of assumption 4 should not be underestimated. If speculative demand arises from expected utility maximisation, then it need not be nondecreasing in next period's prices; compare the comments concerning risk aversion following assumption 7.
6 It is easy to show that, with speculators, prices arbitrarily close to $\bar{p}$ will be observed infinitely often, regardless of the initial conditions, as long as $\underline{\theta}$ is visited infinitely often (is positively recurrent from every other state) and transition from $\underline{\theta}$ to $\bar{\theta}$ in a single step has positive probability. (This will be true in all our examples.) This is because, however large are initial stocks, if there are sufficiently many high-demand states in a row (an event that will occur eventually), these stocks will be exhausted and prices will rise to $\bar{p}$. (In this regard, note our assumption that the level of stocks is bounded above).
7 Wright and Williams (1982) note the difficulty of determining whether speculation is stabilising given a change in the average price.

BIBLIOGRAPHY

Baumol, William J., 1957, "Speculation, Profitability, and Stability", *Review of Economics and Statistics* 39, August, pp. 263–71 reprinted as Chapter 18 of the present volume.

Diamond, Peter A. and Joseph E. Stiglitz, 1974, "Increases in Risk and in Risk Aversion", *Journal of Economic Theory* 8, July, pp. 337–60.

Farrell, Michael J., 1966, "Profitable Speculation", *Economica* 33, May, pp. 183–93 reprinted as Chapter 20 of the present volume.

Friedman, Milton, 1953, *Essays in Positive Economics* (University of Chicago Press).

Goss, B. A. and Basil S. Yamey, 1978, *The Economics of Futures Trading: Readings*, Second edition (London: Macmillan).

Johnson, Harry G., 1976, "Destabilising Speculation: A General Equilibrium Approach", *Journal of Political Economics* 84, February, pp. 101–8.

Kleidon, Allan W., 1986, "Bias in Small Sample Tests of Stock Price Rationality", *Journal of Business* 59, pp. 237–61.

Kohn, Meir, 1978, "Competitive Speculation", *Econometrica* 46, September, pp. 1061–76.

Krasker, William S., 1980, "The 'Peso Problem' in Testing the Efficiency of Forward Exchange Markets", *Journal of Monetary Economy* 6, April, pp. 269–76.

Mill, John S., 1921, *Principles of Political Economy, with Some of Their Applications to Social Philosophy* (London: Longmans, Green).

Newbery, David M. G. and Joseph E. Stiglitz, 1984, "Pareto Inferior Trade", *Review of Economic Studies* 51, January, pp. 1–12.

Rothschild, Michael and Joseph E. Stiglitz, 1970, "Increasing Risk: 1. A Definition", *Journal of Economic Theory* 2, September, pp. 225–43.

Samuelson, Paul A., 1971, "Stochastic Speculative Price", *Proc. Nat. Acad. Sci.* 68, February, pp. 335–37.

Scheinkman, José A. and Jack Schechtman, 1983, "A Simple Competitive Model with Production and Storage", *Review of Economic Studies* 50, July, pp. 427–41.

Shiller, Robert J., 1981a, "Do Stock Prices Move Too Much to Be Justified by Subsequent Changes in Dividends?", *American Economic Review* 71, June, pp. 421–36.

Shiller, Robert J., 1981b, "The Use of Volatility Measures in Assessing Market Efficiency", *Journal of Finance* 36, May, pp. 291–304.

Smith, Adam, 1937, *An Inquiry into the Nature and Causes of the Wealth of Nations*, Fifth edition 1789, Reprint, edited by Edwin Cannan (New York: Modern Library).

Telser, Lester G., 1959, "A Theory of Speculation Relating Profitability and Stability", *Review of Economics and Statistics* 41, August, pp. 295–301; reprinted as Chapter 19 of the present volume.

Wright, Brian D. and Jeffrey C. Williams, 1982, "The Economic Role of Commodity Storage", *Economic Journal* 92, September, pp. 596–614.

5

RETURNS TO SPECULATORS AND THE COSTS OF HEDGING

Introduction
Lester G. Telser

This section contains a sequence of articles, beginning with Houthakker (1957) and ending with Hartzmark (1991), all seeking to measure the profits of speculators and hedgers with a view to assessing the validity of the neoclassical theory of risk transfer from hedgers to speculators. Houthakker claims the evidence supports this theory. Rockwell is sceptical. Dusak uses modern portfolio theory and claims her estimates do not imply a positive return on average to speculators. Lastly, Hartzmark, using data closest to the actual outcome of speculation, shows that the returns to speculators are so highly skewed that describing them as being in the business of selling insurance to hedgers is misleading.

Houthakker presents highly ingenious analyses of the available data in order to draw inferences about the profits and losses of hedgers and speculators. At the outset it is necessary to draw attention to the fact that these figures refer only to reporting traders who had open positions at the end of a trading session large enough to require them to report these to the Federal regulatory agency that then was the Commodity Exchange Authority. Positions of the non-reporting traders are calculated by subtracting the figures for the reporting traders from the known total open interest. Therefore, there are two, not three, degrees of freedom in these estimates. "The profit or loss of that group was then found by multiplying the end-of-month position by the change in the average price". However, futures prices change a good deal within a trading session, not to mention during a whole month. Yet Houthakker does not try to measure the reliability of his estimates by reckoning the highly variable character of the prices although the price data for so doing were readily available. Houthakker's estimates of gains and losses make no provision for transaction costs. The total cost of the transactions is proportional to the volume of trade, not to the open interest. The volume of trade is several times larger than the open interest. How much profit would remain after subtracting transaction costs which apply to each trade is not known. Hence, omission of estimates of the transaction costs is a potentially major shortcoming of Houthakker's analysis. It would be very hazardous to follow his advice and try to earn a positive return by a steadfast adherence to a long position in futures as a seller of price insurance to short hedgers.

Like Working, Houthakker seems troubled by Blair Stewart's results in his study of gains and losses (*An Analysis of Speculative Trading in Grain Futures*, 1949), which is based on nearly 9,000 traders' accounts of a large national brokerage firm in Chicago covering the period from 1925 to 1934. Stewart summarises the main results of his study in his Table 26 (p. 57). There are 2,184 profit traders and 6,598 loss traders. The net profit per winner is US$945 and the net loss per loser was US$1,812. There are 59 hedgers in the study with profits, and the profit per hedger is US$16,442. There are 81 hedgers in the study with losses, and the loss per hedger is US$17,037. It is also important to note sizeable differences among commodities. Thus the profit per hedger is much bigger than the loss per hedger in corn and oats. In wheat the results are nearly even and in rye the loss per hedger is much bigger than the profit per hedger. These results, admittedly for another time, differ so much from Houthakker's estimates that he is impelled to offer explanations. Of his two, the second, that the brokerage firm went bankrupt in 1934, is especially unpersuasive. Many firms went bankrupt in 1934. The early period of Stewart's study, 1925 to 1929, was one of the most prosperous in US history.

Of greater importance than the average result per trader is the shape of the frequency distribution of the returns. Stewart presents the frequency distributions of profits and losses separately for speculators, Table 28, and hedgers, Table 29, in each of the four commodities in his study. Skewness is the salient feature for the speculators. Only a handful of speculators made significant profits. Most speculators had small losses. Of those with profits, 39% had profits below US$100, and 84% had profits below US$1,000. Stewart concludes his description of the frequency distributions with the following statement: "It is apparent that a very large percentage of the traders in the sample operated on a small scale, and also that many of them discontinued trading before either large profits or large losses had been accumulated" (p. 61).

The next chapter by Charles S. Rockwell covers a large number of commodities and a longer period than Houthakker's while using the same kinds of estimates and approach. It is, therefore, all the more noteworthy that Rockwell asserts that his more comprehensive evidence leads him strongly to reject the theory of normal backwardation and to embrace the view that either skill or luck is the main explanation of positive speculative profits.

The third chapter by Katherine Dusak introduces a new approach to the search for a risk premium. It does so by asking whether adding positions in futures changes the risk of a well-diversified portfolio of wealth. This chapter includes a concise exposition of modern portfolio theory. Dusak also presents evidence bearing on an important aspect of the problem by plotting the cumulative distribution function of the sample returns on probability paper and thereby offers clear evidence of non-normality. The reasons for this and their implications form the subject of the two articles by Mandelbrot in Section 6. Dusak concludes that her results for wheat, corn and soybeans do not support the normal backwardation theory.

The fourth chapter in this quartet by Michael L. Hartzmark, turns to data in the hands of the Commodity Futures Trading Commission, the current regulatory agency of futures trading. In order to estimate directly the profits and losses of reporting traders, it uses daily figures for the period July 1, 1977 to December 31, 1981 covering nine futures markets, including two financial futures, T-bonds on the Chicago Board of Trade and 90-day T-Bills on the Chicago Mercantile Exchange. Hartzmark abandons the distinction between hedgers and speculators and uses instead two classes, commercial and non-commercial traders. It should be emphasised that at any stage of his study at which he had to make a choice, it was to bias the results against the hypothesis that traders lack skill. Even so, applying a battery of tests to different aspects of his sample, Hartzmark concludes there is almost no evidence of skill and that luck predominates. It appears on this evidence that normal backwardation as a theory of futures should be respectfully interred.

22
Can Speculators Forecast Prices?*

H. S. Houthakker
Harvard University

The role of speculation in the economic system is still a matter of controversy. In popular parlance the word has acquired an unfavourable connotation; most economists would probably say speculation is at best a necessary evil, though some would regard it as an unnecessary source of instability. One of the main issues in evaluating speculation is no doubt the degree of skill with which speculators can forecast prices: the more accurately prices are forecast, the less they will fluctuate, and the easier therefore the adjustments which interested parties have to make. Thus formulated the question leaves open to what extent the prices that actually emerge are in some sense optimal, for steadiness is only a minor characteristic of optimality. The very difficulty of defining optimality in a dynamic context, however, is a sufficient reason for separately considering speculators' success in predicting prices as they are. For this purpose we shall consider data concerning three important American commodity markets.[1]

In commodity futures markets a measure of the forecasting ability of speculators is not hard to find, for it is immediately reflected in their profits and losses. Except for hedgers, whose futures commitments are offset by commitments in the cash market, the buying and selling of futures contracts have no purpose other than to profit from changes in futures prices. The problem, then, consists in estimating and analysing speculators' profits.

The best source of information on this subject would be the actual trading records of speculators, but these are rarely available. An important study based on data of this type was presented by Blair Stewart,[2] who

---

*This paper was first published in the *Review of Economics and Statistics* 39(2), pp. 143–51 (1957) and is reprinted with the permission of the MIT Press.

made a detailed analysis of the accounts of about 9,000 customers of a nationwide brokers' firm during the period 1925–1934. These accounts reflected almost exclusively speculative transactions in grain futures, mainly by non-professional traders. The most striking results were that nearly 75% of the speculators lost money and that in the entire sample total losses were about six times as large as total gains. Since in the futures markets as a whole gains and losses cancel out (apart from commissions, which in futures trading are small), the question arises by whom corresponding profits were made. Although the coverage of Stewart's material was not wide enough to give much information on this point, he seems to have thought it difficult to account for these heavy losses and to have suspected some unknown bias in his sample.

There were, in fact, two possible sources of bias. In the first place, prices in 1934 were much lower than in 1925, while the customers tended to prefer the long side. This effect, however, does not explain a great deal, since the trading experience of the shorts in the sample was not much less disastrous than that of the longs. A second source of bias may have been that the firm with which the accounts were held went bankrupt, which casts some doubt on the reliability of the advice it presumably gave to its customers.

If no actual trading accounts are available, estimates of gains and losses must be made from price movements and assumptions about commitments. This was done for speculators by Working[3] and for hedgers by Yamey[4] and others. The technique of the present paper is basically similar to theirs, but we were able to replace some assumptions about commitments by observed data and to consider a much longer period.

The method of estimating profits is based on monthly figures of open commitments and futures prices. The commitments are divided into three groups: (large) hedging, (large) speculative, and non-reporting. This division corresponds to the reporting requirements under the Commodity Exchange Act. Traders whose commitments in any one futures contract exceed the reporting limit (200,000 bushels in the case of wheat and corn and 5,000 bales for cotton) have to communicate their entire position to the Commodity Exchange Authority, which classifies futures commitments into hedging or speculative.[5] The remaining commitments are those of small traders, and it is commonly assumed that they are predominantly speculative in nature. It also seems clear that the reporting traders (both hedgers and speculators) are almost exclusively professionals, and that the figures for non-reporting traders are representative of the small non-professional speculators.

To estimate profits and losses it was assumed that the commitments of a group of traders that existed at the end of a month were opened at the average price during that month and closed out at the average price during the following month. The profit or loss of that group was then found by multiplying the end-of-month position by the change in the average price. Thus if large speculators were long 10 million bushels of May wheat on March 31, and the average price of May wheat was US$1.60 per bushel during March and US$1.55 during April, then their loss on that position was put at US$500,000. Commission charges have been ignored throughout. It need hardly be said that this estimation procedure is no more than approximate and could be improved in various ways, but it should be accurate enough for the purpose of this paper.

In the case of wheat and corn the calculation just described could be performed for each futures contract (ie, delivery month) separately, thanks to a recent analysis of the Commodity Exchange Authority[6] which cross-classifies open contracts by future and group of traders. Total profit or loss for each group was then found as the sum of the profits or losses in each futures contract, calculated by multiplying the position in a future by the change in the average price of that future. This procedure will be referred to as Method A.

For cotton, Method A could not be applied because a cross-classification is not available. It therefore had to be assumed that the percentage distribution of open commitments between futures was the same for all groups of traders, and hence the same as the distribution of total open commitments between futures, which is known from Department of Agriculture data.[7] The price change used was a weighted average of the changes in the average price of each future, the weights being given again by the percentage distribution of total open contracts between futures. This procedure, to be called Method B, was also applied to corn and wheat as a check. As may be seen from Table 5 the results from Methods A and B are grossly different, although there are systematic discrepancies which will be discussed below.

The price data used were monthly averages of daily closing prices in Chicago (for grains) and New York (for cotton), obtained courtesy of the Commodity Exchange Authority. Results are given by crop years, which start on July 1 for wheat, August 1 for cotton, and October 1 for corn. Open contract data for grains in the crop years starting in 1937–1939 refer to the Chicago Board of Trade only, for 1946–1951 to all United States markets combined. The first six months of the crop year 1946–1947 had to be omitted in wheat because futures trading was still restricted by the aftermath of wartime measures. Open contract data for cotton are based on New York and New Orleans together in crop years beginning in 1937–1944; for the remaining years they also include the significant cotton futures market in Chicago.

Despite the considerable variability of the entries in Table 1 certain broad conclusions may be drawn. In all three commodities the large hedgers lost and the large speculators gained. The small traders lost in the grains but did quite well in cotton, although it will be noted that of their

Table 1 Net profits (+) or losses (−) of three categories of traders in commodity futures[a] (US$ million)

Crop year[d]	Corn[b]			Wheat[b]			Cotton[c]		
	Large hedgers	Large spec's	Small traders	Large hedgers	Large spec's	Small traders	Large hedgers	Large spec's	Small traders
1937-38	+0.46	+0.22	−0.68	+21.93	+0.36	−22.30	−3.43[e]	+0.44[e]	+2.99[e]
1938-39	+1.68	−0.81	−0.88	+5.91	−0.45	−5.46	−3.80	+0.58	+3.22
1939-40	−1.67	+0.56	+1.11	−2.59	+1.70	−0.90	−8.04	+1.59	+6.45
Sub-total	+0.47	−0.02	−0.45	+25.26	+1.61	−26.87	−15.27	+2.61	+12.65
1940-41	—	—	—	—	—	—	−20.98	+2.04	+18.95
1941-42	—	—	—	—	—	—	−9.39	+1.80	+7.59
1942-43	—	—	—	—	—	—	−7.14	+0.82	+6.33
1943-44	—	—	—	—	—	—	−1.84	+1.12	+0.72
1944-45	—	—	—	—	—	—	−3.59	+1.41	+2.19
1945-46	—	—	—	—	—	—	−79.77	+15.06	+64.71
Sub-total	—	—	—	—	—	—	−122.72	+22.24	+100.48
1946-47	−0.20	+6.12	−5.92	+6.77[f]	+1.43[f]	−8.20[f]	+11.00	+1.87	−12.86
1947-48	−0.36	+1.28	−0.92	−22.86	+13.39	+9.46	−12.80	+3.35	+9.46
1948-49	+3.58	−0.55	−3.03	−0.34	+1.56	−1.22	+2.18	+1.85	−4.02
1949-50	−6.06	+2.56	+3.50	−5.44	+5.10	+0.34	−12.93	+7.28	+5.65
1950-51	−5.52	+2.50	+3.02	−0.47	−0.19	+0.66	−34.11	+9.25	+24.86
1951-52	+2.00	−0.27	−1.73	−9.19	+4.24	+4.95	+1.20	+4.13	−5.33
Sub-total	−6.56	+11.65	−5.08	−31.53	+25.54	+5.99	−45.47	+27.73	+17.75
Grand total	−6.09	+11.62	−5.53	−6.28	+27.16	−20.88	−183.45	+52.58	+130.88

[a] Figures may not check downward or across because of rounding.
[b] Computed by Method A (see text). Prewar years Chicago Board of Trade only; postwar years all markets combined.
[c] Computed by Method B (see text). Until August 1, 1945 New York and New Orleans only; thereafter all markets combined.
[d] Crop years start October 1 for corn, July 1 for wheat, August 1 for cotton.
[e] Excluding first two months.
[f] Excluding first six months.

total computed profit of US$130.9 million no less than US$100.5 million was made during the period 1940–1946, which was excluded in the grains because of lack of data. In the case of the hedgers, only profits and losses on futures commitments are shown, which have to be offset against profits and losses in the cash market.

Most conspicuous in these results is the consistent profitability of the large speculators' transactions. In cotton they made a net profit in every year observed, and although in corn and wheat they lost in a few years, they never lost much. A tabulation of the monthly figures underlying Table 1 is shown in Table 2. It will be seen that the large speculators had net profits in 59% of all months for corn, 61% of all months for wheat, and 68% of all months for cotton. If, to make the period for cotton comparable to the period for the grains, the crop years beginning in 1940 through 1945 are omitted, the percentage for cotton becomes 65%. These scores are sufficiently different from 50% to provide *prima facie* evidence of forecasting skill; some tests of this hypothesis will be presented below.

Less forecasting ability is apparent from the results of the small traders. They gained in 55% of all months for corn, 46% of all months for wheat, and 64% of all months for cotton. Again leaving out the period 1940–1946 the score for cotton drops to 61%.

The main purpose of Table 2 is to show to what extent gains and losses are connected with a net long or net short position. Both large speculators and small traders are net long most of the time and therefore stand to gain when prices go up. During the period of observation cotton prices rose fairly steadily; wheat and corn prices declined on balance during each of the two sub-periods, though in corn the number of months with price rises exceeded the number with price falls. This behaviour of prices explains a good deal of the discrepancy between small traders' results for grains and for cotton, especially when it is considered that in each of the three commodities small traders were net short about 20% of the time. The latter figure, incidentally, shows that the traditional picture of the small speculator as an incurable bull, too ignorant to understand short selling, is incorrect. In fact, small traders do not appear to be less inclined to the short side than the large professional speculators. In cotton small traders were net short in 38 months as against only 11 for the large speculators. In grains the pattern, though opposite to that for cotton, is not very marked (20 against 25 for wheat, 21 against 24 for corn).

On the other hand it is clear that the small traders are rather less successful when net short than the large speculators in similar circumstances. Thus in wheat, although prices fell in 53 out of 102 months, the small traders were short mostly in months when prices were rising, whereas the large speculators in that market were remarkably accurate in their choice of the short side. There is some evidence, particularly from the early postwar years, that small traders were unduly cycle-conscious

Table 2 Number of months with profits and losses

Months with:	Corn			Wheat			Cotton		
	Prices rising	Prices falling	Total	Prices rising	Prices falling	Total	Prices rising	Prices falling	Total
Large speculators' net profit	52	12	64	43	19	62	116	5	121
Large speculators' net loss	12	32	44	6	34	40	6	48	54
Small traders' net profit	51	8	59	38	9	47	99	15	114
Small traders' net loss	13	36	49	11	44	55	23	39	62
Total months	64	44	108	49	53	102	122	54[a]	178[b]

[a] Including one month in which large speculators broke even.
[b] Including two months in which prices did not change.

and therefore unwilling to believe that high prices could last for long. In the end this Cassandra attitude often turned out to be correct, but by then the initial losses had sometimes so undermined the small traders' courage or their margins that they were no longer able to reap the fruits of their badly-timed foresight. This happened for instance in the corn market during the boom of 1947. In the wheat market of 1947, too, small traders were initially speculating against the rapid price rise, but after a long period of losses they reversed themselves and made large profits from the tail end of the boom, only to lose again when prices broke early in 1948. If it is correct to explain the small speculators' actions by a belief that price rises will always be followed by falls, then the usual arguments about the destabilising influence of speculation may require reconsideration.

In Table 3 the totals from Table 1 are analysed by short and long positions. Apart from the difference in small traders' net profits noted previously, the general pattern is the same for the three commodities. The hedgers, who are nearly always net short in the futures markets, are the mainspring of profits for the other traders, who share in proportion to their net long position. In all three commodities the large speculators and small traders lost on balance on their short positions. It cannot be inferred from this that speculators would have done better to stick to the long side, for their short positions are often one half of a spread or straddle (ie, they are offset by a long position in another delivery). Spreading is not only a means of saving on margin requirements[8] but it is helpful in distributing different maturities between speculators according to their preferences.[9]

The essence of futures trading, however, is the transfer of price risks from the hedgers to the speculators in return for a risk premium, and this is clearly illustrated in Table 3. Even in wheat and corn, where prices fell during the period of observation, a risk premium was produced. As it happened the whole premium went to the large speculators, who in addition obtained some of the small traders' funds. In cotton the risk premium went to both large and small traders. Of course a net risk premium accrues to speculators only in the long run, and not necessarily in any given period of time.

The exact mechanism by which the risk premium is transferred cannot be described in this paper. Its principal component is a tendency for the price of a futures contract to rise from the inception of trading to the delivery date. The existence of this tendency, which is implied by Keynes's theory of "normal backwardation",[10] can be statistically demonstrated in various ways.

The main implication for the present analysis is that *in the long run* no great amount of skill is necessary to make a profit in the futures market: all one has to do is to maintain a long position. In this way a trader, if he has enough patience and capital to cover temporary losses, will sooner or

Table 3 Profits (+) and losses (−) of three categories of traders on long and short futures[a] (US$ million)

	Large hedgers			Large speculators			Small traders		
	Long	Short	Net	Long	Short	Net	Long	Short	Net
Corn									
1937–40	−0.84	+1.31	+0.47	−0.47	+0.44	−0.02	−1.67	+1.22	−0.45
1946–52	+11.28	−17.85	−6.56	+27.97	−16.32	+11.65	+34.74	−39.83	−5.08
Total	+10.44	−16.54	−6.09	+27.50	−15.88	+11.63	+33.08	−38.61	−5.53
Wheat									
1937–40	−4.79	+30.04	+25.26	−8.68	+10.29	+1.61	−40.98	+14.11	−26.87
1947–52	+30.82	−62.36	−31.53	+41.99	−16.44	+25.54	+57.20	−51.21	+5.99
Total	+26.04	−32.31	−6.28	+33.31	−6.15	+27.16	+16.22	−37.10	−20.88
Cotton									
1937–40	+9.99	−25.26	−15.27	+5.23	−2.62	+2.61	+22.54	−9.89	+12.65
1940–46	+49.30	−172.01	−122.72	+61.70	−39.46	+22.24	+219.78	−119.31	+100.48
1946–52	+98.59	−144.06	−45.47	+98.83	−71.10	+27.73	+250.77	−233.02	+17.75
Total	+157.88	−341.34	−183.45	+165.75	−113.18	+52.58	+493.09	−361.21	+130.88
Total all commodities	+194.36	−390.19	−195.82	+226.56	−135.21	+91.36	+542.39	−437.92	+104.67

[a] The footnotes of Table 1 apply also to Table 3.

later secure his portion of the risk premium. If, moreover, he can predict short-term price movements more accurately than other speculators, and adjusts his position accordingly, he may make a further profit at their expense. Conversely if he is outguessed by other speculators he may lose his share of the risk premium and more. There are consequently two kinds of skill: general skill, which consists only in being long and requires no information, and special skill, which involves a continuous adjustment to changes in current information. The two types of skill may be positive or negative: a negative general skill means a proclivity for the short side, whereas a negative special skill implies a tendency to be short when prices go up and long when prices go down.

The extent to which a category of traders possesses these two skills may be measured (*ex post*) from the following equation:

$$y_t = a + \beta x_t + \varepsilon_t \qquad (1)$$

in which y_t is the net position of that category, at a certain time t (here, the end of each month); x_t is an index of the change in prices around time t (more particularly the index used to estimate gains and losses by Method B described above) and ε_t is a random disturbance. The general skill is reflected in the constant term a: it is clearly positive when the group tends to be long irrespective of price changes. The coefficient β measures the special skill. What matters for our purpose is not the absolute magnitude of the estimates of a and β, but rather their statistical significance, which can be found by comparing each estimate with its standard error.

It is important to realise that (1) is not a behaviour equation; it is purely an *ex post* relation. Estimates of a and β are given in Table 4, with standard errors in brackets. The number of observations and the correlation coefficient are also given. As an aid in judging significance we note that if a or β is "really" zero, its estimate has a 30% chance of exceeding its standard error and a 5% chance of exceeding twice its standard error. It is hardly necessary to go into further refinements since the results are rather clear-cut.

Table 4 shows that both speculators and small traders possess general skill, since all the estimates of a very considerably exceed their standard errors in all three commodities. A conspicuous difference appears in the measure of special skill, however. The estimates of β for the small traders all fall short of their standard errors and must therefore be regarded as insignificant, with the exception of post-war wheat where the small traders' special skill appears to be significantly negative. The special skill coefficients for the large speculators are significantly positive in wheat and cotton but not in corn, where they are positive but very small.

It seems clear therefore, that there are real differences in the ability of large and small traders to forecast price changes. This implies also that the

Table 4 Estimates of a and β in equation (1), with standard errors (in parentheses) and correlation coefficients

	Number of observations	Large speculators			Small traders		
		a	β	r	a	β	r
Corn[a]							
1937–40	36	4.20 (0.77)	+0.0318 (0.232)	0.0235	9.56 (1.42)	+0.0206 (0.430)	0.0082
1946–52	72	4.72 (0.52)	+0.0416 (0.051)	0.0927	7.58 (1.03)	−0.2220 (0.101)	0.2533
Wheat[a]							
1937–40	36	4.87 (0.72)	+0.2375 (0.111)	0.3451	43.13 (8.18)	−0.3370 (1.250)	0.0462
1947–52	66	6.48 (0.84)	+0.2024 (0.082)	0.2950	10.51 (1.79)	−0.0770 (0.174)	0.0553
Cotton[b]							
1937–45	94	62.2 (5.3)	+31.59 (10.42)	0.3030	621.6 (54.4)	−90.56 (106.99)	0.0879
1945–52	84	121.7 (9.9)	+16.57 (6.67)	0.2646	153.2 (47.1)	+26.01 (31.81)	0.0889

[a] Net position in million of bushels, price changes in cents per bushel.
[b] Net position in thousands of bales, price changes in cents per pound.

differences in profits and losses exhibited in Tables 1–3 are not wholly due to random causes.

We must now consider another aspect of relative skill. So far we have looked only at the total net position of a category of traders, that is to say at the net position in all futures contracts combined. Since, however, the prices of different deliveries do not usually move in an exactly parallel manner, there is also scope for skill in choosing the futures in which to be long or short; this might be called distributive skill to distinguish it from the sort of skill analysed in Table 4.[11]

It is possible to estimate distributive skill by comparing results from the two methods used for estimating profits and losses in Table 1. Method A, used there for corn and wheat, was based on the actual distribution between futures of commitments of the three groups of traders, whereas Method B, used for cotton, was based on the assumption that the distribution between futures was the same for all three groups. By applying Method B to the grains, and subtracting the gains or losses it gives from those estimated by Method A, we will therefore obtain a measure of the gains and losses due to a more or less skilful distribution of a given over-all position between different deliveries.

Table 5 shows that Method B gives small profits (or larger losses) to the large hedgers and large speculators, and larger profits (or smaller losses) to the small traders. This would imply that the large traders have a

Table 5 Analysis of distributive skill (US$ million)

	Large hedgers		Large speculators		Small traders	
	Method B	Method A −Method B	Method B	Method A −Method B	Method B	Method A −Method B
Corn						
1937–40	+0.40	+0.07	−0.07	+0.05	−0.33	−0.12
1946–52	−7.49	+0.93	+11.32	+0.33	−3.83	−1.25
Total	−7.09	+1.00	+11.25	+0.38	−4.16	−1.37
Wheat						
1937–40	+25.04	+0.22	+1.29	+0.32	−26.33	−0.54
1947–52	−32.59	+1.06	+22.99	+2.55	+9.60	−3.61
Total	−7.56	+1.28	+24.29	+2.87	−16.73	−4.15

positive distributive skill. The differences between the results from Methods A and B are not large however, and the question arises whether they are not merely due to an accumulation of random errors. By way of a crude test[12] it was found that the large corn speculators showed evidence of positive distributve skill in 54 months out of 108, negative skill in 51 months, and equal results from Methods A and B in the remaining three months. The small corn traders showed positive distributive skill in 43 months, negative skill in 62 months, and zero skill in three months. In wheat the large speculators showed positive skill 56 months out of 102, negative skill in 41 months, and a tie in five months; for the small traders these figures were respectively 42, 59 and 1. On the basis of these figures the apparent positive distributive skill of the large speculators is not statistically significant; the apparent negative distributive skill of the small traders, on the other hand, cannot plausibly be attributed to random causes only.[13]

It appears, therefore, that the distribution between futures is one of the factors influencing the relative profitability of large and small traders' commitments. Further evidence on this point is provided by an analysis of the monthly profits and losses in corn and wheat for individual futures contracts. For this purpose futures have been grouped together according to their distance from maturity. Thus at the end of February the May future is regarded as three months distant from maturity, the July future as five months distant, and so on. The expiring future (in this case the March future) is consequently treated as one month away. Then the profits and losses on all futures one month distant from maturity, 2 months distant, and so on, were added up. The totals appear in Table 6.

Although the results are not as clear-cut as they might be we can nevertheless find some indication of a difference in success according to the distance from maturity. The large speculators do better in the near futures (those close to maturity) than in the very distant ones, and the opposite is true for the small trader. The exceptions as regards the large speculators are the corn futures three of four months distant from maturity, in which they lose, in common with the small traders, and in which, consequently, the hedgers gain. The last two lines of Table 6 show that small traders lost twice as much in the near futures as they gained in the distant futures. It would be interesting to do the same analysis for cotton, but the data are not available.

It is not difficult to explain these differences. The price behaviour of the near futures depends to a large extent on the magnitude and ownership of deliverable stocks at the relevant terminals (Chicago, Kansas City, and Minneapolis for wheat, Chicago for corn), and this is a matter on which non-professionals cannot easily inform themselves. Price movements in the more distant contracts, on the other hand, are influenced mainly by basic supply and demand factors such as crop prospects, the general

Table 6 Net profits and losses of three categories of traders by distance from maturity of futures contracts (US$ million)

Months from maturity	Corn			Wheat		
	Large hedgers	Large speculators	Small traders	Large hedgers	Large speculators	Small traders
1	−2.59	+4.37	−1.78	−1.04	+4.57	−3.53
2	−1.47	+5.29	−3.82	+8.90	+4.03	−12.93
3	+0.59	−0.23	−0.36	+16.01	+4.13	−20.14
4	+3.85	−1.25	−2.61	−3.32	+5.25	−1.93
5	−0.54	+2.57	−2.02	+5.73	+1.72	−7.44
6	−2.63	+3.13	−0.50	−9.45	+5.00	+4.45
7	−1.87	+0.48	+1.39	−5.68	+1.82	+3.86
8	−0.84	−1.08	+1.92	−8.26	+1.33	+6.93
9	−0.56	−0.60	+1.16	−9.06	+0.67	+8.39
10	−0.03	−0.77	+0.80	−0.16	−0.71	+0.86
11	−0.01	−0.27	+0.28	+0.06	−0.67	+0.61
Total	−6.09	+11.62	−5.53	−6.28	+27.16	−20.88
1–6	−2.79	+13.88	−11.09	+16.82	+24.71	−41.53
7–11	−3.30	−2.26	+5.56	−23.09	+2.45	+20.65

economic outlook, or government policy. In evaluating the latter factors the professionals have no particular comparative advantage. Indeed it is often profitable for them to use their superior knowledge by taking a long or short position in the near futures, at the same time taking an opposite position in the more distant deliveries in order to limit their risks. We have already mentioned that such spreading accounts for a major part of the large speculators' operations. By taking the other side of the distant half of these spreads the small traders may then earn a risk premium from the professionals; the other side of the near half is more likely to be taken by hedgers, who rarely go into distant futures. This type of spreading is quite similar to hedging, which is based on hedgers' superior knowledge of the cash market.

Returning now to the question raised in the title we conclude that large speculators show definite evidence of forecasting skill, both in the long and in the short run. Since these large speculators are professionals whose existence depends on their skill, this finding is hardly revolutionary, edifying though it is to see virtue rewarded. The experience of the small traders indicates that they do quite well when they stick to the long side, where the theory of "normal backwardation" assures them of a profit in the long run. It appears, moreover, that non-professionals would have done well to confine themselves to the more distant futures.

1 These results are part of an investigation of commodity futures undertaken at the Cowles Commission for Research in Economics with the valuable assistance of Lester G. Telser and supported by the Rockefeller Foundation.
2 Blair Stewart, "An Analysis of Speculative Trading in Grain Futures", US Department of Agriculture Technical Bulletin 1001, October 1949.
3 H. Working, "Financial Results of Speculative Holding of Wheat", *Wheat Studies* 7 (July 1931).
4 B. S. Yamey, "Investigation of Hedging on an Organised Produce Exchange", *The Manchester School* 19 (1951).
5 A special category of speculative commitments is "spreading" or "straddling" positions, in which a long position in one or more futures contracts is offset by a short position in one or more other contracts.
6 US Department of Agriculture, "Grain Futures Statistics 1921–1951", Statistical Bulletin 131 (July 1953).
7 US Department of Agriculture, *Cotton Futures Statistics* (three issues covering 1937–1945), and *Commodity Futures Statistics* (annual).
8 Because the differences between the prices of various contracts (also known as "spreads") are less volatile than these prices themselves.
9 J. M. Mehl, formerly Administrator of the Commodity Exchange Authority, ascribes the recent increase in spreading also to income tax considerations, since it permits the transformation of short-term into long-term profits. Cf J. M. Mehl, *Futures Trading Under the Commodity Exchange Act 1946–1954*, (1954), US Department of Agriculture (December), 20.
10 J. M. Keynes, *Treatise on Money*, 1930, Vol. II. pp. 142–44 (London). See also J. R. Hicks, *Value and Capital*, 1939, pp. 137–39 (Oxford).

11 In principle this distributive skill might also be divided into general distributive skill, leading to a long position in those contracts which *on average* tend to go up most, and special distributive skill, consisting in an ability to buy those futures which *in a given period of time* will go up most or sell those which fall most. It does not appear, however, that different deliveries have markedly different rates of average increase in the long run. There would consequently be no scope for general distributive skill, and the distinction between general and special skill would be redundant here.

12 More refined tests could not be applied either here or in Table 2 because the distribution of gains and losses is not of the normal type.

13 If distributive skill were really zero, so that positive and negative skill was equally likely, the standard error for each of the grains would be about five months.

23

Normal Backwardation, Forecasting, and the Returns to Commodity Futures Traders*

Charles S. Rockwell

Two theories are advanced to explain the returns of speculators in commodity futures markets. One, the "theory of normal backwardation", views speculative returns as directly linked to the bearing of risk; the other, which we shall call the "forecasting theory", considers returns to be determined by the ability of speculators to forecast prices accurately. Although competitive, these theories are not mutually exclusive. This paper presents evidence on the extent to which each of these competing explanations may have been operative in the United States commodity futures markets from 1947 to 1965.

The approach used here is similar to that employed by Professor Houthakker (1957a) in his article "Can Speculators Forecast Prices?" That is, the commitments of reporting speculators, hedgers, spreaders and non-reporting traders are obtained from Commodity Exchange Authority (CEA) data and are then multiplied by an appropriate price measure to obtain an estimate of that group's futures market return. The principal difference between this study and that of Professor Houthakker is the much broader coverage obtained here. While Houthakker had available approximately 324 monthly observations on three markets (cotton,[1] wheat, and corn) from 1937 to 1940 and 1946 to 1952, we make use of over 7,900 semi-monthly observations covering 25 markets for the 18 years since 1947. This broader coverage makes possible much more conclusive inferences about the mechanism which determines the returns to speculators and the futures costs of hedging.

*This paper was first published in Food Research Institute Studies 7, pp. 107–30 (1967) and is reprinted with the permission of the Food Research Institute. This research was supported in part by the Office of Naval Research under contract ONR 222(77) with the University of California and by a grant from the Ford Foundation to the Graduate School of Business Administration, University of California, Berkeley, administered by the Center for Research in Management Science. Support was also received from the Committee on Research at Berkeley, the Economic Growth Center at Yale University, and the Yale Computer Center.

The quantitative arguments of this study use only the values of the variables and their first moments. The fact that the sign of the aggregate profit estimates presented in this paper is often critically dependent upon the results of a particular year and market is consistent with the existing evidence that futures prices may be of the stable Paretian type and consequently have infinite variances. Therefore, neither estimates of variances nor significance tests are made. This is not unduly restrictive since the most important findings of the paper are concerned with the sign of variables, and in the fortunate cases where the wrong sign is encountered no measure of dispersion is required. In the less fortunate cases, the persuasiveness of the conclusions concerning the flow of profits must rest upon casual inspections of the consistency over markets and through time of the dollar value of the profit flow and upon the economic significance of these profits as measured by the rate of return on traders' holdings (average annual profits divided by the value of the outstanding contracts which the trader holds). Although it is possible to make significance tests without using second moments, limitations of funds and time prohibited this. Therefore the quantitative breadth of this study is gained at the cost of some statistical sharpness.

The first section of the paper defines the theory of normal backwardation and examines the different assumptions which are made concerning the forecasting ability of speculators. In its simplest form, the theory assumes that speculators: (1) are net long; (2) require positive profits; and (3) are unable to forecast prices. These assumptions may be satisfied if futures prices rise on the average during the lives of each contract, and this is the chief prediction of the theory. If speculators are assumed to be unable to forecast prices, it is appropriate to consider all of their profits to be a reward for risk-bearing and none to be a reward for forecasting. Consequently, advocates of this version of the theory contend that the profit flow between hedgers and speculators is analogous to the flow of insurance premiums between insured and insurer. Speculators, like insurers, are guaranteed an actuarial expectation of gain simply by being long. The amount of their gain depends only upon the size of their position (the amount of risk they bear) and not upon their forecasting ability.

However, because of the third assumption (that speculators are unable to forecast prices) it is possible to construct counter examples showing that the three assumptions are neither necessary nor sufficient to warrant the conclusion of rising prices.[2] More recent formulations of the theory of normal backwardation avoid these counter examples by dropping the third assumption and instead assume that speculators are able to forecast prices. They also contain a corollary that if speculators are net short prices must fall. These improvements in the theory, however, make the interpretation of speculators' profits ambiguous. Since profits depend upon forecasting ability as well as upon the quantity of risk borne, the insurance premium analogy is no longer adequate in itself. Determination of the proportion in which profits divide between a risk premium and a forecasting reward is the principal objective of this paper.

The second section of the paper briefly describes the data and estimation techniques used and presents the estimated profit flows both in terms of dollar values and rates of return. To facilitate reading, the text does not contain data classified by individual markets. Instead, three different market aggregations are used: (1) an "All Markets" total representing aggregation over all 25 markets; (2) a "Large Markets" total which includes only wheat at Chicago, cotton at New York and soybeans; and (3) a "Small Markets" total which excludes the three markets mentioned above. The reason for excluding these three markets is shown graphically on the abscissa of Figure 1(a), p. 587. Wheat at Chicago, cotton at New York and soybeans have such large average values of open interest that these markets can be meaningfully differentiated from all others. A complete set of tables for individual markets may be obtained from the author.

Inspection of the dollar value of profits and the rates of return before commissions indicates that the flow of profits in the three large markets is quite different from that in the 23 small markets. However, in both aggregations, large speculators make substantial and consistent profits. In the large markets small traders make positive but inconsequential profits so that the losses of hedgers become the profits of large speculators. In the small markets, however, it is hedgers who make positive but inconsequential gains with the result that the profits of large speculators come from the pockets of small speculators. It is a general characteristic of the results for all 25 markets that they are determined by what happens in the three large markets. Consequently, the overall 6% profit rate of large speculators is financed by a modest 2% rate of loss by hedgers. It should be noted that the true rate of return on investment for speculators differs grossly from the rate measured here due to the existence of very small (5 or 10%) margin requirements. Finally, the rates of return on the long open interest tend to be symmetrically distributed around zero for the 23 smaller markets, whereas the rates of return in the three markets with the largest average value of the open interest are positive, and quite substantial.

Having measured the profit flow, we next attempt to determine the proportion of this flow which can be attributed to normal backwardation. This is done by defining normal backwardation as the returns which would accrue to a naïve speculator who is long when hedgers are net short and short when hedgers are net long. The magnitude of his positions is assumed to be proportional to the size of the open interest. An inspection of the naïve traders' returns yields two conclusions: first, the corollary to the theory of normal backwardation which states that prices should fall when speculators are net short is false – prices rise consistently under these conditions causing losses to short speculators; second, the rate of gain which accrues to the naïve speculator when he is net long is so small relative to the dispersion of that rate for different markets that we conclude there is no significant tendency toward normal backwardation in the markets investigated here. The dispersion of the rates of return by market for the naïve trader is plotted in Figure 2, p. 591.

This failure to find any consistent evidence of normal backwardation implies the acceptance of the extreme alternative hypothesis that all important profit flows are to be explained in terms of forecasting ability. That is, the proportion of profits attributable to normal backwardation is zero. However, it is possible to define two levels of forecasting skill: first, an elementary ability which is called Basic Forecasting Skill; and second, a more sophisticated ability which is called Special Forecasting Skill. Basic Skill measures the ability of a group to be long in markets where prices rise over the total period of observation and short in markets where prices rise over the total period of observation. Special Skill, therefore, measures a trader's ability to forecast price movements whose duration is shorter than the total period of observation. An examination of the results of this division of profit confirms the conclusion that it is the degree of forecasting ability which controls the flow of profits. We find that hedgers have negative values for both Basic and Special Forecasting Skills; small traders have a positive value for Basic Skill but an equally large negative value for Special Skill; and large speculators have positive values for both measures (1.3% and 4.8% rates of return respectively). Thus large speculators, the only trading group to earn consistent and economically significant profits, acquire three-quarters of these profits because of their ability to forecast short-term price trends and only one-fourth because of their ability to forecast long-term price trends.

In summary, the evidence presented here indicates that it is forecasting ability and not the bearing of risk that determines the profits of speculators. While the theory of normal backwardation may be valid for particular markets under special conditions, it is not adequate as a general explanation of the flow of profits in commodity markets.

The fact that the gross profits of small traders are zero implies that they consistently make substantial net losses after commissions. Since this group is predominantly composed of small speculators, and since this group holds 46% of the value of all contracts, the principal assumptions of the theory of normal backwardation are not met. Small speculators do not require an *ex post* history of profits in order to continue trading. There are at least three possible explanations of this. Small speculators are either risk seekers (and are consequently willing to lose money for the privilege of speculating); comprise a stable population of risk averters who are unable to forecast prices, but do not realise this; or finally, constitute a changing population of risk averters, in which the successful forecasters rise to become large speculators while the unsuccessful withdraw from the market and are replaced with new blood.[3] Unfortunately, we are unable to ascertain the relative validity of these hypotheses.

The implication of these findings with respect to the price effects of speculation may now be stated. The existence of a subset of speculators who are able to forecast price changes causes futures prices on the average to be an unbiased estimate of the ultimate spot price. In a modified form, however, this conclusion readmits a question which the theory of normal backwardation was thought to answer: why are large speculators consistently net long, even when we consider sets of markets where there is clearly no tendency for prices to rise?[4] Since large speculators own only a small fraction of all commitments, it is quite possible for them to be either net short or net long quite independently of the sign of net hedging commitments. The answer to this may be that even the more sophisticated speculators have an irrational preference for the long side. However, it may well be equally true that the distribution of price changes is asymmetric so that skewness and moments other than the mean influence the decisions of speculators to be net long.

That futures prices, on the average, are unbiased estimates of ultimate spot prices need not imply that this result holds either for all markets or for all time periods within a market. As an example of the former, coffee futures prices (a market not covered by this study) have exhibited a strong upward tendency in the post-war period that is quite consistent with the theory of normal backwardation. As an example of the latter, this study shows that if hedgers are net long, futures prices tend to rise. Similar examples of temporary price bias conditional upon special conditions of time and market structure may be found in the papers presented by Lester Telser and Paul Cootner at this Symposium. Perhaps the principal value of this paper is that it puts into better perspective these, as well as other studies, which demonstrate the existence of price bias. The results presented here suggest that this evidence of bias is critically dependent both upon the markets which are selected and upon the special structural characteristics which determine any conditional price forecasts. In contrast, the overall generalisation from the data investigated here is that the futures price is an unbiased estimate of the ultimate spot price.

THE ROLE OF FORECASTING IN THE THEORY OF NORMAL BACKWARDATION

The theory of normal backwardation predicts that under certain assumptions it is necessary on the average for the price of futures contracts to rise. Two of the assumptions of the theory as originally stated by Keynes (1923, pp. 784–86) are that speculators be net long and be risk averters (that is, they require a positive history of profits if they are to continue trading). Under these circumstances, a rising trend in prices is the mechanism that rewards long speculators for the risks they bear.

To Keynes, the possibility that speculators may be better forecasters than hedgers is a "dubious proposition" (Keynes, 1923 p. 785). This contention appears to be reversed in later formulations of the theory of normal backwardation by Hicks (1953, p. 138) and Houthakker (1957b, p. 23). Since forecasting ability, or its absence, is a central theme in this paper, and since Keynes' position as stated in the *Manchester Guardian Commercial* is not very well known, an extensive quote from that source may be helpful (Keynes, 1923, p. 785).

> In most writings on this subject, great stress is laid on the service performed by the professional speculator in bringing about a harmony between short-period and long-period demand and supply, through his action in stimulating or retarding in *good time* the one or the other. This may be the case. But it presumes that the speculator is better informed on the average than the producers and consumers themselves. Which, speaking generally, is rather a dubious proposition. The most important function of the speculator in the great organised "Futures" markets is, I think, somewhat different. He is not so much a prophet (though it may be a belief in his own gifts of prophecy that tempts him into the business), as a risk-bearer ... without paying the slightest attention to the prospects of the commodity he deals in or giving a thought to it, he may, one decade with another, earn substantial remuneration *merely* by running risks and allowing the results of one season to average with those of others: just as an insurance company makes profits . . . .

In Keynes' version of the theory, it is the speculators' inability to forecast accurately that makes them dependent upon the incidental, and probably unanticipated, rising price level to provide a positive history of profits. The assumption that speculators are unable to forecast prices makes it unambiguous to interpret whatever profits they receive as a risk premium paid to them by hedgers and not as a reward for forecasting. The postulation of "no forecasting" ability, however, raises problems concerning the prediction that prices must rise. If the level of the net short position of hedgers is subject to variations, it is possible for speculators to have a positive history of profits without prices rising on the average: for example, prices rise one unit in period one and fall one unit in period two, and speculators are net long two units in period one but only one unit in period two.[5] This, of course, implies that speculators do not correctly forecast the price fall in period two, which is quite consistent with the assumed absence of forecasting ability. The converse may also be shown: that is, a rise in prices need not result in profits for long speculators. Thus, the assumptions of the theory of normal backwardation are neither necessary nor sufficient for the prediction that prices rise.

The principal modification of the theory of normal backwardation made by Hicks and Houthakker is to assume that speculators are able to forecast prices. This distinction may be seen by contrasting the position of Keynes as stated above with that of Hicks (1953, p. 138) in *Value and Capital*.

> Futures prices are therefore nearly always made partly by *speculators*, ... whose action tends to raise the futures price to a more *reasonable level* (last italics mine) .... But it is of the essence of speculation, as opposed to hedging, that the speculator puts himself into a more risky position as a result of his forward trading .... He will therefore only be willing to go on buying futures so long as the futures price remains definitely below the spot price he expects .... The difference between these two prices ... is called by Keynes "normal backwardation."

It seems clear that while both Keynes and Hicks share the same prediction they do not agree upon the underlying model.

One consequence of granting speculators even a modest amount of predictive ability is that it frees the backwardation hypothesis from counter examples (such as the one stated above) that involve speculators being net long during periods when prices fall in a "predictable" manner.[6] Thus, the assumptions of the Hicks–Houthakker version of the theory necessarily imply that prices must rise on the average. However, this improvement in the logic of the theory is gained at a cost: the returns of speculators may no longer be viewed unambiguously as a reward for bearing risk. Rather they represent a mixed payment for forecasting and risk bearing, the proportions of the mixture being determinable only by empirical investigation. The view held by Keynes that the returns of speculators may be interpreted as an insurance premium, will be valid only if the forecasting component of profits is relatively small.

The empirical procedure originally planned for this study was first to measure a normal backwardation component of profits and then define the difference between this amount and the actual returns as the forecasting component. Either component may be negative in value, but a negative backwardation component causes the selection of a distinctly different path of investigation from that caused by a negative forecasting component. For example, a negative value for forecasting profits, which is more than offset by a positive value for backwardation profits, would be quite consistent with the Keynesian version of the theory; negative forecasting ability is an admissible phenomenon. However, if backwardation profits are non-positive, it necessitates the rejection of the theory and, hence, requires a different framework of analysis. In practice, we conclude that the backwardation component is zero and, therefore, adopt the position that forecasting ability is the only important determinant of profits. Although the last section of this paper divides profits into two components representing different degrees of forecasting skill, neither of these components is a measure of the profits attributable to normal backwardation. However, the significance of this division depends upon our first being able both to define an empirically meaningful measure of normal backwardation and show that its value is non-positive.

It is convenient to discuss at this point two problems which arise in defining an empirical estimate of normal backwardation. The first problem arises when hedgers are net long rather than net short. The Keynes and Hicks formulations clearly assume hedgers to be net short. These authors, however, were concerned with the futures markets for international industrial commodities during the 1920s and 1930s when it may well be true that hedgers were consistently net short. On the other hand, the 25 markets covered by this study are predominantly for agricultural commodities, and it will be shown that hedgers are net long for substantial periods of time. It is difficult to see any reason why the theory of normal backwardation in either its Keynesian or Hicks–Houthakker formulation should not be broadened to predict a price fall whenever hedgers are net long. This modification is suggested by both Houthakker (1957b, p. 22) and Cootner (1960, p. 400).

The second problem concerns which weights should be used in aggregating over individual contracts and, a fortiori, commodities. There are at least three possibilities: (1) each contract may be given a weight of one; (2) each contract may be given a weight equal to the average value of the open interest in that contract (taken over all time periods during which that contract trades); and (3) each contract may be weighted by the actual open interest existing on that date. The first alternative, unity weights, gives undue importance to inactive contracts and commodities and need not be considered. The choice between alternatives two and three is more difficult. Numerous arguments can be made for either side. The most important consideration, however, would seem to be protection against misleading results caused by changing market structure. For example, although cotton at New York has the second largest average open interest value of any commodity, trading on this market is almost non-existent by the end of the period. To weight the price performance of these last years with the large open interest that prevailed earlier could cause the same spurious results as applying a weight of one to all contracts and all time periods. Therefore, this study measures normal backwardation as the sum of the return on the total long open interest when hedgers are net short and of the return on the total short open interest when hedgers are net long.

If this measure is to be used, what is its relation to the existing theories of normal backwardation? Normal backwardation describes the profits of marginal speculators who possess no forecasting ability. This is true whether we deal with Keynesian or Hicks–Houthakker versions. We may therefore conceive of normal backwardation as the return earned by a hypothetical speculator who follows a naïve strategy of being constantly long when hedgers are net short and constantly short when hedgers are net long. The naïve strategy used here requires that the hypothetical trader adjusts the size of his positions to maintain them as a constant proportion of the total open interest. In practice, the author's earlier work shows that the results of this strategy do not differ significantly from the results obtained when the trader is assumed to have positions of a fixed size (Rockwell, 1964, p. 114).

RETURNS TO FUTURES TRADERS

Description of the data

Except for the "Commitments of Reporting Traders" for wheat at Chicago, Minneapolis, and Kansas City, all the data are taken from the annual US Department of Agriculture Commodity Exchange Authority publication, *Commodity Futures Statistics*. Before 1962–1963 *Commodity Futures Statistics* presents only aggregate commitments for all wheat markets. Professor Roger Gray, however, made available unpublished CEA statistics on wheat commitments disaggregated into the above three markets which are used in computing wheat profits. For cotton, separate New York and New Orleans prices and open interest are used, but the commitment data for both markets are combined. That is, it is assumed that the proportion of reporting speculators, hedgers, etc., for New York or New Orleans is equal to the aggregate ratio of reporting speculators or hedgers to the aggregate open interest totalled over both markets.

Only the United States commodity futures markets regulated by the CEA are included in the above publication. This study therefore does not cover the unregulated markets such as tin, rubber, coffee, and cocoa which constitute approximately 20% of total futures trading in the United States. Within the set of markets for which statistics are available, the only important market excluded is the one for grain sorghums. This exclusion is for computational convenience since the computer program cannot easily handle a market such as sorghums where reported positions do not continuously exist and where price units change.

The selection of the years covered is influenced by the desire to: maximise the number of post-war observations; have representative trading patterns; and have a reasonably stable general price level. All three of these criteria are satisfied by the period from July 15, 1947 to July 31, 1965. The year 1946–1947 is not included because, as of July 1946, trading had not yet been resumed in some commodities. Therefore, it is not safe to assume that normal trading patterns existed during the period before July of 1947. From 1947 to 1965, the "wholesale price index for farm products" fell at an annual rate of approximately 0.5% per year while the

Dow-Jones "futures price index" fell at an annual rate of 0.7%.[7] This magnitude of price changes is sufficiently small that we may neglect the influence of unanticipated changes in the general level of prices as a determinant of the returns to futures trading.

However, due to differences in coverage, references to general price indexes are not a sufficient indication that unanticipated price changes did not occur. In particular, the Dow-Jones "futures price index" includes a number of industrial commodities not covered by this study. Information on price behaviour in the markets covered here is contained in columns four through seven of Table 1. Columns four and five show the first and last price quotations in the "nearby contract" (that is, the first contract expiring after the first observation; generally, the July contract). Column six is the average annual percentage price change of this contract during the period of observation: that is

$$\frac{(P_{end} - P_{start})/P_{start}}{\text{Number of years}}$$

Since this measure shows that price *levels* fell in 22 out of the 25 markets, and many by substantial amounts, it might be argued that prices during the period under study are not sufficiently stable to support the analysis being made. Column seven is included in order to nullify that contention. This column gives the percentage change in price (of the contract quoted in columns four and five) that occurs between the time the contract matures at the end of one year and the time when it begins trading in the following year. The percentage figure is obtained by summing this difference over all years and dividing by the initial year: that is, $[\Sigma_y(P_{y+1} - P_y)]/P_{start}$ (the percentage change may be greater than 100). This is a measure of the amount of price change that took place between contract years and is therefore "forecasted" by traders in that market. For 14 out of 25 markets the "forecasted" price declines are greater than the actual declines. This means that for over half the markets prices rose, on the whole, during the periods that the contracts actually traded. These figures, of course, use only one contract in each market. A 25-market aggregate index of price changes, in this one contract, using the values of the total open interest in each market as weights, shows that the price level for the entire period falls at an average annual rate of 1.2%. However, the average annual decline in the price level that occurs between the expiration date of the old contracts and the initiation of trading in the new contracts is 1.9%. This evidence is consistent with the predictions of a model which assumes normal backwardation and perfect forecasting ability. While these results are not as convincing as a stable, weighted index of the contracts traded would be, they do indicate that unanticipated price change does not distort importantly the normal profit flow during this period.

Table 1 Description of data and price levels

Commodity and markets[a]	Period of observation From	To	Number of semi-monthly observations	Change in price level of nearby contract[b] (US$) Start	End	Annual percentage price change From start to end	Between maturity years
Wheat, Chicago Board of Trade	7/47	6/65	432	2.39375	1.42250	−2.3	−2.2
Wheat, Kansas City Board of Trade	7/50	6/65	360	2.30375	1.43500	−2.3	−1.6
Wheat, Minneapolis Grain Exchange	7/50	6/65	360	2.36125	1.59750	−2.2	−2.5
Corn, Chicago Board of Trade	7/47	6/65	432	2.30375	1.32250	−2.2	+1.4
Oats, Chicago Board of Trade	7/47	6/65	432	1.02000	0.67750	−1.9	−5.1
Rye, Chicago Board of Trade	7/47	6/65	432	2.52000	1.15750	−3.0	−1.6
Soybeans, Chicago Board of Trade	7/47	6/65	432	2.78000	2.96000	+0.36	−0.7
Soybean meal,[c] Chicago Board of Trade	7/47	6/65	432	87.50	71.10	−1.0	−12.0
Soybean oil,[d] Chicago Board of Trade	7/50	6/65	360	0.1245	0.1008	−1.3	−3.1
Cotton, New York Cotton Exchange	7/47	6/64	408	0.3898	0.3328	−0.9	−2.8
Cotton, New Orleans Cotton Exchange	7/50	6/60	240	0.3569	0.3278	−0.8	−4.3
Cottonseed meal, Memphis Merchants Exchange Clearing Association	7/47	6/60	312	79.90	54.00	−2.5	−9.4
Cottonseed oil, New York Produce Exchange	7/47	6/65	432	0.2350	0.1232	−2.6	−0.6
Lard, Chicago, Board of Trade	7/47	6/62	360	0.1960	0.0870	−3.7	−1.6
Flaxseed, Minneapolis Grain Exchange	7/50	6/62	288	3.7150	3.1900	−1.2	+2.0
Shell eggs, Chicago Mercantile Exchange	7/47	6/65	432	0.5262	0.3490	−1.9	−0.1
Frozen eggs, Chicago Mercantile Exchange	7/61	6/65	96	0.2635	0.2687	+0.5	−4.6
Potatoes, New York Mercantile Exchange	7/47	6/65	432	2.96	2.58	−0.7	+5.0
Wool tops, Wool Association of the New York Cotton Exchange	7/47	6/62	360	1.570	1.666	+0.4	−4.7
Grease wool, Wool Association of the New York Cotton Exchange	5/54	6/63	257	1.413	1.190	−1.7	−0.2
Bran, Kansas City Board of Trade	7/47	6/56	216	58.50	33.20	−4.8	−7.8
Shorts, Kansas City Board of Trade	7/47	6/56	216	60.00	38.90	−3.9	−9.8
Middlings, Kansas City Board of Trade	7/55	6/56	24	37.00	35.15	−5.0	−3.4
Onions, Chicago Mercantile Exchange	9/55	6/59	91	2.10	1.30	−9.5	−1.0
Butter, Chicago Mercantile Exchange	7/47	6/53	144	0.6775	0.6120	−1.6	−2.4

[a] "Large Markets" are wheat at Chicago, cotton at New York, and soybeans.
[b] The nearby contract is the first contract that expires after the first observation; generally it is July.
[c] Soybean meal is for the Memphis Merchants Exchange Association until July 1953.
[d] Soybean oil is for the New York Produce Exchange until July 1950.

Except for grease wool and onions, all markets are covered from July 15 to June 30. In these two markets, data for the first two or three months of the initial year are not available; the alternatives are either to begin with the first available observation or disregard these observations and begin on the next July 15. Since the first available observation is close to July 15, the former procedure is followed.

Estimation of traders' commitments and returns

One of the most important difficulties in using the CEA statistics is that they do not present data on traders' commitments cross-classified according to contract month. That is, we know what the total open interest is and we know how it is divided among trading groups and contract months, but we do not know the joint distribution.

Therefore, we follow Professor Houthakker's example (Houthakker, 1957a, p. 144) and assume that the percentage distribution of the commitments of each trading group by contract month is equal. This is equivalent to assuming that the distribution of the total open interest according to trading groups and according to contract month is statistically independent. To estimate profits we first estimate the mean open interest for each contract as a simple average of the initial and terminal open interest two weeks later; and second, we multiply this quantity measure by the change in price during the period $(P_{t+1} - P_t)$ to obtain an estimate of profits. In addition to this measure of the dollar profits, a measure of the value of the open interest for each contract is obtained and is subsequently used to convert dollar profits into rates of return.

Distribution of the open interest

Table 2 exhibits the percentage distribution of the value of the total open interest aggregated over all available time periods, from 1947 to 1965 according to trading groups. Net spreaders are frequently omitted because the net positions of this group are small and not the concern of this study.

These statistics on the distribution of the open interest may be used for three purposes: first, and of greatest importance to this study, they indicate whether hedgers are net short on the average and whether large speculators tend to be on the same side of the market as hedgers or that of non-reporting traders; second, they show the balance which exists between the long and short positions of any trading group (that is, they provide a measure of the homogeneity of a group's positions); and third, they indicate the "exclusiveness" of the large speculator category. With regard to this last use, if, for example, in one market the sum of the long and short positions of large speculators is one-tenth that of non-reporting traders, it suggests that in this market the large speculators are a more elite group than they are in another market where large speculators'

Table 2 Value of group commitments as a percentage of the value of the total open interest

Trading groups	All markets	All markets	All markets
Non-reporting (small) traders			
Small traders, long	54	51	53
Small traders, short	42	33	39
Reporting (large) traders			
Large speculators, long	9	13	11
Large speculators, short	4	6	5
Spreaders, long	20	14	18
Spreaders, short	20	15	18
Hedgers, long	17	21	18
Hedgers, short	33	45	38
Net commitments			
Small traders	12	18	14
Large speculators	5	7	6
Hedgers	−17	−24	−19
Spreaders[a]	0.1	− 0.4	− 0.1

[a]Used only to maintain accounting balances.

positions are just equal to non-reporting traders' positions. This distinction is important because the CEA reporting level for traders imposes a somewhat arbitrary dichotomy of traders, and it is therefore necessary to remember that the "exclusiveness" of the large speculator category does vary from market to market and also through time. Such a reservation is not so important for hedgers, because there is evidence that most hedgers' positions are above the reporting limit so that nearly all hedging is contained in the reporting hedger category.[8] This also implies that the non-reporting trader category, for practical purposes, may be considered as a small speculator category.[9]

From the statistics presented, it can be forcefully concluded that both non-reporting traders and reporting speculators are net long, and that reporting hedgers are net short. The only important exception to this conclusion is cottonseed meal. Cottonseed meal exactly reverses the normal pattern; but this phenomenon is easily explained as the outcome of a spreading operation whereby speculators offset long positions in soybean meal with short positions in cottonseed meal. Frozen eggs also deviate from the normal pattern, but only for reporting speculators. In this case the net short position of reporting speculators may be explained in terms of the large negative value of the rate of return on the long open interest.

Table 3 Aggregate profits by trading groups: long, short, and net (US$)

Trading group	Large markets		Small markets		All markets[a]	
	Long	Short	Long	Short	Long	Short
Small traders	369.7	−303.6	−68.1	− 1.4	301.6	−305.0
Reporting speculators	114.8	3.1	38.8	22.2	153.5	25.3
Reporting spreaders[b]	159.0	−159.6	5.5	− 3.4	164.5	−163.1
Reporting hedgers	108.1	−291.2	25.4	−18.8	133.5	−310.1
Total long open interest[a]	751.4		1.5		752.9	
Small traders, net		66.1		−69.5		− 3.4
Reporting speculators, net		117.8		61.0		178.8
Reporting hedgers, net		−183.2		6.5		−176.6

[a]Due to rounding, totals are not necessarily exact sums of components shown.
[b]This category is included only for balance purposes. The sum of the net positions is not zero because of its omission.

Although large speculators and small traders are both consistently net long, there is a clear difference between the ratio of short to long positions for non-reporting traders, and for reporting speculators. The short to long ratios for non-reporting traders in the Small, Large and All Markets aggregates are 65, 78, and 74% respectively. While the corresponding ratios for large speculators are 46, 44, and 45, and for hedgers the long to short ratios are 47, 52, and 47. These results are reasonably consistent over markets.

It is to be concluded, therefore, that reporting speculators' and reporting hedgers' positions are more unbalanced than those of non-reporting traders. This, in turn, suggests that the expectations of large speculators are more homogeneous than those of small speculators.

The relatively small proportion of total commitments held by large speculators is evidence that this group is an elite subset of the speculative population. Summing both long and short commitments, large speculators' holdings are less than one-fifth of the value of small traders' holdings. Since the average size of their commitments is much larger than that of small traders (perhaps by a factor of at least 10), the proportion of speculators classified as large is apt to be less than 2% of the total population of speculators.

Aggregate profits

Table 3 presents aggregate profits for the various trading categories according to the Large, Small and All Markets categories. For the All

Markets total, the long position profits of each trading group is positive. About two-fifths of the US$752.9 million total return on the long open interest goes to non-reporting traders, and the remainder divides rather evenly among the other three groups. The distribution of short position losses is similar to the distribution of long position profits in that two-fifths is borne by small traders; it differs in that reporting speculators make profits on the short side as well as the long. Thus, short hedgers bear two-fifths of the short side losses but receive only one-fifth of the long side gains.

For the Small Markets, reporting long hedgers and speculators make about equal amounts: the total of these two groups is equal to the losses of long small traders. Given that long small traders make profits in the 25-market aggregate, it is surprising that they have losses in the 23-market aggregate. On the short side, only large speculators make money and their total is roughly equal to the losses of the short hedgers. The total return on the long open interest in the 23 small markets is essentially zero.

In the Large Markets, almost half of the long side profit goes to non-reporting traders, and the remainder splits about evenly among the other three groups. The short side loss pattern differs from the long side gain pattern in that large speculators make some positive profits and the short position losses of hedgers are more than twice their long position gains. In marked contrast to the inconsequential Small Markets return on the total long open interest, this measure for Large Markets is sizeable (US$751 million) and explains virtually all of the US$753 million total for the 25-market aggregate.

Turning now to the net returns presented in the bottom three rows of Table 3, the most striking feature of the All Markets column is that the short side losses of small traders more than offset their long side gains causing this group to show a small net loss. The gains of large speculators are won, therefore, almost entirely from hedgers. In the 23 small markets, however, this result is reversed. Hedgers show a small net profit and the sizeable gains of the large speculators are made from the small traders.[10]

Consequently, we note again that it is the profit flow in the three large markets that determines the behaviour of the 25-market aggregate. In particular, the losses of hedgers in these three markets are large enough not only to provide profits to large speculators, but also to provide profits to small traders sufficient to offset their losses in the remaining 23 markets.

The temporal consistency of the profit flows may be judged from Tables 4 and 5. These two tables present annual profits on the total long open interest and for net trading groups for the All Markets and Small Markets aggregates respectively. For the 18 years, the All Markets results show

Table 4 Annual profits for all markets (US$)

Year	Small traders, net	Large speculators, net	Hedgers, net	Total
1947/48	16.2	19.5	−34.2	115.9
1948/49	−13.5	− 0.5	13.8	− 48.2
1949/50	7.9	17.0	−24.9	153.3
1950/51	47.5	28.9	−76.1	229.5
1951/52	−10.5	7.8	2.7	126.2
1952/53	−44.3	− 4.3	46.4	−171.9
1953/54	12.8	16.3	−29.7	113.4
1954/55	−17.0	5.1	12.0	− 27.7
1955/56	2.5	12.7	−15.5	73.9
1956/57	− 6.4	6.5	− 0.1	5.8
1957/58	− 9.4	0.9	8.4	− 27.4
1958/59	− 3.6	1.4	2.5	− 3.7
1959/60	−14.6	4.5	10.2	− 38.3
1960/61	63.3	35.2	−98.6	217.8
1961/62	−19.4	− 5.4	24.8	− 95.9
1962/63	− 3.6	2.1	1.6	50.5
1963/64	−24.1	10.0	14.1	− 75.0
1964/65	13.0	21.2	−34.1	155.0
Total	− 3.4	178.8	−176.6	752.9

negative profits for small traders in 11 years, positive profits for large speculators in 15 years, and negative profits for hedgers in 8 years. The consistency of large speculators' profits and small speculators' losses is notable. The same characteristics hold for the 23 small markets.

Rates of return

An economically more meaningful description of the profit flow may be made in terms of the average annual rate of return earned by traders on their invested capital. Ideally, profits should be stated net of commissions and taxes, and invested capital should include "safety reserves" as well as margin requirements. As a proxy for this true rate of return, we use the ratio of gross profits to the dollar value of the contracts held.[11] Omission of commissions causes a serious upward bias in the results for all groups. This bias is apt to be strongest for non-reporting traders who have the greatest relative overlap of long and short positions and who are least likely to own a seat on the exchange. However, the use of the value of the contract in the denominator introduces a gross understatement of the true return. Actual margin requirements are only 5% or 10% of the contract value and, even after allowing for a one-to-one "safety reserve", the true rate of return would be five to ten times larger than that measured here.

Therefore, the principal use of the rate of return variables defined here

Table 5 Annual profits for small markets (US$)

Year	Small traders, net	Large speculators, net	Hedgers, net	Total
1947/48	− 3.3	7.1	− 3.8	46.6
1948/49	−12.5	− 1.2	13.8	−53.7
1949/50	6.7	6.0	−12.8	32.3
1950/51	25.4	13.3	−37.4	73.2
1951/52	−10.6	2.2	8.0	27.1
1952/53	−27.3	− 0.9	26.2	−78.5
1953/54	− 2.3	5.2	− 3.7	14.0
1954/55	−11.4	1.6	9.8	−18.5
1955/56	− 0.3	5.6	− 5.2	7.5
1956/57	− 8.3	1.3	6.9	−24.2
1957/58	− 5.0	0.4	4.3	−10.3
1958/59	− 1.6	1.8	0.4	3.2
1959/60	− 6.2	4.6	1.7	−21.8
1960/61	7.3	3.1	−10.4	39.3
1961/62	−16.8	− 4.8	21.7	−63.8
1962/63	0.9	2.4	− 3.2	23.1
1963/64	−11.2	1.9	9.1	−47.0
1964/65	7.3	11.6	−18.8	53.1
Total	−69.5	61.0	6.5	1.5

must be in comparing the relative profits of different trading groups and not in making judgements about absolute values. If, however, large speculators (who generally may be presumed to have a seat on the exchange and a consequent low commission rate) show a rate of return over, say, 5%, it does suggest their true rate may be 25% a year or more. A figure of this size does indicate a large absolute return. Moreover, we may also say something about the absolute size of hedging costs. Since hedgers are offsetting existing or planned positions in the cash market (which are presumed equal to the dollar value of their futures holding), our rate of return is a direct measure of the gross cost of placing a year-long hedge. It is a gross cost because commissions and the bid-ask differential are omitted. This rate may be compared with the merchandising margins of hedgers to indicate the extent to which the futures cost of placing a hedge is a deterrent to hedging. Rate of return data for trading groups and the three different market aggregates are presented in Table 6.

For All Markets, the rate of return on the long open interest is 4.0% per year, a fairly substantial magnitude. Looking at the long positions of the three trading groups, it is notable that hedgers have virtually the same rate on their long positions, 3.8%, as is earned on the total open interest. Consequently, since the large speculators' return of 7.6% is greater than the return on the total open interest of 4.0%, non-reporting traders receive

Table 6 Aggregate rates of return by trading groups (%)

Groups and positions	Large markets	Small markets	All markets
Total positions			
All groups	6.1	0.0	4.0
Small traders, long	5.6	−2.0	3.0
Small traders, short	−5.8	−0.0	−4.1
Large speculators, long	10.1	4.3	7.6
Large speculators, short	0.5	5.0	2.7
Hedgers, long	5.3	1.7	3.8
Hedgers, short	−7.1	− .6	−4.3
Net positions			
Small traders, net	0.6	−1.2	−0.0
Large speculators, net	7.2	4.6	6.1
Hedgers, net	−3.0	0.1	−1.7

less than that, 3.0%. On short positions, both hedgers and non-reporting traders do slightly worse than the average, enabling large speculators to actually earn a positive rate of profit of 2.7%, even though the trend in prices is against them.

The rates of return on net positions are quite diverse.[12] We noted earlier that: (1) non-reporting traders are consistently net long; (2) prices rise on the average; (3) and, paradoxically, the absolute profits of non-reporting traders are, nevertheless, essentially zero. The answer, of course, must be that their rate of loss on short positions is sufficiently larger than their rate of gain on long positions to nullify any benefit they receive from being net long. In a similar fashion, hedgers do slightly worse on both long and short positions than do traders as a whole. This factor is not as important a contributor to their net rate of loss of 1.7% as is the simple fact that they are net short by a two-to-one margin. The size of hedging cost suggested by a rate of return of −1.7% is not inconsequential, but it is substantially less than many experts have suggested. The most striking feature of the net rates of return is the absolute magnitude of large speculators' returns, 6.1%. Recalling that a multiplier of from five to ten is required to obtain a true rate of return on investment, this suggests that the true rate may be as high as 25% to 50%. This large return is a consequence of two facts: large speculators tend to have high rates of return for both long and short positions; and their ratio of long to short positions is large.

The most significant feature of the rates of return for Large Markets and Small Markets separately is that in total the positive rate of return for All Markets is due solely to conditions in the three large markets where it reaches 6.1%. In the 23 smaller markets it is essentially zero. This, of course, results in there being a nearly zero cost of hedging in the Small

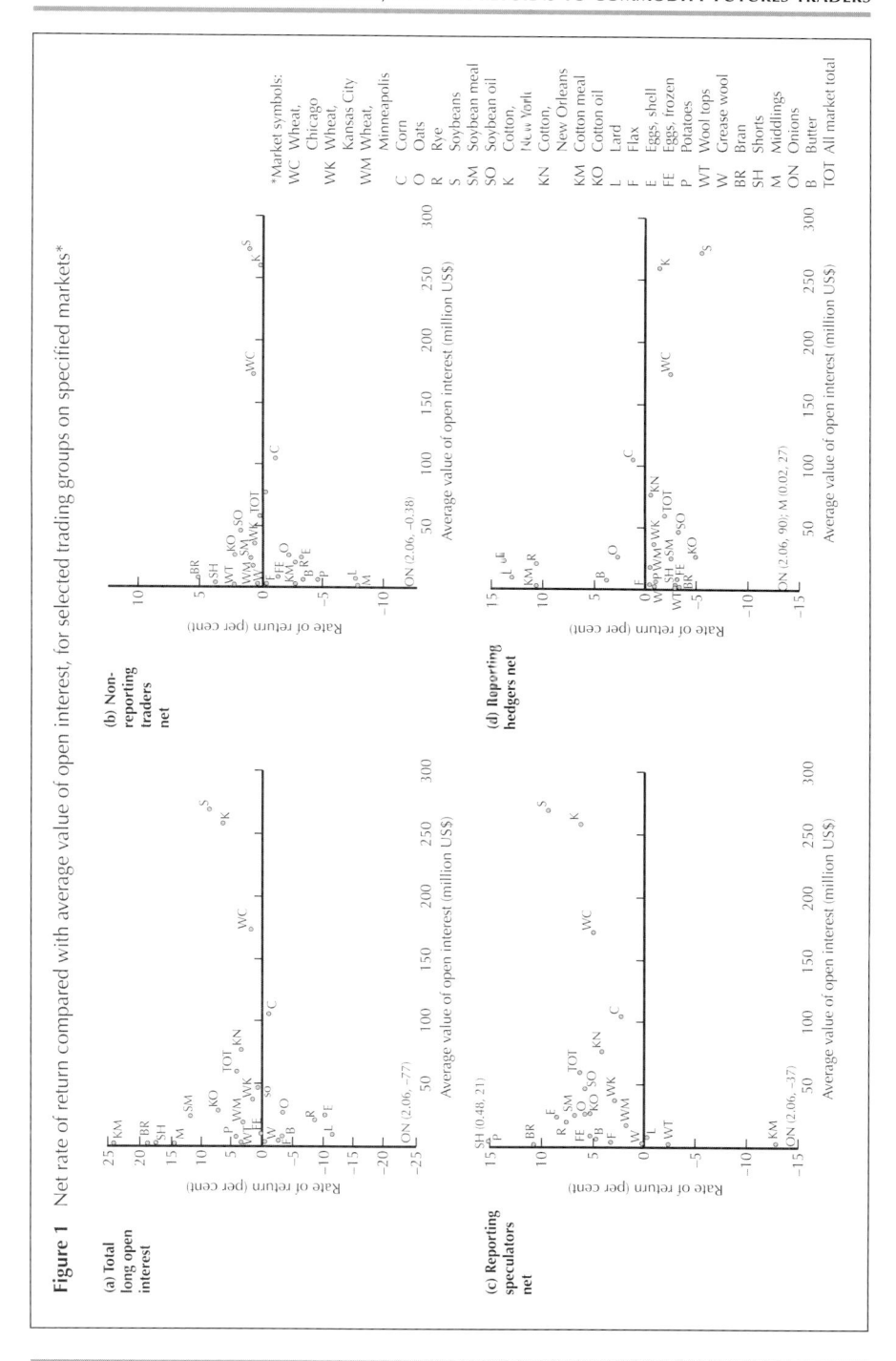

FORECASTING, AND THE RETURNS TO COMMODITY FUTURES TRADERS

Figure 1 Net rate of return compared with average value of open interest, for selected trading groups on specified markets*

(a) Total long open interest

(b) Non-reporting traders net

(c) Reporting speculators net

(d) Reporting hedgers net

*Market symbols:
WC Wheat, Chicago
WK Wheat, Kansas City
WM Wheat, Minneapolis
C Corn
O Oats
R Rye
S Soybeans
SM Soybean meal
SO Soybean oil
K Cotton, New York
KN Cotton, New Orleans
KM Cotton meal
KO Cotton oil
L Lard
F Flax
E Eggs, shell
FE Eggs, frozen
P Potatoes
WT Wool tops
W Grease wool
BR Bran
SH Shorts
M Middlings
ON Onions
B Butter
TOT All market total

587

Markets but a 3% cost in Large Markets. Other features for these two sets of markets are similar to those found for All Markets: non-reporting traders do worse on their long positions than average traders; hedgers do worse on their short positions than average traders; and large speculators do substantially better than average on both their long and short positions.

The final evidence on rates of return is given in Figure 1. These four scatter diagrams plot the rate of return for each of the 25 markets and for the All Markets total as a function of the average value of open interest in those markets. The abscissa values are, therefore, identical for all tables. There are separate plots for the total long open interest, and the net rate of return for small traders, speculators, and hedgers.

Figure 1(a) shows the scatter for long open interest. If the three largest markets are excluded, the rates of return are seen to be distributed around zero with considerable symmetry. The symmetry between positive and negative rates of return is broken when the three largest markets are included: all three have positive values, and the magnitudes for soybeans and cotton at New York are substantial. Thus, different hypotheses may be needed to explain the rate of return in the three largest markets and the 22 smaller markets.

As expected, Figure 1(b), showing the scatter for the net rates of non-reporting traders, is a "squashed" version of Figure 1(a). The rates of return in All Markets tend to be reduced because of the offsetting long and short positions and because of the relatively low rate of return on long positions.[13] Figure 1(b) also tends to be more symmetrical than Figure 1(a) and much less distorted by the effect of the three large markets. This symmetry around zero offers convincing evidence that the lack of profits of small traders cannot be explained by an atypically poor performance in a few markets offsetting satisfactory profits in most markets.

The scatter for reporting speculators, Figure 1(c), is quite different from either of the preceding two scatters. The rate of return for the three largest markets has improved; but what really attracts attention is the lack of markets with significant negative rates of return. Only cottonseed meal and onions are exceptions, and the peculiarities of each of these markets have already been discussed. Thus, positive rates for large speculators are as consistently reflected when we disaggregate over markets as they are when we disaggregate over time.

Figure 1(d) exhibits the results for hedgers. It is essentially a mirror image of Figure 1(a). This is a consequence of hedgers being net short and earning a rate of return on long and short positions approximately equal to that on the total open interest. However, this mirror image characteristic is not satisfied for hedgers' extreme negative returns, although it is well satisfied for extreme positive values. That is, for the four markets with the largest negative rates of return on the total open interest (onions, lard, shell eggs, and rye), the positive rates of return of net hedgers are almost equally as large as the return for all trading groups. On the other hand, for the five markets with the largest positive rates of return on the total open interest (cottonseed meal, bran, shorts, middlings, and soybean meal), the negative rates of return of net hedgers are less than one-fourth of the return for all trading groups. The cause of this can be determined from an inspection of the distribution of long and short positions by markets. For the four markets where prices fell more rapidly, hedgers' net positions are highly unbalanced in favour of the short side. In contrast, for the four markets where prices rose most rapidly, their net positions are highly unbalanced in favour of the long side. This suggests that, although hedgers do no better than average on the whole, they do have a knack for forecasting and profiting from extreme moves in prices.

The results presented in this section may be summarised by two conclusions. First, reporting speculators make significant profits on their long and their short positions. Their net profits are significant both from the point of view of consistency from year to year and market to market, and from the point of view of the magnitude of the rate of return. This cannot be said for small traders, whose returns are essentially zero and are negative if transaction costs are considered. The net costs of hedging are negative, but not large, and the important losses are concentrated in one market, soybeans.

Second, excluding the three largest markets, the rates of return on the total open interest are symmetrically distributed around zero. For the three large markets, however, there is a tendency toward positive returns on the total long open interest. In these large markets there is a 6% average return on the total open interest.

DETERMINANTS OF THE RETURNS TO FUTURES TRADERS
The role of normal backwardation

If normal backwardation is defined as the returns which a naïve speculator earns by keeping his commitments long, in proportion to the total open interest when hedgers are net short, and short, in proportion to the total open interest when hedgers are net long, then the rate of return on the total long open interest, presented above, is closely related to the rate of normal backwardation. However, it is necessary to multiply profits in a given market by minus one for each period that hedgers are net long in that market.

Table 7 presents a comparison of profits on the long open interest, for all periods with profits on the long open interest when hedgers are net long. As explained previously, profits which accrue while hedgers are net long are subtracted from total profits to obtain the measure of profits used

Table 7 Comparison of total returns and returns when hedgers are net short and net long

Item	Large markets	Small markets	All markets
Percentage of periods hedgers net long	25.2	13.5	15.4
Dollar profits on long positions (US$)			
Total	751.4	1.5	752.9
When hedgers net short	619.5	−25.6	593.9
When hedgers net long	132.0	27.0	159.0
Profits due to normal backwardation	487.5	−52.6	434.9
Percentage return on long positions			
When hedgers net short	6.1	− 0.4	3.7
When hedgers net long	6.3	4.0	5.7
Rate of normal backwardation	4.0	− 0.8	2.3

in computing the rate of normal backwardation (or more exactly, twice the profit for periods hedgers are net long must be subtracted from the profit for all periods). The theory of normal backwardation predicts that the subtrahend will be negative so that profits after the subtraction will be larger than they were before. Line four of Table 7, "Dollar profits on long positions when hedgers net long", shows with great force that the theory of normal backwardation is not supported by the data. The profits for both large and small markets are positive, not negative. Given that hedgers are net long only 15% of the time, the magnitude of the profits is sizeable. Indeed, not only is the sign of profits inconsistent with the theory of normal backwardation, but also the rates of return for all three aggregations are greater when hedgers are net long than when they are net short! Thus, an adjustment for the sign of net hedging results in a reduction in the rate of normal backwardation for Large, Small and All Markets to 4.0, −0.8, and 2.3% respectively.[14]

Although we must reject the prediction that prices fall when hedgers are net long, it is still possible that the theory of normal backwardation is supported when we aggregate over all time periods. Since hedgers are net long only 15% of the time, the successful performance of the theory of the remaining 85% could easily lead to a correct overall prediction. Figure 2 plots the rate of normal backwardation against the average value of the open interest for each market. The scatter is similar to Figure 1 except that the mean for All Markets is reduced from 4.0% to 2.3%, and a negative skew is introduced. Although the mean is still positive, both its small magnitude and the fact that normal backwardation is negative for 11 out of 25 markets

Figure 2 Rate of normal backwardation compared with average value of open interest, for specified markets*

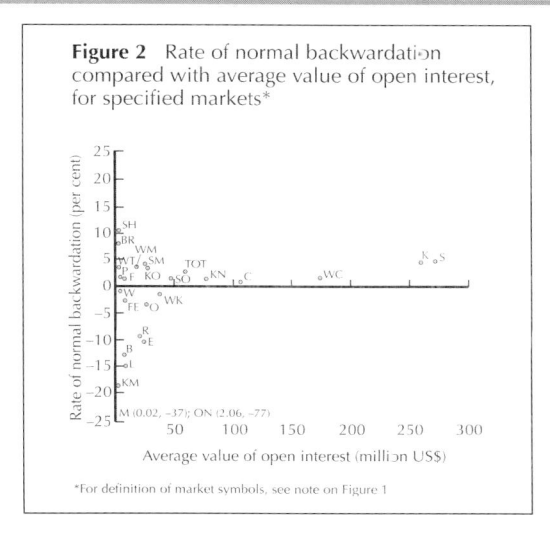

*For definition of market symbols, see note on Figure 1

must lead to the conclusion that a tendency toward normal backwardation is neither a consistent nor an important general characteristic of futures markets. In fact, only one of the 14 markets with positive measures of normal backwardation (shorts at 10.1) has a return in the 10% or more range postulated by Keynes (1930, p. 143). In contrast, there are six markets with negative returns of an absolute magnitude greater than 10%.

The conclusion of this section is that: normal backwardation is not characteristic of the 23 smaller markets either when hedgers are net long or net short; and it is characteristic of the three larger markets only when hedgers are net short. The theory clearly does not have general applicability for all futures markets and it is questionable whether an analysis of variance performed over the 25 markets would indicate a single market with a positive return significantly greater than zero.

The role of basic and special forecasting skills

In this section the rates of return of net trading groups are partitioned into two components: one, a reward defined as Basic Forecasting Skill; and the other, a residual component defined as Special Forecasting Skill.

The decomposition is performed in the following manner. Let V_m^L and V_m^S be the total value of a trading group's long and short commitments, in a single market m, aggregated over all time periods, and let R_m be the rate of return on the long open interest in that market. Then, any net trading group's rate of profit attributable to Basic Forecasting Skill is given by

$$R_m^B = \frac{R_m(V_m^L - V_m^S)}{V_m^L + V_m^S}$$

Denoting the group's actual rate of return by R_m^A, we then obtain the measure of Special Forecasting Skill as a residual of $R_m^F = R_m^A - R_m^B$.

Aggregation over any set of markets is accomplished by computing

$$R^B = \frac{\Sigma_m R_m(V^L - V^S)}{\Sigma_m(V_m^L + V_m^S)} \quad \text{and} \quad R^F = R^A - R^B$$

The measure R_m^B will be positive when R_m is positive and the group is net long on the average ($V_m^L - V_m^S > 0$), or when R_m is negative and the group is net short on the average ($V_m^L - V_m^S < 0$). Thus, this measure of Basic Skill is different from that proposed by Professor Houthakker (1957a, pp. 148–49). He measures the presence of Basic Skill in terms of the intercept coefficients of a regression of the quantity of commitments upon the change in price. If the intercept is positive, that is, if the expected value of a group"s commitments is positive when price change is zero, then a trading group is said to exhibit positive Basic Skill. This, of course, is only proper if the theory of normal backwardation is correct in that prices do rise on the average. However, the conclusion reached in this paper is that there is no important tendency for prices to rise, and, consequently, such a definition of Basic Skill is misleading. The measure used here defines Basic Skill as the ability to be net long on the average in markets where prices rise on the average, and to be net short on the average in markets where prices fall on the average. This measures the long run ability of a trading group to stay on the profitable side of the market. Special Forecasting Skill, defined as a residual, measures the success with which a trading group varies its position, from year to year and period to period, to profit from short run price trends (that is, from price trends whose duration is shorter than the total period of observation).

We may conclude this discussion of definitions by noting that both Houthakker's measure of Basic Skill and the one employed here seek to measure the extent to which traders" returns can be adequately described by a simple naïve strategy of being constantly on one side of a market. The remaining profits may then be interpreted as reflecting the traders' ability to forecast shorter price trends, and this is defined as Special Forecasting Skill. The definitions of Basic Skill differ because this study finds that normal backwardation is not a general characteristic of futures markets: therefore, it is more useful to use the actual trend in prices in defining the naïve trading strategy than to use the hypothetical returns that are predicted by a theory of normal backwardation, a theory which is not consistent with the data.

Table 8 shows that small traders exhibit a consistent negative value for Special Forecasting Skill, R^F. This measure is negative for both the Large and Small Markets aggregates and for 18 out of the 25 individual markets. While the absolute negative magnitude of R^F appears small, this is

Table 8 Division of rate of return according to basic and special forecasting skills by net trading groups (per cent)

Trading and skill groups	Large markets	Small markets	All markets
Small traders net			
R^F	−.1	−1.2	−.4
R^B	.7	.0	.4
R^A	.6	−1.2	−.0
Large speculators net			
R^F	5.0	3.9	4.8
R^B	2.2	.7	1.3
R^A	7.2	4.6	6.1
Hedgers net			
R^F	−.9	.8	−.6
R^B	−2.1	−.7	−1.0
R^A	−3.0	.1	−1.7

partially due to the fact that the positions of this group are almost balanced and consequently the denominator term $V_m^L + V_m^S$ is large. The only important profits for small traders occur in the three large markets where rising prices reward them for being net long so that Basic Skill R^B is positive.[15]

For large speculators, the situation is quite different. They have positive values of R^B and R^F for all three market aggregates. Nearly four-fifths (79%) of their total profits, however, are due to Special Skill and only one-fifth to Basic Skill. The conclusion must be that the substantial profits of large speculators are not an automatic return for simply being on the correct side of the market, but instead a reward for forecasting. This is confirmed on an individual market basis where the R^F variable for reporting speculators is positive for 22 out of the 24 markets.

The profit dichotomy for hedgers is not of great interest due to their offsetting commitments in the cash market. However, the positive value of 0.8 for R^F is consistent with our observation that hedgers are able to adjust the balance of their positions in response to major price movements in order to reduce their losses.[16]

---

1 The data on cotton cover the full period from 1937 to 1952.
2 These counter examples involve fluctuating commitment levels for speculators.

3 This explanation is stressed by Lester Telser (1960, p. 407).
4 Large speculators" long positions account for 11% of the value of the total open interest while their short positions account for only 5%. These statistics are presented in Table 2.
5 A version of this argument is used by Paul Cootner (1960, p. 400) to support a "hedging pressure" theory of price movement.
6 For example, "hedging pressure" theories, where the direction of price change is directly related to the magnitude of short hedges, imply a lack of foresight by speculators who take positions early in the season that are inconsistent with the assumption of forecasting ability.
7 The "wholesale price index" is taken from US President (1966). The Dow-Jones "futures price index" is from the Wall Street Journal for the trading days July 15, 1947 and June 29, 1965. (The respective values are 148.73 and 129.75.)
8 For a discussion of this evidence see Larson (1961).
9 The terms "reporting speculator" and "large speculator" are used synonymously as are the terms "non-reporting trader" and "small trader". The equivalence of the category "small speculator" and "non-reporting trader" (or "small trader"), however, is only approximate.
10 Statistical significance tests performed on the first differences of the annual net profit flows by trading groups from 1947 to 1963 reveal that only the net profits of large speculators are significant at the 0.05 level. This result holds for both the Large and All Market series. The only other potentially significant figure is the amount of the losses of small traders in the Large Markets, and it is only significant at the 0.20 level. For more detail see C. S. Rockwell (1964, p. 84).
11 The dollar value of a contract is obtained from the product of three factors: price per unit, units per futures contract, and the number of futures contracts held. This value is imperfectly linked to total marginal requirements which the exchanges alter only at discreet intervals to reflect major changes in price levels, price volatility, and, at times, trading activity.
12 The rate of return on net positions is defined as the sum of aggregate profits on long positions divided by the total value of long and short positions.
13 To provide partial visual compensation for the effect of offsetting long and short positions, Figure 1(b–d) are plotted with ordinate units one-half of those in Figure 1(a).
14 The sign of net hedging used in all of the above computations is the sign of the difference of the average size of hedgers' long positions less the average size of their short positions for each period and market. That is, $(Q_2^L + Q_1^L)/2 - (Q_2^S + Q_1^S)/2$. This is consistent with the general definition of profits as $\Delta P(Q_2 + Q_1)/2$. However, to test the sensitivity of our conclusions against alternative definitions of what constitutes "when hedgers are net short", we also computed dollar profits according to whether hedgers are net long or short at the beginning of the period. That is, according to the sign of $Q_1^L - Q_1^S$, the results are that the total long open interest profits are US$488 million when hedgers are net short, and US$284.9 million when hedgers are net long. Consequently the profits attributable to normal backwardation are US$203.1 million. The results are sensitive, but even more unfavourable to the theory of normal backwardation.
15 This return occurs in the manner predicted by the theory of normal backwardation; but the variable R^B cannot properly be construed as the rate of normal backwardation since it contains long position profits earned while hedgers are net long.
16 The sum of a group's long and short commitments is used as divisor in Table 8 to obtain the rate of return on net positions. This makes cross-group comparisons of R^B and cross-group comparisons of R^F misleading because of differences in the short to long ratio among groups. Calculations based on the rate of return to the "marginal" traders in each group (using the denominator $/V_m^L + V_m^S/$) yield the following percentage results for all markets:

Trading group	R^A	R^B	R^F
Small traders, net	−0.1	2.1	−2.2
Large speculators, net	12.9	2.8	10.1
Hedgers, net	−4.2	−2.6	−1.6

BIBLIOGRAPHY

Cootner, P., 1960, "Returns to Speculators: Telser versus Keynes", *The Journal of Political Economy*, August.

Hicks, J. R., 1953, *Value and Capital*, Second edition (Oxford).

Houthakker, H. S., 1957a, "Can Speculators Forecast Prices?", *The Review of Economics and Statistics* 39, May, pp. 143–57; reprinted as Chapter 22 of the present volume.

Houthakker, H. S., 1957b, "Restatement of the Theory of Normal Backwardation", *Cowles Foundation Discussion Paper* (44), December.

Keynes, J. M., 1923, "Some Aspects of Commodity Markets", *Manchester Guardian Commercial: European Reconstruction Series*, Section 13, March.

Keynes, J. M., 1930, *A Treatise on Money*, volume 2 (New York).

Larson, A. B., 1961, "Estimation of Hedging and Speculative Positions in Futures Markets", *Food Research Institute Studies*, November.

Telser, L., 1960, "Returns to Speculators: Telser versus Keynes: Reply", *The Journal of Political Economy* , August.

Rockwell, C. S., 1964, "Profits, Normal Backwardation and Forecasting in Commodity Futures", unpublished PhD dissertation. (University of California, Berkeley).

US President, 1966, Economic Report of the President (Washington, DC).

24
*Futures Trading and Investor Returns: An Investigation of Commodity Market Risk Premiums**

Katherine Dusak

The long-standing controversy over whether speculators in a futures market earn a risk premium is analysed within the context of the capital asset pricing model recently developed by Sharpe, Lintner and others. Under that approach the risk premium required on a futures contract should depend not on the variability of prices but on the extent to which the variations in prices are systematically related to variations in the return on total wealth. The systematic risk was estimated for a sample of wheat, corn and soybean futures contracts over the period 1952 to 1967 and found to be close to zero in all three cases. Average realised holding period returns on the contracts over the same period were close to zero.

Considerable controversy exists over the amount and the nature of the returns earned by speculators in commodity futures markets. At one extreme is the position first set forth by J. M. Keynes in his *Treatise on Money* (1930, pp. 135–44) that a futures market is an insurance scheme in which the speculators underwrite the risks of price fluctuation of the spot commodity. The non-speculators or "hedgers" on the other side of the market must expect to pay and, according to Keynes, they do in fact pay, on the average a significant premium to the speculator-insurers for this service. At the other extreme have been those such as C. O. Hardy (1940) who argue that for many speculators a futures market is a gambling casino. Far from demanding and receiving compensation for taking over the risks of price fluctuation from the hedgers, speculators, as a class, are willing to pay for the privilege of gambling in this socially acceptable form (with the losers continually being replaced at the tables by new

*This paper was first published in the *Journal of Political Economy* 81, November–December, pp. 1387–406 (1973) and is reprinted with the permission of the University of Chicago Press. The author wishes to thank Eugene Fama, Charles Nelson, Harry Roberts and especially Merton H. Miller for many helpful comments and suggestions.

arrivals). Despite many empirical studies, the conflict between the insurance interpretation and the gambling interpretation of returns to speculators in futures markets remains unresolved.[1]

This paper offers another and quite different interpretation of the returns to speculators in futures markets. It is argued that futures markets are no different in principle from the markets for any other risky portfolio assets. Futures markets are perhaps more colourful than many other subsegments of the capital market such as the New York Stock Exchange or the bond market, and the terminology of futures markets is perhaps more arcane, but these differences in form should not obscure the fundamental properties that futures market assets share with other investment instruments: in particular, they are all candidates for inclusion in the investor's portfolio.

The portfolio approach, by itself, makes no presumption as to whether returns to speculators are positive, as Keynes hypothesised, or zeroish to negative, as Hardy believed. It says, rather, that returns on any risky capital asset, including futures market assets, are governed by that asset's contribution, positive, negative, or zero, to the risk of a large and well-diversified portfolio of assets (in fact, all assets, in principle). In contrast to this portfolio measure of risk, Keynes and his later followers identify the risk of a futures market asset solely with its price variability.[2] These differences in the proposed measures of risk make it possible to test the portfolio and Keynesian interpretations of futures markets against each other and, in principle, also against the Hardy gambling casino view.

It turns out that for each of the commodity futures studied (wheat, corn, and soybeans) returns and portfolio risk are both close to zero during the sample period even though variability or risk in the Keynesian sense is high. Hence, as far as this set of observations is concerned, the data conform better to the portfolio point of view than to the Keynesian insurance interpretation. The sample did not permit any direct confrontation between the portfolio interpretation and the Hardy view, but some indirect light is thrown on this part of the controversy and some suggestions for further tests are offered.

In the next section the salient points of the equilibrium pricing of portfolio assets are noted, and futures contracts are analysed within this context. Measures of Keynesian and portfolio asset risk are then developed and interpreted in the light of the returns observed.

CAPITAL ASSET PRICING: THE DETERMINATION OF AN EQUILIBRIUM RISK-RETURN RELATION

A model of the equilibrium pricing of portfolio assets was proposed originally by Sharpe (1964) and extended by Lintner (1965), Mossin (1966), and Fama (1971).[3] Sharpe showed that conditions exist under which the equilibrium risk-return relation for any capital asset i can be represented as

$$E(\tilde{R}_i) = R_f + \left[\frac{E(\tilde{R}_w) - R_f}{\sigma(\tilde{R}_w)}\right] \frac{\partial \sigma(\tilde{R}_w)}{\partial x_i} \quad (1)$$

where $\tilde{R}_i$ is the random rate of return on asset i, $E(\tilde{R}_i)$ is its mathematical expectation, and R_f is the pure time return to capital or the so-called riskless rate of interest; $\tilde{R}_w$ is the random rate of return on a representative dollar of total wealth or, equivalently, the return on a portfolio containing all existing assets in the proportions, x_i, in which they are actually outstanding: $E(\tilde{R}_w)$ is the expected rate of return on total wealth, and $\sigma(\tilde{R}_w)$, the standard deviation of the return on total wealth, is a measure of the risk involved in holding a representative dollar of total wealth. The term $[\partial \sigma(\tilde{R}_w)]/\partial x_i$ is the marginal contribution of asset i to the risk of the return on total wealth, $\sigma(\tilde{R}_w)$. Thus expression (1) says that, in equilibrium, the expected rate of return on any asset i will be equal to the riskless rate of interest plus a risk premium proportional to the contribution of the asset to the risk of the return on total wealth.

To see some of the broader implications of this proposition and especially to highlight its fundamental difference from the simple Keynesian approach to risk, note that since

$$\sigma(\tilde{R}_w) = \left[\sum_{i=1}^{N} \sum_{j=1}^{N} x_i x_j \operatorname{Cov}(\tilde{R}_i, \tilde{R}_j)\right]^{1/2}$$

it follows that

$$\frac{\partial \sigma(\tilde{R}_w)}{\partial x_i} = \frac{1}{\sigma(\tilde{R}_w)} \left[\sum_{j=1}^{N} x_j \operatorname{Cov}(\tilde{R}_i, \tilde{R}_j)\right]$$

$$= \frac{1}{\sigma(\tilde{R}_w)} \left[x_i \sigma^2(\tilde{R}_i) + \sum_{j \neq i}^{N} x_j \operatorname{Cov}(\tilde{R}_i, \tilde{R}_j)\right]$$

Thus what governs the riskiness of any asset i is not merely its own variance $\sigma^2(\tilde{R}_i)$ but its weighted covariance with all the other assets making up total wealth. Normally the latter terms can be expected to swamp the former, since there are $N - 1$ terms making up the covariance portion and only one in the variance portion, and that one, moreover, weighted by a very small number, x_i.

Additional insight into the equilibrium risk-return relation is gained by noting that the expression

$$\sum_{j=1}^{N} x_j \operatorname{Cov}(\tilde{R}_i, \tilde{R}_j)$$

can be rewritten as $\text{Cov}(\tilde{R}_i, \tilde{R}_w)$, the covariance of return on asset i with that of total wealth. Hence we can rewrite (1) as

$$E(\tilde{R}_i) = R_f + \left[\frac{E(\tilde{R}_w) - R_f}{\sigma(\tilde{R}_w)}\right] \frac{\text{Cov}(\tilde{R}_i, \tilde{R}_w)}{\sigma(\tilde{R}_w)} \qquad (2)$$

or equivalently as

$$E(\tilde{R}_i) - R_f = [E(\tilde{R}_w) - R_f]\beta_i \qquad (3)$$

where $\beta_i \equiv [\text{Cov}(\tilde{R}_i, \tilde{R}_w)]/\sigma^2(\tilde{R}_w)$. The coefficient β_i can be interpreted as the relative risk of asset i, since it measures the risk of asset i relative to that of total wealth. Equation (3) then says that the risk premium expected on asset i is proportional, in equilibrium, to its systematic risk β_i, the factor of proportionality being the risk premium expected on a representative dollar of total wealth.

Needless to say, the capital asset pricing model rests on a set of fairly strong assumptions. Nevertheless, it has proven to be remarkably robust empirically. Studies by Miller and Scholes (1972), Black, Jensen and Scholes (1972), and Fama and MacBeth (1972) indicate that while simple expressions such as equations (2) and (3) may not be entirely satisfactory descriptions of the relations between return and relative risk, there is a strong connection between them, whereas there seems to be virtually none between the risk premium and measures of nonportfolio risk.[4]

APPLICATION OF THE CAPITAL ASSET PRICING MODEL TO FUTURES CONTRACTS

One difficulty in applying the Sharpe model of capital asset pricing to the risk-return relation on futures contracts is that of defining the appropriate capital asset and its rate of return. Since virtually all futures contracts are bought (and sold short) on margins that typically range from 5 to 10% of the face value of the contract, it might seem at first sight that we can treat the margin as the capital investment and treat the ratio of the net profit at closeout to the initial margin as the rate of return on investment. In fact, one theoretical study, that of Schrock (1971), takes this point of view and makes it the basis of a standard mean-variance portfolio analysis, à la Markowitz (1959), though restricting attention only to futures market assets.

This appealing procedure for computing futures market returns breaks down, however, as soon as we trace the subsequent history of the payment that is turned over to the broker. Unlike other capital assets such as common stocks where the margin is transferred from buyer to seller, the margin on a futures contract is kept in escrow by the broker. Not only does the seller of the futures contract not receive the capital transfer from the buyers, but he actually has to deposit an equivalent amount of his own funds in the broker's escrow account. At closeout, the broker returns the escrowed margin plus or minus any profits or losses (net of commissions in the case of profits and inclusive of commissions in the event of losses) that occurred over the period.

The margin, despite surface appearances, is thus not a portfolio asset in the sense of the Sharpe general-equilibrium model, but merely a good-faith deposit to guarantee performance by the parties to the contract. If the brokers had other ways of ensuring that traders did not make commitments beyond their resources, then no such performance bonds would be required. For example, forward foreign exchange markets, where firms deal through their own banking connections, typically operate without any explicit margins, whereas participants in public futures currency markets are required to post margins.[5]

Although the rate of return on the margin is not a meaningful number from a general-equilibrium point of view, and need not even exist if other types of guarantees could serve, there is another natural candidate which can always be computed: namely, the percentage change in the futures price. We cannot interpret this percentage change as a rate of return comparable to the $\tilde{R}_i$, in equation (2) above, since the holder invests no current resources in the contract. But we can interpret it as essentially the risk premium, $\tilde{R}_i - R_f$, on the spot commodity.[6]

What corresponds to the full return $\tilde{R}_i$ is the return (net of storage costs) that would accrue to the holder of an unhedged spot commodity.[7] That return consists of interest on the capital invested in the commodity plus any return, positive or negative, over and above pure interest due to the unanticipated change in the price of the commodity. If the spot holder chooses to hedge his holding, he thereby converts it to a riskless asset on which he earns only the riskless rate, R_f. The purchaser of the futures contract who takes over the risk has no capital of his own invested and hence earns no interest or pure time return on capital. He receives only the return over and above interest, which is to say, $\tilde{R}_i - R_f$.

This argument can be formalised by restating the Sharpe equilibrium conditions in present-value form. We say that the expected return on any asset i can be expressed as

$$E(\tilde{R}_i) = (1 - \beta_i)R_f + \beta_i E(\tilde{R}_w) \qquad (4)$$

where $\beta_i = \text{Cov}(\tilde{R}_i, \tilde{R}_w)/\sigma^2(\tilde{R}_w)$. Equivalently, since we can represent $E(\tilde{R}_i)$ in terms of period 0 and period 1 prices for the asset as $[E(\tilde{P}_{i,1}) - P_{i,0}]/P_{i,0}$, the equilibrium risk-return relation on asset i can be expressed as

$$P_{i,0} = \frac{E(\tilde{P}_{i,1}) - [E(\tilde{R}_w) - R_f]P_{i,0}\beta_i}{(1 + R_f)} \qquad (5)$$

Expression (5) says that the current price of any asset i is the discounted value (at the riskless rate) of its expected period 1 price, adjusted downward for risk by the factor $[E(\tilde{R}_w) - R_f]P_{i,0}\beta_i$.

Now suppose one were interested in knowing the price of asset i under a contractual agreement to purchase the asset at time 0 but with payment deferred a period to time 1. Clearly the current price for the asset under such an agreement must be given by $P_{i,0}(1 + R_f)$. That is, since the transaction is made at time 0 but consummated at time 1, the purchaser must pay a one-period credit, or borrowing charge of $P_{i,0}R_f$ in addition to the current price $P_{i,0}$. Multiplying both sides of equation (5) by $(1 + R_f)$ we see that

$$P_{i,0}(1 + R_f) = E(\tilde{P}_{i,1}) - [E(\tilde{R}_w) - R_f]P_{i,0}\beta_i \qquad (6)$$

But the contractual agreement just described is a futures contract where asset i refers to the spot commodity. Hence the expression $P_{i,0}(1 + R_f)$ can be interpreted as the current futures price for delivery and payment of the spot commodity one period later, and $E(\tilde{P}_{i,1})$ can be interpreted as the spot price expected to prevail at time 1. The essential point is that buying a futures contract is like buying a capital asset on credit where the capital asset in this case happens to be the spot commodity.[8] The only issue is what is the "discount for cash" or, equivalently, the financing charge. Since the financing is assumed to be riskless, the correct charge is clearly R_f. That is, if $P_{i,0}$ is the current price for immediate payment, $P_{i,0}(1 + R_f)$ must be the price if the buyer buys on one-period credit terms.

Setting $P_{f,0} = P_{i,0}(1 + R_f)$ and rearranging terms, we get

$$\frac{E(\tilde{P}_{i,1}) - P_{f,0}}{P_{i,0}} = \beta_i[E(\tilde{R}_w) - R_f] \qquad (7)$$

Equation (7) can be interpreted as expressing the risk premium on the spot commodity as the change in the futures price divided by the period 0 spot price. Thus once again we see that futures contracts, properly interpreted, pose no problem for capital market theory.

One implication of this analysis is that there are two essentially equivalent ways of calculating the risk premium. On the one hand, we can try to measure the risk premium by taking the percentage change in spot prices (net of storage) over a given interval minus the riskless rate. Alternatively, we can approximate the risk premium as the percentage change in the futures price over the same interval.[9] Of these two approaches, it is the latter that will be adopted here. Data on futures prices are more accessible than spot price data and, of course, use of futures prices also avoids the necessity of having to estimate the storage costs directly. It is important to remember, however, that this choice of measurement is essentially a matter of computational convenience; and that the relevant risk from the general-equilibrium point of view remains the risk inherent in the ownership of the spot commodity itself, regardless of who actually chooses to bear it or what measurement strategy we choose to employ.[10]

THE EMPIRICAL PROPERTIES OF FUTURES MARKET RETURNS

Tests of the risk-return relationship in the futures market are based on a sample of three heavily traded agricultural commodities: wheat, corn and soybeans. There are five different contracts per year for wheat and corn and six for soybeans. For all contracts, semi-monthly price quotations were obtained for a 15-year period from May 15, 1952 through November 15, 1967 – resulting in an approximate sample size of 300 observations per contract.[11] In all cases returns were computed as a simple two-week holding period yield with no allowance made for transaction costs. Following universal practice, returns have been computed separately for each commodity contract (eg, May wheat or September corn).[12] It should be noted that the return series computed in this way is discontinuous, since published price quotations on any one contract are typically available over a 9- or 10-month span.

Since the procedures for computing all subsequent statistical measures assume serial independence of returns, serial correlation coefficients of orders 1–10 have been computed for each contract return series.[13] The results are presented in Table 1. As can be seen, the coefficients fluctuate about zero. Out of a total of 160 correlation coefficients, 132 are less than .10, and only 11 coefficients are more than two standard deviations away from zero. Even the largest in absolute value, moreover, is only .22 and hence accounts for only a trivial portion of the variation in returns on the particular contract. There are about the same number of negative as positive coefficients, with no particular pattern in the signs.[14]

CONSTRUCTION AND INTERPRETATION OF A KEYNESIAN RISK MEASURE

There is nothing in Keynes' essentially heuristic discussion of futures market risk to suggest the use of any one measure of simple variability over another. Subsequent writers have adapted the Keynesian argument to the Markowitz mean-variance framework (eg, Johnson 1960; Schrock 1971). The use of sample variances to measure risk is open to objection, however, if the distribution of returns is stable non-Gaussian, as some have suspected may be true for futures market returns. For such distributions the second and higher-order moments of the distributions do not exist. The variances and standard deviations in any particular sample are always finite, but their behaviour will be erratic and affected by outliers.[15]

Table 1 Semi-monthly serial correlation coefficients for wheat, corn, soybeans ($\tau = 1, 10$)

Contract	1	2	3	4	5	6	7	8	9	10
Wheat:										
July	.14*	.01	−.15*	−.07	.10	−.07	−.02	.03	.12	−.03
Mar	.07	−.07	−.03	.05	.03	.11	.12	−.11	.06	−.22*
May	.17*	.03	−.09	−.11	.06	.10	.10	.06	−.05	−.18*
Sept	.15*	.07	−.02	−.03	.07	−.01	.10	.09	.15	.05
Dec	.14*	.10	.04	−.01	.04	−.14*	−.08	−.01	.03	−.11
$\bar{\rho}$	.13	.03	−.05	−.03	.06	−.00	.00	.01	.06	−.10
Corn:										
July	−.02	−.04	.03	.03	−.05	−.02	.01	−.03	.08	−.06
Mar	.01	.08	−.09	−.00	−.03	.10	.04	.17*	−.07	−.03
May	.02	.08	.00	−.03	−.04	−.02	.00	.05	.05	.05
Sept	.10	.10	.02	−.02	−.04	.03	.08	−.04	.03	−.03
Dec	.02	.06	.02	−.04	−.07	.05	−.07	−.06	−.01	−.04
$\bar{\rho}$	.03	.06	−.00	−.01	−.05	.03	.01	.02	.02	−.02
Soybeans:										
Jan	.02	.05	−.04	−.04	−.11	.09	.03	−.10	−.03	.01
Mar	.03	.18*	.02	.07	−.05	.13	−.05	.05	−.07	.10
May	.09	.17*	.16*	.11	.06	.10	.09	.19*	.06	.10
July	.09	.15*	−.07	.16*	.02	.06	−.02	−.01	.07	−.05
Sept	.06	−.03	.07	.00	−.03	.04	.20*	−.11	−.08	.09
Nov	.03	.06	−.07	−.08	−.09	.01	.13	−.09	.04	−.03
$\bar{\rho}$	.05	.10	.01	.04	−.03	.07	.06	−.03	−.00	.04
SE (r_τ) for $N = 300$	.058	.058	.058	.058	.058	.058	.058	.059	.059	.059

*Coefficient is twice its computed standard error.

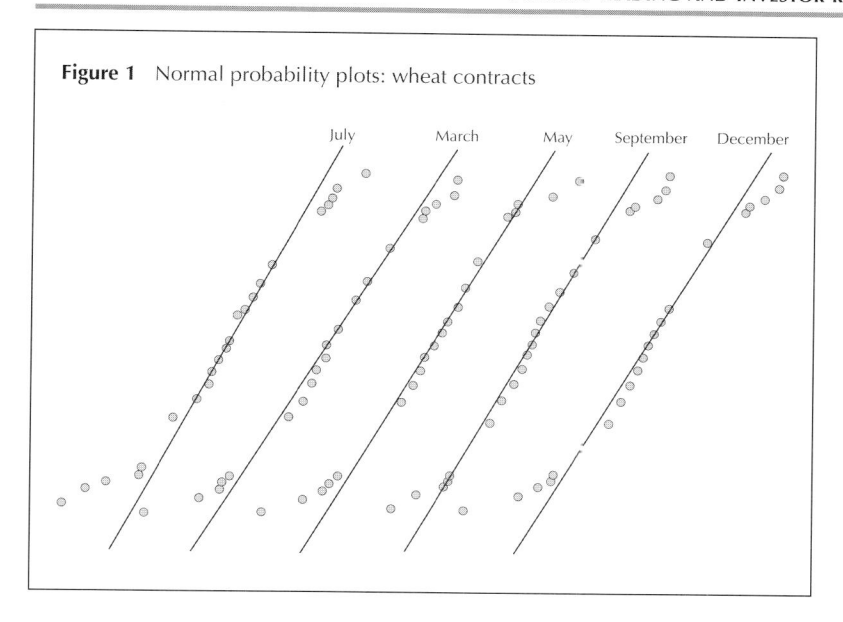

Figure 1 Normal probability plots: wheat contracts

July March May September December

Evidence of non-normality is indicated by the normal probability plots of the cumulative distributions of sample returns in Figures 1–3. To facilitate comparisons among the distributions of contract returns, the normal probability plots have been grouped by commodity. Five observations in the critical upper and lower tails of the distributions have been plotted and every twenty-fifth observation in the less revealing middle range. If the distributions were normal, the plots would closely approximate a straight line with slope $1/s$ and intercept $\bar{x}/s$, where s is the sample standard deviation and $\bar{x}$ is the sample mean (Roberts 1964, p. 13). As can be seen, there are substantial departures from linearity, not only in the tail areas but in the middle range as well, in every graph. The departure from normality in the tails is particularly marked in the case of the six soybean contracts.[16]

It can be shown that any symmetric stable distribution is characterised by three parameters: a shape parameter, α; a location parameter; and a scale or dispersion parameter. For the normal distribution α has the value 2; the first moment or mean serves as a measure of the location parameter, and the standard deviation (divided by $\sqrt{2}$) defines the scale. The fat-tailed distributions encountered in studies of asset pricing have shape parameter α less than 2. For such distributions, the mean can still serve as the location parameter, provided α is greater than one (the case of $\alpha = 1$ being the Cauchy distribution), although it has been shown that a truncated mean has smaller sampling dispersion than the sample mean and thus is a better estimator of the location parameter.[17] But, as noted

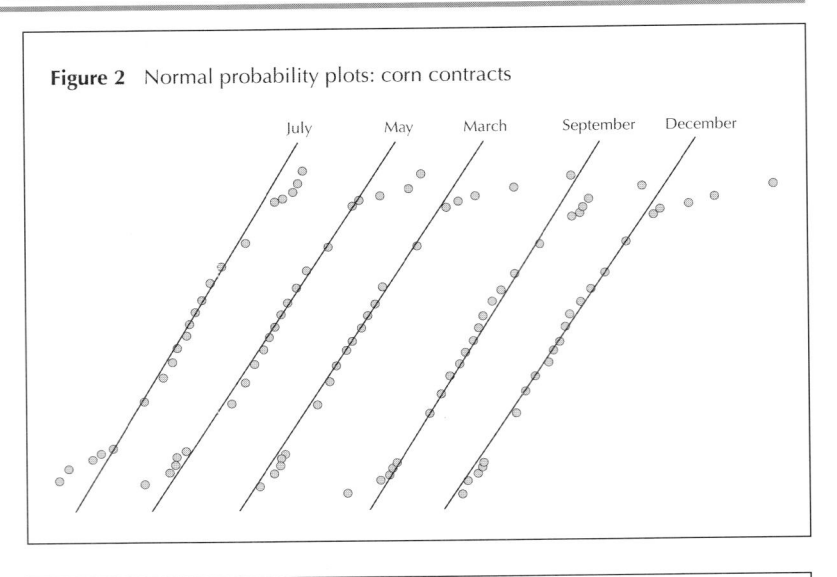

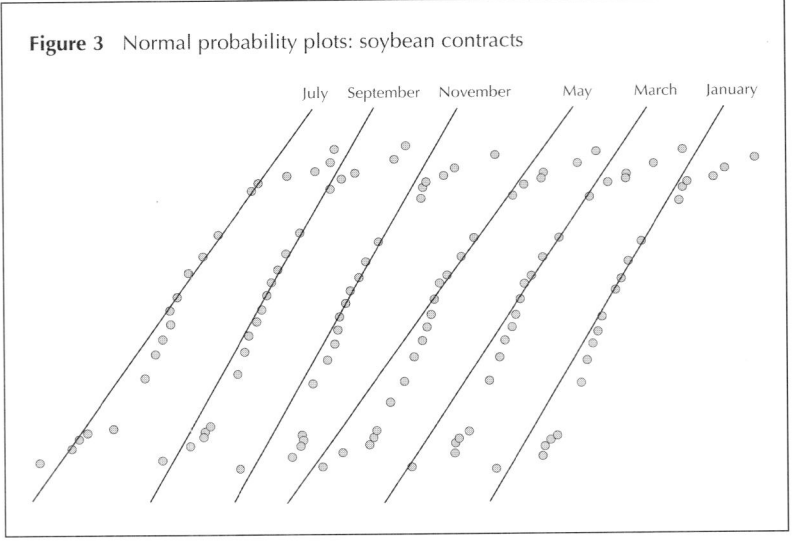

Figure 2 Normal probability plots: corn contracts

July May March September December

Figure 3 Normal probability plots: soybean contracts

July September November May March January

above, the second and higher-order moments, and hence also the standard deviation, are not finite. Interfractile ranges do exist, however, and have been found to serve quite adequately as measures of scale or dispersion.

Estimates of α, the scale factor, the 0.5 truncated mean, and standard errors for the last two estimators are presented in Table 2.[18] Following Fama and Roll (1968, 1971) the 0.28–0.72 interfractile range was used to estimate the scale factor, and a fractile matching procedure (in this case the 0.95) was used to estimate α.

From column 1 of Table 2, it would seem safe to conclude that the distributions of returns on futures contracts conform better to the stable non-Gaussian family than to the normal distribution. The values of α range from 1.44 to 1.84, with half of the estimates below 1.56.

The scale factors, which I shall interpret as measures of Keynesian risk, and their standard errors of estimate are shown in columns 2 and 3 of Table 2.[19] To judge how large these scale factors are, we can compare them to the corresponding dispersion parameter for some other more familiar capital asset such as common stock. The most convenient measure of common stock returns for our purposes is the Standard and Poor Composite Index of 500 industrial common stocks. The estimated dispersion parameter for that index taken semi-monthly over the sample period 1952–1967 is 0.0170, which is the same order of magnitude as the scale factors for the commodities. Since the Standard and Poor Index is, in effect a well-diversified portfolio, we know that the variability of the average stock return will be two or three times as large (see King 1966; Blume 1968) and hence also two or three times that of returns on futures market assets.

These comparisons may surprise those accustomed to thinking of futures markets as especially volatile, and the futures contract as one of the riskiest of capital assets.[20] The impression of substantial return volatility probably arises from the practice of calculating percentage returns on the margin. It should be remembered, though, that the margin is not a capital asset within the economic meaning of that term. Hence in the general-equilibrium context the variability of rates of return on the margin is not a relevant measure of risk.

Since the variability of futures returns is about as great as that of a diversified portfolio of common stock, we should expect that if Keynes was correct in identifying asset risk with simple variability, then the mean return over and above the riskless rate should be about the same for both assets. The mean rate of return (over and above the riskless rate) on the Standard and Poor Index over the period 1952–1967 on a semi-monthly basis was approximately 0.0029 (with a standard error of 0.0012) without allowing for dividends.[21] Had dividends been included, they would probably have added another 0.0017 to bring the total return to 0.0046. This figure is in striking contrast to the point estimates of the truncated means for the commodity returns (Table 2, col. 4). All of the truncated means for corn returns are negative and of roughly the same magnitude. In the case of soybeans, four of the six truncated means are negative but the range between the smallest and largest values is only 0.00143, which is the same order of magnitude as the standard errors of the estimates. The truncated means for wheat returns exhibit somewhat greater varia-

Table 2 Estimates of stable paretian parameters for wheat, corn, and soybeans

Contract*	α† (1)	Scale factor (2)	SE‡ of scale factor (3)	Truncated mean return (4)	SE§ of truncated mean (5)
Wheat:					
July (302)	1.55	.01111	.00085	−.00164	.00126
Mar (302)	1.75	.01228	.00091	.00060	.00139
May (302)	1.70	.01259	.00094	.00096	.00142
Sept (319)	1.56	.01127	.00086	−.00194	.00127
Dec (319)	1.74	.01184	.00088	.00044	.00134
Corn:					
July (301)	1.52	.01027	.00079	−.00158	.00116
Mar (301)	1.65	.01222	.00092	−.00381	.00138
May (301)	1.49	.01062	.00082	−.00268	.00120
Sept (320)	1.65	.01136	.00086	−.00243	.00128
Dec (320)	1.84	.01304	.00092	−.00212	.00147
Soybeans:					
Jan (287)	1.49	.01293	.00100	−.00025	.00146
Mar (287)	1.47	.01347	.00105	−.00029	.00152
May (287)	1.44	.01309	.00102	.00038	.00148
July (287)	1.44	.01399	.00109	.00006	.00158
Sept (287)	1.66	.01391	.00105	−.00105	.00157
Nov (287)	1.50	.01212	.00093	−.00071	.00137

*Numbers of observations are given in parentheses.
†No exact methods have been derived for computing the standard error of $\alpha_{.95}$ or its bias. Using Monte Carlo techniques, Fama and Roll (1971, p. 333) report that for samples of 299 observations the standard deviation of the values of $\hat{\alpha}_{.95}$ in 199 separate replications was about 0.13 when the true value of α was 1.5, about 0.15 when the true value of α was 1.7, and about 0.12 when the true value was 2.0. The mean value of $\alpha_{.95}$ was slightly less than true of α when that value was very close to two; the apparent bias was only 0.04 when α was 1.9 and was beyond detection at a value for α of 1.7.
‡An expression for the variance of the scale factor is given in Fama and Roll (1971, p. 331). Standard errors have been computed from estimates of $\sigma(s)$ for standardised symmetric stable distributions. See Fama and Roll (1971, Table 1, p. 332).
§The standard error of the truncated mean has been computed as: $s\sigma(\bar{x}_{.5,N})$ where s is the scale factor from the underlying distribution of returns and $s\sigma(\bar{x}_{.5,N})$ is the standard deviation for the .5 truncated mean from a standardised normal distribution. For a discussion of this estimator see Roll (1968, p. 30).

tion. The mean returns for the July and September contracts are large and negative whereas the mean returns for the March, May and December contracts are slightly positive. For all but two of the 16 contracts, however, the mean returns are within two standard errors of zero.

These results are a serious blow to the theory of normal backwardation. Using Keynes's definition of asset risk, anyone who invested (ie, sold insurance) in wheat, corn and soybean futures in the period 1952–1967 incurred risk for which he received on average a return very close to zero, if not actually negative. What is even more damaging to the Keynesian theory is the fact that for the same amount of risk (defined as simple return variability) an investment in a diversified portfolio of common stocks over the same period would have yielded a substantial positive return over and above the riskless rate.[22]

CONSTRUCTION AND INTERPRETATION OF THE RISK MEASURES FOR THE CAPITAL MARKET INTERPRETATION

The Sharpe model of capital asset pricing defines asset risk as the contribution the asset makes to the variability of return on a well-diversified portfolio containing, in principle, all assets in the proportions in which they are outstanding. An estimate of the relative risk can be obtained from the linear regression:

$$\tilde{R}_i = \alpha_i + \beta_i \tilde{R}_w + \tilde{\varepsilon}_i \qquad (8)$$

where the usual assumptions of the linear regression model are assumed to hold.

Although equation (8) implies that the independent variable is the return on total wealth, such a variable is virtually impossible to construct, and instead some proxy measure must be utilised. In this study the return on the value-weighted Standard and Poor Index of 500 Common Stocks is used as a proxy for the return on total wealth. Common stocks, after all, represent an important fraction of total wealth, so that even in a more comprehensive index they would be heavily weighted. This has been, moreover, the standard approach followed in most studies of the asset pricing model.

The selection of the Standard and Poor Index to represent common stocks was dictated by the fact that the leading alternative, the Fisher index (1966), is available on a monthly basis, whereas the futures market returns are computed on a semi-monthly basis. The main drawback of the Standard and Poor Index is that it does not include the dividend component of returns on common stock. Dividends, however, are not highly variable in the short run, and their omission is not likely to have any noticeable effects on the regression coefficients that I will be using as measures of risk.[23]

Consistent with the interpretation of the futures return as a risk premium, that is, as a return over and above interest, the market index variable is also stated in risk premium form. As a measure of the riskless rate of interest, I used the 15-day Treasury bill rate.[24]

The estimates of α'_i (where α' denotes regression variables expressed in risk premium form) and β_i from equation (8), their standard errors, the R^2s, and the first-order serial correlation coefficients of the residuals for the sample period 1952–1967 are presented in Table 3 for each of the 16 commodity contracts. The most striking feature of Table 3 is the small size

Table 3 Regression parameters for wheat, corn, and soybeans

Commodity*	$\hat{\alpha}'_i$	SE $(\hat{\alpha}'_i)$	$\hat{\beta}_i$	SE $(\hat{\beta}_i)$	R^2	Autocorrelation coefficient of residuals
Wheat:						
July (302)	−.020	.001	.048	.051	.003	.148
March (302)	.000	.001	.098	.049	.013	.080
May (302)	−.000	.001	.028	.051	.001	.163
Sept (319)	−.002	.001	.068	.051	.006	.149
Dec (319)	−.000	.001	.059	.048	.005	.163
Corn:						
July (301)	−.001	.001	.038	.046	.002	−.041
March (301)	−.003	.001	−.009	.050	.000	.015
May (301)	−.002	.001	−.027	.048	.001	.032
Sept (320)	−.002	.001	.032	.048	.001	.100
Dec (320)	−.001	.001	.007	.047	.000	.017
Soybeans (287 all contracts):						
Jan	.002	.001	.019	.058	.000	.015
March	.003	.002	.100	.065	.008	.018
May	.003	.002	.119	.068	.011	.071
July	.002	.002	.080	.076	.004	.083
Sept	.001	.001	.077	.065	.005	.060
Nov	.002	.001	.043	.058	.002	.023

*Numbers of observations are given in parentheses.

of the regression coefficients, which range from 0.007 to 0.119. With few exceptions the standard errors are approximately the same size, if not somewhat larger, than the regression coefficients. Furthermore, the standard errors in Table 3 may be understated because ordinary least squares is not efficient if the underlying returns are non-Gaussian (see Fama and Babiak (1968) for a discussion of this point). Thus the smallness of the regression coefficients relative to their standard errors is on balance even more pronounced than the figures in the table indicate. In the case of the intercept term, the standard errors are also large (in only two, possibly three, cases are they as small as half the value of the coefficient), which is consistent with expression (3) and a value of the intercept of zero. The low serial correlation of the residuals suggests that the assumption of independence upon which the calculation of the standard error is predicted is a tenable one.

It is clear from Table 3 that relative risk for wheat, corn, and soybeans is very close to zero.[25] To judge how low a level of systematic risk these regression coefficients represent, it is worth while, perhaps, to compare them to the regression coefficients for some well-known common stocks. By construction, of course, the average stock has $\beta = 1$. For American Telephone and Telegraph, considered to be a safe "widows and orphans stock", β was 0.34 over our sample interval. The average regression coefficient for the electric utility industry was 0.41, and the corresponding figure for the gas utility industry was 0.45.[26] Clearly, then, compared to common stocks, the systematic risk measures for wheat, corn and soybeans are low indeed.

Since the mean returns (which are actually risk premiums) are also very close to zero, we may conclude that the data on commodity future returns during our sample period conform better to the capital markets model than to the Keynesian model. In fact, the contest is not even close.[27]

SOME CONCLUDING OBSERVATIONS ON HARDY AND KEYNES

Because both mean returns and systematic risk were zero, the sample evidence permits no direct confrontation between the capital market approach and the Hardy gambling casino theory, which also predicts a mean return of zero. Had we found a commodity for which the β's were substantially and unambiguously positive while the means were zero or negative (or found a commodity with significant negative intercept terms in expression (8)), we could have concluded that for such a case the evidence was more consistent with the gambling than with the portfolio interpretation. We would also have had to conclude that risk-averse investors are apparently not shrewd enough to recognise a bargain. For the existence of a futures asset with a positive value of β when regressed on a stock index and a zero (or negative) mean would make it attractive for risk averters to become "short speculators" in that market. Selling futures short under such conditions would create an asset that was negatively correlated with the rest of the investor's portfolio and yet not reduce the mean return on the portfolio as a whole.

The possibility that other commodities besides these I studied may turn out to have non-zero β's also suggests a way of reconciling Keynesian and capital market views of risk and returns in futures markets. When Keynes wrote *The Treatise on Money* in the late 1920s, the variability that he identified with asset risk may in fact have included a sizeable systematic risk component. In the late 1920s commodity prices were not subject to effective price support.[28] Thus prices could be expected to be more variable and also to be more strongly associated than at present with cyclical swings in the economy. It may well also have been the case that the particular commodities Keynes used as examples of futures market risk – cotton and copper – were strongly associated with the level of activity in British manufacturing in the early 1930s. If such a connection existed, share prices and the prices of raw commodities, including futures, would be related to each other. With cotton as the major input to a large sector of British manufacturers, it would hardly be surprising to observe a high correlation between the returns on cotton futures and the returns on British industrial stocks.

This reinterpretation of Keynes suggests that if my sample were broadened to include commodities more intimately associated with American manufacture, there might perhaps be cases of commodity futures having high positive β's and positive means as well. But such interesting prospects must await future research.

1 Among the most influential papers devoted to the Keynes–Hardy controversy have been those of Telser (1958, 1960, 1967) and Cootner (1960a, 1960b). Other studies of the returns to speculation in futures markets include Houthakker (1957), Gray (1961), Smidt (1965), Rockwell (1967) and Stevenson and Bear (1970).

2 This is at least the conventional interpretation of the Keynesian position as suggested by the following quotation: "It will be seen that, under the present regime of very widely fluctuating prices for individual commodities, the cost of insurance against price changes – which is additional to any charges for interest or warehousing – is very high" (Keynes 1930, p. 144). A somewhat broader interpretation emphasises the insurance premium and tries to relate the size and sign of this premium to variations in the stocks of the commodity over the production cycle (see Cootner 1960a, 1960b).

3 The description of the equilibrium pricing model presented here assumes some familiarity on the part of the reader. For a more complete discussion, see Sharpe (1964) or Lintner (1965). A detailed exposition of the model is also given in Fama and Miller (1972, chapter 7).

4 Recently Black (1972) has generalised the Sharpe model by replacing the riskless asset having return R_f with another asset whose return is a random variable but whose covariance with total wealth is zero. Empirical tests by Black, Jensen and Scholes (1972) and Fama and Macbeth (1972) seem to indicate that the generalised model fits the data somewhat better than the Sharpe version.

5 That entering into a future contract need involve no margin or other specific payment that could be interpreted as an "investment" (and hence that could serve as the basis for computing a "rate of return"), does not mean that the mean-variance portfolio model cannot be applied at the microlevel to analyse an investor's decision process. The price changes on the contracts held will affect *terminal* wealth, just as in the case of any other asset; but the contracts do not appear in the *initial* wealth constraint. For a rigorous treatment in the context of forward foreign exchange, see Leland (1971). A study by Johnson (1960) of futures spot commodity holdings also proceeds in this way. That is, the entire analysis is conducted in terms of price changes and not in terms of rates of return. The fact that the margin does not really represent capital invested in futures contracts, even in those markets where margin is required, might perhaps have been appreciated earlier by analysis of futures markets trading if brokers paid interest on the escrowed funds (or, what amounts to the same thing, if they allowed all traders to deposit or to hypothecate income-earning assets rather than cash). In practice, of course, the brokers presumably do pay interest on the escrowed funds, but only in the hard-to-see form of lower commissions or higher levels of "free" services than would otherwise be the case.

6 I abstract from such complications as transaction costs, basis risk, the business risk of the storage and processing industries, limitations on borrowing, and so on. Or, what amounts to the same thing, I assume that differences in the returns on spot and futures market assets from these sources are so small and so unsystematic relative to the variations in returns on both assets as a consequence of price fluctuations that they can safely be ignored in a first approximation. Some of the main second-order qualifications are indicated at various points in the text and notes in the course of the discussion.

7 Actually total return to the spot commodity holder can be decomposed into three components: a pure time return to capital, a risk premium, and remuneration for storage costs, defined in this context as insurance charges, spoilage and warehousing and administrative costs. Since we are concerned only with the return to capital embodied in the spot commodity, $\tilde{R}_i$, the "full" return on the spot commodity is to be understood as net of storage costs.

8 It does not really matter whether there is a spot commodity in existence yet. That is, just as I can order a car not yet produced, so I can agree to accept delivery next period at a specified price of a commodity still unproduced. The "implicit" spot price, which always exists, is then simply the futures price minus a discount for payment in advance, ie, $P_i = P_f/(1 + R_f)$. Note also that the seller need not actually contemplate producing the spot commodity; ie, he can be a pure short speculator. He merely offers to make delivery to you next period, intending, if necessary, to go out and buy the spot commodity then, if you insist on delivery rather than settlement.

9 I have argued that in equilibrium $P_{f,t} = P_{i,t}(1 + R_f)$, where i refers to the spot commodity. Thus the percentage change in the futures price underestimates the risk premium on the spot commodity by the factor $1/(1 + R_f)$. Given the intervals over which I will be computing returns, the factor $1/(1 + R_f)$ is likely to be very small. Hence for simplicity of exposition I shall refer (somewhat loosely) to the percentage change in the futures price as representing the risk premium on the spot commodity.

10 I have assumed that unanticipated changes in the spot and futures prices are perfectly correlated. (For some evidence on the high degree of correlation between spot and futures prices, see Houthakker (1968).) Where this correlation is not perfect, the spot commodity holder will bear some risk even though hedged, and some compensation for this risk may be impounded in his return. Working (1953) has made this type of risk central to his analysis of futures markets. He regards spot commodity holders not as passive short hedgers but as speculators on the movement of the spot-future price differential, or basis, over time. He argues that professional commodity dealers are better able to predict differentials or relative prices than price levels. Thus they assume a position in both the spot and futures markets in response to expected changes in the basis. Since Working's hypothesis appears to have no testable implications with respect to the risk-return relation in futures markets of the kind that are of main concern in this chapter, I will not pursue it further here.

11 Price quotations were taken from US Department of Agriculture (1952–1967). Lester Telser kindly supplied price data for 1952–1964. The rest were collected independently from the same source. The terminal date of 1967 was the last year for which price data were available at the time this study was started. The initial date of 1952 was chosen to minimise wartime and postwar controls on commodity prices and futures trading.

12 Another possible principle for computing returns would be on the basis of time to contract expiration (eg, wheat contracts with exactly four months to run or corn contracts with two and a half months to run). In a later paper I shall show that the theoretical and empirical justification for looking at returns in this way is weaker than certain treatments of the matter, notably Samuelson's (1965), would suggest.

13 The sample serial correlation coefficient is defined as $\hat{r}_\tau = \text{cov}(\tilde{u}_t, \tilde{u}_{t-\tau})/\text{var}\,\tilde{u}_t$, where in this case $\tau = 1, \ldots, 10$ and u_t is the 2-week rate of return. Even in the case where $\tilde{u}_t$ belongs to the family of distributions for which variance does not exist, it has been shown that $\tilde{r}_\tau$ is an adequate descriptive measure of the serial correlation in the population in the sense that it behaves much the same as its counterpart from a normally distributed sample of observations (see Fama and Babiak 1968, p. 1146). Under the hypothesis that the true serial correlation is zero, the standard error of the sample serial correlation coefficient is given by $\sigma(\hat{r}_t) = \sqrt{1/(N - \tau)}$, where N is the sample size and $\tau = 1, \ldots, 10$.

14 These results are consistent with previous studies of the time series properties of futures prices. Studies by Larson (1960), Houthakker (1961), Smidt (1965), and Stevenson and Bear

(1970) tend to show that although there are occasions when futures prices appear to have exhibited some degree of dependence, there have been no striking cases of large and pervasive price trends or patterns. Computed measures of statistical dependence have usually been small, and the profitability of trading rules has typically been less than that obtained by following a policy of buy and hold.

15 See Fama and Roll (1971, p. 332) for evidence on the sampling variability of the standard deviation when the sample values come from a non-Gaussian distribution.
16 The phenomenon of fat tails (ie, more probability in the tail area of the distribution than in the Gaussian distribution) for distributions of futures returns has previously been noted, but not rigorously investigated, by Smidt (1965), Stevenson and Bear (1970) and Houthakker (1961).
17 "The g truncated sample mean is the average of the middle 100g percent of the ordered observations in the sample. That is, in computing the mean, $100\ (1-g)/2\%$ of the observations in each tail of the data distribution are discarded" (Fama and Roll 1968, p. 826). The optimum degree of truncation depends on the size of α. The lower the value of α, the greater the optimum degree of truncation, reflecting the fact that the more outliers there are in a sample, the greater the number of observations that must be deleted before an efficient estimate of location is obtained. Fama and Roll (1968, p. 832) conclude that "an estimator which performs very well for most values of alpha and N (sample size) is the .5 truncated mean".
18 The procedure for estimating assumes that successive returns are independent – a hypothesis that has already been tested.
19 In principle, any number of interfractile ranges might serve as a measure of risk. For reasons of simplicity and economy we have chosen the same interfractile range used to estimate the dispersion parameter of the distribution.
20 Some indirect evidence on this point is afforded by the refusal of Merrill Lynch, Pierce, Fenner, and Smith to sell futures contracts to women on the grounds that they do not have the psychological stamina to withstand futures market price fluctuations.
21 An ordinary sample mean has been computed for the return on the Standard and Poor Index, since the distribution of stock returns more closely approaches normality than do the distributions of commodity returns (Officer 1971).
22 Note that the evidence presented in Table 2 does not constitute 16 different tests of the Keynesian hypothesis. The similarity in the distribution parameters (and indeed, of the entire distributions, as a glance at Figs 1–3 will testify) suggests that within any commodity group the distribution of returns has the same parameters and that such slight differences as do exist represent only sampling fluctuations. Correlation coefficients between the returns on different contracts of the same commodity have been computed. Out of 35 coefficients, 12 were .90 or higher; another 17 were between .80 and .89; and only six were below .79, the lowest being .72. In some cases, as, for example, the adjacent July and September wheat contracts or the adjacent March and May contracts for corn, the correlation was virtually perfect. As a group, wheat contracts seemed to exhibit the most interdependence, and soybeans the least. This high correlation between returns on different contracts is especially striking in light of the insistence on contract uniqueness in much of the traditional literature on futures markets. The contemporaneous coefficients of correlation between returns for the same contract but different commodities have also been computed. Out of 13 correlation coefficients only two are even as high as .5, and even the highest of these coefficients (.67) is lower than the lowest correlation for returns on the same commodity (.72). Under these circumstances, then, there would appear to be little objection to maintaining a distinction among the three commodities.
23 There are problems posed by the fact that the underlying distributions conform better to stable non-Gaussian than to normal distributions. It can be shown, however, that the ordinary least-squares coefficients are consistent estimators of the corresponding popula-

tion parameters, but not necessarily efficient ones; particularly as α departs further from two. However, the loss of efficiency is not likely to be of much import for samples as large as we will be using (300 observations). See Blattberg and Sargent (1971).

24 Since the variability of the bill rate was small relative to that of the Standard and Poor Index or to that of futures returns during my sample period, the estimates turned out to be virtually identical with those obtained when the index was used in regular return form. Other specifications of the regression equation were also tested, such as the use of logarithms of the price relative, rather than percentage rates of return. There was little difference in explanatory power and no noteworthy change in the absolute or relative sizes of the coefficients.

25 It may strike some readers as paradoxical that there could exist an asset whose return is a random variable and whose β is zero. Remember, however, that a zero β asset has only zero covariance with other assets on *average*. With some assets its return will be positively correlated, and with others, negatively correlated. In fact, since the zero β assets will themselves be part of total wealth, they must be negatively correlated, on balance, with all other assets in the market portfolio. Because of this covariance with other assets, the zero β asset does make a sufficient contribution to the diversification and hence the risk reduction of the total portfolio to justify its inclusion even at a mean return (over and above interest) of zero. For a general treatment of zero β assets within the context of the Sharpe model, see Black (1972).

26 Information supplied by Merton H. Miller and Myron Scholes, from an unpublished manuscript. The regression coefficients have been estimated using annual rates of return over the period 1947–1966.

27 Estimates of the Keynesian risk measure, or scale factor, the truncated means, and the regression coefficients have been computed by 5-year subperiod intervals. Although there is some tendency for both the scale factor and the regression coefficient to be higher in the first 5-year period (especially for wheat), there is no systematic pattern between risk and return. More often than not, high risk, in terms of either simple variability or systematic risk, is associated with negative means, and low risk with positive means.

28 There were, of course, a number of price or output stabilisation schemes in operation during this period, few of which were successful for any length of time.

BIBLIOGRAPHY

Black, Fischer, 1972, "Capital Market Equilibrium with Restricted Borrowing", *Journal of Business* 45, July, pp. 444–55.

Black, Fischer, Michael Jensen and Myron Scholes, 1972, "The Capital Asset Pricing Model: Some Empirical Results", in *Studies in the Theory of Capital Markets*, edited by Michael Jensen (New York: Praeger).

Blattberg, Robert and Thomas Sargent, 1971, "Regression with Non-Gaussian Disturbances: Some Sampling Results", *Econometrica* 39, May, pp. 501–10.

Blume, Marshall E., 1968, "The Assessment of Portfolio Performance: An Application of Portfolio Theory" PhD dissertation (University of Chicago).

Cootner, Paul, 1960a, "Returns in Speculators: Telser versus Keynes", *Journal of Political Economics* 68, August, pp. 396–404(a).

Cootner, Paul, 1960b, "Rejoinder", *Journal of Political Economics* 68, pp. 415–18(b).

Fama, Eugene, 1971, "Risk, Return, and Equilibrium", *Journal of Political Economics* 79, January/February, pp. 30–55.

Fama, Eugene and Harvey Babiak, 1968, "Dividend Policy: An Empirical Analysis", *Journal of the American Statistical Association* 63, December, pp. 1132–61.

Fama, Eugene and James MacBeth, 1972, "Risk, Return, and Equilibrium: Empirical Tests", manuscript (University of Chicago).

Fama, Eugene and Merton H. Miller, 1972, *The Theory of Finance* (New York: Holt, Rinehart & Winston).

Fama, Eugene and Richard Roll, 1968, "Some Properties of Symmetric Stable Distributions", *Journal of the American Statistical Association* 63, September, pp. 817–36.

Fama, Eugene and Richard Roll, 1971, "Parameter Estimates for Symmetric Stable Distributions", *Journal of the American Statistical Association* 66, June, pp. 331–38.

Fischer, Lawrence, 1966, "Some New Stock Market Indexes", *Journal of Business* 39 (supplement), pp. 191–225.

Gray, R. W., 1961, "The Search for a Risk Premium", *Journal of Political Economics* 69, pp. 250–60.

Hardy, Charles O., 1940, *Risk and Risk-Bearing* (Chicago: University of Chicago Press).

Houthakker, H. S., 1957, "Can Speculators Forecast Prices?" *Review of Economics and Statistics* 39, May, pp. 143–51; reprinted as Chapter 22 of the present volume.

Houthakker, H. S., 1961, "Systematic and Random Elements in Short-Term Price Movements", *American Economic Review* 51, May, pp. 164–72.

Houthakker, H. S., 1968, "Normal Backwardation", in *Value, Capital, and Growth: Papers in Honour of Sir John R. Hicks*, edited by J. N. Wolfe (Chicago: Aldine).

Johnson, Leland, 1960, "The Theory of Hedging and Speculation in Commodity Futures", *Review of Economic Studies* 27, June, pp. 139–51.

Keynes, J. M., 1930, *Treatise on Money*, Volume 2 (London: Macmillan).

King, Benjamin, F., 1966, "Market and Industry Factors in Stock Price Behavior", *Journal of Business* 39, January, pp. 139–90.

Larsen, Arnold, 1960, "Measurement of a Random Process in Future Prices", *Food Research Institute Studies*, Stanford University 1, November, pp. 313–24.

Leland, Hayne F., 1971, "Optimal Forward Exchange Positions", *Journal of Political Economics* 89, pp. 257–69.

Lintner, John, 1965, "Security Prices, Risk, and Maximal Gains from Diversification", *Journal of Finance* 20, December, pp. 587–615.

Markowitz, Harry M., 1959, *Portfolio Selection: Efficient Diversification of Investments* (New York: Wiley).

Miller, Merton H. and Myron S. Scholes, 1972, "Rates of Return in Relation to Risk: A Re-examination of Some Recent Findings", in *Studies in the Theory of Capital Markets*, edited by Michael Jensen (New York: Praeger).

Mossin, Jan., 1966, "Equilibrium in a Capital Asset Market", *Econometrica* 37, pp. 763–68.

Officer, Robert R., 1971, "A Time Series Examination of the Market Factor of the New York Stock Exchange". PhD dissertation (University of Chicago).

Roberts, Harry V., 1964, *Statistical Inference and Decision*, Lithographed, (University of Chicago).

Rockwell, Charles S., 1967, "Normal Backwardation, Forecasting and the Returns to Commodity Futures Traders", *Food Research Institute Studies* 8 (supplement), Stanford University; reprinted as Chapter 23 of the present volume.

Roll, Richard, 1968, "The Efficient Market Model Applied to US Treasury Bill Rates". PhD dissertation, Graduate School of Business (University of Chicago).

Samuelson, P. A., 1965, "Proof that Properly Anticipated Prices Fluctuate Randomly", *Industrial Management Review* 8, pp. 41–9.

Schrock, Nichols W., 1971, "The Theory of Asset Choice; Simultaneous Holding of Short and Long Positions in the Futures Market", *Journal of Political Economics* 79, March/April, pp. 270–93.

Sharpe, William, 1964, "Capital Asset Prices: A Theory of Market Equilibrium under Conditions of Risk", *Journal of Finance* 19, pp. 425–42.

Smidt, Seymour, 1965, "A Test of Serial Independence of Price Changes in Soybean Futures", *Food Research Institute Studies* 5, pp. 117–36.

Stevenson, Richard A. and Robert M. Bear, 1970, "Commodity Futures: Trend or Random Walks?" *Journal of Finance* 25, pp. 65–81.

Telser, Lester, 1958, "Futures Trading and the Storage of Cotton and Wheat", *Journal of Political Economics* 66, June, pp. 233–55; reprinted as Chapter 6 of the present volume.

Telser, Lester, 1960, "Returns to Speculators: Telser versus Keynes: Reply", *Journal of Political Economics* 67, pp. 404–15.

Telser, Lester, 1967, "The Supply of Speculative Services in Wheat, Corn and Soybeans", *Food Research Institute Studies* 7, supplement, pp. 131–76.

US Department of Agriculture, Commodity Exchange Authority, 1952–1967, *Commodity Futures Statistics* (Washington: Government Printing Office).

Working, Holbrook, 1953, "Futures Trading and Hedging", *American Economic Review* 18, June, pp. 314–43.

25
*Luck versus Forecast Ability: Determinants of Trader Performance in Futures Markets**

Michael L. Hartzmark
Cragar Industries

> *Statistical techniques are used to demonstrate that the fortunes of individual futures traders are determined by luck, not forecast ability. Even though a large number of traders appear to exhibit significantly superior forecast ability, the investigation strongly supports three conclusions: there are fewer participants with significantly superior skill than expected if participants trade randomly, there are more traders exhibiting no skill than expected if participants trade randomly, and forecast ability is not correlated over time – superior forecasters in the early period are only average forecasters in the later period. Therefore it is luck that determines trader performance.*

The empirical evidence presented in this article lends little support for the hypothesis that futures traders possess the ability or skill to consistently earn positive profits. The statistical analysis utilises the techniques introduced to study the performance of mutual and commodity futures fund managers (Jensen 1968; Kon and Jen 1978, 1979; Henriksson and Merton 1981; Merton 1981; Chang and Lewellen 1984; Henriksson 1984; Jagannathan and Korajczyk 1986; Cumby and Modest 1987). The most striking difference between this article and the previous studies is in the use of highly detailed daily transactions data on individual investors. On a daily basis the traders' *ex ante* predictions and *ex post* realisations

---

*This paper was first published in the *Journal of Business* 64, pp. 49–74 (1991) and is reprinted with the permission of the University of Chicago Press. The author wishes to acknowledge financial support from the Olin Foundation's Faculty Fellowship Program, the University of Michigan Business School Summer Support Program, and the Center for the Study of the Economy and the State at the University of Chicago is gratefully acknowledged. Special thanks to Lester Telser for his insightful comments. Comments from the anonymous referee, Eugene Fama, and the members of the Economic and Legal Organisation Workshop at the University of Chicago were also helpful. Research assistance from David Barker and Jeffrey Pontiff has been helpful. Computer support from M. Steven Potash and Frank Luby is also acknowledged. All errors are my responsibility.

are directly observed. Using this information and employing the non-parametric statistical procedures developed by Henriksson and Merton (1981, hereafter HM) and modified by Cumby and Modest (1987, hereafter CM), it is possible to determine the actual forecast ability of the individual traders.

Two different types of forecast ability or market timing are examined. The first type is called "consistent ability". A trader possessing this skill performs well because he is able systematically and consistently to predict the correct *direction* of future price movements. In other words, he establishes long (short) positions more often than not prior to an increase (decrease) in the futures price. The other type of forecast skill is called "big hit" ability. A trader possessing this ability is able to predict both the *magnitude* and the *direction* of price changes and will thus establish his largest positions (make his biggest bets) when the highest returns (largest absolute price movements) are anticipated.

DESCRIPTION OF THE DATA BASE

The data used in this article come directly from the Commodity Futures Trading Commission (CFTC) reports on the end-of-day commitments of large traders. In all futures markets, those traders, who either at the beginning or the end of a trading day hold commitments exceeding certain levels, must (as specified by CFTC regulations) report their trading activity, indicating their speculative and hedge, long and short positions separately for each contract maturity month.

Nine markets are analysed covering the period from July 1, 1977, to December 31, 1981.[1] Included in this sample of nine markets are the three US wheat markets. In the empirical analysis the positions held by individual traders in these three markets are aggregated.[2] The nine markets include: (1) oats traded on the Chicago Board of Trade (CBT), (2) wheat traded on the CBT, (3) wheat traded on the Minneapolis Grain Exchange (MGE), (4) wheat traded on the Kansas City Board of Trade (KBT), (5) pork bellies traded on the Chicago Mercantile Exchange (CME), (6) live cattle traded on the CME, (7) feeder cattle traded on the CME, (8) US T-bonds traded on the CBT, and (9) 90-day T-bills traded on the International Monetary Market (IMM).

The motivation for participating in the futures market will likely differ for commercial and non-commercial traders.[3] Therefore, traders are categorised depending on the nature of the positions they report (ie, hedge versus speculative). Traders reporting only hedge positions over the full $4\frac{1}{2}$-year period are classified as commercial traders (or pure hedgers). Traders reporting only speculative positions are designated as non-commercial traders (or pure speculators). For those traders who report both hedge and speculative positions, the confidential files kept by the CFTC are consulted to determine if the trader's business is directly

Table 1 Descriptive statistics on trader profits and average net position values for all markets combined (in millions of dollars)

	All traders	Commercial traders	Non-commercial traders
Total	2,229	607	1,622
Position value:			
Mean	−1.04	−7.20**	1.26**
Median	0.02	−1.84	0.19
Standard deviation	30.43	52.99	14.26
Skewness	−16.82	−11.89	9.09
Kurtosis	440.48	167.08	191.72
Range	1,163.35	1,038.10	460.91
Profits:			
Total	1,046.7	763.40	283.38
Mean	0.47**	1.26**	0.17**
Median	0.02	0.01	0.02
Standard deviation	6.16	10.93	2.68
Skewness	18.25	11.55	7.28
Kurtosis	541.33	195.37	161.81
Normality test[a]	0.32**	0.31**	0.25**

[a]For the test of normality the Kolomogorov D-statistic is used.
**Significant at a 1% level.

related to the market for the underlying commodity. If the trader is in a closely related business (eg, farmer, government securities dealer, cattle breeder) he is placed in the commercial category.[4]

For statistical reasons individual traders making less than 25 separate transactions are not included in the analysis. This results in the exclusion of 2,338 traders who were in Hartzmark (1984, 1987). Even so, the total number of traders analysed is 2,229. The average number of transactions per trader ranges from 48 in the oats market to 483 in the live cattle market (see Table 1).[5]

The average net position value[6] held by commercial traders is substantially larger than the non-commercial position. This might be explained because commercial participants are less constrained by the rules limiting position size. In addition, the commercial traders hedge very large cash positions that have values that are directly related to the level of prices in the futures market. The very largest positions are net short on average, causing the size distribution of the individual positions to be negatively skewed. Commercial traders of interest rate futures hold much larger net positions than traders in the other commodities. As one would expect, a large proportion of the traders in the sample establish relatively small positions.

Hartzmark (1984, 1987) shows that the commercial traders earn the largest dollar profits. The one significant difference in the overall performance of the non-commercial traders examined in this article as compared with the results for the 4,567 traders reported in Hartzmark (1984, 1987) is that the aggregate profits of non-commercial traders are significantly different from zero.[7] In Hartzmark (1987) the returns to non-commercial traders are shown to be insignificantly different from zero. This indicates that the 2,338 traders discarded here are mostly non-commercial traders who earn small negative returns, on average. Therefore, if the selection criteria induces any bias, it is toward rejection of the luck hypothesis since the "noise" traders are removed.

In general, the return distributions in each market and for all markets combined are highly skewed and have large peaks around zero dollars (see Table 1). Since the return distribution is a combination of the different position sizes and the price change distribution, a highly skewed return distribution is not evidence of certain traders possessing skill. For example, in the interest rate markets the commercial traders are net short, on average. Furthermore, the commercial traders with the largest positions (in absolute value) are also net short. If negative price changes are observed (as they were over most of the period), then the returns from a naïve strategy of buying a short position and holding it over the period would have offered significant positive profits. Moreover, the returns would be positively skewed since the largest traders in the interest rate markets (who are net short) would be the biggest winners. Non-commercial interest rate traders with their smaller net long positions would have small losses.

Overall, the general direction of the price movements and the varying magnitude of positions held by traders generate the profit distributions. In the following sections an empirical attempt is made to determine whether anything more than "riding the tide" explains the performances of the individual traders.

STATISTICAL METHODS EMPLOYED TO DETERMINE FORECAST ABILITY

Testing for consistent ability

A statistical procedure introduced by HM and modified by CM is used to test for "consistent" forecast ability. To begin, the number of "correct" forecasts that each individual trader makes is observed. A trader is correct when he is long (short) and the subsequent price movement is up (down). A calculation can then be made of the probability of observing that number of correct predictions assuming forecasts are made randomly. To calculate these probability levels for each trader one needs to observe over the whole period (1) the number of correct predictions that prices fall (ie, number of times the trader is short and the price goes down); (2) the number of upticks; (3) the number of downticks; and (4) the number of predictions made. With this information, a calculation can be made of the expected number of correct predictions when a trader is short.[8] The predicted and actual numbers of correct predictions are compared to determine if they are statistically different from each other. The magnitude of the difference and the associated significance level will indicate whether the individual trader possesses superior, inferior, or no forecast ability.

The statistical method introduced by CM is used in this article.[9] The binary variable $Z(t)$ will indicate the direction of the *actual* price movement between time t and $t + 1$. The variable $Z(t)$ is equal to one if the price goes up (the dollar return, $R(t)$, is greater than zero), and $Z(t)$ is equal to zero otherwise. The binary variable $U(t)$ indicates the trader's prediction at time t. The variable $U(t)$ is equal to one if the trader takes a long position – indicating that he thinks the price is going up – otherwise, $U(t)$ is equal to zero. The log odds that an individual trader is long at time t and the price goes up between t and $t + 1$ is given as

$$\log\left\{\frac{\text{probability }[Z(t) = 1]}{\text{probability }[Z(t) = 0]}\right\} = a + BU(t)$$

Given observed positions and price changes, one can test directly whether the trader possesses forecast ability. When $Z(t)$ is independent of $U(t)$, then B equals zero and the trader possesses no forecast ability. If B is significantly greater than zero, the trader possesses superior forecast ability. If B is significantly less than zero, the trader possesses inferior ability.[10]

A logit equation is specified to determine the sign and magnitude of B. Because the standard errors and the degrees of freedom differ dramatically across traders, the relative magnitudes of the parameter estimates and the t-statistics examined alone cannot be used to make inferences about the ability of a single trader. The probability significance level associated with the parameter estimate, B, offers the necessary information to implement the analysis to follow.

Testing for big hit ability

It is likely that a trader does more than simply predict the direction of price movements. He adjusts the magnitude of his position depending on the strength of his conviction. The tests for big hit forecast ability take into account both the magnitude of the trader's net position and the magnitude of the actual price change. Simply because the number of correct predictions an individual trader makes is not above some statistically significant level does not mean that the forecasts are poor. It may be the trader is better able to predict big price changes rather than small changes. To include this additional information the assumption that the magnitude of the price change is independent of the probability of a correct prediction must be relaxed.

To measure big hit ability CM assumes that the magnitude of the price change or dollar return, $R(t)$, depends linearly on the forecast, or, in other words, that the probability of a correct forecast is greater for larger price changes. In this article, because we have even more information than CM, it is assumed that $R(t)$ depends linearly on the net position held by the trader. Big hit ability is indicated if the trader holds his largest positions when there are the largest price movements in a favourable direction.

Two effects are being combined using this measure. First, we are determining whether the probability of correctly predicting the price change is linearly related to the size of the position. In addition, we are testing to see if this probability is greater the larger the subsequent price change.

Define $LS(t)$ as the net position (long minus short contracts) at time t, such that $LS(t)$ is greater than zero if the trader is net long, and $LS(t)$ is less than zero if the trader is net short.[11] The regression equation combining the two effects is then

$$R(t) = a' + B'\, LS(t) + e(t)$$

Testing whether the trader possesses big hit ability is identical to determining whether B' equals zero. If B' is significantly greater than zero, then, as before, the trader possesses superior ability. While if B' is significantly less than zero, the trader exhibits inferior big hit ability.

Deriving forecast coefficients from the regressions

The magnitude of the parameter estimate alone gives little information about whether the individual trader possesses significant forecast ability. For example, observing B_i equal to 0.35 and B_j equal to 0.55 does not indicate whether trader j is a better forecaster than trader i. Each trader is in the market for a different amount of time (ie, the degrees of freedom are different), and the standard errors of the parameter estimates may also differ dramatically. Term B_i may be significantly different from zero, while B_j is not.

To derive comparable measures of ability across all traders, the probability significance levels for each trader are transformed into forecast coefficients (FC_i). These measures incorporate information on the sign of the parameter estimates, standard errors, and degrees of freedom into one aggregate measure. In all tests that follow the forecast coefficient for the ith trader is defined as

$$FC_i = (1 - \text{probability level}_i) \times (\text{sign of parameter estimate}_i).$$

For example, if the probability significance level from the logit or ordinary least squares (OLS) equation is 0.25 and B_i is equal to -0.90, then

$$FC_i = (-1) - (1 - 0.25) = -0.75$$

Therefore, the range $(-1.0 <= FC_i <= 1.0)$ encompasses the universe of traders: those with statistically significant inferior ability, those with no ability, and those with statistically significant ability.

The expected distribution of the forecast coefficients

Even in the case when profits are randomly generated one would expect to observe a certain proportion of individuals with forecast coefficients with extreme values (eg, less than -0.90 or greater than 0.90). Therefore, observing forecast coefficients above 0.90 is not sufficient to indicate that there is significant forecast ability in the market as a whole.[12] Therefore, the null hypothesis to be tested is:

Hypothesis. Returns are generated by a stochastic process, thus the individual forecast coefficients are uniformly distributed over an interval spanning -0.999 to 0.999.

If the positive tail of the distribution of the forecast coefficients is fatter than expected, one can conclude that a greater than expected number of traders possess significant forecast ability, and the null hypothesis is rejected. By the same reasoning, if the negative tail of the distribution is fatter than expected, one can conclude that there is a significant number of inferior forecasters. If both tails are thinner than expected, one of two conclusions can be reached: either (1) the standard errors of the regression parameters are somehow biased upward causing the associated probability levels to be biased toward one, or (2) there is some dependence across traders. This latter explanation is plausible if the traders are communicating with one another or using similar trading strategies.

EMPIRICAL TESTS FOR FORECAST ABILITY

Consistent forecast ability

Table 2 presents the descriptive statistics for the distributions of consistent forecast coefficients. The Kolomogorov D-statistic and the chi-square tests are used to determine whether the distributions are uniform.[13] If a trader always positions himself on one side of the market, a consistent forecast coefficient cannot be calculated.[14] Therefore, the number of traders having consistent forecast coefficients differs from the number having big hit coefficients. For example, nine of 48 oat traders have 25 or more transactions but hold net positions only on one side of the market. Comparing the "All Markets" category in Tables 1 and 2 indicates that 240 traders take positions on one side only.

625

Table 2 Descriptive statistics: consistent forecast coefficients ($\times 10^{-2}$, except traders and χ^2)

	Number of traders	Mean coefficient	Median coefficient	Standard deviation	Skewness	D-statistic	χ^2 statistic
Oats:							
All traders	39	−3.88	−2.58	51.55	7.62	13.53	1.90
Commercial	20	−10.58	−18.56	53.09	31.78	20.46	1.00
Non-commercial	19	3.17	−0.19	50.32	−15.87	14.48	3.89
Wheat:							
All traders	308	−3.70	−0.84	51.46	4.22	8.86	49.66**
Commercial	142	−5.31	−0.84	52.78	1.71	9.62	34.90**
Non-commercial	166	−2.32	−0.87	50.42	7.50	9.42	42.43**
Pork bellies:							
All traders	301	2.62	0.66	55.96	−2.35	6.38	22.92
Commercial	26	15.13*	1.13	43.31	32.53	25.58*	17.08**
Non-commercial	275	1.44	0.66	56.92	−1.01	5.71	23.84
Live cattle:							
All traders	425	4.45*	0.65	55.66	−5.78	7.46*	56.44**
Commercial	82	−0.65	−1.06	52.51	10.49	9.27	30.20*
Non-commercial	343	5.67*	0.91	56.39	−9.77	8.19*	42.36**
Feeder cattle:							
All traders	151	2.80	0.61	57.98	−8.02	7.62	37.48**
Commercial	48	5.32	6.91	55.19	−27.22	12.05	12.83
Non-commercial	103	1.63	−0.60	59.47	−0.32	6.36	32.73*
T-bonds:							
All traders	411	3.18	0.58	53.15	−7.25	6.72*	38.20**
Commercial	86	8.29	15.07	55.59	−18.58	8.29	10.28
Non-commercial	325	1.82	0.00	52.49	−4.74	6.87*	40.85**
T-bills:							
All traders	354	−2.02	−1.26	53.60	8.49	5.32	20.24
Commercial	83	2.68	−1.12	52.12	9.91	7.68	16.71
Non-commercial	271	−3.46	−2.93	54.06	8.79	5.71	14.39
All markets:							
All traders	1989	1.21	0.00	54.30	−.89	4.38**	122.85**
Commercial	487	1.16	−0.64	52.99	−1.26	5.85*	51.36**
Non-commercial	1502	1.22	0.00	54.74	−0.78	4.34**	94.91**

*Significant at a 10% level.
**Significant at a 1% level.

Across the markets the means of the forecast coefficients are almost always indistinguishably different from zero. Only the commercial traders of pork bellies show significant, positive forecast ability, on average. Given the selection procedures used to develop this sample, it is not necessary for the mean forecast coefficient to be zero. Many market participants are not included (ie, scalpers and small traders).[15] If the large reporting traders represent an elite subset of successful survivors in the market, then one would expect a positive mean forecast coefficient. Alternatively, the large degree of turnover in these markets could explain negative average forecast coefficients. With constant exit and entry of traders possessing poor forecast ability (but who establish large positions and make at least 25 transactions), one would expect the mean forecast coefficient to be negative. If the mean coefficient is significantly positive or negative, the null hypothesis is rejected since implicit in each result is that traders have differing skills.

With the exception of the "All Markets" distribution (which is symmetric) the distributions are almost all negatively skewed. None of the observed standard deviations are significantly different from those expected from a uniform distribution spanning an interval from -1.0 to 1.0.[16]

The D-statistics indicate that uniformity is rejected in only one of seven markets for commercial traders and in only two of seven markets for non-commercial traders. The fact that uniformity is accepted in the individual markets and rejected for all markets combined is likely the result of the increased sample size and thus the increased precision of the test.

Figure 1 offers a clear illustration of why uniformity is rejected for "All Markets". The bars in this Chart show the percentage of traders that are observed in each of 20 equal-sized intervals. The midpoints of the intervals are indicated on the Charts. For example, the 0.95 interval includes coefficients between 0.90 and 1.00. The horizontal lines represent the percentage of traders expected in each interval if skill is uniformly distributed.

Uniformity is not rejected because there are more outliers than expected. On the contrary, it results because there are more traders with forecast coefficients close to zero than expected. The "All Markets" Chart indicates that, if anything, there are more poor forecasters than expected. But again, it is the central portion that dominates. Only in pork bellies, live cattle, and feeder cattle are there more coefficients above 90% than expected from chance. In none of the individual markets does one observe more coefficients below negative 90% than would emerge from a random draw.

In general, examination of the *individual* markets shows that the forecast coefficients are randomly distributed. Therefore, the null hypothesis is

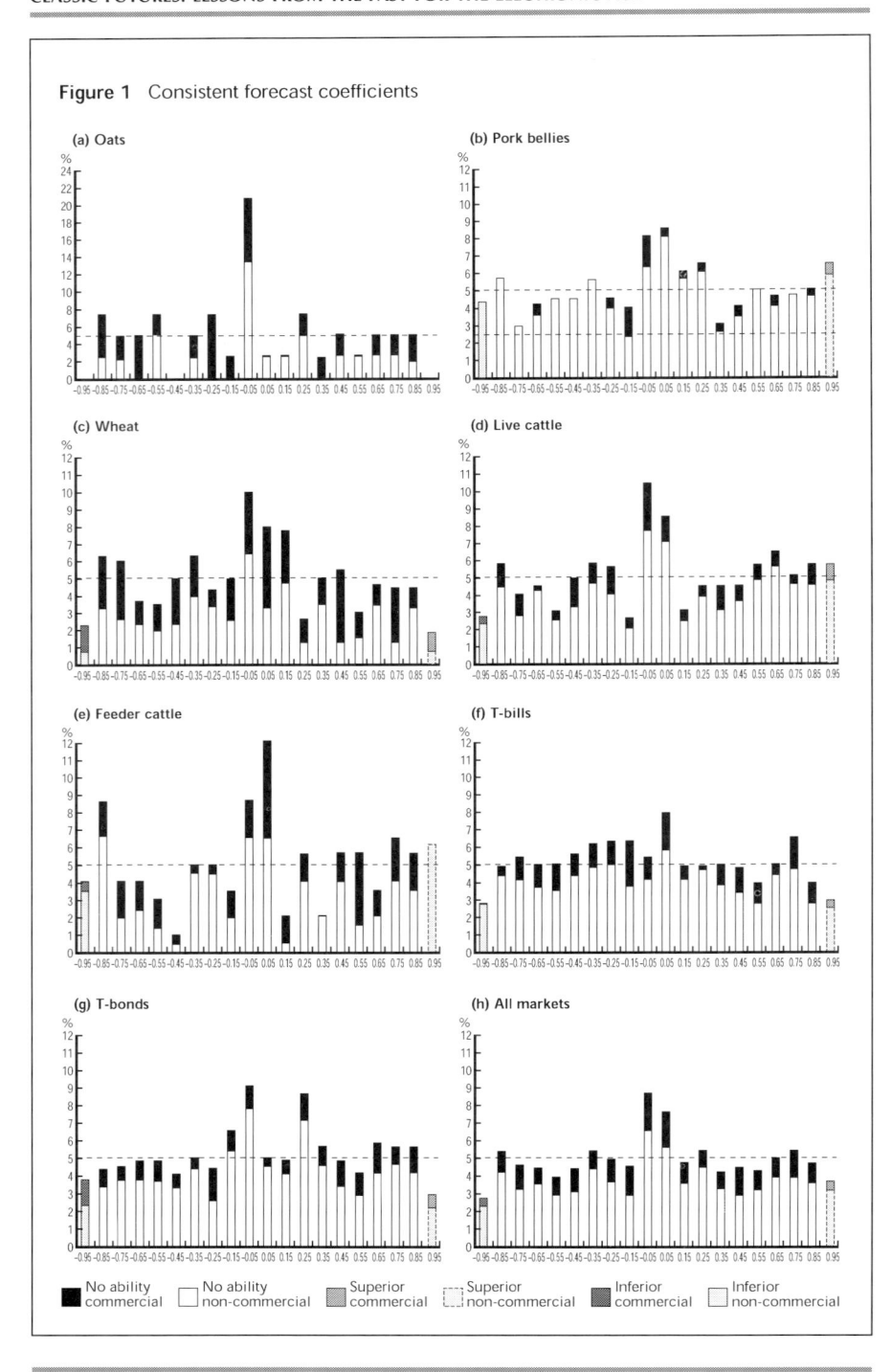

Figure 1 Consistent forecast coefficients

supported. When all markets are combined, forecast ability appears nonrandom because of dependence among traders. If a substantial proportion of the traders in these markets follow the same technical strategies or respond in unison to the suggestions made by the newsletters or advisory services, one would observe such a bunching.

Why traders with apparently poor skills remain in these markets and achieve such large size is puzzling. One often-cited suggestion is that these poor forecasters are using these markets to offset risks they have in other related markets. Therefore, the traders look like poor forecasters in a one-dimensional sense since they are losing money in the futures market. However, in a multidimensional approach, since there is a trade-off between profits and risks, the traders are actually increasing their expected utilities. If this is the case, one would expect to observe a significant difference between the performance of commercial and non-commercial traders. The commercial traders are more likely to use these markets to hedge their cash market price risks. They can "afford" to look like bad forecasters in the futures market since they will have the opposite performance in the cash markets, or at least reduce their overall business risks. The non-commercial traders do not have the same opportunities to use futures markets directly to reduce their price risks. Most academic studies have demonstrated that futures markets do not reduce systematic price risks (Dusak 1973; Bodie and Rosansky 1980; Baxter, Conine, and Tamarkin 1985; Ehrhardt, Jordan, and Walkling 1987; Elton, Gruber, and Rentzler 1987).

In general, there are no significant differences between the commercial and non-commercial distributions. If anything, the non-commercial traders exhibit more poor forecast skill. The risk hypothesis as an explanation for the observed distributions of forecast coefficients is not supported.

Big hit ability

To calculate the big hit forecast coefficients, all of the available information on individual positions and market price movements is used. The regression of the magnitude of the price change between time t and $t+1$ on the size of the position at time t, indicates whether the individual trader makes his biggest bets when he expects the largest price changes.[17]

The results of the big hit regressions are presented in Table 3. One major difference between these and the consistent forecast coefficients is that three of the markets have significant negative mean coefficients for non-commercial traders. The mean coefficient for the non-commercial traders in "All Markets" is also negative and significant. Except for the oat market, all the signs on the means of the non-commercial distributions are negative.

Table 3 Descriptive statistics: big hit coefficients ($\times 10^{-2}$, except traders and χ^2)

	Number of traders	Mean coefficient	Median coefficient	Standard deviation	Skewness	D-statistic	χ^2 statistic
Oats:							
All traders	48	14.14*	7.91	54.72	−3.01	10.29	5.96
Commercial	27	6.81	−1.39	60.93	20.71	11.18	9.85*
Non-commercial	21	23.59*	7.98	45.22	−23.31	26.69*	14.00**
Wheat:							
All traders	341	−4.37*	−5.50	48.97	12.77	12.05**	63.52**
Commercial	168	−1.03	0.97	53.74	6.50	8.17	20.33
Non-commercial	173	−7.62	−6.03	43.75	12.45	18.09**	69.89**
Pork bellies:							
All traders	342	−2.89	−0.95	55.17	6.48	7.68*	30.28*
Commercial	28	−4.63	−2.34	59.52	22.94	13.55	1.64
Non-commercial	314	−2.73	−0.95	54.86	4.94	7.59*	28.29**
Live cattle:							
All traders	483	−0.95	−1.50	54.59	2.06	5.48	36.54**
Commercial	121	−2.58	−9.03	62.63	3.03	9.24	22.31
Non-commercial	362	−0.41	−0.32	51.71	2.92	8.08	55.24**
Feeder cattle:							
All traders	193	−3.28	−0.69	56.96	−0.73	6.99	24.10
Commercial	69	0.28	0.27	61.18	−3.62	5.52	6.22
Non-commercial	124	−5.25	−1.10	54.62	−1.26	10.89	27.61*
T-bonds:							
All traders	439	−2.48	−0.78	51.25	−0.27	10.21**	57.54**
Commercial	96	1.80	−0.64	52.96	10.29	8.89	13.37
Non-commercial	343	−3.68	−1.09	50.78	−4.13	11.55**	57.06**
T-bills:							
All traders	383	−8.31**	−7.11	50.09	10.92	13.85**	63.42**
Commercial	98	−6.63	−8.28	53.76	12.67	9.63	17.51*
Non-commercial	285	−8.88**	−5.94	48.86	9.45	16.22**	62.30**
All markets:							
All traders	2229	−3.21	−2.87	52.71	5.57	8.22**	171.44**
Commercial	607	−1.46	−3.91	56.76	7.03	3.92	16.29
Non-commercial	1622	−3.87**	−2.66	51.11	3.91	10.13**	211.14**

*Significant at a 10% level.
**Significant at a 1% level.

In almost all of the markets the big hit distributions are positively skewed. The standard deviations are those expected from a uniform distribution. The *D*-statistics for the big hit forecast coefficients are similar to those for the consistent forecast coefficient distributions. One major difference is that the distributions for commercial traders are all uniformly distributed. In five of seven markets for non-commercial traders the distributions are not uniform.

Similar results hold for the chi-square tests. Only the oat commercial distribution is not uniform. Uniformity is rejected for all seven of the non-commercial distributions.

In Figure 2 the percentage bar charts for the seven individual markets and "All Markets" are shown. In the T-bond and feeder cattle markets there are more traders with forecast coefficients less than negative 90% than expected. In the interest rate markets it is interesting to note that there is a reduction of traders with superior big hit coefficients when compared to the consistent coefficients. This is somewhat surprising given the massive profits earned by the traders in these markets. However, given the observed large downward price trends over the period analysed, it did not take a genius to earn large profits.

In the oats market, there are a large number of traders with big hit coefficients greater than 90% and the mean is positive and significant. This is in contrast to the consistent forecast results where no oats traders have coefficients greater than 90%. There are certain commercial traders who take their largest positions immediately prior to the biggest price moves. This may be the result of their possessing inside information or their hands-on grasp of the elements that drive the oat market. This is also consistent with selective hedging operations. Commercial traders may be hedging and consistently taking small losses on their futures positions (thus their negative consistent forecast coefficients). However, when they expect a major price move, they speculate in a big way by adjusting their position sizes accordingly.

Overall, Figures 1 and 2 describe similar relationships between profit and performance. Coefficients bunching around zero result in the rejection of uniformity. If this bunching is due to a statistical anomaly, such as heteroscedasticity, one would expect that the bias would have its greatest impact on the big hit results where nonstationarity in price changes can have adverse effects on the efficiency of the big hit parameter estimates. However, this is not the case, and statistical anomalies do not appear to drive the results. There must be dependence across traders.

EX ANTE TESTS OF FORECAST ABILITY

Do the traders who display superior (or inferior) forecast ability in an early period continue to demonstrate it in a later period? A significant

Figure 2 Big hit forecast coefficients

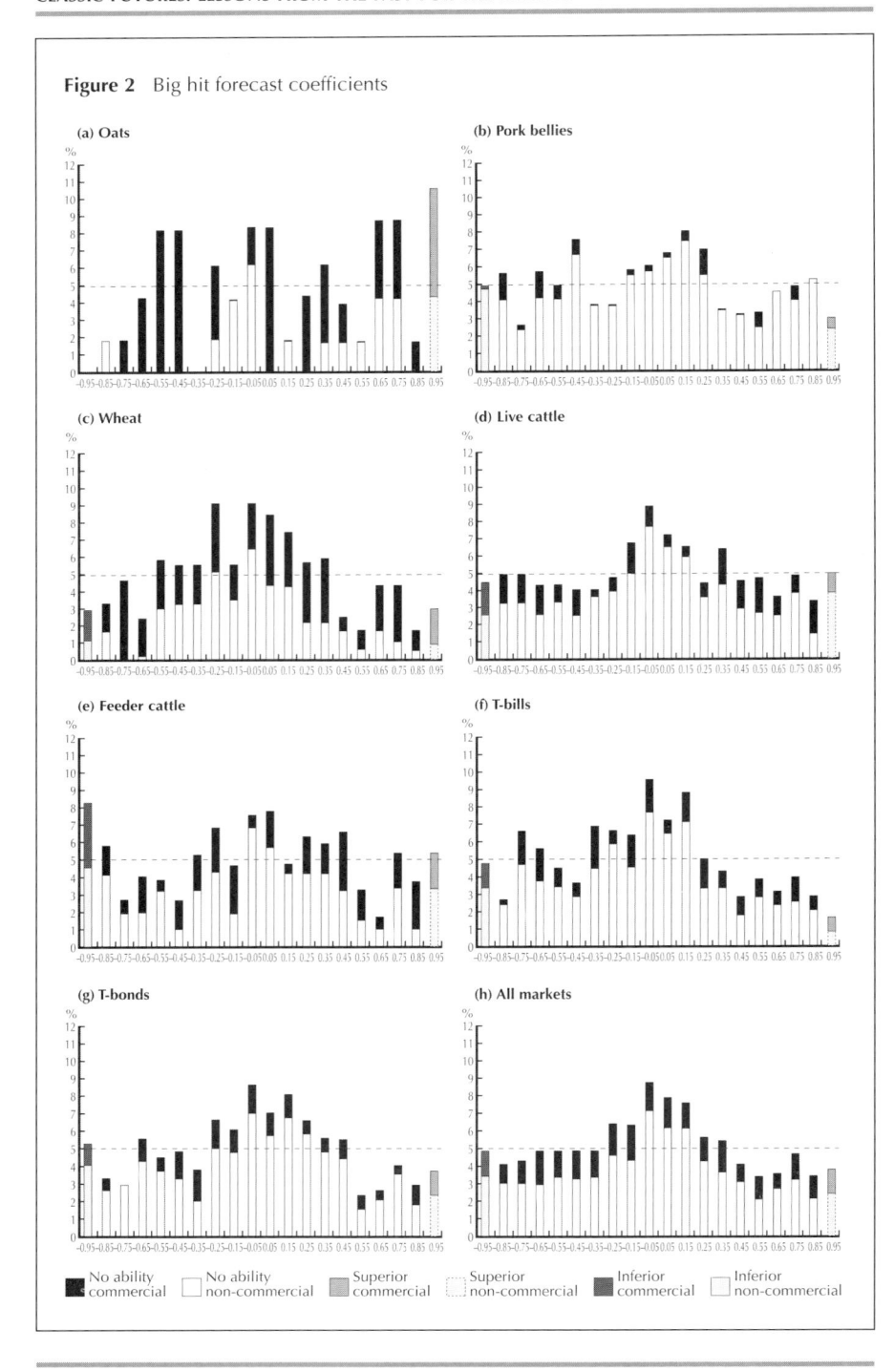

(a) Oats

(b) Pork bellies

(c) Wheat

(d) Live cattle

(e) Feeder cattle

(f) T-bills

(g) T-bonds

(h) All markets

■ No ability commercial □ No ability non-commercial ▨ Superior commercial ▨ Superior non-commercial ■ Inferior commercial □ Inferior non-commercial positive relationship between performance in the two periods supports the skill hypothesis.

To analyse this intertemporal relationship, traders with at least 25 transactions in both an early and a later period are used. The early period extends from January 1, 1977, to September 30, 1979. The later period covers October 1, 1979–December 31, 1981.

Since the commercial traders are mostly hedging (by definition), it is unclear how to interpret the "life-cycle" results for the commercial group.[18] In the previous section this group was analysed to serve as a benchmark with which to compare the performances of non-commercial traders. Since the results in the early period can now be used as the benchmark, I focus only on the performance of non-commercial traders.

Correlation statistics

In Table 4 three variables are correlated for each market to determine if there is a relationship between performance in the early and late periods. The correlations are given for all traders and for the non-commercial traders.

The correlations between dollar returns earned in the two periods are significant when all traders are pooled. For the individual markets the

Table 4 Correlation statistics across periods

	All traders			Non-commercial traders		
	Return	Consistent	Big hit	Return	Consistent	Big Hit
Oats	0.89*	0.01	0.01	0.76*	0.10	−0.06
	26	18	26	9	9	9
Wheat	0.23**	0.23**	0.01	−0.18	0.39**	0.29*
	161	131	161	60	58	60
Pork bellies	0.68**	0.16	0.22*	0.15	0.16	0.26**
	117	99	117	103	88	103
Live cattle	−0.58*	0.04	−0.10	−0.05	0.04	−0.06
	151	126	151	104	96	104
Feeder cattle	0.05	0.16	0.17	−0.44*	0.21	0.18
	50	39	50	27	22	27
T-bonds	0.08	−0.06	−0.13	0.06	−0.02	−0.04
	72	64	72	47	42	47
T-bills	0.33**	−0.00	0.13	0.85**	0.04	0.13
	87	82	87	59	57	59
All markets	−0.01	0.10	0.04	0.20**	0.12*	0.12**
	664	559	664	409	372	409

Note: The number of observations is below the correlation coefficient. Early period is July 1977–September 1979. Late period is October 1979–December 1981. Oats periods are January 1978–June 1979 and July 1979–December 1980.
*Significant at a 10% level.
**Significant at a 1% level.

most puzzling result is in the live cattle market where the correlation is significant but negative.[19] For the non-commercial traders alone there are positive significant correlations only in the T-bill market and for all markets combined. In two markets the correlations are negative and significant. In general, for the non-commercial traders, it does not appear as if dollar performance in one period is positively related to dollar performance in the other period.

As for the forecast coefficients, there does not appear to be any strong correspondence between a trader's observed abilities in the first and second periods. For the non-commercial traders in the wheat markets there are some small significant correlations. In addition, there are significant correlations when all markets are pooled. Even so, these few significant correlations are quite low. Overall, the correlations provide little evidence in support of the skill hypothesis.

Traders broken down by decile of early period forecast ability

In Table 5 the individual early period forecast coefficients are divided up into deciles depending on their relative magnitudes.[20] For each decile, the means of variables describing performance in the later period are calculated. Overall, traders participating during both periods have slightly positive and significant means of early period consistent forecast coefficients. One might expect traders remaining in the market for a second period would have done better than average in the first period, whether positive forecast coefficients are the result of skill or luck. Interestingly, this is not true for the big hitters where the early period mean forecast coefficient is not different from zero (see Table 6). Overall, the second-half consistent forecast coefficient is zero, while the big hit average is negative.

Scanning across deciles it is clear that forecast ability regresses toward zero. Traders exhibiting superior skill in the first half appear to have average (or no) skill in the second half. Traders with inferior skill in the first half improve slightly in the second half.

Only for the traders in the top decile of the consistent forecast coefficients is there some weak evidence supporting the skill hypothesis. This decile is composed of traders who almost all have early period coefficients significantly different from zero at the 10% level. In the second half the significance levels fall, on average, but still remain slightly above the average for the group as a whole.

Other second-period measures of performance are less supportive for the group in the top decile of traders with early period superior consistent forecast ability. In the first half, 84% of these traders earn positive dollar profits (ie, are successful). In the second half, this percentage falls significantly, to 65%. In fact, for the deciles where the early period consistent forecast coefficients are positive (ie, deciles 5–10), the percentage

Table 5 Descriptive statistics by deciles of early period forecast coefficients. Rank by early consistent coefficient (profits in millions of dollars)

Statistic	All traders	Decile 1	Decile 2	Decile 3	Decile 4	Decile 5	Decile 6	Decile 7	Decile 8	Decile 9	Decile 10
Number of traders	372	37	37	37	38	37	37	38	37	37	37
Consistent coefficient:											
First half	0.10**	−0.82**	−0.59**	−0.33**	−0.14**	0.03**	0.19**	0.38**	0.58**	0.76**	0.91**
	(0.03)	(0.01)	(0.01)	(0.01)	(0.01)	(0.01)	(0.01)	(0.01)	(0.01)	(0.01)	(0.10)
Second half	−0.02	−0.07	0.05	−0.18*	−0.07	−0.10	−0.09	0.00	0.01	0.10	0.18*
	(0.03)	(0.09)	(0.11)	(0.08)	(0.09)	(0.09)	(0.09)	(0.10)	(0.09)	(0.10)	(0.09)
Overall	0.09**	−0.57**	−0.28**	−0.28**	−0.02	0.01	0.04	0.26**	0.36**	0.57**	0.79**
	(0.03)	(0.07)	(0.08)	(0.07)	(0.08)	(0.09)	(0.07)	(0.08)	(0.08)	(0.08)	(0.05)
Overall big hit coefficient	−0.04*	−0.01*	−0.12	−0.10	−0.09	−0.09	−0.14*	0.02	−0.01	0.11	0.05
	(0.02)	(0.05)	(0.08)	(0.06)	(0.06)	(0.06)	(0.06)	(0.07)	(0.07)	(0.16)	(0.07)
Profits:											
First half	0.62*	−0.07	0.73	0.38	0.74*	0.61**	0.21	0.67*	0.57*	1.10**	1.26**
	(0.10)	(0.22)	(0.53)	(0.27)	(0.31)	(0.22)	(0.21)	(0.30)	(0.25)	(0.29)	(0.29)
Second half	0.15	0.30*	0.46	0.18	0.25	0.01	−0.24*	1.08	−0.25	−0.21	0.52
	(0.14)	(0.16)	(0.30)	(0.38)	(0.35)	(0.22)	(0.14)	(1.09)	(0.21)	(0.28)	(0.44)
% successful:											
First half	63**	30*	54	46	55	65*	76**	76**	68**	76*	84**
Second half	48	43	51	46	53	46	30	53	41	54	65

Note: Standard errors are in parentheses.
*Significant at a 10% level.
**Significant at a 1% level.

Table 6 Descriptive statistics by deciles of early period forecast coefficients. Rank by early big hit coefficient (profits in millions of dollars)

Statistic	All traders	Decile 1	Decile 2	Decile 3	Decile 4	Decile 5	Decile 6	Decile 7	Decile 8	Decile 9	Decile 10
Number of traders	409	41	41	41	41	41	41	41	41	41	40
Big hit coefficient:											
First half	0.01	−0.85**	−0.53**	−0.32**	−0.16**	−0.04**	0.06**	0.17**	0.34**	0.56**	0.84**
	(0.02)	(0.02)	(0.01)	(0.01)	(0.01)	(0.00)	(0.00)	(0.01)	(0.01)	(0.01)	(0.01)
Second half	0.06**	−0.13	−0.25**	−0.12	−0.12*	0.08	−0.01	−0.09	0.09	−0.08	0.01
	(0.02)	(0.09)	(0.08)	(0.07)	(0.07)	(0.07)	(0.08)	(0.07)	(0.08)	(0.08)	(0.09)
Overall	−0.02	−0.40**	−0.30**	−0.17**	−0.09*	0.09*	0.03	0.02	0.16**	0.19**	0.30**
	(0.02)	(0.08)	(0.06)	(0.06)	(0.05)	(0.05)	(0.06)	(0.04)	(0.05)	(0.06)	(0.09)
Overall consistent coefficient	0.08**	0.06	−0.10	−0.06	−0.00	−0.11	0.01	0.22**	0.23**	0.29**	0.30**
	(0.03)	(0.09)	(0.08)	(0.09)	(0.10)	(0.10)	(0.10)	(0.10)	(0.09)	(0.09)	(0.09)
Profits:											
First half	0.64**	0.90*	0.92*	0.81**	0.09	0.00	0.46**	1.26**	0.78**	0.76**	1.50**
	(0.10)	(0.43)	(0.48)	(0.28)	(0.20)	(0.15)	(0.16)	(0.18)	(0.24)	(0.26)	(0.54)
Second half	0.04	0.17	1.10	−0.17	0.27	−0.03	−0.03	0.11	−0.26	−0.36	−0.09
	(0.14)	(0.16)	(1.03)	(0.23)	(0.31)	(0.13)	(0.40)	(0.16)	(0.31)	(0.26)	(0.20)
% successful:											
First half	65**	63*	68**	59	54	59**	68*	46*	78**	78*	72**
Second half	47	51	41	51	44	29	51	59	49	37	45

Note: Standard errors are in parentheses.
*Significant at a 10% level.
**Significant at a 1% level.

of late period winners is always significantly below the percentage of early period winners. Furthermore, except for decile 6, the second-half percentages are all indistinguishably different from 50%. This is exactly what would be expected from random trading. Only in the bottom decile does this percentage increase significantly. The percentages of winners in the second half fall and approach 50% in the big hit forecast coefficients deciles 5–10 as well. This again is strong evidence supporting a regression to the mean or luck hypothesis.

The mean dollar profits earned in the first half by all the traders who remain in the market during the second period are significantly different from zero. This probably explains what induces the traders to remain in the market during the second period. However, in the second period the mean dollar profits are not different from zero. Scanning the big hit forecast coefficient deciles, one observes a dramatic decrease in profits during the second period in deciles 6–10. In fact, the traders in the top decile earn US$1.5 million, on average, in the early period and lose US$90,000, on average, in the later period. The results are similar for the consistent deciles as well. Significant positive profits earned in the first half turn into nonpositive profits in the second half.

Summarising, the second-half performances of the traders with the lowest and highest early period big hit coefficients are indistinguishable. This is not quite true when the early period consistent forecast coefficients are ordered. In this case, the second-half coefficients, profits, and percentage of successful traders are all greater for the most successful early period forecasters than those for the other deciles.

Examination of the early period outliers

It may be the case that only a small number of traders make up an elite subset of superior forecasters. Most traders may have no significant forecast ability, with only a few outliers consistently exhibiting skill. The traders with big hit and consistent early period forecast coefficients greater than 0.8 (in absolute value) are examined in detail in Table 7.

There are 25 traders with early period big hit coefficients greater than 0.8.[21] If second-period forecast coefficients are determined by luck, then 10% of the 25 (or 2.5) traders should be lucky enough to have coefficients above 0.8 in both the early and late periods. If luck determines the outcomes, then 50% of the 25 (or 12.5) traders should have coefficients less than zero in the second period. And finally, if luck is important, the mean forecast coefficients should be symmetrically distributed and insignificantly different from zero (which they are). Overall, Table 7 shows for the early period big hit outliers that the later period forecast coefficients look like they are generated by a stochastic process.

The same cannot be said unequivocally about the traders with outlying consistent coefficients. The number of outlying traders with the opposite

637

Table 7 Traders with outlying early period forecasting coefficients

	First-half big hit coefficient				First-half consistent coefficient			
Statistic	≥0.8	≥0.9	≤−0.8	≤−0.9	≥0.8	≥0.9	≤−0.8	≤−0.9
Number of traders	25	13	28	16	42	18	20	7
Number with same significance second half	3	2	5	1	6	1	4	1
Number with opposite sign second half	14	5	12	8	14	4	7	2
First-half coefficient	0.90	0.95	−0.91	−0.95	−0.90	0.96	−0.89	−0.96
Second-half coefficient	−0.02	0.21	−0.06	−0.06	0.13	0.30*	−0.16	−0.38
Maximum second-half coefficient	0.98	0.98	1.00	1.00	1.00	1.00	0.94	0.94
Minimum second-half coefficient	−0.99	−0.76	−0.92	−0.91	−0.99	−0.93	−0.94	−0.94

*Significant at the 10% level.

sign in the second period is smaller than expected. At the same time it is not far enough away from the level expected by chance to offer strong evidence supporting the skill hypothesis. The mean second-period consistent coefficient for the 0.9 outliers is positive and significant. This corresponds with the results for decile 10 in Table 5 but is still only weak support for the skill hypothesis.

CONCLUSION

The empirical evidence presented in this article strongly supports the contention that the returns to traders of futures are randomly generated. The support for the luck hypothesis comes from two sources: (1) the observed distributions of forecast coefficients that are either uniform or peaked at zero; and (2) the fact that the abilities in the first period of both the superior and inferior traders regress toward the mean in the second period.

There are two questions emerging from the analysis presented here. First, it is not clear why there is a massive bunching of traders with no ability. It is suggested that this dependence is due to the fact that many individuals use very similar trading strategies or information sources. Second, it is not clear why these large reporting traders, a subset of all participants, earn significant positive returns. If the performances of all traders, whether large or small, are due to luck, one would not expect the large traders to consistently perform any better than the small traders. Yet in all studies to date the small traders are the big losers and the large traders are the big winners (Stewart 1949; Houthakker 1957; Rockwell 1964, 1977; Hartzmark 1984, 1987).

There is little support for the hypothesis that futures traders holding large positions possess the ability to consistently earn profits. It does appear that commercial traders show slightly better forecast ability than non-commercial traders. These commercial particpants are the traders with access to the most timely information that they may be able to profit from. This is dramatically displayed in the oats market, where commercial traders do not posses significant consistent forecast ability but demonstrate significant big hit ability. The commercial traders also make up a higher proportion of the biggest winners than one would expect if everything were random. There are also more commercial traders with superior forecast ability than expected. In the intertemporal analysis, there is some weak support for the skill hypothesis when observing the results of the very few superior outliers with consistent forecast ability. They stand out, although without observing the underlying characteristics of this elite group it is impossible to determine if skill plays any part in determining performance.

What motivates new traders to continually enter these markets and old traders to remain? A sophisticated investment strategy that results in persistent losses in one financial market? An "irrational" belief that they possess superior skill? The desire to gamble on their beliefs and the consumption they derive from the activity? These are important questions, the answers to which will provide insight in to the overall performance of futures markets.

1 Data for oats run from January 1, 1978, to December 31, 1980. Data for T-bonds run from August 22, 1977, to December 31, 1981.
2 This aggregation is appropriate since the contract specifications are all comparable. This is not true for any of the other closely related contracts such as live and feeder cattle.
3 Commercial traders are those participants whose main line of business is focused on the underlying cash commodity.
4 For a more detailed explanation of the decomposition, see Hartzmark (1984, 1987). About 30% of all traders report both hedge and speculative positions. Of these traders, 49% are classified as commercial traders.
5 Transactions are defined as either purchases or sales. Tests for oats and pork bellies were performed in which an observation for each day the trader was in the market was used, not just observations when a transaction was made (ie, even days when the change in position was zero were included). The results were similar, except there were many traders who made one or two transactions and simply held onto their position for more than 25 days. It does not seem appropriate to include these traders in the tests. In addition, the computer costs would grow out of sight if all observations were included in the statistical procedures. If a trader remains on the same side of the market after a transaction, then it still counts as an update. For example, if the trader increases his long position from 100 to 300 contracts, or reduces his long position from 100 to 50 contracts, I assume that he has made a new prediction about the magnitude by which the price will increase. Conversely, it is implicitly

assumed that, if the trader retains the same exact position over a long period of time, his price forecast has not changed, even if the price level has.

6. The net position value is simply the dollar value of all long contacts minus the dollar value of all short contracts held by the individual trader on a given day. The average net position value is the average of the net position value for an individual trader over the period he is in the market.

7. Daily dollar profits for each trader for each contract held are calculated by multiplying the end-of-day positions by the change in the settlement price between the current day and the following day. This is the same procedure used by the central clearinghouse to mark each trader's account to the market price at the end of each trading session. The total dollar profits earned by the trader are then used to measure performance. A percentage rate of return is not used because this measure has little meaning as a performance measure in the futures market. Since the net supply of contracts in futures markets is zero, there is no meaningful way to determine the magnitude of total investment in the market (ie, the denominator for any percentage rate of return would equal zero). In addition, the opportunity cost of investing is quite small (Telser 1981; Hartzmark 1986).

8. One can do all calculations using the same information and get identical results using the long positions as predictions.

9. The probability of correctly predicting the direction of the price movements is assumed to be independent of the magnitude of the subsequent price movements, therefore using CM there is no need to rely on any of the equilibrium models of security valuation. This is especially helpful in a study examining futures markets since researchers still disagree on the appropriate model or market proxy to use.

10. The HM test is equivalent to testing whether B is significantly different from zero.

11. For $LS(t)$ the number of contracts, not the dollar position value, is used. This avoids any problem if price changes and price levels (which are part of the position value) are related. Net zero positions are excluded, even if profits are earned on a spread.

12. See Denton (1985) for a detailed application of this principle using a fair coin toss game.

13. The D-statistic is sensitive to departures in the shape of the actual distribution from uniformity. The chi-square goodness of fit test is better for finding any irregularities in the actual distribution (Sachs 1984).

14. In this case in which traders are always long (predicting the price is going up) or always short (predicting the price is going down), the hypergeometric distribution collapses into a binomial distribution. A unique maximum-likelihood estimate using the logit procedure cannot be found.

15. In most markets the combined holdings of the sample traders total more than 50% of the open interest.

16. The variance of a uniform distribution is $(a - b)^2/12$, where a and b are the end points.

17. An adjustment for heteroscedasticity is made to account for expected differences in the variance of the price changes over different time intervals. If it is assumed that daily price changes have a constant variance, then holding period price changes do not have a constant variance. Since there are usually several days between each of the trader's transactions, the square root of the number of days between each transaction is used as a weight in the adjustment procedure. There is one other source of heteroscedasticity that may be important, but is not corrected for in the regressions. In some of the markets (especially in T-bills and T-bonds) there are large changes in the daily variance of the price changes over time. Given the large number of regressions run and the fact that the traders were in for different periods, this cannot easily be adjusted for. However, few traders participate for long enough for this type of heteroscedasticity to be a problem.

18. In an early period a commercial trader may perform well in the futures market and poorly in the cash market. In the later period he may have the opposite results. However, the trader may have met his goals in each period.

19 All types of questions have been asked about activity in the live cattle market. It appears to be an anomaly. The results here support this. The negative correlation may also be due to price trends. The early period is one where prices trend upward, while in the later period the trend is slightly downward or flat.

20 Deciles for individual markets were also examined. The results were similar. In addition to the statistics presented in Tables 5 and 6, average duration size, and serial correlation measures were calculated. There are no significant differences in any of these measures across the deciles. The correlations related past performance (cumulative profits up to and including month $t-1$) to current performance (profits in month t). Most correlations averaged about -0.25. This suggests that current performance is negatively related to the past record.

21 This is slightly less than one would expect from chance given that there are 409 traders who participate in both periods, and 10% of these (or 41 traders) should be in the interval from 0.8 to 1.0.

BIBLIOGRAPHY

Baxter, J., T. Conine and M. Tamarkin, 1985, "On commodity market risk premiums", *Journal of Futures Markets* 5, Spring, pp. 121–25.

Bodie, Z. and V. I. Rosansky, 1980, "Risk and return in commodity futures", *Financial Analysts Journal* 36, May/June, pp. 27–39.

Chang, E. C. and W. G. Lewellen, 1984, "Market timing and mutual fund investment performance", *Journal of Business* 57, January, pp. 57–72.

Cumby, R. E. and D. M. Modest, 1987, "Testing for market timing ability: A framework for forecast evaluation", *Journal of Financial Economics* 19, September, pp. 169–90.

Denton, F., 1985, "The effect of professional advice on the stability of a speculative market", *Journal of Political Economy* 93, October, pp. 977–93.

Dusak, K., 1973, "Futures trading and investor returns: An investigation of commodity market risk premiums", *Journal of Political Economy* 81, November, pp. 1387–406; reprinted as Chapter 24 of the present volume.

Ehrhardt, M., J. Jordan and R. Walking, 1987, "An application of arbitrage pricing theory to futures markets: Tests of normal backwardation", *Journal of Futures Markets* 7, February, pp. 21–34.

Elton, E., M. Gruber and J. Rentzler, 1987, "Professionally managed, publicly traded commodity funds", *Journal of Business* 60, April, pp. 175–200.

Hartzmark, M. L., 1984, "The distribution of large trader returns in futures markets: Theory and evidence", PhD dissertation (University of Chicago).

Hartzmark, M. L., 1986, "The effects of changing margin levels on futures market activity, the composition of traders in the market, and price performance", *Journal of Business* 59, April, pp. S147–S80.

Hartzmark, M. L., 1987, "Returns to individual traders of futures: Aggregate results", *Journal of Political Economy* 95, December, pp. 1292–306.

Henriksson, R. D., 1984, "Market timing and mutual fund performance: An empirical investigation", *Journal of Business* 57, January, pp. 73–96.

Henriksson, R. D. and R. C. Merton, 1981, "On market timing and investment performance II: Statistical procedures for evaluating forecasting skills", *Journal of Business* 54, October, pp. 513–33.

Houthakker, H., 1957, "Can speculators forecast prices?", *Review of Economics and Statistics* 39, February, pp. 143–57; reprinted as Chapter 22 of the present volume.

Jagannathan, R. and R. A. Korajczyk, 1986, "Assessing the market timing performance of managed portfolios", *Journal of Business* 59, April, pp. 217–35.

Jensen, M. C., 1968, "The performance of mutual funds in the period 1945–1964", *Journal of Finance* 23, May, pp. 389–416.

Kon, S. and F. C. Jen, 1978, "Estimation of time-varying systematic risk and performance for mutual fund portfolios: An application of switching regression", *Journal of Finance* 33, May, pp. 457–76.

Kon, S. and F. C. Jen, 1979, "The investment performance of mutual funds: An empirical investigation of timing, selectivity, and market efficiency", *Journal of Business* 52, April, pp. 263–89.

Merton, R. C., 1981, "On market timing and investment performance I: An equilibrium theory of value for market forecasts", *Journal of Business* 54, July, pp. 363–406.

Rockwell, C. S., 1964, "Profits, normal backwardation, and forecasting in commodity futures", PhD dissertation (University of California at Berkeley).

Rockwell, C. S., 1977, "Normal backwardation, forecasting, and the returns to commodity futures traders", in A. E. Peck (ed.), *Selected Writings on Futures Markets*, volume 2 (Chicago: Board Trade of the City of Chicago). Reprinted as Chapter 23 of the present volume.

Sachs, L., 1984, *Applied Statistics*, second edition (New York: Springer Verlag).

Stewart, B., 1949, "An analysis of speculative trading in grain futures", Technical Bulletin 1001. (Washington, DC: US Department of Agriculture).

Telser, L. G., 1981, "Margins and futures contracts", *Journal of Futures Markets* 73, Summer, pp. 225–53.

6

NEW RESULTS ON THE RANDOM PROPERTIES OF FUTURES PRICES

Introduction

Lester G. Telser

The two chapters by Benoit Mandelbrot, published long before he attained world fame, remain on the frontier of research on the probability distributions of futures and securities prices. Few observers of futures and securities prices do not know that very large and very small price changes are much more common than would be true for random draws from a Gaussian normal distribution. Dusak's diagrams (Chapter 24), show this clearly. Nevertheless, the widely used Black–Scholes formula for stock option pricing and Black's formula for pricing futures' options (Chapter 9 in Section 2) assumes that futures' price changes or percentage price changes are random draws from a normal distribution. Such formulas underestimate the probability of extremes. To emphasise this point, consider the behaviour of the Dow-Jones Index of Industrials. From the first trading session in January 1929 to the last in December 1988, the daily percentage change of the index was at least 2% in 950 sessions, about 6% of the total number of trading sessions in that time, much more than would be expected from a normal distribution with the observed mean and standard deviation. There were three sessions with enormous percentage drops: October 19, 1987, 22.61% and October 28, 1929 and October 29, 1929, with a combined drop of 23.17%. These drops were more than six standard deviations from the mean daily percentage change. For a normal distribution the probability would be less than one chance in 10 billion. Yet these large changes occurred three times in about 15,000 trading sessions. This raises a rhetorical question: is it better to use a theory that you know is wrong but is easier to learn and manipulate, or is it better to study and use a more accurate theory, which is harder to learn and manipulate? Mandelbrot's two chapters should help one do the latter – both articles focus on the probabilistic aspects of the prices, not the economic theory.

The normal distribution applies to price changes taken as reflecting the sum of many independent random effects of about equal size, many of which offset each other. That is, some effects tend to raise and others to reduce prices, but the combined effect tends to nullify this. While adhering to the hypothesis that price changes do reflect the sum of many independent effects, normality does not describe the relatively high frequency of price changes for which some effects are occasionally large relative to the myriad effects. For these conditions mathematicians have derived a more general central limit theorem, giving the distribution for the sum of a large number of random effects that embraces the normal as a special case. There is a practical problem of describing these limiting distributions in closed form. This is possible only in a few cases, the Cauchy distribution, the Arc Sine distribution and the Gaussian (normal) distribution. Therefore, one is driven to descriptions of the empirical distribution of price changes in terms of the observed frequencies of the price changes. If the more general central limit theorem does apply, then the observed frequencies would converge to a limiting distribution. While the mean and standard deviations would not give a reliable picture, the median and the interquartile range would.

Mandelbrot's chapter written in 1963, explains the basic ideas of his new approach to the study of the variation of speculative prices. It includes numerical results for cotton prices, spanning 1816 to 1958, to illustrate the potency of these new ideas. Among the more important lessons from this analysis is an improved estimate of the probability of extreme price changes. To put it another way, these methods give a better measure of the risk. Mandelbrot emphasises the dangers of relying on sample estimates of the standard deviation of price changes in order to calculate the probability of extremes.

If prices take a random walk and price changes are independent, then the curve traced by the prices will be continuous but nowhere differentiable. This means it would not be possible to predict the next price on the basis of knowing the whole path of past prices even though the path is continuous. However, a random walk generated by a Paretian distribution is not even continuous so it displays an even less regular pattern than one from a normal distribution.

Mandelbrot's second chapter in this section endorses these remarks. It starts with some explanations of aspects of his approach in response to misunderstandings and criticisms of his 1963 article. The analogy between an actuarial table of life expectancy and the study of medicine may help clarify Mandelbrot's position. A company in the business of selling life insurance uses a table of mortality rates more or less detailed with respect to certain pertinent variables such as age, sex, previous health and the like, to calculate the premiums it will charge its customers. The company relies on scientific results in order to decide which aspects of any individual policyholder's state are likely to influence their life expectancy; often a physical examination is required to aid its decision. However, scientific research on the causes of death and disease are no part of the business of a life insurance company. A detailed inquiry into the reasons for a policyholder's death apart from a death certificate is highly unusual. In contrast a scientist may consider the facts about an individual's death as highly relevant to his scientific investigation. Someone who buys or sells options wants data about the distribution of price changes analogous to that desired by a life insurance company, while an economist studies price episodes analogous to the way that a scientist may seek data about the causes of death. For example, the study of "speculative bubbles" is an attempt to explain huge price changes as the result of a bursting bubble.

The 1967 chapter offers additional evidence to buttress Mandelbrot's views about Paretian distributions. It shows distributions of price changes for wheat, rail stock prices and interest rates. It closes with remarks about the difficulties of estimating parameters for stable Paretian distributions. Mandelbrot has continued his research in finance and obtained many new results, which have been published in his book *Fractals and Scaling in Finance: Discontinuity, Concentration Risk*, 1997, (Springer–Verlag).

26

*The Variation of Certain Speculative Prices**

Benoit Mandelbrot
Yale University

The name of Louis Bachelier is often mentioned in books on diffusion process. Until very recently, however, few people realised that his early (1900) and path-breaking contribution was the construction of a random-walk model for security and commodity markets.[1] Bachelier's simplest and most important model goes as follows: let $Z(t)$ be the price of a stock, or of a unit of a commodity, at the end of time period t. Then it is assumed that successive differences of the form $Z(t + T) - Z(t)$ are independent, Gaussian or normally distributed, random proportional to the differencing interval T.[2]

Despite the fundamental importance of Bachelier's process, which has come to be called "Brownian motion", it is now obvious that it does not account for the abundant data accumulated since 1900 by empirical economists, simply because *the empirical distributions of price changes are usually too "peaked" to be relative to samples from Gaussian populations.*[3] That is, the histograms of price changes are indeed unimodal and their central "bells" remind one of the "Gaussian ogive". But there are typically so many "outliers" that ogives fitted to the mean square of price changes are much lower and flatter than the distribution of the data themselves (see,

---

*This paper was first published in the *Journal of Business* 36, pp. 394–419 (1963) and is reprinted with the permission of the University of Chicago Press. The theory developed in this paper is a natural continuation of my study of the distribution of income. I was still working on the latter when Hendrik S. Houthakker directed my interest toward the distribution of price changes. The present model was thus suggested by Houthakker's data; it was discussed with him all along and was first publicly presented at his seminar. I therefore owe him a great debt of gratitude. The extensive computations required by this work were performed on the 7090 computer of the IBM Research Center and were mostly programmed by N. J. Anthony, R. Coren, and F. l. Zarnfaller. Many of the data which I have used were most kindly supplied by F. Lowenstein and J. Donald of the Economic Statistics section of the United States Department of Agriculture. Some stages of the present work were supported in part by the Office of Naval Research, under contract number Nonr-3775(00), NR-047040.

Figure 1

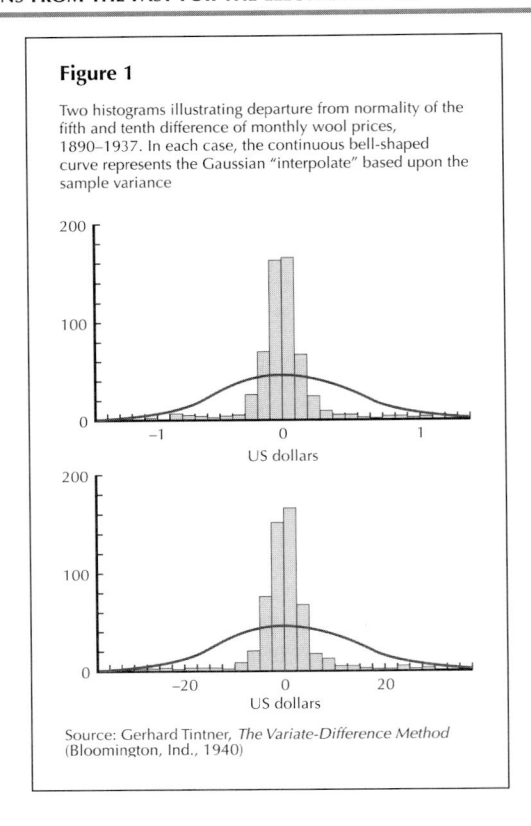

Two histograms illustrating departure from normality of the fifth and tenth difference of monthly wool prices, 1890–1937. In each case, the continuous bell-shaped curve represents the Gaussian "interpolate" based upon the sample variance

Source: Gerhard Tintner, *The Variate-Difference Method* (Bloomington, Ind., 1940)

eg, Fig. 1). The tails of the distributions of price changes are in fact so extraordinarily long that the sample second moments typically vary in an erratic fashion. For example, the second moment reproduced in Figure 2 does not seem to tend to any limit even though the sample size is enormous by economic standards, and even though the series to which it applies is presumably stationary.

It is my opinion that these facts warrant a radically new approach to the problem of price variation.[4] The purpose of this paper will be to present and test such a new model of price behaviour in speculative markets. The principal feature of this model is that starting from the Bachelier process as applied to $\log_e Z(t)$ instead of $Z(t)$, I shall replace the Gaussian distributions throughout by another family of probability laws, to be referred to as "stable Paretian", which were first described in Paul Lévy's classic *Calcul des probabilites* (1925). In a somewhat complex way, the Gaussian is a limiting case of this new family, so the new model is actually a generalisation of that of Bachelier.

Since the stable Paretian probability laws are relatively unknown, I shall

Figure 2

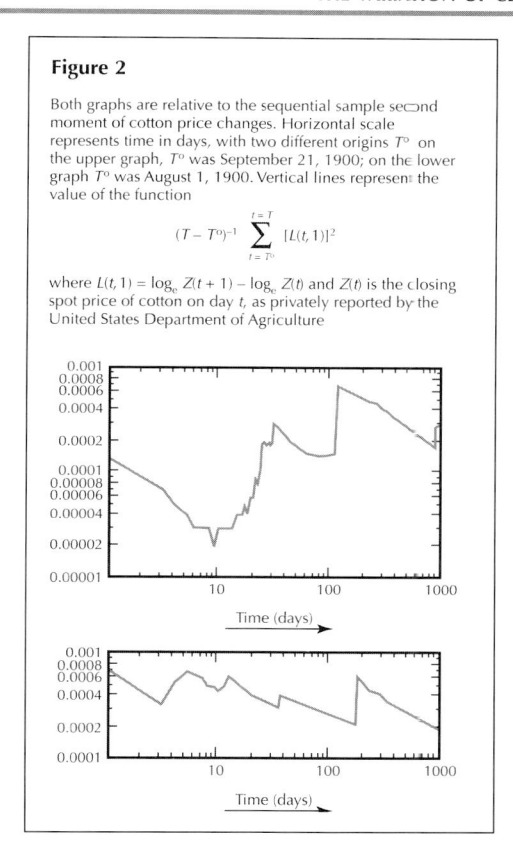

Both graphs are relative to the sequential sample second moment of cotton price changes. Horizontal scale represents time in days, with two different origins T^o on the upper graph, T^o was September 21, 1900; on the lower graph T^o was August 1, 1900. Vertical lines represent the value of the function

$$(T - T^o)^{-1} \sum_{t=T^o}^{t=T} [L(t,1)]^2$$

where $L(t, 1) = \log_e Z(t + 1) - \log_e Z(t)$ and $Z(t)$ is the closing spot price of cotton on day t, as privately reported by the United States Department of Agriculture

begin with a discussion of some of the more important mathematical properties of these laws. Following this, the results of empirical tests of the stable Paretian model will be examined. The remaining sections of the paper will then be devoted to a discussion of some of the more sophisticated mathematical and descriptive properties of the stable Paretian model. I shall, in particular, examine its bearing on the very possibility of implementing the stop-loss rules of speculation (the sixth section).

MATHEMATICAL TOOLS: PAUL LEVY'S STABLE PARETIAN LAWS
Property of "stability" of the Gaussian law and its generalisation

One of the principal attractions of the modified Bachelier process is that the logarithmic relative

$$L(t, T) = \log_e Z(t + T) - \log_e Z(t)$$

is a Gaussian random variable for *every* value of T; the only thing that changes with T is the standard deviation of $L(t, T)$. This feature is the consequence of the following fact;

Let G' and G'' be two independent Gaussian random variables, of zero means and of mean squares respectively equal to σ'^2 and σ''^2. Then, the sum $G' + G''$ is also a Gaussian variable, of mean square equal to $\sigma'^2 + \sigma''^2$. In particular, the "reduced" Gaussian variable, with zero mean and unit mean square, is a solution to

$$s'U + s''U = sU \tag{S}$$

where s is a function of s' and s'' given by the auxiliary relation

$$s^2 = s'^2 + s''^2 \tag{A$_2$}$$

It should be stressed that, from the viewpoint of equation (S) and relation (A$_2$), the quantities s', s'', and s are simply scale factors that "happen" to be closely related to the root-mean-square in the Gaussian case.

The property (S) expresses a kind of stability or invariance under addition, which is so fundamental in probability theory that it came to be referred to simply as "stability". The Gaussian is the only solution of equation (S) for which the second moment is finite – or for which the relation (A$_2$) is satisfied. When the variance is allowed to be infinite, however, (S) possesses many other solutions. This was shown constructively by Cauchy, who considered the random variable U for which

$$Pr(U > u) = Pr(U < -u) = \tfrac{1}{2} - (1/\pi)\tan^{-1}(u)$$

so that its density is of the form

$$dPr(U < u) = [\pi(1 + u^2)]^{-1}$$

For this law, integral moments of all orders are finite, and the auxiliary relation takes the form

$$s = s' + s'' \tag{A1}$$

where the scale factors s', s'', and s are not defined by any moment.

As to the general solution of equation (S), discovered by Paul Levy,[5] the logarithm of its characteristic function takes the form

$$\log \int_{-\infty}^{\infty} \exp(iuz)\, dPr(U < u) = i\delta z - \gamma |z|^\alpha [1 + i\beta(z/|z|)\tan(\alpha\pi/2)] \tag{PL}$$

It is clear that the Gaussian law and the law of Cauchy are stable and that they correspond to the cases ($\alpha = 2$) and ($\alpha = 1; \beta = 0$), respectively.

Equation (PL) determines a family of distribution and density functions $Pr(U < u)$ and $dPr(U < u)$ that depend continuously upon four parameters which also happen to play the roles usually associated with the first four moments of U, as, for example, in Karl Pearson's classification.

First of all, the α is an index of "peakedness" that varies from 0 (excluded) to 2 (included); if $\alpha = 1$, β must vanish. This α will turn out to be intimately related to Pareto's exponent. The β is an index of "skewness" that can vary from -1 to $+1$. If $\beta = 0$, the stable densities are symmetric.

One can say that α and β together determine the "type" of a stable random variable, and such a variable can be called "reduced" if $\gamma = 1$ and $\delta = 0$. It is easy to see that, if U is reduced, sU is a stable variable having the same values for α, β and δ and having a value of γ equal to s^α: this means that the third parameter, γ, is a scale factor raised to the power α. Suppose now that U' and U'' are two independent stable variables, reduced and having the same values for α and β; since the characteristic function of $s'U' + s''U''$ is the product of those of $s'U'$ and $s''U''$, the equation (S) is readily seen to be accompanied by the auxiliary relation

$$s^\alpha = s'^\alpha + s''^\alpha \tag{A}$$

If on the contrary U' and U'' are stable with the same values of α, β and of $\delta = 0$, but with different values of γ (respectively, γ' and γ''), the sum $U' + U''$ is stable with the parameters α, β, $\gamma = \gamma' + \gamma''$ and $\delta = 0$. Thus the familiar additivity property of the Gaussian "variance" (defined by a mean-square) is now played by either γ or by a scale factor raised to the power α.

The final parameter of (PL) is δ; strictly speaking, equation (S) requires that $\delta = 0$, but we have added the term $i\delta z$ to (PL) in order to introduce a location parameter. If $1 < \alpha \leq 2$ so that $E(U)$ is finite, one has $\delta = E(U)$; if $\beta = 0$ so that the stable variable has a symmetric density function, δ is the median or modal value of U; but δ has no obvious interpretation when $0 < \alpha < 1$ with $\beta \neq 0$.

Addition of more than two stable random variables

Let the independent variables U_n satisfy the condition (PL) with values of α, β, γ, and δ equal for all n. Then, the logarithm of the characteristic function of

$$S_N = U_1 + U_2 + \ldots U_n + \ldots U_N$$

is N times the logarithm of the characteristic function of U_n, and it equals

$$i\delta Nz - N\gamma|z|^\alpha[1 + i\beta(z/|z|)\tan(\alpha\pi/2)]$$

so that S_N is stable with the same α and β as U_n, and with parameters δ and γ multiplied by N. It readily follows that

$$U_n - \delta \quad \text{and} \quad N^{-1/\alpha} \sum_{n=1}^{N}(U_n - \delta)$$

have identical characteristic functions and thus are identically distributed random variables. (This is, of course, a most familiar fact in the Gaussian case, $\alpha = 2$.)

The generalisation of the classical "$T^{1/2}$ Law". In the Gaussian model of Bachelier, in which daily increments of $Z(t)$ are Gaussian with the standard deviation $\sigma(1)$, the standard deviation of the change of $Z(t)$ over T days is equal to $\sigma(T) = T^{1/2}\sigma(1)$.

The corresponding prediction of my model is the following: consider any scale factor such as the intersextile range, that is, the difference between the quantity U^+ which is exceeded by one-sixth of the data, and the quantity U^- which is larger than one-sixth of the data. It is easy to find that the expected range satisfies

$$E[U^+(T) - U^-(T)] = T^{1/\alpha}E[U^+(1) - U^-(1)]$$

We should also expect that the deviations from these expectations exceed those observed in the Gaussian case.

Differences between successive means of $Z(t)$. In all cases, the average of $Z(t)$ over the time span $t^0 + 1$ to $t^0 + N$ can then be written as

$$(1/N)[Z(t^0 + 1) + Z(t^0 + 2) + \ldots Z(t^0 + N)] = (1/N)\{NZ(t^0 + 1)$$
$$+ (N-1)[Z(t^0 + 2) - Z(t^0 + 1)] + \ldots$$
$$+ (N-n)[Z(t^0 + n + 1) - Z(t^0 + n)]$$
$$+ \ldots [Z(t^0 + N) - Z(t^0 + N - 1)]\}$$

On the contrary, let the average over the time span $t^0 - N - 1$ to t^0 be written as

$$(1/N)[NZ(t^0) - (N-1)[Z(t^0) - Z(t^0 - 1)] - \ldots - (N-n)[Z(t^0 - n + 1)$$
$$- Z(t^0 - n)] - \ldots [Z(t^0 - N + 2) - Z(t^0 - N + 1)]\}$$

Thus, if the expression $Z(t + 1) - Z(t)$ is a stable variable $U(t)$ with $\delta = 0$, the difference between successive means of values of Z is given by

$$U(t^0) + [(N - 1)/N][U(t^0 + 1) + U(t^0 - 1)] + [(N - n)/N][U(t^0 + n) + U(t^0 - n)] + \ldots [U(t^0 + N - 1) + U(t^0 - N + 1)]$$

This is clearly a stable variable, with the same α and β as the original U, and with a scale parameter equal to

$$\gamma^0(N) = [1 + 2(N - 1)^\alpha N^{-\alpha} + \ldots 2(N - n)^\alpha N^{-\alpha} + \ldots + 2]\gamma(U)$$

As $N \to \infty$, one has

$$\gamma^0(N)/\gamma(U) \to 2N/(\alpha + 1)$$

whereas a genuine monthly change of $Z(t)$ has a parameter $\gamma(N) = N\gamma(U)$; thus the effect of averaging is to multiply γ by the expression $2/(\alpha + 1)$, which is smaller than 1 if $\alpha > 1$.

Stable distributions and the law of Pareto

Except for the Gaussian limit case, the densities of the stable random variables follow a generalisation of the asymptotic behaviour of the Cauchy law. It is clear for example that, as $u \to \infty$, the Cauchy density behaves as follows:

$$uPr(U > u) = uPr(U < -u) \to 1/\pi$$

More generally, Lévy has shown that the tails of *all* non-Gaussian stable laws follow an asymptotic form of the law of Pareto, in the sense that there exist two constants $C' = \sigma'^\alpha$ and $C'' = \sigma''^\alpha$, linked by $\beta = (C' - C'')/(C' + C'')$, such that, when $u \to \infty$, $u^\alpha Pr(U > u) \to C' = \sigma'^\alpha$ and $u^\alpha Pr(U < -u) \to C'' = \sigma''^\alpha$.

Hence *both* tails are Paretion if $|\beta| \neq 1$, a solid reason for replacing the term "stable non-Gaussian" by the less negative one of "*stable Paretian*". The two numbers σ' and σ'' share the role of the standard deviation of a Gaussian variable and will be designated as the "standard positive deviation" and the "standard and negative deviation".

In the extreme cases where $\beta = 1$ and hence $C'' = 0$ (respectively, where $\beta = -1$ and $C' = 0$), the negative tail (respectively, the positive tail) decreases faster then the law of Pareto of index α. In fact, one can prove[6] that it withers away even faster than the Gaussian density so that the extreme cases of stable laws are practically J-shaped. They play an important role in my theory of the distributions of personal income or of city sizes. A number of further properties of stable laws may therefore be found in my publications devoted to these topics.[7]

Stable variables as the only possible limits of weighted sums of independent identically distributed addends

The stability of the Gaussian law may be considered as being only a matter of convenience, and it is often thought that the following property is more important.

Let the U_n be independent, identically distributed, random variables, with a finite $\sigma^2 = E[U_n - E(U)]^2$. Then the classical central limit theorem asserts that

$$\lim_{N \to \infty} N^{-1/2} \sigma^{-1} \sum_{n=1}^{N} [U_n - E(U)]$$

is a reduced Gaussian variable.

This result is of course the basis of the explanation of the presumed occurrence of the Gaussian law in many practical applications relative to sums of a variety of random effects. But the essential thing in all these aggregate arguments is not that $\Sigma[U_n - E(U)]$ is weighted by any special factor, such as $N^{-1/2}$, but rather that the following is true:

There exist two functions, $A(N)$ and $B(N)$, such that, as $N \to \infty$, the weighted sum

$$A(N) \sum_{n=1}^{N} U_n - B(N) \qquad (L)$$

has a limit that is finite and is not reduced to a non-random constant.

If the variance of U_n is not finite, however, condition (L) may remain satisfied while the limit ceases to be Gaussian. For example, if U_n is stable non-Gaussian, the linearly weighted sum

$$N^{-1/\alpha} \Sigma (U_n - \delta)$$

was seen to be *identical in law* to U_n, so that the "limit" of that expression is already attained for $N = 1$ and is a stable non-Gaussian law. Let us now suppose that U_n is asymptotically Paretian with $0 < \alpha < 2$, but not stable; one can show that the limit exists in a real sense, and that it is the stable Paretian law having the same value of α. Again the function $A(N)$ can be chosen equal to $N^{-1/\alpha}$. These results are crucial but I had better not attempt to rederive them here. There is little sense in copying the readily available full mathematical arguments, and experience shows that what was intended to be an illuminating heuristic explanation often looks like another instance in which far-reaching conclusions are based on loose thoughts. Let me therefore just quote the facts:

The problem of the existence of a limit for $A(N)\Sigma U_n - B(N)$ can be solved by introducing the following generalisation of the asymptotic law of Pareto:[8]

The conditions of Pareto–Doeblin–Gnedenko. Introduce the notations

$$Pr(U > u) = Q'(u)u^{-\alpha}$$

$$Pr(U < -u) = Q''(u)u^{-\alpha}$$

The conditions of P–D–G require that (a) *when* $u \to \infty$, $Q'(u)/Q''(u)$ *tends to a limit* C'/C'', (b) *there exists a value of* $\alpha > 0$ *such that for every* $k > 0$, *and for* $u \to \infty$, *one has*

$$\frac{Q'(u) + Q''(u)}{Q'(ku) + Q''(ku)} \to 1$$

These conditions generalise the law of Pareto, for which $Q'(u)$ and $Q''(u)$ themselves tend to limits as $u \to \infty$. With their help, and unless $\alpha = 1$, the problem of the existence of weighting factors $A(N)$ and $B(N)$ is solved by the following theorem:

If the U_n are independent, identically distributed random variables, there may exist no functions $A(N)$ and $B(N)$ such that $A(N)\Sigma U_n - B(N)$ tends to a proper limit. But, if such functions $A(N)$ and $B(N)$ exist, one knows that the limit is one of the solutions of the stability equation (S). *More precisely, the limit is Gaussian if and only if the U_n has finite variance; the limit is stable non-Gaussian if and only if the conditions of Pareto–Doeblin–Gnedenko are satisfied for some $0 < \alpha < 2$. Then $\beta = (C' - C'')/(C' + C'')$ and $A(N)$ is determined by the requirement that*

$$NPr[U > uA^{-1}(N)] \to C'u^{-\alpha}$$

(Whichever the value of α, the P–D–G condition (b) also plays a central role in the study of the distribution of the random variable max U_n.)

As an application of the above definition and theorem, let us examine the product of two independent, identically distributed Paretian (but not stable) variables U' and U''. First of all, for $u > 0$, one can write

$$Pr(U'U'' > u) = Pr(U' > 0; U'' > 0 \quad \text{and} \quad \log U' + \log U'' > \log u)$$
$$+ Pr(U' < 0; U'' < 0 \quad \text{and} \quad \log|U'| + \log|U''| > \log u)$$

But it follows from the law of Pareto that

$$Pr(U > e^z) \sim C' - \exp(-\alpha z) \quad \text{and} \quad Pr(U < -e^z) \sim C'' \exp(-\alpha z)$$

657

where U is either U' or U''. Hence, the two terms P' and P'' that add up to $Pr(U'U'' > u)$ satisfy

$$P' \sim C'^2 \alpha z \exp(-\alpha z) \quad \text{and} \quad P'' \sim C''^2 \alpha z \exp(-\alpha z)$$

Therefore

$$Pr(U'U'' > u) \sim \alpha(C'^2 + C''^2)(\log_e u) u^{-\alpha}$$

Similarly

$$Pr(U'U'' < -u) \sim \alpha 2 C'C''(\log_e u) u^{-\alpha}$$

It is obvious that the Pareto–Doeblin–Gnedenko conditions are satisfied for the functions $Q'(u) \sim (C'^2 + C''^2)\alpha \log_e u$ and $Q''(u) \sim 2C'C'' \alpha \log_e u$. Hence the weighted expression

$$(N \log N)^{-1/\alpha} \sum_{n=1}^{N} U'_n U''_n$$

converges toward a stable Paretian limit with the exponent α and the skewness

$$\beta = (C'^2 + C''^2 - 2C'C'')/(C'^2 + C''^2 + 2C'C'') = [(C' - C'')/(C' + C'')]^2 \geq 0$$

In particular, the positive tail should always be bigger than the negative.

Shape of stable Paretian distributions outside asymptotic range

The result of the third part of the second section should not hide the fact that the asymptotic behaviour is seldom the main thing in the applications. For example, if the sample size is N, the orders of magnitude of the largest and smallest item are given by

$$N Pr[U > u^+(N)] = 1$$

and

$$N Pr[U < -u^-(N)] = 1$$

and the interesting values of u lie between $-u^-$ and u^+. Unfortunately, except in the cases of Gauss and of Cauchy and the case ($\alpha = \frac{1}{2}$; $\beta = 1$), there are no known closed expressions for the stable densities and the theory only says the following: (a) the densities are always unimodal; (b) the densities depend continuously upon the parameters; (c) if $\beta > 0$, the positive tail is the fatter – hence, if the mean is finite (ie, if $1 < \alpha < 2$), it is greater than the most probable value and greater than the median.

To go further, I had to resort to numerical calculations. Let us, however, begin by interpolative arguments.

The symmetric cases, $\beta = 0$. For $\alpha = 1$, one has the Cauchy law, whose density $[\pi(1 + u^2)]^{-1}$ is always *smaller* than the Paretian density $(1/\pi)u^2$ toward which it tends in relative value as $u \to \infty$. Therefore,

$$Pr(U > u) < (1/\pi)u$$

and it follows that for $\alpha = 1$ the doubly logarithmic graph of $\log_e[Pr(U > u)]$ is entirely on the left side of its straight asymptote. By continuity, the same shape must apply when α is only a little higher or a little lower than 1.

For $\alpha = 2$, the doubly logarithmic graph of the Gaussian $\log_e[Pr(U > u)]$ drops down very fast to negligible values. Hence, again by continuity, the graph for $\alpha = 2 - \varepsilon$ must also begin by a rapid decrease. But, since its ultimate slope is close to 2, it must have a point of inflection corresponding to a maximum slope greater than 2, and it must begin by "overshooting" its straight asymptote.

Interpolating between 1 and 2, we see that there exists a smallest value of α, say α^0, for which the doubly logarithmic graph begins by overshooting its asymptote. In the neighbourhood of α^e, the asymptotic α can be measured as a slope even if the sample is small. If $\alpha < \alpha^0$, the asymptotic slope will be underestimated by the slope of small samples; for $\alpha > \alpha^0$ it will be overestimated. The numerical evaluation of the densities yields a value of α^0 in the neighbourhood of 1.5. A graphical presentation of the results of this section is given in Figure 3.

The skew cases. If the positive tail is fatter than the negative one, it may well happen that its doubly logarithmic graph begins by overshooting its asymptote, while the doubly logarithmic graph of the negative tail does not. Hence, there are two critical values of α^0, one for each tail; if the skewness is slight, α is between the critical values and the sample size is not large enough, the graphs of the two tails will have slightly different over-all apparent slopes.

Joint distribution of independent stable paretian variables

Let $p_1(u_1)$ and $p_2(u_2)$ be the densities of U_1 and of U_2. If both u_1 and u_2 are large, the joint probability density is given by

$$p^0(u_1, u_2) = \alpha C_1' u_1^{-(\alpha+1)} \alpha C_2' u_2^{-(\alpha+1)} = \alpha^2 C_1' C_2' (u_1 u_2)^{-(\alpha+1)}$$

Hence, the lines of equal probability are portions of the hyperbolas

$$u_1 u_2 = \text{constant}$$

Figure 3

The various lines are doubly logarithmic plots of the symmetric stable Paretian probability distributions with $\delta = 0$, $\gamma = 1$, $\beta = 0$ and various values of α. Horizontally, $\log_e u$; vertically, $\log_e Pr(U > u) = \log_e Pr(U < -u)$.

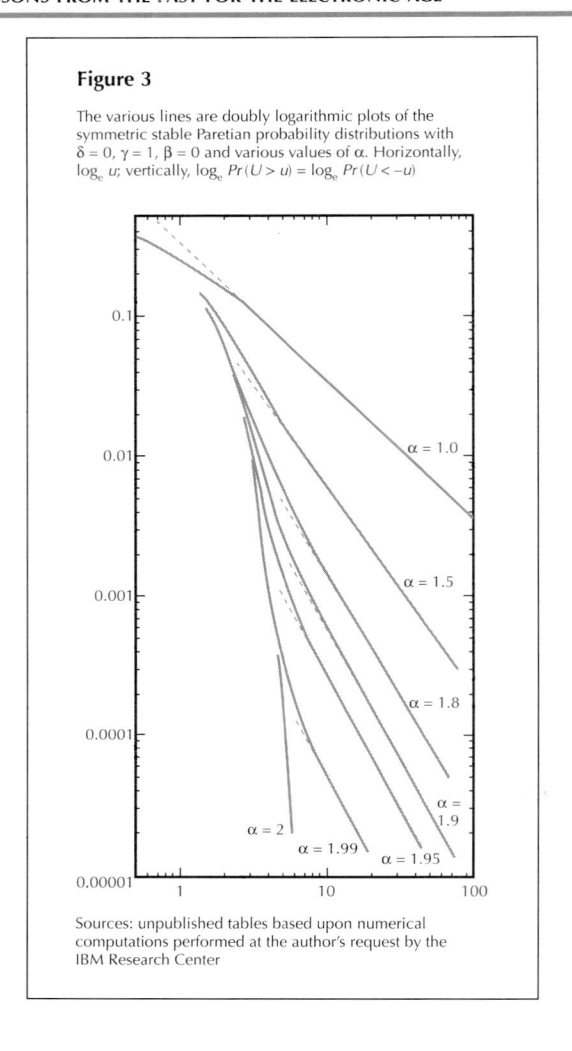

[Plot showing curves for $\alpha = 1.0$, $\alpha = 1.5$, $\alpha = 1.8$, $\alpha = 1.9$, $\alpha = 1.95$, $\alpha = 1.99$, $\alpha = 2$]

Sources: unpublished tables based upon numerical computations performed at the author's request by the IBM Research Center

In the regions where either U_1 or U_2 is large (but not both), these bits of hyperbolas are linked together as in Figure 4. That is, the isolines of small probability have a characteristic "plus-sign" shape. On the contrary, when both U_1 and U_2 are small, $\log_e p_1(u_1)$ and $\log_e p_2(u_2)$ are near their maxima and therefore can be locally approximated by $a_1 - (u_1/b_1)^2$ and $a_2 - (u_2/b_2)^2$. Hence, the probability isolines are ellipses of the form

$$(u_1/b_1)^2 + (u_2/b_2)^2 = \text{constant}$$

The transition between the ellipses and the "plus signs" is, of course, continuous.

Figure 4

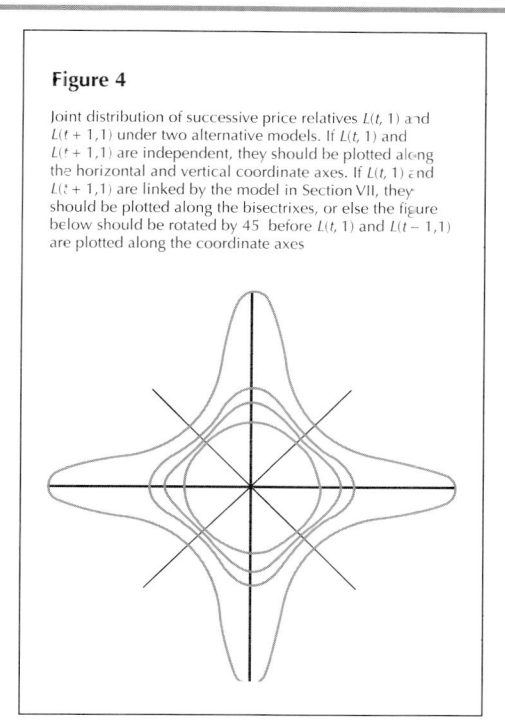

Joint distribution of successive price relatives $L(t, 1)$ and $L(t + 1,1)$ under two alternative models. If $L(t, 1)$ and $L(t + 1,1)$ are independent, they should be plotted along the horizontal and vertical coordinate axes. If $L(t, 1)$ and $L(t + 1,1)$ are linked by the model in Section VII, they should be plotted along the bisectrixes, or else the figure below should be rotated by 45 before $L(t, 1)$ and $L(t - 1,1)$ are plotted along the coordinate axes.

Distribution of U_1, when U_1 and U_2 are independent stable Paretian variables and $U_1 + U_2 = U$ is known

This conditional distribution can be obtained as the intersection between the surface that represents the joint density $p^0(u_1, u_2)$ and the plane $u_1 + u_2 = u$. Hence the conditional distribution is unimodal for small u. For large u, it has two sharply distinct maxima located near $u_1 = 0$ and near $u_2 = 0$.

More precisely, the conditional density of U_1 is given by $p_1(u_1)p_2(u - u_1)/q(u)$, where $q(u)$ is the density of $U = U_1 + U_2$. Let u be positive and very large; if u_1 is small, one can use the Paretian approximations for $p_2(u_2)$ and $q(u)$, obtaining

$$p_1(u_1)p_2(u - u_1)/q(u) \sim [C_1'/(C_1' + C_2')]p_1(u_1)$$

If u_2 is small, one similarly obtains

$$p_1(u_1)p_2(u - u_1)/q(u) \sim [C_2'/C_1' + C_2')]p_2(u - u_1)$$

In other words, the conditional density $p_1(u_1)p_2(u - u_1)/q(u)$ looks as if two unconditioned distributions scaled down in the ratios $C_1'/(C_1' + C_2')$

and $C_2'/(C_1' + C_2')$ had been placed near $u_1 = 0$ and $u_1 = u$. If u is negative but $|u|$ is very large, a similar result holds with C_1'' and C_2'' replacing C_1' and C_2'.

For example, for $\alpha = 2 - \varepsilon$ and $C_1' = C_2'$, the conditioned distribution is made up of two almost Gaussian bells, scaled down to one-half of their height. But, as α tends toward 2, these two bells become smaller and a third bell appears near $u_1 = u/2$. Ultimately, the two side bells vanish and one is left with a central bell which corresponds to the fact that when the sum $U_1 + U_2$ is known, the conditional distribution of a Gaussian U_1 is itself Gaussian.

EMPIRICAL TESTS OF THE STABLE PARETIAN LAWS: COTTON PRICES

This section will have two main goals. First, from the viewpoint of statistical economics, its purpose is to motivate and develop a model of the variation of speculative prices based on the stable Paretian laws discussed in the previous section. Second, from the viewpoint of statistics considered as the theory of data analysis, I shall use the theorems concerning the sums ΣU_n to build a new test of the law of Pareto. Before moving on to the main points of the section, however, let us examine two alternative ways of treating the excessive numbers of large price changes usually observed in the data.

Explanation of large price changes by causal or random "contaminators"
One very common approach is to note that, *a posteriori*, large price changes are usually traceable to well-determined "causes" that should be eliminated before one attempts a stochastic model of the remainder. Such preliminary censorship obviously brings any distribution closer to the Gaussian. This is, for example, what happens when one restricts himself to the study of "quiet periods" of price change. There need not be any observable discontinuity between the "outliers" and the rest of the distribution, however, and the above censorship is therefore usually indeterminate.

Another popular and classical procedure assumes that observations are generated by a mixture of two normal distributions, one of which has a small weight but a large variance and is considered as a random "contaminator". In order to explain the sample behaviour of the moments, it unfortunately becomes necessary to introduce a larger number of contaminators, and the simplicity of the model is destroyed.

Introduction of the law of Pareto to represent price changes
I propose to explain the erratic behaviour of sample moments by assuming that the population moments are infinite, an approach that I have used with success in a number of other applications and which I have explained and demonstrated in detail elsewhere.

This hypothesis amounts practically to the law of Pareto. Let us indeed assume that the increment

$$L(t, 1) = \log_e Z(t + 1) - \log_e Z(t)$$

is a random variable with infinite population moments beyond the first. This implies that its density $p(u)$ is such that $\int p(u) u^2 du$ diverges but $\int p(u) u du$ converges (the integrals being taken all the way to infinity). It is of course natural, at least in the first stage of heuristic motivating argument, to assume that $p(u)$ is somehow "well behaved" for large u; if so, our two requirements mean that as $u \to \infty$, $p(u) u^3$ tends to infinity and $p(u) u^2$ tends to zero.

In words: $p(u)$ must somehow decrease faster than u^{-2} and slower than u^{-3}. From the analytical viewpoint, the simplest expressions of this type are those with an asymptotically Paretian behaviour. *This was the first motivation of the present study.* It is surprising that I could find no record of earlier application of the law of Pareto to two-tailed phenomena.

My further motivation was more theoretical. Granted that the facts impose a revision of Bachelier's process, it would be simple indeed if one could at least preserve the convenient feature of the Gaussian model that the various increments,

$$L(t, T) = \log_e Z(t + T) - \log_e Z(t)$$

depend upon T only to the extent of having different scale parameters. From all other viewpoints, price increments over days, weeks, months, and years would have the same distribution, which would also rule the fixed-base relatives. This naturally leads directly to the probabilists' concept of stability examined in the previous section.

In other terms, the facts concerning moments, together with a desire to have a simple representation, suggested a check as to whether the logarithmic price relatives for unsmoothed and unprocessed time series relative to very active speculative markets are stable Paretian. Cotton provided a good example, and the present paper will be limited to the examination of that case. I have, however, also established that my theory applies to many other commodities (such as wheat and other edible grains), to many securities (such as those of the railroads in their 19th-century heyday), and to interest rates such as those of call or time money.[9] On the other hand, there are unquestionably many economic phenomena for which much fewer "outliers" are observed, even though the available series are very long; it is natural in these cases to favour Bachelier's Gaussian model – known to be a limiting case in my theory as well as its prototype. I must, however, postpone a discussion of the limits of validity of my approach to the study of prices.

Pareto's graphical method applied to cotton-price changes

Let us begin by examining in Figure 5 the doubly logarithmic graphs of various kinds of cotton price changes as if they were independent of each other. The theoretical log $Pr(U > u)$, relative to $\delta = 0$, $\alpha = 1.7$, and $\beta = 0$, is plotted (*solid curve*) on the same graph for comparison. If the various cotton prices followed the stable Paretian law with $\delta = 0$, $\alpha = 1.7$ and $\beta = 0$, the various graphs should be horizontal translates of each other, and a cursory examination shows that the data are in close conformity with the predictions of my model. A closer examination suggests that the positive tails contain systematically fewer data than the negative tails, suggesting that β actually takes a small negative value. This is also confirmed by the fact that the negative tails alone begin by slightly "overshooting" their asymptotes, creating the bulge that should be

Figure 5

Composite of doubly logarithmic graphs of positive and negative tails for three kinds of cotton price relatives, together with cumulated density function of a stable distribution. Horizontal scale *u* of lines 1a, 1b and 1c is marked only on lower edge, and horizontal scale *u* of lines 2a, 2b and 2c is marked along uppr edge. Verticle scale gives the following relative frequencies:

(1a) $Fr[\log_e Z(t + \text{one day}) - \log_e Z(t) > u]$, (2a) $Fr[\log_e Z(t + \text{one day}) - \log_e Z(t) < -u]$, both for the daily closing prices of cotton in New York, 1900–5 (source: private communication from the United States Department of Agriculture).

(1b) $Fr[\log_e Z(t + \text{one day}) - \log_e Z(t) > u]$, (2b) $Fr[\log_e Z(t + \text{one day}) - \log_e Z(t) < -u]$, both for an index of daily closing prices of cotton in the United States, 1944–58 (source: private communication from Hendrik S. Houthakker).

(1c) $Fr[\log_e Z(t + \text{one month}) - \log_e Z(t) > u]$, (2c) $Fr[\log_e Z(t + \text{one month}) - \log_e Z(t) < -u]$, both for the closing prices of cotton on the 15th of each month in New York, 1880–1940 (source: private communication from the United States Department of Agriculture).

The reader is advised to copy on a transparency the horizontal axis and the theoretical distribution and to move both horizontally until the theoretical curve is superimposed on either of the empirical graphs; the only discrepancy is observed for line 2b; it is slight and would imply an even greater departure from normality.

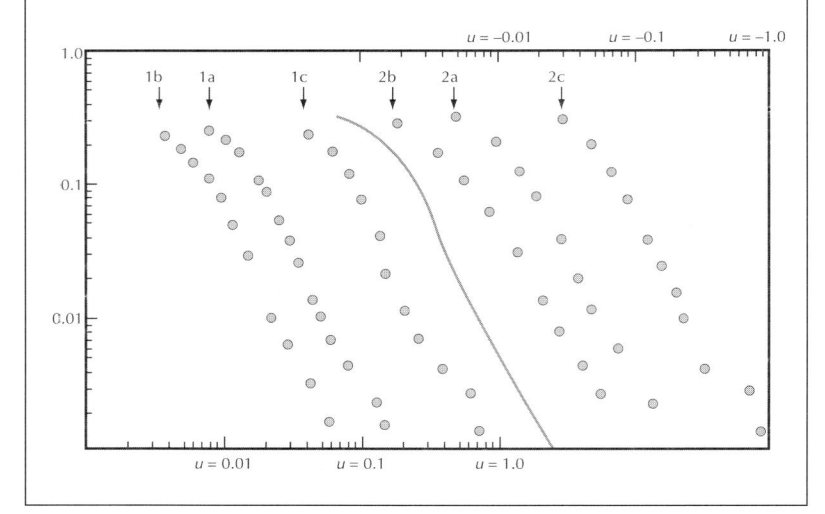

expected α when is greater than the critical α^0 value relative to one tail but not to the other.

Application of the graphical method to the study of changes in the distribution across time

Let us now look more closely at the labels of the various series examined in the previous section. Two of the graphs refer to daily changes of cotton prices, near 1900 and near 1950, respectively. It is clear that these graphs do not coincide but are horizontal translates of each other. This implies that between 1900 and 1950 the generating process has changed only to the extent that the scale γ has become much smaller.

Our next test will be relative to monthly price changes over a longer time span. It would be best to examine the actual changes between, say, the middle of one month to the middle of the next. A longer sample is available, however, when one takes the reported monthly averages of the price of cotton; the graphs of Figure 6 were obtained in this way.

If cotton prices were indeed generated by a stationary stochastic process, our graphs should be straight, parallel, and uniformly spaced. However, each of the 15-year subsamples contains only 200-odd months, so that the separate graphs cannot be expected to be as straight as those relative to our usual samples of 1,000-odd items. The graphs of Figure 6 are, indeed, not quite as neat as those relating to longer periods; but, in the absence of accurate statistical tests, they seem adequately straight and uniformly spaced, except for the period 1880–1896.

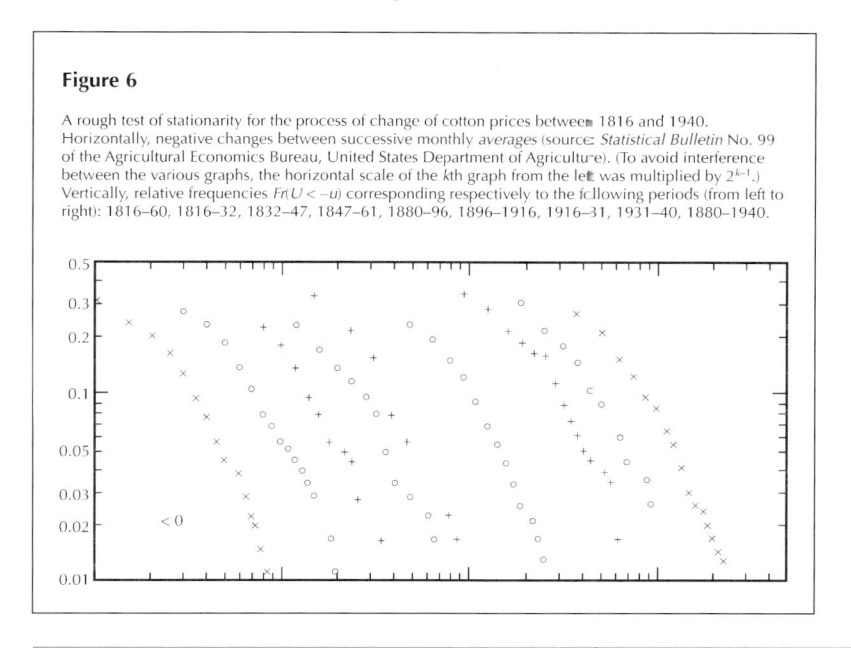

Figure 6

A rough test of stationarity for the process of change of cotton prices between 1816 and 1940. Horizontally, negative changes between successive monthly *averages* (source: *Statistical Bulletin* No. 99 of the Agricultural Economics Bureau, United States Department of Agriculture). (To avoid interference between the various graphs, the horizontal scale of the *k*th graph from the left was multiplied by 2^{k-1}.) Vertically, relative frequencies $Fr(U < -u)$ corresponding respectively to the following periods (from left to right): 1816–60, 1816–32, 1832–47, 1847–61, 1880–96, 1896–1916, 1916–31, 1931–40, 1880–1940.

I conjecture therefore that, since 1816, the process generating cotton prices has changed only in its scale, with the possible exception of the Civil War and of the periods of controlled or supported prices. Long series of monthly price changes should therefore be represented by *mixtures* of stable Paretian laws; such mixtures remain Paretian.[10]

Application of the graphical method to study effects of averaging

It is, of course, possible to derive mathematically the expected distribution of the changes between successive monthly means of the highest and lowest quotation; but the result is so cumbersome as to be useless. I have, however, ascertained that the empirical distribution of these changes does not differ significantly from the distribution of the changes between the monthly means obtained by averaging all the daily closing quotations within months; one may therefore speak of a single average price for each month.

We then see on Figure 7 that the greater part of the distribution of the averages differs from that of actual monthly changes by a horizontal translation to the left, as predicted in the third part of the second section (actually, in order to apply the argument of that section, it would be necessary to rephrase it by replacing $Z(t)$ by $\log_e Z(t)$ throughout; however, the geometric and arithmetic averages of daily $Z(t)$ do not differ much in the case of medium-sized over-all monthly changes of $Z(t)$).

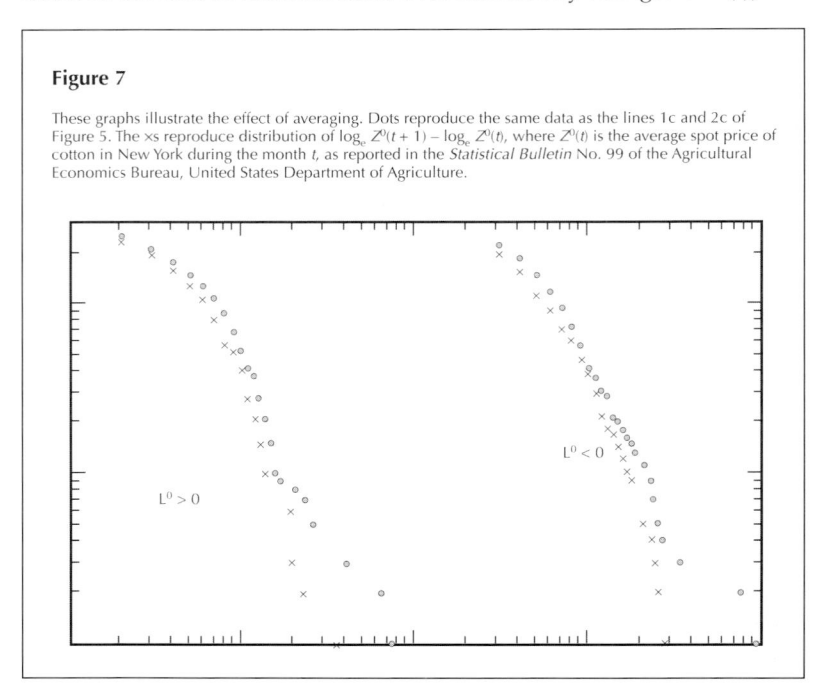

Figure 7

These graphs illustrate the effect of averaging. Dots reproduce the same data as the lines 1c and 2c of Figure 5. The xs reproduce distribution of $\log_e Z^0(t+1) - \log_e Z^0(t)$, where $Z^0(t)$ is the average spot price of cotton in New York during the month t, as reported in the *Statistical Bulletin* No. 99 of the Agricultural Economics Bureau, United States Department of Agriculture.

However, the largest changes between successive averages are smaller than predicted. This seems to suggest that the dependence between successive daily changes has less effect upon actual monthly changes than upon the regularity with which these changes are performed.

A new presentation of the evidence

Let me now show that my evidence concerning daily changes of cotton price strengthens my evidence concerning monthly changes and conversely.

The basic assumption of my argument is that successive daily changes of log (price) are independent. (This argument will thus have to be revised when the assumption is improved upon.) Moreover, the population second moment of $L(t)$ seems to be infinite and the monthly or yearly price changes are patently not Gaussian. Hence the problem of whether any limit theorem whatsoever applies to $\log_e Z(t + T) - \log_e Z(t)$ can also be answered *in theory* by examining whether the daily changes satisfy the Pareto–Doeblin–Gnedenko conditions. *In practice*, however, it is impossible to ever attain an infinitely large differencing interval T or to ever verify any condition relative to an infinitely large value of the random variable u. Hence one must consider that a month or a year is infinitely long, and that the largest observed daily changes of $\log_e Z(t)$ are infinitely large. Under these circumstances, one can make the following inferences.

Inference from aggregation. The cotton price data concerning daily changes of $\log_e Z(t)$ surely appear to follow the weaker condition of Pareto–Doeblin–Gnedenko. Hence, from the property of stability and according to the second section, one should expect to find that, as T increases,

$$T^{-1/\alpha}\{\log_e Z(t + T) - \log_e Z(t) - TE[L(t, 1)]\}$$

tends toward a stable Paretian variable with zero mean.

Inference from disaggregation. Data seem to indicate that price changes over weeks and months follow the same law up to a change of scale. This law must therefore be one of the possible non-Gaussian limits, that is, it must be a stable Paretian. As a result, the inverse part of the theorem of the fourth part of the second section shows that the daily changes of log $Z(t)$ must satisfy the conditions of Pareto–Doeblin–Gnedenko.

It is pleasant to see that the inverse condition of P–D–G, which greatly embarrassed me in my work on the distribution of income, can be put to use in the theory of prices.

A few of the difficulties involved in making the above two inferences will now be discussed.

Disaggregation. The P–D–G conditions are weaker than the asymptotic law of Pareto because they require that limits exist for $Q'(u)/Q''(u)$ and for

$[Q'(u) + Q''(u)]/[Q'(ku) + Q''(ku)]$, but not for $Q'(u)$ and $Q''(u)$ taken separately. Suppose, however, that $Q'(u)$ and $Q''(u)$ still vary a great deal in the useful range of large daily variations of prices. If so, $A(N)\Sigma U_n - B(N)$ will not approach its own limit until *extremely* large values of N are reached. Therefore, if one believes that the limit is rapidly attained, the functions $Q'(u)$ and $Q''(u)$ of daily changes must vary very little in the regions of the tails of the usual samples. In other words, it is necessary after all that the asymptotic law of Pareto apply to daily price changes.

Aggregation. Here, the difficulties are of a different order. From the mathematical viewpoint, the stable Paretian law should become increasingly accurate as T increases. Practically, however, there is no sense in even considering values of T as long as a century, because one cannot hope to get samples sufficiently long to have adequately inhabited tails. The year is an acceptable span for certain grains, but only if one is not worried by the fact that the long available series of yearly prices are ill known and variable averages of small numbers of quotations, not prices actually quoted on some market on a fixed day of each year.

From the viewpoint of economics, there are two much more fundamental difficulties with very large T. First of all, the model of independent daily L's eliminates from consideration every "trend", except perhaps the exponential growth or decay due to a non-vanishing δ. Many trends that are negligible on the daily basis would, however, be expected to be predominant on the monthly or yearly basis. For example, weather might have upon yearly changes of agricultural prices an effect different from the simple addition of speculative daily price movements.

The second difficulty lies in the "linear" character of the aggregation of successive L's used in my model. Since I use natural logarithms, a small $\log_e Z(t + T) - \log_e Z(t)$ will be indistinguishable from the relative price change $[Z(t + T) - Z(t)]/Z(t)$. The addition of small L's is therefore related to the so-called "principle of random proportionate effect"; it also means that the stochastic mechanism of prices readjusts itself immediately to any level that $Z(t)$ may have attained. This assumption is quite usual, but very strong. In particular, I shall show that, if one finds that $\log Z(t +$ one week$) - \log Z(t)$ is very large, it is very likely that it differs little from the change relative to the single day of most rapid price variation (see the fifth part of the fifth section); naturally, this conclusion only holds for independent L's. As a result, the greatest of N successive daily price changes will be so large that one may question both the use of $\log_e Z(t)$ and the independence of the L's.

There are other reasons (see the second part of the fourth section) to expect to find that a simple addition of speculative daily price changes predicts values too high for the price changes over periods such as whole months.

Given all these potential difficulties, I was frankly astonished by the quality of the predictions of my model concerning the distribution of the changes of cotton prices between the fifteenth of one month and the fifteenth of the next. The negative tail has the expected bulge, and even the most extreme changes are precise extrapolates from the rest of the curve. Even the artificial exision of the Great Depression and similar periods would not affect the general results very greatly.

It was therefore interesting to check whether the ratios between the scale coefficients, $C'(T)/C'(1)$ and $C''(T)/C''(1)$, were both equal to T, as predicted by my theory whenever the ratios of standard deviations $\sigma'(T)/\sigma'(s)$ and $\sigma''(T)/\sigma''(s)$ follow the $T^{1/\alpha}$ generalisation of the "$T^{1/2}$ Law" referred to in the second part of the second section. If the ratios of the C parameter are different from T, their value may serve as a measure of the degree of dependence between successive $L(t, 1)$.

The above ratios were absurdly large in my original comparison between the daily changes near 1950 of the cotton prices collected by Houthakker and the monthly changes between 1880 and 1940 of the prices communicated by the USDA. This suggested that the supported prices around 1950 varied less than their earlier counterparts. Therefore I repeated the plot of daily changes for the period near 1900, chosen haphazardly but not actually at random. The new values of $C'(T)/C'(1)$ and $C''(T)/C'(1)$ became quite reasonable, equal to each other and to 18. In 1900, there were seven trading days per week, but they subsequently decreased to five. Besides, one cannot be too dogmatic about estimating $C'(T)/C'(1)$. Therefore the behaviour of this ratio indicated that the "apparent" number of trading days per month was somewhat smaller than the actual number.

WHY ONE SHOULD EXPECT TO FIND NONSENSE MOMENTS AND NONSENSE PERIODICITIES IN ECONOMIC TIME SERIES

Behaviour of second moments and failure of the least-squares method of forecasting

It is amusing to note that the first known non-Gaussian stable law, namely, Cauchy's distribution, was introduced in the course of a study of the method of least squares. In a surprisingly lively argument following Cauchy's 1853 paper, J. Bienaymé[11] stressed that a method based upon the minimisation of the sum of squares of sample deviations cannot be reasonably used if the expected value of this sum is known to be infinite. The same argument applies fully to the problem of least-squares smoothing of economic time series, when the "noise" follows a stable Paretian law other than that of Cauchy.

Similarly, consider the problem of least-squares forecasting, that is, of the minimisation of the expected value of the square of the error of extrapolation. In the stable Paretian case this expected value will be finite for every forecast, so that the method is, at best, extremely questionable. One can perhaps apply a method of "least ζ-power" of the forecasting error, where $\zeta < \alpha$, but such an approach would not have the formal simplicity of least squares manipulations; the most hopeful case is that of $\zeta = 1$, which corresponds to the minimisation of the sum of absolute values of the errors of forecasting.

Behaviour of the kurtosis and its failure as a measure of "peakedness"
Pearson's index of "kurtosis" is defined as

$$-3 + \frac{\text{fourth moment}}{\text{square of the second moment}}$$

If $0 < \alpha < 2$, the numerator and the denominator both have an infinite expected value. One can, however, show that the kurtosis behaves proportionately to its "typical" value given by

$$\frac{(1/N)(\text{most probable value of } \Sigma L^4)}{[(1/N)(\text{most probable value of } \Sigma L^2)]^2} = \frac{\text{const. } N^{-1+4/\alpha}}{[\text{const. } N^{-1+2/\alpha}]^2} = \text{const. } N$$

In other words, the kurtosis is expected to increase without bound as $N \to \infty$. For small N, things are less simple but presumably quite similar.

Let me examine the work of Cootner in this light.[12] He has developed the tempting hypothesis that prices vary at random only as long as they do not reach either an upper or a lower bound, that are considered by well-informed speculators to delimit an interval of reasonable values of the price. If and when ill-informed speculators let the price go too high or too low, the operations of the well-informed speculators will induce this price to come back within a "penumbra" *à la* Taussig. Under the circumstances, the price changes over periods of, say, 14 weeks should be smaller than would be expected if the contributing weekly changes were independent.

This theory is very attractive *a priori* but could not be generally true because, in the case of cotton, it is not supported by the facts. As for Cootner's own justification, it is based upon the observation that the price changes of certain securities over periods of 14 weeks have a much smaller kurtosis than one-week changes. Unfortunately, his sample contains 250-odd weekly changes and only 18 14-week periods. Hence, on the basis of general evidence concerning speculative prices, I would have expected *a priori* to find a small kurtosis for the longer time increment, and Cootner's evidence is not a proof of his theory; other methods must be used in order to attack the still very open problem of the possible dependence between successive price changes.

Method of spectral analysis of random time series

Applied mathematicians are frequently presented these days with the task of describing the stochastic mechanism capable of generating a given time series $u(t)$, known or presumed to be random. The response to such questions is usually to investigate first what is obtained by applying the theory of the "second-order random processes". That is, assuming that $E(U) = 0$, one forms the sample covariance

$$r(\tau) = \left(\frac{1}{N-\tau}\right) \sum_{t=T^0+1}^{t=T^0+N-\tau} u(t)u(t+\tau)$$

which is used, somewhat indirectly, to evaluate the population covariance

$$R(\tau) = E[U(t)U(t+\tau)]$$

Of course, $R(\tau)$ is always assumed to be finite for all; its Fourier transform gives the "spectral density" of the "harmonic decomposition" of $U(t)$ into a sum of sine and cosine terms.

Broadly speaking, this method has been very successful, though many small-sample problems remain unsolved. Its applications to economics have, however, been questionable even in the large-sample case. Within the context of my theory, there is unfortunately nothing surprising in such a finding. The expression $2E[U(t)U(t-\tau)]$ equals indeed $E[U(t) + U(t+\tau)]^2 - E[U(t)]^2 - E[U(t+\tau)]^2$; these three variances are all infinite for time series covered by my model, so that spectral analysis loses its theoretical motivation. I must, however, postpone a more detailed examination of this fascinating problem.

SAMPLE FUNCTIONS GENERATED BY STABLE PARETIAN PROCESSES; SMALL-SAMPLE ESTIMATION OF THE MEAN "DRIFT" OF SUCH A PROCESS

The curves generated by stable Paretian processes present an even larger number of interesting formations than the curves generated by Bachelier's Brownian motion. If the price increase over a long period of time happens *a posteriori* to have been usually large, in a stable Paretian market, one should expect to find that this change was mostly performed during a few periods of especially high activity. That is, one will find in most cases that the majority of the contributing daily changes are distributed on a fairly symmetric curve, while a few especially high values fall well outside this curve. If the total increase is of the usual size, the only difference will be that the daily changes will show no "outliers".

In this section these results will be used to solve one small-sample statistical problem, that of the estimation of the mean drift δ, when the other parameters are known. We shall see that there is no "sufficient statistic" for this problem, and that the maximum likelihood equation does not necessarily have a single root. This has severe consequences from the viewpoint of the very definition of the concept of "trend".

Certain properties of sample paths of Brownian motion

As noted by Bachelier and (independently of him and of each other) by several modern writers,[13] the sample paths of the Brownian motion very much "look like" the empirical curves of time variation of prices or of price indexes. At closer inspection, however, one sees very well the effect of the abnormal number of large positive and negative changes of $\log_e Z(t)$. At still closer inspection, one finds that the differences concern some of the economically most interesting features of the generalised central-limit theorem of the calculus of probability. It is therefore necessary to discuss this question in detail, beginning with a reminder of some classic facts concerning Gaussian random variables.

Conditional distribution of a Gaussian $L(t)$, *knowing* $L(t, T) = L(t, 1) + \ldots + L(t + T - 1, 1)$. Let the probability density of $L(t, T)$ be

$$(2\pi\sigma^2 T)^{-1/2} \exp[-(u - \delta T)^2 / 2T\sigma^2]$$

It is then easy to see that – if one knows the value u of $L(t, T)$ – the density of any of the quantities $L(t + \tau, 1)$ is given by

$$[2\pi\sigma^2(T-1)/T]^{-1/2} \exp\left[-\frac{(u' - u/T)^2}{2\sigma^2(T-1)/T}\right]$$

We see that each of the contributing $L(t + \tau, 1)$ equals u/T plus a Gaussian error term. For large T, that term has the same variance as the unconditioned $L(t, 1)$; one can in fact prove that the value of u has little influence upon the size of the largest of those "noise terms". One can therefore say that, whichever its value, u is roughly uniformly distributed over the T time intervals, each contributing negligibly to the whole.

Sufficiency of u *for the estimation of the mean drift* δ *from the* $L(t + \tau, 1)$. In particular, δ has vanished from the distribution of any $L(t + \tau, 1)$ conditioned by the value of u. This fact is expressed in mathematical statistics by saying that u is a "sufficient statistic" for the estimation of δ from the values of all the $L(t + \tau, 1)$. That is, whichever method of estimation a statistician may favour, his estimate of δ must be a function of u alone. The knowledge of intermediate values of $\log_e Z(t + \tau)$ is of no help to him. Most methods recommend estimating δ by u/T and extrapolating the future linearly from the two known points, $\log_e Z(t)$ and $\log_e Z(t + T)$.

Since the causes of any price movement can be traced back only if it is ample enough, the only thing that can be explained in the Gaussian case is the mean drift interpreted as a trend, and Bachelier's model, which assumes a zero mean for the price changes, can only represent the movement of prices once the broad causal parts or trends have been removed.

Sample from a process of independent stable Paretian increments

Returning to the stable Paretian case, suppose that one knows the values of γ and β (or of C' and C'') and of α. The remaining parameter is the mean drift δ, which one must estimate starting from the known $L(t, T) = \log_e Z(t + T) - \log_e Z(t)$.

The unbiased estimate of δ is $L(t, T)/T$, while the maximum likelihood estimate matches the observed $L(t, T)$ to its *a priori, most probable* value. The "bias" of the maximum likelihood is therefore given by an expression of the form $\gamma^{1/\alpha} f(\beta)$, where the function $f(\beta)$ must be determined from the numerical tables of the stable Paretian densities. Since β is mostly manifested in the relative sizes of the tails, its evaluation requires very large samples, and the quality of one's predictions will depend greatly upon the quality of one's knowledge of the past.

It is, of course, not at all clear that anybody would wish the extrapolation to be unbiased with respect to the mean of the change of the *logarithm* of the price. Moreover, the bias of the maximum likelihood estimate comes principally from an underestimate of the size of changes that are so large as to be catastrophic. The forecaster may therefore very well wish to treat such changes separately and to take account of his private feelings about many things that are not included in the independent-increment model.

Two samples from a stable Paretian process

Suppose now that T is even and that one knows $L(t, T/2)$ and $L(t + T/2, T/2)$ and their sum $L(t, T)$. We have seen in the seventh part of the second section that, when the value $u = L(t, T)$ is given, the conditional distribution of $L(t, T/2)$ depends very sharply upon u. This means that the total change u is not a sufficient statistic for the estimation of δ; in other words, the estimates of δ will be changed by the knowledge of $L(t, T/2)$ and $L(t + T/2, T/2)$.

Consider for example the most likely value δ. If $L(t, T/2)$ and $L(t + T/2, T/2)$ are of the same order of magnitude, this estimate will remain close to $L(t, T)/T$, as in the Gaussian case, But suppose that *the actually observed* values of $L(t, T/2)$ and $L(t + T/2, T/2)$ are very unequal, thus implying that at least one of these quantities is very different from their common mean and median. Such an event is most likely to occur when δ is close to the observed value either of $L(t + T/2, T/2)/(T/2)$ or of $L(t, T/2)/(T/2)$.

We see that as a result, the maximum likelihood equation for δ has two roots, respectively close to $2L(t, T/2)/T$ and to $2L(t + T/2, T/2)/T$. That is, the maximum-likelihood procedure says that one should neglect one of the available items of information, any weighted mean of the two recommended extrapolations being worse than either; but nothing says which item one should neglect.

It is clear that few economists will accept such advice. Some will stress that the most likely value of δ is actually nothing but the most probable value in the case of a uniform distribution of *a priori* probabilities of δ. But it seldom happens that *a priori* probabilities are uniformly distributed. It is also true, of course, that they are usually very poorly determined; in the present problem, however, the economist will not need to determine these *a priori* probabilities with any precision: it will be sufficient to choose the most likely *for him* of the two maximum-likelihood estimates.

An alternative approach to be presented later in this paper will argue that successive increments of $\log_e Z(t)$ are not really independent, so that the estimation of δ depends upon the order of the values of $L(t, T/2)$ and $L(t + T/2, T/2)$ as well as upon their sizes. This may help eliminate the indeterminacy of estimation.

A third alternative consists in abandoning the hypothesis that δ is the same for both changes $L(t, T/2)$ and $L(t + T/2, T/2)$. For example, if these changes are very unequal, one may be tempted to believe that the trend δ is not linear but parabolic. Extrapolation would then approximately amount to choosing among the two maximum-likelihood estimates the one which is chronologically the latest. This is an example of a variety of configurations which would have been so unlikely in the Gaussian case that they should be considered as non-random and would be of help in extrapolation. In the stable Paretian case, however, their probability may be substantial.

Three samples from a stable Paretian process

The number of possibilities increases rapidly with the sample size. Assume now that T is a multiple of 3, and consider $L(t, T/3)$, $L(t + T/3, T/3)$, and $L(t + 2T/3, T/3)$. If these three quantities are of comparable size, the knowledge of $\log Z(t + T/3)$ and $\log Z(t + 2T/3)$ will again bring little change to the estimate based upon $L(t, T)$.

But suppose that one datum is very large and the other are of much smaller and comparable sizes. Then, the likelihood equation will have two local maximums, having very different positions and sufficiently equal sizes to make it impossible to dismiss the smaller one. The absolute maximum yields the estimate $\delta = (3/2T)$ (sum of the two small data); the smaller local maximum yields the estimate $\delta = (3/T)$ (the large datum).

Suppose finally that the three data are of very unequal sizes. Then the maximum likelihood equation has *three* roots.

This indeterminacy of maximum likelihood can again be lifted by one of the three methods of the third part of the fifth section. For example, if the middle datum only is large, the method of non-linear extrapolation will suggest a logistic growth. If the data increase or decrease – when taken chronologically – one will rather try a parabolic trend. Again the probability of these configurations arising from chance under my model will be much greater than in the Gaussian case.

A large number of samples from a stable Paretian process

Let us now jump to a very large number of data. In order to investigate the predictions of my stable Paretian model, we must first re-examine the meaning to be attached to the statement that, in order that a sum of random variables follow a central limit of probability, it is necessary that each of the addends be negligible relative to the sum.

It is quite true, of course, that one can speak of limit laws only if the value of the sum is not *dominated* by any single addend known in advance. That is, to study the limit of $A(N)\Sigma U_n - B(N)$, one must assume that (for every n) $Pr|A(N)U_n - B(N)/N|>\varepsilon)$ tends to zero with $1/N$.

As each addend decreases with $1/N$, their number increases, however, and the condition of the preceding paragraph does not by itself insure that the largest of the $|A(N)U_n - B(N)/N|$ is negligible in comparison with their sum. As a matter of fact, the last condition is true only if the limit of the sum is Gaussian. In the Paretian case, on the contrary, the following ratios,

$$\frac{\max |A(N)U_n - B(N)/N|}{A(N)\Sigma U_n - B(N)}$$

and

$$\frac{\text{sum of } k \text{ largest } |A(N)U_n - B(N)/N|}{A(N)\Sigma U_n - B(BN)}$$

tend to non-vanishing limits as N increases.[14] If one knows moreover that the sum $A(N)\Sigma U_n - B(N)$ happens to be large, one can prove that the above ratios should be expected to be close to one.

Returning to a process with independent stable Paretian $L(t)$, we may say the following: if, knowing α, β, γ, and δ, one observes that $L(t, T = \text{one month})$ is *not* large, the contribution of the day of largest price change is likely to be non-negligible in relative value, but it will remain small in absolute value. For large but finite N, this will not differ too much from the Gaussian prediction that even the largest addend is negligible.

Suppose however that $L(t, T = \text{one month})$ is *very* large. The Paretian theory then predicts that the sum of a few largest daily changes will be very close to the total $L(t, T)$; if one plots the frequencies of various values of $L(t, 1)$, conditioned by a known and very large for $L(t, T)$, one should expect to find that the law of $L(t + \tau, 1)$ contains a few widely "outlying" values. However, if the outlying values are taken out, the conditioned distribution of $L(t + \tau, 1)$ should depend little upon the value of the conditioning $L(t, T)$. I believe this last prediction to be very well satisfied by prices.

Implications concerning estimation. Suppose now that δ is unknown and that one has a large sample of $L(t + \tau, 1)$'s. The estimation procedure consists in that case of plotting the empirical histogram and translating it horizontally until one has optimised its fit to the theoretical density curve. One knows in advance that this best value will be very little influenced by the largest outliers. Hence "rejection of the outliers" is fully justified in the present case, at least in its basic idea.

Conclusions concerning estimation

The observations made in the preceding sections seem to confirm some economists' feeling that prediction is feasible only if the sample size is both very large and stationary, or if the sample size is small but the sample values are of comparable sizes. One can also predict when the sample size is one, but here the unicity of the estimator is only due to ignorance.

Causality and randomness in stable Paretian processes

We mentioned in the first part of the fifth section that, in order to be "causally explainable", an economic change must at least be large enough to allow the economist to trace back the sequence of its causes. As a result, the only causal part of a Gaussian random function is the mean drift δ. This will also apply to stable Paretian random functions when their changes happen to be roughly uniformly distributed.

Things are different when $\log_e Z(t)$ varies greatly between the times t and $t + T$, changing mostly during a few of the contributing days. Then, these largest changes are sufficiently clear-cut, and are sufficiently separated from "noise", to be traced back and explained causally, just as well as the mean drift.

In other words, a careful observer of a stable Paretian random function will be able to extract causal parts from it. But, if the total change of $\log_e Z(t)$ is neither very large nor very small, there will be a large degree of arbitrariness in this distinction between causal and random. Hence one could not tell whether the predicted proportions of the two kinds of effects are empirically correct.

To sum up, the distinction between the causal and the random areas is sharp in the Gaussian case and very diffuse in the stable Paretian case. This seems to me to be a strong recommendation in favour of the stable Paretian process as a model of speculative markets. Of course, I have not the slightest idea why the large price movements should be represented in this way by a simple extrapolation of movements of ordinary size. I came to believe, however, that it is very desirable that both "trend" and "noise" be aspects of the same deeper "truth", which may not be explainable today, but which can be adequately described. I am surely not antagonistic to the ideal of economics: eventually to decompose even the "noise" into parts similar to the trend and to link various series to each other. But, until we can approximate this ideal, we can at least represent some trends as being similar to "noise".

Causality and randomness in aggregation "in parallel"

Borrowing a term from elementary electrical circuit theory, the addition of successive daily changes of a price may be designated by the term "aggregation in series", the term "aggregation in parallel" applying to the operation

$$L(t, T) = \sum_{i=1}^{I} L(i, t, T) = \sum_{i=1}^{I} \sum_{\tau=0}^{T-1} L(i, t + \tau, 1)$$

where i refers to "events" that occur simultaneously during a given time interval such as T or 1.

In the Gaussian case, one should, of course, expect any occurrence of a large value for $L(t, T)$ to be traceable to a rare conjunction of large changes in all or most of the $L(i, t, T)$. In the stable Paretian case, one should on the contrary expect large changes $L(t, T)$ to be traceable to one or a small number of the contributing $L(i, t, T)$. It seems obvious that the Paretian prediction is closer to the facts.

To add up the two types of aggregation in a Paretian world, a large $L(t, T)$ is likely to be traceable to the fact that $L(i, t + \tau, 1)$ happens to be very large for one or a few sets of values of i and of τ. These contributions would stand out sharply and be causally explainable. But, after a while, they should of course rejoin the "noise" made up by the other factors. The next rapid change of $\log_e Z(t)$ should be due to other "causes". If a contribution is "trend-making" in the above sense during a large number of time-increments, one will, of course, doubt that it falls under the same theory as the fluctuations.

PRICE VARIATION IN CONTINUOUS TIME AND THE THEORY OF SPECULATION

The main point of this section is to show that certain systems of speculation, which would have been advantageous if one could implement them, cannot in reality be followed in the case of price series generated by a Paretian process.

Infinite divisibility of stable Paretian laws

Whichever N, it is possible to consider that a stable Paretian increment

$$L(t, 1) = \log_e Z(t + 1) - \log_e Z(t)$$

is the sum of N independent, identically distributed, random variables, and that those variables differ from $L(t)$ only by the value of the constants γ, C' and C'', which are N times smaller.

In fact, it is possible to interpolate the process of independent stable Paretian increments to continuous time, assuming that $L(t, dt)$ is a stable Paretian variable with a scale coefficient $\gamma(dt) = dt\gamma(1)$. This interpolated process is a very important "zeroth" order approximation to the actual price changes. That is, its predictions are surely modified by the mechanisms of the market, but they are very illuminating nonetheless.

Path functions of a stable process in continuous time

It is almost universally assumed, in mathematical models of physical or of social sciences, that all functions can safely be considered as being continuous and as having as many derivatives as one may wish. The functions generated by Bachelier are indeed continuous ("almost surely almost everywhere", but we may forget this qualification"; although they have no derivatives ("almost surely almost nowhere"), we need not be concerned because price quotations are always rounded to simple fractions of the unit of currency.

In the Paretian case things are quite different. If my process is interpolated to continuous t, the paths which it generates become everywhere discontinuous (or rather, they become "almost surely almost everywhere discontinuous"). That is, most of their variation is performed through non-infinitesimal "jumps", the number of jumps larger than u and located within a time increment T, being given by the law $C'T|\acute{a}(u^{-\alpha})|$.

Let us examine a few aspects of this discontinuity. Again, very small jumps of $\log_e Z(t)$ could not be perceived, since price quotations are always expressed in simple fractions. It is more interesting to note that there is a non-negligible probability that a jump of price is so large that "supply and demand" cease to be matched. In other words, the stable Paretian model may be considered as predicting the occurrence of phenomena likely to force the market to close. In a Gaussian model such large changes are so extremely unlikely that the occasional closure of the markets must be explained by non-stochastic considerations.

The most interesting fact is, however, the large probability predicted for medium-sized jumps by the stable Paretian model. Clearly, if those medium-sized movements were oscillatory, they could be eliminated by market mechanisms such as the activities of the specialists. But if the movement is all in one direction, market specialists could at best transform a discontinuity into a change that is rapid but progressive. On the other hand, very few transactions would then be expected at the intermediate smoothing prices. As a result, even if the price Z^0 is quoted transiently, it may be impossible to act rapidly enough to satisfy more than a minute fraction of orders to "sell at Z^0". In other words, a large number of intermediate prices are quoted even if $Z(t)$ performs a large jump in a short time; but they are likely to be so fleeting, and to apply to so few transactions, that they are irrelevant from the viewpoint of actually enforcing a "stop loss order" of any kind. In less extreme cases – as, for example, when borrowings are oversubscribed – the market may have to resort to special rules of allocation.

These remarks are the crux of my criticism of certain systematic methods: they would perhaps be very advantageous if only they could be enforced, but in fact they can only be enforced by very few traders. I shall be content here with a discussion of one example of this kind of reasoning.

The fairness of Alexander's game

S. S. Alexander has suggested the following rule of speculation: "if the market goes up 5%, go long and stay long until it moves down 5%, at which time sell and go short until it again goes up 5%".[15]

This procedure is motivated by the fact that, according to Alexander's interpretation, data would suggest that "in speculative markets, price changes appear to follow a random walk over time; but ... if the market has moved up x%, it is likely to move up more than x% further before it moves down x%". He calls this phenomenon the "persistence of moves". Since there is no possible persistence of moves in any "random walk" with zero mean, we see that if Alexander's interpretation of facts were confirmed, one would have to look at a very early stage for a theoretical improvement over the random walk model.

In order to follow this rule, one must of course watch a price series continuously in time and buy or sell whenever its variation attains the perceived value. In other words, this rule can be strictly followed if and only if the process $Z(t)$ generates continuous path functions, as for example in the original Gaussian process of Bachelier

Alexander's procedure cannot be followed, however, in the case of my own first-approximation model of price change in which there is a probability equal to one that the first move *not smaller* than 5% is *greater* than 5% and *not equal* to 5%. It is therefore mandatory to modify Alexander's scheme to suggest buying or selling when moves of 5% are *first exceeded*. One can prove that the stable Paretian theory predicts that this game also is fair. Therefore, the evidence – as interpreted by

Alexander – would again suggest that one must go beyond the simple model of independent increments of price.

But Alexander's inference was actually based upon the discontinuous series constituted by the closing prices on successive days. He assumed that the intermediate prices could be interpolated by some continuous function of continuous time – the actual form of which need not be specified. That is, whenever there was a difference of *over* 5% between the closing price on day F' and day F'', Alexander implicitly assumed that there was at least one instance between these moments when the price had gone up *exactly* 5%. He recommends buying at this instant, and he computes the empirical returns to the speculator as if he were able to follow this procedure.

For price series generated by my process, however, the price actually paid for a stock will almost always be greater than that corresponding to a 5% rise; hence the speculator will almost always have paid more than assumed in Alexander's evaluation of the returns. On the contrary, the price received will almost always be less than suggested by Alexander. Hence, at best, Alexander overestimates the yield corresponding to his method of speculation and, at worst, the very impression that the yield is positive may be a delusion due to overoptimistic evaluation of what happens during the few most rapid price changes.

One can of course imagine contracts guaranteeing that the broker will charge (or credit) his client the actual price quotation nearest by excess (or default) to a price agreed upon, irrespective of whether the broker was able to perform the transaction at the price agreed upon. Such a system would make Alexander's procedure advantageous to the speculator; but the money he would be making on the average would come from his broker and not from the market; and brokerage fees would have to be such as to make the game at best fair in the long run.

A MORE REFINED MODEL OF PRICE VARIATION

Broadly speaking, the predictions of my main model seem to me to be reasonable. At closer inspection, however, one notes that large price changes are not isolated between periods of slow change; they rather tend to be the result of several fluctuations, some of which "overshoot" the final change. Similarly, the movement of prices in periods of tranquillity seem to be smoother than predicted by my process. In other words, large changes tend to be followed by large changes – of either sign – and small changes tend to be followed by small changes, so that the isolatines of low probability of $[L(t, 1), L(t - 1, 1)]$ are X-shaped. In the case of daily cotton prices, Hendrik S. Houthakker stressed this fact in several conferences and private conversation.

Such an X shape can be easily obtained by rotation from the "plus-sign shape" which was observed in Figure 4 to be applicable when $L(t, 1)$ and

$L(t-1,1)$ are statistically independent and symmetric. The necessary rotation introduces the two expressions:

$$S(t) = (1/2)[L(t,1) + L(t-1,1)] = (1/2)[\log_e Z(t+1) - \log_e Z(t-1)]$$

and

$$D(t) = (1/2)[L(t,1) - L(t-1,1)] = (1/2)[\log_e Z(t+1) - 2\log_e Z(t) + \log_e Z(t-1)]$$

It follows that in order to obtain the X shape of the empirical isolines, it would be sufficient to assume that the first and second finite differences of $\log_e Z(t)$ are two stable Paretian random variables, independent of each other and naturally of $\log_e Z(t)$ (see Figure 4). Such a process is invariant by time inversion.

It is interesting to note that the distribution of $L(t,1)$, conditioned by the known $L(t-1,1)$, is asymptotically Paretian with an exponent equal to $2\alpha + 1$.[16] Since, for the usual range of α, $2\alpha + 1$ is greater than 4, it is clear that no stable Paretian law can be associated with the conditioned $L(t,1)$. In fact, even the kurtosis is finite for the conditioned $L(t,1)$.

Let us then consider a Markovian process with the transition probability I have just introduced. If the initial $L(T^0, 1)$ is small, the first values of $L(t,1)$ will be weakly Paretian with a high exponent $2\alpha + 1$, so that $\log_e Z(t)$ will begin by fluctuating much less rapidly than in the case of independent $L(t,1)$. Eventually, however, a large $L(t^0, 1)$ will appear. Thereafter, $L(t,1)$ will fluctuate for some time between values of the orders of magnitude of $L(t^0, 1)$ and $-L(t^0, 1)$. This will last long enough to compensate fully for the deficiency of large values during the period of slow variation. In other words, the occasional sharp changes of $L(t,1)$ predicted by the model of independent $L(t,1)$ are replaced by oscillatory periods, and the periods without sharp change are less fluctuating than when the $L(t,1)$ are independent.

We see that, for the correct estimation of α, it is mandatory to avoid the elimination of periods of rapid change of prices. One *cannot* argue that they are "causally" explainable and ought to be eliminated before the "noise" is examined more closely. If one succeeded in eliminating all large changes in which way, one would have a Gaussian-like remainder which, however, would be devoid of any significance.

---

1 The present text is a modified version of my "Research Note", NC-87, issued on March 26, 1962 by the Research Center of the International Business Machines Corporation. I have been careful to avoid any change in substance, but certain parts of that exposition have been clarified, and I have omitted some less essential sections, paragraphs, or sentences. The first and second sections correspond roughly to chapters 1 and 2 of the original, the following

two sections correspond to chapters 4 and 5, the fifth and sixth sections to chapter 6, and the last section to chapter 7.

2 The simple Bachelier model implicitly assumes that the variance of the differences $Z(t + T) - Z(t)$ is independent of the level of $Z(t)$. There is reason to expect, however, that the standard deviation of $\Delta Z(t)$ will be proportional to the price level, and for this reason many modern authors have suggested that the original assumption of independent increments of $Z(t)$ be replaced by the assumption of independent and Gaussian increments of $\log_e Z(t)$.

Since Bachelier's original work is fairly inaccessible, it is good to mention more than one reference: "Théorie de la spéculation" (Paris Doctoral Dissertation in Mathematics, March 29, 1900) *Annales de l'Ecole Normale Supérieure*, ser. 3, 17 (1900), 21–86; Théorie mathématique du jeu," *Annales de l'Ecole Normale Supérieure*, ser. 3, 18 (1901), 143–210; *Calcul des probabilités* 1912 (Paris: Gauthier-Villars); *Le jeu, la chance et le hasard* 1914 (Paris, reprinted up to 1929 at least).

3 To the best of my knowledge, the first to note this fact was Wesley C. Mitchell, "The Making and Using of Index Numbers", *Introduction to Index Numbers and Wholesale Prices in the United States and Foreign Countries* (published in 1915 as Bulletin 173 of the US Bureau of Labor Statistics, reprinted in 1921 as Bulletin 284). But unquestionable proof was only given by Maurice Olivier in "Les Nombres indices de la variation des prix" 1926 (Paris doctoral dissertation), and Frederick C. Mills in *The Behavior of Prices* 1927 (New York: National Bureau of Economic Research). Other evidence, referring either to $Z(t)$ or to $\log_e Z(t)$ and plotted on various kinds of coordinates, can be found in Arnold Larson, "Measurement of a Random Process in Future Prices, *Food Research Institute Studies* (1960), 1, pp. 313–24; M. F. M. Osborne, "Brownian Motion in the Stock Market", *Operations Research* 7 (1959), pp. 145–73, pp. 807–11; S. S. Alexander, "Price Movements in Speculative Markets: Trends of Random Walks?" 1961, *Industrial Management Review of MIT* 2(2), pp. 7–26.

4 Such an approach has also been necessary – and successful – in other contexts; for background information and many additional explanations see my "New Methods in Statistical Economics", 1963, *Journal of Political Economy* 71, October.

I believe, however, that each of the applications should stand on their own feet and have minimised the number of cross references.

5 Paul Lévy, 1925, *Calcul des probabilités* (Paris: Gauthier-Villars); Paul Lévy, 1937, *Théorie de l'addition des variables aléatoires*, Second edition, 1954, (Paris: Gauthier-Villars). The most accessible source on these problems is, however, B. V. Gnedenko and A. N.Kolmogoroff, 1954, *Limit Distributions for Sums of Independent Random Variables*, translated by K. L. Chung (Reading: Addison-Wesley Press).

6 A. V. Skorohod, 1954, "Asymptotic Formulas for Stable Distribution Laws", *Dokl. Ak. Nauk SSSR*, 98, pp. 731–35, or *Select. Tranl. Math. Stat. Proba. Am. Math. Soc.*, 1961, pp. 157–61.

7 Benoit Mandelbrot 1960, "The Pareto–Lévy Law and the Distribution of Income", *International Economic Review* 1, pp. 79–106, as amended in "The Stable Paretian Income Distribution, When the Apparent Exponent Is near Two", *International Economic Review* 4, 1963, pp. 111–15; see also my "Stable Paretian Random Functions and the Multiplicative Variation of Income", *Econometrica* 29, 1961, pp. 517–43, and "Paretian Distributions and Income Maximisation" 1962, *Quarterly Journal of Economics* 76, pp. 57–85.

8 See Gnedenko and Kolmogoroff, *op. cit.*, n. 4, p. 175, who use a notation that does not emphasise, as I hope to do, the relation between the law of Pareto and its present generalisation.

9 These examples were mentioned in my 1962 "Research Note" (*op. cit.*, n. 1). My presentation, however, was too sketchy and could not be improved upon without modification of the substance of that "Note" as well as its form. I prefer to postpone examination of all the other examples as well as the search for the point at which my model of cotton prices ceases to predict the facts correctly.

10 See my "New Methods in Statistical Economics", 1963, *Journal of Political Economy*, October.

11 J. Bienaymé, 1853, "Considérations a l'appui de la découverte de Laplace sur la loi de probabilté dans la méthode des moindres carrés", *Comptes rendus, Académ.ie des Sciences de Paris* 37, pp. 309–24 (esp. pp. 321–23).

12 Paul H. Cootner, 1962, "Stock Prices: Random Walks vs. Finite Markov Chains", *Industrial Management Review of MIT* 3, pp. 24–45.

13 See esp. Holbrook Working, 1934, "A Random-Difference Series for Use in the Analysis of Time Series", *Journal of the American Statistical Association* 29, pp. 11–24; Maurice Kendall, 1953, "The Analysis of Economic Time-Series – Part 1: Prices", *Journal of the Royal Statistical Society*, Ser. A, 116, pp. 11–34; M. F. M. Osborne, 1959, "Brownian Motion in the Stock Market", *op. cit.*; Harry V. Roberts, 1959, "Stock-Market "Patterns" and Financial Analysis: Methodological Suggestions", *Journal of Finance* 14, pp. 1–10; and S. S. Alexander, "Price Movements in Speculative Markets: Trends or Random Walks", *op. cit.*, n. 3.

14 Donald Darling, 1952, "The Influence of the Maximum Term in the Addition of Independent Random Variables", *Transactions of the American Mathematical Society* 70, pp. 95–107; and D. Z. Arov and A. A. Bobrov, 1960, "The Extreme Members of Samples and Their Role in the Sum of Independent Variables", *Theory of Probability and Its Applications* 5, pp. 415–35.

15 S. S. Alexander, *op. cit.* n. 3.

16 Proof: $Pr[L(t, 1) > u$, when $w < L(t - 1, 1) < w + dw]$ is the product by $(1/dw)$ of the integral of the probability density of $[L(t - 1, 1)L(t, 1)]$, over a strip that differs infinitesimally from the zone defined by

$$S(t) > (u + w)/2$$
$$w + S(t < D(t) < w + S(t) + dw$$

Hence, if u is large as compared to w, the conditional probability in question is equal to the following integral, carried from $(u + w)/2$ to ∞.

$$\int C' \alpha s^{-(\alpha+1)} C' \alpha (s + w)^{-(\alpha+1)} ds \sim (2\alpha + 1)^{-1} (C')^2 \alpha^2 2^{-(2\alpha+1)} u^{-(2\alpha+1)}$$

27

*The Variation of Some Other Speculative Prices**

Benoit Mandelbrot
Yale University

The present article continues my earlier work, "The Variation of Certain Speculative Prices" (VCSP) (Mandelbrot, 1963c). There, it was argued that the description of time series of prices requires probability models less special than the widely used Gaussian, because the price relatives of *certain* price series have a variance so large that it may in practice be assumed infinite.

The second section of the present work restates the theoretical argument of VCSP, with little mathematics, but stress upon the motivation of my generalisation of the Gaussian model. I trust that this will (implicitly) show certain responses to VCSP to have been unwarranted.

Because very similar reservations about VCSP were often voiced by different authors, and because I hope that they will be withdrawn and do not want to preserve them through controversy, the text will name neither the friendly nor the unfriendly commentators of VCSP, though many are listed in the bibliography.

The third section is devoted to additional empirical evidence in favour of my "stable Paretian" model, relative to wheat, railroad securities, and rates of exchange or of interest. Moreover, some unpublished figures concerning cotton prices are incorporated in the second section. Much of the newly published empirical evidence was already quoted in the larger unpublished work from which VCSP is excerpted (Mandelbrot, 1963c, n. 9).[1] The evidence now available is so extensive that only a small fraction of it can be reported below.

The fourth section is a token contribution to the statistical problems raised by VCSP. The statistics of the stable processes has raised many exciting and very new questions and attracts increasing attention. The

---

*This paper was first published in the *Journal of Business* 40, pp. 393–413 (1967) and is reprinted with the permission of the University of Chicago Press. The author wishes to thank Professor Eugene F. Fama.

practical applicability of the findings of VCSP is naturally much dependent upon the development of statistics.

The preparation of this paper having been very slow, it would by itself give an outdated idea of the status of the theory started in VCSP. Much progress has been made since, (Fama, 1965; Fama and Marshall, 1966; Mandelbrot, 1966; Mandelbrot and Taylor) and in papers by Eugene Fama and myself. Mandelbrot and Taylor touches upon a currently active issue, being devoted to various relations existing between, on the one hand, price changes over fixed time intervals (such as days) and, on the other hand, price changes between successive transactions. Though the distribution of the latter changes is necessarily very short tailed (for institutional and other reasons), the number of transactions within a day is sufficiently variable to account for the long-tailedness of the distribution of daily price changes.

THE STABLE PARETIAN MODEL OF PRICE VARIATION
Bachelier's theory of speculation

Consider a time series of prices, $Z(t)$, and designate by $L(t, T)$ its logarithmic relative

$$L(t, T) = \log_e Z(t, T) - \log_e Z(t)$$

The basic model of price variation, a modification of one proposed in 1900 in Louis Bachelier's theory of speculation (Bachelier, 1900), assumes that successive increments $L(t, T)$ have the following properties: they are (a) random, (b) statistically independent, (c) identically distributed, and (d) their marginal distribution is Gaussian with zero mean. Such a process is called a "stationary Gaussian random walk" or "Brownian motion".

Although his model continues to be extremely important, all four of Bachelier's assumptions are working approximations that should not be made into dogmas. In fact, writing in 1914,[2] Bachelier (1914) himself made no mention of his earlier claims of the existence of empirical evidence in favour of Brownian motion, and noted that his original model diverges from the evidence in at least two ways: first of all, the sample variance of $L(t, T)$ varies in time. He attributed this to variability of the population variance, interpreted the sample histograms as being relative to mixtures, and noted that the tails of the histogram could be expected to be fatter than in the Gaussian case. Second, Bachelier noted that no reasonable mixture of Gaussian distributions could account for the size of the very largest price changes, and he treated them as "contaminators" or "outliers." Thus, he pioneered, not only in stating the oft-rediscovered Gaussian random-walk model, but also in exposing its oft-rediscovered major weaknesses.

However, new advances in the theory of speculation are still best expressed as improvements upon his 1900 model: the approach to price variation, proposed in VCSP, shows that an appropriate generalisation of hypothesis (d) suffices to "save" (a), (b), and (c) in many cases, and in others greatly postpones the need of amending them. I shall comment upon Bachelier's four hypotheses, then come to the argument of VCSP. Readers acquainted with VCSP may proceed immediately to the third section.

Randomness

I have only a few words about the description of price variation by a random process. To say that a price change is random implies, not that it is irrational, but only that it was unpredictable before the fact *and* is describable by the powerful mathematical theory of probability. Therefore, there are two alternatives to randomness, namely, "predictable behaviour" and "haphazard behaviour", where I see the latter term as meaning "unpredictable and not subject to probability theory". By treating the largest price changes as "outliers", Bachelier implicitly resorted to this concept of "haphazard." This might have been unavoidable in his time, but the power of probability theory has much increased since and should be used to the fullest.

Independence

To assume statistical independence of successive $L(t, T)$ is undoubtedly a simplification of reality.[3] The single but very strong defence of this hypothesis is in the surprising fact that models making this assumption account for many features of price behaviour.

Incidentally, independence implies that no investor can use his knowledge of past data to increase his expected profit. But the converse is *not* true. There exist processes in which the expected profit vanishes, but dependence is of extremely long range, and knowledge of the past may be profitable to those investors whose utility function differs from the market's. See, for example, the "martingale" model (Mandelbrot, 1966), which is developed and generalised in Mandelbrot, (the latter paper also touches on various aspects of the spectral analysis of economic time series, also an active topic whose relations with my work have aroused interest; for example, when a time series is non-Gaussian, its spectral whiteness, that is, lack of correlation, is compatible with great departures from the random-walk hypothesis).

Stationarity

One implication of stationarity is that sample moments vary little from sample to sample, as long as the sample length is sufficient. In fact, it is notorious that price moments often "misbehave" from this viewpoint

Figure 1

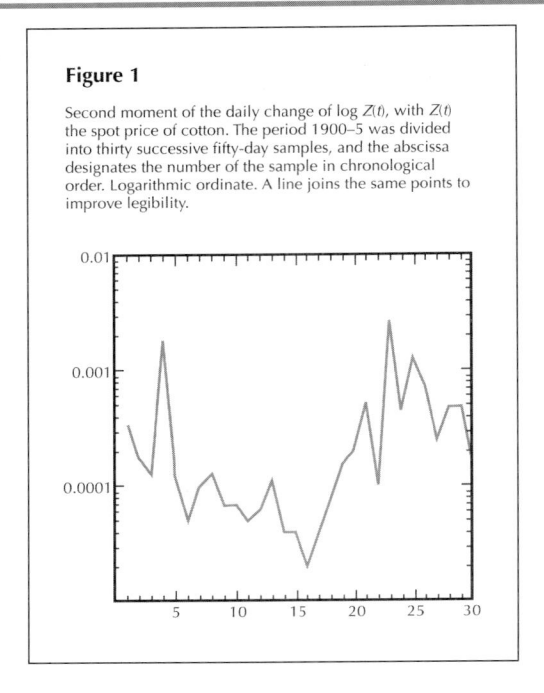

Second moment of the daily change of log $Z(t)$, with $Z(t)$ the spot price of cotton. The period 1900–5 was divided into thirty successive fifty-day samples, and the abscissa designates the number of the sample in chronological order. Logarithmic ordinate. A line joins the same points to improve legibility.

(though this fact is understated in the literature, since "negative" results are seldom published; see, however, F. C. Mills, 1927).

Figure 1 is an example of the enormous variability in time of the sample second moment. The points refer to successive fifty-day sample means of $[L(t, 1)]^2$ for cotton prices in the period 1900–1905 (recall that $L(t, 1)$ is the daily logarithmic price relative). According to Bachelier's model, these sample means should already have stabilised near the population mean. Since no stabilisation is in fact observed, we see conclusively that the price of cotton did not follow a Gaussian stationary random walk.

To account for this, it is usual to say that the mechanism of price variation itself changes in time. We shall, loosely speaking, distinguish systematic, random, and haphazard changes of mechanism.

To refer to systematic changes is especially tempting. Indeed, to explain the temporal changes of the statistical parameter of the process of price variation would constitute a worthwhile first step toward an ultimate explanation of price variation itself. An example of systematic change is given by the yearly seasonal effects, which are strong in the case of agricultural commodities. However, in Figure 1 not all ends of season are accompanied by large price changes, and not all large price changes occur at any prescribed time in the growing season.

The most controversial systematic changes are those due to deliberate changes in the policies of the government and of the exchanges. The existence of long-term changes of this type is unquestionable. For example, the third section of VCSP concluded that various measures of the scale of $L(t, T)$ (such as the interquartile interval) have varied in the case of cotton prices between 1816 and 1958. One evidence is that lines *1a* and *2a* of Figure 5 in VCSP, relative to the 1900s, clearly differ from lines *1b* and *2b*, relative to the 1950s. This decrease in price variability must, at least in part, be a consequence of the deep changes in economic policy that occurred in the early half of this century. However, precisely because it is so easy to read in the facts a proof of the success or failure of changes in economic policy, the temptation to resort to systematic non-stationarity must be carefully controlled.

An example of "random change" is a random-walk process in which the sizes and probabilities of the steps are chosen by some other prices. If this second "master process" is stationary, $Z(t)$ itself is not a random walk but remains a stationary random process.

The final possibility is that the variability of the price mechanism is haphazard, that is, not capable of being treated by probability theory. In practice, it is not very reasonable to resort to the haphazard at this late stage: indeed, why bother to construct complicated statistical models for the behaviour of prices if one expects this behaviour to change before the model has had time to unfold? Moreover, and more important, early resort to the haphazard need not be necessary, as is demonstrated by the smoothness and regularity of the graph of Figure 2, which is the histogram of the data of Figure 1.

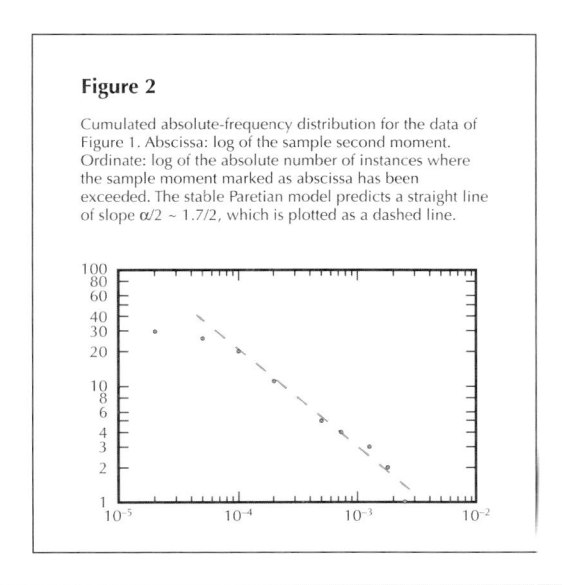

Figure 2

Cumulated absolute-frequency distribution for the data of Figure 1. Abscissa: log of the sample second moment. Ordinate: log of the absolute number of instances where the sample moment marked as abscissa has been exceeded. The stable Paretian model predicts a straight line of slope $\alpha/2 \sim 1.7/2$, which is plotted as a dashed line.

Gaussian hypothesis

Bachelier's assumption, that the marginal distribution of $L(t, T)$ is Gaussian with vanishing expectation, might be convenient, but virtually every student of the distribution of prices has commented on their leptokurtic (ie, very long-tailed) character.[4] As mentioned, Bachelier himself realised this and regarded $L(t, T)$ as a contaminated mixture of Gaussian variables; see Mandelbrot and Howard.

Infinite variance and the stable Paretian distributions

Still other approaches were suggested to take account of the failure of Brownian motion to fit data on price variation. A common feature of all these approaches, however, is that each new fact necessitates an addition to the explanation. Since a new set of parameters is thereby added. I don't doubt that reasonable "curve-fitting" is achievable in many cases.

However, this form of "symptomatic medicine" (a new drug for each complaint) could not be the last word! The effectiveness as well as the beauty of science is that it sometimes evolves central assumptions, from which many independently made observations can be shown to be consequences. These observations are thus organised, and fresh facts can be predicted. The ambition of VCSP was to suggest such a central assumption, the infinite-variance hypothesis, and to show that it accounts for substantial features of price series (of various degrees of volatility) without non-stationarity, without mixture, without master processes, without contamination, and with a choice of increasingly accurate assumptions about interdependence of successive price changes.

When selecting a family of distributions to implement the infinite-variance hypothesis, one must be led by mathematical convenience (eg, the existence of a ready-made mathematical theory) and by simplicity. For a probability distribution, one important criterion of simplicity is the variety of the properties of "invariance" that it possesses: for example, it would be most desirable to have the same distribution (up to some – hopefully linear – weighting) apply to daily, monthly, etc., price changes. Another measure of simplicity is the role that a family of distributions plays in central limit theorems of the calculus of probability.

Following these reasons, VCSP proposed to represent the marginal distribution of $L(t, T)$ by an appropriate member of a family of probability laws called "stable",[5] which measure volatility by a single parameter α[6] ranging between 2 and 0 and whose simplest members are the symmetric probability densities $p_\alpha(u) = (1/\pi) \int_0^\infty \exp(-\gamma s^\alpha) \cos(su) ds$.

Their limit case $\alpha = 2$ is, duly, Gaussian, but my theory also allows non-Gaussian or "stable Paretian"[7] cases $\alpha < 2$, which indeed turn out to represent satisfactorily the data on volatile prices (see VCSP and below).

In assessing the realm of applicability of my theory, one should always understand it as including its classic limit. It is therefore impossible to claim that VCSP was "disproved" when one has pointed out price series for which the Gaussian hypothesis may be tenable.

Now to discuss the main fact, that stable variables with $\alpha < 2$ have an infinite[8] population variance (one says sometimes that they have "no variance"). Concern was expressed at the implication of this feature for statistics, and surprise was expressed at the paradoxically discontinuous change that seems to occur when α becomes exactly 2.

This impression of paradox is unfounded. The population variance itself cannot be measured, and every measurable characteristic of a stable distribution behaves continuously near $\alpha = 2$. We shall present an example later on. Consequently, there is no "black and white" contrast between the Paretian case $\alpha < 2$ and the Gaussian $\alpha = 2$, but a continuous shading of grey,[9] which is less desperate but more sensible and more interesting. The fact that the population second moment is discontinuous at $\alpha = 2$ "only" shows that it is not well suited to a study of price variation.[10]

In particular, the applicability of second-order statistical methods is questionable. "Questionable" does not mean "totally inapplicable", because the statistical methods based upon variances suffer no sudden and catastrophic breakdown as α ceases to equal 2. Therefore, to be unduly concerned with a few specks of "grey" in a price series whose α is near 2 may be as inadvisable as to treat very grey series as white. Moreover, statistics would be unduly restricted if its tools were to be used only where justified. (As a matter of fact, the quality of a statistical method is partly assessed by its "robustness", ie, the quality of its performance when used without justification.) However, one should look for other methods. For example, as predicted, least-squares forecasting (as applied to past data) would have often led to very poor inferences; least-sums-of-absolute-deviations forecasting is always at least as good and usually much superior, and its development should be pressed.

This section will end by two remarks concerning second and fourth moments, respectively.

The behaviour of sample second moments. Define $V(\alpha, N)$ as the variance of a sample of N independent random variables $u_1, \ldots u_n, \ldots u_N$, whose common distribution is stable of exponent α. To obtain a balanced view of the practical properties of such variables, it is best *not* to focus upon mathematical expectations and/or infinite sample sizes. One should rather consider quantiles and samples of large but finite size. Let us therefore select a "finite horizon" by choosing a value of N and a quantile threshold q (such that events whose probability is below q will be considered to be "unlikely"). Save for extreme cases contributing to a "tail" of probability q, the values of $V(\alpha, N)$ will be less than some function $V(\alpha, N, q)$ whose behaviour summarises much of what needs to be known about the sample variance.

As mentioned earlier, when N is finite and $q > 0$, the function $V(\alpha, N, q)$ varies smoothly with α. For example, over a wide range of values of N, the derivative of $V(\alpha, N, q)$ at $\alpha = 2$ is very close to zero, so that $V(\alpha, N, q)$ changes very little if $\alpha = 2.00$ is replaced by, say, $\alpha = 1.99$. This insensitivity is due to the fact that $\alpha = 2.00$ and $\alpha = 1.99$ only differ in the sizes that they predict for some outliers, constituting a small proportion of all cases, whose effects were excluded by the definition of $V(\alpha, N, q)$. By increasing N, or by decreasing q, one decreases the range of exponents in which α is approximable by 2.

If one really objects to infinite variance, while being only concerned with meaningful finite-sample problems, one may "truncate" U so as to attribute to its variance a very large finite value depending upon α, N, and q. The resulting theory may have the asset of familiarity, but the speculation of the value of the truncated variance will be *useless* because it will tell nothing about the "transient" behaviour of $V(\alpha, N, q)$ when N is finite and small. Thus, even when one knows the variance to be finite but very large (as is the case in certain of my more detailed models of price variation; see Mandelbrot), the study of the behaviour of $V(\alpha, N, q)$ is much simplified if one approximates the distribution with finite but very large variance by a distribution with infinite variance. This feature can be illustrated by the following homely example: it is well known that photography is simplest when the object is at an infinite distance from the camera. Therefore, even if the actual distance is known to be finite, the photographer ought to set the distance at infinity if that distance exceeds some finite threshold, depending upon the quality of the lens and its aperture.

The behaviour of sample kurtosis. Pearson's kurtosis, a measure of the peakedness of a distribution defined by $E(U^4)[E(U^2)]^{-2} - 3$, was discussed in VCSP. But the discussion lacked numerical illustration and was called obscure, and additional detail may be useful. If $E(U^2) = \infty$, the value of the kurtosis is indeterminate. One can, however, show that, as $N \to \infty$, and if U is stable with $\alpha < 2$, the random variable $\Sigma_{n=1}^{N} U_n^4 [\Sigma_{n=1}^{n} U_n^2]^{-2}$ tends toward a limit that is different from zero. Therefore, the "expected sample kurtosis", defined as

$$E\left\{ N \sum_{n=1}^{N} U_n^4 \left[\sum_{n=1}^{N} U_n^2 \right]^2 - 3 \right\}$$

is asymptotically proportional to N.

The kurtosis of $L(t, 1)$ was plotted on Figure 3 for the case of cotton, 1900–1905, and is indeed seen to increase steeply with N. Though exact comparison is impossible because the theoretical distribution is not yet

Figure 3

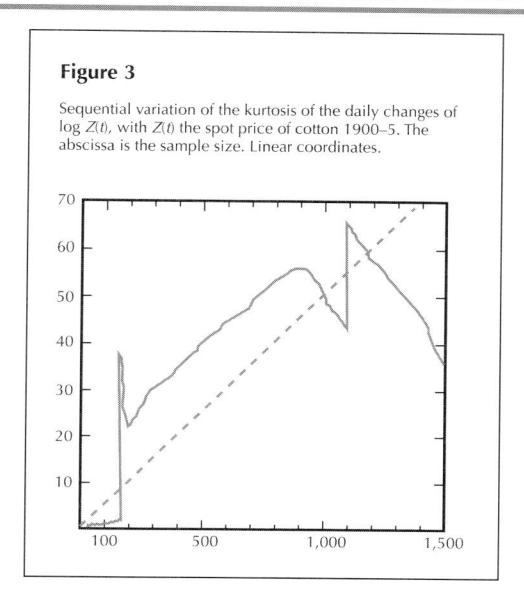

Sequential variation of the kurtosis of the daily changes of log $Z(t)$, with $Z(t)$ the spot price of cotton 1900–5. The abscissa is the sample size. Linear coordinates.

tabulated, this kurtosis indeed fluctuates around a line expressing proportionality to sample size. (For samples less than fifty, the kurtosis was negative.)

Three approximations to a stable distribution: implications for statistics and for the description of the behaviour of prices

It is important that there should exist a single theory of prices that subsumes various degrees of volatility. My theory is unfortunately hard to handle, while simple approximations are available in different ranges of values of α. Thus, given a practical problem with a finite time horizon N, it is best to replace the continuous range of degrees of "greyness" by the following trichotomy (where the boundaries between the categories are dependent upon the problem in question).

The Gaussian $\alpha = 2$ is best known and simplest. Here, not having to worry about a long-tailed marginal distribution, one stands a reasonable chance of rapid progress in the study of dependence. For example, one can use spectral methods and other covariance-oriented approaches. In the immediate neighbourhood of the Gaussian, Gaussian techniques cannot lead one too far astray.

In the zone far away from $\alpha = 2$, another kind of simplicity reigns. Substantial tails of the stable Paretian law are approximable by Pareto's hyperbolic law, with the same α-exponent ruling both tails. The prime example of this zone was provided by the cotton prices studied in VCSP. In this chapter, we shall examine some other price series of similarly high volatility: the prices of some 19th-century rail securities and some exchange and interest rates.

The third and final zone constitutes a transition between the almost Gaussian and the highly Paretian cases and is far more complicated than either. It also provides a test of the meaningfulness and generality of the stable Paretian hypothesis: if that hypothesis holds, the histogram of price changes is expected to plot on bilogarithmetic paper as one of a specific family of inverse-S-shaped curves. (Lévy's α-exponent, therefore, is not to be confused with a "Pareto" slope for a straight bilogarithmic plot (Mandelbrot, 1963a)). If the stable Paretian hypothesis failed, the transition between the almost Gaussian and the highly Paretian cases would be performed in some other way. We shall examine in this light the variation of wheat prices, and we shall find it to conform fully to the stable Paretian prediction concerning the "light grey" zone of low but positive values for $2 - \alpha$ and medium volatility. The first part of the next section, where wheat data are examined, is thus seen to have a purpose similar to Fama's 1964 Chicago thesis (Fama, 1965), which was the first to test further the ideas of VCSP. To minimise "volatility" and maximise the contrast with my original data, Fama chose thirty stocks of large and diversified contemporary corporations and found their stable Paretian "greyness" to be unquestionable although less marked than that of cotton.

ADDITIONAL DATA CONCERNING LOGARITHMIC RELATIVES OF PRICES

The variation of the price of wheat in Chicago, 1883–1936

Introduction. Contrary to the spot prices of cotton, which refer to standardised qualities, wheat cash prices refer to the variable grades of grain. Hence, at any given time (say, at closing time), one can at best speak of a *span* of cash prices, and the closing spans corresponding to successive days very often overlap. As a result, the week is probably the shortest period for which one can reasonably express "the" change of wheat price by a single number rather than by an interval. In any event, the week was chosen in the present work because it is used in H. Working's classic monograph (Working, 1934).[11]

The stable Paretian hypothesis for wheat. It had been suggested (Kendall, 1953) that wheat price relatives follow a Gaussian distribution. Indeed, a casual visual inspection of the histograms of these relatives, as plotted on *natural* coordinates, shows them to be nicely "bell shaped." The importance of the "tails" is, however, notoriously underestimated by plotting the data on natural coordinates. It is, on the contrary, stressed by using probability paper. As seen in Figure 4, probability-paper plots of wheat price relatives are definitely S-shaped (though less so than the plots relative to cotton). As the Gaussian corresponds to $\alpha = 2$, and as I found

Figure 4

Probability-paper plots of the distribution of changes of log $Z(t)$, with $Z(t)$ the spot price of wheat in Chicago, 1883–1934, as reported by Holbrook Working. The scale −0.3, −0.2, −0.1, 0, 0.1, 0.2, 0.3 applies to the weekly changes, marked by dots, and to the yearly charges, marked by +s. The other scale applies to changes over lunar months, marked by ×.

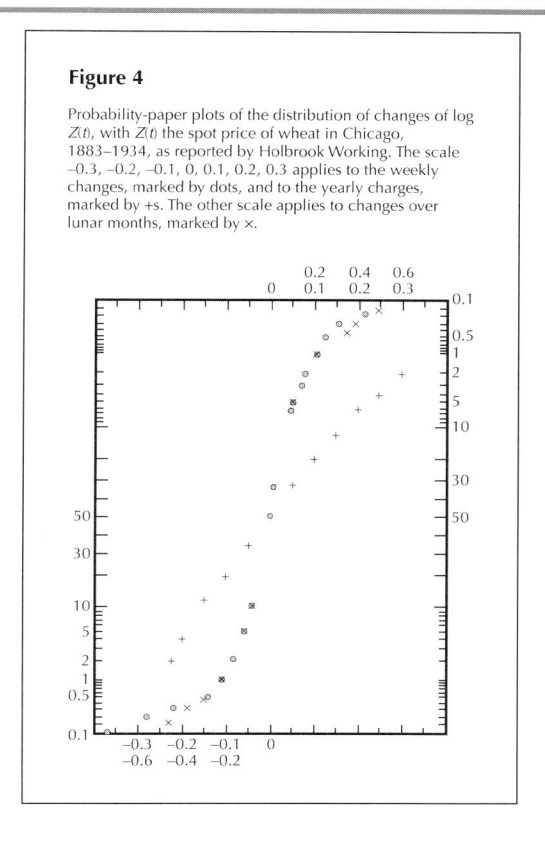

the value $\alpha = 1.7$ for cotton, it is natural to investigate whether wheat is stable Paretian with an α contained between 1.7 and 2.

The evidence of doubly logarithmic graphs. As seen in Figure 5, reproduced from VCSP, this stable Paretian hypothesis would imply that every doubly logarithmic plot of a histogram of wheat price changes should have a characteristic S-shape. It would end with a "Paretian" straight line of slope near 2, but it would start with a region where the local slope increases with u and even begins by exceeding markedly its asymptotic value (Mandelbrot, 1963b).

The above conjecture is indeed verified, as seen in Figures 6 and 7. Moreover, by comparing the data relative to successive subsamples of the period 1883–1936, I found no evidence that the *law* of price variation has changed in kind, despite the erratic behaviour of the outliers.

This is evidence that the stable Paretian hypothesis has *predicted* how a price histogram "should" behave, when it was only known to be intermediate between the highly erratic cotton series and the minimally erratic Gaussian limit.

Figure 5

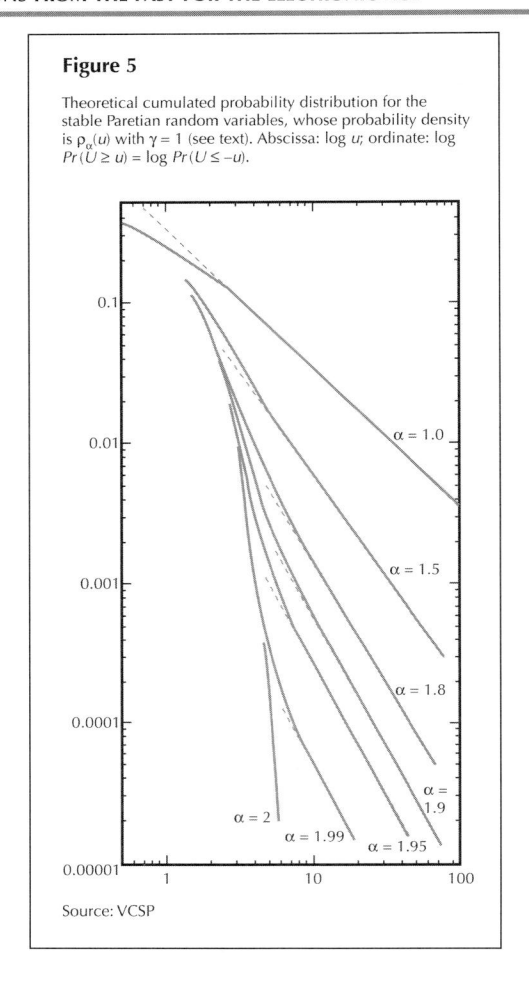

Theoretical cumulated probability distribution for the stable Paretian random variables, whose probability density is $p_\alpha(u)$ with $\gamma = 1$ (see text). Abscissa: log u; ordinate: log $Pr(U \geq u) = \log Pr(U \leq -u)$.

Source: VCSP

To establish the "goodness of fit" of such an S-shaped graph would unfortunately require an even larger sample of data than in the case of the straight graphs characteristic of cotton, while we know the available sample sizes to be rather smaller. Thus the doubly logarithmic evidence is *unavoidably* less clear cut than in the case of cotton.

The evidence of sequential variance. When a series of prices is approximately stationary, a test of whether $\alpha = 2$ or $\alpha < 2$ is provided by the behaviour of the sequential sample second moment. If $\alpha < 2$, the median of the distribution of the sample variance increases as $N^{-1+2/\alpha}$ for "large" N, while it tends to a limit when $\alpha = 2$. *More important*, the variation of the sample variance (about its median value) becomes increasingly erratic as α departs from 2. Thus, the cotton second moment increases very

Figure 6

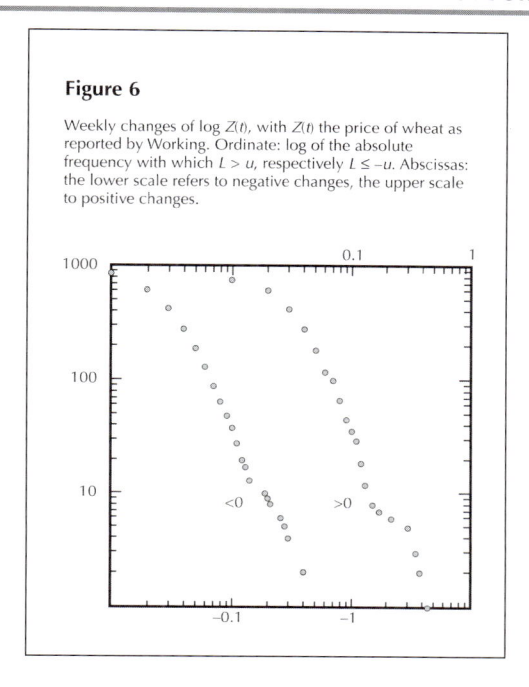

Weekly changes of log $Z(t)$, with $Z(t)$ the price of wheat as reported by Working. Ordinate: log of the absolute frequency with which $L > u$, respectively $L \leq -u$. Abscissas: the lower scale refers to negative changes, the upper scale to positive changes.

erratically, but the wheat second moment should increase more slowly and more regularly. Figure 8 shows that such is indeed the case.

Direct test of the stability. The term "stable" arose from the fact that, when N "stable" random variables U_n are independent and identically distributed, one has

$$Pr\left[N^{-1/\alpha} \sum_{n=1}^{N} U_n \geq u\right] = Pr[U_n \geq u]$$

The stability of Gaussian variables ($\alpha = 2$) is well known and often used in elementary statistics.

I settled on $N = 4$. When the random variables U_n are the weekly price changes, $\sum_{n=1}^{4} U_n$ is the price change over a "lunar month" of four weeks. Since α is expected to be near 2, $4^{-1/\alpha}$ will be near $\frac{1}{2}$.

One can see in Figure 4 that weekly price changes indeed have an S-shaped distribution *indistinguishable* from that of one-half of monthly changes.[12] (The bulk of the graph, corresponding to the central bell containing 80% of the cases, was not plotted for the sake of legibility.) Sampling fluctuations are apparent only at the extreme tails and do not appear systematic.

The combination of Figures 6 and 7 provides another test of stability. They were plotted with absolute, not relative, frequency as the ordinate,

Figure 7

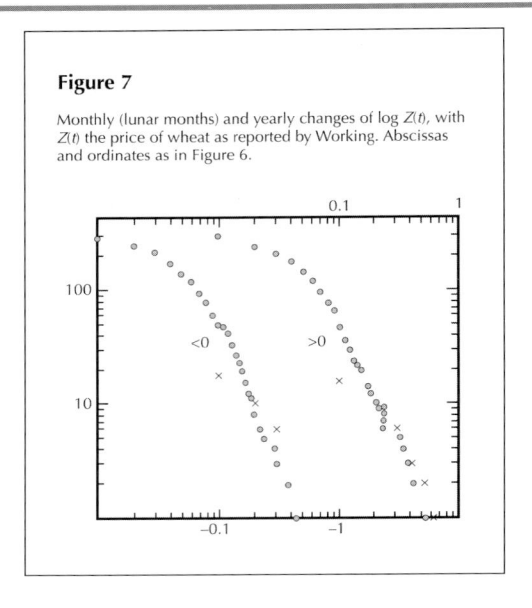

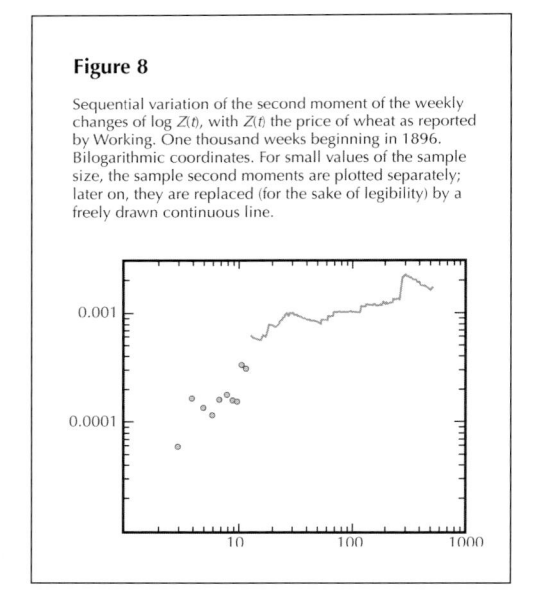

Monthly (lunar months) and yearly changes of log $Z(t)$, with $Z(t)$ the price of wheat as reported by Working. Abscissas and ordinates as in Figure 6.

Figure 8

Sequential variation of the second moment of the weekly changes of log $Z(t)$, with $Z(t)$ the price of wheat as reported by Working. One thousand weeks beginning in 1896. Bilogarithmic coordinates. For small values of the sample size, the sample second moments are plotted separately; later on, they are replaced (for the sake of legibility) by a freely drawn continuous line.

and the stable Paretian theory predicts that such curves should be superposable in their tails, except, of course, for sampling fluctuations. (Roughly, the reason for this prediction is that, in a stable Paretian universe, a large monthly price change is of the same order of magnitude as the largest among the four weekly changes adding to this monthly change.) Clearly, wheat data pass this second test also.

It should be stressed that the two tests, though using the same data, are *distinct* conceptually: Figure 4 compares one-half of a monthly change to the weekly change of the same frequency; Figures 6 and 7 compare monthly and weekly changes of the same size. Stability is thus doubly striking.

The evidence of yearly price changes. Working (1934) also published a table of average January prices of wheat, and Figure 4 also includes the corresponding changes of log $Z(t)$.

Assuming that successive weekly price changes are independent, the evidence of the yearly changes again favours the stable Paretian hypothesis. It is astonishing that the hypothesis of independence of weekly changes can be consistently carried so far, showing no discernible discontinuity between long-term adjustments to follow supply and demand, which would be the subject matter of economics, and the short-term fluctuations that some economists discuss as "mere effect of speculation".

The variation of the prices of railroad stocks, 1857–1936

Railroad stocks were pre-eminent among corporation securities that played for 19th-century speculators a role comparable to that of the basic commodities. Unfortunately, the convenient book of F. R. Macaulay (1936) reports them incompletely: (1) for each of the major stocks, it gives the mean of the highest and lowest quotation during the months of January; (2) for each month, it gives a weighted index of the highest and lowest quotation of every stock.

I began by examining the second series, even though it is averaged too many times for comfort. If one considers that there "should" have been no difference in kind between various 19th-century speculations, one would expect railroad stock changes to be stable Paretian, and averaging would bring an increase in the slope of the corresponding doubly logarithmic graphs, similar to what has been observed in the case of cotton price averages (the third section of VCSP). Indeed, Figure 9, relative to the variation of the monthly averages, yields precisely what one expects for such averages from Paretian processes with an exponent very close to that of cotton.

On yearly data, on the contrary, averaging has little effect. Figure 10 should be regarded as made of two parts, the first five graphs being relative to companies with less-than-average merger activity, the other to companies with above-the-average merger history.

The first five graphs, in my opinion, are a striking confirmation of the ideas suggested by speculation on cotton. Basically, one sees that the

Figure 9

Monthly changes of log $Z(t)$, with $Z(t)$ Macaulay's index of rail stock prices. Abscissas and ordinates as in Figure 6.

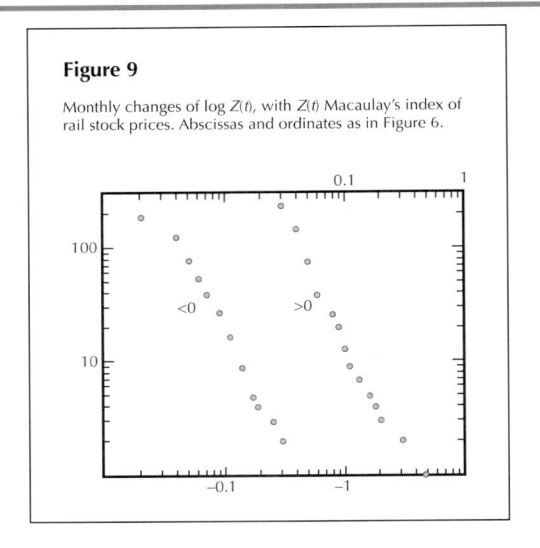

Figure 10

Yearly changes of log $Z(t)$, with $Z(t)$ Macaulay's January index of the price of nine selected rail stocks. Ordinates: absolute frequencies. Abscissas are not marked to avoid confusion: for each graph, they vary from 1 to 10.

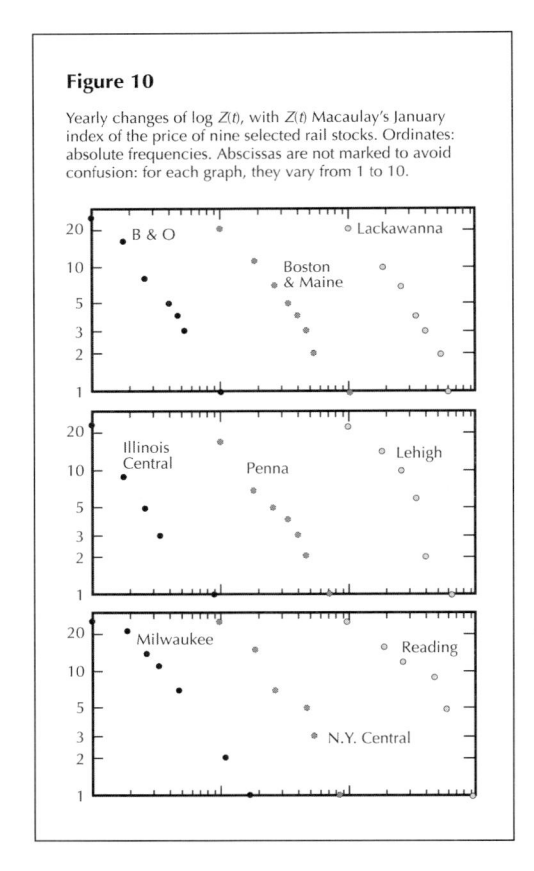

fluctuations of the price of these stocks were all stable Paretian, with the same α-exponent (clearly below the critical value 2). (Moreover, they all had practically the same value of the positive and negative "standard deviations" σ' and σ'', defined in VCSP).

For the companies with an unusual amount of merger activity, the evidence is similar but more erratic.

The variation of interest and exchange rates

Introduction. Various rates of money – and especially the rate of call money in its heyday – are reflections of the overall state of a speculative market. One would therefore expect to find that the behaviour of speculative prices and of speculative rates present strong similarities. But one cannot expect them to be ruled by identical processes. For example, one cannot assume (even as rough approximations) that successive changes of a money rate are statistically independent: such rates would indeed eventually blow up to infinity, or they would vanish. Neither behaviour is admissible. As a result, the distribution of $Z(t)$ itself, which is meaningless when Z is a commodity price, is meaningful when it is a money rate. Moreover, when investigating changes, one will study $Z(t + T) - Z(t)$ rather than $\log Z(t + T) - \log Z(t)$.

The rate of interest on call money. In Figure 11, the abscissa is the excess of Macaulay's rate of call money (Macaulay, 1932) over its "typical" value, 6%. I have not even attempted to plot the distribution of the other tail of the difference "rate minus 6%", since that expression is by definition very short tailed, being bounded by 6%, while the positive value of "rate minus 6%" can go sky high (and occasionally did).

The several lines of Figure 11 correspond, respectively, to the total period 1857–1936 and to three subperiods. They show that call money rates satisfy a single-tailed Paretian law, with an exponent markedly smaller than 2. Scale factors (such as the upper quartile) have changed – a form of non-stationarity – but the exponent seems to have preserved a constant value, lying within the range in which the law of Pareto is known to be invariant under mixing of data from populations have the same α and different γ (Mandelbrot, 1963a).

Other rates of interest. Examine next the distribution of the classic data collected by Erastus B. Bigelow (Fig. 12, dashed line), relative to "street rates of first class paper in Boston" (and New York) at the *end* of each month from January 1836 to December 1860 (Bigelow also reports some rates applicable at the beginnings or middles of the same months, but I disregarded them to avoid the difficulties due to averaging). The dots on Figure 12 again represent the difference between Bigelow's rates and the typical 6%; their behaviour is what we would expect if essentially the same stable Paretian law applied to these rates and those of call money.

Examine, finally, a short sample of rates, reported by L. E. Davis (1960),

Figure 11

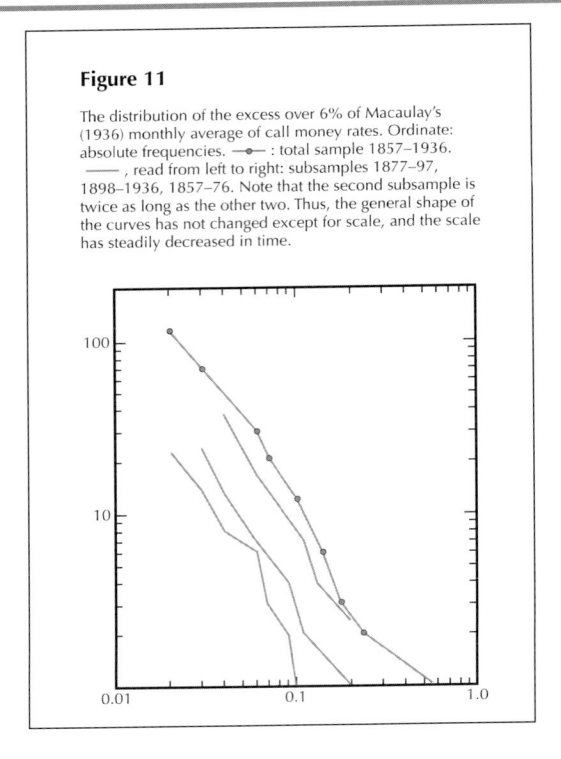

The distribution of the excess over 6% of Macaulay's (1936) monthly average of call money rates. Ordinate: absolute frequencies. —•—: total sample 1857–1936. ———, read from left to right: subsamples 1877–97, 1898–1936, 1857–76. Note that the second subsample is twice as long as the other two. Thus, the general shape of the curves has not changed except for scale, and the scale has steadily decreased in time.

on the basis of the records of New England textile mills. These rates remained much closer to 6% than those of Bigelow and were plotted in such a way that the crosses of Figure 12 represent ten times their excess over 6%. The sample is too short for comfort, but, until further notice, it suggests that the two series have differed mostly by their scales.

The dollar-sterling exchange in the 19th century. The exchange premium or discount in effect on a currency exchange seems to reflect directly the difference between the various "forces" that condition the variations of the values of the two currencies taken separately. This differential quantity has even an advantage over the changes of rates; indeed, one can consider it without resorting to any kind of economic theory, not even the minimal assumption that price changes are more important than price levels. We have therefore plotted the values of the premium or discount between dollar and sterling between 1803 and 1895, as reported by L. E. Davis and J. R. T. Hughes (1960) (Fig. 12). This series is based upon operations which involved credit as well as exchange; in order to eliminate the credit component, the authors used various series of money rates; we also plotted the series based upon Bigelow's rates. One will note

Figure 12

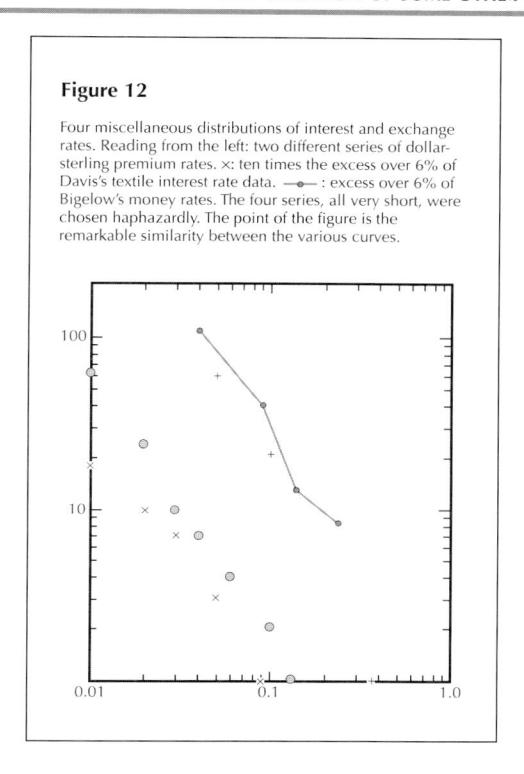

Four miscellaneous distributions of interest and exchange rates. Reading from the left: two different series of dollar-sterling premium rates. ×: ten times the excess over 6% of Davis's textile interest rate data. —•—: excess over 6% of Bigelow's money rates. The four series, all very short, were chosen haphazardly. The point of the figure is the remarkable similarity between the various curves.

that all the graphs of Figure 12 conform strikingly with the expectations generalised from the known behaviour of cotton prices.

STATISTICAL ESTIMATION OF α BY MAXIMUM LIKELIHOOD, WHEN α IS NEARLY 2

Introduction

A handicap for the theory of VCSP is that no closed analytic form is known for the stable Paretian distributions, nor is a closed form likely to be ever discovered. Luckily, the cases where the exponent α is near 1.7 can be dealt with on the basis of an approximating hyperbolic distribution. Now let α be very near 2. To estimate $2 - \alpha$ or to test $\alpha = 2$ against $\alpha < 2$ is extremely important, because $\alpha = 2$ corresponds to the Gaussian law and differs "qualitatively" from other values of α. To estimate such an α is very difficult, however, and the estimate will be intrinsically highly dependent upon the number and the "erratic" sizes of the few most "outlying" values of u_n. I hope to show in the present section that simplifying approximations are fortunately available for certain purposes. The main idea is to represent a stable Paretian density as a sum of two easily manageable expressions, one of which concerns the central "bell", while the other concerns the tails.

A square central "bell" with hyperbolic side portions

Consider the following probability density, which can be defined for $3/2 < \alpha < 2$:

$$p(u) = \alpha - 3/2 \quad \text{if } |u| \leq 1 \text{ (adding up to } 2\alpha - 3)$$
$$p(u) = (2 - \alpha)\alpha |u|^{-(\alpha+1)} \quad \text{if } |u| > 1 \text{ (adding up to } 4 - 2\alpha)$$

When α is near 2, $p(u)$ is a rough first approximation to a stable Paretian density. Its advantage is to lead itself readily to maximum-likelihood estimation.

Let indeed $u_1, \ldots u_n, \ldots, u_N$ be a sample of values of U, ordered by decreasing absolute size, and let M of these have an absolute size greater than 1. Given these sample values u_n, the likelihood of a value of α is defined as being $\Pi p(u_n)$, which equals

$$\left(\alpha - \frac{3}{2}\right)^{N-M} [(2-\alpha)\alpha]^M \left[\prod_{n=1}^{M} |u_n|\right]^{-(\alpha+1)}$$

The logarithm of the likelihood is

$$L(\alpha) = (N - M)\log\left(\alpha - \frac{3}{2}\right) + M\log[(2 - \alpha)\alpha] - (\alpha + 1)\sum_{n=1}^{M} \log|u_n|$$

This $L(\alpha)$ is a continuous function of α. If $M = 0$, it is monotone increasing and attains its maximum for $\alpha = 2$. This is a reasonable answer, since $|U| < 1$ for $\alpha = 2$.

If $M > 0$, on the contrary, $L(\alpha)$ tends to $-\infty$ as α tends to either of the ends of its domain of variation, namely, 3/2 and 2. It has therefore at least one maximum, and the most likely value of α, namely $\hat{\alpha}$, is among the roots of the third-degree algebraic equation.

$$\frac{N - M}{\alpha - \frac{3}{2}} - \frac{M}{2 - \alpha} + \frac{M}{\alpha} - \sum_{n=1}^{M} \log|u_n| = 0$$

Thus, $\hat{\alpha}$ only depends upon M/N and upon $M^{-1}\sum_{n=1}^{M} \log|u_n| = V$, the logarithm of the geometric mean of these u_n's whose absolute value exceeds 1.

Let us examine the latter term closer. The random variable $\log|U|$, conditioned by $\log|U| > 1$, has for distribution

$$Pr(\log|U| > u |\log|U| > 0) = \exp(-\alpha u)$$

Its expected value is $1/\alpha$. Therefore, as the sample sizes M and N tend to infinity, one will have

$$\lim \left[\frac{1}{\alpha} - \frac{1}{M} \sum_{n=1}^{M} \log(u_n)\right] = 0$$

As a result, *in the first approximation*, one can neglect the term

$$\frac{1}{\alpha} - \frac{1}{M} \sum_{n=1}^{M} \log|u_n|$$

when both M and N are large. The equation in $\hat{\alpha}$ simplifies to the first degree and yields

$$\frac{2 - \hat{\alpha}}{\hat{\alpha} - \frac{3}{2}} \sim \frac{M}{N - M}$$

that is,

$$\hat{\alpha} = 2 - M/2N$$

In particular, $\hat{\alpha}$ no longer depends upon the precise values of the u_n but depends only on their relative numbers in the two categories $|U| < 1$ and $|U| > 1$. The ratio M/N may, incidentally, be interpreted as the relative number of outliers for which $|U| > 1$.

For example, if M/N is very small, $\hat{\alpha}$ is very close to 2. (At the other end, if N/M barely exceeds 1, $\hat{\alpha}$ nears 3/2. However, this is a range in which $p(u)$ is a very poor approximation to a stable Paretian probability density.)

It may be observed that, knowing N, M/N is an asymptotically Gaussian random variable. We have thus easily proved that is asymptotically normally distributed for all values of α.

In a second approximation, valid for α near 2, one will insert $\alpha = 2$ in computing the value of

$$W = \frac{1}{\alpha} - \frac{1}{M} \sum \log|u_n|$$

The equation in $\hat{\alpha}$ will thus go down in degree from the third to the second. One of its roots is very large and irrelevant; the other root is such that $\alpha - (2 - M/2N)$ is proportional to W.

Scope of the estimation procedure based upon the counting of outliers

The method of the second part of the fourth section, namely, estimation of $\hat{\alpha}$ from M/N, applies without change under a variety of seemingly generalised conditions:

1. Suppose that the tails are asymmetric, that is,

$$p(u) = (2-\alpha)\alpha p' u^{-(\alpha+1)} \quad \text{if } u > 1$$
$$p(u) = (2-\alpha)\alpha p''|u|^{-(\alpha+1)} \quad \text{if } u < -1$$

where $p' + p'' = 1$. To estimate, one will naturally concentrate upon the random variable $|U|$, which is the same as in the previous part.

2. Further, the results of the previous part remain valid if the conditional density of U, given that $|U| < 1$, is non-uniform but independent of α. Suppose, for example, that for $|u| < 1$, $p(u)$ is equal to $(\alpha - 3/2)$ multiplying the truncated Gaussian density $D\exp(-u^2/2\sigma^2)$, where $1/D(\sigma)$ is defined as equal to $\int_1^{-1} \exp(-s^2/2\sigma^2)ds$. The likelihood of α then equals

$$\left[D(\sigma)2^{-1}\left(\alpha - \frac{3}{2}\right)\right]^{N-M} \exp\left(-\sum_{n=M+1}^{N} \frac{u_n^2}{2\sigma^2}\right)[(2-\alpha)\alpha]^M \left[\prod_{n=1}^{M}|u_n|\right]^{-(\alpha+1)}$$

The maximum likelihood considered as a function of the U_n, is unchanged from the part above.

---

1 This original of VCSP appeared as an IBM Research Note in March, 1962.
2 To my shame, I missed this discussion when sampling this book and privately criticised Bachelier for blind reliance on the Gaussian. Luckily, my criticism was not committed to print.
3 I was surprised to see VCSP criticised for expressing blind belief in independence. For examples of reservations on this account, see its seventh section as well as the final paragraphs of its third and fourth sections.
4 For an old but eminent practitioner's opinion, see Mills (1927); for several recent theorists' opinions see Paul Cootner's anthology (Cootner, 1964).
5 Throughout the long-awaited second volume of William Feller's *Introduction to Probability* (Feller, 1966), one finds a wealth of facts concerning these laws. However, B. V. Gnedenko and A. N. Kolmogoroff's monograph (1954) remains the only up-to-date book discussing these laws in a single chapter. For a compact briefer treatment, see J. Lamperti (1966).
6 α is related to "Pareto's exponent", but it would be extremely dangerous to underestimate the differences between the two concepts (Mandelbrot, 1963b).
7 I first proposed for these cases the term of "Pareto–Lévy laws", then tried to withdraw it. I am now resigned to consider "stable Paretian" and "Pareto–Lévy" as synonymous.
8 I may at this point reassure those who expressed in print the fear that I find $E[L^2]$ to be infinite because I inadvertently took the logarithms of zero.
9 I do not propose the colourful metaphor as a scientific terminology!
10 To my knowledge, $\alpha > 1$ for prices, and the first moment is well suited to the study of log $Z[t]$. ($Z(t)$ itself is another matter.) However, the stable laws with $\alpha < 1$ play a central role in economics (Mandelbrot, 1965, and in *International Economic Review*).

11 It follows that, despite the length of the 1883–1936 record, the number of items in the series of wheat prices is not as large as one might have hoped – although it is naturally very long by the standards of economics.
12 The same method was applied by Fama (1965) to common-stock price changes and he also found that it is a favourable test of their stability.

BIBLIOGRAPHY

Alexander, Sidney S., 1961, "Price Movements in Speculative Markets: Trends of Random Walks", *Industrial Management Review of MIT* 2(2), May, pp. 7–26. Reprinted in *The Random Character of Stock Market Prices* 1964, P. H. Cootner, (ed.), (MIT Press), pp. 199–218.

Alexander, Sidney S., 1964, "Price Movements in Speculative Markets: Trends of Random Walks: No. 2", *ibid.* 4(2), pp. 25–46. Reprinted in *The Random Character of Stock Market Prices* 1964, P. H. Cootner, (ed.), (MIT Press), pp. 338–72.

Bachelier, Louis, 1900, "Théorie de la spéculation", *Annales de l'Ecole Normale Superieure*, Ser. 3. 17 (1900), pp. 21–86. Translated in *The Random Character of Stock Market Prices*, P. H. Cootner, (ed.), (MIT Press, 1964), pp. 17–75.

Bachelier, Louis, 1914, *Le jeu, la chance et le hasard* (Paris: E. Flammarion).

Cootner, Paul H. (ed.), 1964, *The Random Character of Stock Market Prices* (Cambridge: MIT Press).

Davis, L. E., 1960, "The New England Textile Mills and the Capital Market: A Study of Industrial Borrowing 1840–1860", *Journal of Economic History* 20, pp. 1–30.

Davis, L. E. and J. R. T. Hughes, 1960, "A Dollar-Sterling Exchange, 1803–1895", *Economic History Review*, 13, pp. 52–78.

Fama, Eugene F., 1963, "Mandelbrot and the Stable Paretian Hypothesis", *Journal of Business* 36, October, pp. 420–29.

Fama, Eugene F., 1965, "The Behavior of Stock-Market Prices", *ibid.* 38, January, pp. 34-105.

Fama, Eugene F. and Marshall Blume, 1966, "Filter Rules and Stock-Market Trading", *Journal of Business* 39, January, pp. 226–41.

Feller, William, 1966, *An Introduction to the Theory of Probability and Its Applications*, vol. II. (New York: John Wiley & Sons).

Gnedenko, B. V. and A. N. Kolmogoroff, 1954, *Limit Distributions for Sums of Independent Random Variables*. Translated by K. L. Chung. (Reading: Addison-Wesley Press).

Godfrey, M. D., C. W. J. Granger and Oskar Morgenstern, 1964, "The Random Walk Hypothesis of Stock Market Behavior", *Kyklos* 17, pp. 1–30.

Kendall, Maurice G., 1953, "The Analysis of Economic Time-Series – Part 1: Prices", *Journal of the Royal Statistical Society*, Ser. A. 116, pp. 11–34. Reprinted in *The Random Character of Stock Market Prices*, P. H. Cootner, (ed.), (MIT Press, 1964), pp. 85–99.

Lamperti, J., 1966, *Probability, a Survey of the Mathematical Theory* (New York: W. A. Benjamin, Inc).

Macaulay, F. R., 1932, *The Smoothing of Economic Time Series* (New York: National Bureau of Economic Research).

Macaulay, F. R., 1936, *Some Theoretical Problems Suggested by the Movements of Interest Rates,*

Bond Yields, and Stock Prices in the United States since 1856 (New York: National Bureau of Economic Research).

Mandelbrot, Benoit, 1963a, "New Methods in Statistical Economics", *Journal of Political Economy* 71, October, pp. 421–40.

Mandelbrot, Benoit, 1963b, "The Stable Paretian Income Distribution When the Apparent Exponent is near Two", *International Economic Review* 4, January, pp. 111–15.

Mandelbrot, Benoit, 1963c, "The Variation of Certain Speculative Prices", *Journal of Business* 36, October, pp. 294–419. Reprinted in *The Random Character of Stock Market Prices*, Paul Cootner (ed.), (MIT Press, 1964), pp. 307–32; reprinted as Chapter 26 of the present volume.

Mandelbrot, Benoit, 1965, "Very Long-tailed Probability Distributions and the Empirical Distribution of City Sizes", in F. Massarik and P. Rotoosh (eds.), *Mathematical Explorations in Behavioral Science – the Cambria Pines Conference*. (Homewood: Richard D. Irwin, Inc).

Mandelbrot, Benoit, 1966, "Forecasts of Future Prices, Unbiased Markets, and 'Martingale' Models", *Journal of Business* 39, pp. 242–55: note the "errata".

Mandelbrot, Benoit, "Long-Run Linearity in Economic Systems; Roles of J-shaped Spectra and of Infinite Variance", *International Economic Review*.

Mandelbrot, Benoit and Howard M. Taylor, "On the Distribution of Stock Price Differences", *Operations Research*.

Mills, Frederick C., 1927, *The Behavior of Prices* (New York: National Bureau of Economic Research).

Working, Holbrook, 1934, "Prices of Cash Wheat and Futures at Chicago since 1883", *Wheat Studies of the Stanford Food Institute* 2, pp. 75–124.

Notes on Illustrations

1
Amsterdam is regarded as the birthplace of futures and option markets. In the 18th century, the Dutch engaged in some of the earliest organised "time sales", "to arrive" – or put and call contracts – on a range of products related to the whale fisheries and commodities including grain, cocoa and particularly coffee. These contracts were as complex and as sophisticated as those that exist in present-day exchange-traded or over-the-counter derivative markets. The trading in these time dealings – particularly options – is popularly held to be responsible for the speculative mania for tulips that reached its peak in 1634 (trading officially began at the Amsterdam Bourse in 1636).

2
The business interaction between the Dutch and Japanese produced the earliest futures markets in rice circa 1730. The precursor in the 17th century – a Japanese market called **Cho-ai-mai** – resulted from the practice of Japanese feudal lords selling rice for future delivery as protection against weather and war-related price risks.

3
The Chicago Board of Trade (CBOT) is often credited for creating the first formalised rules for engaging in the buying and selling of futures contracts in 1865. These three images from **Harper's Weekly** depict the thriving agrarian economy that existed in Chicago 20 years after the CBOT's original formation in 1848. Because Illinois gaming laws threatened to make such contracts illegal, the physical trade of grain at the exchange was often publicised more than its activities in "time sales", or futures and options contracts. A standardised method for inspecting and grading commodities was established by the Board in 1858 and, aided by the unrivalled network of warehouses and railroads in Chicago, allowed for fungible warehouse receipts and the creation by 1865 of futures and options contracts against which those warehouse receipts could be

used for delivery. By this time, much of the trade in physical commodities and "time sales" was conducted "on change" at the Board, situated just a few blocks north of the modern CBOT at the end of the LaSalle Street "canyon" on Jackson and LaSalle.

4

Chicago and Milwaukee competed heavily to capture the grain export business to the US east coast in the 1850s and 1860s – and that competition involved futures markets. While Milwaukee's once-thriving grain exchange is now a historical landmark rather than a place of mercantile business, this photograph is often shown as proof that the first octagonal trading pit, symbolic of open-outcry futures trading, originated in Milwaukee, not Chicago. In The History of Milwaukee, *published in 1931, John G. Gregory states that "the pit" was a convenience that originated in Milwaukee, citing the Milwaukee Chamber of Commerce's Secretary William J. Langson in 1883 as evidence: "It was in 1870 that I got up the first trading pit. It differed from the present one in not being circular . . . The principle of the thing is very old – it is the Roman amphitheater over again, but on a smaller scale. In 1878, a man in Chicago applied for a patent on the device of a [octagonal] trading pit with steps, and had the cheek to serve us with a notice that if we didn't pay him a royalty he'd sue us. Of course we laughed at the preposterous attempt to obtain money".*

5

The Liverpool Cotton Exchange and Liverpool Corn Exchange dominated European grain and cotton markets in the 19th and early 20th centuries. The participants at these exchanges were often active participants in thriving futures markets that also existed in New York and Chicago – so much so that they were frequently referenced in classic academic study on futures markets, intermarket price differentials, hedging and speculation. Cotton arrivals, according to the Liverpool Cotton Association, reached their highest level in the 1911–1912 season, when no less than 5.2 million bales were imported. The greatest cotton importing market in the world, it provided an unequalled service to spinners in guaranteeing the availability of cotton for forward delivery positions and holding large warehouse stocks. The Liverpool Corn Exchange, built in 1851, still stands, although it is not pictured here. Market historians often credit the Liverpool Cotton Brokers Association and the New York Cotton market, not the CBOT, as having the most plausible claim of pioneering the first highly-organised commodity futures markets.

6

The London Metal Exchange (LME) is the largest metals market in the world. Prior to its incorporation as The London Metal Exchange Company Limited in 1881, the LME operated as The Metal Market and Exchange Company Limited (founded in 1877). Before refurbishment, many of the members' booths were situated in long corridors away from the central trading area. This led to the earliest development of hand signals among traders and clerks, used to relay price information. These LME hand signals, or variations of them, are still used today in all the world's open outcry futures markets; only now, hand signals are used to cope with cacophonous peripheral noise, rather than to negotiate the tricky locations of the LME telephone booths.

7
As demonstrated by this early trading "pit" for gas oil (the first European energy futures contract, launched in 1981), futures trading pits come in all shapes and sizes and range from idiosyncratic rings of phone booths to enormous, tiered octagonal "stadiums" that house thousands of traders. The purpose of these pits, whatever their dimensions: to provide price transparency and liquidity to the broadest range of hedgers and speculators in a particular market as possible.

8
The Chicago Board of Trade (CBOT) had eight previous homes in Chicago before it became a permanent fixture at the Chicago business district intersection of LaSalle and Jackson. This was the first of the exchange's two homes at LaSalle and Jackson, inaugurally opened the evening of April 28, 1885. This etching shows the exterior of the first of the Board's two great trading halls at this location. The building's essence would be captured best by Frank J. Norris in his novel, **The Pit**. Norris wrote: "The lighted office buildings, the murk of rain, the haze of light in the heavens, and raised against it the pile of the Board of Trade Building, black grave, monolithic, crouching on its foundations, like a monstrous sphinx with blind eyes, silent, grave, – crouching there without a sound, without sign of life under the night and the drifting veil of rain".

Worthy of note is the famous light fixture, created by Elmer A. Sperry (top right), that illuminated the building in the evenings and later had to be removed when high winds and unsettled building foundation made the spiralling Tower and light fixture a potential safety hazard. Its electrical power was generated in the Dynamo Room, illustrated in some detail (bottom-left corner). Despite the building's structural flaws, it would house the dominant futures market until a new CBOT building took its place in 1930. Noted the Chicago Tribune on October 9, 1995, the 95th anniversary of Sperry's birth: "In 1885, Sperry contracted to construct and erect one of the most spectacular electric displays seen up to that time. It was a corona of 20 large arc lights 'of the all night variety' about the Board of Trade tower. His company agreed to furnish lights, and to service and operate them for three years. The arcs were mounted on gas pipes carved into a circle of 29 feet in diameter, and hoisted by pulleys and drums". Added the **Tribune** on the first illumination of Sperry's electric light. "The early beacon glowed with 40,000 candle power, compared to a maximum of 3,200 candle power seen up to that time in Chicago. Sperry and his light became known widely and other cities copied Chicago and lit up their tall buildings".

9

Between 1885 and 1905, the year the US Supreme ruled that futures contracts traded on the Chicago Board of Trade were not gambling transactions and sustained the exchange's property right over price quotations being imprudently used by bucketshops. Some of the greatest market speculators who ever traded made their presence known within this great trading room. Included were Benjamin P. Hutchinson, later known as "Old Hutch", Big Jim Patten, a legendary speculator in the early 1900s and Joseph Leiter (son of Chicago millionaire Levi Z. Leiter), who failed as dramatically in his 1898 speculations as Old Hutch succeeded in 1888 and Patten in 1913. So passionate about this room were many legendary speculators that one of them, Arther Cutten reportedly purchased two of the 17 stain glass windows that surrounded the trading floor, when the building was scheduled to be demolished, only to later install them at his estate. These windows reportedly had dimensions of 8 feet by 32 feet and contained allegorical figures of Agriculture, Commerce, Fortune and Order. This etching shows the immaculate fresco and marble interior of the hall, which opened for exchange business on April 29, 1885. The etching also displays a sketch of a communist demonstration that occurred, showing that the Board's special branch of capitalism – speculative futures markets – was not always well-received. Farmers often regarded members of the CBOT as "wind wheat" speculators.

10

Any sketch of "Old Hutch" is hard to come by and, of the few that exist, this is perhaps the most beautifully and accurately rendered. Regarded by some as the only speculator to ever truly corner the Chicago Wheat market in 1888, he began "scalping" as a trader. In his early days, he reported turned quick profits on large volume, often buying and selling in the middle of other traders for as little as a $\frac{1}{4}$-cent profit. He built his wealth that way, reportedly getting the news about foreign markets from reporters in Chicago before the newspapers hit the stands. Once his net worth grew to an estimated US$10–$20 million, his speculations in wheat and other commodities at the CBOT became ever more spectacular – and damaging – to the overall reputation of futures markets in the 1880s. Old Hutch reportedly lost his entire fortune the same way he had made it. In 1891, in only four months, he lost one of his legendary battles with trader John Cudahy. When told his son, the philanthropist and CBOT president Charles, had purchased a painting for the Art Institute of Chicago for US$2,500, Old Hutch issued one of his famous quips: "Look at him. Think about him. A son of mine! He paid $500 apiece for five painted sheep and he could get the real article for $2 a head". Noted an unidentified broker in the Chicago Tribune *on September 26, 1888, during Old Hutch's fabled corner: "He's a cunning fellow, and I'd bet two to one that he sent one set of brokers on the floor to sell and another to buy, simply to draw in the lambs and squeeze them".*

NOTES ON ILLUSTRATIONS

11

The Beijing Commodity Exchange was one of 14 active futures exchanges in China, all formed after 1990. Once an active futures market for such commodities as corn, mung beans (Chinese green beans) and plywood, and also financial futures such as Chinese Treasury bonds, it is now defunct. Due to consolidation, only three futures markets are actively trading contracts in China. Noteworthy about this photograph is the fact that all markets participants were housed centrally in the same room, even though the market traded electronically. Trade could also be conducted verbally in the centre ring area in the middle of the room and was reportedly allowed in periods of fast markets when trade through computer was not sufficient. Throughout the 1990s, many within the Chicago futures industry were known to predict that trading pits and electronic trading platforms might someday 'morph into a "hybrid" trading environment that resembled this. Time will tell.

12

There is no more endearing symbol of futures markets – and the Chicago Board of Trade's important role in their history – than John Warner Norton's mural, "Ceres", that is now displayed in the Chicago Board of Trade's 12th floor all-glass atrium (see cover). Most within the futures industry – except for the lucky few who may have witnessed the construction of the Holabird & Root-designed CBOT building erected in 1930 – will ever have seen this Ceres, a smaller $40\frac{1}{8}$ inch by $11\frac{1}{2}$ inch version. It is the final of 10 original designs that Norton created before painting the final mural, which was first mounted as a centrepiece above the entrance to the North Trading Room (see 13) at the Chicago Board of Trade. For 44 years, until 1974, the mural remained on the trading floor until it ahd to be removed, when the trading floor's ceiling was lowered – to accommodate a new trading hall for the Chicago Board Options Exchange. It is interesting to compare the subtle differences between Norton's final design for the mural and the actual mural itself.

Notes art historian Jim Zimmer: "Norton's design for the towering figure of Ceres in the Exchange room of the Chicago Board of Trade, completed in 1930, was initially vaguely Greek archaic, and then vaguely African with cubist overtones. Dismissing historical and contemporary art as a source, Norton turned, as he usually did to the studio model. He composed an exctic and erotic image of Ceres as a towering symbol of fertility, dramatically ligated from the back and side, striding out of the fields of stylized wheat, corn, and oats – the skyline of the city serving as a backdrop".

The final design of the mural is directly linked to the other, more prominent "Ceres", the 31-foot statue atop the Chicago Board of Trade, also commissioned specifically for the Holabird & Root building. Norton drew inspiration from Storrs' sculpture, though he masterfully reinterpreted Ceres as an artistic subject.

Writes art historian Diane Homan in her 1985 dissertation, "The Ceres Mural at the Chicago Board of Trade": "The Ceres commission must have had an extra personal meaning for John Norton. His father, John Lyman Norton, was at one time 'the richest man in Lockport', but met with financial ruin when assisting his brother, John's uncle, who had severe financial problems while trading commodities at the Board of Trade. John's father was financially devastated which resulted in John Having to leave college early to return home to get a job. Thus, the irony of his commission in the place where his family lost a great deal of money could not have escaped him. One cannot help but view the Ceres mural in light of John Norton's background as a personal statement as well as a public one . . .".

". . . Ceres holds two attributes in either hand. In the right is a sheaf of golden wheat. Stylized in a semi-circular form with decorative forms within it (they are reversals of the oat grains), the sheaf nestles into the bountiful curve of her hip. With her left hand she performs the only true action in the painting by scattering six seeds into the crop surrounding her. As they part from her open hand, they spread out and look like golden coins. Knowing what we do about Norton's family history, it may not be a false interpretation to imagine he means for the grains to look like coins to symbolize the bountiful goddess bestowing riches to the frantically gestering traders below her in the pits who were frequently in need of luck and money".

13–17

There have been many beautiful trading halls built around the world, but perhaps none so spectacular as the Art Deco North Trading Room (see 13) and the Old South Trading Room (see 14) of the Chicago Board of Trade. Both are linked, in a very real sense, to the entire history of derivative markets. And both, before electronic modernisation and an explosion in options trading and interest rate futures overran their priceless attention to detail, were architectural masterpieces.

In 1930, with a blank mural panel still waiting for John Warner Norton's Ceres mural to become the room's centrepiece, the North Trading Room would be the world's greatest trading hall for futures in agricultural commodities such as wheat, corn and oats and rye. Using the soft northern light, CBOT members also engaged in the trade of physical commodities using the black marble tables near the northern windows for careful inspection. The floor was made of hard cherry wood, to resist the scuffing and daily wear and tear that would result from treading on spilled commodities placed on the tables for inspection. In the four corners of the room, images of agricultural, commerce and transportation decorated every wall of the room, even down to the handles that opened the great northern windows. The greatest of these symbolic images of agriculture and the CBOT itself, may have been the six-ton octagonal Art Deco chandelier (see 15) that towered above the trading hall. Images of the CBOT outlined the chandelier's border and the white light from the chandelier reflected another towering silhouette of the Board as it might be seen looking south on LaSalle Street. The chandelier was above the seed trading pit and exchange – all as one unified whole. From 1936 to 1951, the North Room would also be home to new contract launches in futures on soybeans, soy oil and soy meal. These three contracts – innovations in their day – would later account for 40% of all commodity futures traded during the 1950s and 1960s. History would be made in the Old South Room as well. The room would first become a securities exchange, launched on September 16, 1929, that began trading 11 stocks (see 14). The new securities exchange at the Chicago Board of Trade was in response to a decline in futures trading activity in various commodity futures contracts resulting from increased government regulation. Later converted from securities exchange to smoker's lounge in the 1960s, this room would be the first home of the Chicago Board Options Exchange (see 17), which Board members had funded and organised. On April 26, 1973, CBOE opened for business in this very space. Trading was conducted in call options on 16 underlying securities at the outset, and later puts options, once the Securities and Exchange Commission became comfortable with the idea in 1977. This market quickly took off, thanks in large part to the pioneering option pricing formula first developed by Fischer Black and Myron Scholes, and later tweaked to perfection by Robert Merton. Said the market's founder and first president, Joseph Sullivan, before the option market's launch: "Its potential, we believe, is every bit as great as that of the commodity futures markets; and the basic principles we have employed are analogous to those employed to put commodity futures trading on an organized basis about a century ago".

CBOE, in 1999, traded an average of over 1 million contracts per day for the year. Back in the 1970s, the CBOE would move from the CBOT's Old South Trading Room to a new trading hall (see 17) in its earliest construction phase in this photo. Soon after, the Old South Trading Room would once again become home to the CBOT's next "interest rate futures revolution". This revolution – representing a bold move from tangible agricultural commodities to less tangible "commodities" such as the movement of interest rates – would begin with the Government National Mortgage Association futures on October 20, 1975. The successful launch of Treasury bond futures in the Old South Room came soon after in August 1977. By 1981, the bond futures contract would become the world's largest, trading 13.9 million contracts. With the ceiling lowered to accommodate CBOE, and space in the T-bond pit in short supply in the Old South Room, the full-blown interest rate future and options revolution would then finally migrate to the North Trading Room once more, though the room would never again look as pristine.

18

The CBOT's cross-town rival, the Chicago Mercantile Exchange, was also busy at work as a commodities futures market and financial futures market inventor in the 1970s. Once known as the Chicago Butter and Egg Board, the exchange became known as the Chicago Mercantile Exchange on October 6, 1919. The exchange, in its earlier days, had tried to trade everything from apples, potatoes, shrimp, cheese, turkeys, cow hides and onions, to scrap metal futures. In 1953, it would be wild price gyrations in onion futures on the CME's trading floor (pictured) that grabbed the attention of politicians and regulators. Up until then, onions had not been considered a regulated commodity. Because so many market participants claimed to be adversely impacted by the irregular price movements in the onion market in March 1953, the word onion would be added to the list of commodities regulated by the US Commodity Exchange Act. By 1958, the CME would face a serious blow when Congress banned trading in onion futures, its second most actively traded futures contract. It was the first time that futures trading in any commodity was made illegal in the United States. The CME would turn itself around in short order – to become one of the world's largest futures markets. From 1961 to 1964, it established a dominant market in perishable (refrigerated) commodities such as pork bellies and live cattle futures and came to be regarded as an upstart challenger to the CBOT. In 1972, the CME made the bold move of trading currency futures on the same floor that had once been used to trade eggs, butter and onions – though its appearance would undergo a radical transformation. As Bob Tamarkin, author of **The Merc: The Emergence of a Global Financial Powerhouse** noted: "On 16 May 1972 the old trading floor at 110 North Franklin took on the cachet of a United Nations festival. There were seven young women for seven currencies; each wore a native costume as they led a group of dignitaries that included William S. Dale, executive director for the US to the International Monetary Fund, to the IMM area. There, Dale cut a ribbon made of the IMM's seven currencies". Ultimately, stock index futures contracts on broad-based indices such as the S&P 500 and Eurodollar futures would become exchange's bellwether contracts – and some of the world's largest.

19

The London International Financial Futures and Options Exchange began its history using the "Chicago open-outcry" model. On September 30, 1982, LIFFE opened. Its success throughout the 1980s and 1990s as an open outcry exchange was demonstrated in December 1991 when it moved to its new trading floor at Cannon Bridge, which covered some 25,000 square feet and included 614 booths, 1,000 screens, and 440 dealer board consoles. By this time, the size and scope and activity of the trading pits in London rivalled those in Chicago and far exceeded the available trading space at the Royal Exchange. LIFFE would later be one of the first traditional open-outcry exchanges to embrace an all-electronic platform.

20

With many futures markets now operating through all-electronic platforms and computer terminals, old trading cards from yesteryear now seem as rare and valuable to collectors as baseball cards. Seen here is a Chicago Board of Trade trading card from 1918, which includes a series of "futures tax" stamps since such transactions were taxed at the time.

21

An electronic trading card used at the Chicago Mercantile Exchange is one small part of the technological innovation that many open-outcry traders believe will allow futures markets in Chicago to remain a hybrid, rather than all-electronic marketplace. This hand-held technology demonstrates the interaction between man and machine that now exists in open-outcry markets, and between the history and future of futures markets in open-outcry trading pits.

22

It's all in the hands. Nothing is more mysterious to observers in the visitor's gallery of any open-outcry futures market than hand signals. Traders are often forced to rely on them exclusively when the volatility of a fast market renders their shouting and screaming useless in the tight confines of the congested trading pit. This photograph captures the top stair gesturing at the London International Financial Futures and Options Exchange before the exchange converted to an all-electronic trading platform, LIFFE Connect.

23

The announcement by The Chicago Mercantile Exchange and Reuters in September 1987 to create an after-hours electronic trading platform for futures trading – Globex – prompted a dramatic change in the competitive landscape. Other major open-outcry futures markets developed plans for their own trading systems, the earliest example of this arguably the Chicago Board of Trade's proposed Aurora project. This abortive venture with Apple Computer, Tandem Computers and Texas Instruments died once the CBOT joined the first Globex alliance. At the time of the Aurora promotional brochure was published in 1989, it was thought that the best way to recreate the price transparency and liquidity of an open-outcry trading pit would be to create a microcosm of a pit on a computer screen. This photograph of an Aurora prototype screen included icons for trader's badges, just as they would appear in an open-outcry pit. Automated Pit Trading System (APT), LIFFE's first successful "after hours" electronic trading platform in the late 1980s, later borrowed some concepts from Aurora, including the posting of trading badges on the screen, and later proved quite successful. Aurora mutated into the CBOT's first electronic trading platform, Project A, after it left the Globex alliance, in the mid-1990s. All these earlier electronic platforms would later be displaced by more state-of-the-art electronic platforms presently used by the world's largest futures markets.

24

Not all trading pits had the vibrancy of Chicago's and London's in the 1990s. The *original open-outcry trading ring of what was known as the Sydney Greasy Wool Futures Exchange is shown during the 1960s in this photo. This exchange later became known as the Sydney Futures Exchange (SFE) and adopted a more contemporary version of open-outcry trading to trade a whole host of financial futures and options contracts successfully until November 12, 1999, when the SFE flipped the switch to an all-electronic trading platform known as Sycom. Despite these technological changes, the answer is yes: some things never change. Greasy wool futures are still traded in Sydney.*

Index

A

Alchian (1950) 405, 510
Alexander (1961) 204
Alexander's game 679–80
American Economic Review 49
American Elevator and Grain Trade 309
Amsterdam 247
An Analysis of Speculative Trading in Grain Futures 550
Andreas (1894) 295
anticipatory hedging 199–200
anticipatory reliability of prices 203–4
anti-trust laws 410
arbitrage 92, 93
 in the forward market 58–9
 in modern produce markets 252–4
Aristotle 29

B

Bachelier 672
 (1900) 649
 (1914) 686
Bachelier process 651
Bachelier's theory of speculation 686–7
backwardation 89
bank clearinghouses 311–12
Bank of England, establishment 267
bank rate policy 77–80
banking system, origins and development 262–7
Baumol
 (1957) 510, 521, 522
 (1959) 510, 518
 criticisms of counter-examples 503–6
 reply to criticisms of counter-examples 505–7

Baxter, Conine and Tamarkin (1985) 629
bears 60, 61, 265, 460, 461, 463
Becker (1976) 405
Berlin Produce Exchange
 average excess of corn prices 434
 comparison of average price with other markets 431–40
 suspension and effect on corn prices 427–65
Bienaymé (1853) 669
big hit ability 623–4, 629–31
Bigelow 701
bills of lading 248, 249–51
Bisbee and Simonds (1884) 291
Black (1976) 357
Black, Jensen and Scholes (1972) 600
Blau, Gerda 105
Blume (1968) 607
Bodie and Rosansky (1980) 629
bonds
 and speculation 56–7
 stability of yields 67
Boyle (1931) 298
bran, futures market 158, 166, 168
Brinegar (1954) 204
Broomhall 442
Brownian motion 649, 672–3, 686
Buffalo Board of Trade 279, 286
Bull deal in cotton 244
bulls 60, 61, 257
Bureau of Labor Statistics, Wholesale Price Index 133, 143
business conditions, information on 23
business recession, and inverse carrying charges 106
business risk, and insurance 15–17

C

call loans 266
call money, rate of interest 701
capital asset pricing, equilibrium risk-return relation 598–600
capital asset pricing model, application to futures contracts 600–603
capital goods
 and income goods 70–71, 73
 and speculative stocks 75–6
capital market interpretation, risk measures 609–12
Capper–Tincher bill, (Grain Futures Act) 307–8, 311
Cargill wheat corner 406, 409
Carlton
 (1982) 366, 370
 (1983a) 366, 370
 (1983b) 366
 (1984) 405
Carlton and Fischel (1983) 375
carrying charges 89–91, 101–2
carrying costs 55
carrying-charge hedging 197
cash markets 75
 and futures markets 107–8
 independence 94
cash prices 91, 92–3
 and spot prices 461
cash-future price differences 91
cattle contracts 274
Cauchy, law of 653
Cauchy (1853) 669
Chang and Lewellen (1984) 619
Chicago Board of Trade 143, 194, 248, 271, 296, 431
 adoption of complete clearing 304–12
 clearinghouse 296–7
 data on grain trading 367
 importance 380
 number of futures markets (1921–83) 381
 rules 325
Chicago Board of Trade Clearing Corporation (CBOTCC) 277, 311, 313
Chicago corn futures, average seasonal course of prices 95–7
Chicago Cotton Clearing Corporation 309
Chicago *Daily Tribune* 308
Chicago Journal of Commerce 363
Chicago Mercantile Exchange (CME) 282
 importance 380
Chicago Tribune 296
Chicago wheat futures 99–100
Chicago wheat market 90–91, 102–3
 temporal price structure 90
Christie case 278
clearing system
 evolution 303–4
 France 301–2
clearinghouse operations 280–81
clearinghouses 282–3
 direct settlements 282–3
 history 284–7
 settlement with complete clearing 298–302
Coase (1937) 279, 311, 312
coffee exchanges, Europe 302
commercial traders 621
commissions, for commodity futures 346
commodities
 characteristics of commodities traded on futures markets 360–62
 correlation between log volume, log open and price variability, means and standard variation 342
 different commodities traded on futures markets since 1921 365
 log open interest and volume of trade 344
 number, volume and types traded on futures markets 363–8
 speculative interest 167
 standard deviation of selected characteristics related to price

variability 339
volume of trade and open interest 341
commodity contracts, pricing 221–39
Commodity Credit Corporation (CCC) 140, 145, 156, 370
Commodity Exchange, Inc. (COMEX), importance 380, 381
Commodity Exchange Act 168, 554
Commodity Exchange Authority 145, 199, 207, 554, 555, 569
commodity futures, commission and margins 346
Commodity Futures Statistics 577
Commodity Futures Trading Commission Act (1974) 410
Commodity Futures Trading Commission (CFTC) 274, 620
commodity holdings 229
commodity options, pricing 230–32
complete clearing 298–302, 304
 adoption at Chicago Board of Trade 304–12
 contrast with ring clearing 299
 origin 301
Conant 218
 (1947) 199
confidence 468, 474, 475
consolidation, and uncertainty 17
constructive speculation 29
 on a wheat exchange 33–5
"consumers' risks" 28
contango 89
contingency 242
contract price, average for commodity futures contracts 345, 346
convenience yield 103, 122, 139, 141
convenience yield theory 131
Cootner 670
 (1960) 206, 576
corn
 contracts, normal probability plots 606

 estimates of stable Paretian parameters 608
 futures, actual and predicted course 202
 prices
  Liverpool 442–3
  New York 442
  and suspension of Berlin Produce Exchange 429–31
 regression parameters 611
 wheat and soybeans correlation coefficients 604
"Corn Trade News" 442
corporations 18–19, 31
 and commodities 229–30
cotton
 annual average commitments of reporting traders 130
 average stocks and average adjusted spreads 137
 classification of open contracts 156
 costs of storage 144
 futures 132–3
 futures price increases 134–5
 interseasonal stock spread regressions 138
 open contract data 555
 price changes 651
 prices, and stable Paretian laws 662–9
 speculative index 164
 spreads 141, 142
 statistical estimates of firms' stockholding schedule 135–42
Cowles and Jones (1937) 203
Craigie, P.G. 462
"credit deadlock" 80
Cumby and Modest (1987) 619, 620
currency futures 372
current price 57, 62

D
Davis (1960) 701
Davis and Hughes (1960) 702
De Lavergne (1931) 301
delivery orders 250, 251

derived consumers' stock schedule 120–21, 124–6
destabilising speculation 523
  example 524–8
  a general model 528–38
  equilibrium 531–3
Diamond and Stiglitz (1974) 535
differential equation model, speculative destabilisation 486–8
diffusion, and risk 18
diffusion process 649
distributive skill, analysis 562–4
dock warrants 248, 249
doctrine of the Multiplier 72–3
dollar-sterling exchange rate, in the 19th century 702–3
Dornbusch 442
Dow Jones "futures price index" 578
downward bias, in futures prices 94–5, 108, 131
Duluth Board of Trade 303
Dumbell (1927) 284, 285
Dunham, R.W. 463, 465
Dusak (1973) 629
Duvel, J.W.T. 193

E
East India Company 284
economic information 23–4
economic time series, nonsense moments and nonsense periodicities 669–71
economics, empirical research in 209–10
The Economist 66
Edwards (1925) 282
Edwards and Edwards (1984) 404
eggs
  classification of open contracts 156
  futures market 169
  non-classified futures contracts 157
  open futures contracts 155–6
  speculative index 164
Ehrhardt, Jordan and Walking (1987) 629

elasticity of expectations 63, 69
Ellison (1905) 284, 285, 292, 294
Elton, Gruber and Bentzler (1987) 629
Emery (1896) 278, 284, 287, 292, 297, 298, 301, 302, 304
equilibrium
  definition 497–8
  of stocks and prices 124–6
equilibrium futures price 127, 128, 129, 130
equilibrium risk–return relation, and capital asset pricing 598–600
Europe, coffee exchanges 302
European commodity options 232
"Evening Corn Trade List" 442
exchange rates
  stability of unpegged flexible exchange rates 490–91
  variation 701–3
exchanges
  and competition 378–86
  fraction of contracts accounted for by the largest exchange 384
  success of type of good by exchange 385
"Exchanges Bill" 428
expected price 57, 61, 62
  and futures prices 124

F
Fama
  1965) 686, 694
  (1971) 598
Fama and MacBeth (1972) 600
Fama and Marshall (1966) 686
Farrell
  (1954) 515
  (1966) 521, 522
Federal Farm Board 146
Federal Reserve Board 144
Federal Reserve System (1985) 402, 414
federal stamp tax 297
Federal Trade Commission 93, 94, 95

Report on the Grain Trade 97
financial futures 282–3, 370–75
  insider trading 375
  margins 374
  regulation and industrial structure 372–3
  tax treatment 373–4
  uncertainty 371
  variation on price 371
firms' stockholding schedule 121–2, 124–6
  in cotton and wheat, statistical estimates 135–42
  and futures trading 128–9
Fischel (1986) 403
Fischel and Grossman (1984) 414
Fisher (1981) 366
Fisher index (1966) 609
Flour-mill hedging 196
Flux, A.W. 427
forecasting, role in theory of normal backwardation 573–7
forecasting ability
  consistent forecast coefficients 626, 628
  correlation statistics 633–4
  empirical tests 625–31
  ex ante tests 631–3
  statistical methods to determine 622–5
  of traders 620, 622–35
forecasting skills 572
  futures traders 591–3
forecasting theory 569
Forrester
  (1931) 294, 295
  (1932) 298
forward contracts 223, 325
  pricing 230–32
  vs. futures contracts 356–8
France, clearing system 301–2
free enterprise 12–13
Friedman 479
  (1953) 509, 518, 521
FTC (1920) 302
futures 32–3, 34
  distribution between, and profitability 564

and money, isomorphism between 261, 262, 272–3
futures commission merchants (FCMs) 280
futures contracts 223–4, 245, 268, 325–6
  application of capital asset pricing model 600–603
  clearing
    offsetting 272–3
    registration 280
    settlement 281
  cumulative frequency of contract life by group (1921–83) 377
  defined 278–9, 312
  distribution of lifetimes (1921–83) 376
  provision for delivery 270–71
  versus forward contracts 356–8
  volume (1954–83) 369
futures dealing, and price of commodities 428
Futures Industry Association Inc. 367
futures markets 396–7
  characteristics of commodities traded 360–62
  and commodity markets 261
  composition (1921–83) 366–7
  composition of trade 369
  concepts concerning 192
  cumulative death rate by commodity group 378
  development 266–8
  different commodities traded since 1921 365
  growth in number and use 368–75
  and hedging 178–9
  industrial structure 362
  and inverse carrying charges 89–112
  Keynes–Hicks theory of futures markets 130–31
  longevity 375–8
  manipulation 397–401, 402, 406–9

futures markets (*continued*)
  monopoly, responses to costs of monopoly 401–6
  monopoly in 397–401
  number (1921–83) 364
  number, volume and types of commodities traded 363–8
  overview 356–60
  price correlations 361–2
  rates of return 584–9
  regulation 355–6
    and the history of manipulation 406–9
    methods and limits 409–14
  research 191
  returns, empirical properties 603
  success for selected exchanges 382
  survival and death rates 377
  trader performance in 619–22
  and uncertainty 123, 361
  value 358–60, 362
futures prices 61, 91, 92–3
  behaviour 225–8
  of commodities 222
  downward bias 94–5
  and expected price 120
  and expected price change 98
  and general price level 132–3
  probability distribution 645
  and spot prices 228–30
  trends 131–5
futures traders
  estimation of commitments and returns 580
  forecasting skills 591–3
  returns to 577–89
  data 577–80
  determinants 589–93
futures trading
  activity and price variability 343
  and firms' stockholding schedule 128–9
  and government regulation 366
  and stockholding 126–30

G
gambling 29
  and speculation 151–2, 242, 246–7
Gaussian distribution 649–50, 672, 686
Gaussian law, stability and its generalisation 651–3
Germany
  futures markets 423–4
  wheat, imports and exports (1892–1900) 439
  wheat and rye, production and total supply (1893–1900) 436
Giffin, Sir Robert 457
Ginsburg, B.W. 460
Glover, Sir John 458
Gorton (1985) 293
Gorton and Mullineaux (1987) 293, 311, 312
Government National Mortgage Association (GNMA) 274, 358
government price-support programme, effect on spreads 138–40
government regulation, and futures trading 366
government stocks 140
Graf (1953) 195
grain elevator 359
Grain and Feed Statistics through 1954 149
grain freight rates (1892–1900) 435
Grain Futures Act, (Capper–Tincher bill) 307–8, 311
Grain Futures Act (1922) 403, 410
Grain Futures Administration 193, 194
  see also Commodity Exchange Authority
grain futures contracts, volume (1921–82) 368
Gray
  (1960) 205
  (1961) 205, 206

H
Hamada and Scholes (1983) 373
Hardy (1940) 597
Hardy gambling casino theory 612

Harper Deal 406
Hartzmark
  (1984) 621, 622
  (1987) 621, 622
  (1991) 549
hedgers, in futures markets 126, 127, 130, 131
hedging 108, 330
  in commodities 158–9
  continuous response of speculation 169–78
  contracts 20
  estimation of totals 180–82
  in the forward market 58, 59
  in modern produce markets 254–6
  multipurpose concept 195–200
  ratio 160, 161
  and speculation 152–62
hedging–market concept 194–5
Henriksson (1984) 619
Henriksson and Merton (1981) 619, 620
Hicks (1953) 574, 575
Hicks, J.R. 80, 119
Hieronymus (1977) 359
Hirschstein (1931) 301
Hoffman (1941) 194
Hoffman and Duvel (1941) 152
Hooker, R.H., discussion of paper 457–65
Hoos and Working (1940) 152
Houthakker
  (1957) 206, 549
  (1957a) 569, 580
  (1957b) 574, 576
  (1959) 324
  (1961) 204, 206
Huebner (1911) 298
Hutchinson wheat fraud 406

I
income goods, and capital goods 70–71, 73
income-stability, and speculation 70–77
independence assumption 512–13, 515, 519

infinite variance, and stable Paretian distributions 690–93
information, on business conditions 23
insider trading, in financial futures 375
instability, and speculation 476
insurance 14–15, 29–31
  and business risk 15–17
interest rates, variation 701–3
inverse carrying charges 89, 94, 99, 103, 104, 107, 108–9, 113–14
  and business recession 106
inverse carrying prices, theory 92
investment 73–5
  precariousness 471–2
  in private business 469
Irwin 194
  (1935) 152, 194
  (1954) 195

J
Jagannathan and Korajczyk (1986) 619
Jeans (1948) 210
Jensen (1968) 619
Jevons (1903) 292
Johnson
  (1960) 603
  (1976) 522
Journal of Farm Economics 47
judgement 10–12

K
Kaldor, Nicholas 103
  Speculation and Economic Stability 45
Kansas City, wheat futures 153
Kansas City Board of Trade 303
Kansas City exchange 301
Kemp (1963) 510
Kendall (1953) 694
Keynes, J.M. 93, 119, 269, 612
  (1923) 573, 574
  (1930) 597
  General Theory 75

Keynes (*continued*)
  theory of normal backwardation 94, 205
Keynes–Hicks theory of futures markets 130–31
Keynesian risk measure 603–9
King (1966) 607
Knight, Frank H., *Risk, Uncertainty and Profit* 3
Kohn (1978) 522, 523, 532, 533
Kon and Jen (1978, 1979) 619
Krasker (1980) 527

L
Larson (1960) 204
Law of Demand 513, 515
Leeman's Act (1866) 460
Leiter corner 447, 448
Leiter deal in wheat 257–8
Lévy (1925) 650
Lintner (1965) 598
liquidity 472
liquidity–preference 67–8
Liverpool
  basis for calls and closing price of futures (1896) 449
  corn prices 442–3
Liverpool Cotton Association 284, 285, 292, 294
Liverpool wheat market 100, 105, 152
  carrying charges 103
Loch, C.S. 464
London, maritime trade 247
London Corn Exchange 298, 459
London Metal Exchange 282, 298
London Produce Exchange 461
London Stock Exchange, clearinghouse 292
long hedging 269–70
  values for eleven commodities 163
long speculation
  in 11 commodities 162
  and unbalanced short hedging 190
long-term expectation 467–78
long-term rate of interest 66–7, 76

Loss (1983) 410

M
Macaulay
  (1932) 701
  (1936) 699
Manchester Guardian Commercial 574
Mancke 450, 451
  (1898) 440
Mandelbrot
  (1963a) 694, 701
  (1963b) 695
  (1963c) 685
  (1966) 686, 687
Mandelbrot and Howard 690
Mandelbrot and Taylor 686
manipulative speculation 35–7
margins
  for commodity futures 346, 347
  financial futures 374
Marine Code (1681) 250
market clearing prices, distribution 326–9
market expectation 98
"market judgement" 178
market-balance concept vs. risk premiums 205–9
marketing 27–8
markets, in the absence in speculators 481–2
Markowitz (1959) 600
Marshall, Alfred
  Industry and Trade 3
  Principles of Economics (1922) 209, 210
"martingale" model 687
mean daily movement (M.D.M.) 443–5
  (1892–1900) 452
Mehl (1931) 198
members, of an organised exchange 324–5
Merton (1981) 619
Michelson–Morley experiment (1887) 210
Middle Ages
  markets 248–9

speculation 241, 243–4
middleman, in modern markets 245
Mill, John Stuart
  (1898) 211
  (1900) 209
  (1921) 521
Miller and Scholes (1972) 600
Mills, F.C. (1927) 688
Minneapolis, wheat futures 114–16
Minneapolis Grain Exchange 302
modern marketing, technique 247
modern markets, transactions 252–8
monetary policy, and stability 77–80
money, and futures, isomorphism between 261, 262
money wages 65
monopoly
  in futures markets 397–401
  market responses to costs 401–6
Mossin (1966) 598
Multiplier 72–3

N
National Bank Act (United States) 267
New York, corn prices 442
New York Coffee Exchange 298
New York Cotton Exchange 143
New York Produce Exchange 297
Newbery and Stiglitz (1984) 524, 540
Non-performance
  problems 281–2
  risk, and margins 293–4
non-speculative cycle models 488–90
non-financial futures, uncertainty and price correlations 368–70
normal backwardation 59, 60, 65, 94, 130, 559, 569, 570
  and profit flow 571–2
  and returns to futures traders 589–91
  role of forecasting in theory of normal backwardation 573–7

"normal price" 64
  in the securities market 66

O
Oil City Exchange 286
oil futures contracts 286–7
onions
  futures market 168–9
  futures trading 149
open contracts 193–4
  values for 11 commodities 163
open futures contracts, eggs 155–6
open interest, distribution 580–82
operational hedging 197–8
option contracts 224–5, 271–2
organised futures markets 321–4
  cross-section data and sources 349–52
  effects of price variability 340–48
  margins 333, 336
  properties 324–6
  sources of price variability 336–40
  theory of net benefit 329–36
  traders 327–8
organised markets, characteristics 31–3
organised speculation, function and nature 241–7
own-interest rates 270
Ownership Fraud 398–9, 409

P
Paretian model 651–62
partnerships 17–18
Patten wheat deal 246, 257–8
Pearson 653
Pearson's index of kurtosis 670, 692
perfect markets 55
peso problem 527
Phillips, Georges, bankruptcy 296
Pillman, J.C. 459
Position Fraud 398–9
Posner (1976) 397
potatoes, futures market 168–9

price correlations, futures markets 361–2
price expectations, grain market 97–101
price forecasts, and speculators 553–67
price stabilising, and speculation 76–7
price of storage 101–7
price variability
 effects on open interest and volume of trade 340–48
 and futures trading activity 343
 sources in organised futures markets 336–40
price variation
 in continuous time 677–80
 a model 680–81
 stable Paretian model 686–95
price-of-storage concept 201–2
pricing
 commodity options 230–32
 forward contracts 230–32
privileges 271
production 539
profits 571
 definition 496–7
 estimating 554
 and speculators 495
Prussian Chambers of Agriculture 431

Q
quantity risk 339

R
railroad stock, variation of prices (1857–1936) 699–701
random walk 202–3, 646
random walk model 649
rate of interest, on call money 701
rates of return, futures markets 584–9
reliably anticipatory prices concept 202–5
Report on the Grain Trade 281
representative expectations 57
Rew, R.H. 462

ring clearing, contrasted with complete clearing 299
ring settlements 287–98
ring system 304
risk, and uncertainty 5
risk measure, for capital market interpretation 609–12
risk premium concept 205
risk premiums 559
 vs. the market–balance concept 204–9
risk-avoidance hedging 200
Robertson, Professor 66
Rockwell (1964) 577
Rothschild and Stiglitz (1970) 535
Rudnick and Carlisle (1983) 373

S
Salop, Scheffman and Schwartz (1986) 406
Samuelson (1971) 521, 540
Savary 249
scalpers 176, 177
Scheinkman and Shechtman (1983) 523, 539, 540
Scherer (1980) 386
Schrock (1971) 600, 603
Scotch iron warrants 252
securities market, "normal price" 66
Select Pleas in the Court of Admiralty 249
selective hedging 198–9
settlement with complete clearing 298–302
shares
 instability 69–70
 and speculation 56–7
Sharpe (1964) 598
Shiller (1981a) 527
short hedging 176–7, 265–6, 268–9
 contracts in 11 commodities 159
 values for 11 commodities 163
short-term rate of interest 75, 77–9
Shriver (1978) 358
Silber (1981) 382
Smith, Adam (1789/1937) 521

Wealth of Nations 51
Smith and Warner (1979) 283, 304
Sowell (1980) 405
soybean contracts, normal
   probability plots 606
soybean meal, hedging 160
soybeans
   estimates of stable Paretian
     parameters 608
   regression parameters 611
   wheat and corn, correlation
     coefficients 604
Spahr (1926) 292
specialisation, and uncertainty
   19–21
spectral analysis of random time
   series 671
speculation 19–21, 28–9, 424–5,
   474
   and bonds and shares 56–7
   in commodities 158
   continuous response to hedging
     169–78
   defined 53, 149–51, 522
   destabilising speculation 523,
     524–8
   estimation of totals 180–82
   in the forward market 58
   and gambling 151–2, 242, 246–7
   and hedging 152–62
   and income-stability 70–77
   and instability 476
   market in the absence in
     speculators 481–2
   in the Middle Ages 241, 243–4
   in modern produce markets
     256–8
   pre-requisites for 55–62
   and price stabilising 76–7
   and price stability 62–70
   profitable 479, 509–20
   defined 510
   and stability 479–81
   and storage 522, 523
   theory of speculation 677–80
   traditional theory 53–5
   and transactions 154
speculative behaviour 482–5

speculative destabilisation,
   differential equation model
   486–8
speculative holding 179
speculative index 160, 161, 162–9,
   179, 182
   five-year average speculative
     indexes, for 11 commodities
     166
   meaning in terms of market
     functions 164
speculative interest, in
   commodities 167
speculative prices
   model of variation 662–9
   variation 649–708
speculative ratio 65, 160
speculative stocks 56, 57, 70
   and capital goods 75–6
   elasticity 61–2, 63
speculators
   earnings 206–7
   influence on the cycle 485–6
   and price forecasts 553–67
   and profits 495, 555, 557
   a model 498–503
   returns to 597–8
   and stability 521–2
spot prices 128, 129, 130–31, 201
   and cash prices 461
   of commodities 221–2
   and futures prices 108, 228–30
spreads 144
   and consumption 142
   cotton 141, 142
   defined 143
   effect of government
     price-support programme
     139–40
   effects of change in stock 137–8
   wheat 141, 144
"squeeze" 257
Sraffa 269
stabilisation, defined 511–12
stabilising speculation, a model
   533–8
stability
   definition 497

stability (continued)
   and monetary policy 77–80
   and speculators 521–2
stable Paretian distributions, and infinite variance 690–93
stable Paretian laws, and cotton prices 662–9
stable Paretian model of price variation 686–95
stable Paretian processes
   causality and randomness 676–7
   sample functions 671–77
standard contracts 322–3, 357
standard deviation
   of the distribution of market clearing prices 322, 328
   of selected characteristics related to price variability 339
standard deviations (1892–1900) 452
Statistics on Cotton and Related Data 144
Stein (1961) 510
Stewart (1949) 200
Stewart, Blair 49
Stigler
   (1964) 386
   (1975) 406
stock index futures 273–4
stock-carrying demand curve 208
stockholding theory 119
storage
   charges 89, 101–7, 113
   costs 122–3, 245
   and speculation 522, 523
   supply curve 102, 108
straddling, in the corn market 463
superstition 7
supply price 64–5

T
Telser 328
   (1958) 206
   (1959) 510, 521, 522
temporal interdependence 518–19
term contracts 247, 248
theory of discount on the future 94

theory of speculation 677–80
Thiselton–Dyer, Sir W. 458
"to arrive" contracts 247, 248
total welfare 540
trader performance, in futures markets 619–22
trader profits 621
traders
   by early period forecasting ability 634–9
   forecasting ability 620, 622–35
   net profits or losses 556, 558, 560, 565
trading groups, aggregate profits 582–4
transactions
   costs 516–17
   and speculation 154

U
Udney Yule, G. 462
unbalanced short hedging, in 11 commodities 162
uncertainty 5–7, 11–12
   in economics 8–10, 22–3
   and futures markets 119, 361
   methods of dealing with 13, 22–3
   and specialisation 19–21
United States
   wheat markets 90–93
   wheat prices 105
   wheat stocks 99
unpegged flexible exchange rates 490–91
US Department of Agriculture 144, 154
US Federal Trade Commission 198
Usher, Abbott Payson, *The History of Mechanical Invention* 237

V
Vaile, Professor Robert S. 113–17
Value and Capital 575
Vance, Lawrence L. 91, 93
   theory of discount on the future 94

INDEX

variation, of interest and exchange rates 701–3
Viner (1956) 185, 490

W

wagering 152, 242
Wall Street Journal 363, 375, 380, 385
warehouse receipts 250–51, 264, 265, 266–7
warrant system 250–51
warrants 284
Welton, T.A. 463, 465
wheat
  annual average commitments of reporting traders 129
  average stocks and average adjusted spreads 137
  changes in amounts of hedging contracts 172, 173
  changes in long speculation and unbalanced short hedging 196
  contracts, normal probability plots 605
  corn and soybeans correlation coefficients 604
  crops, effects of reports of damage 105–6
  estimates of stable Paretian parameters 608
  imports and exports, Germany (1892–1900) 439
  interseasonal stock spread regressions 138
  monthly prices, Berlin and Chicago (1891–92) 453
  non-classified futures contracts 157
  prices
    average prices Berlin and Chicago (1882–1900) 432
    and mean daily movements, Berlin and Chicago (1893–99) 454
  regression parameters 611
  speculation 553–4
  spot and futures prices (1896–97) 450
  spreads 141, 144
  statistical estimates of firms' stockholding schedule 135–42
  stocks
    measurement 145–6
    United States 99
  variation of price in Chicago (1883–1936) 694–9
wheat exchange, constructive speculation on 33–5
wheat futures 132, 133–4
  Kansas City 153
  Minneapolis 114–16
  prices 95
wheat market
  amounts of hedging and speculation 170
  hedging 169–78
  United States 90–92
wheat prices 458–60
  steadiness 440–54
  United States 105
wheat and rye
  average annual spot prices (1892–1900) 433
  production and total supply, Germany (1893–1900) 436
Wiese (1952) 195
Williams (1982) 279, 286
"wind wheat" 437
wool prices, variation 650
wool tops
  classification of open contracts 156
  futures market 156
  non-classified futures contracts 157
Working
  (1934) 694, 699
  (1949b) 203
  (1949c) 201
  (1953, 1967) 324
  (1953a) 195
  (1953b) 195, 196, 210
  (1954) 153
  (1954a) 195
  (1956 and 1958) 204

731

Working (*continued*)
　(1958) 203
　(1960b) 195, 200
　on downward bias of futures prices 131
Working and Hoos (1938) 152
working stocks 104

Y
yield
　of assets 467
　of goods 55–6
　of stocks 103, 104
Young, Sydney 461